A robot manipulator is a movable chain of links interconnected by joints. One end is fixed to the ground, and a hand or end effector that can move freely in space is attached at the other end.

This book begins with an introduction to the subject of robot manipulators. Next, it describes in detail a forward and reverse analysis for serial robot arms. Most of the text focuses on closed-form solution techniques applied to a broad range of manipulator geometries, from typical industrial robot designs (relatively simple geometries) to the most complicated case of seven general links serially connected by six revolute joints. A unique feature of this text is its detailed analysis of 6R-P and 7R mechanisms. Case studies show how the techniques described in the book are used in real engineering applications.

The book meets the need for a thorough, up-to-date analysis of the structure and mobility of serial manipulators and will be useful to both graduate students and engineers working in the field of robotics.

Kinematic analysis of robot manipulators

Kinematic analysis of robot manipulators

CARL D. CRANE III AND JOSEPH DUFFY
University of Florida

CAMBRIDGE
UNIVERSITY PRESS

CAMBRIDGE UNIVERSITY PRESS
Cambridge, New York, Melbourne, Madrid, Cape Town, Singapore, São Paulo

Cambridge University Press
The Edinburgh Building, Cambridge CB2 8RU, UK

Published in the United States of America by Cambridge University Press, New York

www.cambridge.org
Information on this title: www.cambridge.org/9780521570633

First published 1998
This digitally printed version 2008

A catalogue record for this publication is available from the British Library

Library of Congress Cataloguing in Publication data

Crane, Carl D. (Carl David), 1956–
Kinematic analysis of robot manipulators / Carl D. Crane III,
Joseph Duffy.
p. cm.
Includes bibliographical references and index.
ISBN 0-521-57063-8 (hc)
1. Robot – Kinematics. 2. Manipulators (Mechanism) I. Duffy,
Joseph, 1937– . II. Title.
TJ211.412.C7 1998 97-13161
629.8′92 – dc21 CIP

ISBN 978-0-521-57063-3 hardback
ISBN 978-0-521-04793-7 paperback

To Sherry and Anne

Contents

Preface

This text provides a first-level understanding of the structure, mobility, and analysis of serial manipulators. A serial manipulator is an unclosed or open movable polygon consisting of a series of links and joints. One end is fixed to ground, and attached to the open end is a hand or end effector that can move freely in space.

The structure of a serial manipulator is established by labeling the skeletal form (the sequence of joints and links) with appropriate twist angles and perpendicular distances that define the relative positions of sequences of pairs of skew lines. In this way the geometry of the manipulator is defined, and subsequently it is possible to apply various coordinate transformations for points located on the links. Such transformations readily provide a so-called forward analysis, that is, they can be used to provide the position of some point on the end effector together with the orientation of the end effector measured relative to a coordinate system fixed to ground for a specified set of joint variables.

A more difficult problem is the solution of the so-called reverse or inverse analysis. Here, the position of a point on the end effector together with the end effector's orientation is specified. It is required to determine a corresponding set of six joint variables that will position and orient the end effector as desired. There are multiple solution sets of the six joint variables, in contrast to the forward analysis, where only one solution exists. The method of solution presented in this text is to join the end effector to ground by a hypothetical link. The six parameters that specify the position and orientation of the end effector can easily be transformed into the corresponding six parameters that model the hypothetical link. In this way, the reverse analysis of the manipulator is transformed into the analysis of a corresponding closed-loop spatial mechanism, that is, the computation of multiple solution sets of the joint variables of the closed-loop spatial mechanism. The analysis of closed-loop spatial mechanisms with more than four links has proved to be a difficult subject. A brief history is now presented.

As far as the authors are aware, the first attempt to analyze spatial mechanisms was by Dimentberg (1948), who applied screw algebra. Following this, various works on the subject were published by Weckert (1952); Denavit (1958); Wörle (1962); Yang (1963); Yang and Freudenstein (1964); Uicker, Denavit, and Hartenberg (1964); and Pelecudi (1972).

In the late 1960s and early 1970s much attention was focused on the analysis of five-link 3R-2C* spatial mechanisms. All such mechanisms were successfully analyzed; see Yang (1969), Yuan (1970, 1971), Soni and Pamidi (1971), and Duffy and Habib-Olahi (1971,

* Throughout this text R, C, P, E, and S denote respectively revolute, cylindric, prismatic, planar, and spherical (ball and socket) kinematic pairs.

1971a, 1972). However, the results highlighted the difficulties in the analysis of spatial mechanisms. Wallace (1968), Wallace and Freudenstein (1970, 1975), and Tor Fason and Sharma (1973) had also experienced major difficulties in the analysis of five-link RRERR and RRSRR spatial mechanisms.

It had become clear to the second author of this text that some form of shorthand or concise notation was required to write down the lengthy loop equations for spatial five-, six-, and seven-link mechanisms. This was accomplished by Rooney (1974), Keen (1974), and Duffy and Rooney (1975). The last provided the foundation for Duffy (1980). Various six-link 4R-P-C and seven-link 5R-2P spatial mechanisms were analyzed by Duffy and Rooney (1974) and by Duffy (1977). The remaining linkages were analyzed in Duffy (1980).

A significant result was the analysis of the spatial six-link 5R-C mechanism by Duffy and Rooney (1974a). The difficulty was of an order of magnitude two times greater than the analysis of spatial six-link 4R-P-C and seven-link 5R-2P mechanisms. A sixteenth degree polynomial in the tan-half-angle of an output angular displacement was determined by eliminating a pair of variables in a single operation from a set of four equations.

Duffy and Crane (1980) obtained a thirty-second–degree polynomial for the general seven-link 7R spatial mechanism, "the Mount Everest" (Freudenstein [1973]). Recently, Lee and Liang (1987, 1988) obtained sixteenth degree polynomials for the seven-link 6R-P and 7R spatial mechanisms by using and extending the unified notation developed by Duffy (1980), which was translated into the Chinese language by Professor Liang. More recently, the 7R spatial mechanism was solved by Raghavan and Roth (1993) using matrix notation. All these results demonstrate the high degree of difficulty in the solution of closed-loop spatial mechanisms, and Lee, Liang, Raghavan, and Roth are to be congratulated on their results.

An analysis of seven-link 6R-P and 7R mechanisms is given in this text using the unified and extended notation in collaboration with Lee and Liang. Detailed derivations are presented that we hope will greatly assist any reader who wishes to analyze seven-link 6R-P and 7R spatial mechanisms.

Finally, the shorthand recursive notation presented in Duffy (1980) using spherical trigonometry is developed in this text using rotation matrices. The results throughout the text are highlighted by examples.

Acknowledgments

The authors wish to thank P. Adsit, D. Armstrong, W. Abbasi, S. Ridgeway, D. Novick, and J. Wit for their valuable comments and for their assistance in proof reading the text. The authors are indebted to Ms. Florence Padgett, Editor, Cambridge University Press for her patience, advice, and guidance in the preparation of the text.

1

Introduction

This book stems from a first graduate course taught at the University of Florida on robot geometry. It describes in detail a forward and reverse analysis for serial robot manipulators, and a displacement analysis for closed-loop spatial mechanisms.

In the forward analysis, the variable joint angles are given, together with the constant parameters that describe the geometry of the manipulator. The goal is to determine the location (position and orientation) of the robot's end effector. This problem is relatively simple. A single solution for the location of the end effector exists for a given set of joint angle parameters.

The reverse analysis is more difficult because multiple solution sets exist. Here, the desired location of the robot's end effector is specified, and the goal is to obtain all the sets of joint variables for the specified location. In other words, the manipulator has a multiple of distinct configurations for a specified location of the end effector. Here, it is required to compute all these multiple sets of joint variables that determine each distinct configuration.

One method of performing a reverse analysis is to use an iterative technique. In this approach, a multidimensional search is performed employing a minimization of some specified error function. Often, one component of the error function is the square of the distance between the end effector location for the current set of joint parameters and the desired end effector location. The other component of the error function will usually measure the difference in orientation of the end effector from the desired orientation. Two problems arise with the use of an iterative technique. The first is that only one set of joint variables will be calculated. There is no guarantee that the iterative solutions for a pair of neighboring end effector locations will yield the same robot configuration. For example, a planar three-link revolute manipulator has in general two distinct configurations for a specified location of the end effector. When the end effector is in the first quadrant, these two configurations are referred to as elbow up and elbow down. A problem occurs when the manipulator performing a task is in an elbow-down position and the iterative technique yields an elbow-up configuration. The second problem is that the objective function to be minimized in an iterative approach often has mixed units such as $(\text{length})^2 + (\text{radians})^2$, which stems from an error equation that is a combination of position and orientation errors. Such functions are devoid of any geometrical meaning and they are not invariant with a change of units.

The majority of the book will focus on closed-form techniques for solving the reverse analysis problem. In this closed-form approach, all the possible sets of joint parameters that locate (position and orient) the robot's end effector as desired will be found by firstly solving a polynomial in the tan-half-angle of one of the joint variables. Admittedly it is necessary to iterate to solve for the roots of a polynomial of degree greater than four. This is, however, clearly different from performing the multidimensional search. The remaining joint variables are solved for sequentially using appropriate loop equations.

Chapter 2 begins with a definition of position and orientation. Coordinate systems are attached to each of a series of rigid bodies. Following this, transformations are derived that relate the coordinates of a point in one coordinate system to another.

Chapter 3 proceeds to define a link and to describe the different types of joints that can interconnect these links. A coordinate system is attached to each link of a serial robot manipulator, and the transformation that relates these coordinate systems is derived.

The forward analysis is discussed in Chapter 4. The transformations developed in Chapter 3 are used to determine the overall transformation that relates the coordinate system of the last link, the end effector link, to ground. This overall transformation will be used to transform the coordinates of a point in the end effector coordinate system (i.e., a tool point) to its coordinates in the ground coordinate system. The transformation will also define the orientation of the robot's end effector relative to ground.

Chapter 5 presents the detailed problem statement for the reverse analysis, and initially, iterative solution techniques are discussed. Following this discussion, a framework for obtaining a closed-form solution is established by adding a hypothetical link to the free end of the manipulator. This hypothetical link acts to connect the free end to ground and effectively converts the open or unclosed serial manipulator into a closed-loop spatial mechanism.

Chapter 6 introduces spherical closed-loop mechanisms. It is shown that an equivalent corresponding spherical mechanism can be constructed for a serial robot manipulator with a hypothetical closure link. The angular relationships for the equivalent spherical mechanism and the actual spatial manipulator are the same.

After closing the loop, the reverse analysis problem is converted to that of solving for the joint angles for the closed-loop mechanism when one of the joint angles is known. If the newly formed closed-loop mechanism has one degree of freedom, then this problem is solvable. The angular values that solve the closed-loop mechanism will also position and orient the original robot's end effector as desired. The solutions of virtually all closed-loop spatial mechanisms of one degree of freedom are presented in Chapters 7 through 10.

Chapter 11 presents useful reverse analyses for 6R manipulators with special geometry that can be analyzed directly rather than by simplifying the general 7R mechanism analysis. Five examples are given: the Puma 560, Cincinnati Milacron T3-776, and GE P60 industrial robots and two conceptualized by NASA, that is, the space station remote manipulator system and the modified flight telerobotic servicer manipulator system. These five examples demonstrate how the techniques developed in Chapters 5 through 10 are applied to solve real manipulators.

At the conclusion of Chapter 11, the reader should understand the forward and reverse position analyses of serial robot manipulators. These two analyses constitute the first step required for robot control.

Quaternions are introduced in Chapter 12 as an alternative (or supplement) to the coordinate transformation methods discussed in Chapter 2. This material is presented for completeness (it is not required for a basic understanding of the forward and reverse analysis procedure). Many papers have been published describing various applications of quaternions. These papers are difficult to understand without a knowledge of the basics that are presented in this chapter, and the development follows Brand (1947). A more recent and advanced text on rotation operators has been published by Altmann (1986).

Summarizing, this text provides a first-level understanding of the structure and analysis of serial manipulators. It is clear that a manipulator is an unclosed or open movable polygon consisting of a series of joints and links. A geometric description of joints and links is presented that provides a proper means of analysis using appropriate coordinate transformations for points and orientations. It is also clear that any open serial manipulator can be intimately related to a corresponding closed-loop spatial mechanism simply by joining the free end to ground by a hypothetical link. In this way, the reverse position analysis of the serial manipulator is essentially obtained from the solution of the input–output equation of this corresponding closed-loop spatial mechanism.

2

Coordinate transformations

This chapter relates the position and orientation of a coordinate system B in three-dimensional space to a reference coordinate system A. Once this has been accomplished, it is possible to transform the coordinates of any point in coordinate system B to coordinate system A.

2.1 Relative position and orientation of two coordinate systems

Figure 2.1 shows the pair of coordinate systems A and B. The position and orientation of system B relative to A are defined by the vector $\mathbf{V}_{\mathrm{A0}\rightarrow\mathrm{B0}}$, which gives the position of the origin of the B coordinate system relative to the origin of the A system, and the three unit vectors $\mathbf{x}_\mathrm{B}$, $\mathbf{y}_\mathrm{B}$, and $\mathbf{z}_\mathrm{B}$, which point along the coordinate axes of the B coordinate system. Knowledge of these four vectors as measured in the A coordinate system (written as ${}^{\mathrm{A}}\mathbf{V}_{\mathrm{A0}\rightarrow\mathrm{B0}}$, ${}^{\mathrm{A}}\mathbf{x}_\mathrm{B}$, ${}^{\mathrm{A}}\mathbf{y}_\mathrm{B}$, ${}^{\mathrm{A}}\mathbf{z}_\mathrm{B}$) completely defines the position and orientation of the B coordinate system measured with respect to the A coordinate system.

The three unit vectors ${}^{\mathrm{A}}\mathbf{x}_\mathrm{B}$, ${}^{\mathrm{A}}\mathbf{y}_\mathrm{B}$, ${}^{\mathrm{A}}\mathbf{z}_\mathrm{B}$, each of which has three scalar components, represent a total of nine scalar quantities. However, these are not independent because the vectors are unit vectors and they are also mutually perpendicular. Thus, the following constraint equations may be written:

$$\left|{}^{\mathrm{A}}\mathbf{x}_\mathrm{B}\right| = 1, \tag{2.1}$$

$$\left|{}^{\mathrm{A}}\mathbf{y}_\mathrm{B}\right| = 1, \tag{2.2}$$

$$\left|{}^{\mathrm{A}}\mathbf{z}_\mathrm{B}\right| = 1, \tag{2.3}$$

$${}^{\mathrm{A}}\mathbf{x}_\mathrm{B} \cdot {}^{\mathrm{A}}\mathbf{y}_\mathrm{B} = 0, \tag{2.4}$$

$${}^{\mathrm{A}}\mathbf{x}_\mathrm{B} \cdot {}^{\mathrm{A}}\mathbf{z}_\mathrm{B} = 0, \tag{2.5}$$

$${}^{\mathrm{A}}\mathbf{y}_\mathrm{B} \cdot {}^{\mathrm{A}}\mathbf{z}_\mathrm{B} = 0. \tag{2.6}$$

The unit vectors ${}^{\mathrm{A}}\mathbf{x}_\mathrm{B}$, ${}^{\mathrm{A}}\mathbf{y}_\mathrm{B}$, ${}^{\mathrm{A}}\mathbf{z}_\mathrm{B}$ thus represent $9 - 6 = 3$ independent scalar quantities that specify the orientation of the coordinate system B relative to A.

Consider now that the coordinate system B is attached to a rigid body. The vectors ${}^{\mathrm{A}}\mathbf{V}_{\mathrm{A0}\rightarrow\mathrm{B0}}$, ${}^{\mathrm{A}}\mathbf{x}_\mathrm{B}$, ${}^{\mathrm{A}}\mathbf{y}_\mathrm{B}$, and ${}^{\mathrm{A}}\mathbf{z}_\mathrm{B}$, which define the position and orientation of the B coordinate system with respect to the A system and which consist of six independent parameters, can be used to locate the rigid body in space with respect to the A reference frame. Because six independent parameters must be specified to define position and orientation, it is said that a rigid body in space possesses six degrees of freedom.

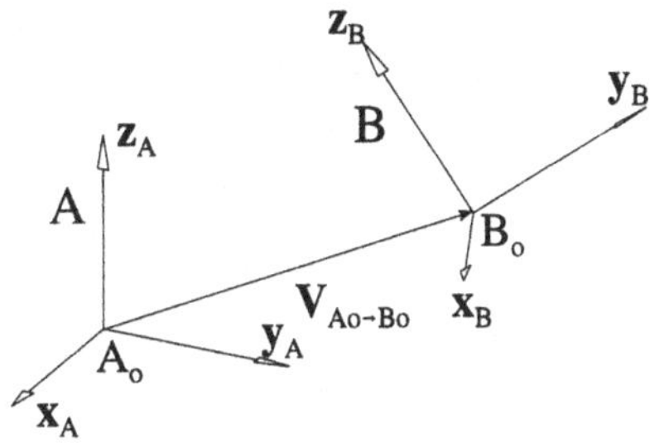

Figure 2.1. Two coordinate systems.

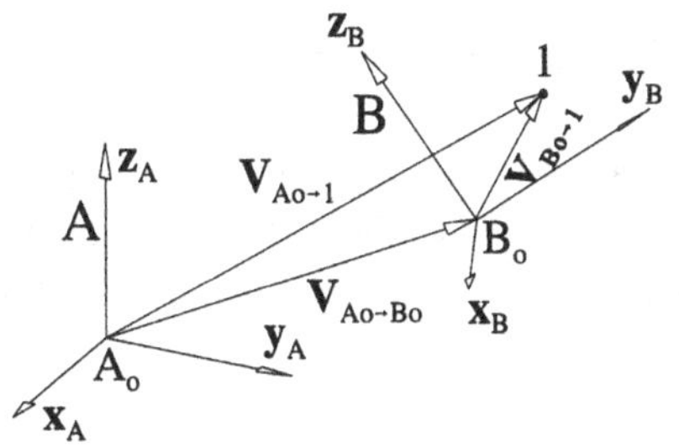

Figure 2.2. Depiction of point transformation problem.

2.2 Point transformations

From here on, the notation ${}^{I}\mathbf{P}_j$ is used to indicate the coordinates of a point j as measured in a coordinate system I. As such, ${}^{I}\mathbf{P}_j$ is a vector that begins at the origin of the I coordinate system and ends at point j and is thus equivalent to ${}^{I}\mathbf{V}_{I0\to j}$.

In many kinematic problems, the position of a point is known in terms of one coordinate system, and it is necessary to determine the position of the same point measured in another coordinate system. The problem statement is presented as follows (see Figure 2.2):

given: ${}^{B}\mathbf{P}_1$, the coordinates of point 1 measured in the B coordinate system (i.e., ${}^{B}\mathbf{V}_{B0\to}1$),
${}^{A}\mathbf{P}_{B0}$, the location of the origin of the B coordinate system measured with respect to the A coordinate system (i.e., ${}^{A}\mathbf{V}_{A0\to B0}$),
${}^{A}\mathbf{x}_B$, ${}^{A}\mathbf{y}_B$, ${}^{A}\mathbf{z}_B$, the orientation of the B coordinate system measured with respect to the A coordinate system,

find: ${}^{A}\mathbf{P}_1$, the coordinates of point 1 measured in the A coordinate system (i.e., ${}^{A}\mathbf{V}_{A0\to 1}$).

From triangle A_0-B_0-1 in Figure 2.2, it may be written that

$$\mathbf{V}_{A0\to 1} = \mathbf{V}_{A0\to B0} + \mathbf{V}_{B0\to 1}. \tag{2.7}$$

Evaluating all the vectors in terms of the A coordinate system gives

$${}^{A}\mathbf{V}_{A0\to 1} = {}^{A}\mathbf{V}_{A0\to B0} + {}^{A}\mathbf{V}_{B0\to 1}. \tag{2.8}$$

It is thus necessary to solve Eq. (2.8) for ${}^{A}\mathbf{V}_{A0\to 1}(= {}^{A}\mathbf{P}_1)$. The first term on the right side of Eq. (2.8) is a given quantity, that is, the coordinates of the origin of the B coordinate system as measured with respect to the A system. The second term on the right side of Eq. (2.8) is yet to be obtained.

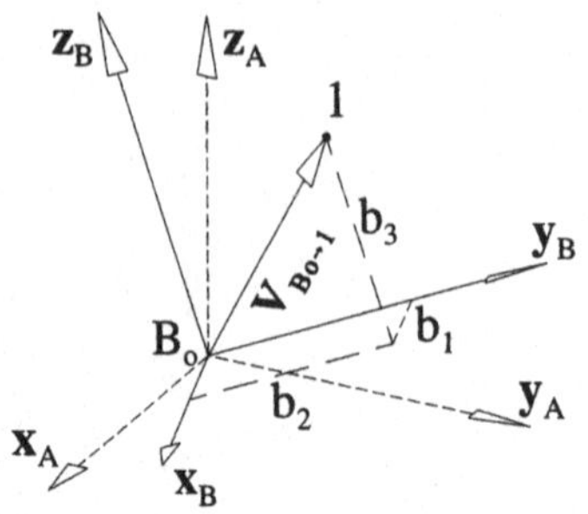

Figure 2.3. Point 1 projected onto the B coordinate system.

Figure 2.3 shows the vector $\mathbf{V}_{B0\to1}$ projected onto the coordinate axes of the B coordinate system. The directions of the coordinate axes of the A coordinate system are also shown as intermittent lines. The vector $\mathbf{V}_{B0\to1}$ may be written in terms of the B coordinate system as

$$^{B}\mathbf{V}_{B0\to1} = b_1{}^{B}\mathbf{x}_B + b_2{}^{B}\mathbf{y}_B + b_3{}^{B}\mathbf{z}_B, \tag{2.9}$$

where the components of $^{B}\mathbf{x}_B$, $^{B}\mathbf{y}_B$, and $^{B}\mathbf{z}_B$ are respectively $[1, 0, 0]^T$, $[0, 1, 0]^T$, and $[0, 0, 1]^T$. Because the vector $^{B}\mathbf{V}_{B0\to1}$ is a given quantity, the values of the scalars b_1, b_2, and b_3 are known. Further, the vector $\mathbf{V}_{B0\to1}$ can now be expressed in terms of the A coordinate system, and thus

$$^{A}\mathbf{V}_{B0\to1} = b_1{}^{A}\mathbf{x}_B + b_2{}^{A}\mathbf{y}_B + b_3{}^{A}\mathbf{z}_B. \tag{2.10}$$

Finally, substituting Eq. (2.10) into Eq. (2.8) gives

$$^{A}\mathbf{V}_{A0\to1} = {}^{A}\mathbf{V}_{A0\to B0} + b_1{}^{A}\mathbf{x}_B + b_2{}^{A}\mathbf{y}_B + b_3{}^{A}\mathbf{z}_B, \tag{2.11}$$

which can be further arranged in matrix form as

$$^{A}\mathbf{V}_{A0\to1} = {}^{A}\mathbf{V}_{A0\to B0} + \left[{}^{A}\mathbf{x}_B \quad {}^{A}\mathbf{y}_B \quad {}^{A}\mathbf{z}_B\right] \begin{bmatrix} b_1 \\ b_2 \\ b_3 \end{bmatrix}, \tag{2.12}$$

where $[{}^{A}\mathbf{x}_B\, {}^{A}\mathbf{y}_B\, {}^{A}\mathbf{z}_B]$ represents a 3×3 matrix that will be designated as

$$^{A}_{B}\mathbf{R} = \left[{}^{A}\mathbf{x}_B \quad {}^{A}\mathbf{y}_B \quad {}^{A}\mathbf{z}_B\right]. \tag{2.13}$$

Substituting $^{B}\mathbf{P}_1 = [b_1, b_2, b_3]^T$, $^{A}\mathbf{P}_1 = {}^{A}\mathbf{V}_{A0\to1}$, and $^{A}\mathbf{P}_{B0} = {}^{A}\mathbf{V}_{A0\to B0}$ yields

$$^{A}\mathbf{P}_1 = {}^{A}\mathbf{P}_{B0} + {}^{A}_{B}\mathbf{R}\,{}^{B}\mathbf{P}_1. \tag{2.14}$$

All terms on the right-hand side of Eq. (2.14) were given directly in the problem statement, where the vectors $^{A}\mathbf{x}_B$, $^{A}\mathbf{y}_B$, and $^{A}\mathbf{z}_B$ are the columns of the matrix $^{A}_{B}\mathbf{R}$.

2.3 4 × 4 transformation matrices

Equation (2.14) expresses the transformation of any point in one coordinate system to a reference coordinate system when the relative position and orientation of the pair of coordinate systems are known. The notation will be slightly modified, however, by introducing homogeneous coordinates.

In homogeneous coordinates, a three-dimensional point given by X, Y, and Z is represented by four scalar values, that is, x, y, z, and w. The three-dimensional and homogeneous coordinates are related by

$$X = x/w, \tag{2.15}$$

$$Y = y/w, \tag{2.16}$$

$$Z = z/w. \tag{2.17}$$

Thus, when $w = 1$, the first three components of the homogeneous coordinates of a point are the same as the three-dimensional coordinates of the point. Points at infinity occur when w equals zero, but these will not be encountered in this book.

By using homogeneous coordinates, Eq. (2.14) may be written as

$$\begin{bmatrix} {}^{A}\mathbf{P}_1 \\ 1 \end{bmatrix} = \begin{bmatrix} & {}^{A}_{B}\mathbf{R} & & {}^{A}\mathbf{P}_{B0} \\ 0 & 0 & 0 & 1 \end{bmatrix} \begin{bmatrix} {}^{B}\mathbf{P}_1 \\ 1 \end{bmatrix}, \tag{2.18}$$

where the matrix ${}^{A}_{B}\mathbf{R}$ and the vector ${}^{A}\mathbf{P}_{B0}$ form the first three rows of a 4 × 4 matrix. The equivalency of Eqs. (2.14) and (2.18) is most easily seen by representing the components of ${}^{A}_{B}\mathbf{R}$, ${}^{A}\mathbf{P}_{B0}$, and ${}^{B}\mathbf{P}_1$ symbolically and then performing the indicated multiplications and additions. The results will be the same for both equations.

The notation ${}^{A}_{B}\mathbf{T}$ will be used to represent the 4 × 4 matrix as

$${}^{A}_{B}\mathbf{T} = \begin{bmatrix} & {}^{A}_{B}\mathbf{R} & & {}^{A}\mathbf{P}_{B0} \\ 0 & 0 & 0 & 1 \end{bmatrix}. \tag{2.19}$$

The point transformation problem can now be written as

$${}^{A}\mathbf{P}_1 = {}^{A}_{B}\mathbf{T}\,{}^{B}\mathbf{P}_1. \tag{2.20}$$

It should be noted that in Eq. (2.14) all the vectors such as ${}^{A}\mathbf{P}_1$ are three dimensional, whereas in Eq. (2.20) each vector is expressed in homogeneous coordinates with $w = 1$.

2.4 Inverse of a transformation

Quite often during robot analyses, it will be necessary to obtain the inverse of a 4 × 4 transformation. In other words, given ${}^{A}_{B}\mathbf{T}$ it will be necessary to obtain ${}^{B}_{A}\mathbf{T}$. The definition of ${}^{A}_{B}\mathbf{T}$ was presented in Eq. (2.19). The matrix ${}^{A}_{B}\mathbf{R}$ is a 3 × 3 matrix whose columns are ${}^{A}\mathbf{x}_B$, ${}^{A}\mathbf{y}_B$, ${}^{A}\mathbf{z}_B$, that is, the coordinates of the unit axis vectors of the B coordinate system measured in the A coordinate system. The vector ${}^{A}\mathbf{P}_{B0}$ represents the coordinates of the origin of the B coordinate system measured with respect to the A coordinate system.

It should be clear that the inverse of ${}^{A}_{B}\mathbf{T}$ can be obtained from Eq. (2.19) by interchanging the letters A and B and that

$$ {}^{B}_{A}\mathbf{T} = \begin{bmatrix} & {}^{B}_{A}\mathbf{R} & & {}^{B}\mathbf{P}_{A0} \\ 0 & 0 & 0 & 1 \end{bmatrix}. \tag{2.21} $$

The inverse will be defined once the matrix ${}^{B}_{A}\mathbf{R}$ and the coordinates of the point ${}^{B}\mathbf{P}_{A0}$ are determined.

The matrix ${}^{A}_{B}\mathbf{R}$ can be written in the form

$$ {}^{A}_{B}\mathbf{R} = \begin{bmatrix} {}^{A}\mathbf{x}_B \cdot {}^{A}\mathbf{x}_A & {}^{A}\mathbf{y}_B \cdot {}^{A}\mathbf{x}_A & {}^{A}\mathbf{z}_B \cdot {}^{A}\mathbf{x}_A \\ {}^{A}\mathbf{x}_B \cdot {}^{A}\mathbf{y}_A & {}^{A}\mathbf{y}_B \cdot {}^{A}\mathbf{y}_A & {}^{A}\mathbf{z}_B \cdot {}^{A}\mathbf{y}_A \\ {}^{A}\mathbf{x}_B \cdot {}^{A}\mathbf{z}_A & {}^{A}\mathbf{y}_B \cdot {}^{A}\mathbf{z}_A & {}^{A}\mathbf{z}_B \cdot {}^{A}\mathbf{z}_A \end{bmatrix}, \tag{2.22} $$

where the components of ${}^{A}\mathbf{x}_A$, ${}^{A}\mathbf{y}_A$, and ${}^{A}\mathbf{z}_A$ are respectively $[1, 0, 0]^T$, $[0, 0, 1]^T$, and $[0, 0, 1]^T$. Each of the nine scalar terms of the 3×3 matrix ${}^{A}_{B}\mathbf{R}$ has been expressed in terms of a scalar product. A scalar product is an invariant operator that can be physically interpreted as being the cosine of the angle between the two unit vectors. The value of the scalar product will remain constant no matter what coordinate system the two vectors are expressed in. Thus,

$$ \begin{aligned} {}^{A}\mathbf{x}_B \cdot {}^{A}\mathbf{x}_A &= {}^{B}\mathbf{x}_B \cdot {}^{B}\mathbf{x}_A, \\ {}^{A}\mathbf{y}_B \cdot {}^{A}\mathbf{y}_A &= {}^{B}\mathbf{y}_B \cdot {}^{B}\mathbf{y}_A, \\ {}^{A}\mathbf{z}_B \cdot {}^{A}\mathbf{z}_A &= {}^{B}\mathbf{z}_B \cdot {}^{B}\mathbf{z}_A. \end{aligned} \tag{2.23} $$

Applying this to all the terms of ${}^{A}_{B}\mathbf{R}$ yields

$$ {}^{A}_{B}\mathbf{R} = \begin{bmatrix} {}^{B}\mathbf{x}_B \cdot {}^{B}\mathbf{x}_A & {}^{B}\mathbf{y}_B \cdot {}^{B}\mathbf{x}_A & {}^{B}\mathbf{z}_B \cdot {}^{B}\mathbf{x}_A \\ {}^{B}\mathbf{x}_B \cdot {}^{B}\mathbf{y}_A & {}^{B}\mathbf{y}_B \cdot {}^{B}\mathbf{y}_A & {}^{B}\mathbf{z}_B \cdot {}^{B}\mathbf{y}_A \\ {}^{B}\mathbf{x}_B \cdot {}^{B}\mathbf{z}_A & {}^{B}\mathbf{y}_B \cdot {}^{B}\mathbf{z}_A & {}^{B}\mathbf{z}_B \cdot {}^{B}\mathbf{z}_A \end{bmatrix}. \tag{2.24} $$

It can be seen that the rows of the 3×3 matrix ${}^{A}_{B}\mathbf{R}$ are ${}^{B}\mathbf{x}_A$, ${}^{B}\mathbf{y}_A$, ${}^{B}\mathbf{z}_A$ by recognizing that ${}^{B}\mathbf{x}_B = [1, 0, 0]^T$, ${}^{B}\mathbf{y}_B = [0, 1, 0]^T$, and ${}^{B}\mathbf{z}_B = [0, 0, 1]^T$. Thus, ${}^{A}_{B}\mathbf{R}$ can be written as

$$ {}^{A}_{B}\mathbf{R} = \begin{bmatrix} {}^{A}\mathbf{x}_B & {}^{A}\mathbf{y}_B & {}^{A}\mathbf{z}_B \end{bmatrix} = \begin{bmatrix} {}^{B}\mathbf{x}_A{}^{T} \\ {}^{B}\mathbf{y}_A{}^{T} \\ {}^{B}\mathbf{z}_A{}^{T} \end{bmatrix}. \tag{2.25} $$

The transpose of Eq. (2.25) is

$$ {}^{A}_{B}\mathbf{R}^{T} = \begin{bmatrix} {}^{B}\mathbf{x}_A & {}^{B}\mathbf{y}_A & {}^{B}\mathbf{z}_A \end{bmatrix}. \tag{2.26} $$

The columns of the 3×3 matrix in Eq. (2.26) are the unit vectors of the A coordinate system measured in terms of the B coordinate system. This is precisely the definition of ${}^{B}_{A}\mathbf{R}$. Thus, it can be concluded that

$$ {}^{B}_{A}\mathbf{R} = {}^{A}_{B}\mathbf{R}^{T}. \tag{2.27} $$

The remaining term to be determined is ${}^{B}\mathbf{P}_{A0}$. This term can readily be calculated from ${}^{A}\mathbf{P}_{B0}$ now that ${}^{B}_{A}\mathbf{R}$ is known. First, the vector ${}^{A}\mathbf{P}_{B0}$ will be transformed to the B coordinate system by utilizing Eq. (2.14) as

$$^{B}\mathbf{P}_{B0} = {}^{B}_{A}\mathbf{R}\,^{A}\mathbf{P}_{B0} + {}^{B}\mathbf{P}_{A0}. \tag{2.28}$$

Now $^{B}\mathbf{P}_{B0} = [0, 0, 0]^{T}$, which are the coordinates of the origin of the B coordinate system measured in the B system. Substituting this result into Eq. (2.28) and rearranging yields

$$^{B}\mathbf{P}_{A0} = -{}^{B}_{A}\mathbf{R}\,^{A}\mathbf{P}_{B0} = -{}^{A}_{B}\mathbf{R}^{T}\,^{A}\mathbf{P}_{B0}. \tag{2.29}$$

Substituting Eqs. (2.27) and (2.29) into Eq. (2.21) yields the final result

$$^{B}_{A}\mathbf{T} = \begin{bmatrix} & {}^{A}_{B}\mathbf{R}^{T} & & -{}^{A}_{B}\mathbf{R}^{T}\,^{A}\mathbf{P}_{B0} \\ 0 & 0 & 0 & 1 \end{bmatrix}. \tag{2.30}$$

2.5 Compound transformations

In Figure 2.4, the coordinates of point 1 are known in terms of the C coordinate system. The position and orientation of the C coordinate system is known relative to the B coordinate system. The position and orientation of the B coordinate system is known in terms of the A coordinate system. The objective is to determine the coordinates of point 1 in terms of the A coordinate system.

From this problem description, it should be apparent that the transformations ${}^{B}_{C}\mathbf{T}$ and ${}^{A}_{B}\mathbf{T}$ are known. Thus, the problem can be solved in two steps. First, the coordinates of point 1 in the B coordinate system can be found from

$$^{B}\mathbf{P}_{1} = {}^{B}_{C}\mathbf{T}\,^{C}\mathbf{P}_{1}. \tag{2.31}$$

Then, the final answer can be obtained from

$$^{A}\mathbf{P}_{1} = {}^{A}_{B}\mathbf{T}\,^{B}\mathbf{P}_{1}. \tag{2.32}$$

Combining these two equations yields

$$^{A}\mathbf{P}_{1} = {}^{A}_{B}\mathbf{T}\,{}^{B}_{C}\mathbf{T}\,^{C}\mathbf{P}_{1}. \tag{2.33}$$

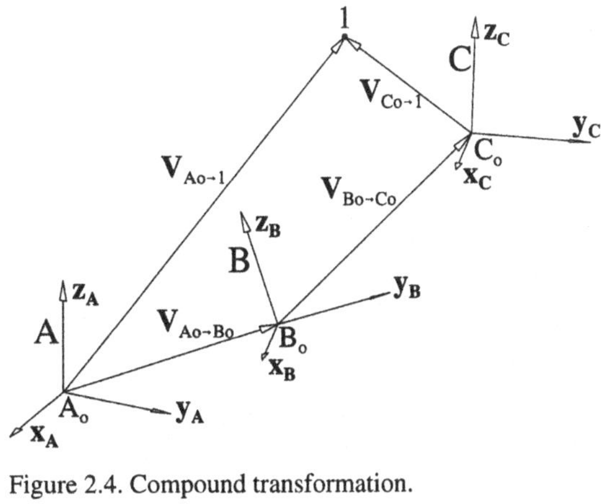

Figure 2.4. Compound transformation.

The term ${}^{A}_{B}\mathbf{T}\,{}^{B}_{C}\mathbf{T}$ transfers a point directly from the C coordinate system to the A coordinate system. Thus, it can be inferred that

$$ {}^{A}_{C}\mathbf{T} = {}^{A}_{B}\mathbf{T}\,{}^{B}_{C}\mathbf{T}. \tag{2.34} $$

The ability to perform matrix multiplication to yield compound transformations is the primary reason why the 4×4 transformation notation is used.

2.6 Standard transformations

In many problems, the relationship between coordinate systems will be defined in terms of rotations about the X, Y, or Z axes. A typical problem statement would be as follows:

given: (1) Coordinate system B is initially aligned with coordinate system A.
(2) Coordinate system B is then rotated α degrees about the X axis.
find: ${}^{A}_{B}\mathbf{R}$ (often written as $\mathbf{R}_{X,\alpha}$) .

Figure 2.5 shows the A and B coordinate systems. By projection, it can be seen that

$$ {}^{A}\mathbf{x}_{B} = \begin{bmatrix} 1 \\ 0 \\ 0 \end{bmatrix}, \tag{2.35} $$

$$ {}^{A}\mathbf{y}_{B} = \begin{bmatrix} 0 \\ \cos\alpha \\ \sin\alpha \end{bmatrix}, \tag{2.36} $$

$$ {}^{A}\mathbf{z}_{B} = \begin{bmatrix} 0 \\ -\sin\alpha \\ \cos\alpha \end{bmatrix}. \tag{2.37} $$

Thus,

$$ {}^{A}_{B}\mathbf{R} = \mathbf{R}_{X,\alpha} = \begin{bmatrix} 1 & 0 & 0 \\ 0 & \cos\alpha & -\sin\alpha \\ 0 & \sin\alpha & \cos\alpha \end{bmatrix}. \tag{2.38} $$

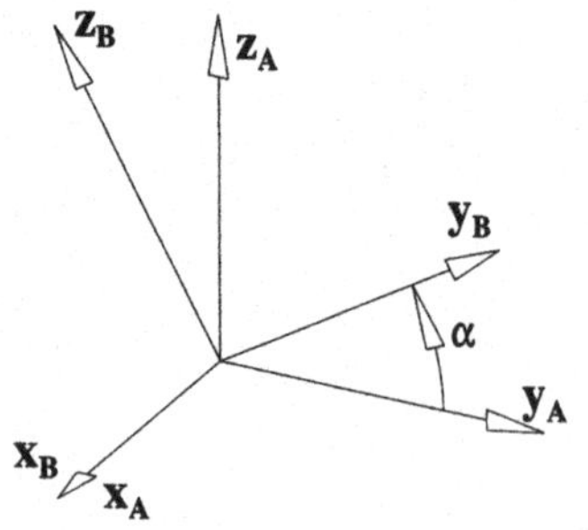

Figure 2.5. Rotation about the X axis.

The problem will now be repeated for rotations about the Y and Z axes.

given: (1) Coordinate system B is initially aligned with coordinate system A.
(2) Coordinate system B is then rotated β degrees about the Y axis.
find: ${}^{A}_{B}\mathbf{R}$ (often written as $\mathbf{R}_{y,\beta}$) .

Again by projection it can be shown that

$$ {}^{A}_{B}\mathbf{R} = \mathbf{R}_{y,\beta} = \begin{bmatrix} \cos\beta & 0 & \sin\beta \\ 0 & 1 & 0 \\ -\sin\beta & 0 & \cos\beta \end{bmatrix} . \tag{2.39} $$

Lastly,

given: (1) Coordinate system B is initially aligned with coordinate system A.
(2) Coordinate system B is then rotated γ degrees about the Z axis.
find: ${}^{A}_{B}\mathbf{R}$ (often written as $\mathbf{R}_{z,\gamma}$).

By projection it can be shown that

$$ {}^{A}_{B}\mathbf{R} = \mathbf{R}_{z,\gamma} = \begin{bmatrix} \cos\gamma & -\sin\gamma & 0 \\ \sin\gamma & \cos\gamma & 0 \\ 0 & 0 & 1 \end{bmatrix} . \tag{2.40} $$

2.7 Example problem

Coordinate system B is initially aligned with coordinate system A. It is translated to the point $[5, 4, 1]^T$ and then rotated 30 degrees about its X axis. Lastly, the coordinate system is rotated 60 degrees about an axis that passes through the point $[2, 0, 2]^T$, measured in the current coordinate system, which is parallel to the Y axis. Find ${}^{A}_{B}\mathbf{T}$.

Figure 2.6 shows the A coordinate system and the B coordinate system after all the operations have been performed. Figure 2.7 also shows four other intermediate coordinate systems labeled C, D, E, and F.

Coordinate system C was initially aligned with coordinate system A and was then translated to the point $[5, 4, 1]^T$. Thus,

$$ {}^{A}_{C}\mathbf{T} = \begin{bmatrix} 1 & 0 & 0 & 5 \\ 0 & 1 & 0 & 4 \\ 0 & 0 & 1 & 1 \\ 0 & 0 & 0 & 1 \end{bmatrix} . \tag{2.41} $$

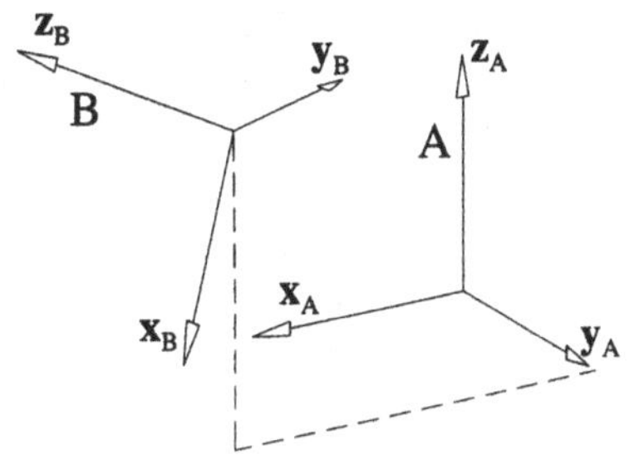

Figure 2.6. Initial and final coordinate systems.

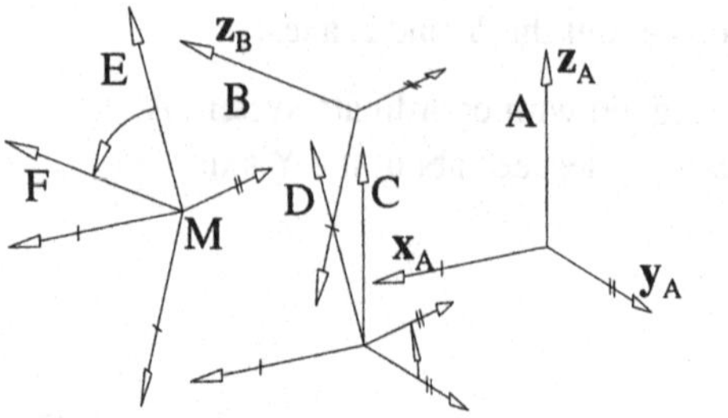

Figure 2.7. Intermediate coordinate systems.

Coordinate system D was initially aligned with coordinate system C and was then rotated 30 degrees about the X axis. Thus,

$$ {}^{C}_{D}\mathbf{T} = \begin{bmatrix} 1 & 0 & 0 & 0 \\ 0 & \cos 30 & -\sin 30 & 0 \\ 0 & \sin 30 & \cos 30 & 0 \\ 0 & 0 & 0 & 1 \end{bmatrix}. \tag{2.42} $$

The last modification to the coordinate system was a rotation of 60 degrees about an axis parallel to the Y axis, which passes through the point $[2, 0, 2]^T$, measured in terms of the D coordinate system. The point that the axis of rotation passes through will be called point M. From observation (see Figure 2.7), it is apparent that ${}^{D}\mathbf{P}_M = {}^{B}\mathbf{P}_M = [2, 0, 2]^T$. Because of this fact, the transformation that relates coordinate system D and coordinate system B will be formed by translating to the point $[2, 0, 2]^T$ (see coordinate system E), rotating 60 degrees about the current Y axis (see coordinate system F), and then translating $[-2, 0, -2]^T$ to obtain coordinate system B. The transformations that accomplish this are

$$ {}^{D}_{E}\mathbf{T} = \begin{bmatrix} 1 & 0 & 0 & 2 \\ 0 & 1 & 0 & 0 \\ 0 & 0 & 1 & 2 \\ 0 & 0 & 0 & 1 \end{bmatrix}, \tag{2.43} $$

$$ {}^{E}_{F}\mathbf{T} = \begin{bmatrix} \cos 60 & 0 & \sin 60 & 0 \\ 0 & 1 & 0 & 0 \\ -\sin 60 & 0 & \cos 60 & 0 \\ 0 & 0 & 0 & 1 \end{bmatrix}, \tag{2.44} $$

$$ {}^{F}_{B}\mathbf{T} = \begin{bmatrix} 1 & 0 & 0 & -2 \\ 0 & 1 & 0 & 0 \\ 0 & 0 & 1 & -2 \\ 0 & 0 & 0 & 1 \end{bmatrix}. \tag{2.45} $$

The overall transformation ${}^{A}_{B}\mathbf{T}$ can be calculated as

$$ {}^{A}_{B}\mathbf{T} = {}^{A}_{C}\mathbf{T}\,{}^{C}_{D}\mathbf{T}\,{}^{D}_{E}\mathbf{T}\,{}^{E}_{F}\mathbf{T}\,{}^{F}_{B}\mathbf{T}. \tag{2.46} $$

The numerical value of ${}^{A}_{B}\mathbf{T}$ is

$$ {}^{A}_{B}\mathbf{T} = \begin{bmatrix} 0.5 & 0 & 0.866 & 4.268 \\ 0.433 & 0.866 & -0.25 & 2.634 \\ -0.75 & 0.5 & 0.433 & 3.366 \\ 0 & 0 & 0 & 1 \end{bmatrix}. \tag{2.47} $$

2.8 General transformations

Two types of additional problems dealing with transformations often occur. In the first, coordinate system B is initially aligned with coordinate system A. An axis and angle of rotation are given about which coordinate system B will be rotated. The objective is to determine ${}^{A}_{B}\mathbf{R}$. The second problem is the opposite. A rotation matrix ${}^{A}_{B}\mathbf{R}$ is given, and it is desired to determine the axis and angle of rotation that is represented by the matrix. Solutions to both problems will be presented in this section.

2.8.1 Determination of equivalent rotation matrix

In this problem, it is assumed that an axis of rotation represented by the unit vector $\mathbf{m} = [m_x, m_y, m_z]^T$ and an angle of rotation, θ, are known. A coordinate system B is initially aligned with a coordinate system A. It is then rotated by an angle θ about the axis $\mathbf{m}$ which passes through the origin (see Figure 2.8). It is desired to find the rotation matrix, ${}^{A}_{B}\mathbf{R}$.

The problem will be solved by first introducing a coordinate system C whose Z axis is parallel to the vector $\mathbf{m}$. The relationship between the A and C coordinate systems can be written as

$$ {}^{A}_{C}\mathbf{R} = \begin{bmatrix} a_x & b_x & m_x \\ a_y & b_y & m_y \\ a_z & b_z & m_z \end{bmatrix}. \tag{2.48} $$

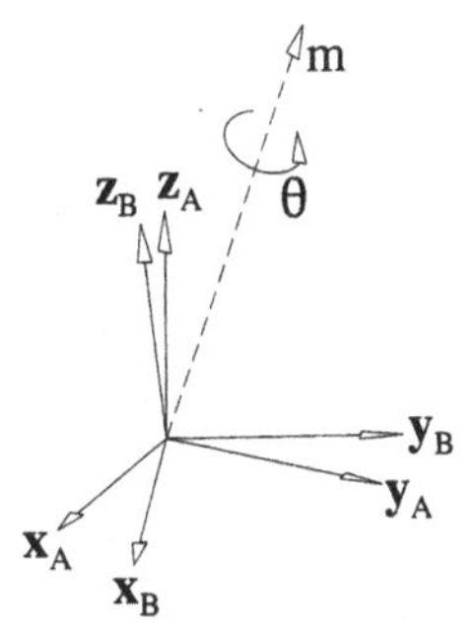

Figure 2.8. Rotation of angle θ about axis $\mathbf{m}$.

Only the terms m_x, m_y, and m_z are known in this equation. A second coordinate system, D, is now aligned with coordinate system C. It is then rotated by an angle θ about its Z axis. The relationship between coordinate systems C and D is

$$ {}^{C}_{D}\mathbf{R} = \begin{bmatrix} \cos\theta & -\sin\theta & 0 \\ \sin\theta & \cos\theta & 0 \\ 0 & 0 & 1 \end{bmatrix}. \tag{2.49}$$

It should be noted that the A and C coordinate systems are essentially rotated together as a single rigid body. In this way the A coordinate system is transformed into the B coordinate system while the C coordinate system is transformed into the D coordinate system. Hence, ${}^{A}_{C}\mathbf{R} = {}^{B}_{D}\mathbf{R}$ and therefore ${}^{D}_{B}\mathbf{R} = {}^{A}_{C}\mathbf{R}^{T}$. The rotation matrix that relates the A and B coordinate system may now be written as

$$ {}^{A}_{B}\mathbf{R} = {}^{A}_{C}\mathbf{R}\,{}^{C}_{D}\mathbf{R}\,{}^{D}_{B}\mathbf{R}. \tag{2.50}$$

Expanding the right side of Eq. (2.50) yields

$$ {}^{A}_{B}\mathbf{R} = \begin{bmatrix} a_x & b_x & m_x \\ a_y & b_y & m_y \\ a_z & b_z & m_z \end{bmatrix} \begin{bmatrix} \cos\theta & -\sin\theta & 0 \\ \sin\theta & \cos\theta & 0 \\ 0 & 0 & 1 \end{bmatrix} \begin{bmatrix} a_x & a_y & a_z \\ b_x & b_y & b_z \\ m_x & m_y & m_z \end{bmatrix}. \tag{2.51}$$

Performing the matrix multiplication and substituting $s = \sin\theta$ and $c = \cos\theta$ yields

$$ {}^{A}_{B}\mathbf{R} = \begin{bmatrix} a_x c + b_x s & -a_x s + b_x c & m_x \\ a_y c + b_y s & -a_y s + b_y c & m_y \\ a_z c + b_z s & -a_z s + b_z c & m_z \end{bmatrix} \begin{bmatrix} a_x & a_y & a_z \\ b_x & b_y & b_z \\ m_x & m_y & m_z \end{bmatrix}. \tag{2.52}$$

$$ {}^{A}_{B}\mathbf{R} = \begin{bmatrix} c(a_x^2+b_x^2)+m_x^2 & c(a_y a_x+b_y b_x)+s(a_y b_x - b_y a_x)+m_x m_y & c(a_z a_x+b_z b_x)+s(a_z b_x - b_z a_x)+m_x m_z \\ c(a_x a_y+b_x b_y)+s(a_x b_y - b_x a_y)+m_x m_y & c(a_y^2+b_y^2)+m_y^2 & c(a_z a_y+b_z b_y)+s(a_z b_y - b_z a_y)+m_y m_z \\ c(a_x a_z+b_x b_z)+s(a_x b_z - b_x a_z)+m_x m_z & c(a_y a_z+b_y b_z)+s(a_y b_z - b_y a_z)+m_y m_z & c(a_z^2+b_z^2)+m_z^2 \end{bmatrix}. \tag{2.53}$$

It must be pointed out that the terms in Eq. (2.53) are not all known. The terms a_x, a_y, a_z, b_x, b_y, and b_z have not been specified. Three facts should be remembered, however. First, the columns (and rows) of ${}^{A}_{C}\mathbf{R}$ are unit vectors. Second, the columns (and rows) of ${}^{A}_{C}\mathbf{R}$ are orthogonal to one another. Third, the last column of ${}^{A}_{C}\mathbf{R}$ can be calculated as the cross product of the first two columns. These facts will be used to simplify Eq. (2.53) and eliminate the unknown terms.

${}^{A}_{B}\mathbf{R}_{1,1}$ represents the element in the first row and first column of Eq. (2.53). This term is written as

$$ {}^{A}_{B}\mathbf{R}_{1,1} = c\left(a_x^2 + b_x^2\right) + m_x^2. \tag{2.54}$$

Because the first row of the matrix in Eq. (2.48) is a unit vector,

$$a_x^2 + b_x^2 + m_x^2 = 1. \quad (2.55)$$

Substituting for $(a_x^2 + b_x^2)$ reduces Eq. (2.54) to the following

$${}^A_B\mathbf{R}_{1,1} = c(1 - m_x^2) + m_x^2 = c + m_x^2(1 - c). \quad (2.56)$$

The element in the second row, first column of the matrix in Eq. (2.53) is written as

$${}^A_B\mathbf{R}_{2,1} = c(a_x a_y + b_x b_y) + s(a_x b_y - b_x a_y) + m_x m_y. \quad (2.57)$$

Because the first two rows of the matrix in Eq. (2.48) are orthogonal, the following equation may be written

$$a_x a_y + b_x b_y + m_x m_y = 0. \quad (2.58)$$

Further, because the third column of the matrix in Eq. (2.48) can be generated as the cross product of the first two columns, the following expression can be written

$$m_z = a_x b_y - b_x a_y. \quad (2.59)$$

Regrouping Eq. (2.58) and then substituting it and Eq. (2.59) into Eq. (2.57) yields

$${}^A_B\mathbf{R}_{2,1} = c(-m_x m_y) + s(m_z) + m_x m_y = m_x m_y(1 - c) + m_z s. \quad (2.60)$$

Similar substitutions may be made on the remaining elements of the matrix ${}^A_B\mathbf{R}$ to eliminate the unknown terms. The final result for the matrix ${}^A_B\mathbf{R}$ is

$${}^A_B\mathbf{R} = \begin{bmatrix} m_x m_x v + c & m_x m_y v - m_z s & m_x m_z v + m_y s \\ m_x m_y v + m_z s & m_y m_y v + c & m_y m_z v - m_x s \\ m_x m_z v - m_y s & m_y m_z v + m_x s & m_z m_z v + c \end{bmatrix}, \quad (2.61)$$

where s and c represent the sine and cosine of θ, and v represents $(1-\cos\theta)$. It is interesting to note that Eq. (2.61) reduces to the matrices expressed in Eqs. (2.38), (2.39), and (2.40) when the axis vector **m** is aligned with the X, Y, and Z coordinate axes respectively.

2.8.2 Determination of axis and angle of rotation

For this problem, it is assumed that a rotation matrix, ${}^A_B\mathbf{R}$, is given, and it is desired to calculate the axis vector, **m**, and the angle of rotation about this axis that would rotate coordinate system A so as to align it with coordinate system B. The rotation matrix may be written as

$${}^A_B\mathbf{R} = \begin{bmatrix} r_{11} & r_{12} & r_{13} \\ r_{21} & r_{22} & r_{23} \\ r_{31} & r_{32} & r_{33} \end{bmatrix}. \quad (2.62)$$

Eq. (2.61) shows how the elements of the rotation matrix can be written in terms of the axis vector **m** and the rotational angle θ. Summing the diagonal elements of the matrices in Eqs. (2.61) and (2.62) and equating the results gives

$$r_{11} + r_{22} + r_{33} = (1 - \cos\theta)(m_x^2 + m_y^2 + m_z^2) + 3\cos\theta. \tag{2.63}$$

Because the axis vector **m** can be considered to be a unit vector, Eq. (2.63) reduces to

$$r_{11} + r_{22} + r_{33} = 1 + 2\cos\theta. \tag{2.64}$$

Solving for $\cos\theta$ gives

$$\cos\theta = \frac{r_{11} + r_{22} + r_{33} - 1}{2}. \tag{2.65}$$

The angle θ is not uniquely defined by Eq. (2.65). Two distinct values of θ in the range of $-\pi$ to $+\pi$ exist that will satisfy this equation. The value of θ that lies in the range of 0 to π will be selected, however, and the unique corresponding axis of rotation will be computed. (Had the value of θ in the range of $-\pi$ to 0 been selected, the resulting rotation axis would point in the opposite direction to the one that will be computed.)

Subtracting the off-diagonal elements of the matrices of Eqs. (2.61) and (2.62) and equating the results yields

$$r_{21} - r_{12} = 2m_z \sin\theta, \tag{2.66}$$

$$r_{13} - r_{31} = 2m_y \sin\theta, \tag{2.67}$$

$$r_{32} - r_{23} = 2m_x \sin\theta. \tag{2.68}$$

The components of the axis vector **m** can be readily computed from these equations.

When the rotation angle θ is very small, the axis vector **m** is not well defined, because the ratios used to compute the vector components all approach $\frac{0}{0}$. When the rotation angle approaches π, the ratios again approach $\frac{0}{0}$. In this case, however, the axis vector is well defined and the problem can be reformulated to obtain an accurate solution.

Equating the diagonal elements of the matrices of Eqs. (2.61) and (2.62) yields

$$r_{11} = m_x^2(1 - \cos\theta) + \cos\theta, \tag{2.69}$$

$$r_{22} = m_y^2(1 - \cos\theta) + \cos\theta, \tag{2.70}$$

$$r_{33} = m_z^2(1 - \cos\theta) + \cos\theta. \tag{2.71}$$

Solving these equations for m_x, m_y, and m_z gives

$$m_x = \pm\sqrt{\frac{r_{11} - \cos\theta}{1 - \cos\theta}}, \tag{2.72}$$

$$m_y = \pm\sqrt{\frac{r_{22} - \cos\theta}{1 - \cos\theta}}, \tag{2.73}$$

$$m_z = \pm\sqrt{\frac{r_{33} - \cos\theta}{1 - \cos\theta}}. \tag{2.74}$$

Because the angle θ has been determined, the remaining issue for each term is whether the positive or negative sign should be used in the equations. Equations (2.66) through (2.68) will be used to determine this information. Because the angle θ was selected to be in the range of 0 to π, $\sin\theta$ will be greater than zero. From Eq. (2.68), it is now obvious that m_x will be positive if the term $r_{32} - r_{23}$ is positive. The sign for the terms m_y and m_z can be deduced in a similar fashion from Eqs. (2.67) and (2.66).

In theory, the axis vector **m** has been determined for the case when θ approaches π. Experience has shown, however, that a numerically more accurate answer results if only the largest-magnitude component of **m** is calculated from Eqs. (2.72) through (2.74). The remaining components can be determined from the following equations, which are obtained by summing the off-diagonal elements of the matrices in Eqs. (2.61) and (2.62) and equating them:

$$r_{12} + r_{21} = 2m_x m_y(1 - \cos\theta), \tag{2.75}$$

$$r_{13} + r_{31} = 2m_x m_z(1 - \cos\theta), \tag{2.76}$$

$$r_{23} + r_{32} = 2m_y m_z(1 - \cos\theta). \tag{2.77}$$

Thus, if the absolute value of m_x as calculated in Eq. (2.72) is larger than the absolute values of m_y and m_z as calculated in Eqs. (2.73) and (2.74), then a more accurate answer for m_y can be obtained by using Eq. (2.75), and a more accurate answer for m_z can be obtained by using Eq. (2.76).

2.9 Summary

This chapter addressed the problem of how to describe the position and orientation of one coordinate system relative to another. It was shown that a convenient representation for position is the specification of the location of the origin of the second coordinate system relative to the first. Orientation can be defined by specifying the coordinates of the unit axis vectors of the second coordinate system measured in the first coordinate system.

It was also shown that the selected method of describing relative position and orientation could be used to easily transform a point between coordinate systems. Homogeneous coordinates were introduced, and the point transformation matrix was expressed as a compact 4×4 matrix.

The point transformation methods introduced in this chapter will be used extensively in the analysis of robot manipulators that follows. Thus, the material presented in this chapter will form the foundation on which the forthcoming three-dimensional kinematic analyses will be based.

2.10 Problems

1. Under what conditions will ${}^{A}_{B}\mathbf{R}$ equal ${}^{B}_{A}\mathbf{R}$?

2. A coordinate system {B} is initially coincident with coordinate system {A}. It is rotated by an angle θ about the X axis and then subsequently rotated by an angle β

about its new Y axis. Determine the orientational relationship of {B} with respect to {A}, ${}^{A}_{B}\mathbf{R}$.

3. The following transformation definitions are given:

$$
{}^{B}_{A}\mathbf{T} = \begin{bmatrix} \frac{\sqrt{3}}{2} & 0 & 0.5 & 20 \\ 0 & -1 & 0 & 0 \\ 0.5 & 0 & -\frac{\sqrt{3}}{2} & 0 \\ 0 & 0 & 0 & 1 \end{bmatrix}, \quad {}^{C}_{A}\mathbf{T} = \begin{bmatrix} \frac{\sqrt{2}}{2} & \frac{\sqrt{2}}{2} & 0 & 0 \\ \frac{\sqrt{2}}{2} & -\frac{\sqrt{2}}{2} & 0 & 0 \\ 0 & 0 & -1 & 10 \\ 0 & 0 & 0 & 1 \end{bmatrix},
$$

$$
{}^{D}_{B}\mathbf{T} = \begin{bmatrix} 1 & 0 & 0 & 0 \\ 0 & \frac{\sqrt{3}}{2} & 0.5 & 10 \\ 0 & -0.5 & \frac{\sqrt{3}}{2} & 0 \\ 0 & 0 & 0 & 1 \end{bmatrix}.
$$

(a) Determine the transformation ${}^{C}_{D}\mathbf{T}$.

(b) The coordinates of point number 1 are $[20, -30, 5]^T$ measured in the D coordinate system. Determine the coordinates of this point as measured in the A, B, and C coordinate systems.

4. Coordinate systems A and B are initially coincident. Coordinate system B is then rotated sixty degrees about a vector parallel to $[2, 4, 7]^T$, which passes through the point $[3, 4, -2]^T$. Determine the transformation ${}^{A}_{B}\mathbf{T}$.

5. The origins of coordinate systems A and B are coincident. You are given the coordinates of three points in the A and B coordinate systems, that is, ${}^{A}\mathbf{P}_1$, ${}^{A}\mathbf{P}_2$, ${}^{A}\mathbf{P}_3$, ${}^{B}\mathbf{P}_1$, ${}^{B}\mathbf{P}_2$, ${}^{B}\mathbf{P}_3$. Determine the rotation matrix ${}^{A}_{B}\mathbf{R}$.

6. The coordinates of point 1 as seen from the A coordinate system are $\begin{bmatrix} 2 \\ 8 \\ 8 \end{bmatrix}$. The coordinates of the same point as seen from the B coordinate system are $\begin{bmatrix} 12 \\ 20 \\ -8 \end{bmatrix}$. The B coordinate system can be obtained by initially aligning it with the A system, translating to a point, and then rotating forty degrees about the Z axis.

(a) Determine the coordinates of the origin of the B coordinate system measured in the A coordinate system on the basis of the given information.

(b) Determine the coordinates of the origin of the A coordinate system measured in the B coordinate system.

7. The transformation that relates the A and B coordinate systems is given as

$$ {}^{A}_{B}\mathbf{T} = \begin{bmatrix} 0.866025 & 0 & 0.5 & 0.26795 \\ 0 & 1 & 0 & 0 \\ -0.5 & 0 & 0.866025 & 1 \\ 0 & 0 & 0 & 1 \end{bmatrix}. $$

Coordinate system B can be obtained from coordinate system A by initially aligning it with A and then rotating coordinate system B about an axis **m** by an angle γ where the rotation axis passes through a point **p**. Determine **m**, γ, and **p**.

8. Coordinate system B is initially aligned with coordinate system A. It is then rotated thirty degrees about an axis that is parallel to the X axis but that passes through the point $[10, 20, 10]^T$.

Coordinate system C is initially aligned with coordinate system A. It is then rotated sixty degrees about an axis $[2, 4, 6]^T$ that passes through the origin.

Determine the transformation that relates the C and B coordinate systems, that is, ${}^{B}_{C}\mathbf{T}$.

9. Coordinate systems A and B are initially aligned and coincident. Coordinate system B is then rotated by an angle of thirty-five degrees about its X axis. It is then rotated 120 degrees about its new Y axis. You wish to return coordinate system B to its origin orientation (aligned with coordinate system A) by performing one rotation. About what axis and by what angle should B be rotated?

10. Write two computer functions named matmult and vecmult that will perform matrix multiplication and matrix and vector multiplication. The C language prototypes for these functions are as follows:

```
void matmult (double ans[4][4], double matrix1[4][4], double matrix2[4][4]);
void vecmult (double ans[4], double matrix1[4][4], double vector1[4]);
```

The function matmult will accept as input two 4×4 matrices, that is, matrix1 and matrix2. The product of matrix1 times matrix2 will be calculated, and the resulting 4×4 matrix will be returned by the function via the parameter ans.

The function vecmult will accept as input one 4×4 matrix and one 4×1 vector, that is, matrix1 and vector1. The product of the matrix times the vector will be calculated, and the resulting 4×1 vector will be returned by the function via the parameter ans. Test your functions by calling them from a main program.

11. Write a computer function named invert_transform that will calculate the inverse of a 4×4 transformation matrix. The C language prototype for this function is as follows:

```
void invert_transform (double result[4][4], double tran[4][4]);
```

The parameter tran will be a 4×4 transformation matrix that is input to the function. The inverse of tran will be calculated and returned via the parameter result. Test your function by calling it from a main program.

3

Manipulator kinematics

3.1 Introduction

The previous chapter introduced point-to-point transformations, that is, the coordinates for a point expressed in some coordinate system were expressed in a second coordinate system. These transformations will now be applied to serial robot manipulators. In this chapter, a spatial link will be defined. Then, different types of joints that can interconnect these spatial links will be discussed. Finally, a standard method of specifying a coordinate system for each link will be introduced together with the transformations that relate these coordinate systems.

3.2 Spatial link

In this text, the term "robot manipulator" will be defined as a serial assemblage (or chain) of links and joints. One end is connected to ground, and at the free end is attached an end effector or gripping device. It will be assumed that a link is a rigid body. Figure 3.1 illustrates a link connecting a pair of consecutive joint axes that are in general skew, labeled with unit directional vectors $\mathbf{S}_i$ and $\mathbf{S}_j$. Two scalar parameters, the link length a_{ij} and the twist angle α_{ij}, define the relative position of this pair of skew axes. The link length is the mutual perpendicular distance between the axes, and the twist angle is the angle between the vectors $\mathbf{S}_i$ and $\mathbf{S}_j$. The unit vector $\mathbf{a}_{ij}$ is defined by $\mathbf{S}_i \times \mathbf{S}_j = \mathbf{a}_{ij} \sin\alpha_{ij}$ as shown in Figure 3.1. Clearly, the choice of the directions of the unit vectors $\mathbf{S}_i$ and $\mathbf{S}_j$ is arbitrary, that is, either $\mathbf{S}_i$ or $\mathbf{S}_j$ can be drawn in the opposite direction. However the cross product $\mathbf{S}_i \times \mathbf{S}_j$ (or $\mathbf{S}_j \times \mathbf{S}_i$) will always determine the direction of $\mathbf{a}_{ij}$, and then α_{ij} is measured in a right-hand sense about $\mathbf{a}_{ij}$.

A kinematic model of a serial manipulator is made by replacing each physical link of the robot with a link drawn along the vector $\mathbf{a}_{ij}$. This is because the physical shape of the actual link is of no geometrical importance. Rather, the geometry of a link is defined by the directions of the vectors $\mathbf{a}_{ij}$, $\mathbf{S}_i$, and $\mathbf{S}_j$ together with the link length $\mathbf{a}_{ij}$ and the twist angle α_{ij}. Figure 3.2 shows the kinematically equivalent link for the physical link shown in Figure 3.1.

Figures 3.3 and 3.4 show two special cases. The first occurs when the perpendicular distance between the vectors $\mathbf{S}_i$ and $\mathbf{S}_j$ is zero, that is, $\mathbf{S}_i$ and $\mathbf{S}_j$ intersect at a finite point. This is called a spherical link. The second occurs when the twist angle, α_{ij}, is zero or π. In this case the link is planar and the vectors $\mathbf{S}_i$ and $\mathbf{S}_j$ are parallel or antiparallel, that is, $\mathbf{S}_i$ and $\mathbf{S}_j$ intersect at a point at infinity.

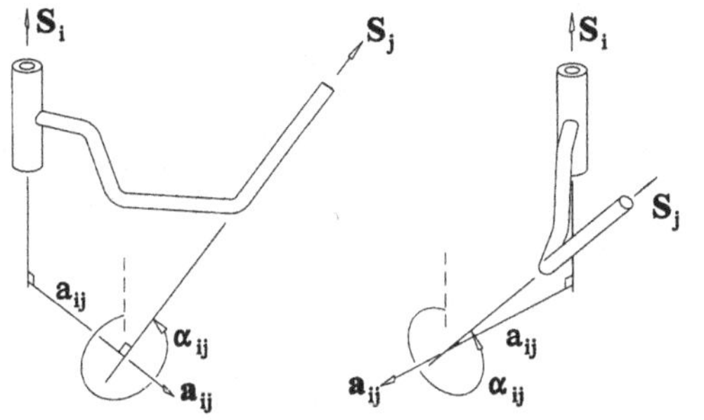

Figure 3.1. Two views of a spatial link.

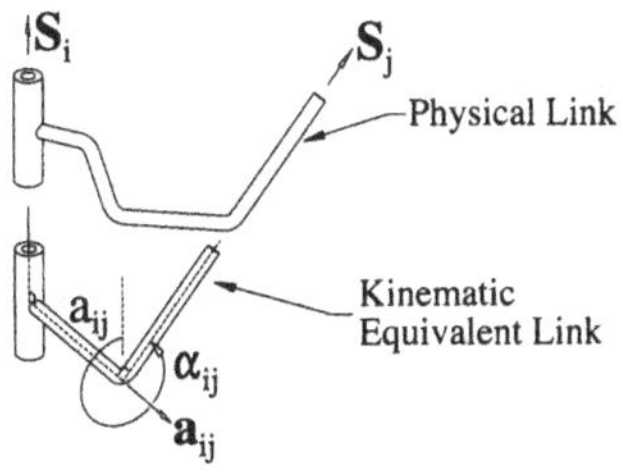

Figure 3.2. Kinematic link.

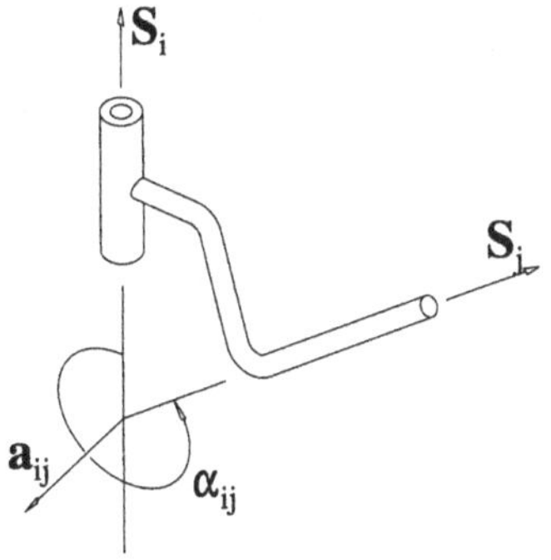

Figure 3.3. Spherical link.

In summary, a link is defined by two scalar parameters, the link length, a_{ij}, and the twist angle, α_{ij}. The direction for the unit vector $\mathbf{a}_{ij}$ is determined by $\mathbf{S}_i \times \mathbf{S}_j = \mathbf{a}_{ij} \sin \alpha_{ij}$, which automatically defines the twist angle.

3.3 Joints

The nature of the relative motion between a pair of successive links is determined by the type of connecting joint.

3.3.1 Revolute joint (R)

One of the simplest and most common joints is the revolute joint, denoted by the letter R. This joint connects two links as shown in Figure 3.5. Link jk is able to rotate relative to link ij about the vector $\mathbf{S}_j$ (it is assumed that the vector $\mathbf{S}_j$ of link ij and the vector $\mathbf{S}_j$

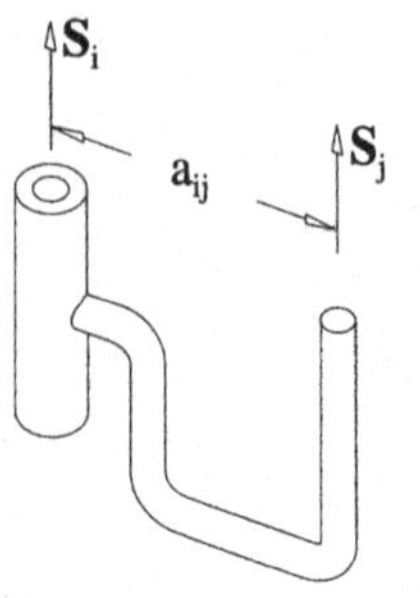

Figure 3.4. Planar link.

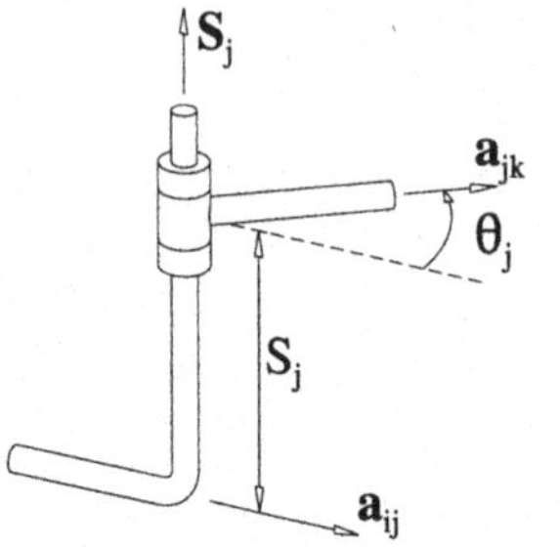

Figure 3.5. Revolute joint.

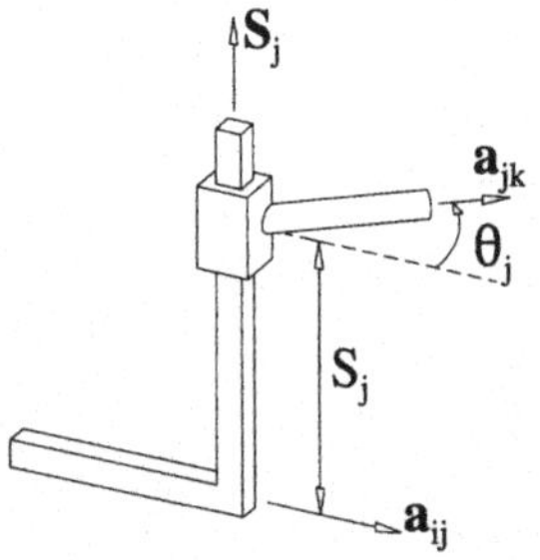

Figure 3.6. Prismatic joint.

of link jk will always be selected so as to be parallel and not antiparallel when the joint is assembled). Link jk thus has one degree of freedom with respect to link ij. The joint angle θ_j measures the relative rotation of the two links and is defined as the angle between the unit vectors $\mathbf{a}_{ij}$ and $\mathbf{a}_{jk}$, measured in a right-hand sense with respect to the unit vector $\mathbf{S}_j$, i.e., $\mathbf{a}_{ij} \times \mathbf{a}_{jk} = \mathbf{S}_j \sin\theta_j$.

Because link jk can rotate only relative to link ij, the distance S_j is a constant. This parameter is called the joint offset distance. It is the mutual perpendicular distance between the vectors $\mathbf{a}_{ij}$ and $\mathbf{a}_{jk}$. In summary, a revolute joint can be completely described by the variable joint angle θ_j and the constant offset value S_j.

3.3.2 Prismatic joint (P)

A prismatic joint, which is denoted by the letter P, allows link jk to translate parallel to the vector $\mathbf{S}_j$ with one degree of freedom relative to link ij (see Figure 3.6). The angle θ_j

is a constant, and it is measured in the same way as for the revolute joint, that is, it is the angle between the vectors $\mathbf{a}_{ij}$ and $\mathbf{a}_{jk}$ measured in a right-hand sense about the vector $\mathbf{S}_j$. The offset distance S_j is a variable for the prismatic joint.

3.3.3 Cylindric joint (C)

A cylindric joint, represented by the letter C, allows link jk to rotate about and translate parallel to the vector $\mathbf{S}_j$ relative to link ij as shown in Figure 3.7. Link jk thus has two independent degrees of freedom relative to link ij. The joint angle θ_j and the offset distance S_j are both variables.

3.3.4 Screw joint (H)

The screw joint, which is denoted by the letter H, is shown in Figure 3.8. For this joint, the offset distance S_j is related to the joint angle θ_j by the linear equation

$$S_j = p_j \theta_j, \tag{3.1}$$

where p_j is the pitch of the screw. Clearly, p_j is a constant; it has units of length/radian, and it may be positive or negative accordingly as the screw has a right- or left-handed thread.

Because the offset distance is a function of the joint angle, link jk has one degree of freedom relative to link ij.

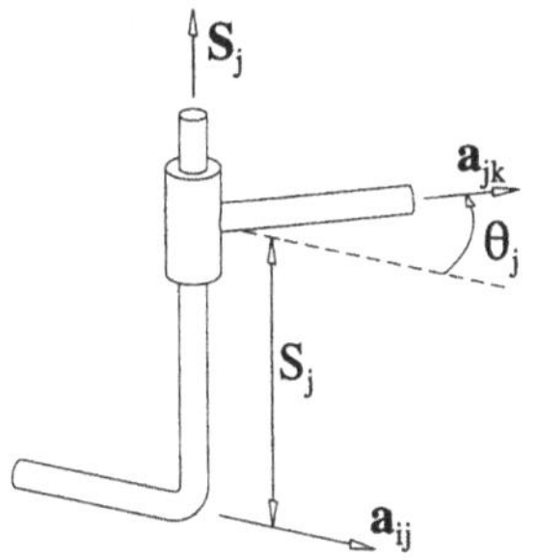

Figure 3.7. Cylindric joint.

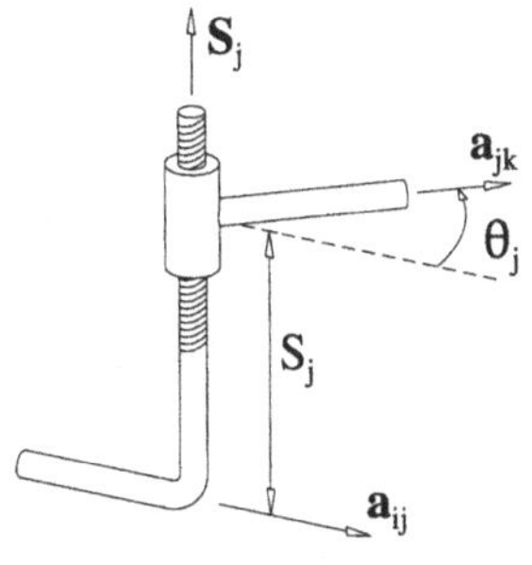

Figure 3.8. Screw joint.

3.3.5 Plane joint (E)

The plane pair (E), which is illustrated in Figure 3.9, permits three independent degrees of freedom between links hi and ij. These freedoms can be considered as a pair of linear displacements in the plane of motion together with a rotation perpendicular to the plane of motion. The three freedoms can be measured, for example, by the pair of coordinates for the origin of the second coordinate system measured in terms of the first coordinate system together with the orientation angle γ, which measures the angle between the direction of $\mathbf{x}_1$ and $\mathbf{x}_2$ measured in a right-hand sense about the direction $\mathbf{z}_1$.

It is not possible to actuate the planar pair in this form in an open loop. However, the plane pair is kinematically equivalent to a combination of two prismatic joints and one revolute joint. The axis of the revolute joint must be perpendicular to the plane formed by the two prismatic joints. Figures 3.10 and 3.11 illustrate such cases. It is important to note that for the plane joint simulated by the PRP shown in Figure 3.10, the following special geometry exists:

$$\alpha_{ij} = \pi/2, \quad a_{ij} = 0,$$
$$\alpha_{jk} = \pi/2, \quad a_{jk} = 0,$$
$$S_j = 0, \quad \theta_i = 0, \quad \theta_k = 0.$$

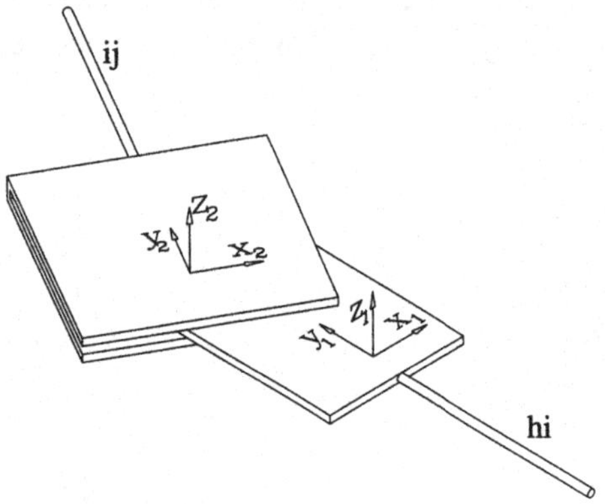

Figure 3.9. Plane joint.

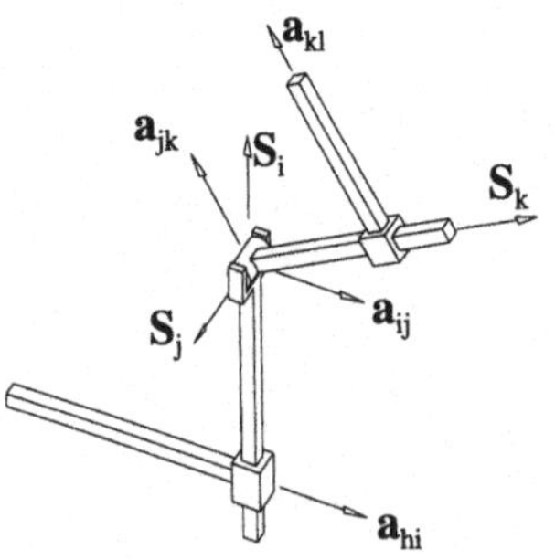

Figure 3.10. Simulation of plane pair (PRP).

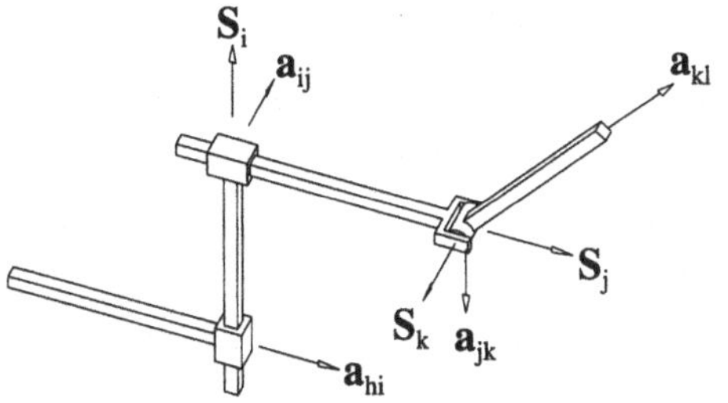

Figure 3.11. Simulation of plane pair (PPR).

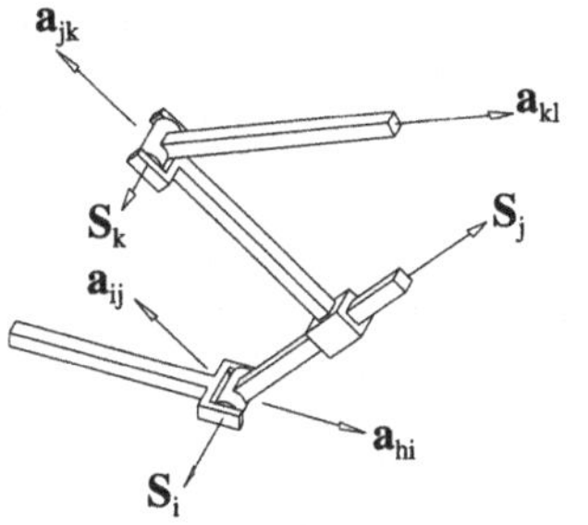

Figure 3.12. Simulation of plane pair (RPR).

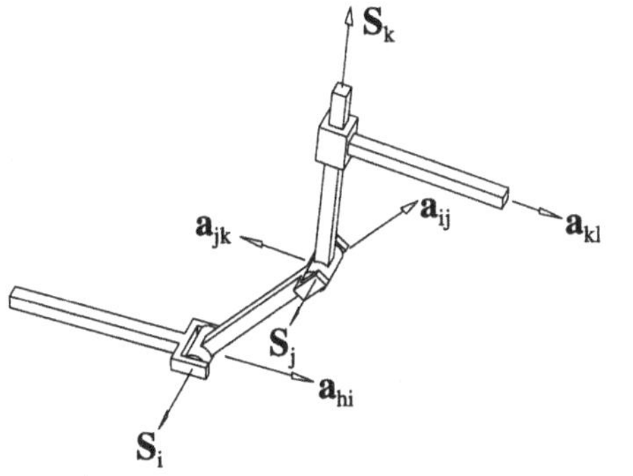

Figure 3.13. Simulation of plane pair (RRP).

For the PPR combination shown in Figure 3.11, the special geometry is as follows:

$$\alpha_{ij} = \pi/2, \quad a_{ij} = 0,$$
$$\alpha_{jk} = \pi/2, \quad a_{jk} = 0,$$
$$S_k = 0, \quad \theta_i = 0, \quad \theta_j = 3\pi/2.$$

The plane pair is kinematically equivalent to a combination of two revolute joints and one prismatic joint as illustrated in Figures 3.12 and 3.13. Lastly, Figure 3.14 illustrates the most practical form of the plane joint. In this case three revolute joints whose axes are parallel will simulate the plane pair. It is left to the reader to deduce the special geometry for these cases.

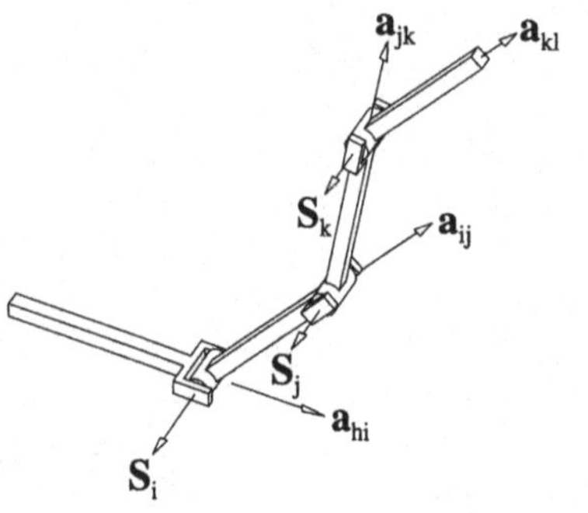

Figure 3.14. Simulation of plane pair (RRR).

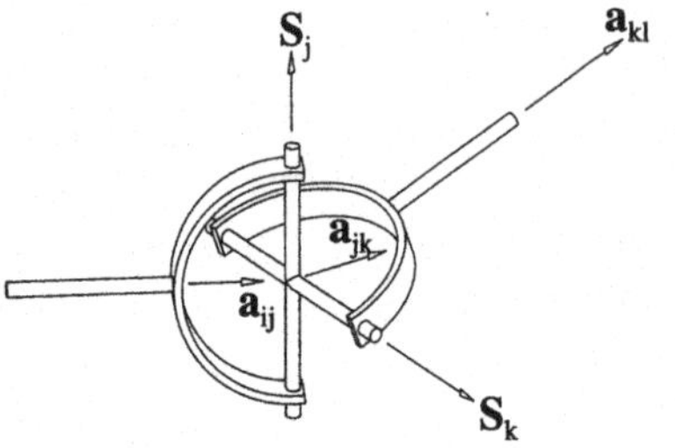

Figure 3.15. Hooke joint.

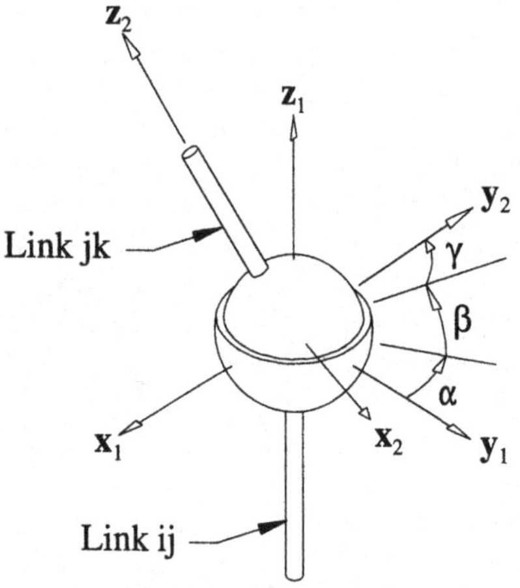

Figure 3.16. Spherical joint (S).

3.3.6 Hooke joint (T)

The Hooke joint is simply two revolute joints whose axes S_j and S_k intersect. It should be apparent that link kl possesses two degrees of freedom relative to link ij. In Figure 3.15, the axes are mutually perpendicular. They can, however, be drawn at any angle.

3.3.7 Spherical joint (S)

The spherical joint, or ball and socket joint, is illustrated in Figure 3.16. Link jk has three degrees of freedom relative to link ij. These three freedoms can be considered as

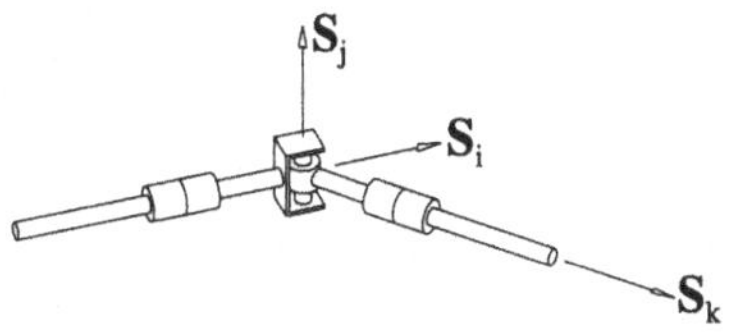

Figure 3.17. Simulation of spherical joint.

three rotations that align a coordinate system attached to link ij with a coordinate system attached to link jk. In Figure 3.16, the origins of the coordinate systems that are attached to each of the links are both located at the center of the spherical joint.

Coordinate system 2, which is attached to link jk, can be obtained from coordinate system 1, which is attached to link ij, by performing the following three rotations:

- a rotation of α about the Z axis
- a rotation of β about the modified X axis
- a rotation of ϕ about the modified Z axis.

The transformation that converts a point known in the second coordinate system to the first coordinate system is simply given by

$$ {}^{1}_{2}\mathbf{R} = \mathbf{R}_{z\alpha}\mathbf{R}_{x\beta}\mathbf{R}_{z\phi}, \tag{3.2}$$

where

$$\mathbf{R}_{z\alpha} = \begin{bmatrix} \cos\alpha & -\sin\alpha & 0 \\ \sin\alpha & \cos\alpha & 0 \\ 0 & 0 & 1 \end{bmatrix}, \quad \mathbf{R}_{x\beta} = \begin{bmatrix} 1 & 0 & 0 \\ 0 & \cos\beta & -\sin\beta \\ 0 & \sin\beta & \cos\beta \end{bmatrix},$$

$$\mathbf{R}_{z\phi} = \begin{bmatrix} \cos\phi & -\sin\phi & 0 \\ \sin\phi & \cos\phi & 0 \\ 0 & 0 & 1 \end{bmatrix}. \tag{3.3}$$

The relationship between the first and second coordinate system could be defined in many different ways, that is, the order and corresponding angles of rotation could be changed.

The design and implementation of a spherical joint can be a complicated process. It is especially involved because one needs to actuate the three freedoms of the joint. The spherical joint, however, can be modeled by three noncoplanar cointersecting revolute joints as shown in Figures 3.17 and 3.18. This method of modeling a spherical joint is commonly used in industrial manipulators.

3.4 Labeling of a kinematic chain

A kinematic chain is shown in Figure 3.19. One body is attached to ground. The present objective is to

(1) select the directions for the joint axis vectors,
(2) select the directions for the link vectors,
(3) label the joint angles and twist angles,

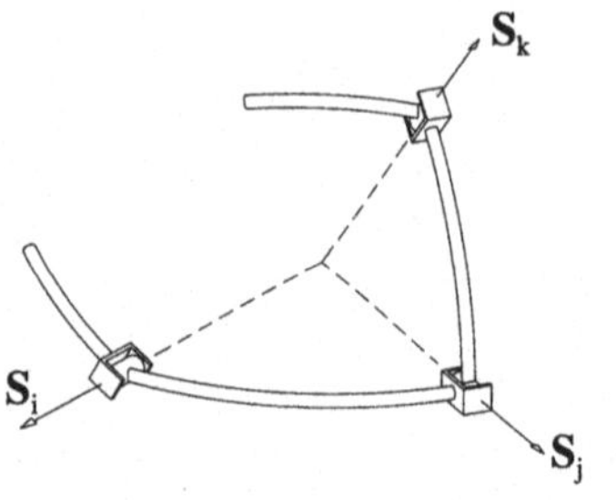

Figure 3.18. Simulation of spherical joint.

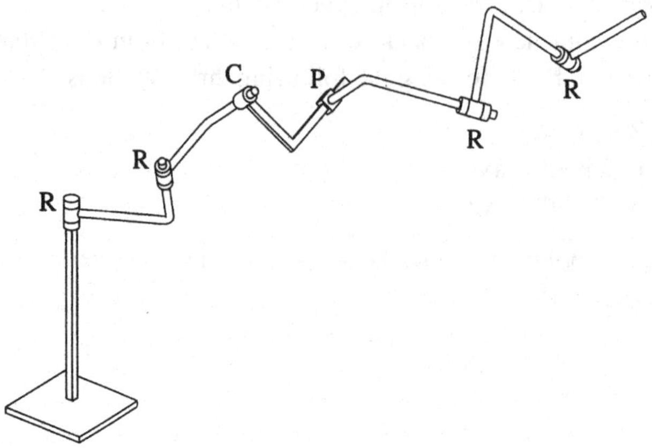

Figure 3.19. Kinematic chain.

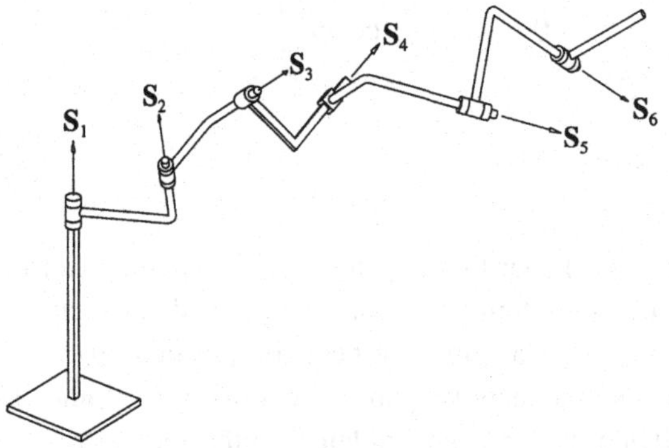

Figure 3.20. Joint vectors labeled.

(4) label the offset and link length distances, and
(5) compile the mechanism parameters in a table listing the constant values and identifying which parameters are variable.

These five steps will be carried out for the kinematic chain shown in Figure 3.19.

Step 1: Label the joint axis vectors.

The first step is to label the joint axes. This is shown in Figure 3.20. One may draw the vector in either direction along the joint axis. However, once directions are selected it is

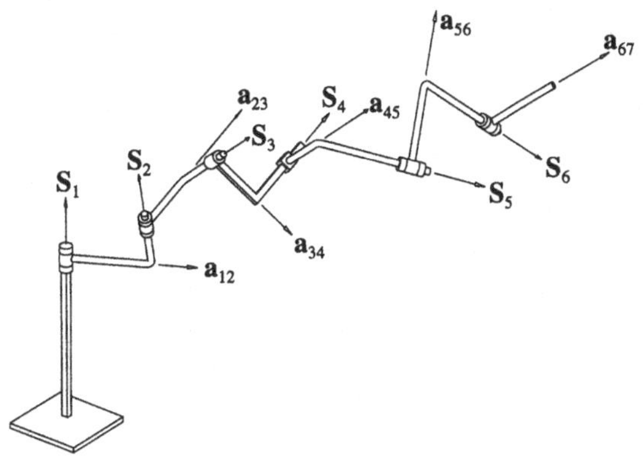

Figure 3.21. Link vectors labeled.

important that they be documented for use in all future analyses. The simplest means of labeling the joint axes is to remember that for a revolute, cylindric, or screw pair, the joint axis is along the line of rotation. A prismatic joint has no particular axis because all points in one body undergo the same relative parallel sliding motion. It is, however, convenient to label this sliding motion by a unit vector drawn on the centerline of the joint.

Step 2: Label the link vectors.

Once the joint axis vectors are specified, the link vectors can be labeled. The link vectors lie along the line that is perpendicular to both of the joint axis vectors that the link connects (see Figure 3.21). The line perpendicular to two joint axes will be unique unless the two joint axes are parallel. If the two joint axes are parallel, then the location but not the direction of the link vector is arbitrarily selected.

The selection of the direction of the vector $\mathbf{a}_{67}$ is somewhat arbitrary because a seventh joint axis, that is, $\mathbf{S}_7$, does not physically exist in this example. The direction of $\mathbf{a}_{67}$ must be selected so that it is perpendicular to $\mathbf{S}_6$, and it must pass through a point on the line of the sixth joint axis. We will later show that the selection of $\mathbf{a}_{67}$ will define a coordinate system attached to the last link of the manipulator and that tool points (points to be positioned in the work space) will be defined in terms of this coordinate system.

Step 3: Label the joint angles and twist angles.

Once the joint axis vectors and the link vectors are specified, the joint angles and twist angles are uniquely defined. For example, in Figure 3.22, θ_3 is defined as the angle between $\mathbf{a}_{23}$ and $\mathbf{a}_{34}$ measured in a right-hand sense about $\mathbf{S}_3$, that is, $\mathbf{a}_{23} \times \mathbf{a}_{34} = \mathbf{S}_3 \sin\theta_3$. Similarly, in Figure 3.23 α_{23} is defined as the angle between $\mathbf{S}_2$ and $\mathbf{S}_3$ measured in a right-hand sense about $\mathbf{a}_{23}$, that is, $\mathbf{S}_2 \times \mathbf{S}_3 = \mathbf{a}_{23} \sin\alpha_{23}$.

The joint angle values that are defined in this manner may not be the same joint parameters that have been defined by the robot manufacturer. The relationship between the kinematic joint angles and the robot manufacturer's joint angles will be linear, as shown by the following equation:

$$\theta_{\mathrm{i(manufacturer)}} = \mathrm{K}_1\theta_{\mathrm{i}} + \mathrm{K}_2. \tag{3.4}$$

The values for K_1 and K_2 can be easily determined by moving each joint of the robot to two positions and recording the kinematic joint angles and the manufacturer's joint angles. K_1 and K_2 can then be obtained by solving two equations with two unknowns.

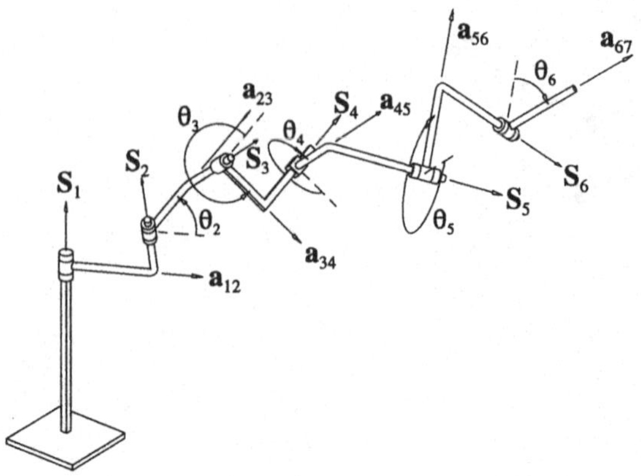

Figure 3.22. Joint angles labeled.

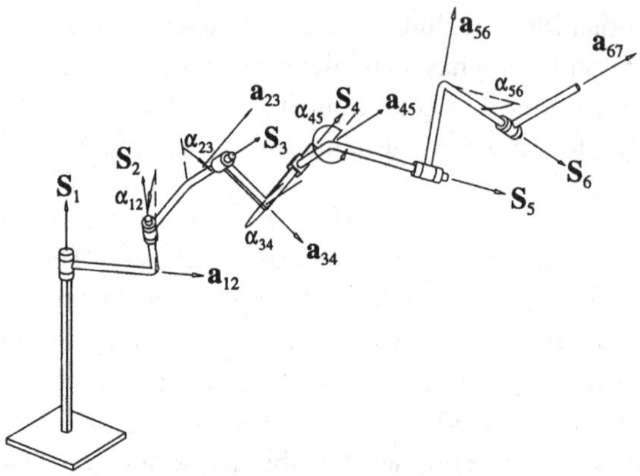

Figure 3.23. Twist angles labeled.

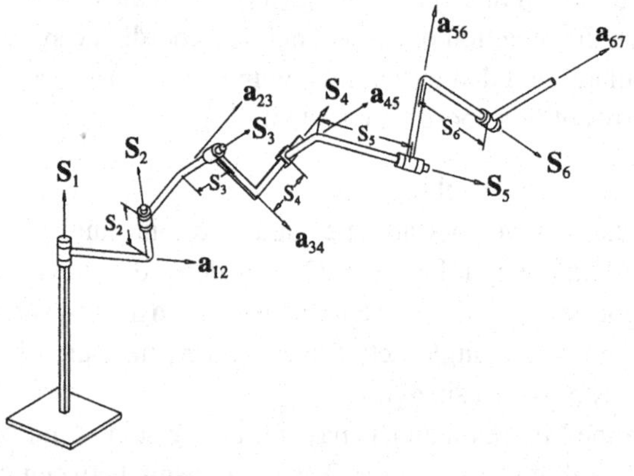

Figure 3.24. Offset lengths labeled.

Step 4: Label the offset and the link length distances.

The offset lengths and link lengths are uniquely defined. For example, S_3 is the distance between the vectors $\mathbf{a}_{23}$ and $\mathbf{a}_{34}$, and a_{23} is the distance between the vectors $\mathbf{S}_2$ and $\mathbf{S}_3$. The offset and link lengths are shown in Figures 3.24 and 3.25. The offset and link lengths may have negative values. For example, the offset distance S_3 will be positive if the direction

Table 3.1. *Mechanism parameters for kinematic chain shown in Figure 3.21.*

Link length, in.	Twist angle, deg.	Joint offset, in.	Joint angle, deg.
$a_{12} = 3.25$	$\alpha_{12} = 30$		$\phi_1 =$ variable
$a_{23} = 2.25$	$\alpha_{23} = 30$	$S_2 = 2.75$	$\theta_2 =$ variable
$a_{34} = 2.125$	$\alpha_{34} = 270$	$S_3 =$ variable	$\theta_3 =$ variable
$a_{45} = 3.5$	$\alpha_{45} = 210$	$S_4 =$ variable	$\theta_4 = 270$
$a_{56} = 3.25$	$\alpha_{56} = 40$	$S_5 = 3.75$	$\theta_5 =$ variable
		$S_6 = 4.75$	$\theta_6 =$ variable

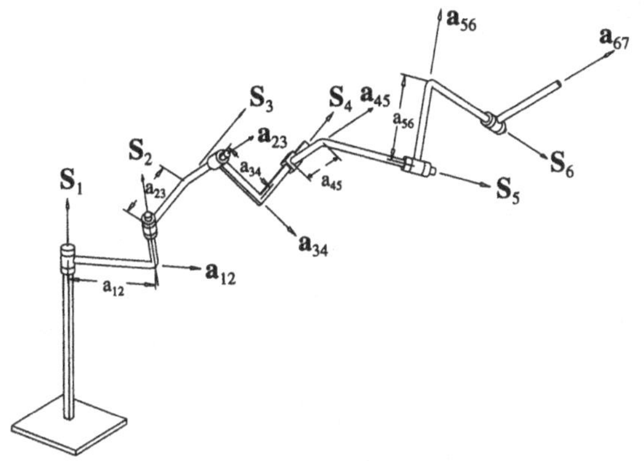

Figure 3.25. Link lengths labeled.

of travel from $\mathbf{a}_{23}$ to $\mathbf{a}_{34}$ is along the direction of $\mathbf{S}_3$. The offset distance S_3 will be negative if moving from $\mathbf{a}_{23}$ to $\mathbf{a}_{34}$ is opposite to the direction of vector $\mathbf{S}_3$.

It is important to note that offset distance S_1 is not defined. According to the labeling convention being used, S_1 would be the distance between the vectors $\mathbf{a}_{01}$ and $\mathbf{a}_{12}$. Because $\mathbf{a}_{01}$ is not defined, the offset distance S_1 is not defined either.

Step 5: Compile the mechanism parameters.

The values for the constant parameters for the kinematic chain must be recorded. The mechanism parameters for the kinematic chain shown in Figure 3.21 are listed in Table 3.1.

The first joint angle must be measured with respect to ground and not relative to another link, as is the case for all the other joint angles. A coordinate system, named the fixed coordinate system, is attached to ground. Its origin is located at the intersection of the vectors $\mathbf{S}_1$ and $\mathbf{a}_{12}$. The Z axis of the fixed coordinate system is along $\mathbf{S}_1$ (see Figure 3.26). The first joint angle, labeled ϕ_1, is defined as the angle between the X axis of the fixed coordinate system and the $\mathbf{a}_{12}$ vector, measured in a right-hand sense about the vector $\mathbf{S}_1$.

The Puma robot is used as a second example of labeling a kinematic chain. Figure 3.27 shows the Puma robot, and Figure 3.28 shows its kinematic drawing. The joint axis vectors are labeled in Figure 3.29, and the link vectors are labeled in Figure 3.30. The twist angles and joint angles as well as the offset lengths and link lengths are now uniquely defined.

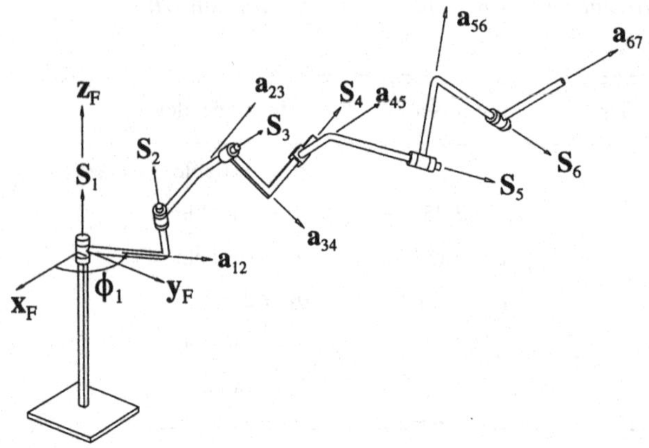

Figure 3.26. Definition of fixed coordinate system.

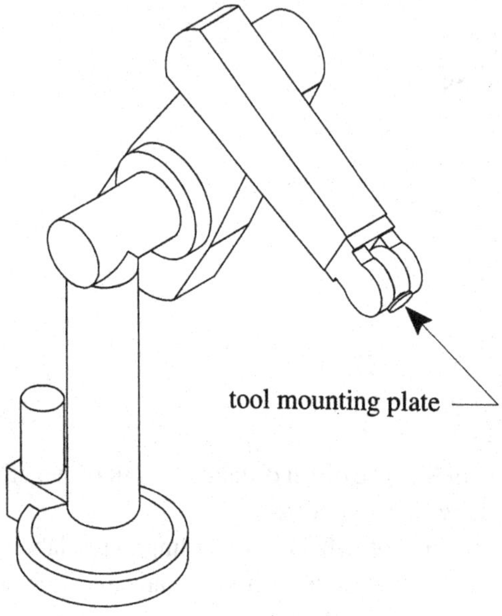

Figure 3.27. Puma robot.

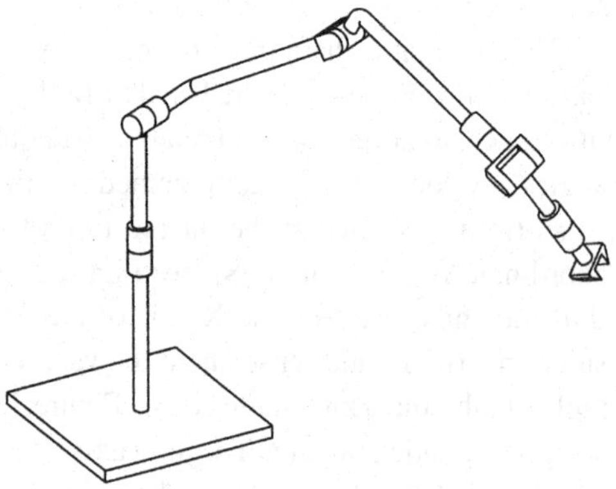

Figure 3.28. Kinematic diagram of Puma robot.

Table 3.2. *Mechanism parameters for Puma robot.*

Link length, in.	Twist angle, deg.	Joint offset, in.	Joint angle, deg.
$a_{12} = 0$	$\alpha_{12} = 90$		ϕ_1 = variable
$a_{23} = 17$	$\alpha_{23} = 0$	$S_2 = 5.9$	θ_2 = variable
$a_{34} = 0.8$	$\alpha_{34} = 270$	$S_3 = 0$	θ_3 = variable
$a_{45} = 0$	$\alpha_{45} = 90$	$S_4 = 17$	θ_4 = variable
$a_{56} = 0$	$\alpha_{56} = 90$	$S_5 = 0$	θ_5 = variable
			θ_6 = variable

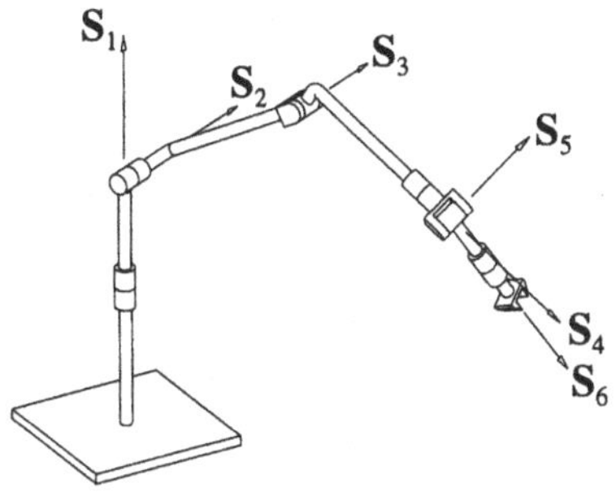

Figure 3.29. Joint vectors labeled.

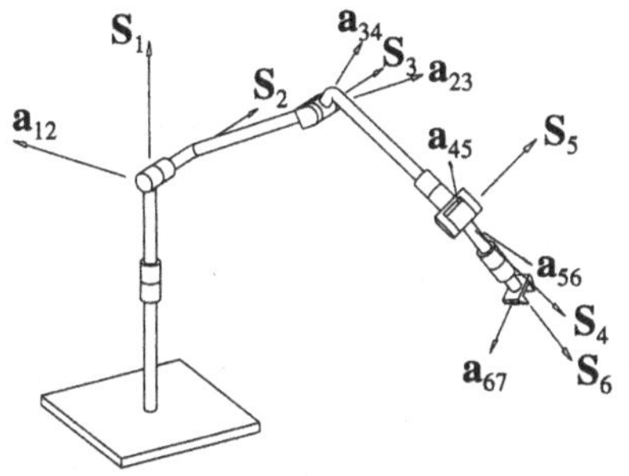

Figure 3.30. Link vectors labeled.

Labeling of these terms is left as an exercise for the reader. Table 3.2 lists the mechanism dimensions for the Puma robot.

3.5 Standard link coordinate systems

For the analysis of manipulator links it is necessary to attach a coordinate system to each rigid body. The selection of the coordinate system for each link will be done systematically. The coordinate system attached to a link ij (see Figure 3.31) will have its origin located at the intersection of $\mathbf{S}_i$ and $\mathbf{a}_{ij}$. The Z axis of the coordinate system will be parallel to $\mathbf{S}_i$. The X axis will be parallel to $\mathbf{a}_{ij}$.

For a serial manipulator, the coordinate system attached to link 12 will be called the first coordinate system. Similarly, the coordinate system attached to link 23 will be called

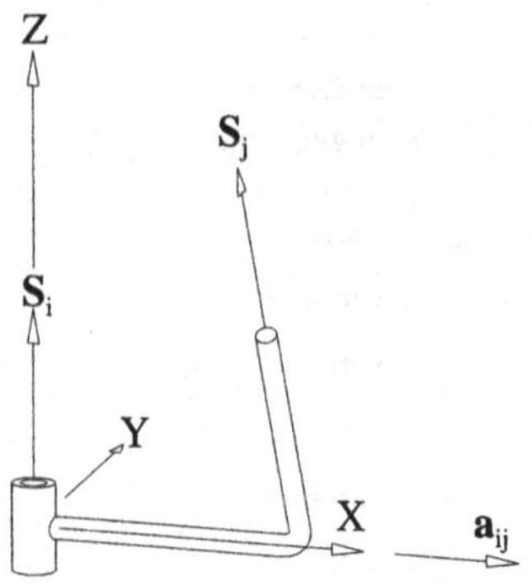

Figure 3.31. Standard link coordinate system.

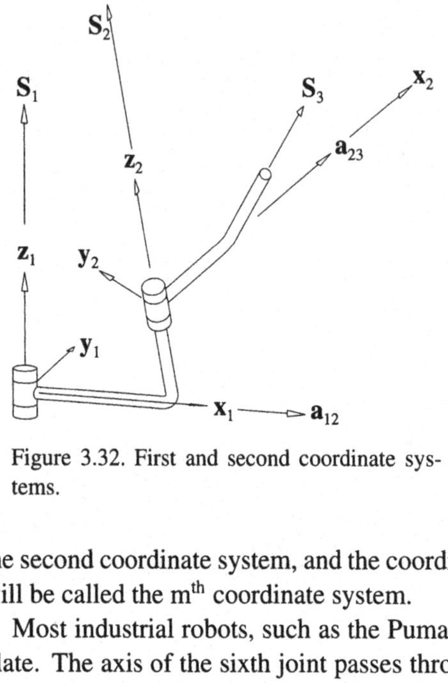

Figure 3.32. First and second coordinate systems.

the second coordinate system, and the coordinate system attached to link mn (n = m + 1) will be called the mth coordinate system.

Most industrial robots, such as the Puma shown in Figure 3.27, have a tool mounting plate. The axis of the sixth joint passes through the center of this plate. In this case, the location and the direction of the vector $\mathbf{a}_{67}$, which is the X axis of the sixth coordinate system, can be arbitrarily selected. Typically, the vector $\mathbf{a}_{67}$ will be placed in the plane of the tool mounting plate. A line will be drawn on the tool mounting plate to signify the $\mathbf{a}_{67}$ vector. Once the vector $\mathbf{a}_{67}$ is specified, the sixth coordinate system as well as the parameters S_6 and θ_6 are defined.

3.6 Transformations between standard coordinate systems

Figure 3.32 shows link a_{12} and link a_{23} of a serial manipulator together with the first and second coordinate systems. It is desired to determine the transformation that relates these two coordinate systems.

The second coordinate system can be obtained by starting with a coordinate system that is initially aligned with the first coordinate system. This new coordinate system is

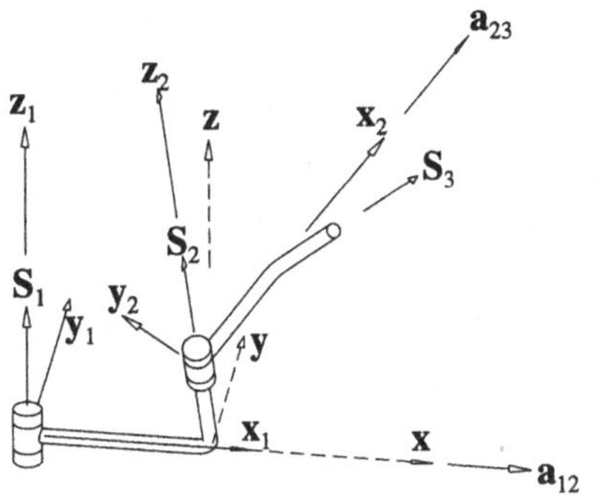

Figure 3.33. First coordinate system translated along $\mathbf{a}_{12}$.

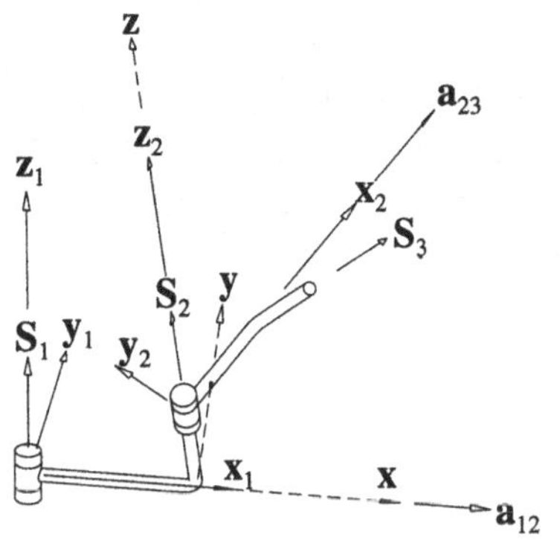

Figure 3.34. Coordinate system rotated about $\mathbf{a}_{12}$ by the angle α_{12}.

then translated by the distance a_{12} along the X axis (see Figure 3.33). Next, it is rotated by the angle α_{12} about the X axis (see Figure 3.34). Following this, the coordinate system is translated along the Z axis by a distance S_2. Lastly, the coordinate system is rotated by the angle θ_2 to align it with the second coordinate system. The transformation that relates the second and first coordinate system can be written as

$$ {}^{1}_{2}\mathbf{T} = \begin{bmatrix} 1 & 0 & 0 & a_{12} \\ 0 & 1 & 0 & 0 \\ 0 & 0 & 1 & 0 \\ 0 & 0 & 0 & 1 \end{bmatrix} \begin{bmatrix} 1 & 0 & 0 & 0 \\ 0 & c_{12} & -s_{12} & 0 \\ 0 & s_{12} & c_{12} & 0 \\ 0 & 0 & 0 & 1 \end{bmatrix} \begin{bmatrix} 1 & 0 & 0 & 0 \\ 0 & 1 & 0 & 0 \\ 0 & 0 & 1 & S_2 \\ 0 & 0 & 0 & 1 \end{bmatrix} \begin{bmatrix} c_2 & -s_2 & 0 & 0 \\ s_2 & c_2 & 0 & 0 \\ 0 & 0 & 1 & 0 \\ 0 & 0 & 0 & 1 \end{bmatrix}, \tag{3.5} $$

or

$$ {}^{1}_{2}\mathbf{T} = \begin{bmatrix} c_2 & -s_2 & 0 & a_{12} \\ s_2c_{12} & c_2c_{12} & -s_{12} & -s_{12}S_2 \\ s_2s_{12} & c_2s_{12} & c_{12} & c_{12}S_2 \\ 0 & 0 & 0 & 1 \end{bmatrix}, \tag{3.6} $$

where s_2 and c_2 represent the sine and cosine of θ_2, and s_{12} and c_{12} represent the sine and cosine of α_{12}. In general, the relationship between the jth and the ith coordinate systems is given by

$$ {}^{i}_{j}\mathbf{T} = \begin{bmatrix} c_j & -s_j & 0 & a_{ij} \\ s_j c_{ij} & c_j c_{ij} & -s_{ij} & -s_{ij} S_j \\ s_j s_{ij} & c_j s_{ij} & c_{ij} & c_{ij} S_j \\ 0 & 0 & 0 & 1 \end{bmatrix}. \tag{3.7} $$

The inverse of this transformation will often be used and is given by

$$ {}^{j}_{i}\mathbf{T} = \begin{bmatrix} c_j & s_j c_{ij} & s_j s_{ij} & -c_j a_{ij} \\ -s_j & c_j c_{ij} & c_j s_{ij} & s_j a_{ij} \\ 0 & -s_{ij} & c_{ij} & -S_j \\ 0 & 0 & 0 & 1 \end{bmatrix}. \tag{3.8} $$

One additional transformation will be presented for completeness. This is the relationship between the first coordinate system and the fixed system. The transformation is simply a rotation about the Z axis by the angle ϕ_1 because the origins of the fixed coordinate system and the first coordinate system are coincident. The transformation is given by

$$ {}^{F}_{1}\mathbf{T} = \begin{bmatrix} \cos\phi_1 & -\sin\phi_1 & 0 & 0 \\ \sin\phi_1 & \cos\phi_1 & 0 & 0 \\ 0 & 0 & 1 & 0 \\ 0 & 0 & 0 & 1 \end{bmatrix}. \tag{3.9} $$

3.7 Summary

In this chapter, the rigid body link was defined and quantified by the link length a_{ij} and the twist angle α_{ij}. Seven types of joints that can interconnect these links were then defined, the most common of which were the revolute joint, R, and the prismatic joint, P. Compound joints such as the Hooke joint, the planar pair, and the ball and socket joint must be simulated by an appropriate series of R and P pairs for actuation in a serial robot manipulator.

The steps for labeling a kinematic chain and defining the mechanism parameters were discussed. Finally, a standard coordinate system was attached to each link and the transformation between coordinate systems was developed.

In the next chapter, the forward kinematic analysis of a robot manipulator will be presented. For this analysis, it is assumed that the constant mechanism parameters and the variable mechanism parameters are all specified. It is necessary to determine the position and orientation of the robot end effector. This analysis will be developed by using the general transformation between standard coordinate systems that was developed in Section 3.6.

3.8 Problems

1. An open chain of links shown in Figure 3.35. Label the joint vectors, link vectors, joint angles, and twist angles on the figure.

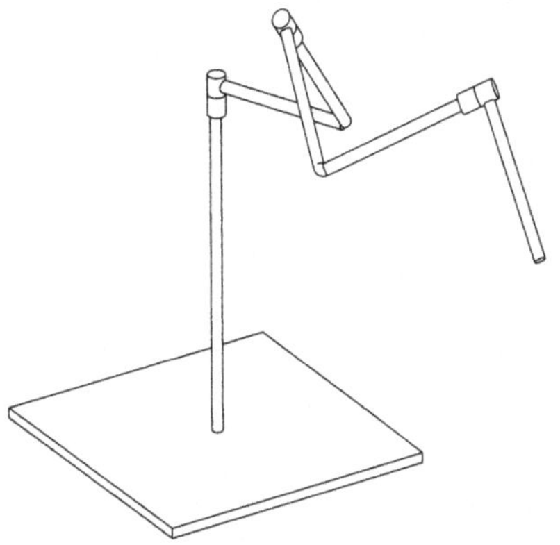

Figure 3.35. Kinematic diagram.

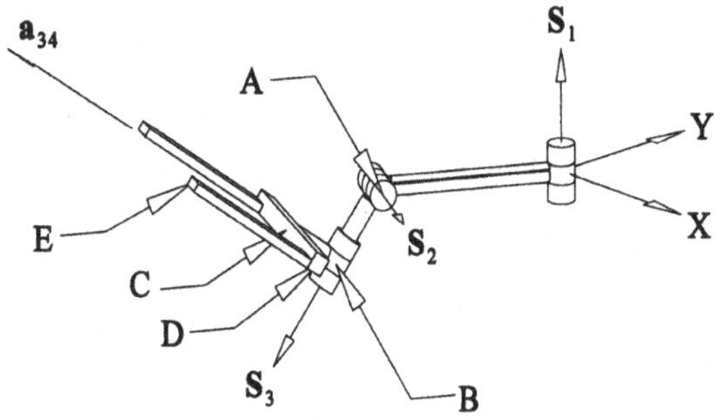

Figure 3.36. RRR manipulator.

2.

(a) Draw a sketch of four general links connected by three revolute joints. Label the twist angles, link lengths, offset lengths, and joint angles.

(b) Specify the special mechanism dimensions necessary to make this mechanism equivalent to a planar pair.

(c) Specify the special mechanism dimensions necessary to make this mechanism equivalent to a ball and socket joint.

3. A 3R manipulator is shown in the Figure 3.36. The vectors $\mathbf{S}_1$, $\mathbf{S}_2$, and $\mathbf{S}_3$ are shown together with a fixed coordinate system. The following information is also known:

Distance between the origin of the fixed coordinate system and point A is ninety-five inches.
Distance between point A and point B is twenty-five inches.
Distance between point B and point C is thirty-five inches.
Distance between point C and point D is thirty inches.
Distance between point D and point E is sixty inches.

(a) Draw the vectors $\mathbf{a}_{12}$ and $\mathbf{a}_{23}$ on the figure, assuming that the twist angles α_{12} and α_{23} are equal to ninety degrees.

(b) Tabulate the mechanism dimensions and give their values.

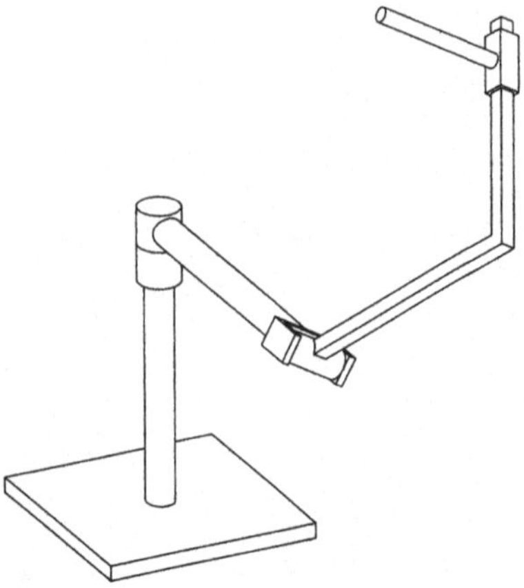

Figure 3.37. RRP manipulator.

(c) Write down the coordinates of point E in terms of the third coordinate system.

(d) At the instant shown in the figure, the joint angle parameters are known to be

$$\phi_1 = 240° \qquad \theta_2 = 120° \qquad \theta_3 = 160°.$$

Determine the coordinates of point E in terms of the fixed coordinate system. Also determine the direction cosines of vectors $\mathbf{S}_3$ and $\mathbf{a}_{34}$ in terms of the fixed coordinate system.

4. An RRP kinematic chain is shown in Figure 3.37. Label all joint and link axes. Label all joint angles and twist angles. What are the variable parameters for this manipulator?

5. The following information is given for a robot manipulator:

$$\alpha_{12} = 50° \qquad a_{12} = 0 \qquad S_2 = 50\,\text{cm} \qquad \phi_1 = 70°$$
$$\alpha_{23} = 90° \qquad a_{23} = 20\,\text{cm} \qquad S_3 = 35\,\text{cm} \qquad \theta_2 = 120°$$
$$\theta_3 = 90°$$

The coordinates of a point measured in terms of the third standard coordinate system are $[8, 2, 0]^T$ cm. Determine the coordinates of this point in terms of the standard fixed coordinate system.

4

Forward kinematic analysis

4.1 Problem statement

A forward analysis of a serial manipulator determines the unique location (position and orientation) for a specified set of joint variables. In practice, the joint variables are monitored continuously as the end effector performs a task. A forward analysis thus monitors continuously the actual location of the end effector, which may of course not be precisely the desired location. The difference is used for location control.

Specifically, for a manipulator comprising six revolute joints, the variable parameters ϕ_1, θ_2, θ_3, θ_4, θ_5, and θ_6 would be known. The goal of the analysis is to determine the coordinates of a tool point that is attached to the last link of the manipulator, that is, link 67 for this case, together with its orientation. Specifically, the forward analysis problem statement for a 6R manipulator is

given: (1) the constant mechanism parameters (link lengths a_{12} through a_{56}, twist angles α_{12} through α_{56}, and joint offsets S_2 through S_5),
(2) the joint offset distance S_6 and the direction of the vector $\mathbf{a}_{67}$ relative to the vector $\mathbf{S}_6$ (to establish the sixth coordinate system),
(3) the variable mechanism parameters (ϕ_1, θ_2, θ_3, θ_4, θ_5, and θ_6), and
(4) the location of a tool point measured in the last coordinate system, ${}^6\mathbf{P}_{\text{tool}}$,

find: (1) the location of the tool point in the fixed coordinate system, ${}^F\mathbf{P}_{\text{tool}}$, and
(2) the orientation of the last coordinate system measured with respect to the fixed system (${}^F_6\mathbf{R}$ for a six-axis robot).

4.2 Forward analysis

The forward analysis is a relatively straightforward problem. Clearly, there is a unique pose for a specified set of six joint variables. The first step of the solution is to obtain the transformation that relates the end effector coordinate system with the fixed coordinate system. Assuming that we have a six-axis robot and using the transformations developed in Section 3.6, the transformation ${}^F_6\mathbf{T}$ can be obtained from

$$ {}^F_6\mathbf{T} = {}^F_1\mathbf{T}\,{}^1_2\mathbf{T}\,{}^2_3\mathbf{T}\,{}^3_4\mathbf{T}\,{}^4_5\mathbf{T}\,{}^5_6\mathbf{T}. \tag{4.1} $$

The orientation of the sixth coordinate system with respect to the fixed system is given by ${}^F_6\mathbf{R}$, the upper left 3×3 matrix of ${}^F_6\mathbf{T}$. As a reminder, the first column of ${}^F_6\mathbf{R}$ is the vector ${}^F\mathbf{a}_{67}$, and the third column is the vector ${}^F\mathbf{S}_6$ by Eq. (2.13) and Section 3.5.

Table 4.1. *Mechanism parameters for Puma robot.*

Link length, in.	Twist angle, deg.	Joint offset, in.	Joint angle, deg.
$a_{12} = 0$	$\alpha_{12} = 90$		$\phi_1 =$ variable
$a_{23} = 17$	$\alpha_{23} = 0$	$S_2 = 5.9$	$\theta_2 =$ variable
$a_{34} = 0.8$	$\alpha_{34} = 270$	$S_3 = 0$	$\theta_3 =$ variable
$a_{45} = 0$	$\alpha_{45} = 90$	$S_4 = 17$	$\theta_4 =$ variable
$a_{56} = 0$	$\alpha_{56} = 90$	$S_5 = 0$	$\theta_5 =$ variable
			$\theta_6 =$ variable

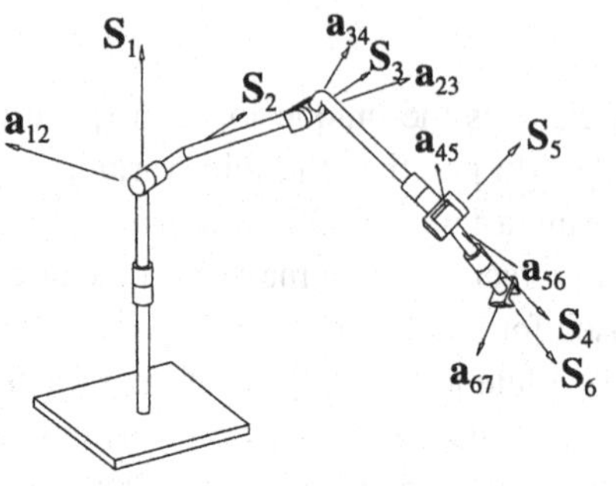

Figure 4.1. Kinematic model of Puma robot.

With the transformation ${}^F_6\mathbf{T}$ now known, the position of the tool point in the fixed coordinate system can be simply found from

$$^F\mathbf{P}_{tool} = {}^F_6\mathbf{T}\,{}^6\mathbf{P}_{tool}. \tag{4.2}$$

The Puma robot, described in detail in Section 3.4, will be used as an example, and Figure 3.28 and Table 3.2 are repeated as Figure 4.1 and Table 4.1. The numerical values listed in Table 4.1 represent all the constant mechanism parameters for the Puma manipulator. It is important to note, however, that the location of the origin of the sixth coordinate system is not yet defined. A value for the offset distance S_6 must be specified in order for the origin of the sixth coordinate system to be defined. The offset S_6 represents the distance between the vectors $\mathbf{a}_{56}$ and $\mathbf{a}_{67}$ measured along the $\mathbf{S}_6$ axis (see Figure 4.1). The vector $\mathbf{a}_{67}$, however, does not physically exist, and thus a unique constant value for S_6 is not automatically defined. This problem is addressed by having the user arbitrarily select a value for S_6. Once this value is chosen, the origin of the sixth coordinate system is uniquely defined. A value of S_6 equal to four inches will be used in this case for the Puma robot.

The next problem is to select a direction of the vector $\mathbf{a}_{67}$. It is known that the vector $\mathbf{a}_{67}$ will pass through the origin of the sixth coordinate system and that $\mathbf{a}_{67}$ must be perpendicular to $\mathbf{S}_6$. However, a planar pencil of lines passes through the origin of the sixth coordinate system and is perpendicular to $\mathbf{S}_6$. The user must align $\mathbf{a}_{67}$ with one of these lines. In a typical application, the value of S_6 may be chosen so that the origin of the sixth coordinate system is located at the center of the tool mounting plate for the robot. A line is then drawn on the face of the tool mounting plate. This line will represent the vector $\mathbf{a}_{67}$. The sixth coordinate system is now completely defined because its origin has been determined by the selection of S_6 and the orientation vectors $\mathbf{S}_6$ and $\mathbf{a}_{67}$ are physically defined. The user would then measure the coordinates of the tool point in terms of this sixth coordinate system, and the forward analysis problem can be completed.

The following numerical data were given for the forward analysis of the Puma robot:

$$S_6 = 4.0\,\text{in.},$$
$$\phi_1 = 5\pi/4, \quad \theta_2 = 5\pi/6, \quad \theta_3 = -\pi/3,$$
$$\theta_4 = \pi/4, \quad \theta_5 = \pi/3, \quad \theta_6 = -\pi/6,$$
$$^6\mathbf{P}_{\text{tool}} = \begin{bmatrix} 5 \\ 3 \\ 7 \end{bmatrix} \text{in.}$$

where the joint angles are given in units of radians. The transformation ${}^F_6\mathbf{T}$ is given as

$$^F_6\mathbf{T} = \begin{bmatrix} 0.997 & -0.002 & 0.079 & 18.577 \\ 0.064 & 0.614 & -0.787 & 23.457 \\ -0.047 & 0.789 & 0.612 & 11.750 \\ 0 & 0 & 0 & 1 \end{bmatrix}, \tag{4.3}$$

and thus the orientation of the sixth coordinate system is known.

The location of the tool point in the fixed coordinate system is calculated from

$$^F\mathbf{P}_{\text{tool}} = {}^F_6\mathbf{T}\,{}^6\mathbf{P}_{\text{tool}}. \tag{4.4}$$

Therefore,

$$^F\mathbf{P}_{\text{tool}} = \begin{bmatrix} 0.997 & -0.002 & 0.079 & 18.577 \\ 0.064 & 0.614 & -0.787 & 23.457 \\ -0.047 & 0.789 & 0.612 & 11.750 \\ 0 & 0 & 0 & 1 \end{bmatrix} \begin{bmatrix} 5 \\ 3 \\ 7 \\ 1 \end{bmatrix}, \tag{4.5}$$

and hence

$$^F\mathbf{P}_{\text{tool}} = \begin{bmatrix} 24.112 \\ 20.113 \\ 18.167 \\ 1 \end{bmatrix} \text{in.} \tag{4.6}$$

4.3 Problems

1. Write a computer function that will perform a forward kinematic analysis of the GE P60 robot. Write a second function that will perform the forward analysis of the Cincinnati Milacron T3-776 robot. Use the kinematic diagrams and mechanism dimensions presented in Sections 11.3 and 11.4 respectively.

The value for the offset S_6 must be selected in order to define the location of the sixth coordinate system of the robot. Use a value of 6.0 cm for the GE P60 robot and 8.0 inches for the Cincinnati Milacron robot.

For both cases, assume that the coordinates of the tool point as measured in the sixth coordinate system are $[12, 8, 5]^T$. Also, the values for the joint angles of the robot are

$$\begin{aligned} \phi_1 &= 50°, \\ \theta_2 &= 120°, \\ \theta_3 &= 295°, \\ \theta_4 &= 30°, \\ \theta_5 &= 190°, \\ \theta_6 &= 100°. \end{aligned}$$

Determine the position of the tool point in terms of the fixed coordinate system. Also determine the orientation of the end effector in terms of the fixed coordinate system, that is, ${}^F\mathbf{a}_{67}$ and ${}^F\mathbf{S}_6$.

A prototype for the function may be written as

```
void forward_ge (double phi1, double th2, double th3,
   double th4, double th5, double th6,
   double S6, double P_tool_6[3],
   double P_tool_F[3], double S6_F[3], double a67_F[3]);
```

2. A three-axis robot is shown in Figure 4.2.

(a) Label all vectors along the links and joint offsets on the accompanying diagram. Label all joint angles and twist angles.

(b) Tabulate the mechanism dimensions and give their values.

(c) Assume you are given the following angular data:

$$\psi_1 = 30°, \qquad \theta_2 = 90°, \qquad \theta_3 = 40°$$

where in this case, the angle ψ_1 is measured between the fixed Y axis and the link vector that is perpendicular to the first two joint axes.

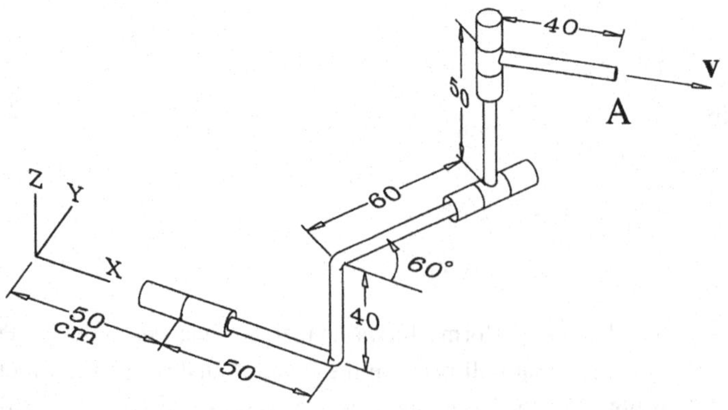

Figure 4.2. RRR manipulator.

Determine the coordinates of point A and the direction cosines of vector **v** in terms of the fixed coordinate system shown in the figure.

3. Suppose you are given an RP manipulator with the following dimensions:

$$a_{12} = 5\text{ in.},\quad \alpha_{12} = 270°,$$
$$\theta_2 = 135°.$$

(a) Draw the manipulator and label the vectors $\mathbf{a}_{12}$, $\mathbf{a}_{23}$, $\mathbf{S}_1$, and $\mathbf{S}_2$. Label the fixed coordinate system and the second coordinate system.

(b) A tool point is given as $[3.0, 1.0, 2.0]^T$ measured in terms of the second coordinate system. Determine the location of the tool point in terms of the fixed coordinate system when $\phi_1 = 45°$ and $S_2 =$ six inches. In addition, determine the direction of the vectors $\mathbf{S}_2$ and $\mathbf{a}_{23}$ in terms of the fixed coordinate system.

5

Reverse kinematic analysis problem statement

A reverse analysis for a 6R serial manipulator determines all possible sets of the six joint variables for any specified end effector location. Each set of six joint variables defines a particular pose for the given end effector location. This analysis is especially important when the end effector must move through a number of finite locations when performing some specified task. This analysis is clearly more difficult than the forward analysis described in Chapter 4. The analysis begins in this chapter by introducing the concept of closing the loop, where a hypothetical link is inserted between the end effector and ground to form a closed-loop spatial mechanism. The analyses of these closed-loop mechanisms are presented in detail in Chapters 7 through 10.

5.1 Problem statement

The problem statement for the reverse analysis of a 6R manipulator is as follows (see Figure 5.1):

given: (1) the constant mechanism parameters (link lengths a_{12} through a_{56}, twist angles α_{12} through α_{56}, and joint offsets S_2 through S_5),

(2) the joint offset distance S_6 and the direction of the vector $\mathbf{a}_{67}$ relative to the vector $\mathbf{S}_6$ (to establish the sixth coordinate system),

(3) the desired position and orientation of the end effector, that is, ${}^F\mathbf{P}_{\text{tool}}$, ${}^F\mathbf{S}_6$, ${}^F\mathbf{a}_{67}$, and

(4) the location of the tool point in the sixth coordinate system, that is, ${}^6\mathbf{P}_{\text{tool}}$,

find: $\phi_1, \theta_2, \theta_3, \theta_4, \theta_5, \theta_6$.

It is assumed that ${}^F\mathbf{S}_6$ and ${}^F\mathbf{a}_{67}$ are unit vectors. From this information, the transformation that relates the sixth coordinate system to ground is given by

$$ {}^F_6\mathbf{T} = \begin{bmatrix} & {}^F_6\mathbf{R} & & {}^F\mathbf{P}_{6\,\text{orig}} \\ 0 & 0 & 0 & 1 \end{bmatrix}, \tag{5.1} $$

where

$$ {}^F_6\mathbf{R} = \left[{}^F\mathbf{a}_{67},\ {}^F\mathbf{S}_6 \times {}^F\mathbf{a}_{67},\ {}^F\mathbf{S}_6\right] \tag{5.2} $$

and

$$ {}^F\mathbf{P}_{6\,\text{orig}} = {}^F\mathbf{P}_{\text{tool}} - \left({}^6\mathbf{P}_{\text{tool}} \cdot \mathbf{i}\right){}^F\mathbf{a}_{67} - \left({}^6\mathbf{P}_{\text{tool}} \cdot \mathbf{j}\right){}^F\mathbf{S}_6 \times {}^F\mathbf{a}_{67} - \left({}^6\mathbf{P}_{\text{tool}} \cdot \mathbf{k}\right){}^F\mathbf{S}_6. \tag{5.3} $$

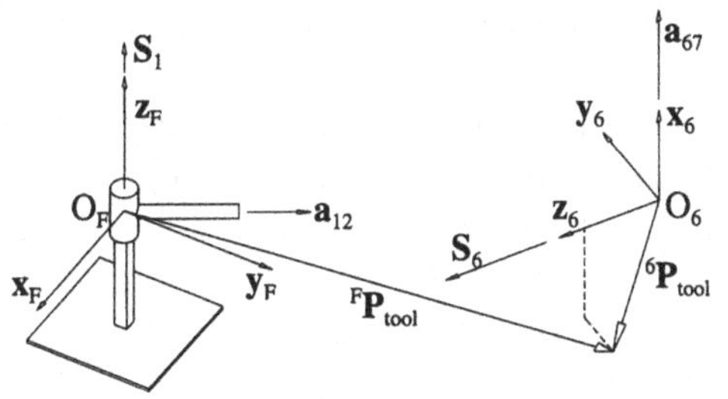

Figure 5.1. Reverse analysis known information.

The 3 × 3 matrix ${}^F_6\mathbf{R}$ is easily determined because ${}^F\mathbf{a}_{67}$ and ${}^F\mathbf{S}_6$ are the X and Z axes of the sixth coordinate system measured with respect to the fixed coordinate system. The location of the origin of the sixth system is determined by projection.

5.2 Iterative solution techniques

Iterative techniques represent one method of solution for the reverse-analysis problem. In these techniques, an initial guess for the joint parameters (ϕ_1 through θ_6 for a 6R robot) is made. A forward analysis is performed to determine the position and orientation of the tool point for the selected joint parameters. The difference between the position and orientation calculated with the forward analysis and the desired position and orientation represent an error that is to be minimized.

In typical iterative techniques, the error must be reduced to or represented by a single scalar value. An objective function, $F(\phi_1, \theta_2, \theta_3, \theta_4, \theta_5, \theta_6)$, is formulated, and search techniques are used to obtain the set of design parameters $(\phi_1, \theta_2, \theta_3, \theta_4, \theta_5, \theta_6)$ that will minimize F. A typical objective function would be the sum of the squares of the position errors in the X , Y, and Z directions, plus the sum of the squares of the orientation errors measured by X-Y-Z fixed angles or some other orientation measurement system. Thus, the objective function could be written as

$$F(\phi_1, \theta_2, \theta_3, \theta_4, \theta_5, \theta_6) = (X_f - X_d)^2 + (Y_f - Y_d)^2 + (Z_f - Z_d)^2 + (\alpha_f - \alpha_d)^2 + (\beta_f - \beta_d)^2 + (\gamma_f - \gamma_d)^2, \tag{5.4}$$

where the subscript f refers to a position or orientation value calculated by performing a forward analysis using the current design parameters, and the subscript d refers to a desired value specified at the start of the reverse analysis problem. The optimal solution for this problem is to obtain a set of values for the design parameters that cause the objective function to equal zero.

A problem exists with Eq. (5.4). It contains terms with different units. The first three terms have units of dimension length^2, whereas the last three terms are dimensionless (radians^2). Such an objective function is without any geometrical meaning. Many times this problem is simply incorrectly ignored or some arbitrary constants that have units of length^2 are used to multiply the orientation differences. Any such multiplication destroys the geometric meaning of orientation. If the specified end effector position and orientation is in the reachable work space of the robot, however, the iterative solution can yield correct results because sets of joint angles will exist that will position and orient the end effector as desired. The objective function will thereby attain its optimal value of zero.

A second problem with the iterative solution is that only one set of joint parameters will be calculated that position and orient the end effector as desired. It will be shown later that up to sixteen sets of joint parameters may exist that will position and orient the end effector of a 6R manipulator as desired.

Some texts might point out that there are two "advantages" to an iterative solution technique. Firstly, a single computer program can be used for virtually any manipulator geometry. Only the values of the mechanism dimensions need be changed for application to a different robot. Secondly, the iterative solution will normally converge with only a few iterations. Usually, a desired position and orientation for the end effector will be very close to the current position and orientation. The current joint parameter values can be used as the initial guess for the design parameters, and the iterative technique should converge to a solution rapidly.

Using the iterative solution technique has serious disadvantages. The problem of mixed units is significant, and if the end effector is commanded to move to a pose that is not within its reachable work space, different "optimal" solutions will result for different units used in the problem. Further, it can be argued that the objective function with mixed units is meaningless. Aside from this argument, the iterative solution cannot guarantee that all solution sets will be determined. If it is important to compute all the sets of joint parameters that can position and orient the robot's end effector as desired (as is often the case), then a closed-form analytical solution must be obtained. This closed-form solution will be the subject of the remainder of this text.

5.3 Closed-form solution technique – hypothetical closure link

A hypothetical closure link will now be added that connects an imaginary joint axis labeled $\mathbf{S}_7$ to the first joint axis, labeled $\mathbf{S}_1$. The direction and location of the vector $\mathbf{S}_7$ must be selected first, where by definition the vector $\mathbf{S}_7$ must be perpendicular to $\mathbf{a}_{67}$. The direction of $\mathbf{S}_7$ will be defined by selecting a value for the angle α_{67}, which will be arbitrarily chosen as 90 degrees. With this selection of α_{67}, $\mathbf{S}_7$ can be calculated from

$$^{\mathrm{F}}\mathbf{S}_7 = {^{\mathrm{F}}\mathbf{a}_{67}} \times {^{\mathrm{F}}\mathbf{S}_6}. \tag{5.5}$$

The vector $\mathbf{S}_7$ will be located so that it passes through the point O_6, the origin of the sixth coordinate system. Thus, the distance $a_{67} = 0$.

Figure 5.2 shows the hypothetical joint axis $\mathbf{S}_7$ and the hypothetical link 71. It will be shown that unique values can be found for the following parameters, which are shown in the figure

$$a_{71}, S_7, S_1, \alpha_{71}, \theta_7, \gamma_1.$$

It is interesting to note that the offset distance S_1 is now defined for the new closed-loop mechanism. S_1 is the distance between the vectors $\mathbf{a}_{71}$ and $\mathbf{a}_{12}$, measured along $\mathbf{S}_1$.

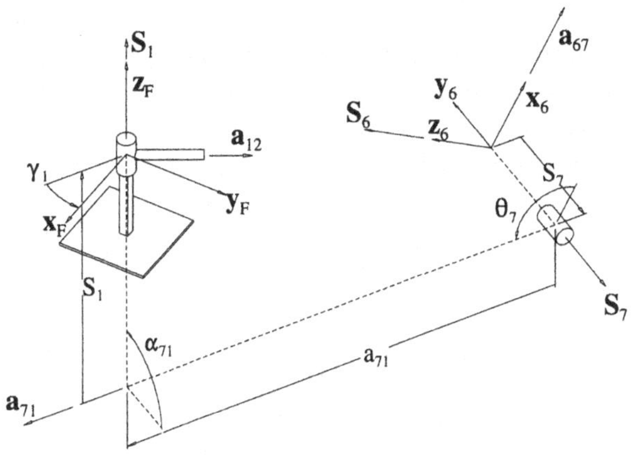

Figure 5.2. Hypothetical closure link.

5.4 General solutions for α_{ij} and θ_j

Expressions for the twist angles α_{ij} and joint angles θ_j will now be presented. The twist angle α_{ij} is defined as the angle between the vectors $\mathbf{S}_i$ and $\mathbf{S}_j$ measured in a right-hand sense about $\mathbf{a}_{ij}$. Because it is assumed that $\mathbf{S}_i$ and $\mathbf{S}_j$ are unit vectors, $\cos\alpha_{ij} = \mathbf{S}_i \cdot \mathbf{S}_j$, where $(\cdot)$ denotes the usual scalar product of a pair of vectors. Knowledge of the cosine of α_{ij} is not sufficient, however, to uniquely determine α_{ij}. There are two distinct angles between 0 and 2π that will have the same cosine value. To uniquely determine α_{ij}, it is also necessary to determine $\sin\alpha_{ij}$. Now, $\mathbf{S}_i \times \mathbf{S}_j = \mathbf{a}_{ij} \sin\alpha_{ij}$, where $(\times)$ denotes the usual vector or cross product of the pair of vectors S_i and S_j. Further, $(\mathbf{S}_i \times \mathbf{S}_j) \cdot \mathbf{a}_{ij} = \sin\alpha_{ij}$. In summary, the expressions for the cosine and sine of α_{ij} are

$$c_{ij} = \mathbf{S}_i \cdot \mathbf{S}_j, \tag{5.6}$$

$$s_{ij} = (\mathbf{S}_i \times \mathbf{S}_j) \cdot \mathbf{a}_{ij}, \tag{5.7}$$

where s_{ij} and c_{ij} represent the sine and cosine of α_{ij} respectively.

Similarly, it can be shown that the cosine and sine of θ_j can be expressed as

$$c_j = \mathbf{a}_{ij} \cdot \mathbf{a}_{jk}, \tag{5.8}$$

$$s_j = (\mathbf{a}_{ij} \times \mathbf{a}_{jk}) \cdot \mathbf{S}_j, \tag{5.9}$$

where s_j and c_j represent the sine and cosine of θ_j respectively.

5.5 Determination of the close-the-loop parameters

The solution for the parameters a_{71}, S_7, S_1, α_{71}, θ_7, and γ_1 begins by determining a direction for the vector $\mathbf{a}_{71}$ measured with respect to the fixed coordinate system. Because $\mathbf{a}_{71}$ must, by definition, be perpendicular to $\mathbf{S}_7$ and $\mathbf{S}_1$, then

$${}^{F}\mathbf{a}_{71} = \frac{{}^{F}\mathbf{S}_7 \times {}^{F}\mathbf{S}_1}{\left|{}^{F}\mathbf{S}_7 \times {}^{F}\mathbf{S}_1\right|}. \tag{5.10}$$

The vector ${}^{F}\mathbf{S}_7$ is given by Eq. (5.5), and because $\mathbf{S}_1$ is parallel to the Z axis of the fixed coordinate system, ${}^{F}\mathbf{S}_1 = [0, 0, 1]^T$. It should be noted that the denominator of Eq. (5.10) will equal zero if the vectors $\mathbf{S}_1$ and $\mathbf{S}_7$ are parallel. This special case is identified if ${}^{F}\mathbf{S}_1 \cdot {}^{F}\mathbf{S}_7 = \pm 1$ and is discussed in Section 5.6.

The twist angle α_{71} can now be calculated by using Eqs. (5.6) and (5.7) and

$$c_{71} = {}^{F}\mathbf{S}_7 \cdot {}^{F}\mathbf{S}_1, \tag{5.11}$$

$$s_{71} = \left({}^{F}\mathbf{S}_7 \times {}^{F}\mathbf{S}_1\right) \cdot {}^{F}\mathbf{a}_{71}. \tag{5.12}$$

The joint angle θ_7 can be found by applying Eqs. (5.8) and (5.9) as

$$c_7 = {}^{F}\mathbf{a}_{67} \cdot {}^{F}\mathbf{a}_{71}, \tag{5.13}$$

$$s_7 = \left({}^{F}\mathbf{a}_{67} \times {}^{F}\mathbf{a}_{71}\right) \cdot {}^{F}\mathbf{S}_7. \tag{5.14}$$

The angle γ_1 is the angle between the vector $\mathbf{a}_{71}$ and the X axis of the fixed coordinate system (see Figure 5.2). The sine and cosine of γ_1 can be determined in a manner similar to that for θ_j; see Eqs. (5.8) and (5.9). It can be shown that

$$\cos\gamma_1 = {}^{F}\mathbf{a}_{71} \cdot \begin{bmatrix} 1 \\ 0 \\ 0 \end{bmatrix}, \tag{5.15}$$

$$\sin\gamma_1 = \left({}^{F}\mathbf{a}_{71} \times \begin{bmatrix} 1 \\ 0 \\ 0 \end{bmatrix} \right) \cdot {}^{F}\mathbf{S}_1. \tag{5.16}$$

At this point, the values for α_{71}, θ_7, and γ_1 have been determined. The remaining parameters to be solved for are the distances S_7, a_{71}, and S_1. These distances will be determined by first writing the vector loop equation

$${}^{F}\mathbf{P}_{6\,\mathrm{orig}} + S_7\,{}^{F}\mathbf{S}_7 + a_{71}\,{}^{F}\mathbf{a}_{71} + S_1\,{}^{F}\mathbf{S}_1 = \mathbf{0}. \tag{5.17}$$

Because all the vectors in Eq. (5.17) are known, the vector equation represents three scalar equations in the three unknowns S_7, a_{71}, and S_1.

The distance S_7 is obtained by forming a cross product of the left and right sides of Eq. (5.17) with ${}^{F}\mathbf{S}_1$ and recognizing that $\mathbf{S}_1 \times \mathbf{S}_1 = \mathbf{0}$. This gives

$$\left({}^{F}\mathbf{P}_{6\,\mathrm{orig}} \times {}^{F}\mathbf{S}_1\right) + S_7\left({}^{F}\mathbf{S}_7 \times {}^{F}\mathbf{S}_1\right) + a_{71}\left({}^{F}\mathbf{a}_{71} \times {}^{F}\mathbf{S}_1\right) = \mathbf{0}. \tag{5.18}$$

Because $({}^{F}\mathbf{S}_7 \times {}^{F}\mathbf{S}_1) = s_{71}\,{}^{F}\mathbf{a}_{71}$, Eq. (5.18) reduces to

$$\left({}^{F}\mathbf{P}_{6\,\mathrm{orig}} \times {}^{F}\mathbf{S}_1\right) + S_7 s_{71}\,{}^{F}\mathbf{a}_{71} + a_{71}\left({}^{F}\mathbf{a}_{71} \times {}^{F}\mathbf{S}_1\right) = \mathbf{0}. \tag{5.19}$$

Forming the scalar product of the left and right sides of Eq. (5.19) with ${}^{F}\mathbf{a}_{71}$ yields

$$\left({}^{F}\mathbf{P}_{6\,\mathrm{orig}} \times {}^{F}\mathbf{S}_1\right) \cdot {}^{F}\mathbf{a}_{71} + S_7 s_{71} = 0. \tag{5.20}$$

Clearly, $({}^{F}\mathbf{a}_{71} \times {}^{F}\mathbf{S}_{1}) \cdot {}^{F}\mathbf{a}_{71} = 0$ and ${}^{F}\mathbf{a}_{71} \cdot {}^{F}\mathbf{a}_{71} = 1$, so Eq. (5.20) can be rearranged to yield

$$S_7 = \frac{\left({}^{F}\mathbf{S}_1 \times {}^{F}\mathbf{P}_{6\,\text{orig}}\right) \cdot {}^{F}\mathbf{a}_{71}}{s_{71}}. \tag{5.21}$$

Similarly, it can be shown that the distances a_{71} and S_1 are given by

$$a_{71} = \frac{\left({}^{F}\mathbf{P}_{6\,\text{orig}} \times {}^{F}\mathbf{S}_1\right) \cdot {}^{F}\mathbf{S}_7}{s_{71}}, \tag{5.22}$$

$$S_1 = \frac{\left({}^{F}\mathbf{P}_{6\,\text{orig}} \times {}^{F}\mathbf{S}_7\right) \cdot {}^{F}\mathbf{a}_{71}}{s_{71}}. \tag{5.23}$$

5.6 Special cases

5.6.1 S_1 and S_7 parallel

Equations (5.10) and (5.21) through (5.23) yield infinite values when $s_{71} = 0$. This occurs when $\mathbf{S}_7$ and $\mathbf{S}_1$ are parallel or antiparallel and when there is no unique vector $a_{71}\mathbf{a}_{71}$ that is mutually perpendicular to $\mathbf{S}_7$ and $\mathbf{S}_1$. This condition is easily identified because from Eq. (5.11), $c_{71} = \pm 1$. It is possible to obtain a solution by selecting $S_7 = 0$, for which Eq. (5.17) reduces to (see also Figure 5.3)

$${}^{F}\mathbf{P}_{6\,\text{orig}} + a_{71}{}^{F}\mathbf{a}_{71} + S_1{}^{F}\mathbf{S}_1 = \mathbf{0}. \tag{5.24}$$

Forming a scalar product of Eq. (5.24) with ${}^{F}\mathbf{S}_1$ and solving for S_1 yields

$$S_1 = -{}^{F}\mathbf{P}_{6\,\text{orig}} \cdot {}^{F}\mathbf{S}_1. \tag{5.25}$$

Rearranging Eq. (5.24) gives

$$a_{71}{}^{F}\mathbf{a}_{71} = -\left({}^{F}\mathbf{P}_{6\,\text{orig}} + S_1\,{}^{F}\mathbf{S}_1\right). \tag{5.26}$$

The right side of Eq. (5.26) is known. However, both a_{71} and ${}^{F}\mathbf{a}_{71}$ are unknown. The

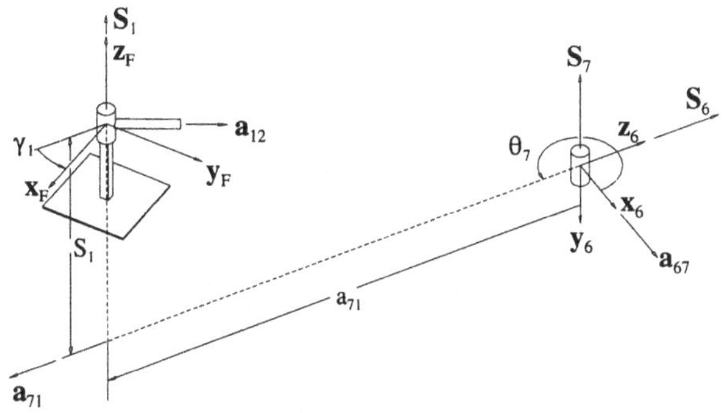

Figure 5.3. Special case where S_7 and S_1 are parallel.

distance a_{71} is easily computed because from (5.26)

$$a_{71} = \left|-({}^{F}\mathbf{P}_{6\,orig} + S_1{}^{F}\mathbf{S}_1)\right|. \tag{5.27}$$

Dividing Eq. (5.26) by a_{71} yields the unknown vector

$${}^{F}\mathbf{a}_{71} = \frac{-({}^{F}\mathbf{P}_{6\,orig} + S_1{}^{F}\mathbf{S}_1)}{a_{71}}. \tag{5.28}$$

Finally, θ_7 and γ_1 are computed using Eqs. (5.13) through (5.16).

5.6.2 S_1 and S_7 collinear

A second special case occurs when $\mathbf{S}_7$ and $\mathbf{S}_1$ are collinear (see Figure 5.4). This is identified when Eq. (5.27) yields $a_{71} = 0$. The direction of the vector $\mathbf{a}_{71}$ in the plane normal to $\mathbf{S}_1$ is now arbitrary. In this case, the angle θ_7 will be chosen as zero, thereby making $\mathbf{a}_{71}$ parallel to $\mathbf{a}_{67}$. The angle γ_1 can now be calculated from Eqs. (5.15) and (5.16) as before.

5.7 Example

The close-the-loop parameters will be calculated for the case of the Puma robot. At the start of the problem, one must select a value for the link offset distance S_6. Following this selection, the location of the origin of the sixth coordinate system on the robot is defined. The position of the tool point in terms of this sixth coordinate system, ${}^{6}\mathbf{P}_{tool}$, must next be specified.

At this point, the reverse kinematic analysis proceeds by the user's specifying the desired position and orientation for the tool point in terms of the fixed coordinate system, that is, ${}^{F}\mathbf{P}_{tool}$, ${}^{F}\mathbf{S}_6$, and ${}^{F}\mathbf{a}_{67}$. The close-the-loop parameters are then calculated as described in Sections 5.5 and 5.6.

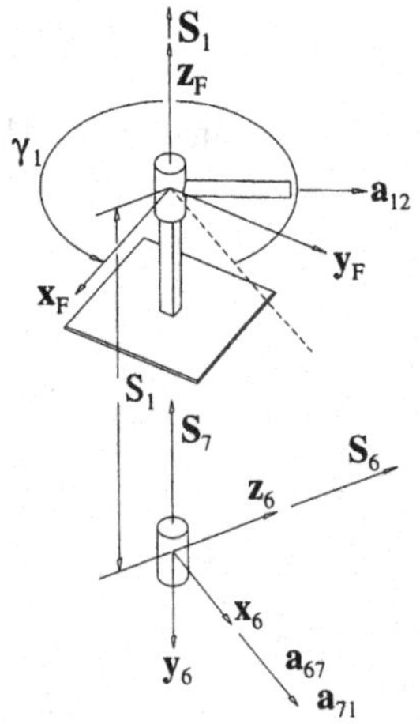

Figure 5.4. Special case where S_7 and S_1 are collinear.

As a numerical example, the following specifications are made:

$$ {}^{6}\mathbf{P}_{\text{tool}} = \begin{bmatrix} 5 \\ 3 \\ 7 \end{bmatrix} \text{ in.}, $$

$$ {}^{F}\mathbf{P}_{\text{tool}} = \begin{bmatrix} 25 \\ 23 \\ 24 \end{bmatrix} \text{ in.}, \quad {}^{F}\mathbf{S}_{6} = \begin{bmatrix} 0.177 \\ 0.884 \\ -0.433 \end{bmatrix}, \quad {}^{F}\mathbf{a}_{67} = \begin{bmatrix} -0.153 \\ 0.459 \\ 0.875 \end{bmatrix}. $$

The following values were calculated by using the close-the-loop procedure and are drawn in Figure 5.5:

$$ \begin{aligned} a_{71} &= -16.68 \text{ in.}, & \alpha_{71} &= 102.50^\circ, \\ S_7 &= 20.67 \text{ in.}, & \theta_7 &= 63.69^\circ, \\ S_1 &= -17.53 \text{ in.}, & \gamma_1 &= -84.79^\circ. \end{aligned} $$

The hypothetical closure link has in effect transformed the open-loop Puma manipulator into a closed-loop spatial mechanism. The parameters for this closed-loop mechanism are listed in Table 5.1. The next chapter will show that this closed-loop spatial mechanism has one degree of freedom. Thus, if one of the variable joint angles of the closed-loop mechanism is known, then the remaining joint variables can be calculated. The angle θ_7 was determined during the close-the-loop process and will serve as the input angle for the one-degree-of-freedom spatial mechanism.

Chapters 7 through 11 will detail how to solve for the variable joint parameters for the vast majority of closed-loop spatial mechanisms. These joint parameters will be the values that are required to position and orient the end effector of corresponding robot manipulators as desired and thus constitute the solution for the reverse kinematic analysis problem.

Table 5.1. *Mechanism parameters for the closed-loop Puma mechanism.*

Link length, in.	Twist angle, deg.	Joint offset, in.	Joint angle, deg.
$a_{12} = 0$	$\alpha_{12} = 90$	$S_1 = -17.53$	$\phi_1 =$ variable
$a_{23} = 17$	$\alpha_{23} = 0$	$S_2 = 5.9$	$\theta_2 =$ variable
$a_{34} = 0.8$	$\alpha_{34} = 270$	$S_3 = 0$	$\theta_3 =$ variable
$a_{45} = 0$	$\alpha_{45} = 90$	$S_4 = 17$	$\theta_4 =$ variable
$a_{56} = 0$	$\alpha_{56} = 90$	$S_5 = 0$	$\theta_5 =$ variable
$a_{67} = 0$	$\alpha_{67} = 90$	$S_6 = 2.0$	$\theta_6 =$ variable
$a_{71} = -16.68$	$\alpha_{71} = 102.50$	$S_7 = 20.67$	$\theta_7 = 63.69$
Robot parameter	User-specified value	Close-the-loop variable	

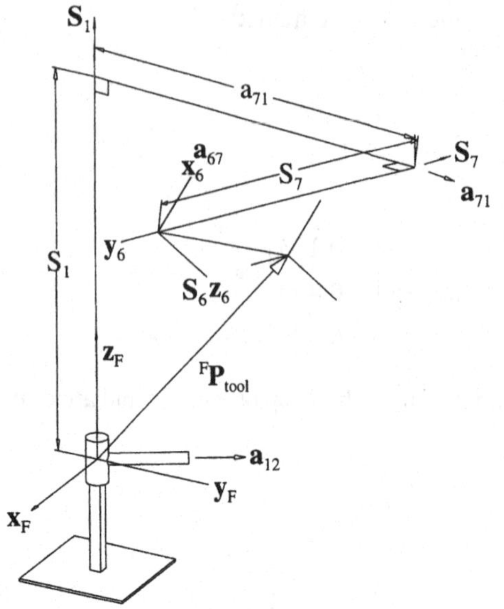

Figure 5.5. Close-the-loop parameters.

5.8 Problems

1. Write a computer function that will perform the close-the-loop analysis. Your program should solve the standard case as well as the two special cases, that is, when $\mathbf{S}_7$ and $\mathbf{S}_1$ are parallel and when $\mathbf{S}_7$ and $\mathbf{S}_1$ are collinear. The C language prototype for your subroutine may be written as

void close_loop (double P_tool_6[3], double P_tool_f[3], double S6_f[3], double a67_f[3], double *a71, double *S7, double *S1, double *al71, double *th7, double *gam1).

Test your subroutine by passing in the values for ${}^6\mathbf{P}_{\text{tool}}$, ${}^F\mathbf{P}_{\text{tool}}$, ${}^F\mathbf{S}_6$, ${}^F\mathbf{a}_{67}$ listed in Section 5.7.

2. The origin of the standard coordinate system attached to an end effector of a robot manipulator is located at the position $\mathbf{r} = 0\mathbf{i} + 0\mathbf{j} + 0\mathbf{k}$. The orientation vector $\mathbf{S}_6$ is parallel to the direction $-\mathbf{j} + \mathbf{k}$, and the vector $\mathbf{a}_{67}$ is parallel to $-\mathbf{j} - \mathbf{k}$. Determine the six close-the-loop parameters for this case.

3. The origin of the standard coordinate system attached to an end effector of a robot manipulator is located at the position $\mathbf{r} = 3\mathbf{j}$. The orientation of the vectors $\mathbf{S}_6$ and $\mathbf{a}_{67}$ as measured in the fixed coordinate system are respectively $[0, 0, -1]^T$ and $[1, 0, 0]^T$.

(a) For the case described, determine the six close-the-loop parameters. Show the angle γ_1 on a drawing, and also indicate the direction of vector $\mathbf{a}_{71}$.

(b) Determine the close-the-loop parameters if the end effector was moved to the origin of the fixed coordinate system while the orientation remained the same.

6

Spherical closed-loop mechanisms

6.1 Equivalent closed-loop spherical mechanism

In the previous chapter, it was shown that any serial manipulator can be transformed into a closed-loop spatial mechanism by constructing a hypothetical closure link. This chapter will focus on the geometry of the new closed-loop mechanism.

A new closed-loop mechanism called the equivalent spherical mechanism will be formed from the original spatial closed-loop mechanism. The first step in creating the equivalent spherical mechanism is to give all the unit joint vectors, $\mathbf{S}_i$, which label revolute or cylindric joint axes, self-parallel translations so that they all meet in a common point O and so that they all point outward from O (see Figure 6.1). Thus the directions of the $\mathbf{S}_i$ vectors are the same for the original spatial mechanism and the cointersecting arrangement.

Consider now that a unit sphere is drawn, centered at point O. The unit vectors $\mathbf{S}_i$ will meet this sphere at a sequence of points, i = 1, 2, 3, . . . , and so forth, as shown in Figure 6.2. Links (arcs of great circles) can be drawn on the unit sphere joining adjacent points, 12, 23, 34, . . . , and so forth. For example, Figure 6.2 illustrates a spherical link joining points 1 and 2 such that the angle between $\mathbf{S}_1$ and $\mathbf{S}_2$ is α_{12}, that is, the same angle as between $\mathbf{S}_1$ and $\mathbf{S}_2$ in the original spatial mechanism. It should be noted that the length of the link connecting points 1 and 2 is $\ell_{12} = r\alpha_{12}$, where α is measured in radians and r is the radius of the unit sphere. For example, if $r = 1$ ft., then $\ell_{12} = \alpha_{12}$ ft.

Finally, the equivalent spherical mechanism is formed by connecting adjacent links $(\alpha_{12}, \alpha_{23})$, $(\alpha_{23}, \alpha_{34})$, . . . , and so forth, with joints. If the joint connecting a pair of adjacent links, for example a_{ij} and a_{jk}, of the original spatial mechanism is a revolute or cylindric joint, then the corresponding adjacent links, α_{ij} and α_{jk}, of the spherical mechanism are joined by a revolute joint. A spherical mechanism can permit relative rotation only between adjacent links. The linear displacement of a cylindric joint is not reflected in the equivalent spherical mechanism. A prismatic joint joining links a_{ij} and a_{jk} in the original spatial mechanism is modeled by a unit vector drawn from point O parallel to the prismatic joint displacement and a solid connection between links α_{ij} and α_{jk}, which preserves the constant angle θ_j between these links. Following this method for connecting adjacent links, the angles α_{ij} and θ_j are defined so as to be the same for the equivalent spherical mechanism and the original spatial mechanism. Thus, *any equations that relate the twist angles and joint angles of the equivalent spherical mechanism will also be valid for the corresponding spatial mechanism.* Figure 6.3 shows a spatial closed-loop five-link mechanism together with its equivalent spherical mechanism.

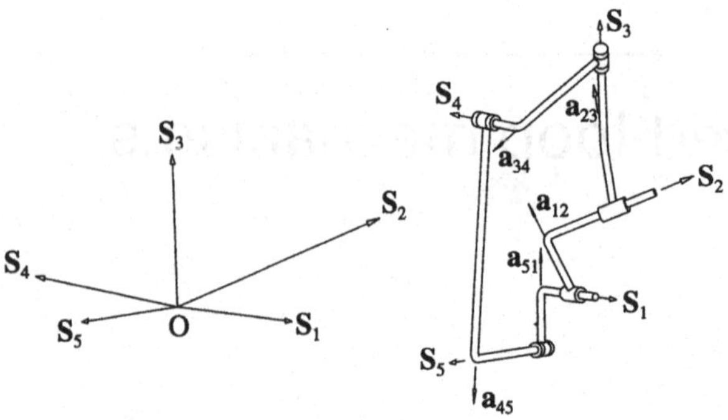

Figure 6.1. Joint axis vectors translated to intersect at a point.

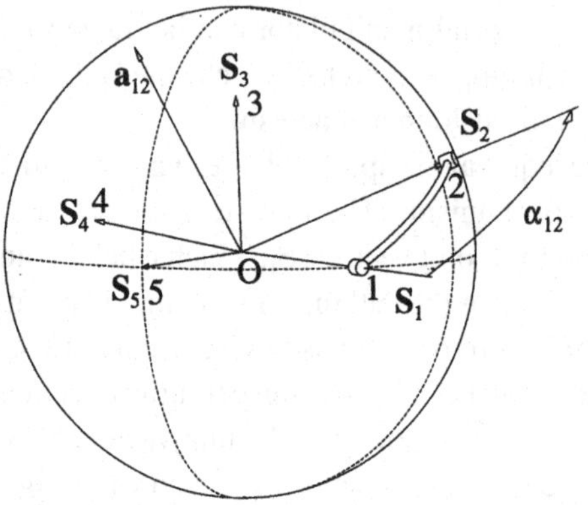

Figure 6.2. Spherical link α_{12} placed between S_1 and S_2.

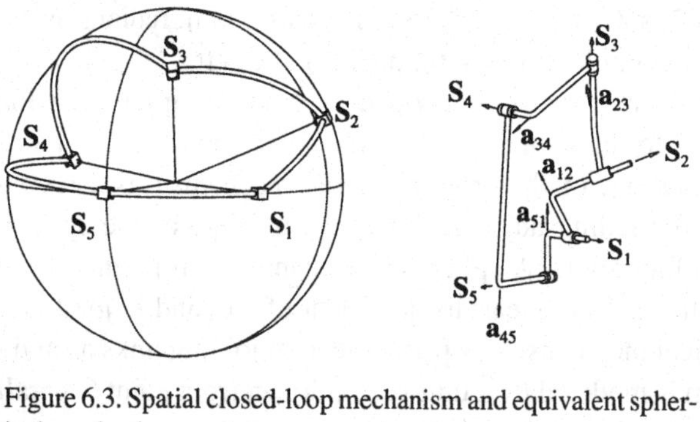

Figure 6.3. Spatial closed-loop mechanism and equivalent spherical mechanism.

6.2 Degrees of freedom

Before generating expressions that contain the twist angles and joint angles of the equivalent spherical mechanism, a method for calculating the number of degrees of freedom of spatial and spherical mechanisms will be presented.

6.2.1 Spatial manipulators and closed-loop mechanisms

In Chapter 2, it was shown that a body in three-dimensional space has six degrees of freedom. Six independent quantities, three related to position and three related to orientation, are needed to completely describe the position and orientation of an object in space. Consider now that there are n unconnected rigid body links. The number of degrees of freedom measured relative to a fixed body or ground, or mobility M, of this set of links is given by

$$M = 6\,n. \tag{6.1}$$

Consider, for example, a system of two unconnected links. This system possesses twelve degrees of freedom. When the pair of links is connected by a revolute joint, the number of degrees of freedom of the system is reduced to seven. Although one of the links possesses six degrees of freedom measured relative to ground, the second link is constrained to rotate about an axis relative to the first link and thus possesses only one additional degree of freedom. The net mobility M of the system of two links is seven. The revolute joint, which allows one relative degree of freedom, has in effect reduced the total mobility of the system by five.

In general, a joint i that connects two links hi and ij will reduce the total mobility of the system by $(6 - f_i)$, where f_i is the number of relative degrees of freedom permitted by joint i. Thus, the net mobility of a system of n links, one of which is connected to ground, is $6(n - 1)$. Further, when they are interconnected by j joints (no two bodies are connected by more than one joint), the net mobility, M, is*

$$M = 6(n - 1) - \sum_{i=1}^{j}(6 - f_i). \tag{6.2}$$

For a single-chain closed-loop spatial mechanism, the number of links will equal the number of joints. Eq. (6.2) reduces to

$$M = \sum_{i=1}^{n} f_i - 6. \tag{6.3}$$

For a serial robot manipulator, the number of joints is one less than the number of links. Eq. (6.2) reduces to

$$M = \sum_{i=1}^{j} f_i. \tag{6.4}$$

If the resulting mobility of a system of links is equal to zero, then the system is a simple structure. If the mobility is less than zero, the system is a redundant structure. If the mobility is equal to one, the overall system has one degree of freedom. Specification of one variable is all that is required to completely position all the links of the system. From Eq. (6.3), a single-chain closed-loop spatial mechanism will have one degree of freedom ($M = 1$) when $\sum_{i=1}^{n} f_i = 7$, that is, the sum of the relative freedoms of the joints equals seven. The spatial mechanism shown in Figure 6.3 has three revolute joints and two cylindric joints and thus has one degree of freedom. In other words, a single input angular displacement will constrain the loop.

* This is the general mobility equation. Special geometry may exist that increases the mobility of the system.

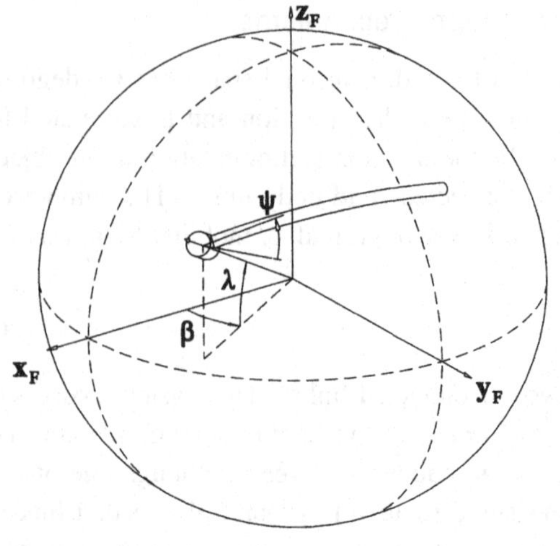

Figure 6.4. Positioning of a spherical link.

6.2.2 Spherical mechanisms

A link of a spherical mechanism has three degrees of freedom. It is shown in Figure 6.4 that the angles β, λ, and ψ completely specify where a link is positioned and oriented on the sphere. By analogy with Eq. (6.2), the mobility of n spherical links, one of which is connected to ground, that are interconnected by j joints may be written as

$$M = 3(n-1) - \sum_{i=1}^{j}(3 - f_i), \tag{6.5}$$

where again f_i represents the relative degrees of freedom of the ith joint.

The number of links for a single-chain closed loop spherical mechanism will equal the number of joints, and for this case Eq. (6.5) reduces to

$$M = \sum_{i=1}^{j} f_i - 3. \tag{6.6}$$

Clearly, when $j = 3$, the mobility for a spherical 3R triangle is zero, $M = 0$. Further, for $j = 4, 5, 6$, and 7 the mobility for a spherical 4R quadrilateral, 5R pentagon, 6R hexagon, and 7R heptagon are respectively $M = 1, 2, 3$, and 4. As an example, the spherical mechanism shown in Figure 6.3, for which $\sum_{i=1}^{5} f_i = 5$ thus possesses $(5-3) = 2$ degrees of freedom.

6.3 Classification of spatial mechanisms

The current objective of this text is to perform a closed-form reverse kinematic analysis for a spatial manipulator. In Chapter 5 it was shown how a hypothetical link could be added to a spatial manipulator to obtain a single-chain closed-loop spatial mechanism. The joint angle for the hypothetical joint (θ_7 for a 6R manipulator) was calculated during the close-the-loop procedure. *When the resulting closed loop mechanism has one degree of freedom, a value of θ_7 is sufficient to define the system. In other words, it is possible*

Table 6.1. *Classification of spatial kinematic chains.*

Group	Number of links	Mechanism
1	4	R-3C
	5	2R-P-2C
	6	3R-2P-C
	7	4R-3P
2	5	3R-2C
	6	4R-P-C
	7	5R-2P
3	6	5R-C
	7	6R-P
4	7	7R

to develop a procedure to compute all the joint parameters of the manipulator, and the reverse analysis will be complete.

Table 6.1 lists all the single closed-loop spatial polygons or closed spatial kinematic chains of links and joints that possess an overall mobility $M = 1$, assuming one link in the chain is held fixed. The various loops are labeled by the numbers of revolute R, prismatic P, and cylindric C kinematic pairs. The listing does not specify the order or sequence of joints.

Reuleaux (1876) stated: "In itself a kinematic chain does not postulate any definite absolute [displacement*]. One must hold fast or fix in position one link of the chain relatively to the portion of surrounding space assumed to be stationary. The relative displacement of links then becomes absolute. A closed kinematic chain of which one link is made stationary is called a mechanism."

The link that is held fixed is called the frame. A change in the selection of a reference frame is known as kinematic inversion. Here we are concerned only with relative displacements, and the relative displacement between any pair of links is independent of the choice of the frame, that is, the kinematic inversion.

In order to identify kinematic inversions it is necessary to firstly specify the sequence of joints. Clearly for four links there is only a single sequence of the four joints R-3C (see Table 6.1). Inversions can be easily identified by drawing planar polygons, and an R-3C chain can be represented by the planar quadrilateral shown in Figure 6.5. An obvious inversion is the spatial four-link RCCC† mechanism with frame a_{41}, an input

* The term "displacement" has been substituted for the term "motion" used in Reuleaux's original text. The motion of a rigid body relative to a reference frame implies not only displacement but velocity, acceleration, and so on (see Hunt (1978)).

† The terminology "R-3C" identifies the types and number of each type of joint that is in a kinematic chain. The terminology "RCCC," however, identifies a specific inversion of an R-3C chain. The input angle is a revolute joint, identifid by the first letter of the sequence. The frame is located between the revolute joint and the cylindric joint represented by the last letter of the sequence. The terminology, where the first letter indicates the type of input joint and the frame is located between the input joint and the joint identified by the last letter of the sequence, will be used throughout the text.

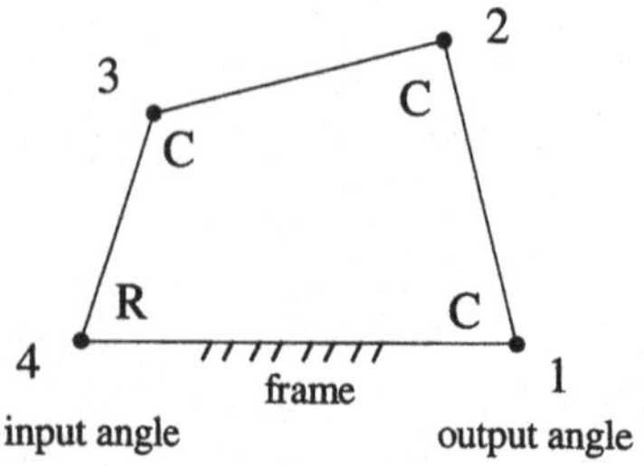

Figure 6.5. Planar representation of an RCCC spatial mechanism.

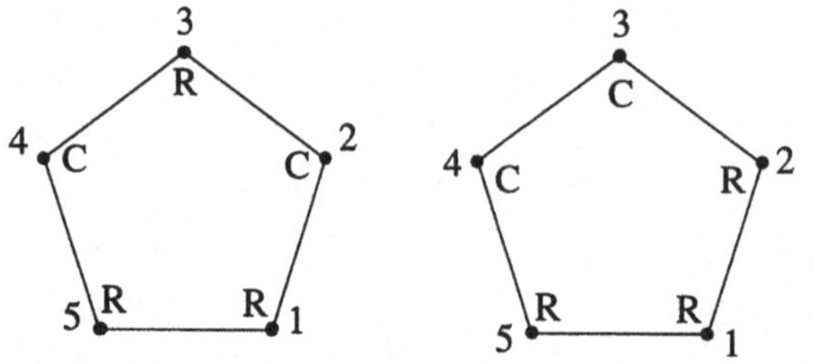

Figure 6.6. Planar representation of two 3R-2C kinematic chains.

joint R, and the output joint C. An electric motor could be used to drive the input joint at constant angular speed. Other inversions are of no practical interest, and from here on only inversions with revolute input joints will be considered.*

A pair of distinct 3R-2C loops is illustrated in Figure 6.6 with kinematic inversions RCRCR, RRCRC, RCRRC, and RCCRR, RRCCR, RRRCC. The process can be continued to identify all inversions of all single-degree-of-freedom closed-loop spatial mechanisms.

Table 6.1 classifies the various kinematic spatial chains, and hence the various inversions, according to group numbers. Each group number is simply the mobility of the equivalent spherical mechanism. For example, the mobility of the equivalent 4R spherical mechanism of the four-link RCCC mechanism is $M = 1$ (see Eq. (6.6)) as are equivalent 4R spherical mechanisms of inversions of the five-link 2R-P-2C, six-link 3R-2P-C, and seven-link 4R-3P kinematic chains. Clearly, all of the inversions of the spatial five-link 3R-2C chains have equivalent 5R spherical mechanisms with mobility $M = 2$.

The grouping of spatial mechanisms according to the mobility M of equivalent spherical mechanisms is important because it can be used to provide a method for the closed-form analysis or the derivation of input–output equations for a given kinematic inversion. This essentially solves the reverse kinematic analysis for serial manipulators. Solutions for the joint parameters of group 1, 2, 3, and 4 mechanisms are presented in detail in Chapters 7, 8, 9, and 10 respectively.

* For a serial manipulator, a hypothetical link is connected between the end link and ground. This hypothetical link is connected to the end link by a revolute joint whose joint angle value is computed. A closed-loop spatial mechanism results where the hypothetical link is the frame and the value of the revolute joint is the input angle. Because closed-loop spatial mechanisms are being analyzed as a means of performing the reverse analysis for a serial manipulator, only closed-loop mechanisms whose input joint is a revolute joint will be considered.

This chapter will continue by providing basic equations that will be used to analyze the various closed-loop mechanisms in Chapters 7 through 10. Specifically, the joint vectors and link vectors will be written in the standard link coordinate systems as defined in Section 3.5 together with other additional coordinate systems. Further equations will be generated that relate the twist angles and joint angles for a spherical triangle, quadrilateral, pentagon, hexagon, and heptagon.

6.4 Generation of expressions for the joint vectors

Expressions for the direction cosines of each of the unit joint vectors ($\mathbf{S}_1$ through $\mathbf{S}_7$) and the unit link vectors ($\mathbf{a}_{12}$ through $\mathbf{a}_{67}$) in each of the standard link coordinate systems are important in the analysis of spherical and spatial mechanisms and manipulators. The derivation of these expressions will begin with the joint vectors expressed in terms of the first standard coordinate system, which has its Z axis along $\mathbf{S}_1$ and its X axis along $\mathbf{a}_{12}$. It should be clear that

$$ {}^1\mathbf{S}_1 = \begin{bmatrix} 0 \\ 0 \\ 1 \end{bmatrix}. \tag{6.7}$$

Analogously, the vector $\mathbf{S}_2$ in the second coordinate system is given by

$$ {}^2\mathbf{S}_2 = \begin{bmatrix} 0 \\ 0 \\ 1 \end{bmatrix}. \tag{6.8}$$

It is important to recognize that all the vectors $\mathbf{S}_i$ and $\mathbf{a}_{ij}$ are drawn from the center of a unit radius sphere and represent the points of penetration on the sphere by the vectors. The coordinates of the point of penetration of the unit vector ${}^2\mathbf{S}_2$ will now be transformed to the first coordinate system by application of the rotational part of the transformation defined by Eq. (3.6). Thus, ${}^2\mathbf{S}_2$ will now be transformed to the first coordinate system by application of the rotational part of the transformation defined by Eq. (3.6). Thus,

$$ {}^1\mathbf{S}_2 = \begin{bmatrix} c_2 & -s_2 & 0 \\ s_2c_{12} & c_2c_{12} & -s_{12} \\ s_2s_{12} & c_2s_{12} & c_{12} \end{bmatrix} \begin{bmatrix} 0 \\ 0 \\ 1 \end{bmatrix} = \begin{bmatrix} 0 \\ -s_{12} \\ c_{12} \end{bmatrix}. \tag{6.9}$$

The process will be extended to obtain the vector $\mathbf{S}_3$ in the first coordinate system as

$$ {}^1\mathbf{S}_3 = {}^1_2\mathbf{R}\,{}^2_3\mathbf{R}\,{}^3\mathbf{S}_3. \tag{6.10}$$

The calculations involved in the matrix multiplication can be reduced by recognizing that ${}^2_3\mathbf{R}^3\mathbf{S}_3 = {}^2\mathbf{S}_3$. The term ${}^2\mathbf{S}_3$ can be obtained simply by an exchange of subscripts $(1 \to 2, 2 \to 3)$ in the right side of Eq. (6.9). Thus, the expression for the direction cosines of $\mathbf{S}_3$ in terms of the first coordinate system can be written as

$$ {}^1\mathbf{S}_3 = \begin{bmatrix} c_2 & -s_2 & 0 \\ s_2c_{12} & c_2c_{12} & -s_{12} \\ s_2s_{12} & c_2s_{12} & c_{12} \end{bmatrix} \begin{bmatrix} 0 \\ -s_{23} \\ c_{23} \end{bmatrix} = \begin{bmatrix} s_{23}s_2 \\ -(s_{12}c_{23} + c_{12}s_{23}c_2) \\ c_{12}c_{23} - s_{12}s_{23}c_2 \end{bmatrix}. \tag{6.11}$$

The right-hand side of Eq. (6.11) is somewhat lengthy, and expressions for ${}^1\mathbf{S}_4$, ${}^1\mathbf{S}_5$, ${}^1\mathbf{S}_6$, and ${}^1\mathbf{S}_7$ are more lengthy and complicated. For convenience a recursive shorthand notation is introduced. This is possible because patterns of combinations of sines and cosines of the twist and joint angles reoccur. The terms $\bar{X}_2$, $\bar{Y}_2$, and $\bar{Z}_2$ are introduced into the right side of Eq. (6.11) and hence

$$^1\mathbf{S}_3 = \begin{bmatrix} \bar{X}_2 \\ \bar{Y}_2 \\ \bar{Z}_2 \end{bmatrix} \tag{6.12}$$

where $\bar{X}_2$, $\bar{Y}_2$, and $\bar{Z}_2$ are defined as

$$\bar{X}_2 = s_{23}s_2, \tag{6.13}$$
$$\bar{Y}_2 = -(s_{12}c_{23} + c_{12}s_{23}c_2), \tag{6.14}$$
$$\bar{Z}_2 = c_{12}c_{23} - s_{12}s_{23}c_2. \tag{6.15}$$

In general, the notation introduced in Eqs. (6.13) to (6.15) can be written as

$$\bar{X}_j = s_{jk}s_j, \tag{6.16}$$
$$\bar{Y}_j = -(s_{ij}c_{jk} + c_{ij}s_{jk}c_j), \tag{6.17}$$
$$\bar{Z}_j = c_{ij}c_{jk} - s_{ij}s_{jk}c_j, \tag{6.18}$$

where the subscript $j = i + 1$ and $k = j + 1$.

Single subscript terms with a different combination of twist angles and joint angles will appear repeatedly. These terms will be defined by X_j, Y_j, and Z_j, where

$$X_j = s_{ij}s_j, \tag{6.19}$$
$$Y_j = -(s_{jk}c_{ij} + c_{jk}s_{ij}c_j), \tag{6.20}$$
$$Z_j = c_{jk}c_{ij} - s_{jk}s_{ij}c_j. \tag{6.21}$$

The definitions $\bar{X}_j$, $\bar{Y}_j$, $\bar{Z}_j$ and X_j, Y_j, Z_j can be related to the geometry of the spherical dyad shown in Figure 6.7 by first writing

$$X = s\,s, \tag{6.22}$$
$$Y = -(s\,c + c\,s\,c), \tag{6.23}$$
$$Z = c\,c - s\,s\,c. \tag{6.24}$$

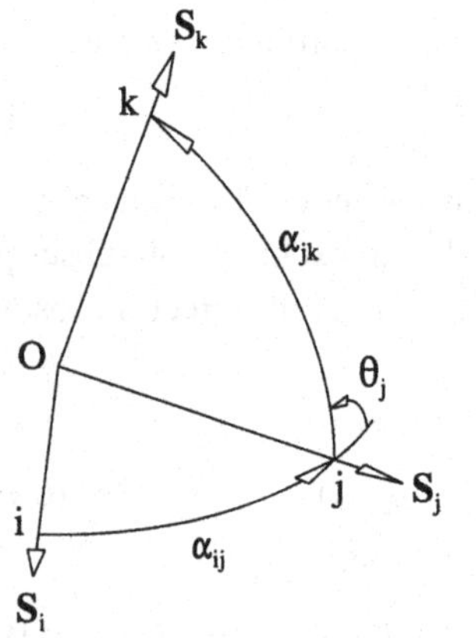

Figure 6.7. Spherical dyad.

The subscript j is an indicator that the expressions contain the joint angle θ_j. Hence,

$$\begin{aligned} \bar{X}_j &= s\, s_j, & X_j &= s\, s_j, \\ \bar{Y}_j &= -(s\, c + c\, s\, c_j), & Y_j &= -(s\, c + c\, s\, c_j), \\ \bar{Z}_j &= c\, c - s\, s\, c_j, & Z_j &= c\, c - s\, s\, c_j. \end{aligned} \tag{6.25}$$

Now, $\bar{X}_j = s_{jk}s_j$ indicates an approach along arc α_{jk} to vertex j, whereas $X_j = s_{ij}s_j$ indicates an approach along arc α_{ij} to vertex j. Hence, one can write

$$\begin{aligned} \bar{X}_j &= s_{jk}\, s_j, & X_j &= s_{ij}\, s_j, \\ \bar{Y}_j &= -(s\, c_{jk} + c\, s_{jk}\, c_j), & Y_j &= -(s\, c_{ij} + c\, s_{ij}\, c_j), \\ \bar{Z}_j &= c\, c_{jk} - s\, s_{jk}\, c_j, & Z_j &= c\, c_{ij} - s\, s_{ij}\, c_j. \end{aligned} \tag{6.26}$$

The remaining unlabeled angle in each of these expressions is simply the other angle completing the dyad, which is respectively α_{ij} and α_{jk}.

Because the vector ${}^1\mathbf{S}_3$ is a unit vector, the sum of the squares of the elements of the vector as expressed in Eqs. (6.13) through (6.15) will equal one. In general, from Eq. (6.16) through Eq. (6.21)

$$\bar{X}_j^2 + \bar{Y}_j^2 + \bar{Z}_j^2 = 1 \tag{6.27}$$

and

$$X_j^2 + Y_j^2 + Z_j^2 = 1. \tag{6.28}$$

Further, a comparison of Eqs. (6.16) through (6.18) and Eqs. (6.19) through (6.21) yields $\bar{X}_j \neq X_j$ and $\bar{Y}_j \neq Y_j$. However,

$$\bar{Z}_j = Z_j. \tag{6.29}$$

The procedure now continues by determining expressions for the vector $\mathbf{S}_4$ in terms of the first coordinate system. The vector ${}^1\mathbf{S}_4$ can be expressed as

$${}^1\mathbf{S}_4 = {}^1_2\mathbf{R}\, {}^2\mathbf{S}_4. \tag{6.30}$$

The term ${}^2\mathbf{S}_4$ can be obtained by an exchange of the coefficients of Eq. (6.12). Thus, ${}^1\mathbf{S}_4$ can be written as

$${}^1\mathbf{S}_4 = \begin{bmatrix} c_2 & -s_2 & 0 \\ s_2c_{12} & c_2c_{12} & -s_{12} \\ s_2s_{12} & c_2s_{12} & c_{12} \end{bmatrix} \begin{bmatrix} \bar{X}_3 \\ \bar{Y}_3 \\ \bar{Z}_3 \end{bmatrix} = \begin{bmatrix} \bar{X}_3c_2 - \bar{Y}_3s_2 \\ c_{12}(\bar{X}_3s_2 + \bar{Y}_3c_2) - s_{12}\bar{Z}_3 \\ s_{12}(\bar{X}_3s_2 + \bar{Y}_3c_2) + c_{12}\bar{Z}_3 \end{bmatrix}, \tag{6.31}$$

where $\bar{X}_3 = s_{34}s_3$, $\bar{Y}_3 = -(s_{23}c_{34} + c_{23}s_{34}c_3)$, and $\bar{Z}_3 = (c_{23}c_{34} - s_{23}s_{34}c_3)$. The terms, "$X_{32}$," "$Y_{32}$," and "$Z_{32}$" will now be defined as

$$X_{32} = \bar{X}_3c_2 - \bar{Y}_3s_2, \tag{6.32}$$

$$Y_{32} = c_{12}(\bar{X}_3s_2 + \bar{Y}_3c_2) - s_{12}\bar{Z}_3, \tag{6.33}$$

$$Z_{32} = s_{12}(\bar{X}_3s_2 + \bar{Y}_3c_2) + c_{12}\bar{Z}_3 \tag{6.34}$$

so that (6.31) can be written as

$$ {}^{1}\mathbf{S}_4 = \begin{bmatrix} X_{32} \\ Y_{32} \\ Z_{32} \end{bmatrix}. \tag{6.35}$$

In general, the terms X_{kj}, Y_{kj}, and Z_{kj}, and a new term, X^*_{kj}, will be defined as

$$X_{kj} = \bar{X}_k c_j - \bar{Y}_k s_j, \tag{6.36}$$

$$X^*_{kj} = \bar{X}_k s_j + \bar{Y}_k c_j, \tag{6.37}$$

$$Y_{kj} = c_{ij}(\bar{X}_k s_j + \bar{Y}_k c_j) - s_{ij}\bar{Z}_k, \tag{6.38}$$

$$Z_{kj} = s_{ij}(\bar{X}_k s_j + \bar{Y}_k c_j) + c_{ij}\bar{Z}_k, \tag{6.39}$$

where $j = i+1$ and $k = j+1$ and where $\bar{X}_k$, $\bar{Y}_k$, and $\bar{Z}_k$ are defined in Eqs. (6.16) through (6.18). Further, the terms X_{ij}, Y_{ij}, Z_{ij}, and X^*_{ij} will be defined for future use as

$$X_{ij} = X_i c_j - Y_i s_j, \tag{6.40}$$

$$X^*_{ij} = X_i s_j + Y_i c_j, \tag{6.41}$$

$$Y_{ij} = c_{jk}(X_i s_j + Y_i c_j) - s_{jk} Z_i, \tag{6.42}$$

$$Z_{ij} = s_{jk}(X_i s_j + Y_i c_j) + c_{jk} Z_i, \tag{6.43}$$

where X_i, Y_i, and Z_i are defined in Eqs. (6.19) through (6.21).

Again, it can be shown that the following equations are true for the double subscripted terms

$$X^2_{kj} + Y^2_{kj} + Z^2_{kj} = 1, \tag{6.44}$$

$$X^2_{ij} + Y^2_{ij} + Z^2_{ij} = 1, \tag{6.45}$$

$$Z_{ij} = Z_{ji}. \tag{6.46}$$

The procedure now continues by determining expressions for the vector $\mathbf{S}_5$ in terms of the first coordinate system. The vector ${}^{1}\mathbf{S}_5$ can be expressed as

$$ {}^{1}\mathbf{S}_5 = {}^{1}_{2}\mathbf{R}\,{}^{2}\mathbf{S}_5. \tag{6.47}$$

The term ${}^{2}\mathbf{S}_5$ can be obtained by an exchange $(1 \to 2, 2 \to 3, 3 \to 4, 4 \to 5)$ of the coefficients of Eq. (6.35). Thus, ${}^{1}\mathbf{S}_5$ can be written as

$$ {}^{1}\mathbf{S}_5 = \begin{bmatrix} c_2 & -s_2 & 0 \\ s_2 c_{12} & c_2 c_{12} & -s_{12} \\ s_2 s_{12} & c_2 s_{12} & c_{12} \end{bmatrix} \begin{bmatrix} X_{43} \\ Y_{43} \\ Z_{43} \end{bmatrix} = \begin{bmatrix} X_{43}c_2 - Y_{43}s_2 \\ c_{12}(X_{43}s_2 + Y_{43}c_2) - s_{12}Z_{43} \\ s_{12}(X_{43}s_2 + Y_{43}c_2) + c_{12}Z_{43} \end{bmatrix}. \tag{6.48}$$

The terms X_{432}, Y_{432}, and Z_{432} will now be defined as

$$X_{432} = X_{43}c_2 - Y_{43}s_2, \tag{6.49}$$

$$Y_{432} = c_{12}(X_{43}s_2 + Y_{43}c_2) - s_{12}Z_{43}, \tag{6.50}$$

$$Z_{432} = s_{12}(X_{43}s_2 + Y_{43}c_2) + c_{12}Z_{43} \tag{6.51}$$

so that the vector ${}^{1}\mathbf{S}_5$ can be expressed as

$$ {}^{1}\mathbf{S}_5 = \begin{bmatrix} X_{432} \\ Y_{432} \\ Z_{432} \end{bmatrix}. \tag{6.52} $$

It should be clear at this point that the expressions for X, Y, and Z with three subscripts are recursive. The definitions for expressions with three subscripts in increasing order are written as

$$ X_{ijk} = X_{ij}c_k - Y_{ij}s_k, \tag{6.53} $$
$$ X^{*}_{ijk} = X_{ij}s_k + Y_{ij}c_k, \tag{6.54} $$
$$ Y_{ijk} = c_{kl}(X_{ij}s_k + Y_{ij}c_k) - s_{kl}Z_{ij}, \tag{6.55} $$
$$ Z_{ijk} = s_{kl}(X_{ij}s_k + Y_{ij}c_k) + c_{kl}Z_{ij}, \tag{6.56} $$

and the definitions for expressions with three subscripts in decreasing order are

$$ X_{kji} = X_{kj}c_i - Y_{kj}s_i, \tag{6.57} $$
$$ X^{*}_{kji} = X_{kj}s_i + Y_{kj}c_i, \tag{6.58} $$
$$ Y_{kji} = c_{hi}(X_{kj}s_i + Y_{kj}c_i) - s_{hi}Z_{kj}, \tag{6.59} $$
$$ Z_{kji} = s_{hi}(X_{kj}s_i + Y_{kj}c_i) + c_{hi}Z_{kj}, \tag{6.60} $$

where $i = h + 1, j = i + 1$ and $k = j + 1$. Further, for a mechanism with n joints, joint $n + 1$ (the joint after joint n) is joint 1.

The process could be repeated two more times to obtain expressions for the vectors $\mathbf{S}_6$ and $\mathbf{S}_7$ in terms of the first coordinate system. However, the notation is recursive and the results are

$$ {}^{1}\mathbf{S}_6 = \begin{bmatrix} X_{5432} \\ Y_{5432} \\ Z_{5432} \end{bmatrix} \tag{6.61} $$

and

$$ {}^{1}\mathbf{S}_7 = \begin{bmatrix} X_{65432} \\ Y_{65432} \\ Z_{65432} \end{bmatrix}, \tag{6.62} $$

where

$$ X_{5432} = X_{543}c_2 - Y_{543}s_2, \tag{6.63} $$
$$ Y_{5432} = c_{12}(X_{543}s_2 + Y_{543}c_2) - s_{12}Z_{543}, \tag{6.64} $$
$$ Z_{5432} = s_{12}(X_{543}s_2 + Y_{543}c_2) + c_{12}Z_{543} \tag{6.65} $$

and

$$X_{65432} = X_{6543}c_2 - Y_{6543}s_2, \tag{6.66}$$
$$Y_{65432} = c_{12}(X_{6543}s_2 + Y_{6543}c_2) - s_{12}Z_{6543}, \tag{6.67}$$
$$Z_{65432} = s_{12}(X_{6543}s_2 + Y_{6543}c_2) + c_{12}Z_{6543}. \tag{6.68}$$

The definitions for expressions with four subscripts in increasing order are written as

$$X_{hijk} = X_{hij}c_k - Y_{hij}s_k, \tag{6.69}$$
$$X^*_{hijk} = X_{hij}s_k + Y_{hij}c_k, \tag{6.70}$$
$$Y_{hijk} = c_{kl}(X_{hij}s_k + Y_{hij}c_k) - s_{kl}Z_{hij}, \tag{6.71}$$
$$Z_{hijk} = s_{kl}(X_{hij}s_k + Y_{hij}c_k) + c_{kl}Z_{hij}, \tag{6.72}$$

and the definitions for expressions with four subscripts in decreasing order are

$$X_{kjih} = X_{kji}c_h - Y_{kji}s_h, \tag{6.73}$$
$$X^*_{kjih} = X_{kji}s_h + Y_{kji}c_h, \tag{6.74}$$
$$Y_{kjih} = c_{gh}(X_{kji}s_h + Y_{kji}c_h) - s_{gh}Z_{kji}, \tag{6.75}$$
$$Z_{kjih} = s_{gh}(X_{kji}s_h + Y_{kji}c_h) + c_{gh}Z_{kji}. \tag{6.76}$$

The definitions for expressions with five subscripts in increasing order are written as

$$X_{hijkl} = X_{hijk}c_l - Y_{hijk}s_l, \tag{6.77}$$
$$X^*_{hijkl} = X_{hijk}s_l + Y_{hijk}c_l, \tag{6.78}$$
$$Y_{hijkl} = c_{lm}(X_{hijk}s_l + Y_{hijk}c_l) - s_{lm}Z_{hijk}, \tag{6.79}$$
$$Z_{hijkl} = s_{lm}(X_{hijk}s_l + Y_{hijk}c_l) + c_{lm}Z_{hijk}, \tag{6.80}$$

and the definitions for expressions with five subscripts in decreasing order are

$$X_{lkjih} = X_{lkji}c_h - Y_{lkji}s_h, \tag{6.81}$$
$$X^*_{lkjih} = X_{lkji}s_h + Y_{lkji}c_h, \tag{6.82}$$
$$Y_{lkjih} = c_{gh}(X_{lkji}s_h + Y_{lkji}c_h) - s_{gh}Z_{lkji}, \tag{6.83}$$
$$Z_{lkjih} = s_{gh}(X_{lkji}s_h + Y_{lkji}c_h) + c_{gh}Z_{lkji}, \tag{6.84}$$

where $h = g + 1, i = h + 1, j = i + 1, k = j + 1, \ell = k + 1$, and $m = \ell + 1$.

As before, the sum of the squares of the multisubscripted X, Y, and Z terms will equal one. Also it is true that

$$Z_{ij...mn} = Z_{nm...ji}. \tag{6.85}$$

In summary, expressions have been found for the joint vectors $\mathbf{S}_1$ through $\mathbf{S}_7$ in terms of the first coordinate system. The results can be summarized as follows:

$$^1\mathbf{S}_1 = \begin{bmatrix} 0 \\ 0 \\ 1 \end{bmatrix}, \quad ^1\mathbf{S}_2 = \begin{bmatrix} 0 \\ -s_{12} \\ c_{12} \end{bmatrix}, \quad ^1\mathbf{S}_3 = \begin{bmatrix} \bar{X}_2 \\ \bar{Y}_2 \\ \bar{Z}_2 \end{bmatrix}, \quad ^1\mathbf{S}_n = \begin{bmatrix} X_{n-1,n-2,...,2} \\ Y_{n-1,n-2,...,2} \\ Z_{n-1,n-2,...,2} \end{bmatrix} \tag{6.86}$$

for $n = 4, 5, 6$, or 7. A similar procedure could be used to obtain the direction cosines for the joint vectors in terms of other standard link coordinate systems.

6.5 Generation of expressions for the link vectors

The direction cosines for the link vectors will be determined in terms of the first coordinate system in a manner similar to that used in the previous section for the joint vectors. The components of the vector $\mathbf{a}_{12}$ measured in the first coordinate system are given by

$$ {}^{1}\mathbf{a}_{12} = \begin{bmatrix} 1 \\ 0 \\ 0 \end{bmatrix}. \tag{6.87} $$

The components of the vector $\mathbf{a}_{23}$ in terms of the second coordinate system are given by

$$ {}^{2}\mathbf{a}_{23} = \begin{bmatrix} 1 \\ 0 \\ 0 \end{bmatrix}. \tag{6.88} $$

Transforming this vector to the first coordinate system via application of the rotation matrix part of Eq. (3.6) gives

$$ {}^{1}\mathbf{a}_{23} = \begin{bmatrix} c_2 & -s_2 & 0 \\ s_2 c_{12} & c_2 c_{12} & -s_{12} \\ s_2 s_{12} & c_2 s_{12} & c_{12} \end{bmatrix} \begin{bmatrix} 1 \\ 0 \\ 0 \end{bmatrix} = \begin{bmatrix} c_2 \\ s_2 c_{12} \\ s_2 s_{12} \end{bmatrix}. \tag{6.89} $$

The vector $\mathbf{a}_{34}$ can be transformed from the third coordinate system to the first system by the following two successive rotations

$$ {}^{1}\mathbf{a}_{34} = {}^{1}_{2}\mathbf{R}\,{}^{2}_{3}\mathbf{R}\,{}^{3}\mathbf{a}_{34}. \tag{6.90} $$

The term ${}^{2}_{3}\mathbf{R}\,{}^{3}\mathbf{a}_{34}$ is equivalent to ${}^{2}\mathbf{a}_{34}$, and this term can be obtained by matrix multiplication or by an exchange of subscripts $(1 \to 2, 2 \to 3, 3 \to 4)$ of Eq. (6.89). The vector ${}^{1}\mathbf{a}_{34}$ is then given by

$$ {}^{1}\mathbf{a}_{34} = \begin{bmatrix} c_2 & -s_2 & 0 \\ s_2 c_{12} & c_2 c_{12} & -s_{12} \\ s_2 s_{12} & c_2 s_{12} & c_{12} \end{bmatrix} \begin{bmatrix} c_3 \\ s_3 c_{23} \\ s_3 s_{23} \end{bmatrix} = \begin{bmatrix} c_2 c_3 - s_2 s_3 c_{23} \\ -s_{12}(s_3 s_{23}) + c_{12}(s_2 c_3 + c_2 s_3 c_{23}) \\ c_{12}(s_3 s_{23}) + s_{12}(s_2 c_3 + c_2 s_3 c_{23}) \end{bmatrix}. \tag{6.91} $$

A shorthand notation for the right-hand side of Eq. (6.91) is introduced using the following definitions

$$ U_{32} = s_3 s_{23}, \tag{6.92} $$

$$ V_{32} = -(s_2 c_3 + c_2 s_3 c_{23}), \tag{6.93} $$

$$ W_{32} = c_2 c_3 - s_2 s_3 c_{23}. \tag{6.94} $$

In general, the notation introduced by Eqs. (6.92) through (6.94) can be written as

$$ U_{ji} = s_j s_{ij}, \tag{6.95} $$

$$ V_{ji} = -(s_i c_j + c_i s_j c_{ij}), \tag{6.96} $$

$$ W_{ji} = c_i c_j - s_i s_j c_{ij}, \tag{6.97} $$

where $j = i + 1$. Analogously, the terms U_{ij}, V_{ij}, and W_{ij} are defined as

$$U_{ij} = s_i s_{ij}, \tag{6.98}$$
$$V_{ij} = -(s_j c_i + c_j s_i c_{ij}), \tag{6.99}$$
$$W_{ij} = c_j c_i - s_j s_i c_{ij}. \tag{6.100}$$

Finally, further abbreviations are introduced as follows

$$U_{321} = U_{32} c_{12} - V_{32} s_{12}, \tag{6.101}$$
$$U^*_{321} = U_{32} s_{12} + V_{32} c_{12} \tag{6.102}$$

so that Eq. (6.91) can now be written in the abbreviated form

$${}^1\mathbf{a}_{34} = \begin{bmatrix} W_{32} \\ -U^*_{321} \\ U_{321} \end{bmatrix}. \tag{6.103}$$

Before proceeding with the determination of the direction cosines of the vector $\mathbf{a}_{45}$ in the first coordinate system, it is instructive to introduce all the definitions for the terms U, U*, V, and W with multiple subscripts. Eqs. (6.95) through (6.100) have defined the expressions for U, V, and W terms with double subscripts in both ascending and descending order. Eqs. (6.101) and (6.102) have introduced some of the triple-subscript terms. All the triple-subscript expressions are defined as follows:

$$U_{ijk} = U_{ij} c_{jk} - V_{ij} s_{jk}, \tag{6.104}$$
$$U^*_{ijk} = U_{ij} s_{jk} + V_{ij} c_{jk}, \tag{6.105}$$
$$V_{ijk} = c_k (U_{ij} s_{jk} + V_{ij} c_{jk}) - s_k W_{ij}, \tag{6.106}$$
$$W_{ijk} = s_k (U_{ij} s_{jk} + V_{ij} c_{jk}) + c_k W_{ij}. \tag{6.107}$$

Expressions for U, U*, V, and W with three or more subscripts are recursive and are valid for both ascending and descending order.

Expressions for four subscripts for U, U*, V, and W are

$$U_{hijk} = U_{hij} c_{jk} - V_{hij} s_{jk}, \tag{6.108}$$
$$U^*_{hijk} = U_{hij} s_{jk} + V_{hij} c_{jk}, \tag{6.109}$$
$$V_{hijk} = c_k (U_{hij} s_{jk} + V_{hij} c_{jk}) - s_k W_{hij}, \tag{6.110}$$
$$W_{hijk} = s_k (U_{hij} s_{jk} + V_{hij} c_{jk}) + c_k W_{hij}, \tag{6.111}$$

expressions for five subscripts for U, U*, V, and W are

$$U_{ghijk} = U_{ghij} c_{jk} - V_{ghij} s_{jk}, \tag{6.112}$$
$$U^*_{ghijk} = U_{ghij} s_{jk} + V_{ghij} c_{jk}, \tag{6.113}$$
$$V_{ghijk} = c_k (U_{ghij} s_{jk} + V_{ghij} c_{jk}) - s_k W_{ghij}, \tag{6.114}$$
$$W_{ghijk} = s_k (U_{ghij} s_{jk} + V_{ghij} c_{jk}) + c_k W_{ghij}, \tag{6.115}$$

and expressions for six subscripts for U, U*, V, and W are

$$U_{fghijk} = U_{fghij}c_{jk} - V_{fghij}s_{jk}, \tag{6.116}$$

$$U^*_{fghijk} = U_{fghij}s_{jk} + V_{fghij}c_{jk}, \tag{6.117}$$

$$V_{fghijk} = c_k(U_{fghij}s_{jk} + V_{fghij}c_{jk}) - s_k W_{fghij}, \tag{6.118}$$

$$W_{fghijk} = s_k(U_{fghij}s_{jk} + V_{fghij}c_{jk}) + c_k W_{fghij}. \tag{6.119}$$

It can be shown for any number of subscripts that

$$W_{ij...mn} = W_{nm...ji} \tag{6.120}$$

and

$$U^2_{ij...mn} + V^2_{ij...mn} + W^2_{ij...mn} = 1. \tag{6.121}$$

Returning now to the problem of transforming the link vectors to the first coordinate system, the vector $\mathbf{a}_{45}$ can be transformed to the first coordinate system as follows

$${}^1\mathbf{a}_{45} = {}^1_2\mathbf{R}\,{}^2\mathbf{a}_{45}. \tag{6.122}$$

The term ${}^2\mathbf{a}_{45}$ may be obtained by an exchange of subscripts in Eq. (6.103), and thus Eq. (6.122) may be written as

$${}^1\mathbf{a}_{45} = \begin{bmatrix} c_2 & -s_2 & 0 \\ s_2c_{12} & c_2c_{12} & -s_{12} \\ s_2s_{12} & c_2s_{12} & c_{12} \end{bmatrix} \begin{bmatrix} W_{43} \\ -U^*_{432} \\ U_{432} \end{bmatrix} = \begin{bmatrix} s_2U^*_{432} + c_2W_{43} \\ -s_{12}U_{432} - c_{12}(c_2U^*_{432} - s_2W_{43}) \\ c_{12}U_{432} - s_{12}(c_2U^*_{432} - s_2W_{43}) \end{bmatrix} \tag{6.123}$$

or

$${}^1\mathbf{a}_{45} = \begin{bmatrix} W_{432} \\ -U^*_{4321} \\ U_{4321} \end{bmatrix}. \tag{6.124}$$

The vector ${}^1\mathbf{a}_{56}$ may be written as

$${}^1\mathbf{a}_{56} = {}^1_2\mathbf{R}\,{}^2\mathbf{a}_{56}. \tag{6.125}$$

The term ${}^2\mathbf{a}_{56}$ may be obtained by the exchange $(1\rightarrow 2, 2\rightarrow 3, 3\rightarrow 4, 4\rightarrow 5,\ 5\rightarrow 6)$ from Eq. (6.124). Eq. (6.125) now becomes

$${}^1\mathbf{a}_{56} = \begin{bmatrix} c_2 & -s_2 & 0 \\ s_2c_{12} & c_2c_{12} & -s_{12} \\ s_2s_{12} & c_2s_{12} & c_{12} \end{bmatrix} \begin{bmatrix} W_{543} \\ -U^*_{5432} \\ U_{5432} \end{bmatrix} = \begin{bmatrix} s_2U^*_{5432} + c_2W_{543} \\ -s_{12}U_{5432} - c_{12}(c_2U^*_{5432} - s_2W_{543}) \\ c_{12}U_{5432} - s_{12}(c_2U^*_{5432} - s_2W_{543}) \end{bmatrix} \tag{6.126}$$

or

$${}^1\mathbf{a}_{56} = \begin{bmatrix} W_{5432} \\ -U^*_{54321} \\ U_{54321} \end{bmatrix}. \tag{6.127}$$

Table 6.2. *Direction cosines expressed in the 1st standard coordinate system.*

$\mathbf{S}_1$	$(0, 0, 1)$	$\mathbf{a}_{12}$	$(1, 0, 0)$
$\mathbf{S}_2$	$(0, -s_{12}, c_{12})$	$\mathbf{a}_{23}$	(c_2, s_2c_{12}, U_{21})
$\mathbf{S}_3$	$(\bar{X}_2, \bar{Y}_2, \bar{Z}_2)$	$\mathbf{a}_{34}$	$(W_{32}, -U^*_{321}, U_{321})$
$\mathbf{S}_4$	(X_{32}, Y_{32}, Z_{32})	$\mathbf{a}_{45}$	$(W_{432}, -U^*_{4321}, U_{4321})$
$\mathbf{S}_5$	$(X_{432}, Y_{432}, Z_{432})$	$\mathbf{a}_{56}$	$(W_{5432}, -U^*_{54321}, U_{54321})$
$\mathbf{S}_6$	$(X_{5432}, Y_{5432}, Z_{5432})$	$\mathbf{a}_{67}$	$(W_{65432}, -U^*_{654321}, U_{654321})$
$\mathbf{S}_7$	$(X_{65432}, Y_{65432}, Z_{65432})$		

Lastly, the vector ${}^1\mathbf{a}_{67}$ may be written as

$$ {}^1\mathbf{a}_{67} = {}^1_2\mathbf{R}\,{}^2\mathbf{a}_{67}. \tag{6.128}$$

The term ${}^1\mathbf{a}_{67}$ may be obtained by the exchange $(1 \to 2, 2 \to 3, 3 \to 4, 4 \to 5, 5 \to 6, 6 \to 7)$ from Eq. (6.127). Equation (6.128) now becomes

$$ {}^1\mathbf{a}_{67} = \begin{bmatrix} c_2 & -s_2 & 0 \\ s_2c_{12} & c_2c_{12} & -s_{12} \\ s_2s_{12} & c_2s_{12} & c_{12} \end{bmatrix} \begin{bmatrix} W_{6543} \\ -U^*_{65432} \\ U_{65432} \end{bmatrix} = \begin{bmatrix} s_2U^*_{65432} + c_2W_{6543} \\ -s_{12}U_{65432} - c_{12}\left(c_2U^*_{65432} - s_2W_{6543}\right) \\ c_{12}U_{65432} - s_{12}\left(c_2U^*_{65432} - s_2W_{6543}\right) \end{bmatrix} \tag{6.129}$$

or

$$ {}^1\mathbf{a}_{67} = \begin{bmatrix} W_{65432} \\ -U^*_{654321} \\ U_{654321} \end{bmatrix}. \tag{6.130}$$

Expressions have now been found for the direction cosines of the unit joint vectors $\mathbf{S}_1$ through $\mathbf{S}_6$ and for the unit link vectors $\mathbf{a}_{12}$ through $\mathbf{a}_{67}$ in terms of the first coordinate system. The results are summarized in Table 6.2. The process may be repeated to determine the direction cosines of the point vectors and the link vectors in any of the standard coordinate systems.

6.6 Spherical triangle

6.6.1 Derivation of fundamental sine, sine–cosine, and cosine laws

The spherical triangle shown in Figure 6.8 has mobility $M = 0$ (see Eq. (6.6)) and is therefore a structure. The objective here is to generate a series of equations that relate the twist angles, α_{ij}, and joint angles, θ_i.

It has been shown that the direction cosines of the vector $\mathbf{S}_3$ in terms of the first coordinate system can be obtained by transforming the vector from the third coordinate system

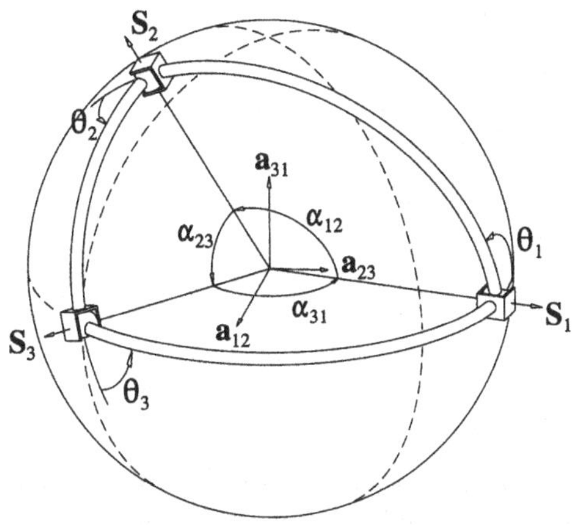

Figure 6.8. Spherical triangle.

to the second and finally to the first coordinate system. The result of the transformations is

$$^1\mathbf{S}_3 = \begin{bmatrix} \bar{X}_2 \\ \bar{Y}_2 \\ \bar{Z}_2 \end{bmatrix}. \tag{6.131}$$

Because the spherical triangle is a closed-loop structure, the vector $\mathbf{S}_3$ can be transformed directly to the first coordinate system. This transformation, ${}^3_1\mathbf{R}$, can be obtained by starting with a coordinate system B initially coincident with the third coordinate system. B is rotated about the vector $\mathbf{a}_{31}$ (the X axis of the third coordinate system) by the angle α_{31}. The Z axis of the B coordinate system is now aligned with the Z axis of the first coordinate system. The B system is now rotated about the Z axis by the angle θ_1. The B coordinate system is now coincident with the first coordinate system. Thus, the rotational transformation that relates the third and first coordinate systems can be written as

$${}^3_1\mathbf{R} = \begin{bmatrix} 1 & 0 & 0 \\ 0 & c_{31} & -s_{31} \\ 0 & s_{31} & c_{31} \end{bmatrix} \begin{bmatrix} c_1 & -s_1 & 0 \\ s_1 & c_1 & 0 \\ 0 & 0 & 1 \end{bmatrix} = \begin{bmatrix} c_1 & -s_1 & 0 \\ c_{31}s_1 & c_{31}c_1 & -s_{31} \\ s_{31}s_1 & s_{31}c_1 & c_{31} \end{bmatrix}. \tag{6.132}$$

Using the inverse of this transformation to obtain the vector $^1\mathbf{S}_3$ gives

$$^1\mathbf{S}_3 = {}^1_3\mathbf{R}\,{}^3\mathbf{S}_3 = \begin{bmatrix} c_1 & c_{31}s_1 & s_{31}s_1 \\ -s_1 & c_{31}c_1 & s_{31}c_1 \\ 0 & -s_{31} & c_{31} \end{bmatrix} \begin{bmatrix} 0 \\ 0 \\ 1 \end{bmatrix} = \begin{bmatrix} s_{31}s_1 \\ s_{31}c_1 \\ c_{31} \end{bmatrix}. \tag{6.133}$$

Equating Eqs. (6.131) and (6.133) gives

$$\bar{X}_2 = s_{31}s_1, \tag{6.134}$$

$$\bar{Y}_2 = s_{31}c_1, \tag{6.135}$$

$$\bar{Z}_2 = c_{31}. \tag{6.136}$$

Equations (6.134) through (6.136) are respectively the sine, sine–cosine, and cosine laws for a spherical triangle.

Table 6.3. *Fundamental sine, sine–cosine, and cosine laws for a spherical triangle.*

$X_1 = s_{23}s_2$	$X_2 = s_{31}s_3$	$X_3 = s_{12}s_1$
$Y_1 = s_{23}c_2$	$Y_2 = s_{31}c_3$	$Y_3 = s_{12}c_1$
$Z_1 = c_{23}$	$Z_2 = c_{31}$	$Z_3 = c_{12}$
$\bar{X}_1 = s_{23}s_3$	$\bar{X}_2 = s_{31}s_1$	$\bar{X}_3 = s_{12}s_2$
$\bar{Y}_1 = s_{23}c_3$	$\bar{Y}_2 = s_{31}c_1$	$\bar{Y}_3 = s_{12}c_2$
$\bar{Z}_1 = c_{23}$	$\bar{Z}_2 = c_{31}$	$\bar{Z}_3 = c_{12}$

A second set of laws can be generated by rotating the vector ${}^2\mathbf{S}_2$ to the third coordinate system in two directions, that is, directly from the second to the third coordinate system and from the second to the first and then to the third. Equating the results of the two transformations yields the second set of sine, sine–cosine, and cosine laws. This result can be obtained simply by exchanging the subscripts in Eqs. (6.134) through (6.136) as follows:

$$\begin{array}{ccc} 1 & 2 & 3 \\ \downarrow & \downarrow & \downarrow . \\ 3 & 1 & 2 \end{array} \tag{6.137}$$

According to this exchange of subscripts, θ_1 is replaced by θ_3, θ_2 is replaced by θ_1, θ_3 is replaced by θ_2, α_{12} is replaced by α_{31}, α_{23} is replaced by α_{12}, and α_{31} is replaced by α_{23}. This yields

$$\bar{X}_1 = s_{23}s_3, \tag{6.138}$$

$$\bar{Y}_1 = s_{23}c_3, \tag{6.139}$$

$$\bar{Z}_1 = c_{23}. \tag{6.140}$$

A total of six sets of sine, sine–cosine, and cosine laws can be generated for a spherical triangle (see Table 6.3). This equates to the six possible ways of reordering the three subscripts. Note that the subscripts appear in increasing order in Eq. (6.137), the same as the original subscripts 1, 2, and 3. For three of the six possible permutations, however, the new subscripts will be in decreasing order. An example of this is the exchange

$$\begin{array}{ccc} 1 & 2 & 3 \\ \downarrow & \downarrow & \downarrow . \\ 3 & 2 & 1 \end{array} \tag{6.141}$$

Whenever an exchange of subscripts occurs, any resulting angle α_{ji} (where $j = i + 1$) should be rewritten as α_{ij}.

6.6.2 Sample problem

Suppose that you have three spherical links. You measure the twist angles and find that $\alpha_{12} = 120°$, $\alpha_{23} = 80°$, and $\alpha_{31} = 135°$ (see Figure 6.9). You wish to determine the joint angles θ_1, θ_2, and θ_3 for the assembled spherical triangle.

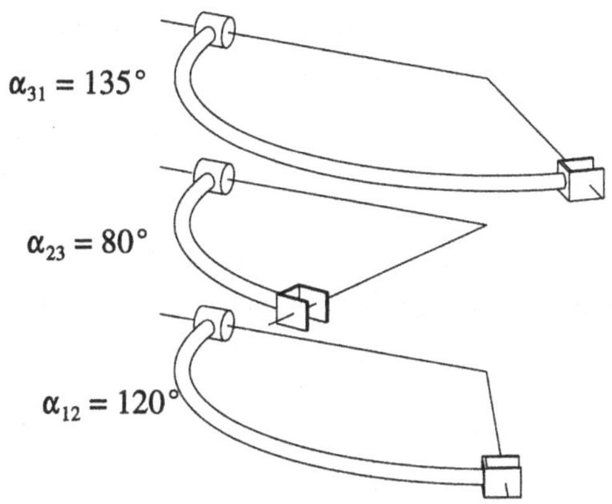

Figure 6.9. Three spherical links.

The solution will begin by examining the equations listed in Table 6.3 in order to find an equation that has only one unknown. Upon examination, it is seen that all the cosine law equations have a single unknown, whereas the sine and sine–cosine law equations have two unknowns. The cosine law $Z_1 = c_{23}$ will be arbitrarily selected. Expanding Z_1 gives

$$c_{12}c_{31} - s_{12}s_{31}c_1 = c_{23}. \tag{6.142}$$

Solving for c_1 gives

$$c_1 = \frac{c_{12}c_{31} - c_{23}}{s_{12}s_{31}}. \tag{6.143}$$

Substituting numerical values gives $c_1 = 0.2938$. There are two distinct angles whose cosines will equal this value. These two angles will be designated as θ_{1A} and θ_{1B}. Thus $\theta_{1A} = 72.92°$ and $\theta_{1B} = 287.08°$. In general, if only cosine θ is known, or only sine θ is known, then two distinct values for θ, where $0 \le \theta \le 2\pi$, exist. However, if both the sine and the cosine of an angle are known, then only one angle is defined that satisfies both the specifications. For the current problem, only c_1 is known. Therefore, two values of θ_1 exist, which will be labeled θ_{1A} and θ_{1B}. This should be expected for this problem because the three links can be assembled in either a clockwise or counterclockwise fashion.

Corresponding values for θ_2 can be computed from the sine and sine–cosine laws (see Table 6.3):

$$X_1 = s_{23}s_2, \tag{6.144}$$

$$Y_1 = s_{23}c_2. \tag{6.145}$$

Expanding X_1 and Y_1 gives

$$s_{31}s_1 = s_{23}s_2, \tag{6.146}$$

$$-(s_{12}c_{31} + c_{12}s_{31}c_1) = s_{23}c_2. \tag{6.147}$$

For the A case, where $\theta_{1A} = 72.92$ degrees, the corresponding values for s_2 and c_2 are

$$s_{2A} = 0.6864, \tag{6.148}$$

$$c_{2A} = 0.7273. \tag{6.149}$$

The unique value for θ_{2A} is 43.34 degrees. For the B case, where $\theta_{1B} = 287.08$ degrees, the corresponding values for s_2 and c_2 are

$$s_{2B} = -0.6864, \tag{6.150}$$

$$c_{2B} = 0.7273. \tag{6.151}$$

The unique value for θ_{2B} is 316.66 degrees.

Corresponding values for θ_3 are computed from the following sine and sine–cosine laws:

$$\bar{X}_1 = s_{23}s_3, \tag{6.152}$$

$$\bar{Y}_1 = s_{23}c_3. \tag{6.153}$$

Expansion of the left sides of Eqs. (6.152) and (6.153) gives

$$s_{12}s_1 = s_{23}s_3, \tag{6.154}$$

$$-(s_{31}c_{12} + c_{31}s_{12}c_1) = s_{23}c_3. \tag{6.155}$$

The terms s_3 and c_3 are the only unknowns in Eqs. (6.154) and (6.155). For the A case, where θ_{1A} equals 72.92 degrees, the corresponding values for s_3 and c_3 are

$$s_{3A} = 0.8406, \tag{6.156}$$

$$c_{3A} = 0.5416. \tag{6.157}$$

The unique value for θ_{3A} is 57.20 degrees. For the B case, where $\theta_{1B} = 287.08$ degrees, the corresponding values for s_3 and c_3 are

$$s_{3B} = -0.8406, \tag{6.158}$$

$$c_{3B} = 0.5416. \tag{6.159}$$

The unique value for θ_{3B} is 302.79 degrees.

Thus, two solution sets have been determined for the three given links. The two assemblies are shown in Figure 6.10.

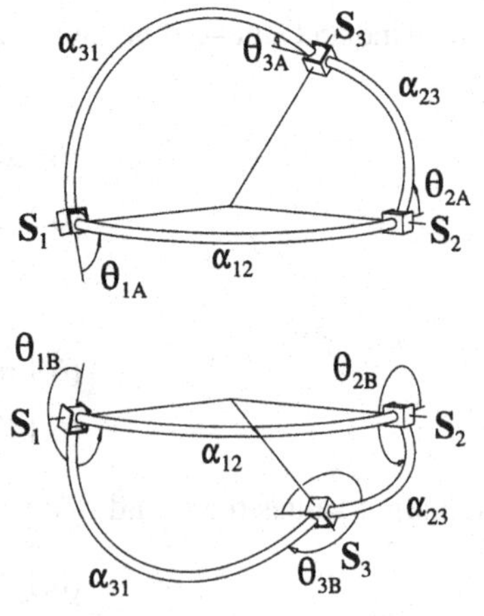

Figure 6.10. Two assemblies of the spherical triangle.

6.6.3 Direction cosines of joint vectors and link vectors for a spherical triangle

The direction cosines for the vectors $\mathbf{S}_1$, $\mathbf{S}_2$, $\mathbf{S}_3$, $\mathbf{a}_{12}$, and $\mathbf{a}_{23}$ expressed in terms of the first coordinate system were determined in Sections 6.4 and 6.5 of this chapter. The vector $\mathbf{a}_{31}$ is yet to be determined. It is known that ${}^3\mathbf{a}_{31} = [1, 0, 0]^T$. Transforming this vector to the second and then to the first coordinate system yields

$$
{}^1\mathbf{a}_{31} = \begin{bmatrix} c_2 & -s_2 & 0 \\ s_2c_{12} & c_2c_{12} & -s_{12} \\ s_2s_{12} & c_2s_{12} & c_{12} \end{bmatrix} \begin{bmatrix} c_3 & -s_3 & 0 \\ s_2c_{23} & c_3c_{23} & -s_{23} \\ s_3s_{23} & c_3s_{23} & c_{23} \end{bmatrix} \begin{bmatrix} 1 \\ 0 \\ 0 \end{bmatrix} = \begin{bmatrix} W_{32} \\ -U^*_{321} \\ U_{321} \end{bmatrix}. \tag{6.160}
$$

The vector ${}^3\mathbf{a}_{31}$ can be rotated directly to the first coordinate system using the transpose of the transformation in Eq. (6.127) as follows:

$$
{}^1\mathbf{a}_{31} = {}^1_3\mathbf{R}\,{}^3\mathbf{a}_{31} = \begin{bmatrix} c_1 & c_{31}s_1 & s_{31}s_1 \\ -s_1 & c_{31}c_1 & s_{31}c_1 \\ 0 & -s_{31} & c_{31} \end{bmatrix} \begin{bmatrix} 1 \\ 0 \\ 0 \end{bmatrix} = \begin{bmatrix} c_1 \\ -s_1 \\ 0 \end{bmatrix}. \tag{6.161}
$$

At this point, the direction cosines of all the link vectors and joint vectors have been calculated in terms of the first coordinate system. The process can be repeated, or an exchange of subscripts can be used to express all the vectors in the second and third coordinate systems. The results of these calculations are presented in the appendix as sets 1 through 3 of the direction cosine table.

The appendix also shows the results of projecting the link and joint vectors onto three additional coordinate systems. For example, set 4 in the appendix projects the vectors onto a coordinate system whose X axis is along the vector $\mathbf{a}_{31}$ and whose Z axis is along $\mathbf{S}_1$. These additional three sets of projections result from performing an exchange of subscripts, where the order of the subscripts is changed from increasing to decreasing.

6.6.4 Polar sine, sine–cosine, and cosine laws for a spherical triangle

In Figure 6.11, the link vectors $\mathbf{a}_{12}$, $\mathbf{a}_{23}$, and $\mathbf{a}_{31}$ have been extended to intersect the unit sphere in three points which can be joined by great circular arcs to form a second triangle

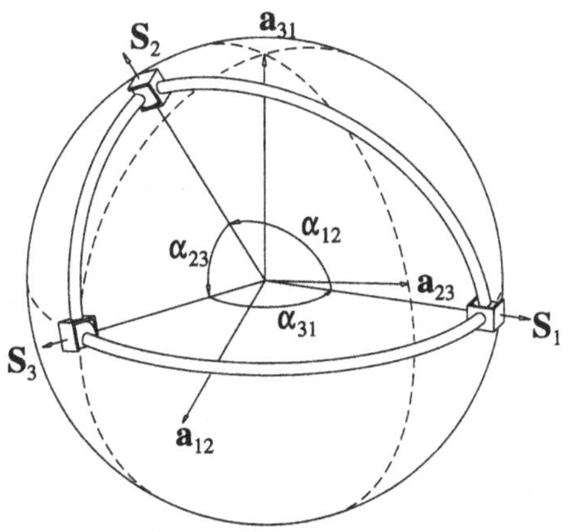

Figure 6.11. Link vectors intersect unit sphere.

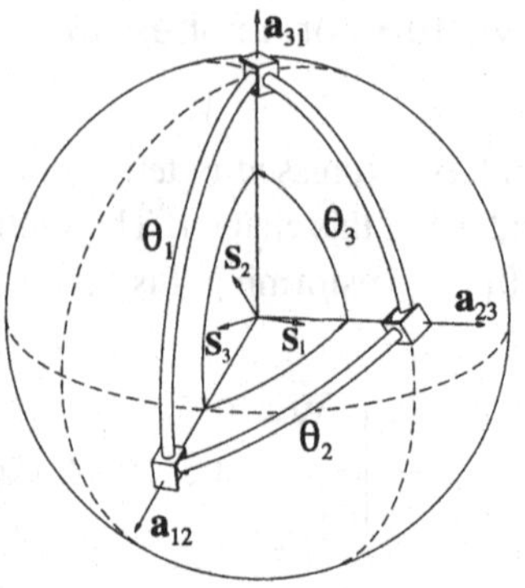

Figure 6.12. Polar triangle.

called the polar triangle. The points of penetration of the unit vectors $\mathbf{a}_{12}$, $\mathbf{a}_{23}$, and $\mathbf{a}_{31}$ are the poles of the planes formed by the pairs of vectors $(\mathbf{S}_1, \mathbf{S}_2)$, $(\mathbf{S}_2, \mathbf{S}_3)$, and $(\mathbf{S}_3, \mathbf{S}_1)$ respectively.

Sets of sine, sine–cosine, and cosine laws can be generated for this polar triangle just as for the spherical triangle. The approach could be similar to that used for the spherical triangle. For example, the vector ${}^3\mathbf{a}_{31}$ could be transformed to the first coordinate system by premultiplying it by ${}^1_2\mathbf{R}{}^2_3\mathbf{R}$. This result would be equated with the vector ${}^3\mathbf{a}_{31}$ premultiplied by ${}^1_3\mathbf{R}$.

A comparison of Figure 6.8 and Figure 6.12 reveals a simpler method of generating the laws for the polar triangle. The link vectors and joint vectors have switched roles. For example, the link vector $\mathbf{a}_{31}$ of the spherical triangle is always perpendicular to the plane containing link α_{31}, which connects the joint vectors $\mathbf{S}_3$ and $\mathbf{S}_1$. The joint vector $\mathbf{S}_3$ of the polar triangle is always perpendicular to the plane containing link θ_3, which connects the link vectors $\mathbf{a}_{23}$ and $\mathbf{a}_{31}$.

Because of the similarity of the spherical triangle and the polar triangle, the laws for the polar triangle can be generated directly by a substitution of variables in the spherical triangle laws. Substituting the definitions of $\bar{X}_2$, $\bar{Y}_2$, and $\bar{Z}_2$ into Eqs. (6.134) through (6.136) yields

$$s_{23}s_2 = s_{31}s_1, \tag{6.162}$$

$$-(s_{12}c_{23} + c_{12}s_{23}c_2) = s_{31}c_1, \tag{6.163}$$

$$c_{12}c_{23} - s_{12}s_{23}c_2 = c_{31}. \tag{6.164}$$

A set of polar sine, sine–cosine, and cosine laws will now be generated by substituting the angles α_{12}, α_{23}, and α_{31} respectively for θ_1, θ_2, and θ_3 and the angles θ_2, θ_3, and θ_1 respectively for α_{12}, α_{23}, and α_{31}. This yields

$$s_3s_{23} = s_1s_{12}, \tag{6.165}$$

$$-(s_2c_3 + c_2s_3c_{23}) = s_1c_{12}, \tag{6.166}$$

$$c_2c_3 - s_2s_3c_{23} = c_1. \tag{6.167}$$

The left side of Eqs. (6.165) through (6.167) can be replaced by the notation U_{32}, V_{32}, and W_{32} based upon the definition of these terms presented in Eqs. (6.92) through (6.94).

Thus, a set of sine, sine–cosine, and cosine laws for a polar triangle may be written as

$$U_{32} = s_1 s_{12}, \tag{6.168}$$

$$V_{32} = s_1 c_{12}, \tag{6.169}$$

$$W_{32} = c_1. \tag{6.170}$$

The substitution of parameters may be applied to each of the six sets of sine, sine–cosine, and cosine laws for the spherical triangle. However, it is simpler to exchange the subscripts 1, 2, 3 in the right and left sides of Eqs. (6.168) through (6.170). Either way, this will result in six sets of polar laws. These six sets are listed in the appendix.

6.7 Spherical quadrilateral

6.7.1 Derivation of fundamental sine, sine–cosine, and cosine laws

A spherical quadrilateral is shown in Figure 6.13. Assuming that one link is attached to ground, the number of degrees of freedom of the mechanism is one. The objective here is to generate a series of equations that relate the twist angles and joint angles of the quadrilateral.

In Section 6.4, it was shown that the vector ${}^4\mathbf{S}_4$ could be rotated from the fourth coordinate system to the third, second, and then first coordinate systems. The result of these transformations (see Eq. (6.35)) was

$${}^1\mathbf{S}_4 = \begin{bmatrix} X_{32} \\ Y_{32} \\ Z_{32} \end{bmatrix}. \tag{6.171}$$

Because the spherical quadrilateral is a closed-loop mechanism, the vector ${}^4\mathbf{S}_4$ can be rotated directly to the first coordinate system via a transformation ${}^1_4\mathbf{R}$. This transformation can be generated in the same manner as was the transformation ${}^1_3\mathbf{R}$ for the spherical triangle,

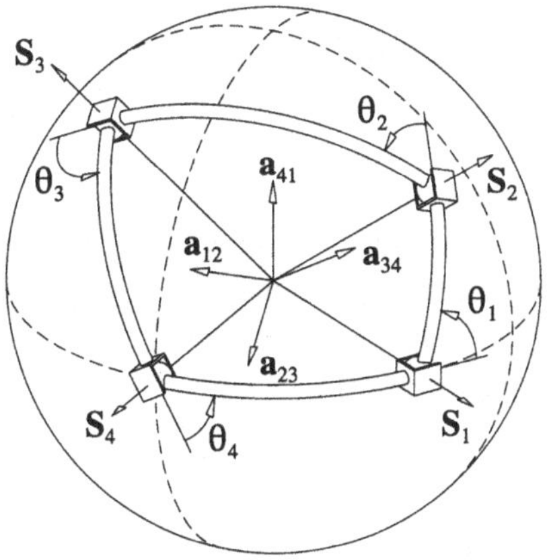

Figure 6.13. Spherical quadrilateral.

that is, by a rotation of α_{41} about the X axis of the fourth coordinate system followed by a rotation of θ_1 about the modified Z axis. The resulting transformation is

$$ {}^{4}_{1}\mathbf{R} = \begin{bmatrix} c_1 & -s_1 & 0 \\ c_{41}s_1 & c_{41}c_1 & -s_{41} \\ s_{41}s_1 & s_{41}c_1 & c_{41} \end{bmatrix}. \tag{6.172} $$

The vector ${}^{1}\mathbf{S}_4$ can now be calculated as

$$ {}^{1}\mathbf{S}_4 = {}^{1}_{4}\mathbf{R}\,{}^{4}\mathbf{S}_4 = \begin{bmatrix} c_1 & c_{41}s_1 & s_{41}s_1 \\ -s_1 & c_{41}c_1 & s_{41}c_1 \\ 0 & -s_{41} & c_{41} \end{bmatrix} \begin{bmatrix} 0 \\ 0 \\ 1 \end{bmatrix} = \begin{bmatrix} s_{41}s_1 \\ s_{41}c_1 \\ c_{41} \end{bmatrix}. \tag{6.173} $$

Equating Eqs. (6.171) and (6.173) yields

$$ X_{32} = s_{41}s_1, \tag{6.174} $$

$$ Y_{32} = s_{41}c_1, \tag{6.175} $$

$$ Z_{32} = c_{41}. \tag{6.176} $$

Equations (6.174) through (6.176) are respectively fundamental sine, sine–cosine, and cosine laws for a spherical quadrilateral. A total of eight fundamental sets of laws can be generated for the quadrilateral by performing an exchange of subscripts. The eight sets of laws are listed in the appendix.

6.7.2 Sample problem

A spherical quadrilateral is formed from four links $\alpha_{12} = 40°$, $\alpha_{23} = 70°$, $\alpha_{34} = 85°$, and $\alpha_{41} = 70°$, where link α_{41} is attached to ground. Because a spherical quadrilateral with revolute joints on the axes $\mathbf{S}_1$, $\mathbf{S}_2$, $\mathbf{S}_3$, and $\mathbf{S}_4$ is a one-degree-of-freedom mechanism, a single joint parameter or input angle must be specified. The angle $\theta_4 = 75°$ is selected as the input angle. The objective of the problem is to obtain values for the joint angles θ_1, θ_2, and θ_3.

(a) Identification of Input/Output Equation

The solution begins by selecting the joint angle to be solved for first, which is typically called the output angle. The angle θ_1 is the output angle for this example because link α_{41} is the frame or ground. The fundamental sine, sine–cosine, and cosine laws in the appendix are then examined in order to identify one equation that contains only one unknown. An advantage of the notation used in this book is that the subscripts of the X, Y, and Z terms identify the joint angles that are contained in the expanded definitions. For example, the term X_{32} contains the angles θ_3 and θ_2. The subscripts 2 or 3 cannot appear in an equation that contains θ_1 as the only unknown. There are apparently two such equations

$$ Z_{41} = c_{23} \tag{6.177} $$

and

$$ Z_{14} = c_{23}, \tag{6.178} $$

which are of course identical because $Z_{41} \equiv Z_{14}$. Expanding the left side of Eq. (6.177) yields

$$s_{12}(X_4 s_1 + Y_4 c_1) + c_{12} Z_4 = c_{23}. \tag{6.179}$$

Equation (6.179) can be regrouped and written as

$$c_1(s_{12} Y_4) + s_1(s_{12} X_4) + (c_{12} Z_4 - c_{23}) = 0. \tag{6.180}$$

It should be recalled that the definitions of the terms X_4, Y_4, and Z_4 are

$$X_4 = s_{34} s_4, \tag{6.181}$$

$$Y_4 = -(s_{41} c_{34} + c_{41} s_{34} c_4), \tag{6.182}$$

$$Z_4 = c_{41} c_{34} - s_{41} s_{34} c_4. \tag{6.183}$$

The right-hand sides of Eqs. (6.181) through (6.183) are all expressed in terms of known parameters. Thus, it can be observed that the terms in parentheses in Eq. (6.180) can be numerically evaluated. The task at hand, therefore, is to solve the equation

$$A c_1 + B s_1 + D = 0 \tag{6.184}$$

where

$$A = s_{12} Y_4, \tag{6.185}$$

$$B = s_{12} X_4, \tag{6.186}$$

$$D = c_{12} Z_4 - c_{23} \tag{6.187}$$

for all values of θ_1. Two solution techniques for this equation will be introduced.

(b) Tan-Half-Angle Solution of $Ac_1 + Bs_1 + D = 0$

In the first solution, the term x_1 will be defined as $\tan(\theta_1/2)$. The following trigonometric identities will be employed

$$s_1 = \frac{2x_1}{1 + x_1^2}, \tag{6.188}$$

$$c_1 = \frac{1 - x_1^2}{1 + x_1^2}. \tag{6.189}$$

Substituting Eqs. (6.188) and (6.189) into Eq. (6.184) yields

$$A \frac{1 - x_1^2}{1 + x_1^2} + B \frac{2x_1}{1 + x_1^2} + D = 0. \tag{6.190}$$

Multiplying Eq. (6.190) by $(1 + x_1^2)$ and regrouping gives

$$(D - A)x_1^2 + (2B)x_1 + (D + A) = 0. \tag{6.191}$$

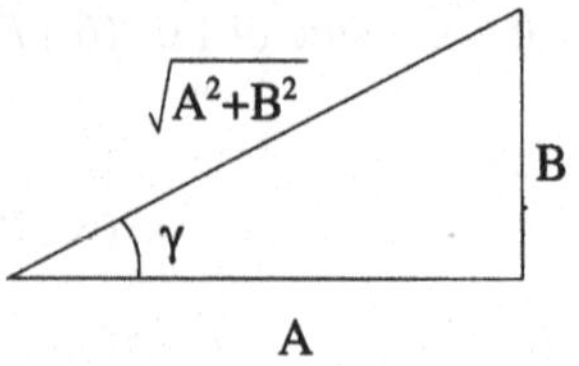

Figure 6.14. Definition of angle γ.

Solving Eq. (6.191) for x_1 yields

$$x_1 = \frac{-B \pm \sqrt{B^2 - (D-A)(D+A)}}{(D-A)}. \tag{6.192}$$

Two real values of x_1 satisfy Eq. (6.191) provided $B^2 - (D-A)(D+A) > 0$. For each of the values of x_1, a unique value for θ_1 can be calculated from the equation

$$\theta_1 = 2\tan^{-1}(x_1). \tag{6.193}$$

This is because $x_1 = \tan(\theta_1/2)$ is single-valued in the range $0 \leq \theta_1 \leq 2\pi$, whereas $\tan(\theta_1)$ is double-valued in the same range.

(c) Trigonometric Solution of $Ac_1 + Bs_1 + D = 0$

A second technique for solving the equation $Ac_1 + Bs_1 + D = 0$ will begin by dividing Eq. (6.184) by the term $\sqrt{A^2+B^2}$ to yield

$$\frac{A}{\sqrt{A^2+B^2}}c_1 + \frac{B}{\sqrt{A^2+B^2}}s_1 + \frac{D}{\sqrt{A^2+B^2}} = 0. \tag{6.194}$$

Using the right-angled triangle shown in Figure 6.14 it is possible to substitute

$$\frac{B}{\sqrt{A^2+B^2}} = \sin\gamma, \tag{6.195}$$

$$\frac{A}{\sqrt{A^2+B^2}} = \cos\gamma. \tag{6.196}$$

Because the sine and cosine of γ are expressed in terms of all known quantities, a unique value for the angle γ can be determined. Substituting Eqs. (6.195) and (6.196) into Eq. (6.194) gives

$$c_\gamma c_1 + s_\gamma s_1 + \frac{D}{\sqrt{A^2+B^2}} = 0. \tag{6.197}$$

Using the trigonometric identity

$$\cos(\alpha - \beta) = \cos(\alpha)\cos(\beta) + \sin(\alpha)\sin(\beta) \tag{6.198}$$

in Eq. (6.197) and regrouping yields

$$\cos(\theta_1 - \gamma) = \frac{-D}{\sqrt{A^2+B^2}}. \tag{6.199}$$

Two values for the quantity $(\theta_1 - \gamma)$ can be found that satisfy Eq. (6.199) as long as the quantity on the right side of Eq. (6.199) is between –1 and 1. Because a unique value for γ has already been determined, two values for θ_1 are now known.

(d) Comparison of Solution Techniques

Three special conditions may occur that will cause one or both of the solution techniques for the equation $Ac_1 + Bs_1 + D = 0$ to yield indeterminate results.

(i) Case 1: $A = D \neq 0, B \neq 0.$

For this first case, the tan-half-angle solution, Eq. (6.192), reduces to

$$x_1 = \frac{-B \pm B}{0}. \tag{6.200}$$

Thus, the two solutions for x_1 are $\frac{-2B}{0}$ and $\frac{0}{0}$. The first value for x_1 that equals infinity corresponds to a value of θ_1 equal to π. The second value for θ_1 cannot be determined.

Two values for θ_1 can be determined by using the trigonometric solution technique for this case. The sine and cosine of γ were previously defined in Eqs. (6.195) and (6.196). The cosine of the difference between θ_1 and γ as expressed in Eq. (6.199) can be written as

$$\cos(\theta_1 - \gamma) = \frac{-A}{\sqrt{A^2 + B^2}}. \tag{6.201}$$

It is apparent from Eqs. (6.196) and (6.201) that for this case, $\cos(\gamma) = -\cos(\theta_1 - \gamma)$. One solution is obviously $\theta_1 = \pi$. The second solution for θ_1 is dependent on the values of the coefficients A and B and can be determined as before by first determining the unique value of the angle γ from Eqs. (6.195) and (6.196) and then obtaining the two values for the quantity $(\theta_1 - \gamma)$ that satisfy Eq. (6.201). The sum of $(\theta_1 - \gamma)$ and γ will yield θ_1.

(ii) Case 2: $A = D \neq 0, B = 0.$

For this case, both solutions of Eq. (6.192) for x_1 will equal $\frac{0}{0}$.

In the trigonometric solution, the sine and cosine of γ will equal 0 and 1 respectively. Thus, the angle γ equals 0. Eq. (6.199) now reduces to

$$\cos\theta_1 = -1. \tag{6.202}$$

The two values of θ_1 that satisfy this equation are a repeated value of π.

(iii) Case 3: $A = D = 0, B \neq 0.$

For the third and final special case, the values for x_1 as determined by the tan-half-angle solution, Eq. (6.192), are $\frac{-2B}{0}$ and $\frac{0}{0}$. The first value corresponds to a value for θ_1 of π. The second value for θ_1 cannot be determined.

In the trigonometric solution, the values for the sine and cosine of γ are respectively 1 and 0. Thus, the angle γ equals $\frac{\pi}{2}$. The value for the cosine of $(\theta_1 - \gamma)$ as defined by

Eq. (6.199) is 0 for this case. Thus, the two values for the angle $(\theta_1 - \gamma)$ are $\frac{\pi}{2}$ and $\frac{3\pi}{2}$. Finally, the two values for θ_1 are 0 and π.

(iv) Summary

Three special cases have been introduced in which two values for θ_1 were not obtained when the tan-half-angle technique was used to solve the equation $Ac_1 + Bs_1 + D = 0$. Proper solutions were obtained, however, when the trigonometric solution technique was used. For this reason, the trigonometric solution is the preferred method for solving this type of equation.

(e) Numerical Solution for θ_1

Upon substituting the numerical values for the twist angles $\alpha_{12}, \alpha_{23}, \alpha_{34}$, and α_{41} and the input angle θ_4 into Eqs. (6.181) through (6.183), Eq. (6.180) can be written as

$$-0.1093c_1 + 0.6185s_1 - 0.5048 = 0. \tag{6.203}$$

The trigonometric solution technique was used to solve this equation for θ_1. The sine and cosine of the angle γ were evaluated from Eqs. (6.195) and (6.196) as 0.9847 and -0.1741. The unique value for γ is 100.02 degrees.

The cosine of $(\theta_1 - \gamma)$ was evaluated from Eq. (6.199) as 0.8037. The two values for $(\theta_1 - \gamma)$ are therefore 36.52 degrees and 323.48 degrees.

The two values for θ_1 can be obtained by summing each value of $(\theta_1 - \gamma)$ with γ. The two values for θ_1 are 136.54 degrees and 63.50 degrees. These two values will be referred to as θ_{1A} and θ_{1B}.

(f) Solution for θ_2

The value for the angle θ_2 may be determined in many ways. However, it is preferred to use the following fundamental sine, sine–cosine, and cosine laws for a spherical quadrilateral:

$$X_{41} = s_{23}s_2, \tag{6.204}$$

$$Y_{41} = s_{23}c_2. \tag{6.205}$$

The left-hand sides of Eqs. (6.204) and (6.205) will be evaluated by substituting θ_4 and θ_{1A} into the definitions of X_{41} and Y_{41}. The value for θ_2 that is associated with θ_4 and θ_{1A} can be uniquely determined because the sine and cosine of θ_2 are computed from Eqs. (6.204) and (6.205). This value for θ_2 will be called θ_{2A}.

Next, θ_4 and θ_{1B} will be substituted into the definitions for X_{41} and Y_{41}. The sine and cosine for θ_{2B} are then determined.

It is important to recognize that θ_{2A} will be the correct value for θ_2 when $\theta_1 = \theta_{1A}$. Similarly, θ_2 will equal θ_{2B} when $\theta_1 = \theta_{1B}$.

The values for θ_{2A} and θ_{2B} for this problem are -38.23 degrees and 38.23 degrees respectively.

(g) Solution for θ_3

As was the case with θ_2, many different equations can be used to calculate θ_3. However, it is preferred to use the following fundamental sine, sine–cosine, and cosine laws for a

Table 6.4. *Solution to spherical quadrilateral sample problem (units in degrees).*

$\theta_4 = 75$	
$\theta_{1A} = 136.54$	$\theta_{1B} = 63.50$
$\theta_{2A} = -38.23$	$\theta_{2B} = 38.23$
$\theta_{3A} = 135.76$	$\theta_{3B} = 87.72$

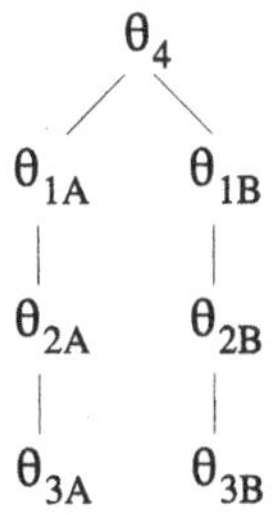

Figure 6.15. Solution tree.

spherical quadrilateral

$$X_{14} = s_{23}s_3, \tag{6.206}$$

$$Y_{14} = s_{23}c_3. \tag{6.207}$$

The definitions of X_{14} and Y_{14} can be determined by substituting the numerical values of θ_4 and θ_{1A}. The corresponding value for θ_3, called θ_{3A} , is then known because the sine and cosine of θ_3 are known.

The process is repeated by substituting θ_4 and θ_{1B} into Eqs. (6.206) and (6.207). The calculated values of s_3 and c_3 determine the corresponding value for the angle θ_{3B}.

For this problem, the values for θ_{3A} and θ_{3B} are 135.76 degrees and 87.72 degrees. The solution is complete, and the results are listed in Table 6.4.

It is often helpful to visualize the solution process by drawing a solution tree (see Figure 6.15). At the top of the tree is the given input angle θ_4. The angles θ_{1A} and θ_{1B} were the first joint angles to be calculated, and therefore they are listed in the tree directly below θ_4. The next angle that was calculated was θ_2. The tree indicates that the angle θ_{2A} was calculated using the values for θ_4 and θ_{1A} in the appropriate equations. The angle θ_{2B} was calculated using θ_4 and θ_{1B} in the same equations. Lastly, the angle θ_{3A} was determined based upon the values of the angles θ_4, θ_{1A}, and θ_{2A}. The value of θ_{3B} was calculated by using θ_4, θ_{1B}, and θ_{2B} in the same equations. The solution tree pictorially shows that there are two solution sets to this problem. For the given input angle, θ_4, the two solution sets are $(\theta_{1A}, \theta_{2A}, \theta_{3A})$ and $(\theta_{1B}, \theta_{2B}, \theta_{3B})$. It is important to remember that a solution comprises an entire set of angles and that care must be taken to correctly organize the angles into the appropriate solution sets.

Figure 6.16 shows the two solutions to the problem. Link 41 is attached to ground.

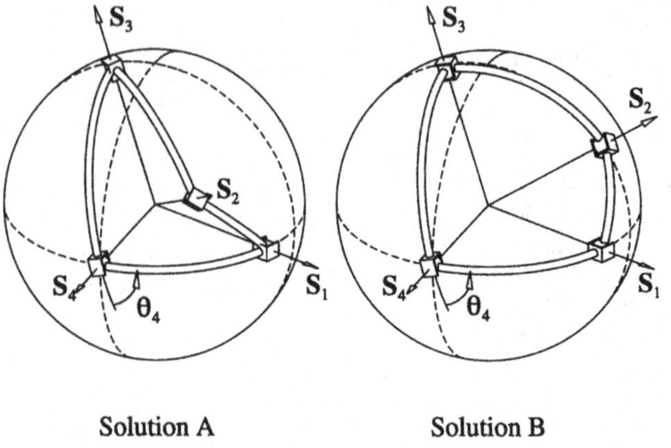

Figure 6.16. Two solutions of a spherical quadrilateral (θ_4 input angle).

The angle θ_4 is the same for both configurations, and thus the position of link 34 is the same for both solutions. Links 12 and 23 are shown in the two solution configurations.

6.7.3 Derivation of subsidiary sine, sine–cosine, and cosine laws

It is possible to generate equations for the spherical quadrilateral in addition to the sine, sine–cosine, and cosine laws derived in Section 6.7.1. These new equations will be referred to as subsidiary sine, sine–cosine, and cosine laws.

One set of subsidiary equations will be generated here by projecting the vector $\mathbf{S}_4$ onto the second coordinate system. This can be done in two ways. Firstly, the vector ${}^4\mathbf{S}_4$ can be rotated to the third and then to the second coordinate sytem. The result of these transformations (see set 2 of the direction cosine table in appendix) is

$$ {}^2\mathbf{S}_4 = \begin{bmatrix} \bar{X}_3 \\ \bar{Y}_3 \\ \bar{Z}_3 \end{bmatrix} . \tag{6.208}$$

The vector ${}^4\mathbf{S}_4$ can also be rotated directly to the first coordinate system and then to the second coordinate system. Using the results from Eq. (6.173), where $\mathbf{S}_4$ has been rotated directly to the first coordinate system,

$$ {}^2\mathbf{S}_4 = {}^2_1\mathbf{R}\,{}^1\mathbf{S}_4 = \begin{bmatrix} c_2 & s_2c_{12} & s_2s_{12} \\ -s_2 & c_2c_{12} & c_2s_{12} \\ 0 & -s_{12} & c_{12} \end{bmatrix} \begin{bmatrix} s_{41}s_1 \\ s_{41}c_1 \\ c_{41} \end{bmatrix} . \tag{6.209}$$

After performing the matrix multiplication and regrouping terms, Eq. (6.209) may be written as

$$ {}^2\mathbf{S}_4 = \begin{bmatrix} X_{12} \\ -X^*_{12} \\ Z_1 \end{bmatrix} . \tag{6.210}$$

Equating Eqs. (6.208) and (6.210) yields the following set of subsidiary sine, sine–cosine, and cosine laws

$$\bar{X}_3 = X_{12}, \tag{6.211}$$

$$\bar{Y}_3 = -X^*_{12}, \tag{6.212}$$

$$\bar{Z}_3 = Z_1. \tag{6.213}$$

A total of eight sets of subsidiary laws can be generated by performing each of the eight possible exchanges of subscripts. The resulting sets of subsidiary laws for the quadrilateral are listed in the appendix.

With regard to the previous sample problem, had the angle θ_2 been selected as the first joint angle to be solved, the following subsidiary cosine law would have been used:

$$Z_4 = \bar{Z}_2. \tag{6.214}$$

The remainder of the solution is left as an exercise for the reader.

6.7.4 Direction cosines of joint vectors and link vectors for a spherical quadrilateral

The direction cosines for all the vectors of the spherical quadrilateral except for $\mathbf{a}_{41}$ were determined with respect to the first coordinate system in Section 6.5. The vector ${}^1\mathbf{a}_{41}$ can be determined by rotating the vector ${}^4\mathbf{a}_{41}$, which equals $[1, 0, 0]^T$, directly to the first coordinate system as follows:

$${}^1\mathbf{a}_{41} = {}^1_4\mathbf{R}\,{}^4\mathbf{a}_{41} = \begin{bmatrix} c_1 & c_{41}s_1 & s_{41}s_1 \\ -s_1 & c_{41}c_1 & s_{41}c_1 \\ 0 & -s_{41} & c_{41} \end{bmatrix} \begin{bmatrix} 1 \\ 0 \\ 0 \end{bmatrix} = \begin{bmatrix} c_1 \\ -s_1 \\ 0 \end{bmatrix}. \tag{6.215}$$

All the vectors of the spherical quadrilateral have now been determined in terms of the first standard coordinate system. The process can be repeated to express all the vectors in the second, third, and fourth coordinate systems. The results of this are listed in the appendix as sets 1 through 4 of the direction cosine table.

The appendix also shows the results of projecting the vectors onto four additional coordinate systems. These results were obtained by performing the four exchanges of subscripts where the order of the indices was changed from increasing to decreasing. The results are listed in the appendix and will be used in future analyses.

6.7.5 Polar sine, sine–cosine, and cosine laws for a spherical quadrilateral

A polar quadrilateral may be formed by extending the link vectors $\mathbf{a}_{ij}$ so that they intersect the unit sphere. New spherical links are placed between the $\mathbf{a}_{ij}$ vectors to maintain their relative orientation. As with the spherical and polar triangle, the roles of the twist angles and joint angles have been interchanged. For example, in the spherical quadrilateral the length of the link that separates the vectors $\mathbf{S}_2$ and $\mathbf{S}_3$ is α_{23}. In the polar quadrilateral, the length of the link that separates the vectors $\mathbf{a}_{23}$ and $\mathbf{a}_{34}$ is θ_3.

Because of the interchange of the roles of the twist angles and the joint angles, polar sine, sine–cosine, and cosine laws and subsidiary polar sine, sine–cosine, and cosine laws can be generated by an appropriate exchange of variables. The results of this variable exchange are presented in the appendix.

6.8 Spherical pentagon

6.8.1 Generation of fundamental, subsidiary, and polar sine, sine–cosine, and cosine laws

A spherical pentagon is shown in Figure 6.17. Link 51 is attached to ground, and the mechanism has two degrees of freedom. Thus, when all the twist angles, α_{ij}, are specified together with two of the joint angles, θ_i, it is possible to solve for the remaining joint angles.

As with the spherical triangle and the spherical quadrilateral, fundamental sine, sine–cosine, and cosine laws can be generated by transforming the vector ${}^5\mathbf{S}_5$ to the first coordinate system in two directions, that is, via the fourth, third, and second coordinate systems, and directly from the fifth to the first coordinate system. A total of ten sets of fundamental laws can be determined by an exchange of subscripts. The resulting ten sets of fundamental laws for the spherical pentagon are listed in the appendix.

Subsidiary formulas can be generated by rotating the vector ${}^5\mathbf{S}_5$ to the second coordinate system via the fourth and third coordinate systems and then equating the result to ${}^5\mathbf{S}_5$ as it is rotated to the second coordinate system via the first coordinate system. Another set of subsidiary laws can be generated by rotating the vector ${}^5\mathbf{S}_5$ to the third coordinate system in two directions. All the resulting subsidiary equations are listed in the appendix.

Similar to the polar triangle and quadrilateral, a polar pentagon can be formed by allowing the link vectors to intersect the unit sphere and then placing spherical links between adjacent link vectors to maintain their relative orientation. An appropriate exchange of variables in the fundamental and subsidiary laws results in the polar fundamental and subsidiary laws. All the polar sine, sine–cosine, and cosine laws for the polar pentagon are listed in the appendix.

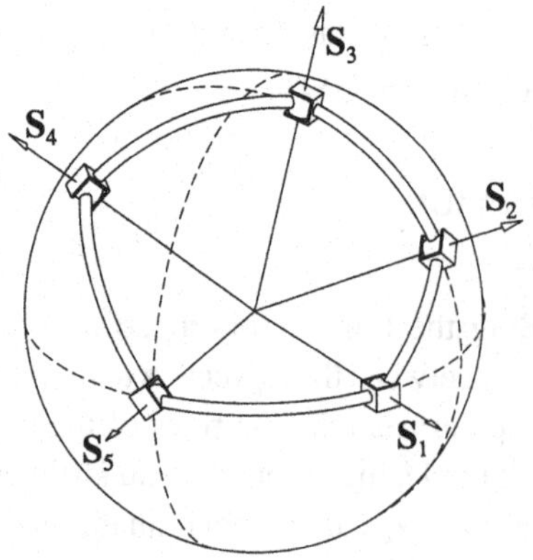

Figure 6.17. Spherical pentagon.

Table 6.5. *Spherical pentagon parameters (units in degrees).*

$\alpha_{12} = 40$	$\theta_1 = ?$
$\alpha_{23} = 35$	$\theta_2 = ?$
$\alpha_{34} = 80$	$\theta_3 = 100$
$\alpha_{45} = 60$	$\theta_4 = ?$
$\alpha_{51} = 90$	$\theta_5 = 65$

6.8.2 Sample problem

Table 6.5 shows the specified angular values for a spherical pentagon where it is assumed that link 51 is attached to ground. Because two joint angles are given, and the spherical pentagon is a two-degree-of-freedom mechanism, it should be possible to determine values for the remaining joint angles.

At the outset, it is necessary to decide which joint angle to solve for first. For this example, the angle θ_1 will be chosen, and it is named the output angle because it connects link 12 to the frame. The task at hand is to obtain the input/output equation, that is, an equation that contains the angle θ_1 as its only unknown. After reviewing all the fundamental and subsidiary sine, sine–cosine, and cosine laws for a spherical pentagon, the following equation was identified:

$$Z_{51} = \bar{Z}_3. \tag{6.216}$$

The definition of the right-hand side is expanded as

$$\bar{Z}_3 = c_{23}c_{34} - s_{23}s_{34}c_3. \tag{6.217}$$

Substituting the given mechanism parameters into Eq. (6.217) yields

$$\bar{Z}_3 = 0.2403. \tag{6.218}$$

Substituting the definition of Z_{51} into Eq. (6.216) yields

$$s_{12}(X_5 s_1 + Y_5 c_1) + c_{12}Z_5 - \bar{Z}_3 = 0. \tag{6.219}$$

The only unknown in Eq. (6.219) is θ_1. Regrouping this equation gives

$$(s_{12}Y_5)c_1 + (s_{12}X_5)s_1 + (c_{12}Z_5 - \bar{Z}_3) = 0. \tag{6.220}$$

This equation is of the form $Ac_1 + Bs_1 + D = 0$, where

$$A = s_{12}Y_5, \tag{6.221}$$

$$B = s_{12}X_5, \tag{6.222}$$

$$D = c_{12}Z_5 - \bar{Z}_3. \tag{6.223}$$

Using the trigonometric solution technique developed in Section 6.7.2, the two values of θ_1 are $\theta_{1A} = 93.01°$ and $\theta_{1B} = 151.99°$.

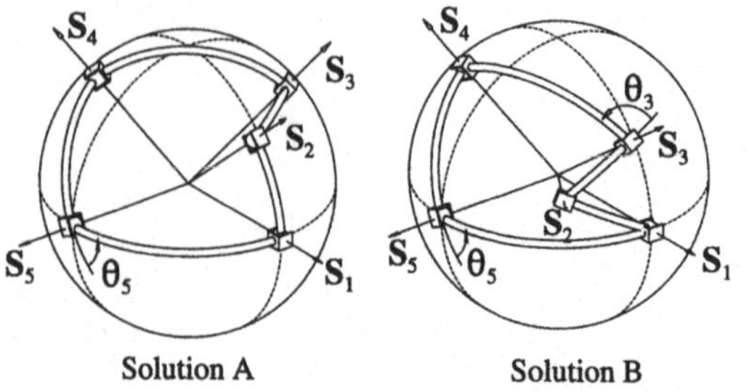

Figure 6.18. Two solutions of a spherical pentagon (θ_5 and θ_3 input angles).

The following two subsidiary equations will be used to determine corresponding values for θ_2:

$$X_{51} = X_{32}, \tag{6.224}$$

$$Y_{51} = -X^*_{32}. \tag{6.225}$$

The given value of θ_5 and the calculated value θ_{1A} will be substituted into the left-hand sides of Eqs. (6.224) and (6.225) to yield numerical values. Substituting the definitions for X_{32} and X^*_{32} yields

$$\bar{X}_3 c_2 - \bar{Y}_3 s_2 = X_{51}, \tag{6.226}$$

$$-(\bar{X}_3 s_2 + \bar{Y}_3 c_2) = Y_{51}. \tag{6.227}$$

Equations (6.226) and (6.227) represent two equations in the two unknowns s_2 and c_2. Substituting θ_5, θ_3, and θ_{1A} into these equations and solving for s_2 and c_2 yields a unique value for θ_2, called θ_{2A}. The angle θ_{2B} will be determined by substituting θ_5 and θ_{1B} into Eqs. (6.226) and (6.227) and then solving for the sine and cosine of θ_{2B}. The calculated values for the angles θ_{2A} and θ_{2B} are -64.23 degrees and -120.55 degrees respectively.

The last variable to be determined is θ_4. This angle will be obtained from the following fundamental sine and sine–cosine laws:

$$X_{123} = s_{45} s_4, \tag{6.228}$$

$$Y_{123} = s_{45} c_4. \tag{6.229}$$

Numerical values for the left-hand sides of these equations will be obtained by substituting the values for θ_3, θ_{1A}, and θ_{2A}. The sine and cosine for the corresponding angle θ_{4A} is readily calculated. Next, the values of θ_3, θ_{1B}, and θ_{2B} are substituted into the equations, and the angle θ_{4B} is determined. The calculated values for the angles θ_{4A} and θ_{4B} are 48.52 degrees and 86.56 degrees respectively. The two solution configurations for the spherical pentagon are shown in Figure 6.18.

6.9 Spherical hexagon and spherical heptagon

6.9.1 Generation of fundamental, subsidiary, and polar sine, sine–cosine, and cosine laws

A spherical hexagon and a spherical heptagon are comprised respectively of six and seven spherical links (see Figures 6.19 and 6.20). The spherical hexagon is a three-degree-

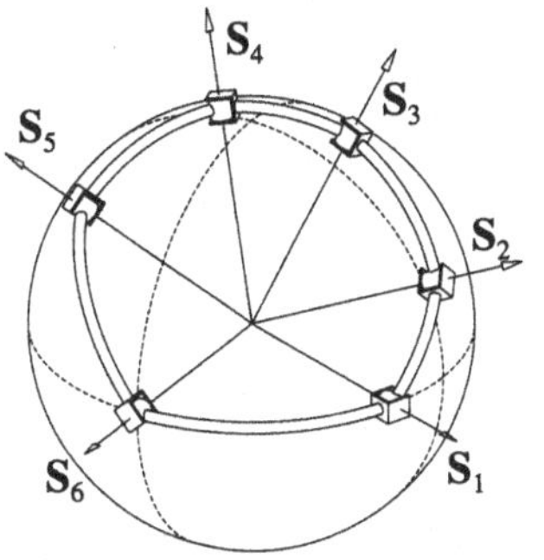

Figure 6.19. Spherical hexagon.

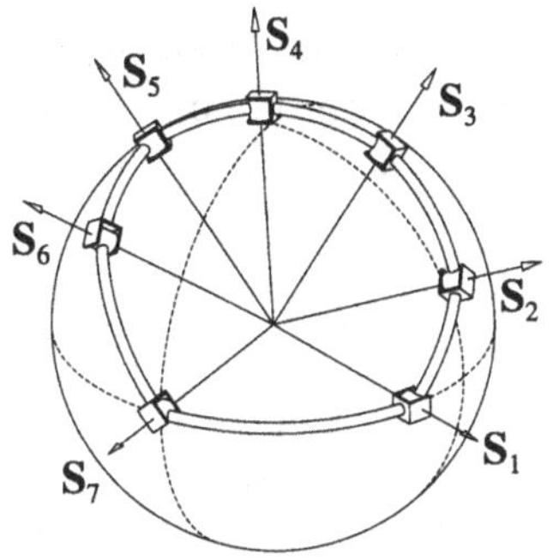

Figure 6.20. Spherical heptagon.

of-freedom mechanism, whereas the spherical heptagon is a four-degree-of-freedom mechanism.

For both of these spherical mechanisms, it is possible to generate fundamental sine, sine–cosine, and cosine laws by transforming the last joint vector to the first coordinate system in two directions. A total of twelve sets can be generated for the spherical hexagon and fourteen sets for the spherical heptagon by applying an exchange of subscripts. All the fundamental laws for the hexagon and the heptagon are listed in the appendix.

Subsidiary laws can be generated by rotating the last joint vector to the second, third, ..., and $(n - 2)$ coordinate systems, where n is six for the hexagon and seven for the heptagon. All the resulting subsidiary laws are listed in the appendix.

As was the case with the spherical triangle, quadrilateral, and pentagon, a polar hexagon and a polar heptagon can be formed by allowing the link vectors to intersect the unit sphere and then placing spherical links between adjacent link vectors to maintain their relative orientation. An appropriate exchange of variables in the fundamental and subsidiary laws results in the polar fundamental and subsidiary laws. All the polar sine, sine–cosine, and cosine laws for the polar hexagon and the polar heptagon are listed in the appendix.

6.9.2 Sample problem

Table 6.6 shows the specified angular values for a spherical heptagon. Four joint angles are given because the mechanism is a four-degree-of-freedom device. The objective is to calculate the remaining three joint angles.

Table 6.6. *Spherical heptagon parameters (units in degrees).*

$\alpha_{12} = 40$	$\theta_1 = \ ?$
$\alpha_{23} = 20$	$\theta_2 = 50$
$\alpha_{34} = 35$	$\theta_3 = \ ?$
$\alpha_{45} = 30$	$\theta_4 = 330$
$\alpha_{56} = 80$	$\theta_5 = \ ?$
$\alpha_{67} = 45$	$\theta_6 = 115$
$\alpha_{71} = 65$	$\theta_7 = 340$

θ_1 is selected as the angle to be solved for first. The fundamental and subsidiary sine, sine–cosine, and cosine laws for a spherical heptagon will be examined in order to find an equation in which θ_1 is the only unknown. The following subsidiary cosine law was identified:

$$Z_{6712} = \bar{Z}_4. \tag{6.230}$$

The right-hand side of this equation can be evaluated because all the angles that comprise the term Z_4 are known. Expanding the left-hand side and regrouping yields

$$s_{23}(X_{671}s_2 + Y_{671}c_2) + c_{23}Z_{671} - \bar{Z}_4 = 0. \tag{6.231}$$

Substituting the definitions of the terms X_{671}, Y_{671}, and Z_{671} and then grouping the sine and cosine of θ_1 terms gives

$$\begin{aligned} &c_1(s_{23}s_2X_{67} + s_{23}c_2c_{12}Y_{67} + c_{23}s_{12}Y_{67}) \\ &\quad + s_1(-s_{23}s_2Y_{67} + s_{23}c_2c_{12}X_{67} + c_{23}s_{12}X_{67}) \\ &\quad + (-s_{23}c_2s_{12}Z_{67} + c_{23}c_{12}Z_{67} - \bar{Z}_4) = 0. \end{aligned} \tag{6.232}$$

Equation (6.232) is of the form $Ac_1 + Bs_1 + D = 0$. Substituting the given values, Eq. (6.232) numerically evaluates to

$$c_1(-0.1039) + s_1(0.8082) + (-0.4346) = 0. \tag{6.233}$$

Using the trigonometric solution technique developed in Section 6.7.2, the two values of θ_1 are $\theta_{1A} = 39.56°$ and $\theta_{1B} = 155.09°$.

The following subsidiary sine and sine–cosine laws are used to solve for corresponding values of θ_3

$$X_{6712} = X_{43}, \tag{6.234}$$

$$Y_{6712} = -X^*_{43}. \tag{6.235}$$

The given values of θ_6, θ_7, and θ_2 together with the angle θ_{1A} are substituted into the left-hand sides of the equations to yield numerical values. Substituting the definitions of

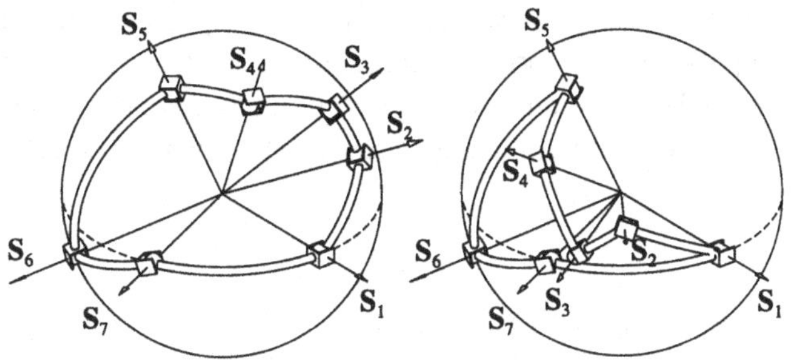

Solution A Solution B

Figure 6.21. Two solutions of a spherical heptagon (θ_7, θ_6, θ_4, θ_2 input angles).

X_{43} and X^*_{43} into the right-hand sides yields

$$X_{6712} = \bar{X}_4 c_3 - \bar{Y}_4 s_3, \qquad (6.236)$$

$$Y_{6712} = -(\bar{X}_4 s_3 + \bar{Y}_4 c_3). \qquad (6.237)$$

Equations (6.236) and (6.237) represent two equations in the two unknowns s_3 and c_3. Solving for these parameters, a unique value of θ_3 can be determined that corresponds to θ_{1A}. This value will be called θ_{3A}. The angle θ_{3B} is determined by substituting θ_{1B} into Eqs. (6.236) and (6.237) and then solving for the sine and cosine of θ_{3B}. The calculated values for the angles θ_{3A} and θ_{3B} are 51.45 degrees and −92.99 degrees respectively.

Finally, corresponding values of θ_5 are determined from the following fundamental sine and sine–cosine laws:

$$X_{71234} = s_{56} s_5, \qquad (6.238)$$

$$Y_{71234} = s_{56} c_5. \qquad (6.239)$$

Numerical values for the left-hand sides of these equations will be obtained by substituting the given values for θ_7, θ_6, θ_4, and θ_2 and the calculated values θ_{1A} and θ_{3A}. The sine and cosine for the corresponding angle θ_{5A} is then determined. Next, the given values and the calculated values θ_{1B} and θ_{3B} are substituted into the equations, and the angle θ_{5B} is determined. The calculated values for the angles θ_{5A} and θ_{5B} are 39.38 degrees and 141.44 degrees. The two solution configurations for the spherical heptagon are shown in Figure 6.21.

6.10 Summary

Chapter 5 showed how an open-loop robot manipulator could be converted to a closed-loop spatial mechanism by solving for the parameters of an imaginary link that closes the loop. In this chapter, the equivalent spherical mechanism for a given spatial mechanism was introduced. All single-degree-of-freedom spatial mechanisms were classified according to the number of degrees of freedom of their equivalent spherical mechanism.

All equations that relate the joint and twist angles of an equivalent spherical mechanism will also be valid for the original spatial mechanism. For this reason, sets of fundamental and subsidiary sine, sine–cosine, and cosine laws were generated for a spherical triangle, quadrilateral. . . heptagon. Polar polygons and corresponding and laws were also generated for each case. The spherical and polar equations are fully listed in the appendix. These equations will serve as a toolbox to be used in future analyses of robot manipulators.

6.11 Problems

1. Completely expand the definitions of the following terms:

   (a) $X_{6543}, Y_{6543}, Z_{6543}$

   (b) $U_{3456}, V_{3456}, W_{3456}$

2. Write a computer function that will use the trigonometric approach to solve an equation of the form Ac + Bs + D = 0. A C language prototype for this function is written as follows:

   ```
   int solve_trig (double A, double B, double D, double *ang_a, double *ang_b);
   ```

   The subroutine should return one if two real solutions were calculated and zero otherwise. Test your subroutine using data that you have checked by hand (or with a program such as MathCad, Mathematica, Maple, etc.)

3. A spherical quadrilateral is to be formed from the following four links:

   $$\alpha_{12} = 75°, \alpha_{23} = 110°, \alpha_{34} = 60°, \alpha_{41} = 80°.$$

   The value of θ_1, the input angle for this case, is 120 degrees. Determine the two sets of solutions for the remaining joint angles of the quadrilateral.

4. Assume that the twist angles (αs) of a spherical pentagon are all known. Further, values for the angles θ_4 and θ_1 are known. Explain how you would obtain values for the remaining joint angles. How many solution sets exist?

5. Write a computer subroutine that will solve two linear equations in two unknowns. The pair of equations may be written as

   $$A_1 x + B_1 y = D_1$$
   $$A_2 x + B_2 y = D_2,$$

   where the coefficients A_1, A_2, B_1, B_2, D_1, and D_2 are known and the parameters x and y are unknown. The C language prototype for your program may be written as

   ```
   int solve_pair (double *x, double *y, double A1, double B1, double D1,
       double A2, double B2, double D2);
   ```

   The function will return a value of one if values of x and y could be obtained. It will return zero if the two equations are linearly dependent and thus unique values of x and y could not be obtained.

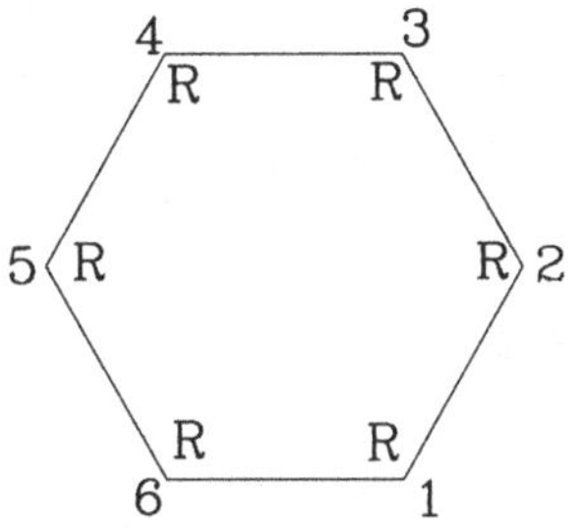

Figure 6.22. Planar representation of a spatial hexagon.

Test your subroutine using data that you have checked by hand (or with a program such as MathCad, Mathematica, Maple, etc.).

6. Assume that the twist angles (αs) of a spherical hexagon are all known. Further, values for the angles θ_6, θ_4, and θ_3 are known. Explain how you would obtain values for the remaining joint angles. How many solution sets exist?

7. Write a computer program that will solve a spherical quadrilateral. In particular, your program will ask the user to enter values for the angles α_{12}, α_{23}, α_{34}, and α_{41}. Next, the user will be prompted to enter a value for the input angle θ_4. With these inputs, your program must print the two set of values for the remaining joint angles.

Your program must identify the case where no solution exists. In other words, no solution for the angles θ_1, θ_2, and θ_3 may exist for the given values of the inputs.

8. You are trying to build a spherical triangle whose angles are $\theta_1 = 120°$, $\theta_2 = 80°$, and $\theta_3 = 140°$. What values should α_{12}, α_{23}, and α_{31} have so you can build your triangle?

9. Assume that you are given numerical values for the angles θ_6, θ_5, and θ_3 for the planar represented of a closed-loop spherical mechanism shown in Figure 6.22 (all the α angles are assumed to be known also).

(a) Write an equation that can be used to solve for θ_1. Factor this equation into the form

$$A(c_1) + B(s_1) + D = 0.$$

(b) Show how to solve an equation of the form

$$A(c_1) + B(s_1) + D = 0$$

for the angle θ_1. How many values of θ_1 may satisfy this equation?

(c) Assuming that θ_1 is known, describe how to solve for θ_4.

(d) Assuming that θ_1 and θ_4 are now known, describe how to solve for θ_2.

7

Displacement analysis of group 1 spatial mechanisms

7.1 Introduction

A group 1 spatial mechanism was defined in Chapter 6 as a one-degree-of-freedom closed-loop kinematic chain with one link, the frame, fixed to the ground and whose equivalent spherical mechanism also has one degree of freedom. All the mechanism dimensions, the link lengths and twist angles, are assumed to be known at the outset. Also, the offset (joint angle) of each revolute (prismatic) joint connecting a pair of links that in general are skew is assumed to be known. The joint angle of a revolute pair (the input pair) connecting a link (the input link) to the frame is assumed to be known.

It will be seen in this chapter that the analysis of all group 1 spatial mechanisms can proceed by first determining the unknown joint parameters from sine, sine–cosine, and cosine laws for the equivalent spherical mechanism, which is essentially a spherical four-link mechanism because there can be no relative motion on a sphere of a pair of links connected by a prismatic joint. The remaining unknown displacements will then be determined from writing the vector loop equation for the mechanism and projecting this equation onto three linearly independent directions, which yields three scalar equations in three unknown displacement values. Two example mechanisms will be presented in this chapter followed by an analysis of the CCC spatial robot manipulator.

7.2 RCPCR mechanism

Figure 7.1 shows an RCPCR group 1 spatial mechanism and its equivalent spherical mechanism. Figure 7.2 shows a planar representation of the same mechanism. Link a_{51} is attached to ground. The specific problem statement is presented as follows:

given: $a_{12}, a_{23}, a_{34}, a_{45}, a_{51}$,
$\alpha_{12}, \alpha_{23}, \alpha_{34}, \alpha_{45}, \alpha_{51}$,
S_1, S_5, θ_3, and
θ_5 (input angle),

find: $\theta_1, \theta_2, \theta_4, S_2, S_3$, and S_4.

It was mentioned in the previous section that the analysis can be decoupled in that the three unknown joint angles can be solved for first from analyzing the equivalent spherical mechanism. The angle θ_1 is identified as the output angle because it is connected to the frame, and this angle will be calculated first.

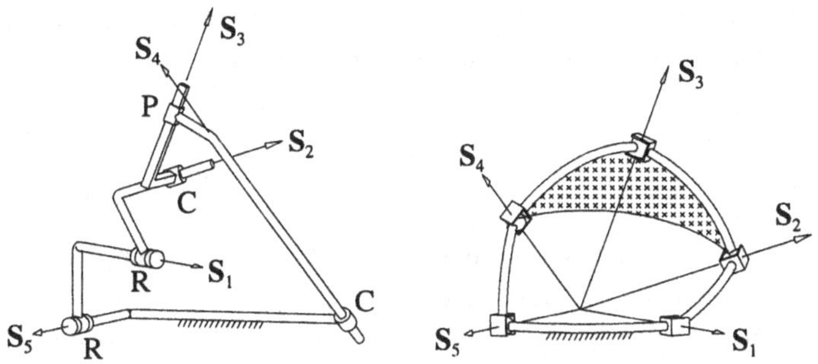

Figure 7.1. RCPCR spatial mechanism and its equivalent spherical mechanism.

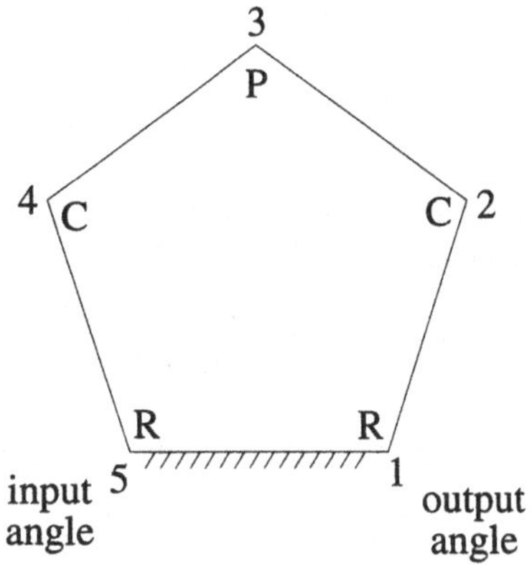

Figure 7.2. Planar representation of RCPCR spatial mechanism.

The objective now is to find a spherical equation that contains only the input angle, θ_5, the output angle, θ_1, and the constant angle, θ_3. All cosine laws for a spherical pentagon contain three joint parameters, and there is a unique subsidiary cosine law that contains θ_5, θ_1, and θ_3. This subsidiary cosine law is

$$Z_{51} = \bar{Z}_3. \tag{7.1}$$

Expanding the left side gives

$$s_{12}(X_5 s_1 + Y_5 c_1) + c_{12} Z_5 - \bar{Z}_3 = 0. \tag{7.2}$$

This equation can be factored into the form of $Ac_1 + Bs_1 + D = 0$ and can be solved for two values of θ_1 by using the solution technique presented in Section 6.7.2(c). These two solutions will be called θ_{1A} and θ_{1B}.

Unique corresponding values for the angle θ_2 can be obtained from the following two spherical equations:

$$X_{51} = X_{32}, \tag{7.3}$$

$$Y_{51} = -X^*_{32}. \tag{7.4}$$

The given value of θ_5 and the calculated value θ_{1A} will be substituted into the left sides of Eqs. (7.3) and (7.4) to yield numerical values. Expanding X_{32} and X^*_{32} yields

$$\bar{X}_3 c_2 - \bar{Y}_3 s_2 = X_{51}, \tag{7.5}$$

$$-(\bar{X}_3 s_2 + \bar{Y}_3 c_2) = Y_{51}. \tag{7.6}$$

Equations (7.5) and (7.6) represent two equations in the two unknowns s_2 and c_2. Solving for these parameters, a unique value for θ_2 can be determined that corresponds to the angle θ_{1A}. This value will be called θ_{2A}. The angle θ_{2B} is found by substituting θ_5 and θ_{1B} into Eqs. (7.5) and (7.6) and then solving for the sine and cosine of θ_{2B}.

The last joint angle to be determined is θ_4. This angle will be determined from the following fundamental sine and sine–cosine laws:

$$X_{123} = s_{45} s_4, \tag{7.7}$$

$$Y_{123} = s_{45} c_4. \tag{7.8}$$

Numerical values can be obtained for the left sides of these equations by substituting the constant joint angle θ_3 and the previously calculated values of θ_{1A} and θ_{2A}. The sine and cosine of the corresponding angle θ_{4A} is readily calculated. The procedure is repeated by substituting θ_3, θ_{1B}, and θ_{2B} into Eqs. (7.7) and (7.8) and then solving for the sine and cosine of θ_{4B}.

The remaining parameters to be determined are the three variable offset distances S_2, S_3, and S_4. These will be determined by first writing the vector loop equation for the mechanism as

$$S_1\mathbf{S}_1 + a_{12}\mathbf{a}_{12} + S_2\mathbf{S}_2 + a_{23}\mathbf{a}_{23} + S_3\mathbf{S}_3 + a_{34}\mathbf{a}_{34} + S_4\mathbf{S}_4 + a_{45}\mathbf{a}_{45} + S_5\mathbf{S}_5 + a_{51}\mathbf{a}_{51} = \mathbf{0}. \tag{7.9}$$

Because all the joint angles are now known for each of the two configurations of the mechanism, the joint and offset vectors can be calculated in terms of any desired coordinate system. Thus, projecting the vector loop equation onto any three linearly independent vectors will yield three scalar equations in the unknowns S_2, S_3, and S_4.

Quite often, a judicious selection of a projection vector will simplify the solution for the three unknowns. For example, projecting Eq. (7.9) onto the vector $\mathbf{a}_{34}$ yields

$$\mathbf{a}_{34} \cdot (S_1\mathbf{S}_1 + a_{12}\mathbf{a}_{12} + S_2\mathbf{S}_2 + a_{23}\mathbf{a}_{23} + S_3\mathbf{S}_3 + a_{34}\mathbf{a}_{34} + S_4\mathbf{S}_4$$
$$+ a_{45}\mathbf{a}_{45} + S_5\mathbf{S}_5 + a_{51}\mathbf{a}_{51}) = 0. \tag{7.10}$$

Using the sets of direction cosines for a spatial pentagon that are listed in the appendix to evaluate the scalar products yields

$$S_1 X_{54} + a_{12} W_{23} + S_2 X_3 + a_{23} c_3 + a_{34} + a_{45} c_4 + S_5 \bar{X}_4 + a_{51} W_{54} = 0. \tag{7.11}$$

This equation contains the parameter S_2 as its only unknown. The calculated values for θ_{1A}, θ_{2A}, and θ_{4A} are substituted into Eq. (7.11) to find the corresponding value for S_2, that is, S_{2A}. The process is repeated by substituting values for θ_{1B}, θ_{2B}, and θ_{4B} to find the corresponding value for S_{2B}.

Table 7.1. *RCPCR mechanism parameters.*

Link length, cm.	Twist angle, deg.	Joint offset, cm.	Joint angle, deg.
$a_{12} = 20$	$\alpha_{12} = 45$	$S_1 = 15$	$\theta_1 =$ variable
$a_{23} = 40$	$\alpha_{23} = 60$	$S_2 =$ variable	$\theta_2 =$ variable
$a_{34} = 30$	$\alpha_{34} = 25$	$S_3 =$ variable	$\theta_3 = 65$
$a_{45} = 20$	$\alpha_{45} = 30$	$S_4 =$ variable	$\theta_4 =$ variable
$a_{51} = 30$	$\alpha_{51} = 70$	$S_5 = 35$	$\theta_5 = 60$ (input)

Table 7.2. *Calculated configurations for the RCPCR spatial mechanism.*

	Solution A	Solution B
θ_1, degrees	223.10	85.48
θ_2, degrees	−100.84	−112.84
θ_4, degrees	113.75	26.37
S_2, cm	−49.13	−112.84
S_3, cm	35.94	197.38
S_4, cm	−58.83	−223.36

The parameter S_4 can be determined by projecting the vector loop equation onto the vector $\mathbf{a}_{23}$. Expansion of the scalar products yields

$$S_1X_2 + a_{12}c_2 + a_{23} + a_{34}c_3 + S_4\bar{X}_3 + a_{45}W_{43} + S_5X_{43} + a_{51}W_{12} = 0. \tag{7.12}$$

The distance S_4 is the only unknown in Eq. (7.12). The calculated values of θ_{1A}, θ_{2A}, and θ_{4A} are substituted into the equation to yield S_{4A}. The values θ_{1B}, θ_{2B}, and θ_{4B} are then input to determine S_{4B}.

The final parameter, S_3, will be obtained by projecting the vector loop equation onto the vector $\mathbf{S}_3$. Expanding the scalar products yields

$$S_1Z_2 + a_{12}U_{23} + S_2c_{23} + S_3 + S_4c_{34} + a_{45}U_{43} + S_5\bar{Z}_4 + a_{51}U_{543} = 0. \tag{7.13}$$

The corresponding value for S_3, that is, S_{3A}, is obtained by substituting S_{2A}, S_{4A}, θ_{1A}, θ_{2A}, and θ_{4A} into Eq. (7.13). Similarly, S_{3B} is obtained by substituting S_{2B}, S_{4B}, θ_{1B}, θ_{2B}, and θ_{4B} into the equation.

At this point the analysis of the group 1 RCPCR mechanism is complete. Two solution configurations were determined. Table 7.1 shows data that were used for a numerical example. The calculated values for the two configurations are listed in Table 7.2.

7.3 RRPRPPR mechanism

Figure 7.3 shows an RRPRPPR group 1 spatial mechanism with its equivalent spherical mechanism. Figure 7.4 shows a planar representation of the same mechanism. Link a_{71} is attached to ground. The specific problem statement is listed as follows:

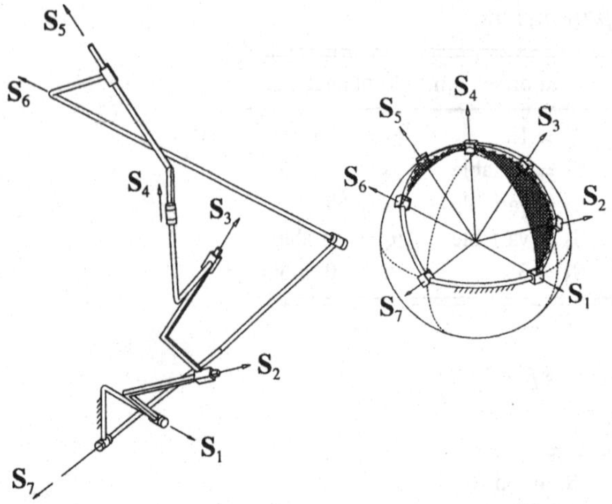

Figure 7.3. RRPRPPR spatial mechanism and its equivalent spherical mechanism.

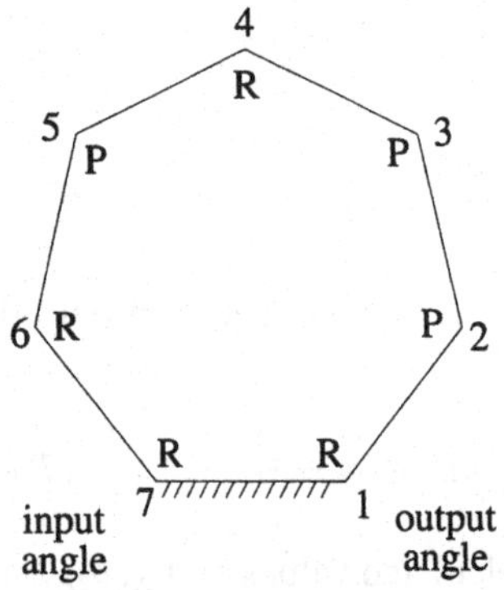

Figure 7.4. Planar representation of RRPRPPR spatial mechanism.

given: $a_{12}, a_{23}, a_{34}, a_{45}, a_{56}, a_{67}, a_{71}$,
$\alpha_{12}, \alpha_{23}, \alpha_{34}, \alpha_{45}, \alpha_{56}, \alpha_{67}, \alpha_{71}$,
$S_1, S_4, S_6, S_7, \theta_2, \theta_3, \theta_5$, and
θ_7 (input angle),

find: $\theta_1, \theta_4, \theta_6, S_2, S_3$, and S_5.

The analysis will proceed as in the previous section, where the three unknown joint angles are obtained first followed by the three unknown joint offsets. The angle θ_1 is identified as the output angle because it is connected to the frame, and θ_1 will be calculated first.

The objective now is to find a spherical equation that contains only the input angle, θ_7, the output angle, θ_1, and the constant angles, θ_2, θ_3, and θ_5. All cosine laws for a spherical heptagon contain five joint parameters, and a unique subsidiary cosine law contains θ_7, θ_1, θ_2, θ_3, and θ_5. This subsidiary cosine law is

$$Z_{3217} = Z_5. \tag{7.14}$$

Expanding the left side of Eq. (7.14) and rearranging gives

$$s_{67}(X_{321}s_7 + Y_{321}c_7) + c_{67}Z_{321} - Z_5 = 0. \tag{7.15}$$

Expanding the definitions of X_{321}, Y_{321}, and Z_{321} and then grouping all the terms containing the sine and cosine of θ_1 yields

$$c_1[X_7X_{32} - Y_7Y_{32}] + s_1[-X_7Y_{32} - Y_7X_{32}] + [Z_7Z_{32} - Z_5] = 0. \tag{7.16}$$

Because all the terms in brackets are defined in terms of the given constants, this equation can be solved for two values of θ_1, that is, θ_{1A} and θ_{1B}, by using the technique described in Section 6.7.2(c).

The angle θ_6 will be solved for next. One solution technique starts by writing the following subsidiary sine and sine–cosine laws for a spherical heptagon:

$$X_{3217} = X_{56}, \tag{7.17}$$

$$Y_{3217} = -X^*_{56}. \tag{7.18}$$

Expanding the definitions for the terms on the right-hand side of these equations gives

$$X_{3217} = X_5c_6 - Y_5s_6, \tag{7.19}$$

$$Y_{3217} = -(X_5s_6 + Y_5c_6). \tag{7.20}$$

Equations (7.19) and (7.20) represent two linear equations in the two unknowns s_6 and c_6. A unique corresponding value for θ_6, called θ_{6A}, can be found by substituting the given joint and twist angle parameters and θ_{1A} into the two equations and then solving for s_{6A} and c_{6A}. Similarly, the given joint and twist angle parameters and θ_{1B} are substituted into the two equations to yield s_{6B} and c_{6B}.

The last remaining joint angle to be computed is θ_4. This angle can be obtained by writing the following two fundamental sine and sine–cosine laws for a spherical heptagon:

$$X_{67123} = s_{45}s_4, \tag{7.21}$$

$$Y_{67123} = s_{45}c_4. \tag{7.22}$$

Substituting the given joint and twist angles and the previously calculated values for θ_{1A} and θ_{6A} into the pair of equations will yield the corresponding values for the sine and cosine of θ_{4A}. The corresponding value for θ_{4B} will be determined in a similar fashion.

At this point, two solution sets exist for the joint angles of the spatial mechanism. Corresponding values for the offset parameters S_2, S_3, and S_5 must yet be determined. These parameters can be calculated as in the previous section by writing the vector loop equation for the spatial mechanism as

$$S_1\mathbf{S}_1 + a_{12}\mathbf{a}_{12} + S_2\mathbf{S}_2 + a_{23}\mathbf{a}_{23} + S_3\mathbf{S}_3 + a_{34}\mathbf{a}_{34} + S_4\mathbf{S}_4 + a_{45}\mathbf{a}_{45}$$
$$+ S_5\mathbf{S}_5 + a_{56}\mathbf{a}_{56} + S_6\mathbf{S}_6 + a_{67}\mathbf{a}_{67} + S_7\mathbf{S}_7 + a_{71}\mathbf{a}_{71} = \mathbf{0}. \tag{7.23}$$

All joint parameters are known for each of the two solution sets. Thus, the vector loop equation can be projected onto any three linearly independent directions to yield three scalar equations in the three unknowns S_2, S_3, and S_5. For example, projecting the vector loop equation onto the vector $\mathbf{a}_{23}$ yields

$$\mathbf{a}_{23} \cdot (S_1\mathbf{S}_1 + a_{12}\mathbf{a}_{12} + S_2\mathbf{S}_2 + a_{23}\mathbf{a}_{23} + S_3\mathbf{S}_3 + a_{34}\mathbf{a}_{34} + S_4\mathbf{S}_4 + a_{45}\mathbf{a}_{45} + S_5\mathbf{S}_5 + a_{56}\mathbf{a}_{56} + S_6\mathbf{S}_6 + a_{67}\mathbf{a}_{67} + S_7\mathbf{S}_7 + a_{71}\mathbf{a}_{71}) = 0.x \quad (7.24)$$

Evaluating the scalar products using the sets of direction cosines for a heptagon listed in the appendix gives

$$S_1X_2 + a_{12}c_2 + a_{23} + a_{34}c_3 + S_4\bar{X}_3 + a_{45}W_{43} + S_5X_{43} + a_{56}W_{543} + S_6X_{543} + a_{67}W_{712} + S_7X_{12} + a_{71}W_{12} = 0. \quad (7.25)$$

Upon substituting the constant mechanism parameters and the calculated values for θ_{1A}, θ_{4A}, and θ_{6A} into this equation, the corresponding value for the joint offset S_5 can be determined, as this is the only unknown in Eq. (7.25). The value for S_5 corresponding to θ_{1B}, θ_{4B}, and θ_{6B} can be found in a similar fashion.

Projecting the vector loop equation onto the vector $\mathbf{a}_{12}$ and expanding the scalar products using the sets of direction cosines in the appendix yields

$$a_{12} + a_{23}c_2 + S_3\bar{X}_2 + a_{34}W_{32} + S_4X_{32} + a_{45}W_{432} + S_5X_{432} + a_{56}W_{671} + S_6X_{71} + a_{67}W_{71} + S_7X_1 + a_{71}c_1 = 0. \quad (7.26)$$

This equation can be used to solve for S_3 for the A case by substituting values for the constant mechanism parameters and the calculated values of θ_{1A}, θ_{4A}, θ_{6A}, and S_{5A} into this equation. The offset value for S_3 for the B case can be found by substituting the corresponding values of θ_{1B}, θ_{4B}, θ_{6B}, and S_{5B} into Eq. (7.26).

The last joint offset to be calculated is S_2. This parameter may be determined by obtaining a scalar equation by projecting the vector loop equation onto any arbitrary direction independent of $\mathbf{a}_{23}$ and $\mathbf{a}_{12}$. Projecting the vector loop equation onto the direction of $\mathbf{S}_2$ and expanding the scalar products yields

$$S_1c_{12} + S_2 + S_3c_{23} + a_{34}U_{32} + S_4\bar{Z}_3 + a_{45}U_{432} + S_5Z_{43} + a_{56}U_{6712} + S_6Z_{71} + a_{67}U_{712} + S_7Z_1 + a_{71}U_{12} = 0. \quad (7.27)$$

This equation can be used to determine values for S_2 for the two solution sets by substituting the constant mechanism parameters and each set of previous solutions for θ_1, θ_4, θ_6, S_3, and S_5.

At this point, the solution of the RRPRPPR mechanism is complete. It was shown that the three unknown joint angles could be determined first by analyzing the equivalent spherical mechanism. The three corresponding unknown joint displacements could then be determined by projecting the vector loop equation onto any three linearly independent directions to yield three scalar equations in the three unknown joint offset variables. Data

Table 7.3. *RRPRPPR mechanism parameters.*

Link length, cm.	Twist angle, deg.	Joint offset, cm.	Joint angle, deg.
$a_{12} = 20$	$\alpha_{12} = 45$	$S_1 = 15$	$\theta_1 =$ variable
$a_{23} = 40$	$\alpha_{23} = 60$	$S_2 =$ variable	$\theta_2 = 80$
$a_{34} = 30$	$\alpha_{34} = 25$	$S_3 =$ variable	$\theta_3 = 65$
$a_{45} = 20$	$\alpha_{45} = 30$	$S_4 = 35$	$\theta_4 =$ variable
$a_{56} = 20$	$\alpha_{56} = 70$	$S_5 =$ variable	$\theta_5 = 20$
$a_{67} = 10$	$\alpha_{67} = 20$	$S_6 = 25$	$\theta_6 =$ variable
$a_{71} = 30$	$\alpha_{71} = 40$	$S_7 = 50$	$\theta_7 = 300$ (input)

Table 7.4. *Calculated configurations for the RRPRPPR spatial mechanism.*

	Solution A	Solution B
θ_1, degrees	235.43	−5.15
θ_4, degrees	101.77	14.25
θ_6, degrees	−91.17	160.51
S_2, cm	−53.98	−110.15
S_3, cm	104.23	110.12
S_5, cm	−134.30	−208.09

used in a numerical example are listed in Table 7.3, and the two solution configurations are listed in Table 7.4.

All group 1 spatial mechanisms follow the pattern demonstrated in this and the previous section. In other words, group 1 spatial mechanisms can always be solved by decoupling the problem. The unknown joint displacements can be calculated first by analyzing the equivalent spherical mechanism, which is, of course, a spherical 4R mechanism with mobility $M = 1$. The unknown joint offsets can then be determined by projecting the vector loop equation onto three linearly independent directions. Group 1 spatial mechanisms thus represent the simplest form of a spatial closed-loop device, and as such further examples will not be developed in this chapter. The next section, however, will present a solution for the simplest spatial manipulator, that is, the CCC manipulator.

7.4 CCC spatial manipulator

Figure 7.5 shows a CCC spatial manipulator. The fixed coordinate system has been attached so that the Z axis is parallel with the vector $\mathbf{S}_1$. It is important to note that the origin of the fixed coordinate system does not always coincide with the origin of the first coordinate system, that is, the intersection of the vectors $\mathbf{S}_1$ and $\mathbf{a}_{12}$, because the first joint of the manipulator is a cylindric joint. The distance from the origin of the fixed coordinate system to the origin of the first coordinate system is defined in the figure as L_1.

The specific problem statement for the reverse analysis of the CCC spatial manipulator is stated as follows:

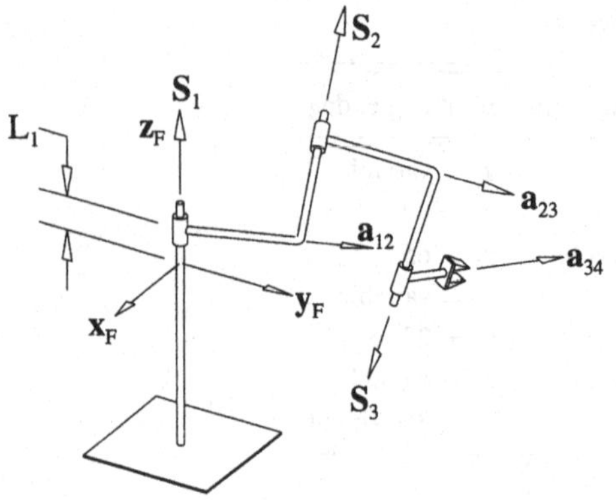

Figure 7.5. CCC spatial manipulator.

given:

constant mechanism parameters:

$a_{12}, a_{23},$

$\alpha_{12}, \alpha_{23},$

position and orientation of end effector:

${}^{F}\mathbf{P}_{tool}, {}^{F}\mathbf{S}_3, {}^{F}\mathbf{a}_{34}$, and

location of tool point in the 3rd coordinate system:

${}^{3}\mathbf{P}_{tool},$

find: $\phi_1, \theta_2, \theta_3, L_1, S_2$, and S_3.

The location of the third coordinate system can be obtained using Figure 7.5. Its origin is located at the intersection of the vectors $\mathbf{S}_3$ and $\mathbf{a}_{34}$. Its Z axis is parallel to the vector $\mathbf{S}_3$, and its X axis is parallel to the vector $\mathbf{a}_{34}$. The location of the tool point in terms of this third coordinate system is given together with the desired position of the tool point and orientation of the vectors $\mathbf{S}_3$ and $\mathbf{a}_{34}$ in the fixed coordinate system. The objective is to determine the values for the variable parameters that will position and orient the end effector as desired.

The first step of the analysis will be to determine the location of the origin of the third coordinate system as measured in the fixed coordinate system. Using the results of Eq. (5.3), the location of the origin of the third coordinate system may be calculated from

$${}^{F}\mathbf{P}_{3orig} = {}^{F}\mathbf{P}_{tool} - ({}^{3}\mathbf{P}_{tool} \cdot \mathbf{i}){}^{F}\mathbf{a}_{34} - ({}^{3}\mathbf{P}_{tool} \cdot \mathbf{j}){}^{F}\mathbf{S}_3 \times {}^{F}\mathbf{a}_{34} - ({}^{3}\mathbf{P}_{tool} \cdot \mathbf{k}){}^{F}\mathbf{S}_3. \quad (7.28)$$

The second step of the analysis will be to form a closed-loop mechanism by calculating the parameters of a hypothetical fourth link as described in Chapter 5. Arbitrary values may be selected for the parameters a_{34} and α_{34}. For this analysis, the link length a_{34} will be selected as zero and the twist angle α_{34} will be selected as ninety degrees. With these two selections, the direction of the vector $\mathbf{S}_4$ is known in terms of the fixed coordinate system. Further, it is known that the vector $\mathbf{S}_4$ passes through the origin of the third coordinate system because the link length a_{34} was chosen as zero.

Table 7.5. *Constant mechanism parameters for the CCC manipulator.*

Link length, cm.	Twist angle, deg.	Joint offset, cm.	Joint angle, deg.
$a_{12} = 12$	$\alpha_{12} = 27$	L_1 = variable	ϕ_1 = variable
$a_{23} = 23$	$\alpha_{23} = 45$	S_2 = variable	θ_2 = variable
		S_3 = variable	θ_3 = variable

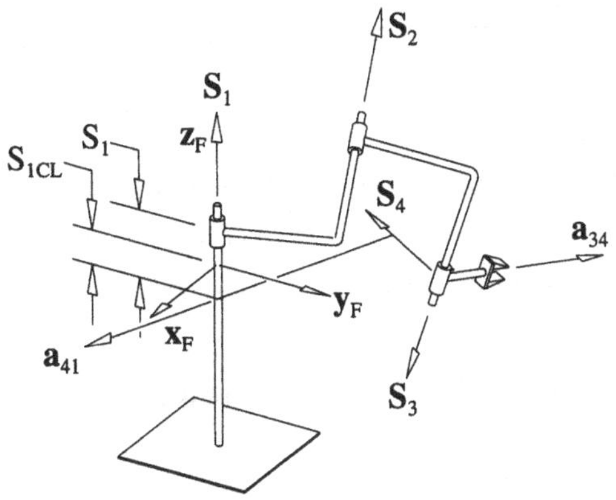

Figure 7.6. Hypothetical closure link for CCC manipulator.

Figure 7.6 shows the hypothetical closure link for the CCC manipulator. The analysis presented in Chapter 5 will be used to determine the close-the-loop parameters, that is, the three distances S_{1CL}, S_4, and a_{41} and the three angles θ_4, α_{41}, and γ_1. It should be noted that the close-the-loop parameter S_{1CL} (see Figure 7.6) is the distance along the vector $\mathbf{S}_1$ from the intersection of the vectors $\mathbf{S}_1$ and $\mathbf{a}_{41}$ to the origin of the fixed coordinate system. The desired output value L_1 (see Figure 7.5) will equal the difference between the distance S_1 (defined in the traditional way as the perpendicular distance between the vectors $\mathbf{a}_{41}$ and $\mathbf{a}_{12}$) and the distance S_{1CL}.

Upon completing the close-the-loop procedure, the CCC manipulator can be analyzed as a closed-loop RCCC spatial mechanism. The equivalent RRRR spherical mechanism is analyzed first as discussed in Section 6.7.2 to yield two solution sets for the joint angles θ_1, θ_2, and θ_3. The two corresponding values for ϕ_1 are then found from the equation

$$\phi_1 = \theta_1 - \gamma_1. \tag{7.29}$$

The corresponding values for the three displacements S_1, S_2, and S_3 are then determined by projecting the vector loop equation onto any three linearly independent directions and then solving the three scalar equations for the three unknown displacements. Finally, the value of the parameter L_1 is found as the difference between the calculated value of S_1 and the close-the-loop parameter S_{1CL}.

At this point the analysis of the CCC manipulator is complete, and a numerical example is now presented. Table 7.5 shows the constant mechanism parameters of a CCC

Table 7.6. *RCCC mechanism parameters.*

Link length, cm.	Twist angle, deg.	Joint offset, cm.	Joint angle, deg.
$a_{12} = 12$	$\alpha_{12} = 27$	$S_1 =$ variable	$\theta_1 =$ variable
$a_{23} = 23$	$\alpha_{23} = 45$	$S_2 =$ variable	$\theta_2 =$ variable
$a_{34}^{*} = 0$	$\alpha_{34}^{*} = 90$	$S_3 =$ variable	$\theta_3 =$ variable
$a_{41}^{**} = -38.6996$	$\alpha_{41}^{**} = 102.77$	$S_4^{**} = -34.3735$	$\theta_4^{**} = 112.10$ (input angle)

* = arbitrary selection
** = close-the-loop parameter

manipulator. The desired position and orientation for the manipulator are given as

$$^F\mathbf{P}_{\text{tool}} = \begin{bmatrix} 45.529 \\ -11.355 \\ 50.125 \end{bmatrix} \text{cm}, \qquad ^F\mathbf{S}_3 = \begin{bmatrix} 0.6672 \\ -0.6482 \\ 0.3671 \end{bmatrix} \qquad ^F\mathbf{a}_{34} = \begin{bmatrix} -0.0900 \\ 0.4188 \\ 0.9036 \end{bmatrix}, \tag{7.30}$$

and the position of the tool point in terms of the third coordinate system is given as

$$^3\mathbf{P}_{\text{tool}} = \begin{bmatrix} 6 \\ 8 \\ 2 \end{bmatrix} \text{cm}. \tag{7.31}$$

The objective is to calculate sets of values for the parameters L_1, S_2, S_3, ϕ_1, θ_2, and θ_3 that will position and orient the end effector as desired. The solution begins by determining the location of the origin of the third coordinate system by using Eq. (7.28). The calculated value was determined to be

$$^F\mathbf{P}_{3\text{orig}} = \begin{bmatrix} 50.6499 \\ -7.4845 \\ 42.2003 \end{bmatrix} \text{cm}. \tag{7.32}$$

Values for the parameters α_{34} and a_{34} were arbitrarily selected as ninety degrees and zero respectively, and the close-the-loop parameters were calculated to be

$$S_{1CL} = -49.8007\,\text{cm}, \quad a_{41} = -38.6996\,\text{cm}, \quad S_4 = -34.3735\,\text{cm}, \quad \theta_4 = 112.10^\circ,$$
$$\alpha_{41} = 102.77^\circ, \quad \gamma_1 = 49.31^\circ. \tag{7.33}$$

Table 7.6 shows the mechanism parameters for the newly formed closed-loop RCCC spatial mechanism. All parameters of the mechanism are known, and the three unknown joint angles can be determined as presented in Section 6.7.2. The three joint offset values are calculated by projecting the vector loop equation onto three different directions.

Table 7.7 shows the resulting two configurations of the manipulator that will position and orient the end effector as desired.

7.5 Summary

Group 1 spatial mechanisms can always be solved by first analyzing the corresponding equivalent spherical mechanism to determine the two configuration sets of the three

Table 7.7. *Calculated configurations for the CCC manipulator.*

	Solution A	Solution B
ϕ_1, degrees	71.63	20.02
θ_2, degrees	−34.94	34.94
θ_3, degrees	92.48	60.02
L_1, cm	41.59	10.04
S_2, cm	−14.97	14.97
S_3, cm	54.31	35.00

variable joint angles. The corresponding values for the three variable joint offsets can next be determined by projecting the vector loop equation for the mechanism onto any three arbitrary directions. This will yield three scalar equations in the three unknown joint offset distances. It was shown in the example problems that a judicious selection of the projection directions can simplify the solution.

A complete example of the reverse analysis of the simplest spatial manipulator, the CCC manipulator, was also presented. This example illustrated the concept of closing the loop with a hypothetical link in order to form a new closed-loop spatial mechanism. The resulting group 1 spatial mechanism was then analyzed using the techniques presented in this chapter.

7.6 Problems

1. A CCC robot has the following dimensions:

$$\alpha_{12} = 60^\circ \qquad a_{12} = 30\,\text{in.}$$
$$\alpha_{23} = -30 \qquad a_{23} = 12.$$

The tool point in terms of the third coordinate system is given as $[6, 8, 2]^T$. Determine the values for the parameters L_1, S_2, S_3, ϕ_1, θ_2, and θ_3 that will position the tool point at $[45, -11, 50]^T$ measured in terms of the fixed coordinate system. The orientation of the end effector is to be specified by

$$S_3 = [1, 2, 2]^T \quad \text{and} \quad a_{34} = [2, -2, 1]^T.$$

2. A planar representation of a spatial RPRPRPR mechanism with link a_{71} fixed to ground is illustrated in Figure 7.7. The link a_{71} is fixed to ground.

(a) What group mechanism is this? Why?

(b) Assuming that the input angle θ_7 is specified together with all constant mechanism dimensions, obtain an input/output equation of the form

$$Ac_1 + Bs_1 + D = 0.$$

Obtain expressions for A, B, and D.

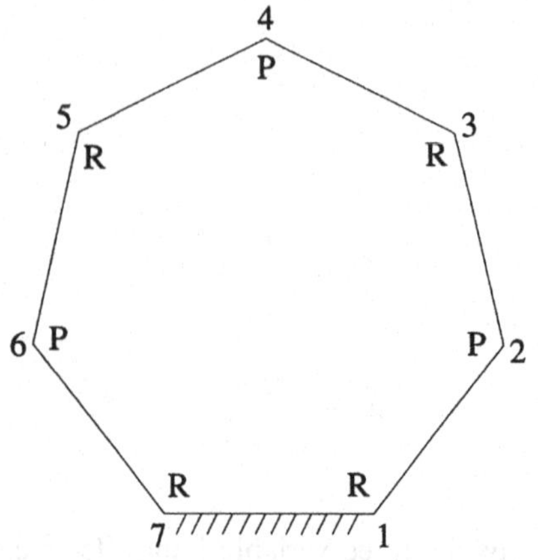

Figure 7.7. RPRPRPR mechanism.

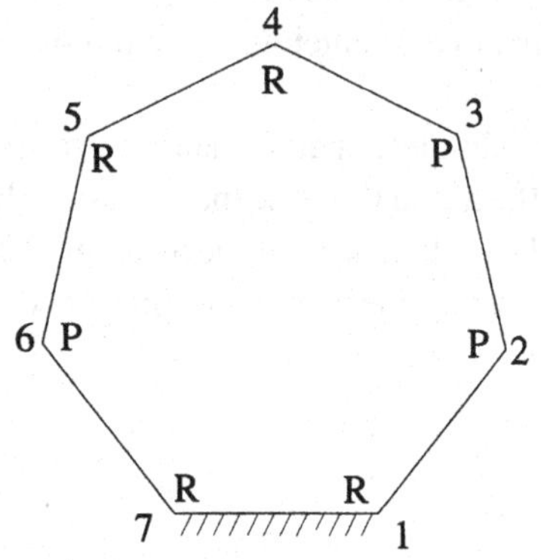

Figure 7.8. RPRRPPR mechanism.

(c) Show how to solve the equation from part (b) for all values of θ_1 in terms of the parameters A, B, and D.

(d) Explain how to solve for corresponding values of θ_3 and θ_5.

(e) Explain how to solve for the three slider displacements S_2, S_4, and S_6.

3. A planar representation of a group 1 spatial closed-loop mechanism with link a_{71} fixed to ground is shown in Figure 7.8.

(a) Assuming that all constant mechanism parameters are known and that the angle θ_7 is given as an input angle, explain how to solve for the angle θ_1. How many values for θ_1 can be found?

(b) Assuming that you have successfully solved for θ_1, explain how you would solve for the angle θ_4.

(c) Assuming that you have successfully solved for θ_1 and θ_4, explain how you would solve for the angle θ_5.

(d) Finally, assuming that you have successfully solved for θ_1, θ_4, and θ_5, explain how you would solve for the slider displacements S_2, S_3, and S_6.

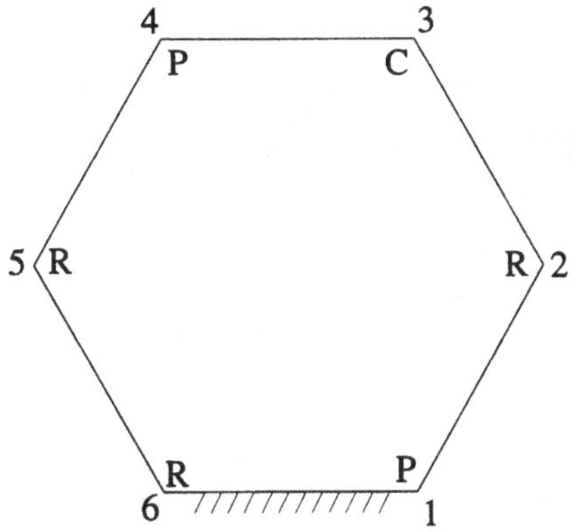

Figure 7.9. RRPCRP mechanism.

4. Shown in Figure 7.9 is a planar representation of a six-link RRPCRP group 1 spatial mechanism with link a_{61} fixed to ground. The input parameter is θ_6, and the parameter to be solved for first is θ_2.

(a) Assuming that all constant mechanism dimensions are known, what link lengths, offsets, twist angles, and joint angles are still unknown?

(b) Write an equation that contains only the input angle and the output angle. Expand the equation into the format:

$$Ac_2 + Bs_2 + D = 0.$$

(c) Describe how you would solve an equation of the form $Ac_2 + Bs_2 + D = 0$. How many values of θ_2 would satisfy this equation?

(d) Assuming that θ_2 is now known, explain how you would determine the remaining angles θ_3 and θ_5.

(e) Assuming that all joint angles are now known, explain how you would solve for the slider displacements S_4, S_3 and S_1.

8

Group 2 spatial mechanisms

8.1 Problem statement

A group 2 spatial mechanism is defined as a one-degree-of-freedom single-loop spatial mechanism whose equivalent spherical mechanism has two degrees of freedom. The simplest group 2 mechanism consists of five links that are interconnected by three revolute joints and two cylindric joints.

Figure 8.1 shows an RCRCR spatial mechanism and its equivalent spherical mechanism (note that the distance S_4 is negative in the drawing). Link a_{51} is fixed to ground, the input angle is θ_5, and it is assumed that all constant mechanism parameters are known, that is, the constant twist angles, link lengths, and joint offset distances. The objective is to determine values for the unknown output angle, θ_1, together with values for θ_2, θ_3, and θ_4 and the unknown joint offset distances S_2 and S_4. The angle θ_1 is the output angle (the unknown parameter to be solved for first).

Because the RCRCR mechanism is a one-degree-of-freedom device, specification of the angle θ_5 is sufficient to calculate values for all the unknown parameters. Examination of the fundamental and subsidiary cosine laws yields that each cosine law contains three joint variables (sine and sine–cosine laws contain four joint variables). Hence, it is possible to write down only a spherical equation containing the input angle θ_5, the output angle θ_1, together with a third joint angle. The equations that contain θ_2, θ_3, and θ_4 as the third angle are respectively

$$Z_{512} = c_{34}, \tag{8.1}$$

$$Z_{51} = \bar{Z}_3, \tag{8.2}$$

and

$$Z_{451} = c_{23}. \tag{8.3}$$

The solution to the problem will be accomplished by deriving a second equation in the input angle θ_5, the output angle θ_1, and the unwanted angle θ_3. It is necessary to eliminate θ_3 from this equation and Eq. (8.2).

The next section describes alternative solutions for a pair of trigonometric equations. Following this, methods for generating additional equations are derived using vector loop equations and dual numbers. Finally, the second equation in θ_3 for the RCRCR mechanism is derived, and a numerical example is presented.

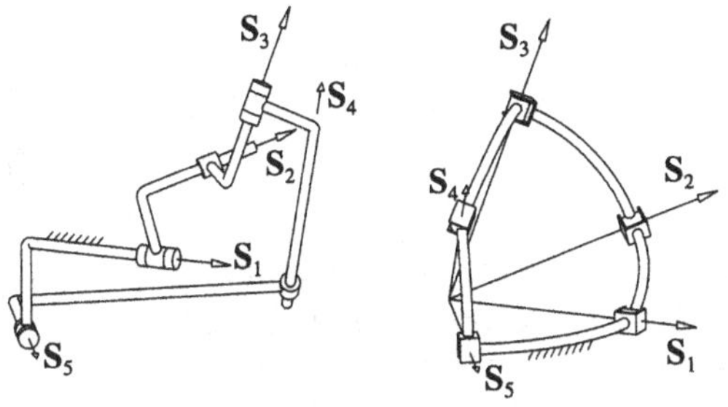

Figure 8.1. Spatial RCRCR group 2 mechanism and its equivalent spherical mechanism.

8.2 Solution of two trigonometric equations in two unknowns

In the example of the RCRCR spatial mechanism, it will be assumed that a second equation can be found that contains θ_3 as the extra unknown. This new equation will be paired with Eq. (8.2), which can be expanded in the form

$$s_{12}(X_5 s_1 + Y_5 c_1) + c_{12} Z_5 = c_{23} c_{34} - s_{23} s_{34} c_3. \tag{8.4}$$

Equation (8.4) and the yet-to-be-determined second equation can be regrouped into the general form

$$c_3(A_i c_1 + B_i s_1 + D_i) + s_3(E_i c_1 + F_i s_1 + G_i) + (H_i c_1 + I_i s_1 + J_i) = 0, \qquad i = 1, 2, \tag{8.5}$$

where the coefficients A_i through J_i can be numerically evaluated as they are all in terms of the given mechanism parameters.

The sine and cosine of an angle, θ_k, can be written in terms of the tan-half-angle, x_k, where $x_k = \tan(\theta_k/2)$ using the following trigonometric identities

$$s_k = \frac{2x_k}{1 + x_k^2}, \tag{8.6}$$

$$c_k = \frac{1 - x_k^2}{1 + x_k^2}. \tag{8.7}$$

Substituting for the sine and cosine of θ_1 and θ_3 and then multiplying throughout by $(1 + x_1^2)(1 + x_3^2)$ in the pair of equations represented by Eq. (8.5) gives

$$\begin{aligned}(1 - x_3^2)&\left[A_i(1 - x_1^2) + B_i(2x_1) + D_i(1 + x_1^2)\right]\\ &+ (2x_3)\left[E_i(1 - x_1^2) + F_i(2x_1) + G_i(1 + x_1^2)\right]\\ &+ (1 + x_3^2)\left[H_i(1 - x_1^2) + I_i(2x_1) + J_i(1 + x_1^2)\right] = 0, \qquad i = 1, 2.\end{aligned} \tag{8.8}$$

Regrouping Eq. (8.8) gives

$$\begin{aligned}x_3^2&\left[x_1^2(A_i - D_i - H_i + J_i) + x_1(2I_i - 2B_i) + (-A_i - D_i + H_i + J_i)\right]\\ &+ x_3\left[x_1^2(-2E_i + 2G_i) + x_1(4F_i) + (2E_i + 2G_i)\right] + \left[x_1^2(-A_i + D_i - H_i + J_i)\right.\\ &\left.+ x_1(2I_i + 2B_i) + (A_i + D_i + H_i + J_i)\right] = 0, \qquad i = 1, 2.\end{aligned} \tag{8.9}$$

Equation (8.9) may be written as

$$x_3^2\left[a_i x_1^2 + b_i x_1 + d_i\right] + x_3\left[e_i x_1^2 + f_i x_1 + g_i\right]\left[h_i x_1^2 + i_i x_1 + j_i\right] = 0, \qquad i = 1, 2, \tag{8.10}$$

where the coefficients a_i through j_i are readily expressed in terms of the coefficients A_i through J_i and are therefore known quantities. These expressions are as follows:

$$\begin{aligned} a_i &= A_i - D_i - H_i + J_i & b_i &= 2(I_i - B_i) & d_i &= -A_i - D_i + H_i + J_i \\ e_i &= 2(G_i - E_i) & f_i &= 4F_i & g_i &= 2(G_i + E_i) \\ h_i &= -A_i + D_i - H_i + J_i & i_i &= 2(I_i + B_i) & j_i &= A_i + D_i + H_i + J_i. \end{aligned} \tag{8.11}$$

The problem of solving for values of θ_1 and θ_3 that satisfy the two equations represented by Eq. (8.5) has now been reduced to determining values for x_1 and x_3 that satisfy the two biquadratic equations represented by Eq. (8.10). Two solution techniques for this problem will be introduced.

8.2.1 Sylvester's solution method

The equations represented by Eq. (8.10) can be written as

$$L_1 x_3^2 + M_1 x_3 + N_1 = 0, \tag{8.12}$$
$$L_2 x_3^2 + M_2 x_3 + N_2 = 0, \tag{8.13}$$

where

$$L_i = a_i x_1^2 + b_i x_1 + d_i, \quad i = 1, 2 \tag{8.14}$$
$$M_i = e_i x_1^2 + f_i x_1 + g_i, \quad i = 1, 2 \tag{8.15}$$
$$N_i = h_i x_1^2 + i_i x_1 + j_i, \quad i = 1, 2. \tag{8.16}$$

The coefficients L_i through N_i are quadratic expressions in the variable x_1.

Equations (8.12) and (8.13) are a pair of quadratic equations in the variable x_3. A quadratic equation clearly has two solutions. The question, however, is what conditions must the coefficients L_i through N_i satisfy in order for one value of x_3 to solve both Eqs. (8.12) and (8.13) simultaneously.

In Sylvester's solution method, Eqs. (8.12) and (8.13) will be rewritten as

$$L_1 t + M_1 u + N_1 v = 0, \tag{8.17}$$
$$L_2 t + M_2 u + N_2 v = 0, \tag{8.18}$$

where $t = x_3^2$, $u = x_3$, and $v = 1$. Equations (8.17) and (8.18) represent two linear homogeneous equations in the three unknowns t, u, and v.

Equations (8.12) and (8.13) are now multiplied by x_3 to yield

$$L_1 x_3^3 + M_1 x_3^2 + N_1 x_3 = 0, \tag{8.19}$$
$$L_2 x_3^3 + M_2 x_3^2 + N_2 x_3 = 0. \tag{8.20}$$

Letting $s = x_3^3$, and using the definitions for t and u, Eqs. (8.19) and (8.20) may be written as

$$L_1 s + M_1 t + N_1 u = 0, \tag{8.21}$$
$$L_2 s + M_2 t + N_2 u = 0. \tag{8.22}$$

Equations (8.17), (8.18), (8.21), and (8.22) represent four linear homogeneous equations in the four unknowns s, t, u, and v. These four equations can be written in matrix form as

$$\begin{bmatrix} 0 & L_1 & M_1 & N_1 \\ 0 & L_2 & M_2 & N_2 \\ L_1 & M_1 & N_1 & 0 \\ L_2 & M_2 & N_2 & 0 \end{bmatrix} \begin{bmatrix} s \\ t \\ u \\ v \end{bmatrix} = \begin{bmatrix} 0 \\ 0 \\ 0 \\ 0 \end{bmatrix}. \tag{8.23}$$

The four linear homogeneous equations must be linearly dependent if there is to be any solution set other than the trivial solution where all the unknowns equal zero. The equations will be linearly dependent if the following determinant equals zero:

$$\begin{vmatrix} 0 & L_1 & M_1 & N_1 \\ 0 & L_2 & M_2 & N_2 \\ L_1 & M_1 & N_1 & 0 \\ L_2 & M_2 & N_2 & 0 \end{vmatrix} = 0. \tag{8.24}$$

The terms L_i through N_i are quadratic in the variable x_1. Expansion of the determinant will thus yield in general an eighth-degree polynomial in the variable x_1.

8.2.2 Bezout's solution method

Equations (8.12) and (8.13) can be thought of as two quadratic equations in the variable x_3. Bezout's solution method proceeds by rewriting these equations as

$$x_3(L_1 x_3 + M_1) + N_1 = 0, \tag{8.25}$$
$$x_3(L_2 x_3 + M_2) + N_2 = 0. \tag{8.26}$$

These two new "linear" equations must be linearly dependent if there is to be a common solution. Equations (8.25) and (8.26) will be linearly dependent if

$$N_2(L_1 x_3 + M_1) - N_1(L_2 x_3 + M_2) = 0. \tag{8.27}$$

This equation can be written as

$$x_3 \begin{vmatrix} L_1 & N_1 \\ L_2 & N_2 \end{vmatrix} + \begin{vmatrix} M_1 & N_1 \\ M_2 & N_2 \end{vmatrix} = 0. \tag{8.28}$$

Equations (8.12) and (8.13) can now be written as

$$x_3^2(L_1) + (M_1 x_3 + N_1) = 0, \tag{8.29}$$
$$x_3^2(L_2) + (M_2 x_3 + N_2) = 0. \tag{8.30}$$

These two new "linear" equations (linear in the term x_3^2) must be linearly dependent if there is to be a common solution. The equations will be linearly dependent if

$$L_1(M_2x_3 + N_2) - L_2(M_1x_3 + N_3) = 0. \tag{8.31}$$

Equation (8.31) can be written as

$$x_3\begin{vmatrix} L_1 & M_1 \\ L_2 & M_2 \end{vmatrix} + \begin{vmatrix} L_1 & N_1 \\ L_2 & N_2 \end{vmatrix} = 0. \tag{8.32}$$

Equations (8.28) and (8.32) are two linear equations in the variable x_3. In order for a common solution for x_3 to exist that simultaneously satisfies Eqs. (8.28) and (8.32), the equations must be linearly dependent and the following expression may be written:

$$\begin{vmatrix} L_1 & M_1 \\ L_2 & M_2 \end{vmatrix}\begin{vmatrix} M_1 & N_1 \\ M_2 & N_2 \end{vmatrix} - \begin{vmatrix} L_1 & N_1 \\ L_2 & N_2 \end{vmatrix}^2 = 0. \tag{8.33}$$

The coefficients L_1 through N_2 are quadratic in the variable x_1. Thus, Eq. (8.33) can be expanded, and in general it will yield an eighth-degree polynomial in the variable x_1, which is the same eighth-degree polynomial obtained by Sylvester's method. The reader may well prefer Bezout's method because the expansion of Eq. (8.33) is simpler than that for Eq. (8.24) and also Eqs. (8.28) and (8.32) give alternate expressions for $x_3 = \tan(\theta_3/2)$. However, special relationships between the coefficients L_1 through N_2 may reduce the degree of this polynomial, and for the RCRCR mechanism it in fact reduces directly to fourth degree.

The corresponding value for x_3 for each value of x_1 can be found from Eq. (8.28) as

$$x_3 = \frac{-\begin{vmatrix} M_1 & N_1 \\ M_2 & N_2 \end{vmatrix}}{\begin{vmatrix} L_1 & N_1 \\ L_2 & N_2 \end{vmatrix}} \tag{8.34}$$

or from (8.32) as

$$x_3 = \frac{-\begin{vmatrix} L_1 & N_1 \\ L_2 & N_2 \end{vmatrix}}{\begin{vmatrix} L_1 & M_1 \\ L_2 & M_2 \end{vmatrix}} \tag{8.35}$$

In summary, it has been shown in this section how two biquadratic equations in two unknowns can be solved. The result is in general an eighth-degree polynomial in one of the variables. Corresponding values for the second variable are readily obtained.

8.3 Generation of additional equations

The previous section showed how two equations, such as Eq. (8.5), that are linear in terms of the sines and cosines of two unknown angles may be solved. The result is, in general, an eighth-degree polynomial in terms of the tan-half-angle of one of the unknowns.

This section will focus on techniques for obtaining the second equation that will be paired with a spherical cosine law to obtain a mechanism's input/output equation. Three types of equations will be introduced and discussed in general terms. After this, specific examples such as the RCRCR mechanism shown in Figure 8.1 will be presented.

8.3.1 Projection of vector loop equation

The vector loop equation for a spatial closed-loop mechanism is formed by summing all the products of the link lengths times the unit link vectors plus the offset distances times the unit joint axis vectors. This summation will equal the zero vector. For the RCRCR mechanism, the vector loop equation is written as

$$S_1\mathbf{S}_1 + a_{12}\mathbf{a}_{12} + S_2\mathbf{S}_2 + a_{23}\mathbf{a}_{23} + S_3\mathbf{S}_3 + a_{34}\mathbf{a}_{34} + S_4\mathbf{S}_4 + a_{45}\mathbf{a}_{45} + S_5\mathbf{S}_5 + a_{51}\mathbf{a}_{51} = \mathbf{0}. \tag{8.36}$$

This vector loop equation can be projected onto any vector, $\mathbf{b}$, to yield a scalar equation. This projection is accomplished by performing a scalar product of each vector in the loop equation with the vector $\mathbf{b}$.

Typically, the vector loop equation is projected onto one of the link vectors or joint axis vectors. This is done because the scalar product can be easily evaluated by use of the direction cosine tables listed in the appendix. For example, for the RCRCR mechanism, the projection of the vector loop equation onto the direction $\mathbf{a}_{34}$ can be written as

$$\begin{aligned} &S_1(\mathbf{S}_1 \cdot \mathbf{a}_{34}) + a_{12}(\mathbf{a}_{12} \cdot \mathbf{a}_{34}) + S_2(\mathbf{S}_2 \cdot \mathbf{a}_{34}) + a_{23}(\mathbf{a}_{23} \cdot \mathbf{a}_{34}) + S_3(\mathbf{S}_3 \cdot \mathbf{a}_{34}) \\ &\quad + a_{34}(\mathbf{a}_{34} \cdot \mathbf{a}_{34}) + S_4(\mathbf{S}_4 \cdot \mathbf{a}_{34}) + a_{45}(\mathbf{a}_{45} \cdot \mathbf{a}_{34}) + S_5(\mathbf{S}_5 \cdot \mathbf{a}_{34}) + a_{51}(\mathbf{a}_{51} \cdot \mathbf{a}_{34}) = 0. \end{aligned} \tag{8.37}$$

The result of a scalar product is independent of the coordinate system that the two vectors are measured in. Thus, each scalar product in Eq. (8.37) may be evaluated using a different set from the direction cosine table for a spherical pentagon, if so desired. For example, using set 8 for the direction cosine table for the first six scalar products and set 3 for the remainder yields the scalar equation

$$S_1X_{23} + a_{12}W_{23} + S_2X_3 + a_{23}c_3 + a_{34} + a_{45}c_4 + S_5\bar{X}_4 + a_{51}W_{54} = 0. \tag{8.38}$$

It will be seen in Chapter 11 that the projection of the vector loop equation will be used often in the reverse-analysis solution of industrial robot manipulators.

8.3.2 Self-scalar product of vector loop equation

A scalar equation can be generated by performing a scalar product of the vector loop equation with itself. For the RCRCR mechanism the self-scalar product may be written as

$$\begin{aligned} &[S_1\mathbf{S}_1 + a_{12}\mathbf{a}_{12} + S_2\mathbf{S}_2 + a_{23}\mathbf{a}_{23} + S_3\mathbf{S}_3 + a_{34}\mathbf{a}_{34} + S_4\mathbf{S}_4 + a_{45}\mathbf{a}_{45} + S_5\mathbf{S}_5 + a_{51}\mathbf{a}_{51}] \\ &\quad \cdot [S_1\mathbf{S}_1 + a_{12}\mathbf{a}_{12} + S_2\mathbf{S}_2 + a_{23}\mathbf{a}_{23} + S_3\mathbf{S}_3 + a_{34}\mathbf{a}_{34} + S_4\mathbf{S}_4 \\ &\quad + a_{45}\mathbf{a}_{45} + S_5\mathbf{S}_5 + a_{51}\mathbf{a}_{51}] = 0. \end{aligned} \tag{8.39}$$

Expanding this equation, dividing by 2, and then regrouping gives

$$
\begin{aligned}
&\frac{1}{2}[S_1^2 + a_{12}^2 + S_2^2 + a_{23}^2 + S_3^2 + a_{34}^2 + S_4^2 + a_{45}^2 + S_5^2 + a_{51}^2] \\
&+ S_1\mathbf{S}_1\cdot\ [a_{12}\mathbf{a}_{12} + S_2\mathbf{S}_2 + a_{23}\mathbf{a}_{23} + S_3\mathbf{S}_3 + a_{34}\mathbf{a}_{34} + S_4\mathbf{S}_4 + a_{45}\mathbf{a}_{45} + S_5\mathbf{S}_5 + a_{51}\mathbf{a}_{51}] \\
&+ a_{12}\mathbf{a}_{12}\cdot [S_2\mathbf{S}_2 + a_{23}\mathbf{a}_{23} + S_3\mathbf{S}_3 + a_{34}\mathbf{a}_{34} + S_4\mathbf{S}_4 + a_{45}\mathbf{a}_{45} + S_5\mathbf{S}_5 + a_{51}\mathbf{a}_{51}] \\
&+ S_2\mathbf{S}_2\cdot\ [a_{23}\mathbf{a}_{23} + S_3\mathbf{S}_3 + a_{34}\mathbf{a}_{34} + S_4\mathbf{S}_4 + a_{45}\mathbf{a}_{45} + S_5\mathbf{S}_5 + a_{51}\mathbf{a}_{51}] \\
&+ a_{23}\mathbf{a}_{23}\cdot [S_3\mathbf{S}_3 + a_{34}\mathbf{a}_{34} + S_4\mathbf{S}_4 + a_{45}\mathbf{a}_{45} + S_5\mathbf{S}_5 + a_{51}\mathbf{a}_{51}] \\
&+ S_3\mathbf{S}_3\cdot\ [a_{34}\mathbf{a}_{34} + S_4\mathbf{S}_4 + a_{45}\mathbf{a}_{45} + S_5\mathbf{S}_5 + a_{51}\mathbf{a}_{51}] \\
&+ a_{34}\mathbf{a}_{34}\cdot [S_4\mathbf{S}_4 + a_{45}\mathbf{a}_{45} + S_5\mathbf{S}_5 + a_{51}\mathbf{a}_{51}] \\
&+ S_4\mathbf{S}_4\cdot\ [a_{45}\mathbf{a}_{45} + S_5\mathbf{S}_5 + a_{51}\mathbf{a}_{51}] \\
&+ a_{45}\mathbf{a}_{45}\cdot [S_5\mathbf{S}_5 + a_{51}\mathbf{a}_{51}] \\
&+ S_5\mathbf{S}_5\cdot\ [a_{51}\mathbf{a}_{51}] = 0.
\end{aligned}
\tag{8.40}
$$

Each of the remaining scalar products can be evaluated in terms of any desired coordinate system to yield a scalar equation. In most applications of the self-scalar product, certain terms of the vector loop equation are moved to the right-hand side of the equal sign. A self-scalar product is then performed for the remaining vectors on the left side of the equation. This is equated to the self-scalar product of the vectors on the right side of the equation. This procedure is used to obtain a scalar equation that does not contain certain unwanted joint angles.

8.3.3 Secondary cosine laws

Secondary cosine laws are scalar equations that have proved to be most useful in the analysis of spatial mechanisms. They can be derived by employing various scalar triple products of vector loop equations for closed polygons.

It is, however, easier and more instructive to derive secondary cosine laws from existing spherical cosine laws, and this is precisely why they have proved to be most useful. They contain the very same joint angles as their corresponding spherical cosine laws, which themselves contain the minimum number of joint variables. Further, this derivation crystallizes the concepts of equivalent spherical and spatial mechanisms (see Section 6.1 and Figure 8.1). Any spherical cosine law that contains certain α_{ij}'s and θ_j's is valid for equivalent spherical and spatial polygons. Dualizing the spherical cosine law provides an additional secondary cosine law for the spatial polygon that contains the same α_{ij}'s and θ_j's together with corresponding a_{ij}'s and S_j's. In order to accomplish this, it is necessary to understand the definitions and various operations of dual numbers and dual angles that were first introduced by Study (1901).

(a) Dual Numbers

A dual number is defined as a pair of real numbers, one of which is associated with the real unit $+1$ and the other of which is associated with the unit ϵ where $\epsilon^2 = \epsilon^3 = \cdots = 0$. For example, a dual number may be written as $5 + 7\epsilon$.

The definition of addition, multiplication, and division of dual numbers is straightforward. The sum of two dual numbers $\hat{a}$ and $\hat{b}$ that are defined as

$$\hat{a} = a + \epsilon a_0, \tag{8.41}$$

$$\hat{b} = b + \epsilon b_0 \tag{8.42}$$

is

$$\hat{a} + \hat{b} = (a + b) + \epsilon (a_0 + b_0). \tag{8.43}$$

The product of $\hat{a}$ and $\hat{b}$ is defined as

$$\hat{a}\hat{b} = (ab) + \epsilon (ab_0 + ba_0). \tag{8.44}$$

The division of dual numbers such as

$$\frac{\hat{a}}{\hat{b}} = \frac{(a + \epsilon a_0)}{(b + \epsilon b_0)} \tag{8.45}$$

is accomplished by multiplying the numerator and denominator by $(b - \epsilon b_0)$ which gives

$$\frac{\hat{a}}{\hat{b}} = \frac{(a + \epsilon a_0)\,(b - \epsilon b_0)}{(b + \epsilon b_0)\,(b - \epsilon b_0)} = \frac{(ab) + \epsilon (a_0 b - b_0 a)}{b^2}. \tag{8.46}$$

Division by a "pure dual number" for which the denominator has no real part is not defined.

Dual numbers may be substituted into functions. For example, suppose that f(x) is defined as

$$f(x) = 3x^2 + 2x + 7. \tag{8.47}$$

Substituting the dual number $(5 + 2\epsilon)$ into the function gives

$$f(5 + 2\epsilon) = 3(5 + 2\epsilon)^2 + 2(5 + 2\epsilon) + 7 = 92 + 64\epsilon. \tag{8.48}$$

The function can also be evaluated by a different manner. The Taylor's series expansion of a function is given by the infinite series

$$f(x + \Delta x) = f(x) + \Delta x \frac{f'(x)}{1!} + (\Delta x)^2 \frac{f''(x)}{2!} + \cdots. \tag{8.49}$$

Substituting the dual number $\hat{a}$ into Eq. (8.49) gives

$$f(a + \epsilon a_0) = f(a) + \epsilon a_0 f'(a). \tag{8.50}$$

Equation (8.50) is not an approximation because the remaining terms of the Taylor's series expansion vanish because $\epsilon^2 = \epsilon^3 = \epsilon^n = 0, n \geq 2$. Using Eq. (8.50) for the function f(x) when $x = (5 + 2\epsilon)$ gives

$$f(5 + 2\epsilon) = f(5) + 2\epsilon f'(5) = 92 + 2\epsilon(32) = 92 + 64\epsilon. \tag{8.51}$$

The Taylor's series expansion will be used extensively when evaluating functions with dual numbers. As a second example, consider the function

$$g(x) = \cos(x) + 3\sin(x). \tag{8.52}$$

Evaluating this function when $x = (2 + 5\epsilon)$ radians gives

$$g(2 + 5\epsilon) = g(2) + 5\epsilon g'(2) = \cos(2) + 3\sin(2) + 5\epsilon(-\sin(2) + 3\cos(2)) = 2.31 - 10.79\epsilon. \tag{8.53}$$

Taylor's series expansion may also be used for a function with more than one variable. For example, if f is a function of three variables, that is, f(x, y, z), then $f(a+\epsilon a_0, b+\epsilon b_0, c+\epsilon c_0)$ is defined by

$$f(a + \epsilon a_0, b + \epsilon b_0, c + \epsilon c_0) = f(a, b, c) + \epsilon a_0 \left.\frac{\partial f}{\partial x}\right|_{\substack{x=a\\y=b\\z=c}} + \epsilon b_0 \left.\frac{\partial f}{\partial y}\right|_{\substack{x=a\\y=b\\z=c}} + \epsilon c_0 \left.\frac{\partial f}{\partial z}\right|_{\substack{x=a\\y=b\\z=c}}. \tag{8.54}$$

Equation (8.54) can be easily extended by the reader for a function with more than three variables.

As an example, consider the function

$$f(x, y) = 4x^2\cos(y) + 3x\sin(y). \tag{8.55}$$

Evaluating this function at $x = 4 + 2\epsilon$ and $y = 3 + 5\epsilon$ yields

$$\begin{aligned} f(4 + 2\epsilon, 3 + 5\epsilon) &= 4(4)^2\cos(3) + 3(4)\sin(3) \\ &\quad + 2\epsilon[8(4)\cos(3) + 3\sin(3)] \\ &\quad + 5\epsilon[-4(4)^2\sin(3) + 3(4)\cos(3)] \\ &= -61.67 - 167.07\epsilon. \end{aligned} \tag{8.56}$$

(b) Dual Angles

A dual angle can be used to measure the relative position of two skew lines in space. For example, the dual angle $\hat{\alpha}_{12} = \alpha_{12} + \epsilon a_{12}$ completely describes the relative position of the lines $\$_1$ and $\$_2$ (the notation $\$_i$ represents the line in space that passes through the ith joint axis of the mechanism). Similarly, the dual angle $\hat{\theta}_2 = \theta_2 + \epsilon S_2$ completely describes the relative position of the lines $\$_{12}$ and $\$_{23}$, where $\$_{12}$ represents the line along link a_{12} and $\$_{23}$ represents the line along link a_{23}.

It can be shown that all the trigonometric identities will be valid for dual numbers. For example, it can be shown that $\sin^2\hat{\theta}_2 + \cos^2\hat{\theta}_2 = 1$ as follows:

$$\sin\hat{\theta}_2 = \sin\theta_2 + \epsilon S_2\cos\theta_2, \tag{8.57}$$

$$\cos\hat{\theta}_2 = \cos\theta_2 - \epsilon S_2\sin\theta_2. \tag{8.58}$$

Squaring and adding the left and right sides of Eqs. (8.57) and (8.58) gives

$$\sin^2 \hat{\theta}_2 + \cos^2 \hat{\theta}_2 = (\sin^2 \theta_2 + \cos^2 \theta_2) + \epsilon[2S_2 \sin\theta_2 \cos\theta_2 - 2S_2 \sin\theta_2 \cos\theta_2] = 1. \tag{8.59}$$

Dual angles will now be inserted into the spherical and polar sine, sine–cosine, and cosine laws, resulting in new equations. These equations were proven to be valid by Kotelnikov (1895) in his Principle of Transference.

(c) Secondary Cosine Laws

Dual angles are most often substituted into spherical cosine laws because these laws have one less joint variable than sine or sine–cosine laws and the objective of most problems is to obtain equations with as few unknowns as possible. The phrase "secondary cosine law" refers to the equation that is formed by equating the dual parts of a spherical cosine law whose twist and joint angles have been replaced by dual angles.

As an example, consider the spherical cosine law

$$Z_{41} = c_{23} \tag{8.60}$$

for an RCCC spatial mechanism. The left-hand side of Eq. (8.60) contains the angles θ_4, θ_1, α_{34}, α_{41}, and α_{12}. The right-hand side contains only the angle α_{23}. Dual angles will now be substituted for all of these angles. Dual angle substitutions will now be made by writing

$$\hat{Z}_{41} = \hat{c}_{23}. \tag{8.61}$$

The right-hand side of Eq. (8.61) is clearly

$$\hat{c}_{23} = c_{23} - \epsilon a_{23} s_{23}. \tag{8.62}$$

The left-hand side of Eq. (8.61) can be written as

$$\hat{Z}_{41} = Z_{41} + \epsilon Z_{041}, \tag{8.63}$$

where

$$Z_{41} = s_{12}(X_4 s_1 + Y_4 c_1) + c_{12} Z_4 \tag{8.64}$$

and

$$Z_{041} = S_4 \frac{\partial Z_{41}}{\partial \theta_4} + S_1 \frac{\partial Z_{41}}{\partial \theta_1} + a_{34} \frac{\partial Z_{41}}{\partial \alpha_{34}} + a_{41} \frac{\partial Z_{41}}{\partial \alpha_{41}} + a_{12} \frac{\partial Z_{41}}{\partial \alpha_{12}}. \tag{8.65}$$

Each of the partial derivatives of Eq. (8.65) must now be evaluated. The partial derivative of Z_{41} with respect to θ_1 is readily evaluated as

$$\frac{\partial Z_{41}}{\partial \theta_1} = s_{12}(X_4 c_1 - Y_4 s_1) = s_{12} X_{41}. \tag{8.66}$$

Similarly, the partial derivative of Z_{41} with respect to α_{12} is readily evaluated as

$$\frac{\partial Z_{41}}{\partial \alpha_1} = c_{12}(X_4 s_1 + Y_4 c_1) - s_{12} Z_4 = Y_{41}. \tag{8.67}$$

The partial derivatives of Z_{41} with respect to θ_4 and α_{34} can more easily be determined by taking a partial derivative Z_{14}, which equals Z_{41}. This is equivalent to performing the following exchange of subscripts on Eqs. (8.66) and (8.67):

$$\begin{matrix} 4 & 1 & 2 & 3 \\ \downarrow & \downarrow & \downarrow & \downarrow \\ 1 & 4 & 3 & 2 \end{matrix}. \tag{8.68}$$

The exchange yields the following two partial derivatives:

$$\frac{\partial Z_{41}}{\partial \theta_4} = s_{34} X_{14}, \tag{8.69}$$

$$\frac{\partial Z_{41}}{\partial \alpha_{34}} = Y_{14}. \tag{8.70}$$

The partial derivative of Z_{41} with respect to α_{41} remains to be evaluated. This partial derivative can be written as

$$\frac{\partial Z_{41}}{\partial \alpha_{41}} = s_{12}\left(\frac{\partial X_4}{\partial \alpha_{41}} s_1 + \frac{\partial Y_4}{\partial \alpha_{41}} c_1\right) + c_{12}\frac{\partial Z_4}{\partial \alpha_{41}}. \tag{8.71}$$

Now,

$$X_4 = s_{34} s_4, \tag{8.72}$$

$$Y_4 = -(s_{41} c_{34} + c_{41} s_{34} c_4), \tag{8.73}$$

$$Z_4 = c_{41} c_{34} - s_{41} s_{34} c_4. \tag{8.74}$$

Taking a partial derivative of each of these terms with respect to α_{41} yields

$$\frac{\partial X_4}{\partial \alpha_{41}} = 0, \qquad \frac{\partial Y_4}{\partial \alpha_{41}} = -Z_4, \qquad \frac{\partial Z_4}{\partial \alpha_{41}} = Y_4. \tag{8.75}$$

Substituting these derivatives into Eq. (8.71) yields

$$\frac{\partial Z_{41}}{\partial \alpha_{41}} = -s_{12} c_1 Z_4 + c_{12} Y_4. \tag{8.76}$$

All the partial derivatives in Eq. (8.64) have been expanded. The secondary cosine law that corresponds to the spherical cosine law $Z_{41} = c_{23}$ may now be written as

$$S_4(s_{34} X_{14}) + S_1(s_{12} X_{41}) + a_{34}(Y_{14}) + a_{41}(-s_{12} c_1 Z_4 + c_{12} Y_4) + a_{12}(Y_{41}) = -a_{23} s_{23}. \tag{8.77}$$

The procedure for generating secondary cosine laws is straightforward. The task of evaluating the partial derivatives can be simplified, however, by use of a computer software package that can perform symbolic manipulations. Several of these types of packages are commercially available.

One example of evaluating a secondary cosine law has been presented in this section. The subsequent sections of this and the next chapter will show how secondary cosine laws can contribute to the determination of the input/output equation for a spatial mechanism.

8.4 Five-link group 2 spatial mechanisms

Five-link group 2 mechanisms will contain three revolute joints and two cylindric joints. These mechanisms are the simplest group 2 mechanisms and will result, in general, in an eighth-degree input/output equation. One exception, however, is the RCRCR mechanism, which will have only a fourth-degree input/output equation because of special reductions in the determinants |LM|, |MN|, and |LN| (see Eq. (8.33)).

Two cases of five-link group 2 mechanisms whose input is a revolute joint will be presented in this section. The two cases represent the situations where the two cylindrical joints are adjacent in the mechanism or they are separated by a revolute joint. All five-link group 2 mechanisms (whose input is one of the revolute joints) may be solved using one of these cases.

8.4.1 Case I: RCRCR spatial mechanism (C joints separated by one R joint)

The RCRCR mechanism is shown in Figure 8.1. A planar representation of the mechanism is shown in Figure 8.2. The problem statement is as follows:

given: constant mechanism parameters, that is,

$a_{12}, a_{23}, a_{34}, a_{45}, a_{51},$

$\alpha_{12}, \alpha_{23}, \alpha_{34}, \alpha_{45}, \alpha_{51},$

$S_1, S_3,$ and S_5 and

input angle, θ_5,

find: $\theta_1, \theta_2, \theta_3, \theta_4, S_2,$ and S_4.

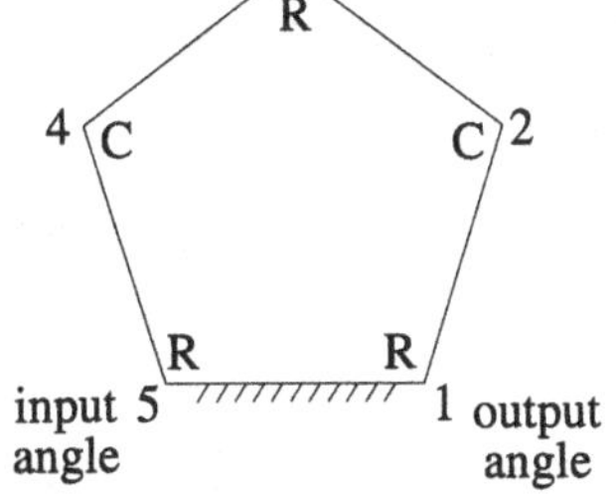

Figure 8.2. Planar representation of RCRCR group 2 spatial mechanism.

The angle θ_1 will be solved for first and is called the output angle. The present objective is to obtain an equation that does not contain the variable joint offsets S_2 or S_4 and that has θ_1 and any one of θ_2, θ_3, or θ_4 as its only unknowns. This equation will be paired with the appropriate spherical cosine law, and the input/output equation will then be obtained as described in Section 8.2.

The necessary equation can be obtained by focusing on the fact that the variable joint offset distances S_2 and S_4 must not appear in the equation. One approach would be to project the vector loop equation onto a vector that is perpendicular to S_2 and S_4, that is, $S_2 \times S_4$. Although this may yield the desired result, the procedure will be complex. As an alternative, dual angles will be substituted into the spherical cosine law

$$Z_{51} = \bar{Z}_3. \tag{8.78}$$

This spherical equation was selected because it does not contain θ_2 or θ_4 and as such its corresponding secondary cosine law will not contain S_2 or S_4.

The secondary cosine law associated with Eq. (8.78) will be written as

$$Z_{051} = \bar{Z}_{03}. \tag{8.79}$$

Now,

$$\bar{Z}_3 = c_{23}c_{34} - s_{23}s_{34}c_3. \tag{8.80}$$

The dual part of this equation, after the dual angles $\hat{\alpha}_{23}$, $\hat{\alpha}_{34}$, and $\hat{\theta}_3$ have been substituted, can be written as

$$\bar{Z}_{03} = a_{23}\frac{\partial \bar{Z}_3}{\partial \alpha_{23}} + a_{34}\frac{\partial \bar{Z}_3}{\partial \alpha_{34}} + S_3\frac{\partial \bar{Z}_3}{\partial \theta_3}. \tag{8.81}$$

Evaluating the partial derivatives yields

$$\bar{Z}_{03} = a_{23}\bar{Y}_3 + a_{34}Y_3 + S_3s_{34}X_3. \tag{8.82}$$

The left side of Eq. (8.79) must now be evaluated. The term Z_{051} can be written as

$$Z_{051} = a_{45}\frac{\partial Z_{51}}{\partial \alpha_{45}} + a_{51}\frac{\partial Z_{51}}{\partial \alpha_{51}} + a_{12}\frac{\partial Z_{51}}{\partial \alpha_{12}} + S_5\frac{\partial Z_{51}}{\partial \theta_5} + S_1\frac{\partial Z_{51}}{\partial \theta_1}. \tag{8.83}$$

After expanding the partial derivatives for the a_{12} and S_1 terms, Eq. (8.83) may be expressed as

$$Z_{051} = a_{12}Y_{51} + S_1s_{12}X_{51} + s_{12}(X_{05}s_1 + Y_{05}c_1) + c_{12}Z_{05}, \tag{8.84}$$

where

$$X_{05} = a_{45}c_{45}s_5 + S_5s_{45}c_5, \tag{8.85}$$

$$Y_{05} = a_{45}(s_{51}s_{45} - c_{51}c_{45}c_5) - a_{51}Z_5 + S_5c_{51}s_{45}s_5, \tag{8.86}$$

$$Z_{05} = a_{45}\bar{Y}_5 + a_{51}Y_5 + S_5s_{51}s_{45}s_5. \tag{8.87}$$

Substituting the results of Eqs. (8.82) and (8.84) into Eq. (8.79) and moving all terms to the left-hand side results in the equation

$$a_{12}Y_{51} + S_1 s_{12} X_{51} + s_{12}(X_{05}s_1 + Y_{05}c_1) + c_{12}Z_{05} - a_{23}\bar{Y}_3 - a_{34}Y_3 - S_3 s_{34} X_3 = 0. \tag{8.88}$$

Equation (8.88) contains the unknown joint angles θ_1 and θ_3, which together with Eq. (8.78) can be used to eliminate θ_3 in order to obtain the input/output equation that contains only θ_1.

Equations (8.88) and (8.78) must be regrouped into the form of Eq. (8.5). Equation (8.78) will be referred to as the first equation, and Eq. (8.88) will be the second equation.

The first equation, is expanded as follows:

$$s_{12}(X_5 s_1 + Y_5 c_1) + c_{12}Z_5 - (c_{23}c_{34} - s_{23}s_{34}c_3) = 0. \tag{8.89}$$

Regrouping this equation yields

$$c_3[s_{23}s_{34}] + s_3[0] + [c_1(s_{12}Y_5) + s_1(s_{12}X_5) + (c_{12}Z_5 - c_{23}c_{34})] = 0. \tag{8.90}$$

Thus, the coefficients for the first equation, when regrouped into the form of Eq. (8.5), are

$$\begin{aligned} &A_1 = 0, && B_1 = 0, && D_1 = s_{23}s_{34}, \\ &E_1 = 0, && F_1 = 0, && G_1 = 0, \\ &H_1 = s_{12}Y_5, && I_1 = s_{12}X_5, && J_1 = c_{12}Z_5 - c_{23}c_{34}. \end{aligned} \tag{8.91}$$

The second equation, is expanded by substituting

$$X_{51} = X_5 c_1 - Y_5 s_1, \tag{8.92}$$

$$Y_{51} = c_{12}(X_5 s_1 + Y_5 c_1) - s_{12}Z_5, \tag{8.93}$$

$$Y_3 = -(s_{34}c_{23} + c_{34}s_{23}c_3), \tag{8.94}$$

$$\bar{Y}_3 = -(s_{23}c_{34} + c_{23}s_{34}c_3), \tag{8.95}$$

$$X_3 = s_{23}s_3 \tag{8.96}$$

and regrouping to yield

$$\begin{aligned} &c_3[a_{23}c_{23}s_{34} + a_{34}c_{34}s_{23}] + s_3[-S_3 s_{34} s_{23}] \\ &\quad + [c_1(a_{12}c_{12}Y_5 + S_1 s_{12} X_5 + s_{12}Y_{05}) + s_1(a_{12}c_{12}X_5 - S_1 s_{12} Y_5 + s_{12}X_{05}) \\ &\quad + (-a_{12}s_{12}Z_5 + c_{12}Z_{05} + a_{23}s_{23}c_{34} + a_{34}s_{34}c_{23})] = 0. \end{aligned} \tag{8.97}$$

Thus, the coefficients for the second equation, when regrouped into the form of Eq. (8.5), are

$$\begin{aligned}
&A_2 = 0, \qquad B_2 = 0, \qquad D_2 = a_{23}c_{23}s_{34} + a_{34}c_{34}s_{23},\\
&E_2 = 0, \qquad F_2 = 0, \qquad G_2 = -S_3 s_{34} s_{23},\\
&H_2 = a_{12}c_{12}Y_5 + S_1 s_{12} X_5 + s_{12} Y_{05},\\
&I_2 = a_{12}c_{12}X_5 - S_1 s_{12} Y_5 + s_{12} X_{05},\\
&J_2 = -a_{12}s_{12}Z_5 + c_{12}Z_{05} + a_{23}s_{23}c_{34} + a_{34}s_{34}c_{23}.
\end{aligned} \tag{8.98}$$

Equations (8.90) and (8.97) represent two equations of the form

$$c_3[A_i c_1 + B_i s_1 + D_i] + s_3[E_i c_1 + F_i s_1 + G_i] + [H_i c_1 + I_i s_1 + J_i] = 0 \qquad i = 1, 2. \tag{8.99}$$

Letting x_3 equal $\tan(\theta_3/2)$, the sine and cosine of θ_3 can be replaced by the following trigonometric identities:

$$s_3 = \frac{2x_3}{1 + x_3^2}, \tag{8.100}$$

$$c_3 = \frac{1 - x_3^2}{1 + x_3^2}. \tag{8.101}$$

Multiplying Eq. (8.99) throughout by $(1 + x_3^2)$ and regrouping gives

$$\begin{aligned}
&x_3^2[(H_i - A_i)c_1 + (I_i - B_i)s_1 + (J_i - D_i)] + x_3[2(E_i c_1 + F_i s_1 + G_i)]\\
&\quad + [(H_i + A_i)c_1 + (I_i + B_i)s_1 + (J_i + D_i)] = 0 \qquad i = 1, 2.
\end{aligned} \tag{8.102}$$

Substituting the coefficients for the first equation, Eq. set (8.91), into Eq. (8.102) yields

$$\begin{aligned}
&x_3^2[(s_{12}Y_5)c_1 + (s_{12}X_5)s_1 + (c_{12}Z_5 - c_{23}c_{34} - s_{23}s_{34})]\\
&\quad + [(s_{12}Y_5)c_1 + (s_{12}X_5)s_1 + (c_{12}Z_5 - c_{23}c_{34} + s_{23}s_{34})] = 0.
\end{aligned} \tag{8.103}$$

Introducing the shorthand notation

$$c_{23\pm34} = \cos(\alpha_{23} \pm \alpha_{34}), \tag{8.104}$$

Eq. (8.103) may be written as

$$\begin{aligned}
&x_3^2[(s_{12}Y_5)c_1 + (s_{12}X_5)s_1 + (c_{12}Z_5 - c_{23-34})]\\
&\quad + [(s_{12}Y_5)c_1 + (s_{12}X_5)s_1 + (c_{12}Z_5 - c_{23+34})] = 0.
\end{aligned} \tag{8.105}$$

Substituting the zero-valued coefficients for the second equation, Eq. set (8.98), into Eq. (8.102) yields

$$x_3^2[H_2 c_1 + I_2 s_1 + (J_2 - D_2)] + x_3[2G_2] + [H_2 c_1 + I_2 s_1 + (J_2 + D_2)] = 0. \tag{8.106}$$

Equations (8.105) and (8.106) represent two equations that are quadratic in the variable x_3.

These two equations may be written as

$$L_i x_3^2 + M_i x_3 + N_i = 0 \qquad i = 1, 2, \tag{8.107}$$

where the coefficients are given as

$$\begin{aligned} L_1 &= (s_{12}Y_5)c_1 + (s_{12}X_5)s_1 + (c_{12}Z_5 - c_{23-34}), \\ M_1 &= 0, \\ N_1 &= (s_{12}Y_5)c_1 + (s_{12}X_5)s_1 + (c_{12}Z_5 - c_{23+34}) \end{aligned} \tag{8.108}$$

and

$$\begin{aligned} L_2 &= H_2c_1 + I_2s_1 + (J_2 - D_2), \\ M_2 &= 2G_2, \\ N_2 &= H_2c_1 + I_2s_1 + (J_2 + D_2). \end{aligned} \tag{8.109}$$

The condition that must exist on the coefficients L_i through M_i of these equations in order for there to be a common root of x_3 was shown in Section 8.2.2 to be

$$\begin{vmatrix} L_1 & M_1 \\ L_2 & M_2 \end{vmatrix} \begin{vmatrix} M_1 & N_1 \\ M_2 & N_2 \end{vmatrix} - \begin{vmatrix} L_1 & N_1 \\ L_2 & N_2 \end{vmatrix}^2 = 0. \tag{8.110}$$

The determinant notation $|S\ \ T|$ is defined as

$$|S\ \ T| = \begin{vmatrix} S_1 & T_1 \\ S_2 & T_2 \end{vmatrix}. \tag{8.111}$$

The determinants $|L\ \ M|$, $|M\ \ N|$, and $|L\ \ N|$ are expanded as

$$|L\ \ M| = L_1M_2 = 2G_2[(s_{12}Y_5)c_1 + (s_{12}X_5)s_1 + (c_{12}Z_5 - c_{23-34})], \tag{8.112}$$

$$|M\ \ N| = -M_2N_1 = -2G_2[(s_{12}Y_5)c_1 + (s_{12}X_5)s_1 + (c_{12}Z_5 - c_{23+34})], \tag{8.113}$$

$$\begin{aligned} |L\ \ N| = {} & 2D_2[(s_{12}Y_5)c_1 + (s_{12}X_5)s_1 + c_{12}Z_5] + [H_2c_1 + I_2s_1 + J_2][c_{23+34} - c_{23-34}] \\ & - D_2[c_{23+34} + c_{23-34}]. \end{aligned} \tag{8.114}$$

Equation (8.114) is simplified by recognizing that $c_{23+34} - c_{23-34} = -2s_{23}s_{34}$ and that $c_{23+34} + c_{23-34} = 2c_{23}c_{34}$ to yield

$$\begin{aligned} |L\ \ N| = {} & 2D_2[(s_{12}Y_5)c_1 + (s_{12}X_5)s_1 + c_{12}Z_5] \\ & + [H_2c_1 + I_2s_1 + J_2](-2s_{23}s_{34}) - D_2(2c_{23}c_{34}). \end{aligned} \tag{8.115}$$

Equations (8.112), (8.113), and (8.115) can be regrouped into the following forms

$$|L\ \ M| = c_1(P_1) + s_1(Q_1) + (R_1), \tag{8.116}$$

$$|M\ \ N| = c_1(P_2) + s_1(Q_2) + (R_2), \tag{8.117}$$

$$|L\ \ N| = c_1(P_3) + s_1(Q_3) + (R_3), \tag{8.118}$$

where

$$P_1 = 2G_2s_{12}Y_5, \quad Q_1 = 2G_2s_{12}X_5, \quad R_1 = 2G_2(c_{12}Z_5 - c_{23-34}),$$
$$P_2 = -2G_2s_{12}Y_5, \quad Q_2 = -2G_2s_{12}X_5, \quad R_2 = -2G_2(c_{12}Z_5 - c_{23+34}),$$
$$P_3 = 2D_2s_{12}Y_5 - 2s_{23}s_{34}H_2,$$
$$Q_3 = 2D_2s_{12}X_5 - 2s_{23}s_{34}I_2,$$
$$R_3 = 2D_2c_{12}Z_5 - 2s_{23}s_{34}J_2 - 2D_2c_{23}c_{34}. \tag{8.119}$$

Each of the three determinants defined in Eqs. (8.16) through (8.18) will be reevaluated by substituting the tan-half-angle trigonometric identities for the sine and cosine of θ_1. Multiplying throughout by $(1 + x_1^2)$ and regrouping yields

$$|L\ M| = \left[x_1^2(R_1 - P_1) + x_1(2Q_1) + (R_1 + P_1)\right]/(1 + x_1^2), \tag{8.120}$$
$$|M\ N| = \left[x_1^2(R_2 - P_2) + x_1(2Q_2) + (R_2 + P_2)\right]/(1 + x_1^2), \tag{8.121}$$
$$|L\ N| = \left[x_1^2(R_3 - P_3) + x_1(2Q_3) + (R_3 + P_3)\right]/(1 + x_1^2). \tag{8.122}$$

The product $|L\ M||M\ N|$ can now be evaluated as

$$|L\ M||M\ N| = \{\left[x_1^2(R_1 - P_1) + x_1(2Q_1) + (R_1 + P_1)\right]$$
$$\times \left[x_1^2(R_2 - P_2) + x_1(2Q_2) + (R_2 + P_2)\right]\}/(1 + x_1^2)^2. \tag{8.123}$$

This equation is expanded to yield

$$|LM||MN| = \{x_1^4[(R_1 - P_1)(R_2 - P_2)] + x_1^3[(R_1 - P_1)(2Q_2) + (R_2 - P_2)(2Q_1)]$$
$$+ x_1^2[(R_1 - P_1)(R_2 + P_2) + (R_2 - P_2)(R_1 + P_1) + 4Q_1Q_2]$$
$$+ x_1[(2Q_1)(R_2 + P_2) + (2Q_2)(R_1 + P_1)]$$
$$+ [(R_1 + P_1)(R_2 + P_2)]\}/(1 + x_1^2)^2. \tag{8.124}$$

The product $|L\,N|^2$ is evaluated as

$$|L\ N|^2 = \{x_1^4\left[(R_3 - P_3)^2\right] + x_1^3[(R_3 - P_3)(4Q_3)] + x_1^2\left[2(R_3 - P_3)(R_3 + P_3) + 4Q_3^2\right]$$
$$+ x_1[(4Q_3)(R_3 + P_3)] + \left[(R_3 + P_3)^2\right]\}/(1 + x_1^2)^2. \tag{8.125}$$

A fourth-order input/output equation for the spatial mechanism is finally obtained by substituting Eqs. (8.124) and (8.125) into Eq. (8.110) and multiplying throughout by $(1 + x_1^2)^2$, which yields the following equation that contains θ_1 as its only unknown:

$$x_1^4\left[(R_1 - P_1)(R_2 - P_2) - (R_3 - P_3)^2\right]$$
$$+ x_1^3[(R_1 - P_1)(2Q_2) + (R_2 - P_2)(2Q_1) - (R_3 - P_3)(4Q_3)]$$
$$+ x_1^2\left[(R_1 - P_1)(R_2 + P_2) + (R_2 - P_2)(R_1 + P_1) + 4Q_1Q_2\right.$$
$$\left.- 2(R_3 - P_3)(R_3 + P_3) - 4Q_3^2\right] + x_1[(2Q_1)(R_2 + P_2) + (2Q_2)(R_1 + P_1)$$
$$- (4Q_3)(R_3 + P_3)] + \left[(R_1 + P_1)(R_2 + P_2) - (R_3 + P_3)^2\right] = 0. \tag{8.126}$$

Equation (8.126) can be solved for up to four distinct values of x_1 and thereby four corresponding values of θ_1. All the coefficients P_1 through R_3 are defined in terms of the constant mechanism dimensions and the given input angle, θ_5.

Corresponding values for the tan-half-angle of θ_3 can be calculated from either Eq. (8.34) or Eq. (8.35). Quite often it is useful when debugging a computer program to compare the two solutions to see if they are the same. Numerical round-off error on the computer may be improved if the two calculated values are averaged together.

The joint angle θ_4 can be obtained by writing the following subsidiary sine and sine–cosine laws for a spherical pentagon:

$$X_{15} = X_{34}, \tag{8.127}$$

$$Y_{15} = -X^*_{34}. \tag{8.128}$$

Expanding the right sides of Eqs. (8.127) and (8.128) gives

$$X_{15} = X_3 c_4 - Y_3 s_4, \tag{8.129}$$

$$Y_{15} = -(X_3 s_4 + Y_3 c_4). \tag{8.130}$$

Equations (8.129) and (8.130) represent two equations in the two unknowns s_4 and c_4. Corresponding values of θ_1 and θ_3 can be substituted into these equations to calculate a unique corresponding value for θ_3.

The remaining joint angle, θ_2, can be determined from the following two fundamental sine and sine–cosine laws for a spherical pentagon:

$$X_{543} = s_{12} s_2, \tag{8.131}$$

$$Y_{543} = s_{12} c_2. \tag{8.132}$$

Corresponding values for the angles θ_1, θ_3, and θ_4 can be substituted into these equations to calculate the corresponding value for θ_2.

The offset distances S_2 and S_4 are the last variables to be determined. These two values will be found by projecting the vector loop equation for the mechanism onto two different directions. Earlier, in Section 8.3.1, the vector loop equation was projected onto the direction a_{34}. The resulting scalar equation was listed as Eq. (8.38) and is repeated here as

$$S_1 X_{23} + a_{12} W_{23} + S_2 X_3 + a_{23} c_3 + a_{34} + a_{45} c_4 + S_5 \bar{X}_4 + a_{51} W_{54} = 0. \tag{8.133}$$

The offset distance S_2 is the only unknown in this equation. The distance S_4 will be determined by projecting the vector loop equation onto the direction $\mathbf{a}_{23}$. This can be written as

$$\begin{aligned} & S_1(\mathbf{S}_1 \cdot \mathbf{a}_{23}) + a_{12}(\mathbf{a}_{12} \cdot \mathbf{a}_{23}) + S_2(\mathbf{S}_2 \cdot \mathbf{a}_{23}) + a_{23}(\mathbf{a}_{23} \cdot \mathbf{a}_{23}) + S_3(\mathbf{S}_3 \cdot \mathbf{a}_{23}) \\ & \quad + a_{34}(\mathbf{a}_{34} \cdot \mathbf{a}_{23}) + S_4(\mathbf{S}_4 \cdot \mathbf{a}_{23}) + a_{45}(\mathbf{a}_{45} \cdot \mathbf{a}_{23}) + S_5(\mathbf{S}_5 \cdot \mathbf{a}_{23}) + a_{51}(\mathbf{a}_{51} \cdot \mathbf{a}_{23}) = 0. \end{aligned} \tag{8.134}$$

Evaluating the scalar products by using the sets of direction cosines listed in the appendix gives

$$S_1 X_2 + a_{12} c_2 + a_{23} + a_{34} c_3 + S_4 \bar{X}_3 + a_{45} W_{43} + S_5 X_{43} + a_{51} W_{12} = 0. \tag{8.135}$$

This equation contains the variable S_4 as its only unknown.

Table 8.1. *RCRCR mechanism parameters.*

Link length, cm.	Twist angle, deg.	Joint offset, cm.	Joint angle, deg.
$a_{12} = 25$	$\alpha_{12} = 60$	$S_1 = 30$	θ_1 = variable
$a_{23} = 30$	$\alpha_{23} = 45$	S_2 = variable	θ_2 = variable
$a_{34} = 40$	$\alpha_{34} = 35$	$S_3 = 25$	θ_3 = variable
$a_{45} = 10$	$\alpha_{45} = 30$	S_4 = variable	θ_4 = variable
$a_{51} = 32$	$\alpha_{51} = 12$	$S_5 = 10$	$\theta_5 = 260$ (input)

Table 8.2. *Calculated configurations for the RCRCR spatial mechanism.*

	Solution			
	A	B	C	D
θ_1, degrees	−104.75	−78.85	3.38	22.84
θ_2, degrees	121.97	−105.96	−120.86	−164.64
θ_3, degrees	134.98	−129.10	−51.22	16.89
θ_4, degrees	−59.73	73.62	−57.70	−120.81
S_2, cm.	46.12	101.88	−1.60	−40.59
S_4, cm.	−92.28	−106.34	−13.25	−26.16

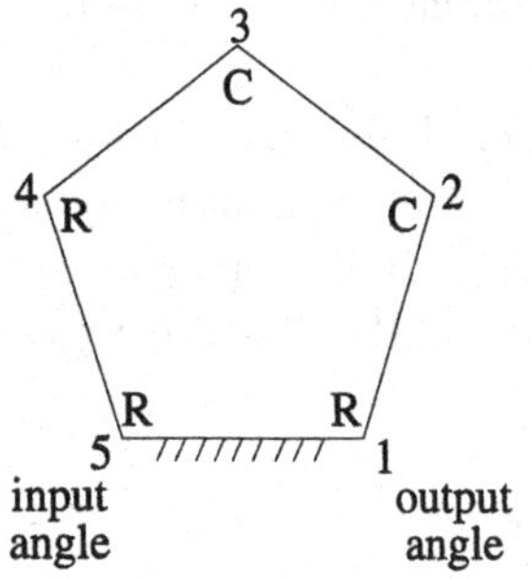

Figure 8.3. Planar representation of RRCCR group 2 spatial mechanism.

This completes the analysis of the RCRCR group 2 spatial mechanism. Four solution configurations were determined. Table 8.1 shows data that were used for a numerical example. The calculated values for the four configurations are listed in Table 8.2.

8.4.2 Case II: RRCCR spatial mechanism (C joints adjacent)

Figure 8.3 shows a planar representation of the RRCCR spatial mechanism. The problem statement for this mechanism is as follows:

given:
constant mechanism parameters, that is,
$a_{12}, a_{23}, a_{34}, a_{45}, a_{51}$,
$\alpha_{12}, \alpha_{23}, \alpha_{34}, \alpha_{45}, \alpha_{51}$,
S_1, S_4, and S_5 and
input angle, θ_5,
find: $\theta_1, \theta_2, \theta_3, \theta_4$, S_2, and S_3.

Because this is a group 2 mechanism, it will be necessary to obtain two equations that contain the input angle, θ_5, the output angle, θ_1, and one additional joint angle. One of these equations will be a spherical cosine law (whether fundamental or subsidiary), and the other will be an equation that contains the constant link length and joint offset parameters. This second equation must not contain the variable offsets S_2 and S_3. One means of obtaining such an equation is to project the vector loop equation onto the direction $\mathbf{a}_{23}$.

The vector loop equation for the mechanism is written as

$$S_1\mathbf{S}_1 + a_{12}\mathbf{a}_{12} + S_2\mathbf{S}_2 + a_{23}\mathbf{a}_{23} + S_3\mathbf{S}_3 + a_{34}\mathbf{a}_{34} + S_4\mathbf{S}_4 + a_{45}\mathbf{a}_{45} + S_5\mathbf{S}_5 + a_{51}\mathbf{a}_{51} = \mathbf{0}. \tag{8.136}$$

Projecting this equation onto the direction $\mathbf{a}_{23}$ yields

$$\begin{aligned} &S_1(\mathbf{S}_1 \cdot \mathbf{a}_{23}) + a_{12}(\mathbf{a}_{12} \cdot \mathbf{a}_{23}) + S_2(\mathbf{S}_2 \cdot \mathbf{a}_{23}) + a_{23}(\mathbf{a}_{23} \cdot \mathbf{a}_{23}) + S_3(\mathbf{S}_3 \cdot \mathbf{a}_{23}) \\ &\quad + a_{34}(\mathbf{a}_{34} \cdot \mathbf{a}_{23}) + S_4(\mathbf{S}_4 \cdot \mathbf{a}_{23}) + a_{45}(\mathbf{a}_{45} \cdot \mathbf{a}_{23}) + S_5(\mathbf{S}_5 \cdot \mathbf{a}_{23}) + a_{51}(\mathbf{a}_{51} \cdot \mathbf{a}_{23}) = 0. \end{aligned} \tag{8.137}$$

The individual scalar products can be evaluated by using the sets of direction cosines listed in the appendix as follows:

$$S_1X_2 + a_{12}c_2 + a_{23} + a_{34}c_3 + S_4X_{512} + a_{45}W_{512} + S_5X_{12} + a_{51}W_{12} = 0. \tag{8.138}$$

The fundamental sine–cosine law

$$Y_{512} = s_{34}c_3 \tag{8.139}$$

can be used to eliminate c_3 from Eq. (8.138) provided s_{34} does not equal zero.* Equation (8.138) can thus be written as

$$S_1X_2 + a_{12}c_2 + a_{23} + a_{34}Y_{512}/s_{34} + S_4X_{512} + a_{45}W_{512} + S_5X_{12} + a_{51}W_{12} = 0. \tag{8.140}$$

Equation (8.140) contains the input angle, θ_5, the output angle, θ_1, and the extra angle, θ_2. This equation will be paired with the fundamental cosine law

$$Z_{512} = c_{34} \tag{8.141}$$

to generate the input/output equation for the mechanism.

* If s_{34} equals zero, then $\mathbf{S}_3$ and $\mathbf{S}_4$ would be parallel or antiparallel. The equivalent spherical mechanism would reduce to a quadrilateral. Spherical equations could then be used to obtain the input/output equation.

Expanding the left side of Eq. (8.141) and regrouping terms yields

$$c_2[s_{23}Y_{51}] + s_2[s_{23}X_{51}] + [c_{23}Z_{51} - c_{34}] = 0. \tag{8.142}$$

Expanding X_{51}, Y_{51}, and Z_{51} and then regrouping gives

$$\begin{aligned} &c_2[c_1(s_{23}c_{12}Y_5) + s_1(s_{23}c_{12}X_5) + (-s_{23}s_{12}Z_5)] + s_2[c_1(s_{23}X_5) + s_1(-s_{23}Y_5)] \\ &\quad + [c_1(c_{23}s_{12}Y_5) + s_1(c_{23}s_{12}X_5) + (c_{23}c_{12}Z_5 - c_{34})] = 0. \end{aligned} \tag{8.143}$$

Expressing Eq. (8.141) in the form of Eq. (8.5) yields†

$$\begin{aligned} &A_1 = s_{23}c_{12}Y_5, \quad B_1 = s_{23}c_{12}X_5, \quad D_1 = -s_{23}s_{12}Z_5, \\ &E_1 = s_{23}X_5, \quad F_1 = -s_{23}Y_5, \quad G_1 = 0, \\ &H_1 = c_{23}s_{12}Y_5, \quad I_1 = c_{23}s_{12}X_5, \quad J_1 = c_{23}c_{12}Z_5 - c_{34}. \end{aligned} \tag{8.144}$$

Expanding X_2, Y_{512}, X_{512}, W_{512}, X_{12}, and W_{12} in Eq. (8.140) and regrouping terms yields

$$\begin{aligned} &c_2[a_{12} + a_{34}c_{23}Y_{51}/s_{34} + S_4X_{51} + a_{45}W_{51} + S_5X_1 + a_{51}c_1] \\ &\quad + s_2[S_1s_{12} + a_{34}c_{23}X_{51}/s_{34} - S_4Y_{51} + a_{45}(U_{51}s_{12} + V_{51}c_{12}) - S_5Y_1 - a_{51}s_1c_{12}] \\ &\quad + [a_{23} - a_{34}s_{23}Z_{51}/s_{34}] = 0. \end{aligned} \tag{8.145}$$

Finally, expanding X_{51}, Y_{51}, Z_{51}, X_1, Y_1, U_{51}, V_{51}, and W_{51} and expressing in the form of Eq. (8.5) yields

$$\begin{aligned} &A_2 = a_{34}c_{23}c_{12}Y_5/s_{34} + S_4X_5 + a_{45}c_5 + a_{51}, \\ &B_2 = a_{34}c_{23}c_{12}X_5/s_{34} - S_4Y_5 - a_{45}s_5c_{51} + S_5s_{51}, \\ &D_2 = a_{12} - a_{34}c_{23}s_{12}Z_5/s_{34}, \\ &E_2 = a_{34}c_{23}X_5/s_{34} - S_4c_{12}Y_5 - a_{45}c_{12}s_5c_{51} + S_5c_{12}s_{51}, \\ &F_2 = -a_{34}c_{23}Y_5/s_{34} - S_4c_{12}X_5 - a_{45}c_{12}c_5 - a_{51}c_{12}, \\ &G_2 = S_1s_{12} + S_4s_{12}Z_5 + a_{45}U_{51}s_{12} + S_5s_{12}c_{51}, \\ &H_2 = -a_{34}s_{23}s_{12}Y_5/s_{34}, \\ &I_2 = -a_{34}s_{23}s_{12}X_5/s_{34}, \\ &J_2 = a_{23} - a_{34}s_{23}c_{12}Z_5/s_{34}. \end{aligned} \tag{8.146}$$

Finally tan-half-angle identities for θ_1 and θ_2 can be used and the pair of equations expressed in the form of Eq. (8.10), for which the coefficients are expressed by Eq. (8.11).

The two equations of the form of Eq. (8.10) are quadratic in the variables x_2 and x_1. These equations can be solved using Bezout's method (Section 8.2.2) to yield an eighth-degree polynomial in the variable x_1. Thus, a maximum of eight values of θ_1 exist for the mechanism for a given value of the input angle θ_5.

† Note that the extra variable in the current equations is θ_2 as opposed to θ_3 as written in Eq. (8.5).

The corresponding value for the parameter x_2 can be found from either

$$x_2 = \frac{-\begin{vmatrix} M_1 & N_1 \\ M_2 & N_2 \end{vmatrix}}{\begin{vmatrix} L_1 & N_1 \\ L_2 & N_2 \end{vmatrix}} \tag{8.147}$$

or

$$x_2 = \frac{-\begin{vmatrix} L_1 & N_1 \\ L_2 & N_2 \end{vmatrix}}{\begin{vmatrix} L_1 & M_1 \\ L_2 & M_2 \end{vmatrix}}, \tag{8.148}$$

where $L_1, M_1,$ and N_1 are defined in Eqs. (8.14) through (8.16) and Eq. set (8.11). Equations (8.147) and (8.148) can be derived in a manner similar to that in which Eqs. (8.34) and (8.35) were derived.

A unique corresponding value of θ_4 can be obtained from the following fundamental sine and sine–cosine laws:

$$X_{215} = s_{34}s_4, \tag{8.149}$$

$$Y_{215} = s_{34}c_4. \tag{8.150}$$

Similarly, a unique corresponding value of θ_3 can then be obtained from the following fundamental sine and sine–cosine laws:

$$X_{512} = s_{34}s_3, \tag{8.151}$$

$$Y_{512} = s_{34}c_3. \tag{8.152}$$

The offset distances S_2 and S_3 are the remaining parameters to be determined. These two values will be found by projecting the vector loop equation for the mechanism onto two different directions. Projecting the vector loop equation onto $\mathbf{a}_{34}$ (see Section 8.3.1) yielded Eq. (8.38), which is repeated here as

$$S_1X_{23} + a_{12}W_{23} + S_2X_3 + a_{23}c_3 + a_{34} + a_{45}c_4 + S_5\bar{X}_4 + a_{51}W_{54} = 0. \tag{8.153}$$

The offset distance S_2 is the only unknown in this equation.

The offset distance S_3 is determined by projecting the vector loop equation onto the direction $\mathbf{a}_{12}$. This can be written as

$$S_1(\mathbf{S}_1 \cdot \mathbf{a}_{12}) + a_{12}(\mathbf{a}_{12} \cdot \mathbf{a}_{12}) + S_2(\mathbf{S}_2 \cdot \mathbf{a}_{12}) + a_{23}(\mathbf{a}_{23} \cdot \mathbf{a}_{12}) + S_3(\mathbf{S}_3 \cdot \mathbf{a}_{12}) + a_{34}(\mathbf{a}_{34} \cdot \mathbf{a}_{12})$$
$$+ S_4(\mathbf{S}_4 \cdot \mathbf{a}_{12}) + a_{45}(\mathbf{a}_{45} \cdot \mathbf{a}_{12}) + S_5(\mathbf{S}_5 \cdot \mathbf{a}_{12}) + a_{51}(\mathbf{a}_{51} \cdot \mathbf{a}_{12}) = 0. \tag{8.154}$$

Evaluating the scalar products by using the sets of direction cosines listed in the appendix gives

$$a_{12} + a_{23}c_2 + S_3\bar{X}_2 + a_{34}W_{32} + S_4X_{32} + a_{45}W_{51} + S_5X_1 + a_{51}c_1 = 0. \tag{8.155}$$

This equation contains the variable S_3 as its only unknown.

Table 8.3. *RRCCR mechanism parameters.*

Link length, cm.	Twist angle, deg.	Joint offset, cm.	Joint angle, deg.
$a_{12} = 12$	$\alpha_{12} = 62$	$S_1 = 80$	$\theta_1 =$ variable
$a_{23} = 34$	$\alpha_{23} = 67$	$S_2 =$ variable	$\theta_2 =$ variable
$a_{34} = 40$	$\alpha_{34} = 73$	$S_3 =$ variable	$\theta_3 =$ variable
$a_{45} = 41$	$\alpha_{45} = 127$	$S_4 = 26$	$\theta_4 =$ variable
$a_{51} = 44$	$\alpha_{51} = 80$	$S_5 = 87$	$\theta_5 = 222$ (input)

Table 8.4. *Calculated configurations for the RRCCR spatial mechanism.*

	Solution							
	A	B	C	D	E	F	G	H
θ_1, deg.	−112.27	−55.92	−11.80	−4.36	14.84	99.08	104.92	150.19
θ_2, deg.	29.06	−154.37	3.24	−3.07	167.77	159.13	−86.11	−138.75
θ_3, deg.	110.96	−163.98	−156.77	−149.63	131.56	59.12	−55.15	−41.78
θ_4, deg.	85.99	−153.67	25.02	25.04	−127.84	−148.26	−40.49	−92.55
S_2, cm.	−135.77	52.99	−2.51	3.18	37.36	−46.10	−54.83	−64.34
S_3, cm.	99.96	−115.49	106.68	105.67	−104.68	−84.69	79.56	−10.82

This completes the analysis of the RRCCR group 2 spatial mechanism. Eight solution configurations were determined. Table 8.3 shows data that were used for a numerical example. The calculated values for the eight configurations are listed in Table 8.4.

It is interesting to plot the calculated joint angles and offsets as the input angle varies in increments between zero and 2π. Figure 8.4 shows the calculated outputs for the RRCCR mechanism whose dimensions are given in Table 8.3 as the angle θ_5 is varied.

8.5 Six-link group 2 spatial mechanisms

All six-link group 2 mechanisms consist of four revolute joints, one cylindric joint, and one prismatic joint, that is, 4R-C-P. The only difference between the various six-link group 2 mechanisms is the order of the types of joints.

The solution method for all the six-link mechanisms is identical to that for the five-link mechanisms of the previous section. That is, two equations will be derived that contain the input angle, the output angle, and one extra joint angle. Elimination of the extra joint angle from the two equations will result in an eighth-degree polynomial in terms of the tan-half-angle of the output angle. One of the two equations will be a subsidiary or fundamental cosine law for a spherical hexagon. The other equation will be either a secondary cosine law (if the C and P joint are not adjacent in the mechanism) or a projection of the vector loop equation onto a link direction vector.

Two examples will be presented in this section, one where the C and P joints are adjacent, and one where they are not. These two examples should be sufficient to demonstrate the solution technique. After completing these examples, the reader should then be able to apply the technique to all other cases.

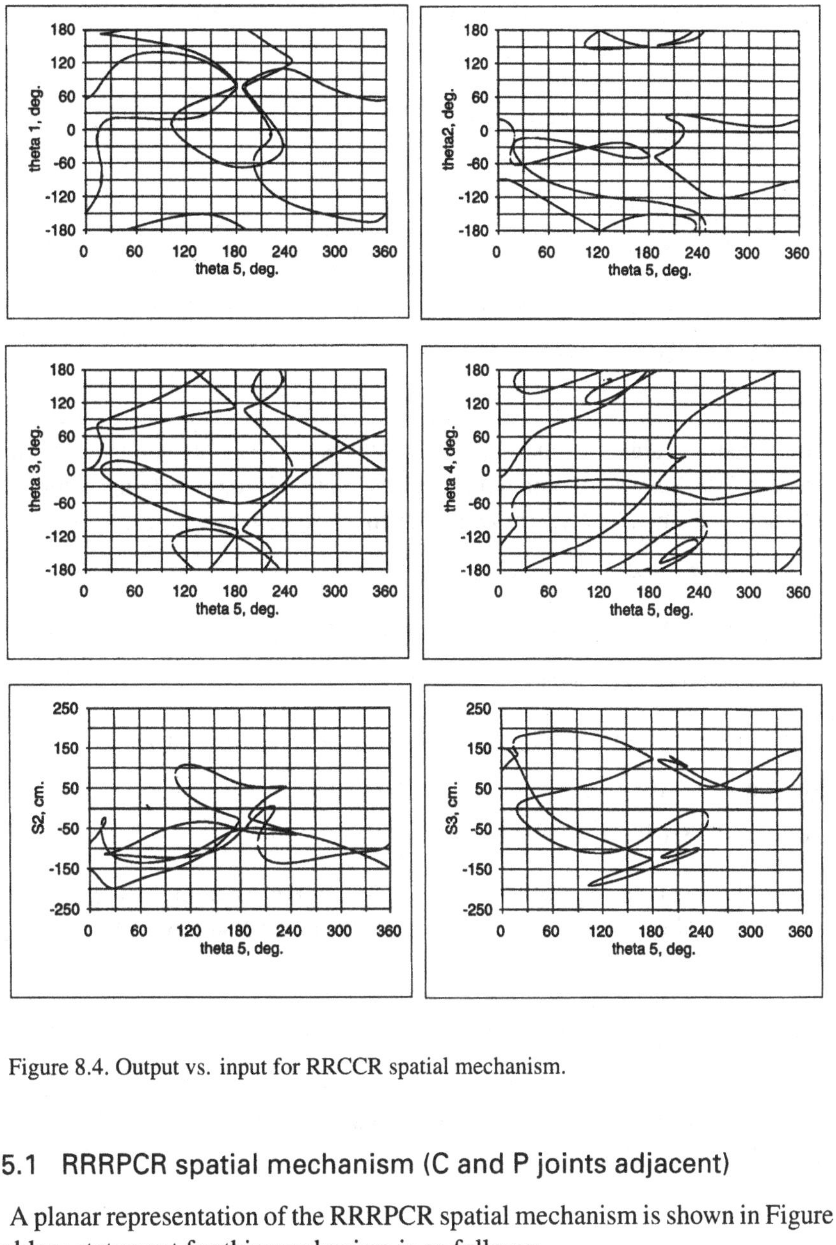

Figure 8.4. Output vs. input for RRCCR spatial mechanism.

8.5.1 RRRPCR spatial mechanism (C and P joints adjacent)

A planar representation of the RRRPCR spatial mechanism is shown in Figure 8.5. The problem statement for this mechanism is as follows:

given: constant mechanism parameters:

$a_{12}, a_{23}, a_{34}, a_{45}, a_{56}, a_{61}$,

$\alpha_{12}, \alpha_{23}, \alpha_{34}, \alpha_{45}, \alpha_{56}, \alpha_{61}$,

S_1, S_4, S_5, S_6, and

θ_3 and

input angle:

θ_6,

find: $\theta_1, \theta_2, \theta_4, \theta_5, S_2$, and S_3.

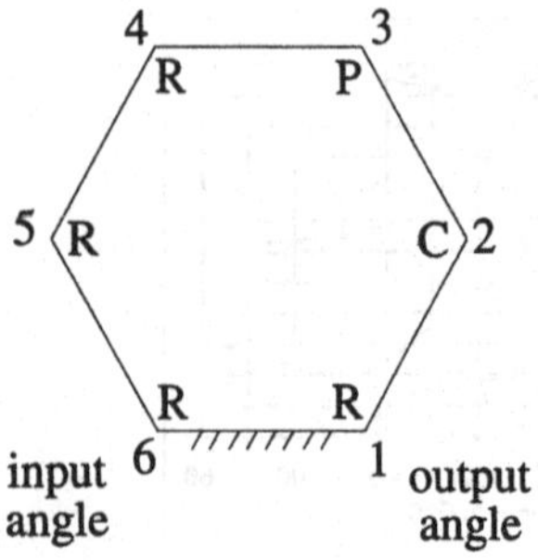

Figure 8.5. Planar representation of RRRPCR spatial mechanism.

The equation that contains the link lengths and offsets must not contain the unknown parameters S_2 and S_3. One way of obtaining such an equation is to project the vector loop equation onto the vector $\mathbf{a}_{23}$. The vector loop equation is written as

$$S_1\mathbf{S}_1 + a_{12}\mathbf{a}_{12} + S_2\mathbf{S}_2 + a_{23}\mathbf{a}_{23} + S_3\mathbf{S}_3 + a_{34}\mathbf{a}_{34} + S_4\mathbf{S}_4 + a_{45}\mathbf{a}_{45} + S_5\mathbf{S}_5 + a_{56}\mathbf{a}_{56} + S_6\mathbf{S}_6 + a_{61}\mathbf{a}_{61} = \mathbf{0}. \tag{8.156}$$

Projecting this equation onto $\mathbf{a}_{23}$ yields

$$S_1(\mathbf{S}_1 \cdot \mathbf{a}_{23}) + a_{12}(\mathbf{a}_{12} \cdot \mathbf{a}_{23}) + S_2(\mathbf{S}_2 \cdot \mathbf{a}_{23}) + a_{23}(\mathbf{a}_{23} \cdot \mathbf{a}_{23}) + S_3(\mathbf{S}_3 \cdot \mathbf{a}_{23}) + a_{34}(\mathbf{a}_{34} \cdot \mathbf{a}_{23}) + S_4(\mathbf{S}_4 \cdot \mathbf{a}_{23}) + a_{45}(\mathbf{a}_{45} \cdot \mathbf{a}_{23}) + S_5(\mathbf{S}_5 \cdot \mathbf{a}_{23}) + a_{56}(\mathbf{a}_{56} \cdot \mathbf{a}_{23}) + S_6(\mathbf{S}_6 \cdot \mathbf{a}_{23}) + a_{61}(\mathbf{a}_{61} \cdot \mathbf{a}_{23}) = \mathbf{0}. \tag{8.157}$$

The sets of direction cosines for a spherical hexagon that are listed in the appendix are used to evaluate the scalar products of the above equation. The resulting equation is

$$S_1X_2 + a_{12}c_2 + a_{23} + a_{34}c_3 + S_4\bar{X}_3 + a_{45}W_{43} + S_5X_{612} + a_{56}W_{612} + S_6X_{12} + a_{61}W_{12} = 0. \tag{8.158}$$

All the terms in this equation, with the exception of W_{43}, contain only the constant mechanism parameters, the input angle, θ_6, the output angle, θ_1, and the extra angle, θ_2. The exception, W_{43}, is defined as follows:

$$W_{43} = c_3c_4 - s_3s_4c_{34}. \tag{8.159}$$

This equation may be modified to yield

$$W_{43} = \frac{(s_{45}c_4)c_3 - c_{34}(s_{45}s_4)s_3}{s_{45}} \tag{8.160}$$

provided s_{45} does not equal zero. Fundamental sine and sine–cosine laws for a spherical hexagon may be used to replace the two terms in parentheses in Eq. (8.160). Thus, this equation may be written as

$$W_{43} = \frac{(Y_{6123})c_3 - c_{34}(X_{6123})s_3}{s_{45}} = \frac{c_{34}Y_{612} - s_{34}Z_{612}c_3}{s_{45}}. \tag{8.161}$$

This expression is now defined in terms of the given parameters, the output angle, and the extra angle, θ_2.

Substituting Eq. (8.161) into Eq. (8.158) yields

$$S_1X_2 + a_{12}c_2 + a_{23} + a_{34}c_3 + S_4\bar{X}_3 + a_{45}\frac{c_{34}Y_{612} - s_{34}Z_{612}c_3}{s_{45}} + S_5X_{612} + a_{56}W_{612} + S_6X_{12} + a_{61}W_{12} = 0. \quad (8.162)$$

This equation contains the output angle, θ_1, and the extra angle, θ_2, as its only unknowns. It will be paired with the fundamental cosine law

$$Z_{6123} = c_{45} \quad (8.163)$$

to yield the appropriate input/output equation for the mechanism.

Both Eqs. (8.162) and (8.163) will be expanded and regrouped into the format of Eq. (8.5). The tan-half-angle substitutions for θ_1 and θ_2 will then be used to yield two biquadratic equations that can be solved for the output angle by using Bezout's method.

Equation (8.163) will be designated as the first equation, and it can be written as

$$s_{34}(X_{612}s_3 + Y_{612}c_3) + c_{34}Z_{612} - c_{45} = 0. \quad (8.164)$$

Expanding X_{612}, Y_{612}, and Z_{612} and regrouping yields

$$c_2[\bar{X}_3X_{61} - \bar{Y}_3Y_{61}] + s_2[-\bar{X}_3Y_{61} - \bar{Y}_3X_{61}] + [\bar{Z}_3Z_{61} - c_{45}] = 0. \quad (8.165)$$

Expanding X_{61}, Y_{61}, and Z_{61} and regrouping yields

$$c_2[c_1(\bar{X}_3X_6 - \bar{Y}_3c_{12}Y_6) + s_1(-\bar{X}_3Y_6 - \bar{Y}_3c_{12}X_6) + (\bar{Y}_3s_{12}Z_6)] + s_2[c_1(-\bar{X}_3c_{12}Y_6 - \bar{Y}_3X_6) + s_1(-\bar{X}_3c_{12}X_6 + \bar{Y}_3Y_6) + (\bar{X}_3s_{12}Z_6)] + [c_1(\bar{Z}_3s_{12}Y_6) + s_1(\bar{Z}_3s_{12}X_6) + (\bar{Z}_3c_{12}Z_6 - c_{45})] = 0. \quad (8.166)$$

The coefficients for the first equation (see Eq. (8.5)) are thus

$$\begin{aligned} A_1 &= \bar{X}_3X_6 - \bar{Y}_3c_{12}Y_6, & B_1 &= -\bar{X}_3Y_6 - \bar{Y}_3c_{12}X_6, & D_1 &= \bar{Y}_3s_{12}Z_6, \\ E_1 &= -\bar{X}_3c_{12}Y_6 - \bar{Y}_3X_6, & F_1 &= -\bar{X}_3c_{12}X_6 + \bar{Y}_3Y_6, & G_1 &= \bar{X}_3s_{12}Z_6, \\ H_1 &= \bar{Z}_3s_{12}Y_6, & I_1 &= \bar{Z}_3s_{12}X_6, & J_1 &= \bar{Z}_3c_{12}Z_6 - c_{45}. \end{aligned} \quad (8.167)$$

The second equation, Eq. (8.162), is expanded and regrouped as follows by substituting for the terms X_2, X_{6123}, Y_{6123}, X_{612}, W_{612}, X_{12}, and W_{12}:

$$c_2\left[a_{12} + S_5X_{61} + a_{56}W_{61} + S_6X_1 + a_{61}c_1 + a_{45}\frac{\bar{Z}_3Y_{61}}{s_{45}}\right] + s_2\left[S_1s_{12} - S_5Y_{61} + a_{56}U^*_{612} - S_6Y_1 - a_{61}s_1c_{12} + a_{45}\frac{\bar{Z}_3X_{61}}{s_{45}}\right] + \left[a_{23} + a_{34}c_3 + S_4\bar{X}_3 + a_{45}\frac{\bar{Y}_3Z_{61}}{s_{45}}\right] = 0. \quad (8.168)$$

This equation is next regrouped into the format of Eq. (8.5) as follows:

$$
\begin{aligned}
&c_2\left[c_1\left(S_5X_6 + a_{56}c_6 + a_{61} + a_{45}\frac{\bar{Z}_3c_{12}Y_6}{s_{45}}\right)\right.\\
&\left.+ s_1\left(-S_5Y_6 - a_{56}s_6c_{61} + S_6s_{61} + a_{45}\frac{\bar{Z}_3c_{12}X_6}{s_{45}}\right) + \left(a_{12} - a_{45}\frac{\bar{Z}_3s_{12}Z_6}{s_{45}}\right)\right]\\
&+ s_2\left[c_1\left(-S_5c_{12}Y_6 - a_{56}c_{12}s_6c_{61} + S_6c_{12}s_{61} + a_{45}\frac{\bar{Z}_3X_6}{s_{45}}\right)\right.\\
&+ s_1\left(-S_5c_{12}X_6 - a_{56}c_{12}c_6 - a_{61}c_{12} - a_{45}\frac{\bar{Z}_3Y_6}{s_{45}}\right)\\
&\left.+ \left(S_1s_{12} + S_5s_{12}Z_6 + a_{56}U_{61}s_{12} + S_6s_{12}c_{61}\right)\right]\\
&+ \left[c_1\left(a_{45}\frac{\bar{Y}_3s_{12}Y_6}{s_{45}}\right) + s_1\left(a_{45}\frac{\bar{Y}_3s_{12}X_6}{s_{45}}\right)\right.\\
&\left.+ \left(a_{23} + a_{34}c_3 + S_4\bar{X}_3 + a_{45}\frac{\bar{Y}_3c_{12}Z_6}{s_{45}}\right)\right] = 0.
\end{aligned}
\tag{8.169}
$$

The coefficients for the second equation (see the format of Eq. (8.5)) are thus

$$
\begin{aligned}
A_2 &= S_5X_6 + a_{56}c_6 + a_{61} + a_{45}\bar{Z}_3c_{12}Y_6/s_{45},\\
B_2 &= -S_5Y_6 - a_{56}s_6c_{61} + S_6s_{61} + a_{45}\bar{Z}_3c_{12}X_6/s_{45},\\
D_2 &= a_{12} - a_{45}\bar{Z}_3s_{12}Z_6/s_{45},\\
E_2 &= -S_5c_{12}Y_6 - a_{56}c_{12}s_6c_{61} + S_6c_{12}s_{61} + a_{45}\bar{Z}_3X_6/s_{45},\\
F_2 &= -S_5c_{12}X_6 - a_{56}c_{12}c_6 - a_{61}c_{12} - a_{45}\bar{Z}_3Y_6/s_{45},\\
G_2 &= S_1s_{12} + S_5s_{12}Z_6 + a_{56}U_{61}s_{12} + S_6s_{12}c_{61},\\
H_2 &= a_{45}\bar{Y}_3s_{12}Y_6/s_{45},\\
I_2 &= a_{45}\bar{Y}_3s_{12}X_6/s_{45},\\
J_2 &= a_{23} + a_{34}c_3 + S_4\bar{X}_3 + a_{45}\bar{Y}_3c_{12}Z_6/s_{45}.
\end{aligned}
\tag{8.170}
$$

Now that both equations have been regrouped into the format of Eq. (8.5), the tan-half-angle identities for the sine and cosine of θ_1 and θ_2 are inserted. The two equations can then be regrouped into the format of Eq. (8.10), where the coefficients of this equation are defined in Eq. set (8.11).

The two equations of the form of Eq. (8.10) are quadratic in the variables x_1 and x_2. These equations can be solved via Bezout's method (Section 8.2.2) to yield an eighth-degree polynomial in the variable x_1.

The corresponding value for the parameter x_2 can be found from either

$$
x_2 = \frac{-\begin{vmatrix} M_1 & N_1 \\ M_2 & N_2 \end{vmatrix}}{\begin{vmatrix} L_1 & N_1 \\ L_2 & N_2 \end{vmatrix}}
\tag{8.171}
$$

or

$$x_2 = \frac{-\begin{vmatrix} L_1 & N_1 \\ L_2 & N_2 \end{vmatrix}}{\begin{vmatrix} L_1 & M_1 \\ L_2 & M_2 \end{vmatrix}}, \tag{8.172}$$

where L_1, M_1, and N_1 are defined in Eqs. (8.14) through (8.16) and Eq. set (8.11). Equations (8.171) and (8.172) can be derived in a manner similar to that in which Eqs. (8.34) and (8.35) were derived.

Two joint angles remained to be solved, that is, θ_4 and θ_5. A unique corresponding value of θ_5 can be obtained from the following fundamental sine and sine–cosine laws:

$$X_{3216} = s_{45}s_5, \tag{8.173}$$

$$Y_{3216} = s_{45}c_5. \tag{8.174}$$

Similarly, a unique corresponding value of θ_4 can then be obtained from the following fundamental sine and sine–cosine laws:

$$X_{6123} = s_{45}s_4, \tag{8.175}$$

$$Y_{6123} = s_{45}c_4. \tag{8.176}$$

The offset distances S_2 and S_3 are the remaining parameters to be determined. These two values will be found by projecting the vector loop equation for the mechanism onto two different directions. Projecting the vector loop equation onto the direction $\mathbf{a}_{34}$ and evaluating the scalar products using the sets of direction cosines provided in the appendix yields

$$S_1X_{23} + a_{12}W_{23} + S_2X_3 + a_{23}c_3 + a_{34} + a_{45}c_4 + S_5\bar{X}_4 + a_{56}W_{54}$$
$$+ S_6X_{54} + a_{61}W_{654} = 0. \tag{8.177}$$

The offset distance S_2 is the only unknown in this equation.

The distance S_3 will be determined by projecting the vector loop equation onto the direction $\mathbf{a}_{12}$. Evaluating the scalar products by using the sets of direction cosines listed in the appendix gives

$$a_{12} + a_{23}c_2 + S_3\bar{X}_2 + a_{34}W_{32} + S_4X_{32} + a_{45}W_{561} + S_5X_{61}$$
$$+ a_{56}W_{61} + S_6X_1 + a_{61}c_1 = 0. \tag{8.178}$$

This equation contains the variable S_3 as its only unknown.

At this point, the analysis of the RRRPCR group 2 spatial mechanism is complete. Eight solution configurations were determined. Table 8.5 shows data that were used for a numerical example. The calculated values for the eight configurations are listed in Table 8.6.

8.5.2 RRPRCR spatial mechanism (C and P joints separated)

A planar representation of the RRPRCR spatial mechanism is shown in Figure 8.6. The problem statement for this mechanism is as follows:

Table 8.5. *RRRPCR mechanism parameters.*

Link length, cm.	Twist angle, deg.	Joint offset, cm.	Joint angle, deg.
$a_{12} = 35$	$\alpha_{12} = 82$	$S_1 = 67$	θ_1 = variable
$a_{23} = 19$	$\alpha_{23} = 78$	S_2 = variable	θ_2 = variable
$a_{34} = 19$	$\alpha_{34} = 34$	S_3 = variable	$\theta_3 = 320$
$a_{45} = 9$	$\alpha_{45} = 93$	$S_4 = 62$	θ_4 = variable
$a_{56} = 22$	$\alpha_{56} = 50$	$S_5 = 71$	θ_5 = variable
$a_{61} = 10$	$\alpha_{61} = 77$	$S_6 = 61$	$\theta_6 = 322$ (input)

Table 8.6. *Calculated configurations for the RRRPCR spatial mechanism.*

	Solution							
	A	B	C	D	E	F	G	H
θ_1, deg.	−107.96	−118.51	−63.47	−22.40	21.05	47.48	−163.33	−173.31
θ_2, deg.	−108.08	62.38	51.38	−145.85	138.65	−76.66	−31.45	134.67
θ_4, deg.	168.94	−95.54	−49.91	92.56	28.47	56.74	−175.62	−98.45
θ_5, deg.	170.53	13.46	46.78	−148.55	156.88	−53.32	87.08	−68.65
S_2, cm.	3.24	−35.43	71.46	143.61	165.13	89.03	−87.61	−68.78
S_3, cm.	−140.58	15.60	26.95	−114.68	−149.83	63.81	−36.81	−89.69

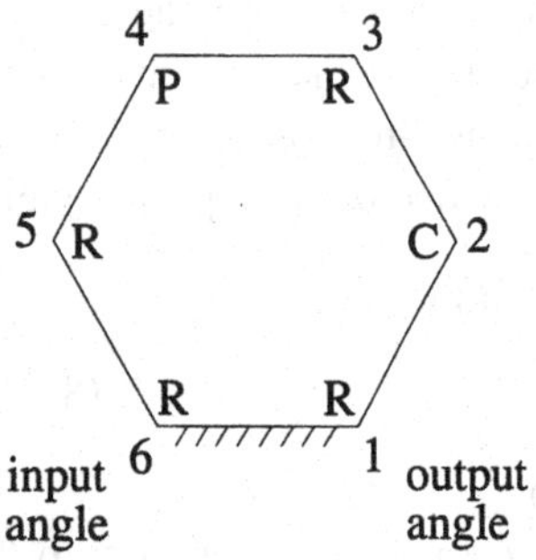

Figure 8.6. Planar representation of the RRPRCR group 2 spatial mechanism.

given:

constant mechanism parameters:

$a_{12}, a_{23}, a_{34}, a_{45}, a_{56}, a_{61}$,

$\alpha_{12}, \alpha_{23}, \alpha_{34}, \alpha_{45}, \alpha_{56}, \alpha_{61}$,

S_1, S_3, S_5, S_6, and

θ_4 and

input angle:

θ_6,

find: $\theta_1, \theta_2, \theta_3, \theta_5, S_2$, and S_4.

Because this is a group 2 mechanism, two equations must be generated that contain only the output angle and one additional angle as unknowns. One of the two equations will be a spherical cosine law. The other equation will contain the link lengths and offsets and must not contain the unknown parameters S_2 and S_4. One way of obtaining this second equation is to write the secondary equation of a cosine law that does not contain the angles θ_2 or θ_4. The equation

$$Z_{561} = \bar{Z}_3 \tag{8.179}$$

will be dualized to yield

$$Z_{0561} = \bar{Z}_{03}. \tag{8.180}$$

The term on the right side of Eq. (8.180) contains the parameters θ_3, α_{23}, and α_{34}. Thus, the right-hand side may be written as

$$\bar{Z}_{03} = S_3\frac{\partial \bar{Z}_3}{\partial \theta_3} + a_{23}\frac{\partial \bar{Z}_3}{\partial \alpha_{23}} + a_{34}\frac{\partial \bar{Z}_3}{\partial \alpha_{34}}. \tag{8.181}$$

Evaluating the partial derivatives yields

$$\bar{Z}_{03} = S_3(s_{23}s_{34}s_3) + a_{23}(\bar{Y}_3) + a_{34}(Y_3). \tag{8.182}$$

The term on the left side of Eq. (8.180) contains the parameters $\theta_5, \theta_6, \theta_1, \alpha_{45}, \alpha_{56}, \alpha_{61}$, and α_{12} and may be written as

$$\begin{aligned} Z_{0561} = {} & S_5\frac{\partial Z_{561}}{\partial \theta_5} + S_6\frac{\partial Z_{561}}{\partial \theta_6} + S_1\frac{\partial Z_{561}}{\partial \theta_1} \\ & + a_{45}\frac{\partial Z_{561}}{\partial \alpha_{45}} + a_{56}\frac{\partial Z_{561}}{\partial \alpha_{56}} + a_{61}\frac{\partial Z_{561}}{\partial \alpha_{61}} + a_{12}\frac{\partial Z_{561}}{\partial \alpha_{12}}. \end{aligned} \tag{8.183}$$

Evaluating the partial derivatives gives

$$\begin{aligned} Z_{0561} = {} & S_5(s_{45}X_{165}) + S_6(-\bar{X}_1X^*_{56} - \bar{Y}_1X_{56}) + S_1(s_{12}X_{561}) + a_{45}(Y_{165}) \\ & + a_{56}(-s_{45}c_5Z_{16} + c_{45}Y_{16}) + a_{61}(-s_{12}c_1Z_{56} + c_{12}Y_{56}) + a_{12}(Y_{561}). \end{aligned} \tag{8.184}$$

Equating the results of Eqs. (8.182) and (8.184) yields the equation

$$\begin{aligned} & S_5(s_{45}X_{165}) + S_6(-\bar{X}_1X^*_{56} - \bar{Y}_1X_{56}) + S_1(s_{12}X_{561}) + a_{45}(Y_{165}) \\ & + a_{56}(-s_{45}c_5Z_{16} + c_{45}Y_{16}) + a_{61}(-s_{12}c_1Z_{56} + c_{12}Y_{56}) + a_{12}(Y_{561}) \\ & + S_3(-s_{23}s_{34}s_3) + a_{23}(-\bar{Y}_3) + a_{34}(-Y_3) = 0. \end{aligned} \tag{8.185}$$

All terms in Eq. (8.185) are expressed in terms of the constant mechanism parameters, the input angle θ_6, the output angle θ_1, and the extra angle θ_5, with the exception of the S_3, a_{23}, and a_{34} terms. The notation $C(S_3)$, $C(a_{23})$, and $C(a_{34})$ is now introduced to represent the coefficients of the S_3, a_{23}, and a_{34} terms. For this case these coefficients are defined as

$$C(S_3) = -s_{23}s_{34}s_3, \tag{8.186}$$

$$C(a_{23}) = -\bar{Y}_3, \tag{8.187}$$

$$C(a_{34}) = -Y_3. \tag{8.188}$$

The coefficient for the S_3 term may be written as

$$C(S_3) = -s_{34}X_3. \tag{8.189}$$

A subsidiary sine law may be used to write this term as follows:

$$C(S_3) = -s_{34}X_{1654}. \tag{8.190}$$

The coefficient for the a_{34} term may be modified by using a subsidiary sine–cosine law to give

$$C(a_{34}) = X^*_{1654}. \tag{8.191}$$

The coefficient for the a_{23} term may be written as

$$C(a_{23}) = s_{23}c_{34} + c_{23}s_{34}c_3. \tag{8.192}$$

Multiplying the right-hand side of Eq. (8.192) by $\frac{s_{23}}{s_{23}}$ yields

$$C(a_{23}) = \frac{s_{23}^2 c_{34} + c_{23}s_{23}s_{34}c_3}{s_{23}}. \tag{8.193}$$

Adding and subtracting the term $c_{23}^2 c_{34}$ from the numerator of Eq. (8.193) and then regrouping terms gives

$$C(a_{23}) = \frac{s_{23}^2 c_{34} + c_{23}^2 c_{34} - c_{23}(c_{23}c_{34} - s_{23}s_{34}c_3)}{s_{23}}. \tag{8.194}$$

Substituting for the definition of Z_3 gives

$$C(a_{23}) = \frac{c_{34} - c_{23}Z_3}{s_{23}}. \tag{8.195}$$

A subsidiary cosine law is used to replace the term Z_3 to give

$$C(a_{23}) = \frac{c_{34} - c_{23}Z_{165}}{s_{23}}. \tag{8.196}$$

All the terms in Eq. (8.185) have now been expressed in terms of the constant mechanism parameters, the input angle θ_6, the output angle θ_1, and the extra angle θ_5. This equation is written as

$$\begin{aligned} &S_5(s_{45}X_{165}) + S_6(-\bar{X}_1X^*_{56} - \bar{Y}_1X_{56}) + S_1(s_{12}X_{561}) + a_{45}(Y_{165}) \\ &\quad + a_{56}(-s_{45}c_5Z_{16} + c_{45}Y_{16}) + a_{61}(-s_{12}c_1Z_{56} + c_{12}Y_{56}) + a_{12}(Y_{561}) \\ &\quad + S_3(-s_{34}X_{1654}) + a_{23}(c_{34} - c_{23}Z_{165})/s_{23} + a_{34}(X^*_{1654}) = 0. \end{aligned} \tag{8.197}$$

This equation will be paired with the fundamental spherical cosine law

$$Z_{4561} = c_{23} \tag{8.198}$$

to yield an eighth-degree input/output equation in the tan-half-angle of θ_1. Both Eqs. (8.197) and (8.198) will be expanded and regrouped into the format of Eq. (8.5) (where the extra angle is now θ_5 instead of θ_2). The tan-half-angle substitutions for θ_1 and θ_5 will then be used to yield two biquadratic equations that can be solved for the output angle by using Bezout's method.

Equations (8.198) and (8.197) will be referred to as the first and second equations for this problem, respectively. These two equations will be regrouped into the format

$$c_5(A_i c_1 + B_i s_1 + D_i) + s_5(E_i c_1 + F_i s_1 + G_i) + (H_i c_1 + I_i s_1 + J_i) = 0 \qquad i = 1, 2. \tag{8.199}$$

The coefficients of the two equations are listed as follows:

$$\begin{aligned}
A_1 &= s_{12}(c_{61}s_6X_4 + Y_4(-s_{61}s_{56} + c_{61}c_{56}c_6)),\\
B_1 &= s_{12}(c_6X_4 - s_6c_{56}Y_4),\\
D_1 &= c_{12}(\bar{X}_6X_4 - \bar{Y}_6Y_4),\\
E_1 &= s_{12}(X_4(-s_{61}s_{56} + c_{61}c_{56}c_6) - c_{61}s_6Y_4),\\
F_1 &= s_{12}(-c_{56}s_6X_4 - c_6Y_4),\\
G_1 &= c_{12}(-\bar{X}_6Y_4 - \bar{Y}_6X_4),\\
H_1 &= s_{12}Y_6Z_4,\\
I_1 &= s_{12}X_6Z_4,\\
J_1 &= c_{12}Z_6Z_4 - c_{23}, \qquad\qquad (8.200)\\
A_2 &= S_5(s_{45}c_{61}s_{12}s_6) + a_{61}(-s_{12}s_{45}\bar{Y}_6) + S_3s_{34}s_{12}[s_4c_{45}(s_{56}s_{61} - c_{56}c_{61}c_6) - c_4c_{61}s_6]\\
&\quad + a_{34}s_{12}[c_{45}c_4(s_{56}s_{61} - c_{56}c_{61}c_6) + s_4s_6c_{61}] + a_{12}c_{12}s_{45}(s_{61}s_{56} - c_{61}c_{56}c_6)\\
&\quad + a_{23}c_{23}s_{45}s_{12}/s_{23}(-s_{56}s_{61} + c_{56}c_{61}c_6) + S_6(c_{61}s_{12}c_{56}s_{45}s_6) + S_1(s_{12}c_{56}s_{45}s_6)\\
&\quad + a_{56}(-s_{12}s_{45}Y_6) + a_{45}s_{12}c_{45}(s_{56}s_{61} - c_{56}c_{61}c_6),\\
B_2 &= S_3s_{34}s_{12}(-c_4c_6 + c_{45}c_{56}s_4s_6) + S_5(s_{45}s_{12}c_6) + a_{34}s_{12}(s_4c_6 + c_{45}c_{56}c_4s_6)\\
&\quad + a_{23}(-c_{23}s_{45}c_{56}s_{12}s_6/s_{23}) + S_1s_{12}s_{45}(-s_{61}s_{56} + c_{61}c_{56}c_6) + a_{45}(c_{45}c_{56}s_{12}s_6)\\
&\quad + a_{56}(-s_{45}s_{56}s_{12}s_6) + S_6(s_{12}c_{56}s_{45}c_6) + a_{12}(c_{12}c_{56}s_{45}s_6),\\
D_2 &= S_3s_{34}c_{12}(s_4c_{45}\bar{Y}_6 - c_4\bar{X}_6) + a_{23}(-c_{23}s_{45}c_{12}\bar{Y}_6/s_{23}) + a_{61}c_{12}s_{45}(s_{61}s_{56} - c_{61}c_{56}c_6)\\
&\quad + a_{34}c_{12}(c_4c_{45}\bar{Y}_6 + s_4\bar{X}_6) + a_{45}(c_{45}c_{12}\bar{Y}_6) + S_5(s_{45}s_{61}c_{12}s_6) + a_{12}(-s_{12}s_{45}\bar{Y}_6)\\
&\quad + S_6(s_{61}c_{12}c_{56}s_{45}s_6) + a_{56}(-s_{45}c_{12}Z_6),\\
E_2 &= a_{61}(-s_{12}s_{61}s_{45}s_6) + S_1(s_{12}s_{45}c_6) + a_{12}(c_{12}c_{61}s_{45}s_6) + S_6(c_{61}s_{12}s_{45}c_6)\\
&\quad + a_{34}s_{12}[c_4c_{45}c_{61}s_6 + s_4(-s_{56}s_{61} + c_{56}c_{61}c_6)] + a_{23}(-c_{23}s_{45}c_{61}s_{12}s_6/s_{23})\\
&\quad + a_{45}(c_{45}c_{61}s_{12}s_6) + S_3s_{12}s_{34}[c_{45}c_{61}s_4s_6 + c_4(s_{56}s_{61} - c_{56}c_{61}c_6)]\\
&\quad + S_5s_{12}s_{45}(-s_{61}s_{56} + c_{61}c_{56}c_6),\\
F_2 &= a_{23}(-c_{23}s_{45}s_{12}c_6/s_{23}) + S_6(-s_{12}s_{45}s_6) + a_{12}(c_{12}s_{45}c_6) + a_{45}(c_{45}s_{12}c_6)\\
&\quad + S_3s_{12}s_{34}(c_{45}s_4c_6 + c_{56}c_4s_6) + S_5(-s_{45}c_{56}s_{12}s_6) + a_{34}s_{12}(c_{45}c_4c_6 - c_{56}s_4s_6)\\
&\quad + S_1(-s_{12}c_{61}s_{45}s_6),
\end{aligned}$$

$$\begin{aligned}
G_2 &= a_{34}c_{12}(-s_4\bar{Y}_6 + c_{45}c_4\bar{X}_6) + a_{23}(-c_{23}s_{45}s_{61}c_{12}s_6/s_{23}) + a_{61}(c_{12}c_{61}s_{45}s_6) \\
&\quad + a_{12}(-s_{12}s_{61}s_{45}s_6) + S_3c_{12}s_{34}(s_4c_{45}\bar{X}_6 + c_4\bar{Y}_6) + a_{45}(c_{45}s_{61}c_{12}s_6) \\
&\quad + S_6(s_{61}c_{12}s_{45}c_6) + S_5(-c_{12}s_{45}\bar{Y}_6), \\
H_2 &= a_{61}(-s_{12}c_{45}Z_6) + a_{45}(-s_{12}s_{45}Y_6) + a_{23}(-s_{12}c_{23}c_{45}Y_6/s_{23}) + a_{12}(c_{12}c_{45}Y_6) \\
&\quad + S_3(-s_{34}s_{12}s_{45}s_4Y_6) + S_6(c_{61}s_{12}s_{56}c_{45}s_6) + S_1(s_{12}s_{56}c_{45}s_6) \\
&\quad + a_{34}(-s_{12}s_{45}c_4Y_6) + a_{56}c_{45}s_{12}(s_{56}s_{61} - c_{56}c_{61}c_6), \\
I_2 &= a_{45}(-s_{45}s_{56}s_{12}s_6) + a_{56}(c_{45}c_{56}s_{12}s_6) + S_6(s_{12}s_{56}c_{45}c_6) + a_{23}(-c_{23}c_{45}s_{56}s_{12}s_6/s_{23}) \\
&\quad + a_{34}(-s_{45}s_{56}s_{12}c_4s_6) + a_{12}(c_{12}s_{56}c_{45}s_6) + S_1(-s_{12}c_{45}Y_6) \\
&\quad + S_3(-s_{34}s_{45}s_{56}s_{12}s_4s_6), \\
J_2 &= S_3(-s_{34}s_{45}c_{12}s_4Z_6) + a_{23}(c_{34}/s_{23}) + a_{23}(-c_{23}c_{12}c_{45}Z_6/s_{23}) + a_{61}(c_{12}c_{45}Y_6) \\
&\quad + a_{34}(-s_{45}c_{12}c_4Z_6) + a_{12}(-c_{45}s_{12}Z_6) + a_{56}(c_{45}c_{12}\bar{Y}_6) \\
&\quad + a_{45}(-s_{45}c_{12}Z_6) + S_6(s_{61}c_{12}s_{56}c_{45}s_6).
\end{aligned} \tag{8.201}$$

Now that both equations have been regrouped into the format of Eq. (8.199), the tan-half-angle identities for the sine and cosine of θ_1 and θ_5 are inserted. The two equations can then be regrouped into the format similar to Eq. (8.10), that is,

$$x_5^2\left[a_ix_1^2 + b_ix_1 + d_i\right] + x_5\left[e_ix_1^2 + f_ix_1 + g_i\right] + \left[h_ix_1^2 + i_ix_1 + j_i\right] = 0, \qquad i = 1, 2, \tag{8.202}$$

where the coefficients of this equation are defined in Eq. set (8.11).

The two equations of the form of Eq. (8.202) are quadratic in the variables x_1 and x_5. These equations can be solved via Bezout's method (Section 8.2.2) to yield an eighth-degree polynomial in the variable x_1.

The corresponding value for the parameter x_5 can be found from either

$$x_5 = \frac{-\begin{vmatrix} M_1 & N_1 \\ M_2 & N_2 \end{vmatrix}}{\begin{vmatrix} L_1 & N_1 \\ L_2 & N_2 \end{vmatrix}} \tag{8.203}$$

or

$$x_5 = \frac{-\begin{vmatrix} L_1 & N_1 \\ L_2 & N_2 \end{vmatrix}}{\begin{vmatrix} L_1 & M_1 \\ L_2 & M_2 \end{vmatrix}}, \tag{8.204}$$

where L_1, M_1, and N_1 are defined in Eqs. (8.14) through (8.16) and Eq. set (8.11). Equations (8.203) and (8.204) can be derived in a manner similar to that in which Eqs. (8.34) and (8.35) were derived.

Two joint angles remain to be solved, that is, θ_2 and θ_3. A unique corresponding value of θ_2 can be obtained from the following fundamental sine and sine–cosine laws:

$$X_{4561} = s_{23}s_2, \tag{8.205}$$

$$Y_{4561} = s_{23}c_2. \tag{8.206}$$

Table 8.7. *RRPRCR mechanism parameters.*

Link length, cm.	Twist angle, deg.	Joint offset, cm.	Joint angle, deg.
$a_{12} = 35$	$\alpha_{12} = 82$	$S_1 = 67$	θ_1 = variable
$a_{23} = 19$	$\alpha_{23} = 78$	S_2 = variable	θ_2 = variable
$a_{34} = 19$	$\alpha_{34} = 34$	$S_3 = 15$	θ_3 = variable
$a_{45} = 9$	$\alpha_{45} = 93$	S_4 = variable	$\theta_4 = -96$
$a_{56} = 22$	$\alpha_{56} = 50$	$S_5 = 71$	θ_5 = variable
$a_{61} = 10$	$\alpha_{61} = 77$	$S_6 = 61$	$\theta_6 = 322$ (input)

Similarly, a unique corresponding value of θ_3 can then be obtained from the following fundamental sine and sine–cosine laws:

$$X_{1654} = s_{23}s_3, \tag{8.207}$$

$$Y_{1654} = s_{23}c_3. \tag{8.208}$$

The offset distances S_2 and S_4 are the remaining parameters to be determined. These two values will be found by projecting the vector loop equation for the mechanism onto two different directions. Projecting the vector loop equation onto the direction $\mathbf{a}_{34}$ and evaluating the scalar products using the sets of direction cosines provided in the appendix yields

$$S_1X_{23} + a_{12}W_{23} + S_2X_3 + a_{23}c_3 + a_{34} \\ + a_{45}c_4 + S_5\bar{X}_4 + a_{56}W_{54} + S_6X_{54} + a_{61}W_{654} = 0. \tag{8.209}$$

The offset distance S_2 is the only unknown in this equation.

The distance S_4 will be determined by projecting the vector loop equation onto the direction a_{12}. Evaluating the scalar products by using the sets of direction cosines listed in the appendix gives

$$a_{12} + a_{23}c_2 + S_3\bar{X}_2 + a_{34}W_{32} + S_4X_{32} \\ + a_{45}W_{561} + S_5X_{61} + a_{56}W_{61} + S_6X_1 + a_{61}c_1 = 0. \tag{8.210}$$

This equation contains the variable S_4 as its only unknown.

At this point, the analysis of the RRPRCR group 2 spatial mechanism is complete. Eight solution configurations were determined. Table 8.7 shows data that were used for a numerical example. The calculated values for the eight configurations are listed in Table 8.8. The eight configurations are shown in Figure 8.7.

8.6 Seven-link group 2 spatial mechanisms

All seven-link group 2 mechanisms comprise five revolute joints and two prismatic joints, that is, 5R-2P. The solution method for these mechanisms is identical to that for the five- and six-link group 2 mechanisms.

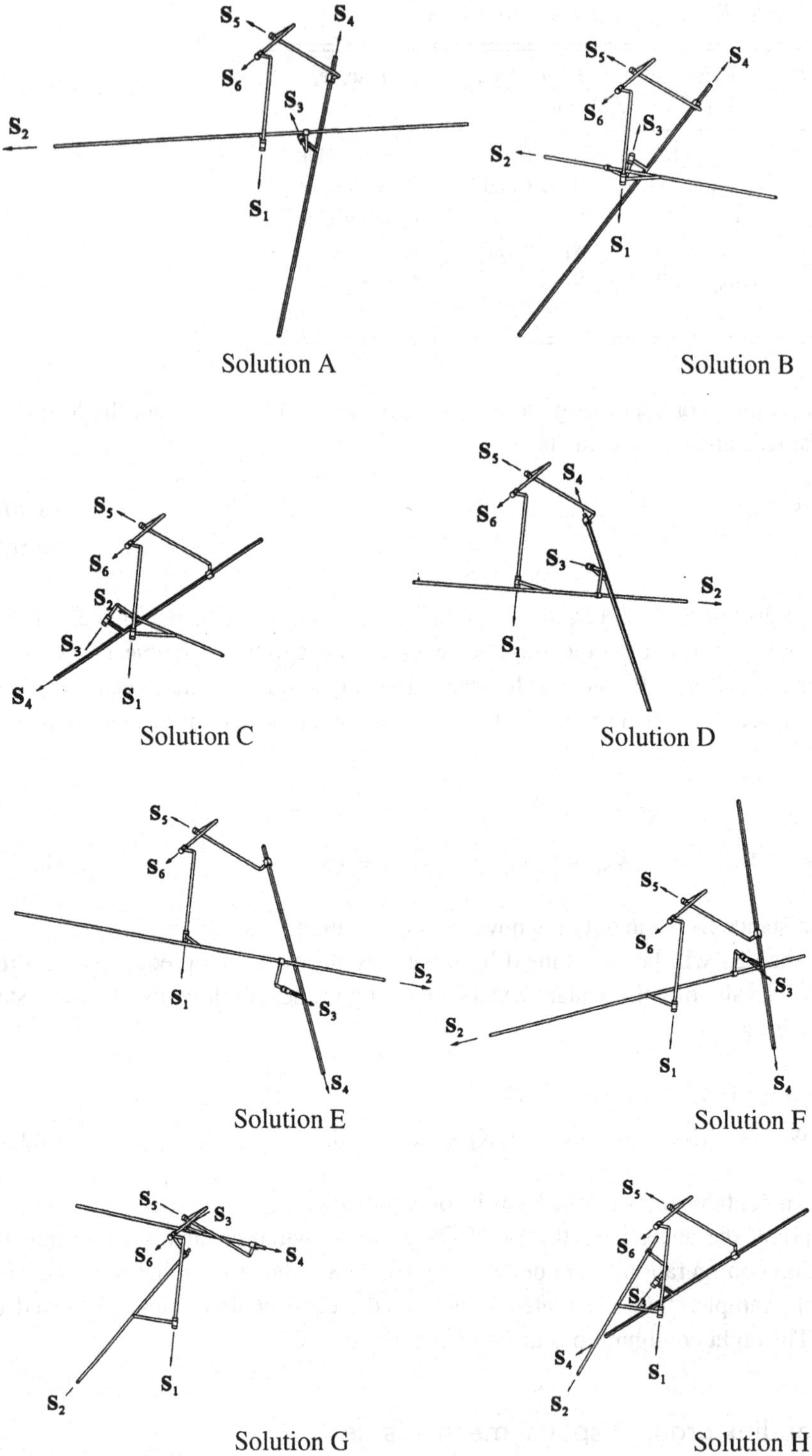

Figure 8.7. Eight configurations of an RRPRCR group 2 spatial mechanism.

Table 8.8. *Calculated configurations for the RRPRCR spatial mechanism.*

	Solution							
	A	B	C	D	E	F	G	H
θ_1, deg.	−118.75	−59.12	−38.23	17.13	45.28	−154.88	169.62	164.59
θ_2, deg.	62.27	24.74	−149.05	−92.70	89.61	−95.70	−49.48	148.25
θ_3, deg.	−39.69	9.66	164.51	115.85	78.19	−116.73	−136.00	−26.77
θ_5, deg.	13.40	56.19	−89.36	−19.39	167.66	174.27	124.31	−87.70
S_2, cm.	−35.60	73.80	138.13	56.12	71.33	−88.82	−150.49	−125.81
S_4, cm.	62.35	61.79	−100.10	72.67	−143.30	−30.80	67.11	−74.25

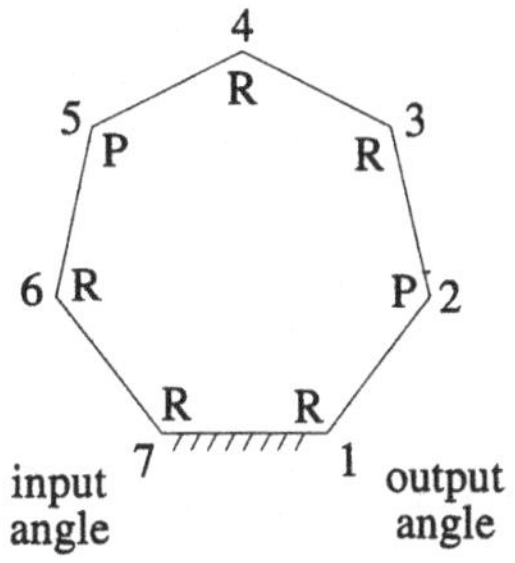

Figure 8.8. Planar representation of the RRPRRPR group 2 spatial mechanism.

Two equations are generated in terms of the input angle, the output angle, one extra joint angle, and the constant mechanism parameters. These two equations are solved simultaneously using Bezout's method to yield an eighth-degree input/output equation.

One seven-link group 2 mechanism will be solved as an example. Figure 8.8 shows a planar representation of an RRPRRPR mechanism. The problem statement for this mechanism is as follows:

given: constant mechanism parameters:
$a_{12}, a_{23}, a_{34}, a_{45}, a_{56}, a_{67}, a_{71}$,
$\alpha_{12}, \alpha_{23}, \alpha_{34}, \alpha_{45}, \alpha_{56}, \alpha_{67}, \alpha_{71}$
S_1, S_3, S_4, S_6, S_7, and
θ_2, θ_5 and
input angle:
θ_7,
find: $\theta_1, \theta_3, \theta_4, \theta_6, S_2$, and S_5.

One equation that will contain the constant link lengths and offsets (but not the unknowns S_2 and S_5) is the secondary cosine law

$$Z_{0671} = Z_{043}. \tag{8.211}$$

The term on the right side of this equation contains the parameters $\theta_3, \theta_4, \alpha_{23}, \alpha_{34}$, and α_{45}

and can thus be written as

$$Z_{043} = S_3\frac{\partial Z_{43}}{\partial \theta_3} + S_4\frac{\partial Z_{43}}{\partial \theta_4} + a_{23}\frac{\partial Z_{43}}{\partial \alpha_{23}} + a_{34}\frac{\partial Z_{43}}{\partial \alpha_{34}} + a_{45}\frac{\partial Z_{43}}{\partial \alpha_{45}}. \tag{8.212}$$

Expanding the partial derivatives gives

$$Z_{043} = S_3(s_{23}X_{43}) + S_4(s_{45}X_{34}) + a_{23}(Y_{43}) + a_{34}(c_{23}\bar{Y}_4 - s_{23}c_3\bar{Z}_4) + a_{45}(Y_{34}). \tag{8.213}$$

The term on the left side of Eq. (8.211) contains the parameters S_6, S_7, S_1, a_{56}, a_{67}, a_{71}, and a_{12}. It can thus be written as follows:

$$\begin{aligned} Z_{0671} = {} & S_6\frac{\partial Z_{671}}{\partial \theta_6} + S_7\frac{\partial Z_{671}}{\partial \theta_7} + S_1\frac{\partial Z_{671}}{\partial \theta_1} \\ & + a_{56}\frac{\partial Z_{671}}{\partial \alpha_{56}} + a_{67}\frac{\partial Z_{671}}{\partial \alpha_{67}} + a_{71}\frac{\partial Z_{671}}{\partial \alpha_{71}} + a_{12}\frac{\partial Z_{671}}{\partial \alpha_{12}}. \end{aligned} \tag{8.214}$$

Expanding the partial derivatives for this equation yields

$$\begin{aligned} Z_{0671} = {} & S_6(s_{56}X_{176}) + S_7(-\bar{X}_1X^*_{67} - \bar{Y}_1X_{67}) + S_1(s_{12}X_{671}) + a_{56}(Y_{176}) \\ & + a_{67}(c_{56}Y_{17} - s_{56}c_6Z_{17}) + a_{71}(c_{12}Y_{67} - s_{12}c_1Z_{67}) + a_{12}(Y_{671}). \end{aligned} \tag{8.215}$$

Substituting the results of Eqs. (8.213) and (8.215) into Eq. (8.211) and moving all terms to the left side of the equation gives

$$\begin{aligned} & S_6(s_{56}X_{176}) + S_7(-\bar{X}_1X^*_{67} - \bar{Y}_1X_{67}) + S_1(s_{12}X_{671}) + a_{56}(Y_{176}) \\ & + a_{67}(c_{56}Y_{17} - s_{56}c_6Z_{17}) + a_{71}(c_{12}Y_{67} - s_{12}c_1Z_{67}) + a_{12}(Y_{671}) + S_3(-s_{23}X_{43}) \\ & + S_4(-s_{45}X_{34}) + a_{23}(-Y_{43}) + a_{34}(-c_{23}\bar{Y}_4 + s_{23}c_3\bar{Z}_4) + a_{45}(-Y_{34}) = 0. \end{aligned} \tag{8.216}$$

The angle θ_6 is selected as the extra joint angle in this equation. Thus, each of the terms in Eq. (8.216) must now be written in terms of the input angle, θ_7, the output angle, θ_1, the extra angle, θ_6, and the two constant angles, θ_2 and θ_5. In other words, the angles θ_3 and θ_4 that appear in the S_3, S_4, a_{23}, a_{34}, and a_{45} terms in Eq. (8.216) must be replaced.

The S_3 term may be written as

$$C(S_3) = -s_{23}X_{43}. \tag{8.217}$$

A subsidiary sine law for a spherical heptagon can be used to rewrite this term as

$$C(S_3) = -s_{23}X_{6712}. \tag{8.218}$$

Similarly, a subsidiary sine law can be used to rewrite the S_4 term as

$$C(S_4) = -s_{45}X_{1765}. \tag{8.219}$$

Subsidiary sine–cosine laws are utilized to rewrite the a_{23} and a_{45} terms as

$$C(a_{23}) = X^*_{6712} \tag{8.220}$$

and

$$C(a_{45}) = X^*_{1765}. \tag{8.221}$$

The last remaining term is a_{34}. This term appears in Eq. (8.216) as

$$C(a_{34}) = -c_{23}\bar{Y}_4 + s_{23}c_3\bar{Z}_4. \tag{8.222}$$

Expanding $\bar{Y}_4$ yields

$$\bar{Y}_4 = -(s_{34}c_{45} + c_{34}s_{45}c_4). \tag{8.223}$$

Multiplying the right side of Eq. (8.223) by $\frac{s_{34}}{s_{34}}$ and regrouping gives

$$\bar{Y}_4 = \frac{-s_{34}^2c_{45} - c_{34}s_{34}s_{45}c_4}{s_{34}}. \tag{8.224}$$

Adding $(c_{34}^2c_{45} - c_{34}^2c_{45})$ to the numerator of Eq. (8.224) and regrouping gives

$$\bar{Y}_4 = \frac{-\left(s_{34}^2 + c_{34}^2\right)c_{45} + c_{34}(c_{34}c_{45} - s_{34}s_{45}c_4)}{s_{34}}. \tag{8.225}$$

This equation can be simplified to yield

$$\bar{Y}_4 = \frac{-c_{45} + c_{34}\bar{Z}_4}{s_{34}}. \tag{8.226}$$

Equation (8.222) can now be written as

$$C(a_{34}) = \frac{c_{23}(c_{45} - c_{34}\bar{Z}_4)}{s_{34}} + s_{23}c_3\bar{Z}_4. \tag{8.227}$$

Regrouping this equation gives

$$C(a_{34}) = \frac{c_{23}c_{45} - \bar{Z}_4Z_3}{s_{34}}. \tag{8.228}$$

Subsidiary cosine laws are used to replace the $\bar{Z}_4$ and Z_3 terms in Eq. (8.228) to give

$$C(a_{34}) = \frac{c_{23}c_{45} - Z_{2176}Z_{1765}}{s_{34}}. \tag{8.229}$$

The coefficient of the a_{34} term has been expressed in terms of the constant mechanism parameters, the input angle, the output angle, and the extra angle, θ_6. However, Eq. (8.229) appears to be of second order in the sines and cosines of the variable joint angles, θ_6 and θ_1. The product $Z_{2176}Z_{1765}$ must be expanded and regrouped in order to reduce it to a linear expression.

The term Z_{1765} is written as follows:

$$Z_{1765} = s_{45}(X_{176}s_5 + Y_{176}c_5) + c_{45}Z_{176}. \tag{8.230}$$

Expanding Y_{176} and Z_{176} gives

$$Z_{1765} = s_{45}s_5X_{176} + s_{45}c_5\left(c_{56}X^*_{176} - s_{56}Z_{17}\right) + c_{45}\left(s_{56}X^*_{176} + c_{56}Z_{17}\right). \tag{8.231}$$

Regrouping this equation yields

$$Z_{1765} = X_5X_{176} - Y_5X^*_{176} + Z_5Z_{17}. \tag{8.232}$$

The term Z_{2176} is written as

$$Z_{2176} = s_{56}X^*_{2176} + c_{56}Z_{217}. \tag{8.233}$$

The results of Eqs. (8.232) and (8.233) may now be used to express the product $Z_{2176}Z_{1765}$ as

$$Z_{2176}Z_{1765} = \left[X_5X_{176} - Y_5X^*_{176} + Z_5Z_{17}\right]\left[s_{56}X^*_{2176} + c_{56}Z_{217}\right]. \tag{8.234}$$

This equation can also be written as

$$Z_{2176}Z_{1765} = s_{56}\left\{X_5X_{176}X^*_{2176} - Y_5X^*_{176}X^*_{2176} + Z_5Z_{17}X^*_{2176}\right\} + c_{56}\{Z_{1765}Z_{217}\}. \tag{8.235}$$

Two new terms, A_{56712} and B_{56712}, will be defined to represent the expressions in the braces of Eq. (8.235). Thus, the following two definitions are introduced

$$A_{56712} = X_5X_{176}X^*_{2176} - Y_5X^*_{176}X^*_{2176} + Z_5Z_{17}X^*_{2176}, \tag{8.236}$$

$$B_{56712} = Z_{1765}Z_{217}, \tag{8.237}$$

and the product $Z_{2176}Z_{1765}$ may be written as

$$Z_{2176}Z_{1765} = s_{56}A_{56712} + c_{56}B_{56712}. \tag{8.238}$$

Equation (8.236) will be simplified by first expanding the following fundamental cosine law for a spherical heptagon:

$$Z_{21765} = c_{34}. \tag{8.239}$$

Expanding Z_{21765} gives

$$s_{45}(X_{2176}s_5 + Y_{2176}c_5) + c_{45}Z_{2176} = c_{34}. \tag{8.240}$$

Expanding Y_{2176} and Z_{2176} gives

$$s_{34}s_5X_{2176} + s_{45}c_5\left(c_{56}X^*_{2176} - s_{56}Z_{217}\right) + c_{45}\left(s_{56}X^*_{2176} + c_{56}Z_{217}\right) = c_{34}. \tag{8.241}$$

Regrouping this equation gives

$$X_5X_{2176} - Y_5X^*_{2176} + Z_5Z_{217} = c_{34}. \tag{8.242}$$

The middle term, $Y_5X^*_{2176}$ (which also appears in Eq. (8.236)), may thus be expressed as

$$Y_5X^*_{2176} = X_5X_{2176} + Z_5Z_{217} - c_{34}. \tag{8.243}$$

This result is now substituted into Eq. (8.237) to give

$$A_{56712} = X_5X_{176}X^*_{2176} - X^*_{176}[X_5X_{2176} + Z_5Z_{217} - c_{34}] + Z_5Z_{17}X^*_{2176}. \tag{8.244}$$

This equation is regrouped as

$$A_{56712} = X_5\left[X_{176}X^*_{2176} - X^*_{176}X_{2176}\right] + Z_5\left[Z_{17}X^*_{2176} - X^*_{176}Z_{217}\right] + c_{34}X^*_{176}. \tag{8.245}$$

The terms X_{176}, X^*_{176}, X_{2176}, and X^*_{2176} are substituted into this equation to give

$$\begin{aligned} A_{56712} = {} & X_5[(X_{17}c_6 - Y_{17}s_6)(X_{217}s_6 + Y_{217}c_6) - (X_{17}s_6 + Y_{17}c_6)(X_{217}c_6 - Y_{217}s_6)] \\ & + Z_5[Z_{17}(X_{217}s_6 + Y_{217}c_6) - (X_{17}s_6 + Y_{17}c_6)Z_{217}] + c_{34}X^*_{176}. \end{aligned} \tag{8.246}$$

Performing the multiplication and substituting for $s_6^2 + c_6^2 = 1$ gives

$$\begin{aligned} A_{56712} = {} & X_5[X_{17}Y_{217} - Y_{17}X_{217}] + Z_5[s_6(X_{217}Z_{17} - Z_{217}X_{17}) \\ & + c_6(Z_{17}Y_{217} - Z_{217}Y_{17})] + c_{34}X^*_{176}. \end{aligned} \tag{8.247}$$

The first term in brackets is written as

$$\begin{aligned} X_{17}Y_{217} - Y_{17}X_{217} = {} & [\bar{X}_1c_7 - \bar{Y}_1s_7][c_{67}(X_{21}s_7 + Y_{21}c_7) - s_{67}Z_{21}] \\ & - [c_{67}(\bar{X}_1s_7 + \bar{Y}_1c_7) - s_{67}\bar{Z}_1][X_{21}c_7 - Y_{21}s_7]. \end{aligned} \tag{8.248}$$

Performing the multiplication and recognizing that $s_7^2 + c_7^2 = 1$ gives

$$\begin{aligned} X_{17}Y_{217} - Y_{17}X_{217} = {} & c_7s_{67}(X_{21}\bar{Z}_1 - Z_{21}\bar{X}_1) \\ & + s_7s_{67}(Z_{21}\bar{Y}_1 - Y_{21}\bar{Z}_1) + c_{67}(Y_{21}\bar{X}_1 - X_{21}\bar{Y}_1). \end{aligned} \tag{8.249}$$

Substituting for $X_{21}, Y_{21}, Z_{21}, \bar{X}_1, \bar{Y}_1$, and $\bar{Z}_1$, regrouping terms, and recognizing that $s_1^2 + c_1^2 = 1$ and $s_{71}^2 + c_{71}^2 = 1$ gives

$$\begin{aligned} X_{17}Y_{217} - Y_{17}X_{217} = {} & c_1[-c_{12}Y_7\bar{X}_2 + X_7(-c_{12}\bar{Y}_2 - s_{12}\bar{Z}_2)] \\ & + s_1[-c_{12}X_7\bar{X}_2 + c_{12}Y_7\bar{Y}_2 + s_{12}Y_7\bar{Z}_2] + [s_{12}Z_7\bar{X}_2]. \end{aligned} \tag{8.250}$$

This term is linear in the sines and cosines of θ_1, θ_2, and θ_7. Equation (8.250) will be simplified by firstly expanding the terms $\bar{X}_2$, $\bar{Y}_2$, and $\bar{Z}_2$ and recognizing that $s_{12}^2 + c_{12}^2 = 1$. The result of this step is

$$X_{17}Y_{217} - Y_{17}X_{217} = s_{23}(c_1c_2X_7 - s_1c_2Y_7 - c_{12}s_1s_2X_7 - c_{12}c_1s_2Y_7 + s_{12}s_2Z_7). \tag{8.251}$$

Regrouping this equation and introducing the terms U_{21}, V_{21}, and W_{21} gives

$$X_{17}Y_{217} - Y_{17}X_{217} = s_{23}(W_{21}X_7 + V_{21}Y_7 + U_{21}Z_7). \tag{8.252}$$

Expanding the terms X_7, Y_7, and Z_7 in this equation and regrouping gives

$$X_{17}Y_{217} - Y_{17}X_{217} = s_{23}[c_{67}(U_{21}c_{71} - V_{21}s_{71}) - s_{67}(c_7(U_{21}s_{71} + V_{21}c_{71}) - s_7W_{21})]. \tag{8.253}$$

Equation (8.253) can now be written as the simplified expression

$$X_{17}Y_{217} - Y_{17}X_{217} = s_{23}U_{2176}. \tag{8.254}$$

The second term of Eq. (8.247) is written as

$$X_{217}Z_{17} - Z_{217}X_{17} = [X_{21}c_7 - Y_{21}s_7][s_{67}(\bar{X}_1s_7 + \bar{Y}_1c_7) + c_{67}\bar{Z}_1] - [s_{67}(X_{21}s_7 + Y_{21}c_7) + c_{67}Z_{21}][\bar{X}_1c_7 - \bar{Y}_1s_7]. \tag{8.255}$$

Multiplying the terms and recognizing that $s_7^2 + c_7^2 = 1$ gives

$$X_{217}Z_{17} - Z_{217}X_{17} = c_7c_{67}(X_{21}\bar{Z}_1 - Z_{21}\bar{X}_1) + s_7c_{67}(Z_{21}\bar{Y}_1 - Y_{21}\bar{Z}_1) + s_{67}(X_{21}\bar{Y}_1 - Y_{21}\bar{X}_1). \tag{8.256}$$

Substituting for $X_{21}, Y_{21}, Z_{21}, \bar{X}_1, \bar{Y}_1$, and $\bar{Z}_1$, regrouping terms, and recognizing that $s_1^2 + c_1^2 = 1$ and $s_{71}^2 + c_{71}^2 = 1$ gives

$$X_{217}Z_{17} - Z_{217}X_{17} = c_1[c_{12}\bar{X}_2(-s_{67}s_{71} + c_{67}c_{71}c_7) - c_{12}c_{67}s_7\bar{Y}_2 - s_{12}c_{67}s_7\bar{Z}_2] + s_1[-c_{12}c_{67}s_7\bar{X}_2 + c_{12}\bar{Y}_2(s_{67}s_{71} - c_{67}c_{71}c_7) + s_{12}\bar{Z}_2(s_{67}s_{71} - c_{67}c_{71}c_7)] + [s_{12}\bar{X}_2\bar{Y}_7]. \tag{8.257}$$

This term is linear in the sines and cosines of θ_1, θ_2, and θ_7. Equation (8.257) will be simplified by first introducing the terms $\bar{X}_2$, $\bar{Y}_2$, and $\bar{Z}_2$ and recognizing that $s_{12}^2 + c_{12}^2 = 1$ to give

$$X_{217}Z_{17} - Z_{217}X_{17} = s_{23}[s_{12}s_2\bar{Y}_7 + c_{67}c_2(s_7c_1 + c_7s_1c_{71}) - s_{67}s_{71}s_1c_2 + c_{12}s_2(-s_{67}s_{71}c_1 + c_{67}c_{71}c_1c_7 - c_{67}s_1s_7)]. \tag{8.258}$$

This equation may be regrouped as

$$X_{217}Z_{17} - Z_{217}X_{17} = s_{23}[U_{21}\bar{Y}_7 + s_{67}s_{71}V_{21} - c_{67}c_{71}c_7V_{21} + c_{67}s_7W_{21}]. \tag{8.259}$$

Expanding $\bar{Y}_7$ and regrouping terms gives

$$X_{217}Z_{17} - Z_{217}X_{17} = s_{23}[-s_{67}(U_{21}c_{71} - V_{21}s_{71}) - c_{67}(c_7(U_{21}s_{71} + V_{21}c_{71}) - s_7W_{21})]. \tag{8.260}$$

The first term in parentheses in Eq. (8.260) is U_{217}, and the second term in parentheses is V_{217}. Thus, Eq. (8.260) may be written simply as

$$X_{217}Z_{17} - Z_{217}X_{17} = -s_{23}U^*_{2176}. \tag{8.261}$$

The third expression in Eq. (8.247) is written as

$$Z_{17}Y_{217} - Z_{217}Y_{17} = [s_{67}(\bar{X}_1 s_7 + \bar{Y}_1 c_7) + c_{67}\bar{Z}_1][c_{67}(X_{21}s_7 + Y_{21}c_7) - s_{67}Z_{21}] - [s_{67}(X_{21}s_7 + Y_{21}c_7) + c_{67}Z_{21}][c_{67}(\bar{X}_1 s_7 + \bar{Y}_1 c_7) - s_{67}\bar{Z}_1]. \quad (8.262)$$

Multiplying the terms on the right side of this equation, regrouping, and recognizing that $s_{67}^2 + c_{67}^2 = 1$ gives

$$Z_{17}Y_{217} - Z_{217}Y_{17} = c_7(\bar{Z}_1 Y_{21} - \bar{Y}_1 Z_{21}) + s_7(\bar{Z}_1 X_{21} - \bar{X}_1 Z_{21}). \quad (8.263)$$

Substituting for $X_{21}, Y_{21}, Z_{21}, \bar{X}_1, \bar{Y}_1$, and $\bar{Z}_1$, regrouping terms, and recognizing that $s_1^2 + c_1^2 = 1$ and $s_{71}^2 + c_{71}^2 = 1$ gives

$$Z_{17}Y_{217} - Z_{217}Y_{17} = c_1[c_{12}c_{71}s_7\bar{X}_2 + c_{12}c_7\bar{Y}_2 + s_{12}c_7\bar{Z}_2] + s_1[c_{12}c_7\bar{X}_2 - c_{12}c_{71}s_7\bar{Y}_2 - s_{12}c_{71}s_7\bar{Z}_2] + [-s_{12}s_{71}s_7\bar{X}_2]. \quad (8.264)$$

This equation can be regrouped to give

$$Z_{17}Y_{217} - Z_{217}Y_{17} = c_{12}(W_{71}\bar{Y}_2 - V_{71}\bar{X}_2) + s_{12}(W_{71}\bar{Z}_2 - U_{71}\bar{X}_2). \quad (8.265)$$

This term is also linear in the sines and cosines of θ_1, θ_2, and θ_7. Equation (8.265) may be simplified by introducing the terms $\bar{X}_2$, $\bar{Y}_2$, and $\bar{Z}_2$ and recognizing that $s_{12}^2 + c_{12}^2 = 1$ to give

$$Z_{17}Y_{217} - Z_{217}Y_{17} = -s_{23}[s_2(U_{71}s_{12} + V_{71}c_{12}) + c_2 W_{71}]. \quad (8.266)$$

The term in brackets in this equation is W_{712}, which equals W_{217}, and therefore

$$Z_{17}Y_{217} - Z_{217}Y_{17} = -s_{23}W_{217}. \quad (8.267)$$

Substituting the results of Eqs. (8.254), (8.261), and (8.267) into (8.247) yields

$$A_{56712} = X_5[s_{23}U_{2176}] + Z_5\left[s_6\left(-s_{23}U^*_{2176}\right) + c_6(-s_{23}W_{217})\right] + c_{34}X^*_{176}. \quad (8.268)$$

Rearranging this equation gives

$$A_{56712} = s_{23}(X_5 U_{2176} - Z_5 W_{2176}) + c_{34}X^*_{176}. \quad (8.269)$$

All that remains to be accomplished is to expand the term B_{56712} as defined in Eq. (8.237) so that it is linear in the sines and cosines of the unknown joint parameters. Expanding $Z_{5671} = Z_{1765}$ in Eq. (8.237) gives

$$B_{56712} = [s_{12}(X_{567}s_1 + Y_{567}c_1) + c_{12}Z_{567}]Z_{217}. \quad (8.270)$$

Expanding Y_{567} and Z_{567} gives

$$B_{56712} = \left[s_{12}\left(X_{567}s_1 + \left(c_{71}X^*_{567} - s_{71}Z_{56}\right)c_1\right) + c_{12}\left(s_{71}X^*_{567} + c_{71}Z_{56}\right)\right]Z_{217}. \quad (8.271)$$

Regrouping terms yields

$$B_{56712} = \bar{X}_1 X_{567} Z_{217} - \bar{Y}_1 X^*_{567} Z_{217} + \bar{Z}_1 Z_{56} Z_{217}. \tag{8.272}$$

The current objective is to express B_{56712} as an expression that is linear in the unknown values θ_1 and θ_6. This task will begin by writing the products $X_{567}Z_{217}$ and $X^*_{567}Z_{217}$ from Eq. (8.272) as

$$X_{567} Z_{217} = s_{67}(X_{21} X_{567} s_7 + Y_{21} X_{567} c_7) + c_{67} Z_{21} X_{567} \tag{8.273}$$

and

$$X^*_{567} Z_{217} = s_{67}\left(X_{21} X^*_{567} s_7 + Y_{21} X^*_{567} c_7\right) + c_{67} Z_{21} X^*_{567}. \tag{8.274}$$

Next, a fundamental cosine law for a spherical heptagon ($Z_{21765} = c_{34}$) is expanded as

$$s_{45}(X_{2176} s_5 + Y_{2176} c_5) + c_{45} Z_{2176} = c_{34}. \tag{8.275}$$

Expanding Y_{2176} and Z_{2176} and regrouping this equation gives

$$X_5 X_{2176} - Y_5 X^*_{2176} + Z_5 Z_{217} = c_{34}. \tag{8.276}$$

Expanding X_{2176} and X^*_{2176} gives

$$X_5(X_{217} c_6 - Y_{217} s_6) - Y_5(X_{217} s_6 + Y_{217} c_6) + Z_5 Z_{217} = c_{34}. \tag{8.277}$$

Rearranging this equation yields

$$X_{56} X_{217} - X^*_{56} Y_{217} + Z_5 Z_{217} = c_{34}. \tag{8.278}$$

Substituting for X_{217}, Y_{217}, and Z_{217} gives

$$\begin{aligned} &X_{56}(X_{21} c_7 - Y_{21} s_7) - X^*_{56}(c_{67}(X_{21} s_7 + Y_{21} c_7) - s_{67} Z_{21}) \\ &\quad + Z_5(s_{67}(X_{21} s_7 + Y_{21} c_7) + c_{67} Z_{21}) = c_{34}. \end{aligned} \tag{8.279}$$

Regrouping this equation yields

$$\begin{aligned} &X_{21}\left(X_{56} c_7 - \left(c_{67} X^*_{56} - s_{67} Z_5\right) s_7\right) - Y_{21}\left(X_{56} s_7\right. \\ &\quad \left. + \left(c_{67} X^*_{56} - s_{67} Z_5\right) c_7\right) + Z_{21}\left(s_{67} X^*_{56} + c_{67} Z_5\right) = c_{34}. \end{aligned} \tag{8.280}$$

This equation may be written as

$$X_{21} X_{567} - Y_{21} X^*_{567} + Z_{21} Z_{56} = c_{34}. \tag{8.281}$$

Multiplying this equation by s_7 and then c_7 and rearranging yields

$$X_{21} X_{567} s_7 = Y_{21} X^*_{567} s_7 + (c_{34} - Z_{21} Z_{56}) s_7 \tag{8.282}$$

and

$$Y_{21} X^*_{567} c_7 = X_{21} X_{567} c_7 - (c_{34} - Z_{21} Z_{56}) c_7. \tag{8.283}$$

Substituting Eq. (8.282) into Eqs. (8.273) and (8.283) into Eq. (8.274) gives

$$X_{567}Z_{217} = s_{67}\left[Y_{21}X^*_{567}s_7 + (c_{34} - Z_{21}Z_{56})s_7 + Y_{21}X_{567}c_7\right] + c_{67}Z_{21}X_{567} \tag{8.284}$$

and

$$X^*_{567}Z_{217} = s_{67}\left[X_{21}X^*_{567}s_7 + X_{21}X_{567}c_7 - (c_{34} - Z_{21}Z_{56})c_7\right] + c_{67}Z_{21}X^*_{567}. \tag{8.285}$$

Expanding the terms X_{567} and X^*_{567} in these two equations yields

$$\begin{aligned} X_{567}Z_{217} = {} & s_{67}[Y_{21}(X_{56}s_7 + Y_{56}c_7)s_7 \\ & + (c_{34} - Z_{21}Z_{56})s_7 + Y_{21}(X_{56}c_7 - Y_{56}s_7)c_7] + c_{67}Z_{21}X_{567} \end{aligned} \tag{8.286}$$

and

$$\begin{aligned} X^*_{567}Z_{217} = {} & s_{67}[X_{21}(X_{56}s_7 + Y_{56}c_7)s_7 + X_{21}(X_{56}c_7 - Y_{56}s_7)c_7 \\ & - (c_{34} - Z_{21}Z_{56})c_7] + c_{67}Z_{21}X^*_{567}. \end{aligned} \tag{8.287}$$

Multiplying and rearranging terms and then recognizing that $s_7^2 + c_7^2 = 1$ gives

$$X_{567}Z_{217} = s_{67}[Y_{21}X_{56} + (c_{34} - Z_{21}Z_{56})s_7] + c_{67}Z_{21}X_{567} \tag{8.288}$$

and

$$X^*_{567}Z_{217} = s_{67}[X_{21}X_{56} - (c_{34} - Z_{21}Z_{56})c_7] + c_{67}Z_{21}X^*_{567}. \tag{8.289}$$

Substituting Eqs. (8.288) and (8.289) into Eq. (8.272) and expanding Z_{217} gives

$$\begin{aligned} B_{56712} = {} & \bar{X}_1[s_{67}(Y_{21}X_{56} + (c_{34} - Z_{21}Z_{56})s_7) + c_{67}Z_{21}X_{567}] \\ & - \bar{Y}_1\left[s_{67}(X_{21}X_{56} - (c_{34} - Z_{21}Z_{56})c_7) + c_{67}Z_{21}X^*_{567}\right] \\ & + \bar{Z}_1Z_{56}[s_{67}(X_{21}s_7 + Y_{21}c_7) + c_{67}Z_{21}]. \end{aligned} \tag{8.290}$$

This equation is linear in the sines and cosines of the input angle, θ_7‡, but is not yet linear in the sines and cosines of the output angle, θ_1. Regrouping this equation gives

$$\begin{aligned} B_{56712} = {} & s_{67}X_{56}(\bar{X}_1Y_{21} - \bar{Y}_1X_{21}) + c_{67}Z_{21}(\bar{Z}_1Z_{56} + \bar{X}_1X_{567} - \bar{Y}_1X^*_{567}) \\ & + s_{67}Z_{56}[s_7(\bar{Z}_1X_{21} - \bar{X}_1Z_{21}) + c_7(\bar{Z}_1Y_{21} - \bar{Y}_1Z_{21})] + c_{34}s_{67}X^*_{17}. \end{aligned} \tag{8.291}$$

The second term in parentheses in Eq. (8.291) may be rewritten as

$$\begin{aligned} \bar{Z}_1Z_{56} + \bar{X}_1X_{567} - \bar{Y}_1X^*_{567} = {} & (c_{71}c_{12} - s_{71}s_{12}c_1)Z_{56} \\ & + (s_{12}s_1)X_{567} + (s_{71}c_{12} + c_{71}s_{12}c_1)X^*_{567}. \end{aligned} \tag{8.292}$$

This equation may be regrouped as

$$\begin{aligned} \bar{Z}_1Z_{56} + \bar{X}_1X_{567} - \bar{Y}_1X^*_{567} = {} & s_{12}s_1X_{567} + s_{12}c_1\left(c_{71}X^*_{567} - s_{71}Z_{56}\right) \\ & + c_{12}\left(s_{71}X^*_{567} + c_{71}Z_{56}\right). \end{aligned} \tag{8.293}$$

‡ It is not important that the equation be linear in the sines and cosines of the given joint parameters. It is only important that it be linear in the sines and cosines of the unknown joint parameters.

Introducing the terms Y_{567} and Z_{567} in the right side of Eq. (8.293) yields

$$\bar{Z}_1 Z_{56} + \bar{X}_1 X_{567} - \bar{Y}_1 X^*_{567} = s_{12}(X_{567} s_1 + Y_{567} c_1) + c_{12} Z_{567}. \tag{8.294}$$

Finally, this term may be simplified by recognizing that the right side of Eq. (8.294) is simply Z_{5671}. Thus,

$$\bar{Z}_1 Z_{56} + \bar{X}_1 X_{567} - \bar{Y}_1 X^*_{567} = Z_{5671}. \tag{8.295}$$

Equation (8.295) is substituted into Eq. (8.291) to yield

$$\begin{aligned} B_{56712} = {} & s_{67} X_{56}(\bar{X}_1 Y_{21} - \bar{Y}_1 X_{21}) + c_{67} Z_{21} Z_{5671} \\ & + s_{67} Z_{56}[s_7(\bar{Z}_1 X_{21} - \bar{X}_1 Z_{21}) + c_7(\bar{Z}_1 Y_{21} - \bar{Y}_1 Z_{21})] + c_{34} s_{67} X^*_{17}. \end{aligned} \tag{8.296}$$

Equation (8.296) is still not linear in the sines and cosines of θ_1. The first term in parentheses in this equation may be written as

$$\begin{aligned} \bar{X}_1 Y_{21} - \bar{Y}_1 X_{21} = {} & [s_{12} s_1][c_{71}(\bar{X}_2 s_1 + \bar{Y}_2 c_1) - s_{71} \bar{Z}_2] \\ & + [s_{71} c_{12} + c_{71} s_{12} c_1][\bar{X}_2 c_1 - \bar{Y}_2 s_1]. \end{aligned} \tag{8.297}$$

Performing the multiplication of terms, regrouping, and recognizing that $s_1^2 + c_1^2 = 1$ yields

$$\bar{X}_1 Y_{21} - \bar{Y}_1 X_{21} = c_1(c_{12} s_{71} \bar{X}_2) + s_1 s_{71}(-c_{12} \bar{Y}_2 - s_{12} \bar{Z}_2) + (s_{12} c_{71} \bar{X}_2). \tag{8.298}$$

Expanding the terms $\bar{X}_2$, $\bar{Y}_2$, and $\bar{Z}_2$ and recognizing that $s_{12}^2 + c_{12}^2 = 1$ gives

$$\bar{X}_1 Y_{21} - \bar{Y}_1 X_{21} = s_{23}[(s_{12} s_2) c_{71} + (s_1 c_2 + c_1 s_2 c_{12}) s_{71}]. \tag{8.299}$$

The term in brackets on the right side of Eq. (8.299) is U_{217}. Thus, Eq. (8.299) may be written as

$$\bar{X}_1 Y_{21} - \bar{Y}_1 X_{21} = s_{23} U_{217}. \tag{8.300}$$

The second term in parentheses in Eq. (8.296) may be expanded as

$$\begin{aligned} \bar{Z}_1 X_{21} - \bar{X}_1 Z_{21} = {} & [c_{71} c_{12} - s_{71} s_{12} c_1][\bar{X}_2 c_1 - \bar{Y}_2 s_1] \\ & - [s_{12} s_1][s_{71}(\bar{X}_2 s_1 + \bar{Y}_2 c_1) + c_{71} \bar{Z}_2]. \end{aligned} \tag{8.301}$$

Multiplying the terms together, regrouping, and recognizing that $s_1^2 + c_1^2 = 1$ yields

$$\bar{Z}_1 X_{21} - \bar{X}_1 Z_{21} = c_1(c_{12} c_{71} \bar{X}_2) + s_1 c_{71}(-c_{12} \bar{Y}_2 - s_{12} \bar{Z}_2) + (-s_{12} s_{71} \bar{X}_2). \tag{8.302}$$

Expanding the terms $\bar{X}_2$, $\bar{Y}_2$, and $\bar{Z}_2$ and recognizing that $s_{12}^2 + c_{12}^2 = 1$ gives

$$\bar{Z}_1 X_{21} - \bar{X}_1 Z_{21} = -s_{23}[(s_{12} s_2) s_{71} - (s_1 c_2 + c_1 s_2 c_{12}) c_{71}]. \tag{8.303}$$

The term in brackets on the right side of Eq. (8.303) is U^*_{217}. Thus, Eq. (8.303) may be written as

$$\bar{Z}_1 X_{21} - \bar{X}_1 Z_{21} = -s_{23} U^*_{217}. \tag{8.304}$$

The third term in parentheses in Eq. (8.296) may be expanded as

$$\begin{aligned} \bar{Z}_1 Y_{21} - \bar{Y}_1 Z_{21} = {} & [c_{71} c_{12} - s_{71} s_{12} c_1][c_{71}(\bar{X}_2 s_1 + \bar{Y}_2 c_1) - s_{71} \bar{Z}_2] \\ & + [s_{71} c_{12} + c_{71} s_{12} c_1][s_{71}(\bar{X}_2 s_1 + \bar{Y}_2 c_1) + c_{71} \bar{Z}_2]. \end{aligned} \tag{8.305}$$

Multiplying the terms together, regrouping, and recognizing that $s_{71}^2 + c_{71}^2 = 1$ yields

$$\bar{Z}_1 Y_{21} - \bar{Y}_1 Z_{21} = c_1(c_{12} \bar{Y}_2 + s_{12} \bar{Z}_2) + s_1(c_{12} \bar{X}_2). \tag{8.306}$$

Expanding the terms $\bar{X}_2$, $\bar{Y}_2$, and $\bar{Z}_2$ and recognizing that $s_{12}^2 + c_{12}^2 = 1$ gives

$$\bar{Z}_1 Y_{21} - \bar{Y}_1 Z_{21} = -s_{23}[c_1 c_2 - s_1 s_2 c_{12}]. \tag{8.307}$$

This equation may be rewritten as

$$\bar{Z}_1 Y_{21} - \bar{Y}_1 Z_{21} = -s_{23} W_{21}. \tag{8.308}$$

The only term in Eq. (8.296) that has not yet been expressed linearly in the sines and cosines of the joint angles is $Z_{21} Z_{5671}$. This term may be written as

$$Z_{21} Z_{5671} = \left[s_{71} X^*_{21} + c_{71} \bar{Z}_2\right] \left[s_{12} X^*_{5671} + c_{12} Z_{567}\right]. \tag{8.309}$$

Performing the multiplication gives

$$Z_{21} Z_{5671} = s_{12} s_{71} X^*_{5671} X^*_{21} + c_{12} s_{71} Z_{567} X^*_{21} + c_{71} \bar{Z}_2 Z_{5671}. \tag{8.310}$$

Expanding X^*_{21} in the first term gives

$$Z_{21} Z_{5671} = s_{12} s_{71} X^*_{5671} (\bar{X}_2 s_1 + \bar{Y}_2 c_1) + c_{12} s_{71} Z_{567} X^*_{21} + c_{71} \bar{Z}_2 Z_{5671}. \tag{8.311}$$

The first term of this equation is the only term that is not linear in the sines and cosines of the joint angles. This term will be modified by first expanding the fundamental cosine law, $Z_{56712} = c_{34}$, as follows:

$$s_{23}(X_{5671} s_2 + Y_{5671} c_2) + c_{23} Z_{5671} = c_{34}. \tag{8.312}$$

Expanding Y_{5671} and Z_{5671} gives

$$s_{23} s_2 X_{5671} + s_{23} c_2 \left(c_{12} X^*_{5671} - s_{12} Z_{567}\right) + c_{23} \left(s_{12} X^*_{5671} + c_{12} Z_{567}\right) = c_{34}. \tag{8.313}$$

Rearranging this equation yields

$$\bar{X}_2 X_{5671} - \bar{Y}_2 X^*_{5671} + \bar{Z}_2 Z_{567} = c_{34}. \tag{8.314}$$

Multiplying Eq. (8.314) by c_1 and rearranging gives

$$\bar{Y}_2 X^*_{5671} c_1 = \bar{X}_2 X_{5671} c_1 - (c_{34} - \bar{Z}_2 Z_{567}) c_1. \tag{8.315}$$

Equation (8.315) is now substituted into Eq. (8.311) to yield

$$\begin{aligned} Z_{21} Z_{5671} = {} & s_{12} s_{71} \left[X^*_{5671} \bar{X}_2 s_1 + \bar{X}_2 X_{5671} c_1 - (c_{34} - \bar{Z}_2 Z_{567}) c_1 \right] \\ & + c_{12} s_{71} Z_{567} X^*_{21} + c_{71} \bar{Z}_2 Z_{5671}. \end{aligned} \tag{8.316}$$

Substituting X^*_{5671} and X_{5671} into this equation, regrouping, and recognizing that $s_1^2 + c_1^2 = 1$ gives

$$Z_{21} Z_{5671} = s_{12} s_{71} [X_{567} \bar{X}_2 - (c_{34} - \bar{Z}_2 Z_{567}) c_1] + c_{12} s_{71} Z_{567} X^*_{21} + c_{71} \bar{Z}_2 Z_{5671}. \tag{8.317}$$

This term is linear in the sines and cosines of the joint angles. Equation (8.317) can now be regrouped as follows:

$$Z_{21} Z_{5671} = s_{71} \left[s_{12} X_{567} \bar{X}_2 + Z_{567} \left(c_{12} X^*_{21} + s_{12} c_1 \bar{Z}_2 \right) \right] + c_{71} \bar{Z}_2 Z_{5671} - s_{12} c_{34} s_{71} c_1. \tag{8.318}$$

The term in parentheses in this equation will be expanded and simplified. This term may be written as

$$c_{12} X^*_{21} + s_{12} c_1 \bar{Z}_2 = c_{12} (\bar{X}_2 s_1 + \bar{Y}_2 c_1) + s_{12} c_1 \bar{Z}_2. \tag{8.319}$$

Expanding $\bar{X}_2$, $\bar{Y}_2$, and $\bar{Z}_2$ and recognizing that $s_{12}^2 + c_{12}^2 = 1$ yields

$$c_{12} X^*_{21} + s_{12} c_1 \bar{Z}_2 = -s_{23} (c_1 c_2 - s_1 s_2 c_{12}). \tag{8.320}$$

Equation (8.320) may now be written as

$$c_{12} X^*_{21} + s_{12} c_1 \bar{Z}_2 = -s_{23} W_{21}. \tag{8.321}$$

Substituting Eq. (8.321) into (8.318) and expanding $\bar{X}_2$ gives

$$Z_{21} Z_{5671} = s_{23} s_{71} (X_{567} U_{21} - Z_{567} W_{21}) + c_{71} \bar{Z}_2 Z_{5671} - s_{12} c_{34} s_{71} c_1. \tag{8.322}$$

Substituting the results of Eqs. (8.300), (8.304), (8.308), and (8.322) into Eq. (8.296) gives the following expression for B_{56712}:

$$\begin{aligned} B_{56712} = {} & s_{67} X_{56} [s_{23} U_{217}] + c_{67} [s_{23} s_{71} (X_{567} U_{21} - Z_{567} W_{21}) + c_{71} \bar{Z}_2 Z_{5671} \\ & - s_{12} c_{34} s_{71} c_1] + s_{67} Z_{56} \left[s_7 \left(-s_{23} U^*_{217} \right) + c_7 (-s_{23} W_{21}) \right] + c_{34} s_{67} X^*_{17}. \end{aligned} \tag{8.323}$$

Now that the terms A_{56712} and B_{56712} have been expressed linearly in terms of the joint angle variables, Eqs. (8.269) and (8.323) may be substituted into Eq. (8.238) to yield an expression for the product $Z_{2176} Z_{1765}$. This product may then be substituted into Eq. (8.229) to yield the following complicated, but linear, expression for $C(a_{34})$:

$$\begin{aligned} C(a_{34}) = \{ & -s_{56}[s_{23}(X_5U_{2176} - Z_5W_{2176}) + c_{34}X^*_{176}] - c_{56}[s_{67}X_{56}(s_{23}U_{217}) \\ & + c_{67}(s_{23}s_{71}(X_{567}U_{21} - Z_{567}W_{21}) + c_{71}\bar{Z}_2Z_{5671} - s_{12}c_{34}s_{71}c_1) \\ & + s_{67}Z_{56}(s_7(-s_{23}U^*_{217}) + c_7(-s_{23}W_{21})) + c_{34}s_{67}X^*_{17}] + c_{23}c_{45}\}/s_{34}. \end{aligned} \tag{8.324}$$

The results of Eqs. (8.218) through (8.221) are now substituted into Eq. (8.216) to give the following result:

$$\begin{aligned} & S_1(s_{12}X_{671}) + S_3(-s_{23}X_{6712}) + S_4(-s_{45}X_{1765}) + S_6(s_{56}X_{176}) \\ & + S_7(-\bar{X}_1X^*_{67} - \bar{Y}_1X_{67}) + a_{12}(Y_{671}) + a_{23}(X^*_{6712}) \\ & + a_{45}(X^*_{1765}) + a_{56}(Y_{176}) + a_{67}(c_{56}Y_{17} - s_{56}c_6Z_{17}) \\ & + a_{71}(c_{12}Y_{67} - s_{12}c_1Z_{67}) + a_{34}[C(a_{34})] = 0, \end{aligned} \tag{8.325}$$

where $C(a_{34})$ is defined in Eq. (8.324).

This equation contains the constant mechanism parameters, the input angle, θ_7, the output angle, θ_1, and the extra angle, θ_6. This equation will be paired with the fundamental cosine law

$$Z_{56712} = c_{34}, \tag{8.326}$$

which also contains the output angle, θ_1, and the extra angle, θ_6, as its only unknowns.

Equation (8.326) will now be referred to as the first equation, and Eq. (8.325) will be now referred to as the second equation for this problem. Both equations will be regrouped into the following format:

$$c_6(A_ic_1 + B_is_1 + D_i) + s_6(E_ic_1 + F_is_1 + G_i) + (H_ic_1 + I_is_1 + J_i) = 0, \qquad i = 1, 2. \tag{8.327}$$

The coefficients of the two equations are listed as follows:

$$\begin{aligned} A_1 &= X_5(c_7\bar{X}_2 - c_{71}s_7\bar{Y}_2) + Y_5(-c_{67}s_7\bar{X}_2 + \bar{Y}_2(s_{67}s_{71} - c_{67}c_{71}c_7)), \\ B_1 &= X_5(-c_{71}s_7\bar{X}_2 - c_7\bar{Y}_2) + Y_5(c_{67}s_7\bar{Y}_2 + \bar{X}_2(s_{67}s_{71} - c_{67}c_{71}c_7)), \\ D_1 &= X_5(\bar{X}_7\bar{Z}_2) + Y_5(-\bar{Y}_7\bar{Z}_2), \\ E_1 &= X_5(-c_{67}s_7\bar{X}_2 + \bar{Y}_2(s_{67}s_{71} - c_{67}c_{71}c_7)) + Y_5(c_{71}s_7\bar{Y}_2 - c_7\bar{X}_2), \\ F_1 &= X_5(c_{67}s_7\bar{Y}_2 + \bar{X}_2(s_{67}s_{71} - c_{67}c_{71}c_7)) + Y_5(c_7\bar{Y}_2 + c_{71}s_7\bar{X}_2), \\ G_1 &= X_5(-\bar{Y}_7\bar{Z}_2) + Y_5(-\bar{X}_7\bar{Z}_2), \\ H_1 &= Z_5(X_7\bar{X}_2 - Y_7\bar{Y}_2), \\ I_1 &= Z_5(-X_7\bar{Y}_2 - Y_7\bar{X}_2), \\ J_1 &= Z_5(Z_7\bar{Z}_2) - c_{34}, \end{aligned} \tag{8.328}$$

$$A_2 = S_1(s_{12}s_{56}c_{67}s_7) + S_3s_{23}s_{56}(-c_2s_7c_{67} + c_{12}s_2(s_{67}s_{71} - c_{67}c_{71}c_7))$$

$$
\begin{aligned}
&+S_4s_{12}s_{45}(-c_5s_7c_{71}+s_5c_{56}(s_{67}s_{71}-c_{67}c_{71}c_7))+S_6(s_{12}s_{56}c_{71}s_7)\\
&+S_7(s_{12}s_{56}c_{67}c_{71}s_7)+a_{12}c_{12}s_{56}(s_{67}s_{71}-c_{67}c_{71}c_7)\\
&+a_{23}s_{56}(s_2s_7c_{67}+c_2c_{12}(s_{67}s_{71}-c_{67}c_{71}c_7))\\
&+a_{45}s_{12}(s_5s_7c_{71}+c_5c_{56}(s_{67}s_{71}-c_{67}c_{71}c_7))\\
&+a_{56}s_{12}c_{56}(s_{67}s_{71}-c_{67}c_{71}c_7)+a_{67}(-s_{12}s_{56}Y_7)\\
&+a_{71}(-s_{12}s_{56}\bar{Y}_7)+a_{34}\big[s_{12}c_{34}s_{56}(-s_{67}s_{71}+c_{67}c_{71}c_7)\\
&+X_5c_{56}\big(s_7c_{67}\big(-s_{12}\bar{Z}_2c_{71}^2+s_{23}s_{71}^2c_2\big)-c_{12}s_{23}s_{67}s_{71}s_2\big)\\
&+Z_5s_{23}s_{56}(c_2c_7-c_{12}c_{71}s_2s_7)+Y_5c_{56}\big(-s_{23}c_{67}s_{71}c_2\bar{Y}_7\\
&+s_{12}c_{67}c_{71}\bar{Z}_2(s_{67}s_{71}-c_{67}c_{71}c_7)+s_{67}^2s_{23}(c_2c_7-s_2s_7c_{12}c_{71})\big)\big]/s_{34},
\end{aligned}
$$

$$
\begin{aligned}
B_2={}&S_1s_{12}s_{56}(-s_{67}s_{71}+c_{67}c_{71}c_7)+S_3s_{23}s_{56}(c_2(s_{67}s_{71}-c_{67}c_{71}c_7)+c_{12}c_{67}s_2s_7)\\
&+S_4s_{12}s_{45}(-c_5c_7+c_{56}c_{67}s_5s_7)+S_6(s_{12}s_{56}c_7)+S_7(s_{12}s_{56}c_{67}c_7)\\
&+a_{12}(c_{12}s_{56}c_{67}s_7)+a_{23}s_{56}(s_2(-s_{67}s_{71}+c_{67}c_{71}c_7)+c_{12}c_{67}c_2s_7)\\
&+a_{45}s_{12}(s_5c_7+c_{56}c_{67}c_5s_7)+a_{56}(s_{12}c_{56}c_{67}s_7)+a_{67}(-s_{12}s_{56}s_{67}s_7)\\
&+a_{34}\big[-s_{12}c_{34}s_{56}c_{67}s_7+Z_5s_{23}s_{56}(-c_{71}c_2s_7-c_{12}s_2c_7)+X_5c_{56}\big(-s_{12}c_{67}c_{71}c_7\bar{Z}_2\\
&-c_{12}s_{23}c_{67}s_{71}^2s_2s_7-s_{23}s_{67}s_{71}c_2\big)+Y_5c_{56}\big(c_{12}s_{23}s_2\big(s_{71}c_{67}\bar{Y}_7-s_{67}^2c_7\big)\\
&+s_7c_{71}\big(s_{12}c_{67}^2\bar{Z}_2-s_{67}^2s_{23}c_2\big)\big)\big]/s_{34},
\end{aligned}
$$

$$
\begin{aligned}
D_2={}&S_3(-s_{12}s_{23}s_{56}s_2\bar{Y}_7)+S_4c_{12}s_{45}(-c_5s_7s_{71}+s_5c_{56}\bar{Y}_7)+S_6(c_{12}s_{56}s_{71}s_7)\\
&+S_7(c_{12}s_{56}c_{67}s_{71}s_7)+a_{12}(-s_{12}s_{56}\bar{Y}_7)+a_{23}(-s_{12}s_{56}c_2\bar{Y}_7)\\
&+a_{45}c_{12}(s_5\bar{X}_7+c_5c_{56}\bar{Y}_7)+a_{56}(c_{12}c_{56}\bar{Y}_7)+a_{67}(-c_{12}s_{56}\bar{Z}_7)\\
&+a_{71}c_{12}s_{56}(s_{67}s_{71}-c_{67}c_{71}c_7)+a_{34}\big[-c_{12}c_{34}s_{56}\bar{Y}_7+Z_5U_{21}s_{23}s_{56}s_{71}s_7\\
&+X_5c_{56}(s_{23}U_{21}\bar{Y}_7-c_{12}c_{67}c_{71}s_{71}s_7\bar{Z}_2)+Y_5c_{56}(U_{21}s_{23}s_{71}s_7\\
&+c_{12}c_{67}c_{71}\bar{Z}_2\bar{Y}_7)\big]/s_{34},
\end{aligned}
$$

$$
\begin{aligned}
E_2={}&S_1(s_{12}s_{56}c_7)+S_3s_{23}s_{56}(-c_2c_7+c_{12}c_{71}s_2s_7)+S_4s_{12}s_{45}(c_5(s_{67}s_{71}-c_{67}c_{71}c_7)\\
&+s_5c_{56}c_{71}s_7)+S_6s_{12}s_{56}(-s_{67}s_{71}+c_{67}c_{71}c_7)+S_7(s_{12}s_{56}c_{71}c_7)+a_{12}(c_{12}s_{56}c_{71}s_7)\\
&+a_{23}s_{56}(s_2c_7+c_{12}c_{71}c_2s_7)+a_{45}s_{12}(s_5(-s_{67}s_{71}+c_{67}c_{71}c_7)+c_5c_{56}c_{71}s_7)\\
&+a_{56}(s_{12}c_{56}c_{71}s_7)+a_{71}(-s_{12}s_{56}s_{71}s_7)+a_{34}\big[-s_{12}c_{34}s_{56}c_{71}s_7\\
&+Z_5s_{23}s_{56}(c_{12}s_2(s_{67}s_{71}-c_{67}c_{71}c_7)-c_{67}c_2s_7)\\
&+Y_5c_{56}\big(s_{23}s_{71}(c_{12}s_{67}s_2-c_{67}s_{71}s_7c_2)+s_{12}c_{67}c_{71}^2s_7\bar{Z}_2\big)\\
&+X_5c_{56}\big(c_{71}(\bar{Z}_2s_{12}c_{67}(s_{67}s_{71}-c_{67}c_{71}c_7)+s_{23}s_{67}(c_{67}s_{71}c_2-s_{67}c_{12}s_2s_7))\\
&+s_{23}c_2c_7\big(s_{71}^2c_{67}^2+s_{67}^2\big)\big)\big]/s_{34},
\end{aligned}
$$

$$
\begin{aligned}
F_2={}&S_1(-s_{12}s_{56}c_{71}s_7)+S_3s_{23}s_{56}(c_{12}s_2c_7+c_{71}c_2s_7)+S_4s_{12}s_{45}(c_{67}c_5s_7+c_{56}s_5c_7)\\
&+S_6(-s_{12}s_{56}c_{67}s_7)+S_7(-s_{12}s_{56}s_7)+a_{12}(c_{12}s_{56}c_7)+a_{23}s_{56}(c_{12}c_2c_7-c_{71}s_2s_7)\\
&+a_{45}s_{12}(c_{56}c_5c_7-c_{67}s_5s_7)+a_{56}(s_{12}c_{56}c_7)+a_{34}\big[-s_{12}c_{34}s_{56}c_7\\
&+Z_5s_{23}s_{56}(c_2(s_{67}s_{71}-c_{67}c_{71}c_7)+c_{12}c_{67}s_2s_7)\\
&+X_5c_{56}\big(s_{23}s_{67}(c_{12}s_2(-s_{67}c_7-c_{67}s_{71}c_{71})-s_{67}c_{71}c_2s_7\big)
\end{aligned}
$$

$$
\begin{aligned}
&- c_{12}s_{23}c_{67}^2s_{71}^2c_7s_2 + s_{12}c_{67}^2c_{71}s_7\bar{Z}_2) + Y_5c_{56}(s_{12}c_{67}c_{71}c_7\bar{Z}_2 \\
&+ s_{23}s_{71}(s_{67}c_2 + c_{12}c_{67}s_{71}s_2s_7))\big]/s_{34}, \\
G_2 &= S_3(-s_{12}s_{23}s_{56}s_{71}s_2s_7) + S_4c_{12}s_{45}(c_5\bar{Y}_7 + c_{56}s_5\bar{X}_7) + S_6(-c_{12}s_{56}\bar{Y}_7) \\
&+ S_7(c_{12}s_{56}s_{71}c_7) + a_{12}(-s_{12}s_{56}s_{71}s_7) + a_{23}(-s_{12}s_{56}s_{71}c_2s_7) \\
&+ a_{45}c_{12}(-s_5\bar{Y}_7 + c_{56}s_{71}c_5s_7) + a_{56}(c_{12}c_{56}s_{71}s_7) + a_{71}(c_{12}s_{56}c_{71}s_7) \\
&+ a_{34}\big[- c_{12}c_{34}s_{56}s_{71}s_7 - Z_5s_{23}s_{56}U_{21}\bar{Y}_7 + X_5c_{56}(s_{23}s_{71}s_7U_{21} + c_{12}c_{67}c_{71}\bar{Z}_2\bar{Y}_7) \\
&+ Y_5c_{56}(c_{12}c_{67}s_{71}c_{71}s_7\bar{Z}_2 - s_{23}U_{21}\bar{Y}_7)\big]/s_{34}, \\
H_2 &= S_1(s_{12}c_{56}s_{67}s_7) + S_3s_{23}c_{56}(-X_7c_2 + c_{12}s_2Y_7) + S_4(-s_{12}s_{45}s_{56}s_5Y_7) \\
&+ S_7(s_{12}c_{56}s_{67}c_{71}s_7) + a_{12}(c_{12}c_{56}Y_7) + a_{23}c_{56}(s_2X_7 + c_{12}c_2Y_7) \\
&+ a_{45}(-s_{12}s_{56}c_5Y_7) + a_{56}(-s_{12}s_{56}Y_7) + a_{67}s_{12}c_{56}(s_{67}s_{71} - c_{67}c_{71}c_7) \\
&+ a_{71}(-s_{12}c_{56}Z_7) + a_{34}\big[-s_{12}c_{34}c_{56}Y_7 + X_5s_{23}s_{56}(c_{12}s_2Y_7 - c_2X_7) \\
&+ Z_5c_{56}c_{67}(-s_{12}c_{71}\bar{Z}_2Y_7 + s_{23}(s_{71}c_2Z_7 + s_{67}(c_7c_2 - s_7s_2c_{12}c_{71})))\big]/s_{34}, \\
I_2 &= S_1(-s_{12}c_{56}Y_7) + S_3s_{23}c_{56}(s_2c_{12}X_7 + c_2Y_7) + S_4(-s_{12}s_{45}s_{56}s_{67}s_5s_7) \\
&+ S_7(s_{12}c_{56}s_{67}c_7) + a_{12}(c_{12}c_{56}s_{67}s_7) + a_{23}c_{56}(c_{12}c_2X_7 - s_2Y_7) \\
&+ a_{45}(-s_{12}s_{56}s_{67}c_5s_7) + a_{56}(-s_{12}s_{56}s_{67}s_7) + a_{67}(s_{12}c_{56}c_{67}s_7) \\
&+ a_{34}\big[-s_{12}c_{34}c_{56}s_{67}s_7 + X_5s_{23}s_{56}(c_2Y_7 + s_2c_{12}X_7) \\
&+ Z_5c_{56}c_{67}(s_2c_{12}s_{23}(-s_{71}Z_7 - s_{67}c_7) + s_{67}c_{71}(-s_{12}s_7\bar{Z}_2 - s_{23}s_7c_2))\big]/s_{34}, \\
J_2 &= S_3(-s_{12}s_{23}c_{56}s_2Z_7) + S_4(-c_{12}s_{45}s_{56}s_5Z_7) + S_7(c_{12}c_{56}s_{67}s_{71}s_7) + a_{12}(-s_{12}c_{56}Z_7) \\
&+ a_{23}(-s_{12}c_{56}c_2Z_7) + a_{45}(-c_{12}s_{56}c_5Z_7) + a_{56}(-c_{12}s_{56}Z_7) + a_{67}(c_{12}c_{56}\bar{Y}_7) \\
&+ a_{71}(c_{12}c_{56}Y_7) + a_{34}[-Z_5\bar{Z}_2Z_7c_{12}c_{56}c_{67}c_{71} - U_{21}X_5Z_7s_{23}s_{56} \\
&+ c_{12}c_{34}c_{56}s_{67}s_{71}c_7 + c_{23}c_{45}]/s_{34}.
\end{aligned}
\tag{8.329}
$$

As in the previous solutions, the tan-half-angle identities for the sine and cosine of θ_1 and θ_6 are now inserted into Eq. set (8.327). These two equations can then be regrouped into the form

$$
x_6^2\left[a_ix_1^2 + b_ix_1 + d_i\right] + x_6\left[e_ix_1^2 + f_ix_1 + g_i\right] + \left[h_ix_1^2 + i_ix_1 + j_i\right] = 0, \qquad i = 1, 2, \tag{8.330}
$$

where the coefficients of this equation are defined in Eq. set (8.11).

The two equations of the form of Eq. (8.330) are quadratic in the variables x_1 and x_6. These equations can be solved via Bezout's method (Section 8.2.2) to yield an eighth-degree polynomial in the variable x_1.

The corresponding value for the parameter x_6 can be found from either

$$
x_6 = \frac{-\begin{vmatrix} M_1 & N_1 \\ M_2 & N_2 \end{vmatrix}}{\begin{vmatrix} L_1 & N_1 \\ L_2 & N_2 \end{vmatrix}} \tag{8.331}
$$

or

$$x_6 = \frac{-\begin{vmatrix} L_1 & N_1 \\ L_2 & N_2 \end{vmatrix}}{\begin{vmatrix} L_1 & M_1 \\ L_2 & M_2 \end{vmatrix}}, \tag{8.332}$$

where L_1, M_1, and N_1 are defined in Eqs. (8.14) through (8.16) and Eq. set (8.11). Equations (8.331) and (8.332) can be derived in a manner similar to that in which Eqs. (8.34) and (8.35) were derived.

Two joint angles remain to be solved, that is, θ_3 and θ_4. A unique corresponding value of θ_3 can be obtained from the following fundamental sine and sine–cosine laws:

$$X_{56712} = s_{34}s_3, \tag{8.333}$$

$$Y_{56712} = s_{34}c_3. \tag{8.334}$$

Similarly, a unique corresponding value of θ_4 can then be obtained from the following fundamental sine and sine–cosine laws:

$$X_{21765} = s_{34}s_4, \tag{8.335}$$

$$Y_{21765} = s_{34}c_4. \tag{8.336}$$

The offset distances S_2 and S_5 are the remaining parameters to be determined. These two values will be found be projecting the vector loop equation for the mechanism onto two different directions. Projecting the vector loop equation onto the direction $\mathbf{a}_{45}$ and evaluating the scalar products using the sets of direction cosines provided in the appendix yields

$$\begin{aligned} &S_1X_{234} + a_{12}W_{234} + S_2X_{34} + a_{23}W_{34} + S_3X_4 + a_{34}c_4 \\ &\quad + a_{45} + a_{56}c_5 + S_6\bar{X}_5 + a_{67}W_{65} + S_7X_{65} + a_{71}W_{765} = 0. \end{aligned} \tag{8.337}$$

The offset distance S_2 is the only unknown in this equation.

The distance S_5 will be determined by projecting the vector loop equation onto the direction $\mathbf{a}_{12}$. Evaluating the scalar products by using the sets of direction cosines listed in the appendix gives

$$\begin{aligned} &a_{12} + a_{23}c_2 + S_3\bar{X}_2 + a_{34}W_{32} + S_4X_{32} + a_{45}W_{432} \\ &\quad + S_5X_{432} + a_{56}W_{671} + S_6X_{71} + a_{67}W_{71} + S_7X_1 + a_{71}c_1 = 0. \end{aligned} \tag{8.338}$$

This equation contains the variable S_5 as its only unknown.

At this point, the analysis of the RRPRRPR group 2 spatial mechanism is complete. Eight solution configurations were determined. Table 8.9 shows data that were used for a numerical example. The calculated values for the eight configurations are listed in Table 8.10. These eight configurations are shown in Figure 8.9.

Table 8.9. *RRPRRPR mechanism parameters.*

Link length, cm.	Twist angle, deg.	Joint offset, cm.	Joint angle, deg.
$a_{12} = 23$	$\alpha_{12} = 120$	$S_1 = 16$	θ_1 = variable
$a_{23} = 13$	$\alpha_{23} = 43$	S_2 = variable	$\theta_2 = 265$
$a_{34} = 40$	$\alpha_{34} = 75$	$S_3 = 65$	θ_3 = variable
$a_{45} = 17$	$\alpha_{45} = 120$	$S_4 = 38$	θ_4 = variable
$a_{56} = 20$	$\alpha_{56} = 92$	S_5 = variable	$\theta_5 = 307$
$a_{67} = 38$	$\alpha_{67} = 111$	$S_6 = 79$	θ_6 = variable
$a_{71} = 24$	$\alpha_{71} = 67$	$S_7 = 83$	$\theta_7 = 256$ (input)

Table 8.10. *Calculated configurations for the RRPRRPR spatial mechanism.*

	Solution							
	A	B	C	D	E	F	G	H
θ_1, deg.	−78.01	25.62	56.45	143.96	165.75	−164.32	57.78	−150.92
θ_3, deg.	−2.91	158.25	−109.73	−73.07	−39.09	65.70	142.42	76.72
θ_4, deg.	−23.85	−46.06	−133.56	129.29	114.33	149.71	0.19	164.22
θ_6, deg.	−53.67	155.97	55.51	12.40	36.28	149.22	−174.72	163.68
S_2, cm.	5.22	33.44	50.99	−51.36	−80.35	−66.98	−73.80	−86.20
S_5, cm.	138.83	36.60	6.50	,41.94	28.63	−47.96	−63.86	−35.09

8.7 Summary

The method of solution for group 2 mechanisms should at this point be apparent. Two equations are generated, each of which contains the input angle, the output angle, the constant mechanism parameters, and one extra joint angle. The two equations are solved simultaneously using Bezout's method to yield, in general, an eighth-degree polynomial in the tan-half-angle of the output variable.

One of the two equations will be either a projection of the vector loop equation or a secondary cosine law. The selection of this equation is guided by the fact that it may not contain either of the unknown joint offsets. Quite often, however, one term of this equation requires additional manipulation in order to be expressed in terms of the required variables (see, for example, Eqs. (8.187) and (8.222)).

Once the equation containing the link lengths and offsets is selected, the extra joint angle is identified. An appropriate cosine law can then be selected that contains the output angle and the extra angle as its only unknowns. The two equations are then factored into the format of Eq. (8.5), and the solution continues as per Bezout's method.

Examples of five-link, six-link, and seven-link group 2 mechanisms have been presented in detail in this chapter. Although a solution for every group 2 mechanism has not been presented, it is hoped that the reader has grasped the solution technique and will be able to solve any group 2 mechanism.

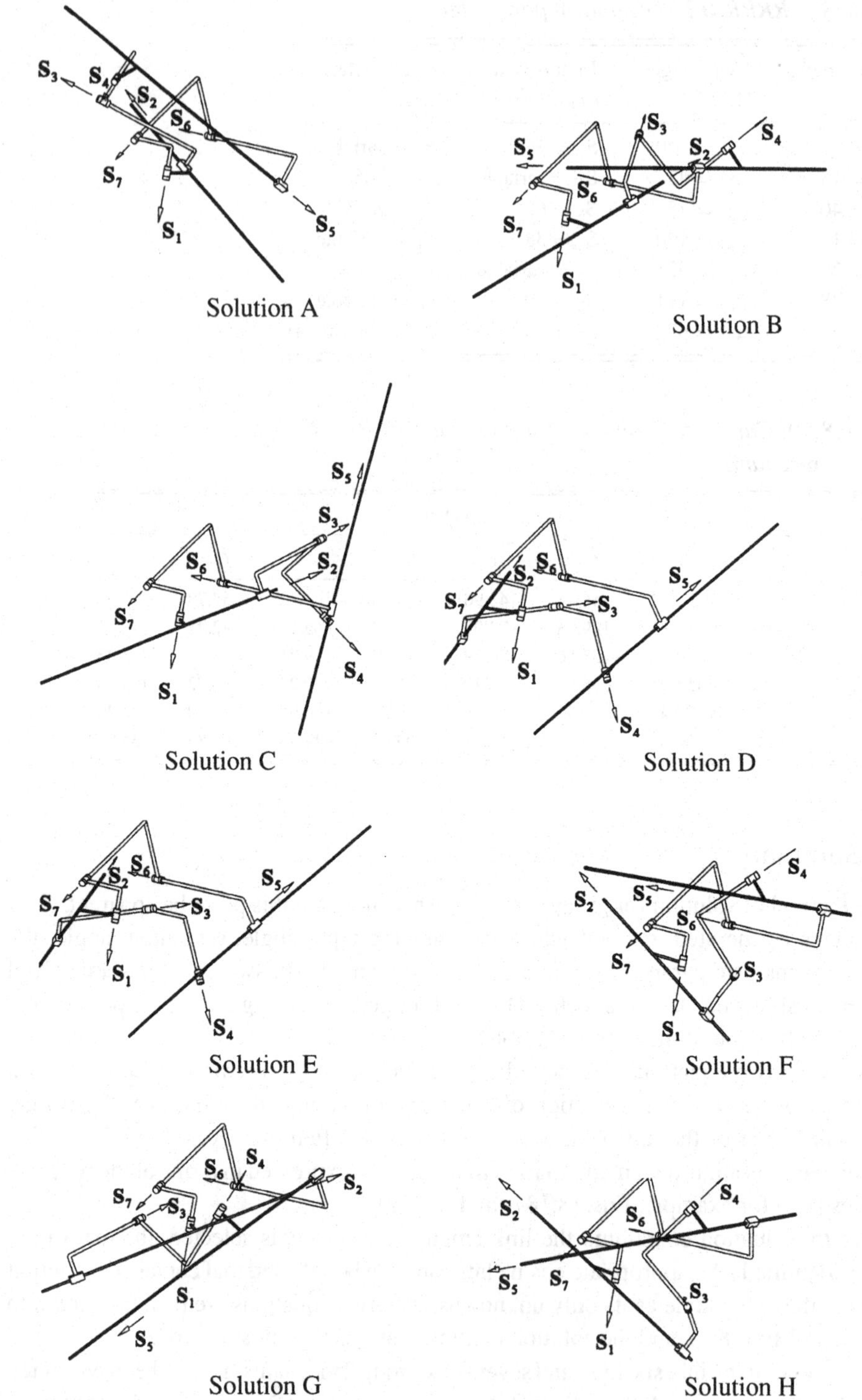

Figure 8.9. Eight configurations of an RRPRCR group 2 spatial mechanism.

8.8 Problems

1. Insert dual angles into the following spherical cosine laws and derive the corresponding secondary cosine laws:

   (a) Spherical pentagon: $Z_{43} = Z_1$

   (b) Spherical hexagon: $Z_{5432} = c_{61}$

   (c) Spherical hexagon: $Z_{12} = Z_{54}$

2. Shown in Figure 8.10 is a planar representation of a five-link RCCRR spatial mechanism. The input parameter is θ_5, and the output parameter is θ_1.

   (a) Assuming that all constant mechanism dimensions are known, what link lengths, offsets, twist angles, and joint angles are still unknown?

   (b) What group mechanism is this? Why?

   (c) Write a secondary cosine law that contains the input angle, the output angle, and only one additional unknown. Expand the secondary law.

   (d) Describe how you would use the equation in part (c) to solve for the output angle θ_1.

3. A spatial six-link RCRPRR mechanism is represented in Figure 8.11. The input parameter is θ_6, and the output parameter is θ_1. It is necessary to obtain an input/output equation for this mechanism. In order to do this it will be necessary to obtain two equations that each have θ_2 as an extra unknown parameter.

   (a) Write a spherical equation that contains the output angle and θ_2 as its only unknowns.

   (b) Write a secondary cosine law that contains the output angle and θ_2 as its only unknowns. Expand your equation as necessary to show that these are the only unknowns in the equation.

   (c) Describe how you would use the equations in parts (a) and (b) in order to solve for θ_1. How many values for θ_1 can be found for each given value of θ_6?

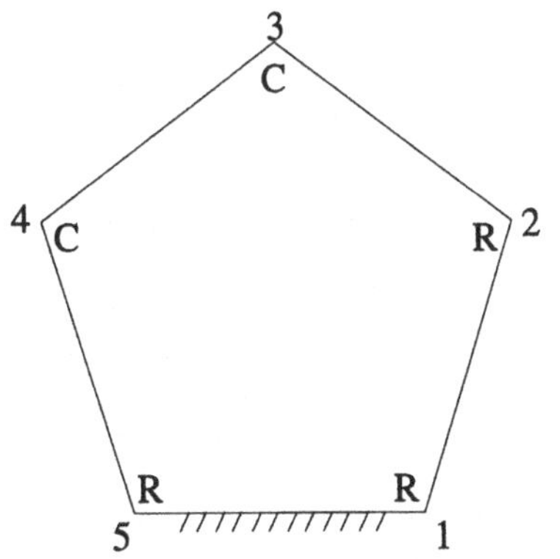

Figure 8.10. RCCRR spatial mechanism.

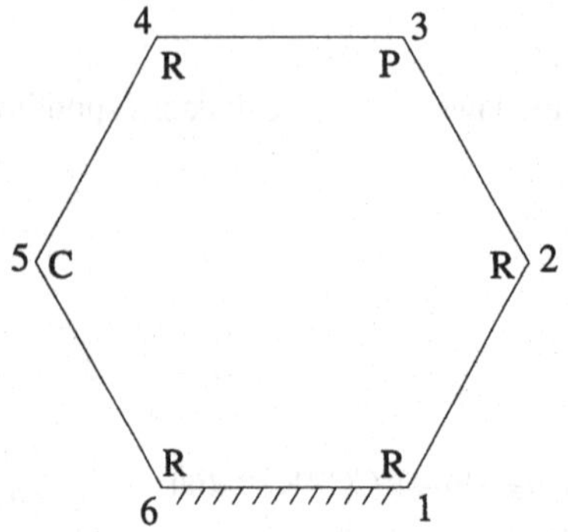

Figure 8.11. RCRPRR spatial mechanism.

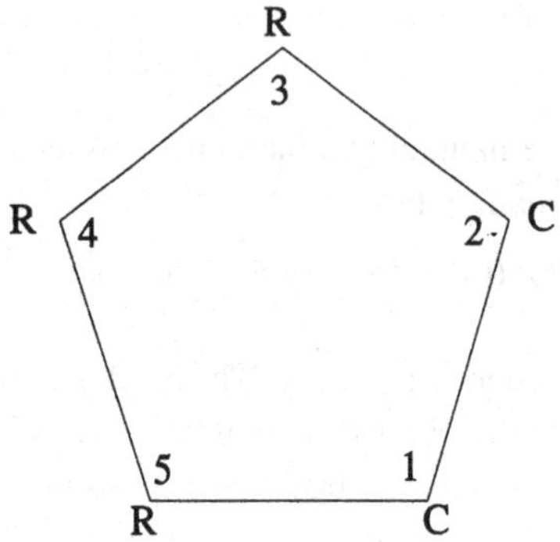

Figure 8.12. RRRCC spatial mechanism.

(d) Assuming that values for θ_1 are now known, describe how you would solve for corresponding values of θ_2.

(e) Describe how you would solve for the remaining unknown mechanism parameters.

4. Evaluate the function

$$f(x, y) = e^x(x^2y + 4x + 6)$$

when $x = 2 + 3\epsilon$ and $y = 1 - 5\epsilon$.

5. Completely expand a secondary cosine law for a spatial quadrilateral that will not contain the offsets S_2 or S_4.

6. A planar representation of a spatial closed-loop mechanism is shown in Figure 8.12.

(a) What group mechanism is this?

(b) Assuming that all constant mechanism parameters are known and that the angle θ_5 is given as an input angle, explain how to solve for the angle θ_4.

(c) Assuming that you have successfully solved for θ_4, explain how you would solve for the angle θ_1.

(d) Assuming that you have successfully solved for θ_4 and θ_1, explain how you would solve for the angle θ_3.

(e) Assuming that you have successfully solved for θ_4, θ_1, and θ_3, explain how you would solve for the angle θ_2.

(f) Assuming that you have successfully solved for θ_4, θ_1, θ_3, and θ_2, explain how you would solve for the slider displacements S_1 and S_2.

9

Group 3 spatial mechanisms

9.1 Introduction

Group 3 mechanisms, namely the six-link 5R-C and seven-link 6R-P mechanisms, all have equivalent spherical mechanisms with mobility three. This means that it is possible to select an appropriate spherical cosine law for the equivalent six-link, six-revolute spherical mechanisms that contains the input angle, the output angle, and two additional angular displacements. The spherical cosine law can be expressed in the form

$$\left(ax_m^2 + bx_m + d\right)x_n^2 + \left(ex_m^2 + fx_m + g\right)x_n + \left(hx_m^2 + ix_m + j\right) = 0, \tag{9.1}$$

where x_m and x_n are the tangents of the half angles (tan-half-angles) of the two additional angles θ_m and θ_n. The coefficients a through j are themselves quadratic in the tan-half-angle of the output angular displacement. It is becoming apparent that it is necessary to eliminate a pair of tan-half-angles in a single operation from a set of equations. This is much more difficult than the elimination problem encountered with group 2 mechanisms, which was the elimination of a single tan-half-angle from a pair of simultaneous equations.

At the outset it appears that it is necessary to form a further two or possibly three equations of the form of Eq. (9.1) and to attempt to apply Sylvester's dialytic method to eliminate x_m and x_n in a single operation. This procedure yields a polynomial in the output tan-half-angle that is of thirty-second degree (or higher). However, it will be shown here that the input–output polynomials for group 3 mechanisms are of sixteenth degree, and they can be derived by generating four simultaneous equations of the form

$$\left(a_i x_m^2 + b_i x_m + d_i\right)x_n + \left(e_i x_m^2 + f_i x_m + g_i\right) = 0, \qquad i = 1\ldots 4, \tag{9.2}$$

where the coefficients a_i through g_i are again quadratic in the tan-half-angle of the output angle. Multiplying these four equations by x_m will yield eight "linear" homogeneous equations in the eight "variables" $x_m^3x_n$, $x_m^2x_n$, x_mx_n, x_m^3, x_m^2, x_m, x_n, and 1. These equations will have a solution only if the equations are linearly dependent. The eight equations will be linearly dependent if the determinant of the coefficient matrix equals zero, and thus

$$\begin{vmatrix} 0 & 0 & a_i & b_i & d_i & e_i & f_i & g_i \\ a_i & e_i & b_i & d_i & 0 & f_i & g_i & 0 \end{vmatrix} = 0. \tag{9.3}$$

Expanding this 8×8 determinant will yield a sixteenth-degree input–output equation because the coefficients a_i through g_i are quadratic in the tan-half-angle of the output angle.

In the next section it will be shown how to generate tan-half-angle laws for spherical mechanisms that can be regrouped into the format shown in Eq. (9.2). After this is completed, the analysis of six-link and seven-link group 3 mechanisms will be presented.

9.2 Tan-half-angle laws

It was shown in Section 6.4 (see Eq. (6.55)) that for an n-sided spherical mechanism, the direction cosines of the vector along the nth joint axis measured in terms of the first coordinate system could be written as

$$ {}^{1}\underline{S}_{n} = \begin{bmatrix} X_{n-1,n-2,\ldots,2} \\ Y_{n-1,n-2,\ldots,2} \\ Z_{n-1,n-2,\ldots,2} \end{bmatrix}. \tag{9.4} $$

The direction of this vector was also calculated by rotating the vector directly from the nth coordinate system to the first coordinate system as

$$ {}^{1}\underline{S}_{n} = \begin{bmatrix} s_{n1}s_{1} \\ s_{n1}c_{1} \\ c_{n1} \end{bmatrix}. \tag{9.5} $$

Equating the components of Eqs. (9.4) and (9.5) resulted in the following fundamental sine, sine–cosine, and cosine laws for a spherical mechanism with n links:

$$ X_{n-1,n-2,\ldots,2} = s_{n1}s_{1}, \tag{9.6} $$

$$ Y_{n-1,n-2,\ldots,2} = s_{n1}c_{1}, \tag{9.7} $$

$$ Z_{n-1,n-2,\ldots,2} = c_{n1}. \tag{9.8} $$

Substituting the half-angle expressions $s_1 = \frac{2x_1}{1+x_1^2}$ and $c_1 = \frac{1-x_1^2}{1+x_1^2}$ where $x_1 = \tan(\frac{\theta_1}{2})$, into Eqs. (9.6) and (9.7) gives

$$ X_{n-1,n-2,\ldots,2} = s_{n1}\left(\frac{2x_1}{1+x_1^2}\right), \tag{9.9} $$

$$ Y_{n-1,n-2,\ldots,2} = s_{n1}\left(\frac{1-x_1^2}{1+x_1^2}\right). \tag{9.10} $$

These two equations may be written as

$$ (X_{n-1,n-2,\ldots,2})x_1^2 + (-2s_{n1})x_1 + (X_{n-1,n-2,\ldots,2}) = 0, \tag{9.11} $$

$$ (Y_{n-1,n-2,\ldots,2} + s_{n1})x_1^2 + (Y_{n-1,n-2,\ldots,2} - s_{n1}) = 0. \tag{9.12} $$

The necessary condition that two quadratic equations of the form $a_i x^2 + b_i x + d_i = 0$ $(i = 1, 2)$ have a common root (see Section 8.2.2) is

$$ \begin{vmatrix} a_1 & b_1 \\ a_2 & b_2 \end{vmatrix} \begin{vmatrix} b_1 & d_1 \\ b_2 & d_2 \end{vmatrix} - \begin{vmatrix} a_1 & d_1 \\ a_2 & d_2 \end{vmatrix}^2 = 0, \tag{9.13} $$

where the common root is evaluated as

$$x_1 = -\frac{\begin{vmatrix} a_1 & d_1 \\ a_2 & d_2 \end{vmatrix}}{\begin{vmatrix} a_1 & b_1 \\ a_2 & b_2 \end{vmatrix}} = -\frac{\begin{vmatrix} b_1 & d_1 \\ b_2 & d_2 \end{vmatrix}}{\begin{vmatrix} a_1 & d_1 \\ a_2 & d_2 \end{vmatrix}}. \quad (9.14)$$

Applying the condition of Eq. (9.13) to Eqs. (9.11) and (9.12) gives

$$X^2_{n-1,n-2,\ldots,2} + Y^2_{n-1,n-2,\ldots,2} = s^2_{n1}. \quad (9.15)$$

Replacing s^2_{n1} by $(1 - c^2_{n1})$ gives

$$X^2_{n-1,n-2,\ldots,2} + Y^2_{n-1,n-2,\ldots,2} + c^2_{n1} = 1. \quad (9.16)$$

Using a fundamental cosine law to substitute for c^2_{n1} yields

$$X^2_{n-1,n-2,\ldots,2} + Y^2_{n-1,n-2,\ldots,2} + Z^2_{n-1,n-2,\ldots,2} = 1. \quad (9.17)$$

This relationship was shown to be true in Section 6.4. Therefore, Eqs (9.11) and (9.12) always have a common root. The value of this common root can be obtained by applying Eq. (9.14), which yields a pair of alternative expressions,

$$x_1 = \frac{X_{n-1,n-2,\ldots,2}}{Y_{n-1,n-2,\ldots,2} + s_{n1}}, \quad x_1 = -\frac{Y_{n-1,n-2,\ldots,2} - s_{n1}}{X_{n-1,n-2,\ldots,2}}. \quad (9.18)$$

These equations represent two new relationships for an n-sided spherical mechanism. These will be called half-angle laws for a spherical n-gon.

Further sets of half-angle laws can be generated from the other fundamental sine and sine–cosine laws. Additionally, more half-angle laws can be generated from pairs of subsidiary sine and sine–cosine laws by following the same procedure. Sets of half-angle laws are presented in the appendix for the spherical quadrilateral through the spherical heptagon.

9.3 Six-link group 3 spatial mechanisms

All six-link group 3 mechanisms contain five revolute joints and one cylindric joint, that is, 5R-C. The only difference between the various six-link group 3 mechanisms is the selection of the frame or fixed link. One example mechanism will be presented in this section. The solution technique developed here is applicable to the various inversions.

Shown in Figure 9.1 is a planar representation of an RCRRRR spatial mechanism. Here, link a_{61} is attached to ground and all the constant mechanism parameters are known together with the input angle, θ_6. The objective is to determine corresponding values for the remaining unknown joint displacements and joint offsets. In particular, the problem statement is as follows:

given: $\alpha_{12}, \alpha_{23}, \alpha_{34}, \alpha_{45}, \alpha_{56}, \alpha_{61}$,
$a_{12}, a_{23}, a_{34}, a_{45}, a_{56}, a_{61}$,
S_1, S_2, S_3, S_4, S_6, and
θ_6 (input angle),

find: $\theta_1, \theta_2, \theta_3, \theta_4$, and θ_5, S_5.

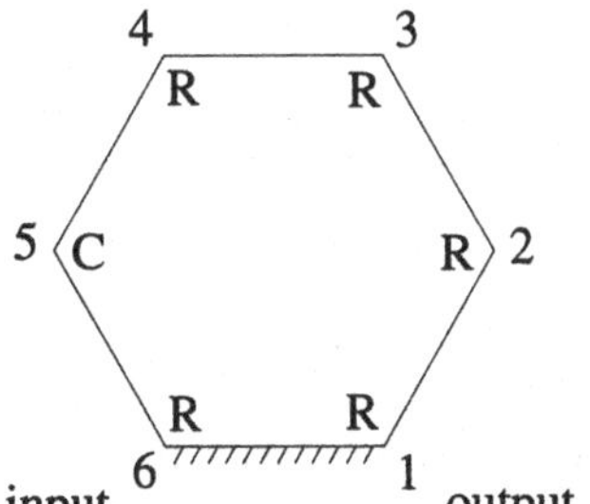

Figure 9.1. Planar representation of an RCRRRR spatial mechanism.

The angle θ_1 is identified as the output angle because it is connected to the frame, and it will be solved for first.

9.3.1 Development of input–output equation (solution for θ_1)

It will be shown that the input–output equation for this mechanism will be generated from four equations of the form of Eq. (9.2). The first pair of equations is derived from the following subsidiary tan-half-angle laws for a spherical hexagon:

$$\bar{X}_4 - X_{612} = (\bar{Y}_4 - Y_{612})x_3, \tag{9.19}$$

$$(\bar{X}_4 + X_{612})x_3 = -(\bar{Y}_4 + Y_{612}). \tag{9.20}$$

It can readily be shown by expanding $\bar{Y}_4$ and $\bar{Z}_4$ that

$$s_{34}\bar{Y}_4 = c_{34}\bar{Z}_4 - c_{45}. \tag{9.21}$$

Using a subsidiary cosine law to substitute for $\bar{Z}_4$ gives

$$s_{34}\bar{Y}_4 = c_{34}Z_{612} - c_{45}. \tag{9.22}$$

Multiplying Eqs. (9.19) and (9.20) by s_{34} and then using Eq. (9.22) to substitute for the quantity $s_{34}\bar{Y}_4$ gives

$$s_{34}(\bar{X}_4 - X_{612}) = (c_{34}Z_{612} - s_{34}Y_{612} - c_{45})x_3, \tag{9.23}$$

$$s_{34}(\bar{X}_4 + X_{612})x_3 = -(c_{34}Z_{612} + s_{34}Y_{612} - c_{45}). \tag{9.24}$$

Corresponding secondary tan-half-angle laws can now be generated by substituting dual angles into Eqs. (9.23) and (9.24). The expansion of the necessary partial derivatives has been demonstrated in Chapter 8 with the exception of the derivative $\frac{dx_3}{d\theta_3}$, where $x_3 = \tan\frac{\theta_3}{2}$. Hence,

$$\frac{dx_3}{d\theta_3} = \frac{1}{2}\left(1 + \tan^2\left(\frac{\theta_3}{2}\right)\right) = \frac{1 + x_3^2}{2}. \tag{9.25}$$

Using this result, the dual of Eq. (9.23) may be written as

$$a_{34}c_{34}(\bar{X}_4 - X_{612}) + s_{34}(\bar{X}_{04} - X_{0612}) = [a_{45}s_{45} + a_{34}(-s_{34}Z_{612} - c_{34}Y_{612}) + c_{34}Z_{0612} - s_{34}Y_{0612}]x_3 + S_3(c_{34}Z_{612} - s_{34}Y_{612} - c_{45})\frac{(1+x_3^2)}{2}. \quad (9.26)$$

It remains to reduce Eq. (9.26) to an equation that is linear in x_3.

Several of the terms of this equation may be substituted by expressions formed from Eqs. (9.23) and (9.24). Multiplying Eq. (9.23) by x_3 and then subtracting the result from Eq. (9.24) gives

$$2s_{34}X_{612}x_3 = -c_{34}Z_{612}(x_3^2+1) + s_{34}Y_{612}(x_3^2-1) + c_{45}(x_3^2+1), \quad (9.27)$$

which can be rearranged in the form

$$(c_{34}Z_{612} - s_{34}Y_{612} - c_{45})x_3^2 = -2s_{34}X_{612}x_3 + (-c_{34}Z_{612} - s_{34}Y_{612} + c_{45}). \quad (9.28)$$

After multiplying Eq. (9.26) throughout by s_{34} and substituting Eqs. (9.28) and (9.23) into Eq. (9.26) yields

$$a_{34}c_{34}(c_{34}Z_{612} - s_{34}Y_{612} - c_{45})x_3 + s_{34}^2(\bar{X}_{04} - X_{0612}) = [a_{45}s_{45} + a_{34}(-s_{34}Z_{612} - c_{34}Y_{612}) + c_{34}Z_{0612} - s_{34}Y_{0612}]s_{34}x_3 + \frac{1}{2}S_3s_{34}(c_{34}Z_{612} - s_{34}Y_{612} - c_{45}) + S_3s_{34}\left(-s_{34}X_{612}x_3 + \frac{1}{2}(-c_{34}Z_{612} - s_{34}Y_{612} + c_{45})\right). \quad (9.29)$$

This equation reduces to

$$a_{34}c_{34}(c_{34}Z_{612} - s_{34}Y_{612} - c_{45})x_3 + s_{34}^2(\bar{X}_{04} - X_{0612}) = [a_{45}s_{45} + a_{34}(-s_{34}Z_{612} - c_{34}Y_{612}) + c_{34}Z_{0612} - s_{34}Y_{0612}]s_{34}x_3 - S_3s_{34}^2Y_{612} - S_3s_{34}^2X_{612}x_3 \quad (9.30)$$

and is finally expressed in the form

$$[a_{34}(Z_{612} - c_{34}c_{45}) - a_{45}s_{34}s_{45} - s_{34}c_{34}Z_{0612} + s_{34}^2Y_{0612} + S_3s_{34}^2X_{612}]x_3 + [s_{34}^2(\bar{X}_{04} - X_{0612}) + S_3s_{34}^2Y_{612}] = 0. \quad (9.31)$$

Analogously, taking partial derivatives, the dual of Eq. (9.24) can be expressed as

$$a_{34}c_{34}(\bar{X}_4 + X_{612})x_3 + S_3s_{34}(\bar{X}_4 + X_{612})\frac{(1+x_3^2)}{2} + s_{34}(\bar{X}_{04} + X_{0612})x_3 = -a_{45}s_{45} + a_{34}(s_{34}Z_{612} - c_{34}Y_{612}) - (c_{34}Z_{0612} + s_{34}Y_{0612}), \quad (9.32)$$

which can be rearranged in the form

$$x_3(\bar{X}_4 + X_{612})\left(a_{34}c_{34} + S_3s_{34}\frac{x_3}{2}\right) + \frac{1}{2}S_3s_{34}(\bar{X}_4 + X_{612}) + s_{34}(\bar{X}_{04} + X_{0612})x_3 = -a_{45}s_{45} + a_{34}(s_{34}Z_{612} - c_{34}Y_{612}) - (c_{34}Z_{0612} + s_{34}Y_{0612}). \quad (9.33)$$

Multiplying by s_{34} and substituting Eq. (9.24) into Eq. (9.33) gives

$$(-c_{34}Z_{612} - s_{34}Y_{612} + c_{45})\left(a_{34}c_{34} + S_3s_{34}\frac{x_3}{2}\right) + \frac{1}{2}S_3s_{34}^2(\bar{X}_4 + X_{612}) + s_{34}^2(\bar{X}_{04} + X_{0612})x_3$$
$$= -a_{45}s_{45}s_{34} + a_{34}s_{34}(s_{34}Z_{612} - c_{34}Y_{612}) - s_{34}(c_{34}Z_{0612} + s_{34}Y_{0612}). \tag{9.34}$$

Equation (9.23) can be rearranged to solve for the product $s_{34}\bar{X}_4$, and

$$s_{34}\bar{X}_4 = (c_{34}Z_{612} - s_{34}Y_{612} - c_{45})x_3 + s_{34}X_{612}. \tag{9.35}$$

This expression may be substituted into Eq. (9.34) to yield

$$(-c_{34}Z_{612} - s_{34}Y_{612} + c_{45})\left(a_{34}c_{34} + S_3s_{34}\frac{x_3}{2}\right) + \frac{1}{2}S_3s_{34}^2X_{612}$$
$$+\frac{1}{2}S_3s_{34}[(c_{34}Z_{612} - s_{34}Y_{612} - c_{45})x_3 + s_{34}X_{612}] + s_{34}^2(\bar{X}_{04} + X_{0612})x_3$$
$$= -a_{45}s_{45}s_{34} + a_{34}s_{34}(s_{34}Z_{612} - c_{34}Y_{612}) - s_{34}(c_{34}Z_{0612} + s_{34}Y_{0612}). \tag{9.36}$$

Rearranging this equation gives

$$\left[-S_3s_{34}^2Y_{612} + s_{34}^2(\bar{X}_{04} + X_{0612})\right]x_3 + \left[a_{34}c_{34}(-c_{34}Z_{612} - s_{34}Y_{612} + c_{45}) + S_3s_{34}^2X_{612}\right.$$
$$\left. + a_{45}s_{34}s_{45} - a_{34}s_{34}(s_{34}Z_{612} - c_{34}Y_{612}) + s_{34}(c_{34}Z_{0612} + s_{34}Y_{0612})\right] = 0. \tag{9.37}$$

Equations (9.31) and (9.37) contain the input angle θ_6, the output angle θ_1, the extra angle θ_2, and the tan-half-angle of θ_3. However, each of these equations also contains the expression $\bar{X}_{04}$, where by definition

$$\bar{X}_{04} = a_{45}c_{45}s_4 + S_4s_{45}c_4. \tag{9.38}$$

This term clearly contains a further unwanted angle θ_4. However, both s_4 and c_4 can be expressed in terms of the angles θ_6, θ_1, and θ_2 using spherical equations. Firstly, multiplying Eq. (9.38) by $s_{34}^2s_{45}$ and substituting $\bar{X}_4 = s_{45}s_4$ gives

$$s_{34}^2s_{45}\bar{X}_{04} = a_{45}s_{34}^2c_{45}\bar{X}_4 + S_4s_{34}s_{45}(s_{34}s_{45}c_4). \tag{9.39}$$

Equation (9.23) may be solved for the product $s_{34}\bar{X}_4$, and the result then substituted into Eq. (9.39). This yields

$$s_{34}^2s_{45}\bar{X}_{04} = a_{45}s_{34}c_{45}[(c_{34}Z_{612} - s_{34}Y_{612} - c_{45})x_3 + s_{34}X_{612}] + S_4s_{34}s_{45}(s_{34}s_{45}c_4). \tag{9.40}$$

Expanding the left side of the subsidiary cosine law $\bar{Z}_4 = Z_{612}$ gives

$$c_{34}c_{45} - s_{34}s_{45}c_4 = Z_{612}. \tag{9.41}$$

This equation may be rearranged to solve for the expression $(s_{34}s_{45}c_4)$, which is then substituted into Eq. (9.40) and yields

$$s_{34}^2 s_{45} \bar{X}_{04} = a_{45} s_{34} c_{45}[(c_{34} Z_{612} - s_{34} Y_{612} - c_{45}) x_3 + s_{34} X_{612}] + S_4 s_{34} s_{45}(c_{34} c_{45} - Z_{612}). \tag{9.42}$$

Substituting Eq. (9.42) into s_{45} times Eq. (9.31) gives

$$\begin{aligned}&\left[a_{34}(Z_{612} - c_{34}c_{45}) - a_{45}s_{34}s_{45} - s_{34}c_{34}Z_{0612} + s_{34}^2 Y_{0612} + S_3 s_{34}^2 X_{612}\right] s_{45} x_3\\ &\quad + a_{45}s_{34}c_{45}[(c_{34}Z_{612} - s_{34}Y_{612} - c_{45})x_3 + s_{34}X_{612}] + S_4 s_{34}s_{45}(c_{34}c_{45} - Z_{612})\\ &\quad - s_{45}s_{34}^2 X_{0612} + S_3 s_{45} s_{34}^2 Y_{612} = 0.\end{aligned} \tag{9.43}$$

This equation may be rearranged as

$$\begin{aligned}&\left\{\left[a_{34}(Z_{612} - c_{34}c_{45}) - s_{34}c_{34}Z_{0612} + s_{34}^2 Y_{0612} + S_3 s_{34}^2 X_{612}\right] s_{45} - a_{45}s_{34}\right.\\ &\quad \left. + a_{45}s_{34}c_{45}(c_{34}Z_{612} - s_{34}Y_{612})\right\} x_3 + a_{45}c_{45}s_{34}^2 X_{612} + S_4 s_{34}s_{45}(c_{34}c_{45}\\ &\quad - Z_{612}) - s_{45}s_{34}^2 X_{0612} + S_3 s_{45}s_{34}^2 Y_{612} = 0.\end{aligned} \tag{9.44}$$

The term $x_3\bar{X}_{04}$ will now be eliminated linearly from Eq. (9.37). Multiplying Eq. (9.39) by x_3 gives

$$s_{34}^2 s_{45}\bar{X}_{04}x_3 = a_{45}s_{34}^2 c_{45}\bar{X}_4 x_3 + S_4 s_{34}s_{45}(s_{34}s_{45}c_4)x_3. \tag{9.45}$$

Equation (9.24) can be rearranged in the form

$$s_{34}\bar{X}_4 x_3 = -(c_{34}Z_{612} + s_{34}Y_{612} - c_{45}) - s_{34}X_{612}x_3. \tag{9.46}$$

Substituting Eq. (9.46) into Eq. (9.45) gives

$$s_{34}^2 s_{45}\bar{X}_{04}x_3 = -a_{45}s_{34}c_{45}(c_{34}Z_{612} + s_{34}Y_{612} - c_{45} + s_{34}X_{612}x_3) + S_4 s_{34}s_{45}(s_{34}s_{45}c_4)x_3. \tag{9.47}$$

Equation (9.47) may now be substituted into s_{45} times Eq. (9.37) to give

$$\begin{aligned}&\left[-S_3 s_{34}^2 Y_{612} + s_{34}^2 X_{0612}\right] s_{45}x_3 - a_{45}s_{34}c_{45}(c_{34}Z_{612} + s_{34}Y_{612} - c_{45} + s_{34}X_{612}x_3)\\ &\quad + S_4 s_{34}s_{45}(s_{34}s_{45}c_4)x_3 + s_{45}\left[a_{34}c_{34}(-c_{34}Z_{612} - s_{34}Y_{612} + c_{45}) + S_3 s_{34}^2 X_{612}\right.\\ &\quad \left. + a_{45}s_{34}s_{45} - a_{34}s_{34}(s_{34}Z_{612} - c_{34}Y_{612}) + s_{34}(c_{34}Z_{0612} + s_{34}Y_{0612})\right] = 0,\end{aligned} \tag{9.48}$$

which can be rearranged in the form

$$\begin{aligned}&\left\{\left[-S_3 s_{34}^2 Y_{612} + s_{34}^2 X_{0612}\right] s_{45} - a_{45}s_{34}^2 c_{45}X_{612}\right.\\ &\quad \left. + S_4 s_{34}s_{45}(s_{34}s_{45}c_4)\right\} x_3 - a_{45}s_{34}c_{45}(c_{34}Z_{612} + s_{34}Y_{612}) + a_{45}s_{34}\\ &\quad + s_{45}\left[a_{34}c_{34}(-c_{34}Z_{612} - s_{34}Y_{612} + c_{45}) + S_3 s_{34}^2 X_{612}\right.\\ &\quad \left. - a_{34}s_{34}(s_{34}Z_{612} - c_{34}Y_{612}) + s_{34}(c_{34}Z_{0612} + s_{34}Y_{0612})\right] = 0.\end{aligned} \tag{9.49}$$

The product $(s_{34}s_{45}c_4)$ can be eliminated using Eq. (9.41), and thus

$$\begin{aligned}&\{[-S_3s_{34}^2Y_{612}+s_{34}^2X_{0612}]s_{45}-a_{45}s_{34}^2c_{45}X_{612}+S_4s_{34}s_{45}(c_{34}c_{45}-Z_{612})\}x_3\\&\quad-a_{45}s_{34}c_{45}(c_{34}Z_{612}+s_{34}Y_{612})+a_{45}s_{34}+s_{45}[a_{34}c_{34}(-c_{34}Z_{612}-s_{34}Y_{612}+c_{45})\\&\quad+S_3s_{34}^2X_{612}-a_{34}s_{34}(s_{34}Z_{612}-c_{34}Y_{612})+s_{34}(c_{34}Z_{0612}+s_{34}Y_{0612})]=0.\end{aligned}\tag{9.50}$$

Simplifying Eq. (9.50) yields

$$\begin{aligned}&\{[-S_3s_{34}^2Y_{612}+s_{34}^2X_{0612}]s_{45}-a_{45}s_{34}^2c_{45}X_{612}+S_4s_{34}s_{45}(c_{34}c_{45}-Z_{612})\}x_3\\&\quad-a_{45}s_{34}c_{45}(c_{34}Z_{612}+s_{34}Y_{612})+a_{45}s_{34}+s_{45}[-a_{34}Z_{612}+a_{34}c_{34}c_{45}\\&\quad+S_3s_{34}^2X_{612}+s_{34}(c_{34}Z_{0612}+s_{34}Y_{0612})]=0.\end{aligned}\tag{9.51}$$

Equations (9.44) and (9.51) are a pair of equations that can be expressed in the format of Eq. (9.2). Two additional equations are now derived from the following secondary cosine law:

$$\bar{Z}_{04}=Z_{0612}.\tag{9.52}$$

Expanding the left side of Eq. (9.52) gives

$$S_4s_{34}\bar{X}_4+a_{34}\bar{Y}_4+a_{45}Y_4=Z_{0612}.\tag{9.53}$$

The term Y_4 may be expressed in terms of Z_4 as follows:

$$s_{45}Y_4=c_{45}Z_4-c_{34}.\tag{9.54}$$

A subsidiary cosine law can be used to substitute for Z_4 to yield

$$s_{45}Y_4=c_{45}Z_{612}-c_{34}.\tag{9.55}$$

Equations (9.55) and (9.22) may now be substituted into $s_{34}s_{45}$ times Eq. (9.53) to give

$$S_4s_{34}^2s_{45}\bar{X}_4+a_{34}s_{45}(c_{34}Z_{612}-c_{45})+a_{45}s_{34}(c_{45}Z_{612}-c_{34})=s_{34}s_{45}Z_{0612}.\tag{9.56}$$

Using Eq. (9.23) to substitute for the term $\bar{X}_4$ gives

$$\begin{aligned}&S_4s_{34}s_{45}[(c_{34}Z_{612}-s_{34}Y_{612}-c_{45})x_3+s_{34}X_{612}]+a_{34}s_{45}(c_{34}Z_{612}-c_{45})\\&\quad+a_{45}s_{34}(c_{45}Z_{612}-c_{34})=s_{34}s_{45}Z_{0612}.\end{aligned}\tag{9.57}$$

This equation can be rearranged as

$$\begin{aligned}&[S_4s_{34}s_{45}(c_{34}Z_{612}-s_{34}Y_{612}-c_{45})]x_3+[S_4s_{34}^2s_{45}X_{612}+a_{34}s_{45}(c_{34}Z_{612}-c_{45})\\&\quad+a_{45}s_{34}(c_{45}Z_{612}-c_{34})-s_{34}s_{45}Z_{0612}]=0.\end{aligned}\tag{9.58}$$

Equation (9.24) may be used to substitute for the term $s_{34}\bar{X}_4x_3$ in the product of x_3 times Eq. (9.56). The result can be written as

$$\begin{aligned}&S_4s_{34}s_{45}[-(c_{34}Z_{612}+s_{34}Y_{612}-c_{45})-s_{34}X_{612}x_3]+[a_{34}s_{45}(c_{34}Z_{612}-c_{45})\\&\quad+a_{45}s_{34}(c_{45}Z_{612}-c_{34})-s_{34}s_{45}Z_{0612}]x_3=0.\end{aligned}\tag{9.59}$$

This equation can be rearranged as

$$\begin{aligned}&\left[-S_4 s_{34}^2 s_{45} X_{612} + a_{34} s_{45}(c_{34} Z_{612} - c_{45}) + a_{45} s_{34}(c_{45} Z_{612} - c_{34}) - s_{34} s_{45} Z_{0612}\right] x_3\\&\quad + [-S_4 s_{34} s_{45}(c_{34} Z_{612} + s_{34} Y_{612} - c_{45})] = 0.\end{aligned} \tag{9.60}$$

Equations (9.58) and (9.60) can be expressed in the form of Eq. (9.2) and these together with (9.44) and (9.50) can be used to obtain a sixteenth degree input/output equation in terms of x_1, the tan-half-angle of the output angle. The four equations can be rewritten in the form

$$\begin{aligned}&\{a_{34} s_{45} Z_{612} - M_1 - M_5 c_{34} Z_{0612} + M_2 Y_{0612} + M_2 S_3 X_{612} + M_4 a_{45}(c_{34} Z_{612} - s_{34} Y_{612})\} x_3\\&\quad + M_3 a_{45} X_{612} + M_5 S_4(c_{34} c_{45} - Z_{612}) + M_2(-X_{0612} + S_3 Y_{612}) = 0,\end{aligned} \tag{9.61}$$

$$\begin{aligned}&\{-M_2 S_3 Y_{612} + M_2 X_{0612} - M_3 a_{45} X_{612} + M_5 S_4(c_{34} c_{45} - Z_{612})\} x_3\\&\quad - M_4 a_{45}(c_{34} Z_{612} + s_{34} Y_{612}) + M_1 + M_2 S_3 X_{612} - a_{34} s_{45} Z_{612}\\&\quad + M_6 Z_{0612} + M_2 Y_{0612} = 0,\end{aligned} \tag{9.62}$$

$$\begin{aligned}&\{M_5 S_4(c_{34} Z_{612} - s_{34} Y_{612} - c_{45})\} x_3 + M_2 S_4 X_{612} + M_7 Z_{612} - a_{34} s_{45} c_{45}\\&\quad - a_{45} s_{34} c_{34} - M_5 Z_{0612} = 0,\end{aligned} \tag{9.63}$$

$$\begin{aligned}&\{-M_2 S_4 X_{612} + M_7 Z_{612} - a_{34} s_{45} c_{45} - a_{45} s_{34} c_{34} - M_5 Z_{0612}\} x_3\\&\quad - M_5 S_4(c_{34} Z_{612} + s_{34} Y_{612} - c_{45}) = 0,\end{aligned} \tag{9.64}$$

where M_1 through M_7 are constants defined as

$$\begin{aligned}&M_1 = a_{34} c_{34} c_{45} s_{45} + a_{45} s_{34}, && M_5 = s_{34} s_{45},\\&M_2 = s_{34}^2 s_{45}, && M_6 = s_{34} c_{34} s_{45},\\&M_3 = s_{34}{}^2 c_{45}, && M_7 = a_{34} c_{34} s_{45} + a_{45} s_{34} c_{45},\\&M_4 = s_{34} c_{45}.\end{aligned} \tag{9.65}$$

Equations (9.61) through (9.64) must be expanded into the format of Eq. (9.2). As a first step the terms X_{0612}, Y_{0612}, and Z_{0612} are expanded as follows:

$$\begin{aligned}X_{0612} &= -S_2 X_{612}^* + X_{061} c_2 - Y_{061} s_2,\\Y_{0612} &= -a_{23} Z_{612} + S_2 c_{23} X_{612} + c_{23}(X_{061} s_2 + Y_{061} c_2) - s_{23} Z_{061},\\Z_{0612} &= a_{23} Y_{612} + S_2 s_{23} X_{612} + s_{23}(X_{061} s_2 + Y_{061} c_2) + c_{23} Z_{061},\end{aligned} \tag{9.66}$$

where

$$\begin{aligned}X_{061} &= -S_1 X_{61}^* + X_{06} c_1 - Y_{06} s_1,\\Y_{061} &= -a_{12} Z_{61} + S_1 c_{12} X_{61} + c_{12}(X_{06} s_1 + Y_{06} c_1) - s_{12} Z_{06},\\Z_{061} &= a_{12} Y_{61} + S_1 s_{12} X_{61} + s_{12}(X_{06} s_1 + Y_{06} c_1) + c_{12} Z_{06},\end{aligned} \tag{9.67}$$

and where

$$\begin{aligned}X_{06} &= a_{56} c_{56} s_6 + S_6 s_{56} c_6,\\Y_{06} &= a_{56}(s_{61} s_{56} - c_{61} c_{56} c_6) - a_{61} Z_6 + S_6 c_{61} s_{56} s_6,\\Z_{06} &= a_{56} \bar{Y}_6 + a_{61} Y_6 + S_6 s_{61} s_{56} s_6.\end{aligned} \tag{9.68}$$

Equations (9.61) through (9.64) may now be regrouped into the form

$$(A_i c_2 + B_i s_2 + D_i)x_3 + (E_i c_2 + F_i s_2 + G_i) = 0, \qquad i = 1 \ldots 4, \tag{9.69}$$

where

$$\begin{aligned} A_i &= A_{i,1}\, c_1 + A_{i,2}\, s_1 + A_{i,3}, \\ &\vdots \\ G_i &= G_{i,1}\, c_1 + G_{i,2}\, s_1 + G_{i,3}. \end{aligned} \tag{9.70}$$

The coefficients $A_{1,1}$ through $G_{4,3}$ are now defined as

$$\begin{aligned}
A_{1,1} = {} & [M_5 c_{34}(a_{12}s_{12}s_{23} - a_{23}c_{12}c_{23}) + M_4 a_{45} c_{12}(s_{23}c_{34} - c_{23}s_{34}) \\
& - M_2(a_{12}s_{12}c_{23} + a_{23}c_{12}s_{23}) + a_{34}c_{12}s_{23}s_{45}]Y_6 + [M_2(S_3 + S_2 c_{23} + S_1 c_{12}c_{23}) \\
& - M_5 s_{23}c_{34}(S_1 c_{12} + S_2)]X_6 + c_{12}[M_2 c_{23} - M_5 s_{23}c_{34}]Y_{06}, \\
A_{1,2} = {} & [M_5 c_{34}(a_{12}s_{12}s_{23} - a_{23}c_{12}c_{23}) + M_4 a_{45} c_{12}(s_{23}c_{34} - c_{23}s_{34}) \\
& - M_2(a_{12}s_{12}c_{23} + a_{23}c_{12}s_{23}) + a_{34}c_{12}s_{23}s_{45}]X_6 - [M_2(S_3 + S_2 c_{23} + S_1 c_{12}c_{23}) \\
& - M_5 s_{23}c_{34}(S_1 c_{12} + S_2)]Y_6 + c_{12}[M_2 c_{23} - M_5 s_{23}c_{34}]X_{06}, \\
A_{1,3} = {} & [M_2(a_{23}s_{12}s_{23} - a_{12}c_{12}c_{23}) + M_4 a_{45} s_{12}(c_{23}s_{34} - s_{23}c_{34}) \\
& + M_5 c_{34}(a_{12}c_{12}s_{23} + a_{23}s_{12}c_{23}) - a_{34}s_{12}s_{23}s_{45}]Z_6 + s_{12}[M_5 s_{23}c_{34} - M_2 c_{23}]Z_{06}, \\
B_{1,1} = {} & [M_4 a_{45}(s_{23}c_{34} - c_{23}s_{34}) - M_2 a_{23}s_{23} - M_5 a_{23}c_{23}c_{34} + a_{34}s_{23}s_{45}]X_6 \\
& - [M_2(S_1 c_{23} + S_3 c_{12} + S_2 c_{12}c_{23}) - M_5 c_{34}(S_1 s_{23} + S_2 c_{12}s_{23})]Y_6 \\
& + (M_2 c_{23} - M_5 s_{23}c_{34})X_{06}, \\
B_{1,2} = {} & -[M_4 a_{45}(s_{23}c_{34} - c_{23}s_{34}) - M_2 a_{23}s_{23} - M_5 a_{23}c_{23}c_{34} + a_{34}s_{23}s_{45}]Y_6 \\
& - [M_2(S_1 c_{23} + S_3 c_{12} + S_2 c_{12}c_{23}) - M_5 c_{34}(S_1 s_{23} + S_2 c_{12}s_{23})]X_6 \\
& + (M_5 s_{23}c_{34} - M_2 c_{23})Y_{06}, \\
B_{1,3} = {} & s_{12}[M_2(S_3 + S_2 c_{23}) - M_5 S_2 s_{23}c_{34}]Z_6, \\
D_{1,1} = {} & [M_5 c_{34}(a_{23}s_{12}s_{23} - a_{12}c_{12}c_{23}) + M_4 a_{45} s_{12}(s_{23}s_{34} + c_{23}c_{34}) \\
& - M_2(a_{12}c_{12}s_{23} + a_{23}s_{12}c_{23}) + a_{34}s_{12}c_{23}s_{45}]Y_6 - s_{12}[M_2 S_1 s_{23} + M_5 S_1 c_{23}c_{34}]X_6 \\
& - s_{12}[M_2 s_{23} + M_5 c_{23}c_{34}]Y_{06}, \\
D_{1,2} = {} & [M_5 c_{34}(a_{23}s_{12}s_{23} - a_{12}c_{12}c_{23}) + M_4 a_{45} s_{12}(s_{23}s_{34} + c_{23}c_{34}) \\
& - M_2(a_{12}c_{12}s_{23} + a_{23}s_{12}c_{23}) + a_{34}s_{12}c_{23}s_{45}]X_6 + s_{12}[M_2 S_1 s_{23} + M_5 S_1 c_{23}c_{34}]Y_6 \\
& - s_{12}[M_2 s_{23} + M_5 c_{23}c_{34}]X_{06}, \\
D_{1,3} = {} & [M_2(a_{12}s_{12}s_{23} - a_{23}c_{12}c_{23}) + M_4 a_{45} c_{12}(s_{23}s_{34} + c_{23}c_{34}) \\
& + M_5 c_{34}(a_{12}s_{12}c_{23} + a_{23}c_{12}s_{23}) + a_{34}c_{12}c_{23}s_{45}]Z_6 \\
& - c_{12}(M_2 s_{23} + M_5 c_{23}c_{34})Z_{06} - M_1, \\
E_{1,1} = {} & [M_2(S_1 + S_2 c_{12} + S_3 c_{12}c_{23}) - M_5 S_4 c_{12}s_{23}]Y_6 + M_3 a_{45}X_6 - M_2 X_{06}, \\
E_{1,2} = {} & [M_2(S_1 + S_2 c_{12} + S_3 c_{12}c_{23}) - M_5 S_4 c_{12}s_{23}]X_6 - M_3 a_{45}Y_6 + M_2 Y_{06},
\end{aligned}$$

$$
\begin{aligned}
E_{1,3} &= s_{12}[M_5S_4s_{23} - M_2(S_2 + S_3c_{23})]Z_6,\\
F_{1,1} &= [M_2(S_2 + S_1c_{12} + S_3c_{23}) - M_5S_4s_{23}]X_6 - [M_2a_{12}s_{12} + M_3a_{45}c_{12}]Y_6\\
&\quad + M_2c_{12}Y_{06},\\
F_{1,2} &= -[M_2(S_2 + S_1c_{12} + S_3c_{23}) - M_5S_4s_{23}]Y_6 - [M_2a_{12}s_{12} + M_3a_{45}c_{12}]X_6\\
&\quad + M_2c_{12}X_{06},\\
F_{1,3} &= [M_3a_{45}s_{12} - M_2a_{12}c_{12}]Z_6 - M_2s_{12}Z_{06},\\
G_{1,1} &= -s_{12}[M_2S_3s_{23} + M_5S_4c_{23}]Y_6,\\
G_{1,2} &= -s_{12}[M_2S_3s_{23} + M_5S_4c_{23}]X_6,\\
G_{1,3} &= -c_{12}[M_2S_3s_{23} + M_5S_4c_{23}]Z_6 + M_5S_4c_{34}c_{45}, \qquad (9.71)\\
A_{2,1} &= [-M_2(S_1 + S_2c_{12} + S_3c_{12}c_{23}) - M_5S_{4c12}s_{23}]Y_6 - M_3a_{45}X_6 + M_2X_{06},\\
A_{2,2} &= [-M_2(S_1 + S_2c_{12} + S_3c_{12}c_{23}) - c_{12}s_{23}]X_6 + M_3a_{45}Y_6 - M_2Y_{06},\\
A_{2,3} &= s_{12}[M_2(S_2 + S_3c_{23}) + M_5S_4s_{23}]Z_6,\\
B_{2,1} &= -[M_2(S_2 + S_1c_{12} + S_3c_{23}) + M_5S_4s_{23}]X_6 + [M_2a_{12}s_{12} + M_3a_{45}c_{12}]Y_6\\
&\quad - M_2c_{12}Y_{06},\\
B_{2,2} &= [M_2a_{12}s_{12} + M_3a_{45}c_{12}]X_6 + [M_2(S_2 + S_1c_{12} + S_3c_{23}) + M_5S_4s_{23}]Y_6\\
&\quad - M_2c_{12}X_{06},\\
B_{2,3} &= [M_2a_{12}c_{12} - M_3a_{45}s_{12}]Z_6 + M_2s_{12}Z_{06},\\
D_{2,1} &= s_{12}[M_2S_3s_{23} - M_5S_4c_{23}]Y_6,\\
D_{2,2} &= s_{12}[M_2S_3s_{23} - M_5S_4c_{23}]X_6,\\
D_{2,3} &= c_{12}[M_2S_3s_{23} - M_5S_4c_{23}]Z_6 + M_5S_4c_{34}c_{45},\\
E_{2,1} &= [M_6(a_{23}c_{12}c_{23} - a_{12}s_{12}s_{23}) - M_2(a_{12}s_{12}c_{23} + a_{23}c_{12}s_{23})\\
&\quad - M_4a_{45}c_{12}(s_{23}c_{34} + c_{23}s_{34}) - a_{34}c_{12}s_{23}s_{45}]Y_6\\
&\quad + [M_2(S_3 + S_2c_{23} + S_1c_{12}c_{23}) + M_6s_{23}(S_2 + S_1c_{12})]X_6\\
&\quad + c_{12}[M_6s_{23} + M_2c_{23}]Y_{06},\\
E_{2,2} &= [M_6(a_{23}c_{12}c_{23} - a_{12}s_{12}s_{23}) - M_2(a_{12}s_{12}c_{23} + a_{23}c_{12}s_{23})\\
&\quad - M_4a_{45}c_{12}(s_{23}c_{34} + c_{23}s_{34}) - a_{34}c_{12}s_{23}s_{45}]X_6\\
&\quad - [M_2(S_3 + S_2c_{23} + S_1c_{12}c_{23}) + M_6s_{23}(S_2 + S_1c_{12})]Y_6\\
&\quad + c_{12}[M_6s_{23} + M_2c_{23}]X_{06},\\
E_{2,3} &= [M_2(a_{23}s_{12}s_{23} - a_{12}c_{12}c_{23}) + M_4a_{45}s_{12}(s_{23}c_{34} + c_{23}s_{34})\\
&\quad - M_6(a_{12}c_{12}s_{23} + a_{23}s_{12}c_{23}) + a_{34}s_{12}s_{23}s_{45}]Z_6 - s_{12}[M_2c_{23} + M_6s_{23}]Z_{06},\\
F_{2,1} &= [M_6a_{23}c_{23} - M_4a_{45}(s_{23}c_{34} + c_{23}s_{34}) - M_2a_{23}s_{23} - a_{34}s_{23}s_{45}]X_6\\
&\quad - [M_2(S_1c_{23} + S_2c_{12}c_{23} + S_3c_{12}) + M_6s_{23}(S_1 + S_2c_{12})]Y_6\\
&\quad + [M_2c_{23} + M_6s_{23}]X_{06},\\
F_{2,2} &= -[M_6a_{23}c_{23} - M_4a_{45}(s_{23}c_{34} + c_{23}s_{34}) - M_2a_{23}s_{23} - a_{34}s_{23}s_{45}]Y_6\\
&\quad - [M_2(S_1c_{23} + S_2c_{12}c_{23} + S_3c_{12}) + M_6s_{23}(S_1 + S_2c_{12})]X_6\\
&\quad - [M_2c_{23} + M_6s_{23}]Y_{06},
\end{aligned}
$$

$$
\begin{aligned}
F_{2,3} &= s_{12}[M_2(S_3 + S_2c_{23}) + M_6S_2s_{23}]Z_6,\\
G_{2,1} &= [M_6(a_{12}c_{12}c_{23} - a_{23}s_{12}s_{23}) + M_4a_{45}s_{12}(s_{23}s_{34} - c_{23}c_{34})\\
&\quad - M_2(a_{12}c_{12}s_{23} + a_{23}s_{12}c_{23}) - a_{34}s_{12}c_{23}s_{45}]Y_6\\
&\quad + s_{12}[M_6S_1c_{23} - M_2S_1s_{23}]X_6 + s_{12}[M_6c_{23} - M_2s_{23}]Y_{06},\\
G_{2,2} &= [M_6(a_{12}c_{12}c_{23} - a_{23}s_{12}s_{23}) + M_4a_{45}s_{12}(s_{23}s_{34} - c_{23}c_{34})\\
&\quad - M_2(a_{12}c_{12}s_{23} + a_{23}s_{12}c_{23}) - a_{34}s_{12}c_{23}s_{45}]X_6\\
&\quad - s_{12}[M_6S_1c_{23} - M_2S_1s_{23}]Y_6 + s_{12}[M_6c_{23} - M_2s_{23}]X_{06},\\
G_{2,3} &= [M_2(a_{12}s_{12}s_{23} - a_{23}c_{12}c_{23}) + M_4a_{45}c_{12}(s_{23}s_{34} - c_{23}c_{34})\\
&\quad - M_6(a_{12}s_{12}c_{23} + a_{23}c_{12}s_{23}) - a_{34}c_{12}c_{23}s_{45}]Z_6 + c_{12}[M_6c_{23} - M_2s_{23}]Z_{06} + M_1,
\end{aligned}
$$

$$
\begin{aligned}
A_{3,1} &= M_5S_4c_{12}Y_6(s_{23}c_{34} - c_{23}s_{34}), \qquad (9.72)\\
A_{3,2} &= M_5S_4c_{12}X_6(s_{23}c_{34} - c_{23}s_{34}),\\
A_{3,3} &= -M_5S_4s_{12}Z_6(s_{23}c_{34} - c_{23}s_{34}),\\
B_{3,1} &= M_5S_4X_6(s_{23}c_{34} - c_{23}s_{34}),\\
B_{3,2} &= -M_5S_4Y_6(s_{23}c_{34} - c_{23}s_{34}),\\
B_{3,3} &= 0,\\
D_{3,1} &= M_5S_4s_{12}Y_6(s_{23}s_{34} + c_{23}c_{34}),\\
D_{3,2} &= M_5S_4s_{12}X_6(s_{23}s_{34} + c_{23}c_{34}),\\
D_{3,3} &= M_5S_4c_{12}Z_6(s_{23}s_{34} + c_{23}c_{34}) - M_5S_4c_{45},\\
E_{3,1} &= [M_5(a_{12}s_{12}s_{23} - a_{23}c_{12}c_{23}) + M_7c_{12}s_{23}]Y_6 + [M_2S_4 - M_5s_{23}(S_2 + S_1c_{12})]X_6\\
&\quad - M_5c_{12}s_{23}Y_{06},\\
E_{3,2} &= -[M_2S_4 - M_5s_{23}(S_2 + S_1c_{12})]Y_6 + [M_5(a_{12}s_{12}s_{23} - a_{23}c_{12}c_{23}) + M_7c_{12}s_{23}]X_6\\
&\quad - M_5c_{12}s_{23}X_{06},\\
E_{3,3} &= [M_5(a_{12}c_{12}s_{23} + a_{23}s_{12}c_{23}) - M_7s_{12}s_{23}]Z_6 + M_5s_{12}s_{23}Z_{06},\\
F_{3,1} &= [M_5s_{23}(S_1 + S_2c_{12}) - M_2S_4c_{12}]Y_6 + [M_7s_{23} - M_5a_{23}c_{23}]X_6 - M_5s_{23}X_{06},\\
F_{3,2} &= -[M_7s_{23} - M_5a_{23}c_{23}]Y_6 + [M_5s_{23}(S_1 + S_2c_{12}) - M_2S_4c_{12}]X_6 + M_5s_{23}Y_{06},\\
F_{3,3} &= [M_2S_4s_{12} - M_5S_2s_{12}s_{23}]Z_6,\\
G_{3,1} &= [M_5(a_{23}s_{12}s_{23} - a_{12}c_{12}c_{23}) + M_7s_{12}c_{23}]Y_6 - M_5S_1s_{12}c_{23}X_6 - M_5s_{12}c_{23}Y_{06},\\
G_{3,2} &= [M_5(a_{23}s_{12}s_{23} - a_{12}c_{12}c_{23}) + M_7s_{12}c_{23}]X_6 + M_5S_1s_{12}c_{23}Y_6 - M_5s_{12}c_{23}X_{06},\\
G_{3,3} &= [M_5(a_{23}c_{12}s_{23} + a_{12}s_{12}c_{23}) + M_7c_{12}c_{23}]Z_6 - a_{34}s_{45}c_{45} - a_{45}s_{34}c_{34}\\
&\quad - M_5c_{12}c_{23}Z_{06}, \qquad (9.73)
\end{aligned}
$$

$$
\begin{aligned}
A_{4,1} &= -[M_5s_{23}(S_2 + S_1c_{12}) + M_2S_4]X_6 + [M_5(a_{12}s_{12}s_{23} - a_{23}c_{12}c_{23}) + M_7c_{12}s_{23}]Y_6\\
&\quad - M_5c_{12}s_{23}Y_{06},\\
A_{4,2} &= [M_5(a_{12}s_{12}s_{23} - a_{23}c_{12}c_{23}) + M_7c_{12}s_{23}]X_6 + [M_5s_{23}(S_2 + S_1c_{12}) + M_2S_4]Y_6\\
&\quad - M_5c_{12}s_{23}X_{06},
\end{aligned}
$$

$$A_{4,3} = [-M_7 s_{12} s_{23} + M_5(a_{12} c_{12} s_{23} + a_{23} s_{12} c_{23})]Z_6 + M_5 s_{12} s_{23} Z_{06},$$
$$B_{4,1} = [M_5 s_{23}(S_1 + S_2 c_{12}) + M_2 S_4 c_{12}]Y_6 + [M_7 s_{23} - M_5 a_{23} c_{23}]X_6 - M_5 s_{23} X_{06},$$
$$B_{4,2} = -[M_7 s_{23} - M_5 a_{23} c_{23}]Y_6 + [M_5 s_{23}(S_1 + S_2 c_{12}) + M_2 S_4 c_{12}]X_6 + M_5 s_{23} Y_{06},$$
$$B_{4,3} = -[M_2 S_4 s_{12} + M_5 S_2 s_{12} s_{23}]Z_6,$$
$$D_{4,1} = [M_5(a_{23} s_{12} s_{23} - a_{12} c_{12} c_{23}) + M_7 s_{12} c_{23}]Y_6 - M_5 S_1 s_{12} c_{23} X_6 - M_5 s_{12} c_{23} Y_{06},$$
$$D_{4,2} = [M_5(a_{23} s_{12} s_{23} - a_{12} c_{12} c_{23}) + M_7 s_{12} c_{23}]X_6 + M_5 S_1 s_{12} c_{23} Y_6 - M_5 s_{12} c_{23} X_{06},$$
$$D_{4,3} = [M_5(a_{12} s_{12} c_{23} + a_{23} c_{12} s_{23}) + M_7 c_{12} c_{23}]Z_6 - a_{34} s_{45} c_{45} - a_{45} s_{34} c_{34} - M_5 c_{12} c_{23} Z_{06},$$
$$E_{4,1} = -M_5 S_4 c_{12} Y_6(c_{23} s_{34} + s_{23} c_{34}),$$
$$E_{4,2} = -M_5 S_4 c_{12} X_6(c_{23} s_{34} + s_{23} c_{34}),$$
$$E_{4,3} = M_5 S_4 s_{12} Z_6(c_{23} s_{34} + s_{23} c_{34}),$$
$$F_{4,1} = -M_5 S_4 X_6(c_{23} s_{34} + s_{23} c_{34}),$$
$$F_{4,2} = M_5 S_4 Y_6(c_{23} s_{34} + s_{23} c_{34}),$$
$$F_{4,3} = 0,$$
$$G_{4,1} = M_5 S_4 s_{12} Y_6(s_{23} s_{34} - c_{23} c_{34}),$$
$$G_{4,2} = M_5 S_4 s_{12} X_6(s_{23} s_{34} - c_{23} c_{34}),$$
$$G_{4,3} = M_5 S_4 c_{12} Z_6(s_{23} s_{34} - c_{23} c_{34}) + M_5 S_4 c_{45}. \tag{9.74}$$

The coefficients defined in Eq. sets (9.71) through (9.74) can all be evaluated numerically because they are expressed in terms of given parameters.

The four equations that are now expressed in the format of (9.69) are next modified by substituting the tan-half-angle expressions for the sines and cosines of θ_1 and θ_2. The equations may be written as follows after multiplying each by the product $(1 + x_1^2)$ $(1 + x_2^2)$:

$$\left(a_i x_2^2 + b_i x_2 + d_i\right)x_3 + \left(e_i x_2^2 + f_i x_2 + g_i\right) = 0, \qquad i = 1 \ldots 4, \tag{9.75}$$

where

$$a_i = a_{i,1} x_1^2 + a_{i,2} x_1 + a_{i,3},$$
$$\vdots \tag{9.76}$$
$$g_i = g_{i,1} x_1^2 + g_{i,2} x_1 + g_{i,3}.$$

The coefficients $a_{i,1}$ through $g_{i,3}$ are defined in terms of $A_{i,1}$ through $G_{i,3}$ as follows:

$$a_{i,1} = D_{i,3} - A_{i,3} - D_{i,1} + A_{i,1},$$
$$a_{i,2} = 2(D_{i,2} - A_{i,2}),$$
$$a_{i,3} = D_{i,3} - A_{i,3} + D_{i,1} - A_{i,1},$$
$$b_{i,1} = 2(B_{i,3} - B_{i,1}),$$
$$b_{i,2} = 4B_{i,2},$$

$$
\begin{aligned}
b_{i,3} &= 2(B_{i,3} + B_{i,1}),\\
d_{i,1} &= D_{i,3} + A_{i,3} - D_{i,1} - A_{i,1},\\
d_{i,2} &= 2(D_{i,2} + A_{i,2}),\\
d_{i,3} &= D_{i,3} + A_{i,3} + D_{i,1} + A_{i,1},\\
e_{i,1} &= G_{i,3} - E_{i,3} - G_{i,1} + E_{i,1},\\
e_{i,2} &= 2(G_{i,2} - E_{i,2}),\\
e_{i,3} &= G_{i,3} - E_{i,3} + G_{i,1} - E_{i,1},\\
f_{i,1} &= 2(F_{i,3} - F_{i,1}),\\
f_{i,2} &= 4F_{i,2},\\
f_{i,3} &= 2(F_{i,3} + F_{i,1}),\\
g_{i,1} &= G_{i,3} + E_{i,3} - G_{i,1} - E_{i,1},\\
g_{i,2} &= 2(G_{i,2} + E_{i,2}),\\
g_{i,3} &= G_{i,3} + E_{i,3} + G_{i,1} + E_{i,1}.
\end{aligned}
\tag{9.77}
$$

Equation set (9.76) represents four equations that are expressed in the format of Eq. (9.2). A sixteenth-degree input/output equation may be obtained from these equations by expanding the 8×8 determinant of Eq. (9.3). This equation is written as

$$
\begin{vmatrix}
0 & 0 & a_1 & b_1 & d_1 & e_1 & f_1 & g_1\\
0 & 0 & a_2 & b_2 & d_2 & e_2 & f_2 & g_2\\
0 & 0 & a_3 & b_3 & d_3 & e_3 & f_3 & g_3\\
0 & 0 & a_4 & b_4 & d_4 & e_4 & f_4 & g_4\\
a_1 & e_1 & b_1 & d_1 & 0 & f_1 & g_1 & 0\\
a_2 & e_2 & b_2 & d_2 & 0 & f_2 & g_2 & 0\\
a_3 & e_3 & b_3 & d_3 & 0 & f_3 & g_3 & 0\\
a_4 & e_4 & b_4 & d_4 & 0 & f_4 & g_4 & 0
\end{vmatrix} = 0.
\tag{9.78}
$$

This determinant may be expanded by using Laplace's theorem, which states that a determinant can be evaluated by taking any m rows of the determinant and forming every possible minor of the mth order from these rows. Each minor is multiplied by its complement, that is, a determinant formed by deleting the rows and columns of the minor from the original matrix. Each product is then given a sign based on whether the sum of the numbers indicating the rows and columns from which the minor is formed is even or odd. The value of the original determinant will equal the sum of the individual products.

As an example, consider the 5×5 matrix

$$
\begin{bmatrix}
a_1 & b_1 & c_1 & d_1 & e_1\\
a_2 & b_2 & c_2 & d_2 & e_2\\
a_3 & b_3 & c_3 & d_3 & e_3\\
a_4 & b_4 & c_4 & d_4 & e_4\\
a_5 & b_5 & c_5 & d_5 & e_5
\end{bmatrix}.
\tag{9.79}
$$

Laplace's theorem may be applied to evaluate the determinant by forming the product of every 2×2 determinant from the first two rows of the matrix times its corresponding 3×3 complement determinant. The determinant can thus be evaluated as

$$\begin{vmatrix} a_1 & b_1 & c_1 & d_1 & e_1 \\ a_2 & b_2 & c_2 & d_2 & e_2 \\ a_3 & b_3 & c_3 & d_3 & e_3 \\ a_4 & b_4 & c_4 & d_4 & e_4 \\ a_5 & b_5 & c_5 & d_5 & e_5 \end{vmatrix} = \begin{vmatrix} a_1 & b_1 \\ a_2 & b_2 \end{vmatrix}\begin{vmatrix} c_3 & d_3 & e_3 \\ c_4 & d_4 & e_4 \\ c_5 & d_5 & e_5 \end{vmatrix} - \begin{vmatrix} a_1 & c_1 \\ a_2 & c_2 \end{vmatrix}\begin{vmatrix} b_3 & d_3 & e_3 \\ b_4 & d_4 & e_4 \\ b_5 & d_5 & e_5 \end{vmatrix}$$

$$+ \begin{vmatrix} a_1 & d_1 \\ a_2 & d_2 \end{vmatrix}\begin{vmatrix} b_3 & c_3 & e_3 \\ b_4 & c_4 & e_4 \\ b_5 & c_5 & e_5 \end{vmatrix} - \begin{vmatrix} a_1 & e_1 \\ a_2 & e_2 \end{vmatrix}\begin{vmatrix} b_3 & c_3 & d_3 \\ b_4 & c_4 & d_4 \\ b_5 & c_5 & d_5 \end{vmatrix} + \begin{vmatrix} b_1 & c_1 \\ b_2 & c_2 \end{vmatrix}\begin{vmatrix} a_3 & d_3 & e_3 \\ a_4 & d_4 & e_4 \\ a_5 & d_5 & e_5 \end{vmatrix}$$

$$- \begin{vmatrix} b_1 & d_1 \\ b_2 & d_2 \end{vmatrix}\begin{vmatrix} a_3 & c_3 & e_3 \\ a_4 & c_4 & e_4 \\ a_5 & c_5 & e_5 \end{vmatrix} + \begin{vmatrix} b_1 & e_1 \\ b_2 & e_2 \end{vmatrix}\begin{vmatrix} a_3 & c_3 & d_3 \\ a_4 & c_4 & d_4 \\ a_5 & c_5 & d_5 \end{vmatrix} + \begin{vmatrix} c_1 & d_1 \\ c_2 & d_2 \end{vmatrix}\begin{vmatrix} a_3 & b_3 & e_3 \\ a_4 & b_4 & e_4 \\ a_5 & b_5 & e_5 \end{vmatrix}$$

$$- \begin{vmatrix} c_1 & e_1 \\ c_2 & e_2 \end{vmatrix}\begin{vmatrix} a_3 & b_3 & d_3 \\ a_4 & b_4 & d_4 \\ a_5 & b_5 & d_5 \end{vmatrix} + \begin{vmatrix} d_1 & e_1 \\ d_2 & e_2 \end{vmatrix}\begin{vmatrix} a_3 & b_3 & c_3 \\ a_4 & b_4 & c_4 \\ a_5 & b_5 & c_5 \end{vmatrix}. \tag{9.80}$$

By using Laplace's theorem, Eq. (9.78) may be written as the summation of products of 4×4 determinants as

$$|aebd||defg| + |aebf||bdfg| - |aebg||bdeg| - |aedf||adfg| + |aedg||adeg| + |aefg||abdg| = 0, \tag{9.81}$$

where the determinant notation $|xyzw|$ is defined by

$$|xyzw| = \begin{vmatrix} x_1 & y_1 & z_1 & w_1 \\ x_2 & y_2 & z_2 & w_2 \\ x_3 & y_3 & z_3 & w_3 \\ x_4 & y_4 & z_4 & w_4 \end{vmatrix}. \tag{9.82}$$

Equation (9.81) represents the sixteenth-degree input/output equation for the mechanism, as each of the 4×4 determinants can be evaluated as an eighth-degree polynomial in terms of the tan-half-angle of θ_1. Although it is possible to generate this equation symbolically, it is far easier to expand the determinants of Eq. (9.81) numerically to produce the input/output equation for the case at hand. This equation can then be solved for the sixteen possible values of θ_1, not all of which may be real-valued.

9.3.2 Determination of θ_2 and θ_3

Expressions for x_2 and x_3 can be obtained by rearranging Eq. set (9.75) as follows:

$$(a_i)x_2^2x_3 + (b_i)x_2x_3 + (e_i)x_2^2 + (f_ix_2 + d_ix_3 + g_i) = 0, \qquad i = 1 \ldots 4. \tag{9.83}$$

Equation set (9.83) represents four homogeneous equations in the four unknowns $x_2^2x_3$, x_2x_3, x_2^2, and 1. A solution will exist only if the equations are linearly dependent. This occurs if the determinant of the coefficients equals zero, that is,

$$\begin{vmatrix} a_1 & b_1 & e_1 & (f_1x_2 + d_1x_3 + g_1) \\ a_2 & b_2 & e_2 & (f_2x_2 + d_2x_3 + g_2) \\ a_3 & b_3 & e_3 & (f_3x_2 + d_3x_3 + g_3) \\ a_4 & b_4 & e_4 & (f_4x_2 + d_4x_3 + g_4) \end{vmatrix} = 0. \tag{9.84}$$

This determinant may be expressed in the form

$$|abef|x_2 + |abed|x_3 + |abeg| = 0. \tag{9.85}$$

Equation set (9.75) may next be rearranged as

$$(a_i)x_2^2x_3 + (d_i)x_3 + (g_i) + (b_ix_3 + e_ix_2 + f_i)x_2 = 0, \qquad i = 1\ldots4. \tag{9.86}$$

This represents four homogeneous equations in the unknowns $x_2^2x_3$, x_3, 1, and x_2. A solution to these equations will exist only if the equations are linearly dependent. Thus, it must be the case that

$$\begin{vmatrix} a_1 & d_1 & g_1 & (b_1x_3 + e_1x_2 + f_1) \\ a_2 & d_2 & g_2 & (b_2x_3 + e_2x_2 + f_2) \\ a_3 & d_3 & g_3 & (b_3x_3 + e_3x_2 + f_3) \\ a_4 & d_4 & g_4 & (b_4x_3 + e_4x_2 + f_4) \end{vmatrix} = 0. \tag{9.87}$$

This determinant may be expressed in the form

$$|adgb|x_3 + |adge|x_2 + |adgf| = 0. \tag{9.88}$$

Solving Eqs. (9.85) and (9.88) for x_2 and x_3 gives

$$x_2 = \frac{-|abeg||adgb| + |adgf||abed|}{|abef||adgb| - |adge||abed|}, \tag{9.89}$$

$$x_3 = \frac{-|abef||adgf| + |adge||abeg|}{|abef||adgb| - |adge||abed|}. \tag{9.90}$$

Equations (9.89) and (9.90) may be used to calculate unique corresponding values for θ_2 and θ_3 for each previously calculated value for θ_1.

9.3.3 Determination of θ_4 and θ_5

Corresponding values for θ_4 and θ_5 may be obtained from the following pairs of fundamental sine and sine–cosine laws for a spherical hexagon:

$$X_{6123} = s_{45}s_4, \tag{9.91}$$

$$Y_{6123} = s_{45}c_4, \tag{9.92}$$

$$X_{3216} = s_{45}s_5, \tag{9.93}$$

$$Y_{3216} = s_{45}c_5. \tag{9.94}$$

Table 9.1. *RCRRRR mechanism parameters.*

Link length, cm.	Twist angle, deg.	Joint offset, cm.	Joint angle, deg.
$a_{12} = 7.2$	$\alpha_{12} = 90$	$S_1 = 2$	$\theta_1 =$ variable
$a_{23} = 0$	$\alpha_{23} = 90$	$S_2 = 2.6$	$\theta_2 =$ variable
$a_{34} = 3.6$	$\alpha_{34} = 90$	$S_3 = 0$	$\theta_3 =$ variable
$a_{45} = 0.8$	$\alpha_{45} = 90$	$S_4 = 0$	$\theta_4 =$ variable
$a_{56} = 2.3$	$\alpha_{56} = 90$	$S_5 =$ variable	$\theta_5 =$ variable
$a_{61} = 9.2$	$\alpha_{61} = 90$	$S_6 = 0.8$	$\theta_6 = 273$ (input)

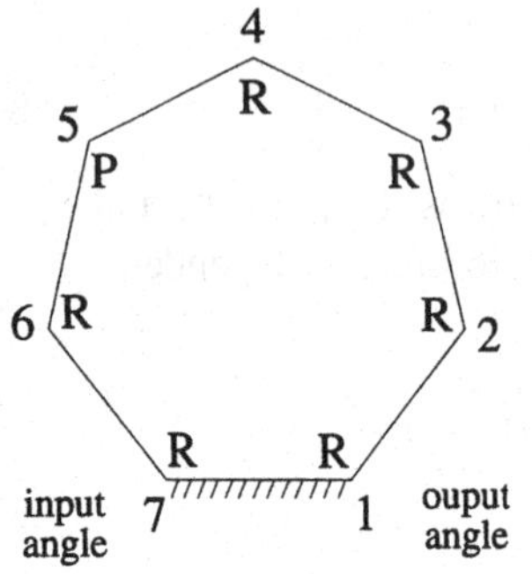

Figure 9.2. Planar representation of an RRPRRRR spatial mechanism.

9.3.4 Determination of S_5

The last parameter to be determined is the offset distance S_5. This may be determined by projecting the vector loop equation onto any direction, resulting in one equation in one unknown. Projecting onto the vector $\mathbf{S}_5$ gives

$$S_1Z_6 + a_{12}U_{165} + S_2Z_{16} + a_{23}U_{345} + S_3Z_4 + a_{34}U_{45} + S_4c_{45} + S_5 + S_6c_{56} + a_{61}U_{65} = 0. \quad (9.95)$$

This equation can readily be solved for the parameter S_5.

9.3.5 Numerical example

Table 9.1 shows the data that were used as input for a numerical example. The calculated values for the sixteen configurations are listed in Table 9.2.

9.4 Seven-link group 3 spatial mechanisms

All seven-link group 3 mechanisms comprise six revolute joints and one prismatic joint, that is, 6R-P. Shown in Figure 9.2 is the planar representation of the closed loop RRPRRRR spatial mechanism.

Table 9.2. *Calculated configurations for the RCRRRR spatial mechanism.*

Solution	θ_1, deg.	θ_2, deg.	θ_3, deg.	θ_4, deg.	θ_5, deg.	S_5, cm.
A	−19.33	87.68	−74.71	−20.04	105.18	16.45
B	13.74	176.81	−13.80	83.90	179.24	13.31
C	14.22	9.41	14.28	−83.88	−177.69	20.48
D	−139.62	79.75	82.62	139.28	−77.75	−0.22
E	28.09	−103.89	119.16	−147.42	58.09	18.80
F	−154.06	4.01	−153.91	−96.60	178.24	5.27
G	−129.76	−98.16	92.90	50.23	98.13	−0.07
H	−5.93	−89.70	−114.52	−173.49	−65.65	16.58
I	12.62	−31.49	−164.81	123.61	−7.88	13.88
J	39.05	79.77	73.26	−41.07	−109.32	18.88
K	−153.71	18.29	27.94	109.31	−8.47	−1.70
L	−174.96	87.74	−81.58	−174.91	81.81	2.22
M	172.26	−93.73	−95.18	7.76	−95.64	2.03
N	−154.71	155.20	−26.88	−109.35	10.95	5.08
O	15.70	−141.46	160.99	−123.92	11.73	19.90
P	−154.38	169.36	154.24	96.59	−175.39	−1.89

There are three distinct inversions of this mechanism,

	Input	Output
RRPRRRR	θ_7	θ_1
RRRPRRR	θ_1	θ_2
RRRRRPR	θ_7	θ_6,

and a detailed analysis of the first of these three inversions will be given here.

It is assumed that all the mechanism dimensions are known together with the input angle θ_7 and that it is required to compute the remaining unknown joint displacements. In particular, the problem statement is as follows:

given: $\alpha_{12}, \alpha_{23}, \alpha_{34}, \alpha_{45}, \alpha_{56}, \alpha_{67}, \alpha_{71}$,
$a_{12}, a_{23}, a_{34}, a_{45}, a_{56}, a_{67}, a_{71}$,
$S_1, S_2, S_3, S_4, \theta_5, S_6, S_7$, and
(θ_7 input angle),

find: $\theta_1, \theta_2, \theta_3, \theta_4, \theta_6$, and S_5.

9.4.1 Development of the input–output equation (solution for θ_1)

It will be shown that the input–output equation for this mechanism will be generated from four equations of the form of Eq. (9.2). Firstly, a pair of equations of this form is generated from projecting the vector loop equation onto two directions that are perpendicular to the direction of S_5, that is, the x and y components of set 5 from the table of direction cosines. Following this, a second pair of equations is generated from secondary tan-half-angle laws.

The projections of the vector loop equation onto the x and y directions of set 5 from the direction cosine table are

$$S_1X_{76} + a_{12}W_{176} + S_2X_{176} + a_{23}W_{2176} + S_3X_{2176} + a_{34}W_{32176} + S_4X_{32176} + a_{45}c_5 + a_{56} + a_{67}c_6 + S_7\bar{X}_6 + a_{71}W_{76} = 0, \quad (9.96)$$

$$S_1Y_{76} - a_{12}U^*_{1765} + S_2Y_{176} - a_{23}U^*_{21765} + S_3Y_{2176} - a_{34}U^*_{321765} + S_4Y_{32176} - a_{45}s_5 - S_6s_{56} + a_{67}s_6c_{56} + S_7Y_6 - a_{71}U^*_{765} = 0. \quad (9.97)$$

These equations can be modified by using fundamental and subsidiary spherical and polar laws to substitute for the terms X_{32176}, Y_{32176}, U^*_{321765}, and W_{32176} to give

$$S_1X_{76} + a_{12}W_{176} + S_2X_{176} + a_{23}W_{2176} + S_3X_{2176} + a_{34}W_{45} + S_4X_5 + a_{45}c_5 + a_{56} + a_{67}c_6 + S_7\bar{X}_6 + a_{71}W_{76} = 0, \quad (9.98)$$

$$S_1Y_{76} - a_{12}U^*_{1765} + S_2Y_{176} - a_{23}U^*_{21765} + S_3Y_{2176} + a_{34}V_{45} + S_4s_{45}c_5 - a_{45}s_5 - S_6s_{56} + a_{67}s_6c_{56} + S_7\bar{Y}_6 - a_{71}U^*_{765} = 0. \quad (9.99)$$

All terms that contain the angle θ_6 are now expanded, and these two equations can be expressed in the forms

$$H_1c_6 - H_2s_6 + M = 0, \quad (9.100)$$

$$c_{56}H_2c_6 + c_{56}H_1s_6 + N = 0, \quad (9.101)$$

where

$$\begin{aligned}
H_1 &= S_1\bar{X}_7 + S_2X_{17} + S_3X_{217} + a_{12}W_{17} + a_{23}W_{217} + a_{67} + a_{71}c_7,\\
H_2 &= S_1\bar{Y}_7 + S_2Y_{17} + S_3Y_{217} - S_7s_{67} - a_{12}U^*_{176} - a_{23}U^*_{2176} + a_{71}c_{67}s_7,\\
M &= S_4X_5 + a_{34}W_{45} + a_{45}c_5 + a_{56}, \qquad (9.102)\\
N &= S_4s_{45}c_5 + a_{34}V_{45} - a_{45}s_5 - H_3s_{56},\\
H_3 &= S_1\bar{Z}_7 + S_2Z_{17} + S_3Z_{217} + S_6 + S_7c_{67} + a_{12}U_{176} + a_{23}U_{2176} + a_{71}U_{76}.
\end{aligned}$$

The last equation in Eq. set (9.102) can be modified by reversing the order of the subscripts on the $Z_{ij...m}$ terms and by substituting for the terms U_{176}, U_{2176}, and U_{76} respectively by the equivalent expressions X_{71}, X_{712}, and X_7. Also, it is noted that the angle θ_4 is contained only in the terms M and N and because of this these terms will be expanded. The Eq. set (9.102) can thus be rewritten in the form

$$\begin{aligned}
H_1 &= S_1\bar{X}_7 + S_2X_{17} + S_3X_{217} + a_{12}W_{17} + a_{23}W_{217} + a_{67} + a_{71}c_7,\\
H_2 &= S_1\bar{Y}_7 + S_2Y_{17} + S_3Y_{217} - S_7s_{67} - a_{12}U^*_{176} - a_{23}U^*_{2176} + a_{71}c_{67}s_7,\\
M &= J_1 + K_{11}c_4 + K_{12}s_4,\\
N &= J_2 + K_{21}c_4 + K_{22}s_4,\\
H_3 &= S_1Z_7 + S_2Z_{71} + S_3Z_{712} + S_6 + S_7c_{67} + a_{12}X_{71} + a_{23}X_{712} + a_{71}X_7, \qquad (9.103)
\end{aligned}$$

where

$$J_1 = a_{56} + a_{45}c_5 + S_4 s_{45} s_5, \qquad K_{11} = a_{34}c_5, \qquad K_{12} = -a_{34}c_{45}s_5,$$

$$J_2 = -s_{56}H_3 + S_4 s_{45} c_5 - a_{45}s_5, \qquad K_{21} = -a_{34}s_5, \qquad K_{22} = -a_{34}c_{45}c_5. \tag{9.104}$$

The solution proceeds by generating two new equations. The first is obtained by subtracting the product of $c_{56}x_6$ times Eq. (9.100) from Eq. (9.101), where $x_6 = \tan(\theta_6/2)$. The second equation is generated by adding the product of c_{56} times Eq. (9.100) to the product of x_6 times Eq. (9.101). These two equations may now be written as

$$c_{56}[H_1(s_6 - c_6x_6) + H_2(c_6 + s_6x_6)] + (N - c_{56}x_6M) = 0, \tag{9.105}$$

$$c_{56}[H_1(s_6x_6 + c_6) + H_2(c_6x_6 - s_6)] + x_6N + c_{56}M. \tag{9.106}$$

The terms $(s_6 - c_6x_6)$ and $(c_6 + s_6x_6)$ will next be expanded by introducing the trigonometric identities of Eqs. (6.157) and (6.158). Firstly,

$$s_6 - c_6x_6 = \frac{2x_6}{1+x_6^2} - \frac{1-x_6^2}{1+x_6^2}x_6. \tag{9.107}$$

Simplifying this expression yields

$$s_6 - c_6x_6 = \frac{x_6(x_6^2+1)}{1+x_6^2} = x_6. \tag{9.108}$$

Secondly,

$$c_6 + s_6x_6 = \frac{1-x_6^2}{1+x_6^2} + \frac{2x_6}{1+x_6^2}x_6 = 1. \tag{9.109}$$

Simplifying Eqs. (9.105) and (9.106) using Eqs. (9.108) and (9.109) gives

$$c_{56}[H_1x_6 + H_2] + (N - c_{56}x_6M) = 0, \tag{9.110}$$

$$c_{56}[H_1 - H_2x_6] + x_6N + c_{56}M, \tag{9.111}$$

which can be rearranged as

$$c_{56}(H_1 - M)x_6 + (c_{56}H_2 + N) = 0, \tag{9.112}$$

$$(c_{56}H_2 - N)x_6 - c_{56}(H_1 + M) = 0. \tag{9.113}$$

Equations (9.112) and (9.113) are not quite in the form of Eq. (9.2). They contain the input angle θ_7, the output angle θ_1, and the two extra angles θ_2 and θ_6 (note that they are linear in x_6). However, they also contain the variable θ_4 in the terms M and N, which has to be eliminated. This final elimination will be performed after the next pair of equations is developed.

A further two equations of the form of Eq. (9.2) are derived from the following subsidiary laws for a spherical heptagon:

$$X_{2176} = X_{45}, \tag{9.114}$$

$$X^*_{2176} = -Y_{45}. \tag{9.115}$$

The following pair of corresponding secondary equations can be obtained by introducing dual angles into Eqs. (9.114) and (9.115):

$$X_{02176} = X_{045}, \tag{9.116}$$

$$X^*_{02176} = -Y_{045}. \tag{9.117}$$

The left side of these equations may be expanded as

$$-S_6(X_{217}s_6 + Y_{217}c_6) + X_{0217}c_6 - Y_{0217}s_6 = X_{045}, \tag{9.118}$$

$$S_6(X_{217}c_6 - Y_{217}s_6) + X_{0217}s_6 + Y_{0217}c_6 = -Y_{045}, \tag{9.119}$$

and regrouping gives

$$(X_{0217} - S_6Y_{217})c_6 - (Y_{0217} + S_6X_{217})s_6 = X_{045}, \tag{9.120}$$

$$(X_{0217} - S_6Y_{217})s_6 + (Y_{0217} + S_6X_{217})c_6 = -Y_{045}. \tag{9.121}$$

The previous two equations have common terms and may be rewritten as

$$\alpha c_6 - \beta s_6 = \lambda_1, \tag{9.122}$$

$$\alpha s_6 + \beta c_6 = \lambda_2, \tag{9.123}$$

where

$$\begin{aligned} \alpha &= X_{0217} - S_6Y_{217}, & \lambda_1 &= X_{045}, \\ \beta &= Y_{0217} + S_6X_{217}, & \lambda_2 &= -Y_{045}. \end{aligned} \tag{9.124}$$

Subtracting x_6 times Eq. (9.122) from Eq. (9.123) yields

$$\alpha(s_6 - c_6x_6) + \beta(c_6 + s_6x_6) = \lambda_2 - \lambda_1x_6. \tag{9.125}$$

Adding Eq. (9.122) to x_6 times Eq. (9.123) yields

$$\alpha(c_6 + s_6x_6) + \beta(c_6x_6 - s_6) = \lambda_1 + \lambda_2x_6. \tag{9.126}$$

Equations (9.125) and (9.126) may be simplified using Eqs. (9.108) and (9.109), and thus

$$\alpha x_6 + \beta = \lambda_2 - \lambda_1x_6, \tag{9.127}$$

$$\alpha - \beta x_6 = \lambda_1 + \lambda_2x_6. \tag{9.128}$$

Regrouping terms gives

$$(\alpha + \lambda_1)x_6 + (\beta - \lambda_2) = 0, \tag{9.129}$$

$$(\beta + \lambda_2)x_6 - (\alpha - \lambda_1) = 0, \tag{9.130}$$

and expanding α and β using Eq. (9.124) gives

$$(X_{0217} - S_6 Y_{217} + X_{045})x_6 + (Y_{0217} + S_6 X_{217} + Y_{045}) = 0, \tag{9.131}$$

$$(Y_{0217} + S_6 X_{217} - Y_{045})x_6 - (X_{0217} - S_6 Y_{217} - X_{045}) = 0. \tag{9.132}$$

It is important to note that each of these equations contains the unwanted slider displacement S_5 in the terms X_{045} and Y_{045}. This displacement will be eliminated by first expanding the definitions of the terms X_{045} and Z_{045} as

$$X_{045} = -S_5(X_4 s_5 + Y_4 c_5) + (X_{04} c_5 - Y_{04} s_5), \tag{9.133}$$

$$Z_{045} = S_5 s_{56}(X_4 c_5 - Y_4 s_5) + a_{56} Y_{45} + s_{56}(X_{04} s_5 + Y_{04} c_5) + c_{56} Z_{04}. \tag{9.134}$$

Part of the unwanted S_5 term can be eliminated by adding $s_{56}c_5$ times Eq. (9.133) to s_5 times Eq. (9.134) and by making the substitution $Z_{045} = Z_{0217}$. This yields

$$s_{56} c_5 X_{045} = -s_5 Z_{0217} - S_5 s_{56} Y_4 + a_{56} s_5 Y_{45} + s_{56} X_{04} + c_{56} s_5 Z_{04}. \tag{9.135}$$

It remains to eliminate $-S_5 s_{56} Y_4$ from this equation. It can readily be shown upon expanding Y_4 and Z_4 that

$$Y_4 = -\frac{1}{s_{45}}(c_{45} Z_4 - c_{34}) \tag{9.136}$$

and hence

$$-S_5 s_{56} Y_4 = -S_5 \frac{s_{56}}{s_{45}} c_{45} Z_4 + S_5 \frac{s_{56}}{s_{45}} c_{34}. \tag{9.137}$$

Now, by projecting the vector loop equation onto the direction of $\mathbf{S}_3$ (using sets 3 and 13 from the table of direction cosines), the following equation may be written:

$$-S_5 Z_4 = S_1 Z_2 + S_2 c_{23} + S_3 + S_4 c_{34} + S_6 Z_{712} + S_7 Z_{12} + a_{12} U_{23} + a_{45} U_{43} \\ + a_{56} U_{543} + a_{67} U_{7123} + a_{71} U_{123}. \tag{9.138}$$

This equation may be modified by substituting for the terms U_{23}, U_{43}, U_{543}, U_{7123}, and U_{123} respectively with $\bar{X}_2$, X_4, X_{45}, X_{217}, and X_{21} to give

$$-S_5 Z_4 = H_4, \tag{9.139}$$

where

$$H_4 = S_1 Z_2 + S_2 c_{23} + S_3 + S_4 c_{34} + S_6 Z_{712} + S_7 Z_{12} + a_{12} \bar{X}_2 + a_{45} X_4 + a_{56} X_{45} \\ + a_{67} X_{217} + a_{71} X_{21}. \tag{9.140}$$

Further, by projecting the vector loop equation onto the direction of $\mathbf{S}_6$ (using sets 6 and 10 from the table of direction cosines), the following equation may be written:

$$-S_5 c_{56} = S_1 Z_7 + S_2 Z_{71} + S_3 Z_{712} + S_4 Z_5 + S_6 + S_7 c_{67} + a_{12} U_{176} + a_{23} U_{2176} \\ + a_{34} U_{456} + a_{45} U_{56} + a_{71} U_{76}. \tag{9.141}$$

This equation may be modified by substituting for the terms U_{176}, U_{2176}, U_{456}, U_{56}, and U_{76} respectively with X_{71}, X_{712}, X_{54}, $\bar{X}_5$, and X_7 to give

$$-S_5 c_{56} = H_5, \tag{9.142}$$

where

$$H_5 = S_1 Z_7 + S_2 Z_{71} + S_3 Z_{712} + S_4 Z_5 + S_6 + S_7 c_{67} + a_{12} X_{71} + a_{23} X_{712} + a_{34} X_{54} + a_{45} \bar{X}_5 + a_{71} X_7. \tag{9.143}$$

The expression for H_5 may also be written as

$$H_5 = H_3 + S_4 Z_5 + a_{34} X_{54} + a_{45} \bar{X}_5, \tag{9.144}$$

where H_3 is defined in Eq. set (9.103).

Substituting Eqs. (9.139) and (9.142) into Eq. (9.137) and then this result into Eq. (9.135) gives the following expression for X_{045} with the slider displacement S_5 completely eliminated:

$$X_{045} = \frac{-s_{45} c_{56} s_5 Z_{0217} + c_{45} s_{56} c_{56} H_4 - c_{34} s_{56} H_5 + s_{45} c_{56} s_5 (a_{56} Y_{45} + c_{56} Z_{04}) + s_{45} s_{56} c_{56} X_{04}}{s_{45} s_{56} c_{56} c_5}. \tag{9.145}$$

The elimination of the slider displacement S_5 from the term Y_{045} is relatively simple. By definition

$$Y_{045} = S_5 c_{56} (X_4 c_5 - Y_4 s_5) - a_{56} \left(s_{56} X^*_{45} + c_{56} Z_4 \right) + c_{56} (X_{04} s_5 + Y_{04} c_5) - s_{56} Z_{04}. \tag{9.146}$$

The unwanted S_5 term can be eliminated by simply subtracting c_{56} times Eq. (9.134) from s_{56} times Eq. (9.146) and by making the substitution $Z_{045} = Z_{0217}$. This yields

$$Y_{045} = \frac{c_{56} Z_{0217} - a_{56} X^*_{45} - Z_{04}}{s_{56}}. \tag{9.147}$$

Substituting the expressions for X_{045} and Y_{045} (Eqs. (9.145) and (9.147)) in Eqs. (9.131) and (9.132) gives two equations that are of the same format as Eqs. (9.112) and (9.113). In other words, these four equations contain the input angle θ_7, the output angle θ_1, the two extra angles θ_2 and θ_6 (note that they are all linear in x_6), and the angle θ_4. It remains to eliminate θ_4 to obtain the format of Eq. (9.2).

All terms in Eqs. (9.131) and (9.132) that contain the angle θ_4 will next be expanded so that the sine and cosine of this angle may be expressed separately. By inspecting these two equations it is apparent that the angle θ_4 is present only in the terms X_{045} and Y_{045}. By inspection of Eqs. (9.145) and (9.147) it is necessary to expand the terms X_4, X_{45}, X^*_{45},

X_{54}, Y_{45}, X_{04}, and Z_{04}, and

$$
\begin{aligned}
X_{45} &= X_4c_5 - Y_4c_5,\\
X^*_{45} &= X_4s_5 + Y_4c_5,\\
Y_{45} &= c_{56}(X_4s_5 + Y_4c_5) - s_{56}Z_4,\\
X_4 &= s_{34}s_4,\\
Y_4 &= -(s_{45}c_{34} + c_{45}s_{34}c_4),\\
Z_4 &= c_{45}c_{34} - s_{45}s_{34}c_4,\\
X_{54} &= \bar{X}_5c_4 - \bar{Y}_5s_4,\\
X_{04} &= a_{34}c_{34}s_4 + S_4s_{34}c_4,\\
Z_{04} &= -a_{34}(c_{45}s_{34} + s_{45}c_{34}c_4) - a_{45}(s_{45}c_{34} + c_{45}s_{34}c_4) + S_4s_{45}s_{34}s_4.
\end{aligned}
\tag{9.148}
$$

After substituting these expansions into Eqs. (9.145) and (9.147), they may be written in the form

$$X_{045} = J_3 + K_{31}c_4 + K_{32}s_4, \tag{9.149}$$

$$Y_{045} = J_4 + K_{41}c_4 + K_{42}s_4, \tag{9.150}$$

where

$$
\begin{aligned}
J_3 &= \big[-s_{45}c_{56}s_5Z_{0217} - c_{34}s_{56}H'_5 + c_{45}s_{56}c_{56}H'_4 - a_{34}s_{34}s_{45}c_{45}c^2_{56}s_5 - a_{45}c_{34}s^2_{45}c^2_{56}s_5\\
&\quad - a_{56}c_{34}s^2_{45}c^2_{56}s_5c_5\big]/(s_{45}s_{56}c_{56}c_5),\\
K_{31} &= \big[S_4s_{34}s_{45}s_{56}c_{56} - a_{34}c_{34}\big(s_{56}\bar{X}_5 + s^2_{45}c^2_{56}s_5\big) - a_{45}s_{34}s_{45}c_{45}c^2_{56}s_5\\
&\quad + a_{56}s_{34}\big(s_{45}c_{56}s_5(s_{45}s_{56} - c_{45}c_{56}c_5) + c^2_{45}s_{56}c_{56}s_5\big)\big]/(s_{45}s_{56}c_{56}c_5),\\
K_{32} &= \big[S_4s_{34}s^2_{45}c^2_{56}s_5 - a_{34}c_{34}c_{45}s^2_{56}c_5 + a_{45}s_{34}c_{45}s_{56}c_{56}\\
&\quad + a_{56}s_{34}c_{56}\big(s_{45}c_{56}s^2_5 + c_{45}s_{56}c_5\big)\big]/(s_{45}s_{56}c_{56}c_5),\\
J_4 &= \frac{c_{56}Z_{0217} + a_{34}s_{34}c_{45} + a_{45}c_{34}s_{45} + a_{56}c_{34}s_{45}c_5}{s_{56}},\\
K_{42} &= \frac{-S_4s_{34}s_{45} - a_{56}s_{34}s_5}{s_{56}},\\
K_{41} &= \frac{a_{34}c_{34}s_{45} + a_{45}s_{34}c_{45} + a_{56}s_{34}c_{45}c_5}{s_{56}},
\end{aligned}
\tag{9.151}
$$

and where

$$
\begin{aligned}
H'_4 &= H_4 - a_{45}X_4 - a_{56}X_{45} = S_1Z_2 + S_2c_{23} + S_3 + S_4c_{34} + S_6Z_{712} + S_7Z_{12}\\
&\quad + a_{12}\bar{X}_2 + a_{67}X_{217} + a_{71}X_{21},\\
H'_5 &= H_5 - a_{34}X_{54} = S_1Z_7 + S_2Z_{71} + S_3Z_{712} + S_4Z_5 + S_6 + S_7c_{67} + a_{12}X_{71}\\
&\quad + a_{23}X_{712} + a_{45}\bar{X}_5 + a_{71}X_7.
\end{aligned}
\tag{9.152}
$$

Substituting the expressions for M and N in Eq. set (9.103) into Eqs. (9.112) and (9.113) and then substituting Eqs. (9.149) and (9.150) into Eqs. (9.131) and (9.132) yields the following four equations:

$$c_{56}(H_1 - J_1 - K_{11}c_4 - K_{12}s_4)x_6 + (c_{56}H_2 + J_2 + K_{21}c_4 + K_{22}s_4) = 0, \quad (9.153)$$

$$(c_{56}H_2 - J_2 - K_{21}c_4 - K_{22}s_4)x_6 - c_{56}(H_1 + J_1 + K_{11}c_4 + K_{12}s_4) = 0, \quad (9.154)$$

$$(X_{0217} - S_6Y_{217} + J_3 + K_{31}c_4 + K_{32}s_4)x_6 + (Y_{0217} + S_6X_{217} + J_4 + K_{41}c_4 + K_{42}s_4) = 0, \quad (9.155)$$

$$(Y_{0217} + S_6X_{217} - J_4 - K_{41}c_4 - K_{42}s_4)x_6 - (X_{0217} - S_6Y_{217} - J_3 - K_{31}c_4 - K_{32}s_4) = 0. \quad (9.156)$$

It is important to recognize that the coefficients in these equations, that is, H_i, J_i, K_{ij}, contain the sines and cosines of the angles θ_1 and θ_2 as their only variables. It now remains to eliminate θ_4 from these four equations without increasing the degree in the sines and cosines of the angles θ_1 and θ_2. This is accomplished by eliminating the four terms c_4x_6, s_4x_6, c_4, and s_4 in a single operation. This operation is derived from the following four spherical equations:

$$(X_{45} + X_{217})x_6 + (Y_{45} + Y_{217}) = 0, \quad (9.157)$$

$$(Y_{45} - Y_{217})x_6 - (X_{45} - X_{217}) = 0, \quad (9.158)$$

$$Z_{45} - Z_{217} = 0, \quad (9.159)$$

$$Z_{45}x_6 - Z_{217}x_6 = 0. \quad (9.160)$$

Note that Eq. (9.160) is equal to the product of x_6 times Eq. (9.159). These four equations can be written in the matrix form

$$\begin{bmatrix} X_{45} \\ -Y_{45} \\ 0 \\ Z_{45} \end{bmatrix} x_6 + \begin{bmatrix} Y_{45} \\ X_{45} \\ Z_{45} \\ 0 \end{bmatrix} = \begin{bmatrix} -X_{217} \\ -Y_{217} \\ 0 \\ Z_{217} \end{bmatrix} x_6 + \begin{bmatrix} -Y_{217} \\ X_{217} \\ Z_{217} \\ 0 \end{bmatrix}. \quad (9.161)$$

All terms that contain the variable angle θ_4, that is, X_{45}, Y_{45}, and Z_{45}, will next be expanded as

$$\begin{aligned} X_{45} &= X_4c_5 - Y_4s_5, \\ Y_{45} &= c_{56}(X_4s_5 + Y_4c_5) - s_{56}Z_4, \\ Z_{45} &= s_{56}(X_4s_5 + Y_4c_5) + c_{56}Z_4, \\ X_4 &= s_{34}s_4, \\ Y_4 &= -(s_{45}c_{34} + c_{45}s_{34}c_4), \\ Z_4 &= c_{45}c_{34} - s_{45}s_{34}c_4. \end{aligned} \quad (9.162)$$

The expression for X_{45} may be expanded as

$$X_{45} = (s_{34}s_4)c_5 + (s_{45}c_{34} + c_{45}s_{34}c_4)s_5 \quad (9.163)$$

and then rearranged as

$$X_{45} = s_{34}\left(X_5' c_4 + c_5 s_4\right) + c_{34} X_5. \tag{9.164}$$

The new term X_5' is defined by

$$X_5' = c_{45} s_5 \tag{9.165}$$

The expression for Y_{45} may be expanded as

$$Y_{45} = c_{56}((s_{34} s_4) s_5 - (s_{45} c_{34} + c_{45} s_{34} c_4) c_5) - s_{56}(c_{45} c_{34} - s_{45} s_{34} c_4). \tag{9.166}$$

This can be rearranged as

$$Y_{45} = s_{34}\left(\bar{Z}_5' c_4 + \bar{X}_5' s_4\right) + c_{34} Y_5, \tag{9.167}$$

where $\bar{X}_5'$ and $\bar{Z}_5'$ are defined here as

$$\bar{X}_5' = c_{56} s_5, \tag{9.168}$$

$$\bar{Z}_5' = s_{45} s_{56} - c_{45} c_{56} c_5. \tag{9.169}$$

Now, $Z_{45} = Z_{54}$ and

$$Z_{54} = s_{34}(\bar{X}_5 s_4 + \bar{Y}_5 c_4) + c_{34} \bar{Z}_5. \tag{9.170}$$

Substituting Eqs. (9.164), (9.166), and (9.170) into Eq. (9.161) and rearranging gives

$$s_{34}\begin{bmatrix} X_5' c_4 + c_5 s_4 \\ -(\bar{Z}_5' c_4 + \bar{X}_5' s_4) \\ 0 \\ \bar{X}_5 s_4 + \bar{Y}_5 c_4 \end{bmatrix} x_6 + c_{34}\begin{bmatrix} X_5 \\ -Y_5 \\ 0 \\ \bar{Z}_5 \end{bmatrix} x_6 + s_{34}\begin{bmatrix} \bar{Z}_5' c_4 + \bar{X}_5' s_4 \\ X_5' c_4 + c_5 s_4 \\ \bar{X}_5 s_4 + \bar{Y}_5 c_4 \\ 0 \end{bmatrix} + c_{34}\begin{bmatrix} Y_5 \\ X_5 \\ \bar{Z}_5 \\ 0 \end{bmatrix}$$

$$= \begin{bmatrix} -X_{217} \\ -Y_{217} \\ 0 \\ Z_{217} \end{bmatrix} x_6 + \begin{bmatrix} -Y_{217} \\ X_{217} \\ Z_{217} \\ 0 \end{bmatrix}. \tag{9.171}$$

This equation can be rearranged as

$$s_{34}\begin{bmatrix} X_5' c_4 + c_5 s_4 \\ -(\bar{Z}_5' c_4 + \bar{X}_5' s_4) \\ 0 \\ \bar{X}_5 s_4 + \bar{Y}_5 c_4 \end{bmatrix} x_6 + s_{34}\begin{bmatrix} \bar{Z}_5' c_4 + \bar{X}_5' s_4 \\ X_5' c_4 + c_5 s_4 \\ \bar{X}_5 s_4 + \bar{Y}_5 c_4 \\ 0 \end{bmatrix} = \begin{bmatrix} -X_{217} \\ -Y_{217} \\ 0 \\ Z_{217} \end{bmatrix} x_6$$

$$+ \begin{bmatrix} -Y_{217} \\ X_{217} \\ Z_{217} \\ 0 \end{bmatrix} - c_{34}\begin{bmatrix} X_5 \\ -Y_5 \\ 0 \\ \bar{Z}_5 \end{bmatrix} x_6 - c_{34}\begin{bmatrix} Y_5 \\ X_5 \\ \bar{Z}_5 \\ 0 \end{bmatrix}. \tag{9.172}$$

The left side of Eq. (9.172) may be regrouped to yield

$$s_{34}\mathbf{M}\begin{bmatrix} c_4x_6 \\ s_4x_6 \\ c_4 \\ s_4 \end{bmatrix} = \begin{bmatrix} -X_{217} \\ -Y_{217} \\ 0 \\ Z_{217} \end{bmatrix} x_6 + \begin{bmatrix} -Y_{217} \\ X_{217} \\ Z_{217} \\ 0 \end{bmatrix} - c_{34}\begin{bmatrix} X_5 \\ -Y_5 \\ 0 \\ \bar{Z}_5 \end{bmatrix} x_6 - c_{34}\begin{bmatrix} Y_5 \\ X_5 \\ \bar{Z}_5 \\ 0 \end{bmatrix}, \tag{9.173}$$

where

$$\mathbf{M} = \begin{bmatrix} X'_5 & c_5 & \bar{Z}'_5 & \bar{X}'_5 \\ -\bar{Z}'_5 & -\bar{X}'_5 & X'_5 & c_5 \\ 0 & 0 & \bar{Y}_5 & \bar{X}_5 \\ \bar{Y}_5 & \bar{X}_5 & 0 & 0 \end{bmatrix}. \tag{9.174}$$

It is important to note that the matrix $\mathbf{M}$ contains only the given constant mechanism parameters. The terms c_4x_6, s_4x_6, c_4, and s_4 can be solved for by inverting $\mathbf{M}$ to obtain the result

$$\begin{bmatrix} c_4x_6 \\ s_4x_6 \\ c_4 \\ s_4 \end{bmatrix} = \frac{\mathbf{M}^{-1}}{s_{34}}\left(\begin{bmatrix} -X_{217} \\ -Y_{217} \\ 0 \\ Z_{217} \end{bmatrix} x_6 + \begin{bmatrix} -Y_{217} \\ X_{217} \\ Z_{217} \\ 0 \end{bmatrix} - c_{34}\begin{bmatrix} X_5 \\ -Y_5 \\ 0 \\ \bar{Z}_5 \end{bmatrix} x_6 - c_{34}\begin{bmatrix} Y_5 \\ X_5 \\ \bar{Z}_5 \\ 0 \end{bmatrix}\right) \tag{9.175}$$

The inverse of $\mathbf{M}$ may be obtained from the equation

$$\mathbf{M}^{-1} = \frac{\text{Adj }\mathbf{M}}{|\mathbf{M}|}, \tag{9.176}$$

where Adj $\mathbf{M}$ is the adjoint matrix of $\mathbf{M}$. In review, Adj $\mathbf{M}$ is the transpose of the matrix of cofactors, that is, the ith row and j^{th} column element of Adj $\mathbf{M}$ is the determinant of the original matrix $\mathbf{M}$ with its jth row and i^{th} column removed. Adj $\mathbf{M}$ is thus calculated as

$$\text{adj }\mathbf{M} =$$

$$\begin{bmatrix} \bar{X}_5\left(\bar{X}'_5\bar{X}_5 - c_5\bar{Y}_5\right) & -\bar{X}_5\left(\bar{Z}'_5\bar{X}_5 - \bar{X}'_5\bar{Y}_5\right) & \bar{X}_5\left(\bar{Z}'_5c_5 - \bar{X}'_5X'_5\right) & -\bar{X}_5\left(c_5X'_5 + \bar{Z}'_5\bar{X}'_5\right) + \bar{Y}_5\left(c_5^2 + \bar{X}'^2_5\right) \\ -\bar{Y}_5\left(X'_5\bar{X}_5 - c_5\bar{Y}_5\right) & \bar{Y}_5\left(\bar{Z}'_5\bar{X}_5 - \bar{X}'_5\bar{Y}_5\right) & -\bar{Y}_5\left(\bar{Z}'_5c_5 - \bar{X}'_5X'_5\right) & \bar{X}_5\left(X'^2_5 + \bar{Z}'^2_5\right) - \bar{Y}_5\left(c_5X'_5 + \bar{Z}'_5\bar{X}'_5\right) \\ \bar{X}_5\left(\bar{Z}'_5\bar{X}_5 - \bar{Y}_5\bar{X}'_5\right) & \bar{X}_5\left(X'_5\bar{X}_5 - \bar{Y}_5c_5\right) & -\bar{X}_5\left(X'_5c_5 + \bar{Z}'_5\bar{X}'_5\right) + \bar{Y}_5\left(c_5^2 + \bar{X}'^2_5\right) & \bar{X}_5\left(X'_5\bar{X}'_5 - \bar{Z}'_5c_5\right) \\ \bar{Y}_5\left(\bar{Y}'_5\bar{X}'_5 - Z'_5\bar{X}_5\right) & \bar{Y}_5\left(\bar{Y}_5c_5 - X'_5\bar{X}_5\right) & \bar{X}_5\left(X'^2_5 + \bar{Z}'^2_5\right) - \bar{Y}_5\left(X'_5c_5 + X'_5\bar{Z}'_5\right) & \bar{Y}_5\left(\bar{Z}'_5c_5 - X'_5\bar{X}'_5\right) \end{bmatrix}. \tag{9.177}$$

This equation can be simplified by expanding the terms in parentheses to yield the intermediate results

$$X'_5\bar{X}_5 - c_5\bar{Y}_5 = -Y_5, \tag{9.178}$$

$$\bar{Z}'_5\bar{X}_5 - Y_5\bar{X}'_5 = X_5, \tag{9.179}$$

$$Z'_5c_5 - \bar{X}'_5X'_5 = -Z_5, \tag{9.180}$$

$$X'_5c_5 + \bar{X}'_5\bar{Z}'_5 = -\bar{X}_5\bar{Y}_5, \tag{9.181}$$

$$X'^2_5 + \bar{Z}'^2_5 = -\bar{Y}^2_5 + 1. \tag{9.182}$$

Using these results, Eq. (9.177) may be rewritten as

$$\text{adj}\,\mathbf{M} = \begin{bmatrix} -\bar{X}_5Y_5 & -\bar{X}_5X_5 & -\bar{X}_5Z_5 & \bar{Y}_5 \\ \bar{Y}_5Y_5 & \bar{Y}_5X_5 & \bar{Y}_5Z_5 & \bar{X}_5 \\ \bar{X}_5X_5 & -\bar{X}_5Y_5 & \bar{Y}_5 & \bar{X}_5Z_5 \\ -\bar{Y}_5X_5 & \bar{Y}_5Y_5 & \bar{X}_5 & -\bar{Y}_5Z_5 \end{bmatrix}. \tag{9.183}$$

The determinant of $\mathbf{M}$ is expanded as

$$|\mathbf{M}| = \bar{X}_5(\bar{X}_5X'^2_5 - 2X'_5\bar{Y}_5c_5 + \bar{X}_5\bar{Z}'^2_5) + \bar{Y}_5(\bar{Y}_5c^2_5 + \bar{Y}_5\bar{X}'^2_5 - 2\bar{X}_5\bar{X}'_5\bar{Z}'_5). \tag{9.184}$$

This can be regrouped as

$$|\mathbf{M}| = X'^2_5\bar{X}^2_5 + \bar{X}'^2_5\bar{Y}^2_5 - 2\bar{X}_5\bar{Y}_5(X'_5c_5 + \bar{X}'_5\bar{Z}'_5) + \bar{X}^2_5\bar{Z}'^2_5 + \bar{Y}^2_5c^2_5. \tag{9.185}$$

The term in parentheses in this equation can be replaced by using Eq. (9.181) to give

$$|\mathbf{M}| = X'^2_5\bar{X}^2_5 + \bar{X}'^2_5\bar{Y}^2_5 - 2\bar{X}_5\bar{Y}_5(-\bar{X}_5\bar{Y}_5) + \bar{X}^2_5\bar{Z}'^2_5 + \bar{Y}^2_5c^2_5. \tag{9.186}$$

This equation is next regrouped as

$$|\mathbf{M}| = X'^2_5\bar{X}^2_5 + \bar{X}'^2_5\bar{Y}^2_5 + \bar{X}^2_5(2\bar{Y}^2_5 + \bar{Z}'^2_5) + \bar{Y}^2_5c^2_5. \tag{9.187}$$

Equation (9.182) can be rearranged so that the quantity $(\bar{Y}^2_5 + \bar{Z}'^2_5)$ can be replaced by $(1 - X'^2_5)$ in the previous equation to yield

$$|\mathbf{M}| = X'^2_5\bar{X}^2_5 + \bar{X}'^2_5\bar{Y}^2_5 + \bar{X}^2_5(\bar{Y}^2_5 + 1 - X'^2_5) + \bar{Y}^2_5c^2_5. \tag{9.188}$$

This equation is now rearranged as

$$|\mathbf{M}| = \bar{X}'^2_5\bar{Y}^2_5 + \bar{X}^2_5(\bar{Y}^2_5 + 1) + \bar{Y}^2_5c^2_5. \tag{9.189}$$

Regrouping this equation again yields

$$|\mathbf{M}| = \bar{X}^2_5 + \bar{Y}^2_5(\bar{X}'^2_5 + \bar{X}^2_5 + c^2_5). \tag{9.190}$$

After substituting the definitions for $\bar{X}'_5$ and $\bar{X}_5$, and recognizing that $(s^2_{56} + c^2_{56}) = 1$ and

$(s_5^2 + c_5^2) = 1$, the determinant of the matrix **M** can be written as

$$|\mathbf{M}| = \bar{X}_5^2 + \bar{Y}_5^2. \tag{9.191}$$

Now that the adjoint and determinant of **M** have been expressed in terms of the constant mechanism parameters, the matrix $\mathbf{M}^{-1}$ as defined in Eq. (9.176) can be substituted into Eq. (9.175) to solve for the terms c_4x_6, s_4x_6, c_4, and s_4. These four terms may be written as

$$\begin{aligned} c_4x_6 &= P_{11}x_6 + P_{12}, \\ s_4x_6 &= P_{21}x_6 + P_{22}, \\ c_4 &= P_{31}x_6 + P_{32}, \\ s_4 &= P_{41}x_6 + P_{42}, \end{aligned} \tag{9.192}$$

where

$$\begin{aligned} P_{11} &= K(\bar{X}_5Y_5X_{217} + \bar{X}_5X_5Y_{217} + \bar{Y}_5Z_{217} - c_{34}\bar{Y}_5Z_5), \\ P_{12} &= K(\bar{X}_5Y_5Y_{217} - \bar{X}_5X_5X_{217} - \bar{X}_5Z_5Z_{217} + c_{34}\bar{X}_5), \\ P_{21} &= K(-\bar{Y}_5Y_5X_{217} - \bar{Y}_5X_5Y_{217} + \bar{X}_5Z_{217} - c_{34}\bar{X}_5Z_5), \\ P_{22} &= K(\bar{Y}_5X_5X_{217} - \bar{Y}_5Y_5Y_{217} + \bar{Y}_5Z_5Z_{217} - c_{34}\bar{Y}_5), \\ P_{31} &= K(-\bar{X}_5X_5X_{217} + \bar{X}_5Y_5Y_{217} + \bar{X}_5Z_5Z_{217} - c_{34}\bar{X}_5), \\ P_{32} &= K(-\bar{X}_5Y_5X_{217} - \bar{X}_5X_5Y_{217} + \bar{Y}_5Z_{217} - c_{34}\bar{Y}_5Z_5), \\ P_{41} &= K(\bar{Y}_5X_5X_{217} - \bar{Y}_5Y_5Y_{217} - \bar{Y}_5Z_5Z_{217} + c_{34}\bar{Y}_5), \\ P_{42} &= K(\bar{Y}_5Y_5X_{217} + \bar{Y}_5X_5Y_{217} + \bar{X}_5Z_{217} - c_{34}\bar{X}_5Z_5), \\ K &= \frac{1}{s_{34}\left(\bar{X}_5^2 + \bar{Y}_5^2\right)}. \end{aligned} \tag{9.193}$$

Substituting Eq. (9.192) into Eqs. (9.153) through (9.156) and regrouping gives

$$[c_{56}(H_1 - K_{12}P_{21} - K_{11}P_{11} - J_1) + K_{21}P_{31} + K_{22}P_{41}]x_6 + [c_{56}(-K_{12}P_{22} + H_2 - K_{11}P_{12}) + K_{21}P_{32} + J_2 + K_{22}P_{42}] = 0, \tag{9.194}$$

$$[c_{56}(H_2 - K_{11}P_{31} - K_{12}P_{41}) - K_{22}P_{21} - K_{21}P_{11} - J_2]x_6 + [c_{56}(-H_1 - K_{11}P_{32} - J_1 - K_{12}P_{42}) - K_{22}P_{22} - K_{21}P_{12}] = 0, \tag{9.195}$$

$$[-S_6Y_{217} + J_3 + K_{32}P_{21} + K_{41}P_{31} + K_{31}P_{11} + K_{42}P_{41} + X_{0217}]x_6 + [K_{42}P_{42} + J_4 + K_{32}P_{22} + Y_{0217} + K_{31}P_{12} + K_{41}P_{32} + S_6X_{217}] = 0, \tag{9.196}$$

$$[S_6X_{217} - J_4 - K_{42}P_{21} + K_{31}P_{31} - K_{41}P_{11} + K_{32}P_{41} + Y_{0217}]x_6 + [K_{32}P_{42} + J_3 - K_{42}P_{22} - X_{0217} - K_{41}P_{12} + K_{31}P_{32} + S_6Y_{217}] = 0. \tag{9.197}$$

The variable parameters in these four equations are the angles θ_1, θ_2, and θ_6. Because these four equations are linear in the tan-half-angle of θ_6, they may be written in the form

$$(A_i c_2 + B_i s_2 + D_i)x_6 + (E_i c_2 + F_i s_2 + G_i) = 0, \qquad i = 1 \ldots 4, \tag{9.198}$$

where

$$\begin{aligned} A_i &= A_{i,1}\, c_1 + A_{i,2}\, s_1 + A_{i,3}, \\ &\vdots \\ G_i &= G_{i,1}\, c_1 + G_{i,2}\, s_1 + G_{i,3}. \end{aligned} \tag{9.199}$$

The following definitions are provided that are used to expand the terms in the four equations that contain the angles θ_1 and θ_2:

$$\begin{aligned}
&X_{217} = X_{21}c_7 - Y_{21}s_7, && Y_{17} = c_{67}(\bar{X}_1 s_7 + \bar{Y}_1 c_7) - s_{67}\bar{Z}_1, \\
&Y_{217} = c_{67}(X_{21}s_7 + Y_{21}c_7) - s_{67}Z_{21}, && Z_{17} = Z_{71}, \\
&Z_{217} = s_{67}(X_{21}s_7 + Y_{21}c_7) + c_{67}Z_{21}, && U_{2176} = U_{217}c_{67} - V_{217}s_{67}, \\
&X_{21} = \bar{X}_2 c_1 - \bar{Y}_2 s_1, && U^*_{2176} = U_{217}s_{67} + V_{217}c_{67}, \\
&Y_{21} = c_{71}(\bar{X}_2 s_1 + Y_2 c_1) - s_{71}\bar{Z}_2, && U_{217} = U_{21}c_{71} - V_{21}s_{71}, \\
&Z_{21} = s_{71}(\bar{X}_2 s_1 + \bar{Y}_2 c_1) + c_{71}\bar{Z}_2, && V_{217} = c_7(U_{21}s_{71} + V_{21}c_{71}) - s_7 W_{21}, \\
&\bar{X}_2 = s_{23}s_2, && W_{217} = s_7(U_{21}s_{71} + V_{21}c_{71}) + c_7 W_{21}, \\
&\bar{Y}_2 = -(s_{12}c_{23} + c_{12}s_{23}c_2), && U_{21} = s_{12}s_2, \\
&\bar{Z}_2 = c_{12}c_{23} - s_{12}s_{23}c_2, && U_{17} = s_{71}s_1, \\
&X_{712} = X_{71}c_2 - Y_{71}s_2, && V_{21} = -(s_1 c_2 + c_1 s_2 c_{12}), \\
&Y_{712} = c_{23}(X_{71}s_2 + Y_{71}c_2) - s_{23}Z_{71}, && W_{21} = c_1 c_2 - s_1 s_2 c_{12}, \\
&Z_{712} = Z_{217}, && U_{176} = U_{17}c_{67} - V_{17}s_{67}, \\
&X_{71} = X_7 c_1 - Y_7 s_1, && U^*_{176} = U_{17}s_{67} + V_{17}c_{67}, \\
&Y_{71} = c_{12}(X_7 s_1 + Y_7 c_1) - s_{12}Z_7, && V_{17} = -(s_7 c_1 + c_7 s_1 c_{71}), \\
&Z_{71} = s_{12}(X_7 s_1 + Y_7 c_1) + c_{12}Z_7, && W_{17} = c_7 c_1 - s_7 s_1 c_{71}. \\
&X_{17} = \bar{X}_1 c_7 - \bar{Y}_1 s_7, &&
\end{aligned} \tag{9.200}$$

The terms X_{0217}, Y_{0217}, and Z_{0217} are expanded as follows:

$$\begin{aligned}
X_{0217} &= -S_7(X_{21}s_7 + Y_{21}c_7) + X_{021}c_7 - Y_{021}s_7, \\
Y_{0217} &= -a_{67}Z_{217} + S_7 c_{67} X_{217} + c_{67}(X_{021}s_7 + Y_{021}c_7) - s_{67}Z_{021}, \\
Z_{0217} &= a_{67}Y_{217} + S_7 s_{67} X_{217} + s_{67}(X_{021}s_7 + Y_{021}c_7) + c_{67}Z_{021}, \\
X_{021} &= -S_1(\bar{X}_2 s_1 + \bar{Y}_2 c_1) + \bar{X}_{02}c_1 - \bar{Y}_{02}s_1, \\
Y_{021} &= -a_{71}Z_{21} + S_1 c_{71} X_{21} + c_{71}(\bar{X}_{02}s_1 + \bar{Y}_{02}c_1) - s_{71}\bar{Z}_{02}, \\
Z_{021} &= a_{71}Y_{21} + S_1 s_{71} X_{21} + s_{71}(\bar{X}_{02}s_1 + \bar{Y}_{02}c_1) + c_{71}\bar{Z}_{02}, \\
\bar{X}_{02} &= a_{23}c_{23}s_2 + S_2 s_{23} c_2, \\
\bar{Y}_{02} &= S_2 c_{12} s_{23} s_2 - a_{12}Z_2 + a_{23}(s_{12}s_{23} - c_{12}c_{23}c_2), \\
\bar{Z}_{02} &= S_2 s_{12} s_{23} s_2 + a_{12}\bar{Y}_2 + a_{23}Y_2.
\end{aligned} \tag{9.201}$$

Using these definitions, the coefficients of Eqs. (9.194) through (9.197) may be written as

$$
\begin{aligned}
A_{1,1} = {} & \bar{X}_5 c_{12} s_{23} K\big[K_{11} c_{56}(-Y_5 \bar{X}_7' - X_5 Z_7') - K_{12} c_{56} Y_7 \\
& + K_{21}(-X_5 \bar{X}_7' + Y_5 Z_7' + Z_5 Y_7)\big] + \bar{Y}_5 c_{12} s_{23} K\big[-K_{11} c_{56} Y_7 \\
& + K_{22}(X_5 \bar{X}_7' - Y_5 Z_7' - Z_5 Y_7) + c_{56} K_{12}(X_5 Z_7' + Y_5 \bar{X}_7')\big] \\
& + c_{56}(a_{23} c_7 + S_3 c_{12} s_{23} c_{71} s_7), \\
A_{1,2} = {} & \bar{X}_5 c_{12} s_{23} K\big[-K_{11} c_{56}(X_5 X_7' + Y_5 c_7) - K_{12} c_{56} X_7 \\
& + K_{21}(-X_5 c_7 + Y_5 X_7' + Z_5 X_7)\big] + \bar{Y}_5 c_{12} s_{23} K\big[-K_{11} c_{56} X_7 \\
& + K_{12} c_{56}(X_5 X_7' + Y_5 c_7) + K_{22}(X_5 c_7 - Y_5 X_7' - Z_5 X_7)\big] \\
& + c_{56}(-a_{23} \bar{X}_7' + S_3 c_{12} s_{23} c_7), \\
A_{1,3} = {} & \bar{X}_5 s_{12} s_{23} K\big[K_{11} c_{56}(X_5 \bar{Y}_7 + Y_5 \bar{X}_7) + K_{12} c_{56} \bar{Z}_7 + K_{21}(X_5 \bar{X}_7 - Y_5 \bar{Y}_7 - Z_5 \bar{Z}_7)\big] \\
& + \bar{Y}_5 s_{12} s_{23} K\big[K_{11} c_{56} \bar{Z}_7 + K_{12} c_{56}(-X_5 \bar{Y}_7 - Y_5 \bar{X}_7) \\
& + K_{22}(-X_5 \bar{X}_7 + Y_5 \bar{Y}_7 + Z_5 \bar{Z}_7)\big] - S_3 s_{12} s_{23} c_{56} \bar{X}_7, \\
B_{1,1} = {} & \bar{X}_5 s_{23} K\big[-K_{11} c_{56}(X_5 X_7' + Y_5 c_7) - K_{12} c_{56} X_7 + K_{21}(-X_5 c_7 + Y_5 X_7' + Z_5 X_7)\big] \\
& + \bar{Y}_5 s_{23} K\big[-K_{11} c_{56} X_7 + K_{12} c_{56}(X_5 X_7' + Y_5 c_7) + K_{22}(X_5 c_7 - Y_5 X_7' - Z_5 X_7)\big] \\
& + c_{56}(-a_{23} c_{12} \bar{X}_7' + S_3 s_{23} c_7), \\
B_{1,2} = {} & \bar{X}_5 s_{23} K\big[K_{11} c_{56}(X_5 Z_7' + Y_5 \bar{X}_7') + K_{12} c_{56} Y_7 + K_{21}(X_5 \bar{X}_7' - Y_5 Z_7' - Z_5 Y_7)\big] \\
& + \bar{Y}_5 s_{23} K\big[K_{11} c_{56} Y_7 + K_{12} c_{56}(-X_5 Z_7' - Y_5 \bar{X}_7') \\
& + K_{22}(-X_5 \bar{X}_7' + Y_5 Z_7' + Z_5 Y_7)\big] - c_{56}(a_{23} c_{12} c_7 + S_3 s_{23} \bar{X}_7'), \\
B_{1,3} = {} & a_{23} s_{12} c_{56} \bar{X}_7, \\
D_{1,1} = {} & \bar{X}_5 s_{12} c_{23} K\big[K_{11} c_{56}(-X_5 Z_7' - Y_5 \bar{X}_7') - K_{12} c_{56} Y_7 \\
& + K_{21}(-X_5 \bar{X}_7' + Y_5 Z_7' + Z_5 Y_7)\big] + \bar{Y}_5 s_{12} c_{23} K\big[-K_{11} c_{56} Y_7 \\
& + K_{12} c_{56}(X_5 Z_7' + Y_5 \bar{X}_7') + K_{22}(X_5 \bar{X}_7' - Y_5 Z_7' - Z_5 Y_7)\big] \\
& + c_{56}(a_{12} c_7 + S_2 s_{12} \bar{X}_7' + S_3 s_{12} c_{23} \bar{X}_7'), \\
D_{1,2} = {} & \bar{X}_5 s_{12} c_{23} K\big[-K_{11} c_{56}(X_5 X_7' + Y_5 c_7) - K_{12} c_{56} X_7 \\
& + K_{21}(-X_5 c_7 + Y_5 X_7' + Z_5 X_7)\big] + \bar{Y}_5 s_{12} c_{23} K\big[-K_{11} c_{56} X_7 \\
& + K_{12} c_{56}(X_5 X_7' + Y_5 c_7) + K_{22}(X_5 c_7 - Y_5 X_7' - Z_5 X_7)\big] \\
& + c_{56}(-a_{12} \bar{X}_7' + S_2 s_{12} c_7 + S_3 s_{12} c_{23} c_7), \\
D_{1,3} = {} & \bar{X}_5 K\big[K_{11} c_{12} c_{23} c_{56}(-X_5 \bar{Y}_7 - Y_5 \bar{X}_7) + K_{12} c_{56}(c_{34} Z_5 - c_{12} c_{23} Z_7) \\
& + K_{21}(-c_{34} + c_{12} c_{23}(-X_5 \bar{X}_7 + Y_5 \bar{Y}_7 + Z_5 Z_7))\big] \\
& + \bar{Y}_5 K\big[K_{11} c_{56}(c_{34} Z_5 - c_{12} c_{23} Z_7) + K_{12} c_{12} c_{23} c_{56}(X_5 \bar{Y}_7 + Y_5 \bar{X}_7) \\
& + K_{22}(c_{34} + c_{12} c_{23}(X_5 \bar{X}_7 - Y_5 \bar{Y}_7 - Z_5 Z_7))\big] \\
& + c_{56}(-J_1 + S_1 \bar{X}_7 + S_2 c_{12} \bar{X}_7 + S_3 c_{12} c_{23} \bar{X}_7 + a_{67} + a_{71} c_7), \\
E_{1,1} = {} & \bar{X}_5 c_{12} s_{23} K\big[K_{11} c_{56}(X_5 \bar{X}_7' - Y_5 Z_7' + Z_5 Y_7) + K_{21}(-X_5 Z_7' - Y_5 \bar{X}_7') + K_{22} Y_7\big] \\
& + \bar{Y}_5 c_{12} s_{23} K\big[K_{12} c_{56}(-X_5 \bar{X}_7' + Y_5 Z_7' - Z_5 Y_7) + K_{21} Y_7 + K_{22}(X_5 Z_7' + Y_5 \bar{X}_7')\big] \\
& + c_{56}(S_3 c_{12} s_{23} Z_7' + a_{23} X_7') - s_{56}(a_{23} X_7 + S_3 c_{12} s_{23} Y_7),
\end{aligned}
$$

$$
\begin{aligned}
E_{1,2} = {} & \bar{X}_5 c_{12} s_{23} K\big[K_{11} c_{56}\big(X_5 c_7 - Y_5 X'_7 + Z_5 X_7\big) - K_{21}\big(X_5 X'_7 + Y_5 c_7\big) + K_{22} X_7\big] \\
& + \bar{Y}_5 c_{12} s_{23} K\big[K_{12} c_{56}\big(-X_5 c_7 + Y_5 X'_7 - Z_5 X_7\big) \\
& + K_{21} X_7 + K_{22}\big(X_5 X'_7 + Y_5 c_7\big)\big] + c_{56}\big(-a_{23} Z'_7 + S_3 c_{12} s_{23} X'_7\big) \\
& + s_{56}\big(a_{23} Y_7 - S_3 c_{12} s_{23} X_7\big), \\
E_{1,3} = {} & \bar{X}_5 s_{12} s_{23} K\big[K_{11} c_{56}(-X_5 \bar{X}_7 + Y_5 \bar{Y}_7 - Z_5 Z_7) + K_{21}(X_5 \bar{Y}_7 + Y_5 \bar{X}_7) - K_{22} Z_7\big] \\
& + \bar{Y}_5 s_{12} s_{23} K\big[K_{12} c_{56}(X_5 \bar{X}_7 - Y_5 \bar{Y}_7 + Z_5 Z_7) - K_{21} Z_7 - K_{22}(X_5 \bar{Y}_7 + Y_5 \bar{X}_7)\big] \\
& + S_3 s_{12} s_{23}(s_{56} Z_7 - c_{56} \bar{Y}_7), \\
F_{1,1} = {} & \bar{X}_5 s_{23} K\big[K_{11} c_{56}\big(X_5 c_7 - Y_5 X'_7 + Z_5 X_7\big) - K_{21}\big(X_5 X'_7 + Y_5 c_7\big) + K_{22} X_7\big] \\
& + \bar{Y}_5 s_{23} K\big[K_{12} c_{56}\big(-X_5 c_7 + Y_5 X'_7 - Z_5 X_7\big) + K_{21} X_7 + K_{22}\big(X_5 X'_7 + Y_5 c_7\big)\big] \\
& + a_{23} c_{12}\big(s_{56} Y_7 - c_{56} Z'_7\big) + S_3 s_{23}\big(-s_{56} X_7 + c_{56} X'_7\big), \\
F_{1,2} = {} & \bar{X}_5 s_{23} K\big[K_{11} c_{56}\big(-X_5 \bar{X}'_7 + Y_5 Z'_7 - Z_5 Y_7\big) + K_{21}\big(X_5 Z'_7 + Y_5 \bar{X}'_7\big) - K_{22} Y_7\big] \\
& + \bar{Y}_5 s_{23} K\big[K_{12} c_{56}\big(X_5 \bar{X}'_7 - Y_5 Z'_7 + Z_5 Y_7\big) - K_{21} Y_7 - K_{22}\big(X_5 Z'_7 + Y_5 \bar{X}'_7\big)\big] \\
& + a_{23} c_{12}\big(s_{56} X_7 - c_{56} X'_7\big) + S_3 s_{23}\big(s_{56} Y_7 - c_{56} Z'_7\big), \\
F_{1,3} = {} & a_{23} s_{12}(c_{56} \bar{Y}_7 - s_{56} Z_7), \\
G_{1,1} = {} & \bar{X}_5 s_{12} c_{23} K\big[K_{11} c_{56}\big(X_5 \bar{X}'_7 - Y_5 Z'_7 + Z_5 Y_7\big) - K_{21}\big(X_5 Z'_7 + Y_5 \bar{X}'_7\big) + K_{22} Y_7\big] \\
& + \bar{Y}_5 s_{12} c_{23} K\big[K_{12} c_{56}\big(-X_5 \bar{X}'_7 + Y_5 Z'_7 - Z_5 Y_7\big) + K_{21} Y_7 + K_{22}\big(X_5 Z'_7 + Y_5 \bar{X}'_7\big)\big] \\
& + S_2 s_{12}\big(-s_{56} Y_7 + c_{56} Z'_7\big) + S_3 s_{12}\big(-c_{23} s_{56} Y_7 + c_{23} c_{56} Z'_7\big) \\
& + a_{12}\big(-s_{56} X_7 + c_{56} X'_7\big), \\
G_{1,2} = {} & \bar{X}_5 s_{12} c_{23} K\big[K_{11} c_{56}\big(X_5 c_7 - Y_5 X'_7 + Z_5 X_7\big) - K_{21}\big(X_5 X'_7 + Y_5 c_7\big) + K_{22} X_7\big] \\
& + \bar{Y}_5 s_{12} c_{23} K\big[K_{12} c_{56}\big(-X_5 c_7 + Y_5 X'_7 - Z_5 X_7\big) + K_{21} X_7 + K_{22}\big(X_5 X'_7 + Y_5 c_7\big)\big] \\
& + S_2 s_{12}\big(-s_{56} X_7 + c_{56} X'_7\big) + S_3 s_{12} c_{23}\big(-s_{56} X_7 + c_{56} X'_7\big) + a_{12}\big(s_{56} Y_7 - c_{56} Z'_7\big), \\
G_{1,3} = {} & \bar{X}_5 K\big[K_{11} c_{56}(-c_{34} + c_{12} c_{23}(X_5 \bar{X}_7 - Y_5 \bar{Y}_7 + Z_5 Z_7)) - K_{21} c_{12} c_{23}(X_5 \bar{Y}_7 \\
& + Y_5 \bar{X}_7) + K_{22}(-c_{34} Z_5 + c_{12} c_{23} Z_7)\big] + \bar{Y}_5 K\big[K_{12} c_{56}(c_{34} + c_{12} c_{23}(-X_5 \bar{X}_7 \\
& + Y_5 \bar{Y}_7 - Z_5 Z_7)) + K_{21}(-c_{34} Z_5 + c_{12} c_{23} Z_7) + K_{22} c_{12} c_{23}(X_5 \bar{Y}_7 + Y_5 \bar{X}_7)\big] \\
& + S_1(c_{56} \bar{Y}_7 - s_{56} Z_7) + S_2 c_{12}(-s_{56} Z_7 + c_{56} \bar{Y}_7) + S_3 c_{12} c_{23}(-s_{56} Z_7 + c_{56} \bar{Y}_7) \\
& + S_4 s_{45} c_5 - S_6 s_{56} - S_7(s_{56} c_{67} + c_{56} s_{67}) - a_{45} s_5 + a_{71}(-s_{56} X_7 + c_{56} X'_7),
\end{aligned}
\tag{9.202}
$$

$$
\begin{aligned}
A_{2,1} = {} & \bar{X}_5 c_{12} s_{23} K\big[K_{11} c_{56}\big(X_5 \bar{X}'_7 - Y_5 Z'_7 - Z_5 Y_7\big) - K_{21}\big(X_5 Z'_7 + Y_5 \bar{X}'_7\big) - K_{22} Y_7\big] \\
& + \bar{Y}_5 c_{12} s_{23} K\big[K_{12} c_{56}\big(-X_5 \bar{X}'_7 + Y_5 Z'_7 + Z_5 Y_7\big) - K_{21} Y_7 + K_{22}\big(X_5 Z'_7 + Y_5 \bar{X}'_7\big)\big] \\
& + a_{23}\big(s_{56} X_7 + c_{56} X'_7\big) + S_3 c_{12} s_{23}\big(s_{56} Y_7 + c_{56} Z'_7\big), \\
A_{2,2} = {} & \bar{X}_5 c_{12} s_{23} K\big[K_{11} c_{56}\big(X_5 c_7 - Y_5 X'_7 - Z_5 X_7\big) - K_{21}\big(X_5 X'_7 + Y_5 c_7\big) - K_{22} X_7\big] \\
& + \bar{Y}_5 c_{12} s_{23} K\big[K_{12} c_{56}\big(-X_5 c_7 + Y_5 X'_7 + Z_5 X_7\big) - K_{21} X_7 + K_{22}\big(X_5 X'_7 + Y_5 c_7\big)\big] \\
& - a_{23}\big(s_{56} Y_7 + c_{56} Z'_7\big) + S_3 c_{12} s_{23}\big(s_{56} X_7 + c_{56} X'_7\big), \\
A_{2,3} = {} & \bar{X}_5 s_{12} s_{23} K\big[K_{11} c_{56}(-X_5 \bar{X}_7 + Y_5 \bar{Y}_7 + Z_5 Z_7) + K_{21}(X_5 \bar{Y}_7 + Y_5 \bar{X}_7) + K_{22} Z_7\big] \\
& + \bar{Y}_5 s_{12} s_{23} K\big[K_{12} c_{56}(X_5 \bar{X}_7 - Y_5 \bar{Y}_7 - Z_5 Z_7) + K_{21} Z_7 - K_{22}(X_5 \bar{Y}_7 + Y_5 \bar{X}_7)\big] \\
& - S_3 s_{12} s_{23}(s_{56} Z_7 + c_{56} \bar{Y}_7),
\end{aligned}
$$

$$\begin{aligned}
B_{2,1} = {} & \bar{X}_5 s_{23} K\big[K_{11} c_{56}\big(X_5 c_7 - Y_5 X_7' - Z_5 X_7\big) - K_{21}\big(X_5 X_7' + Y_5 c_7\big) - K_{22} X_7\big] \\
& + \bar{Y}_5 s_{23} K\big[K_{12} c_{56}\big(-X_5 c_7 + Y_5 X_7' + Z_5 X_7\big) - K_{21} X_7 + K_{22}\big(X_5 X_7' + Y_5 c_7\big)\big] \\
& - a_{23} c_{12}\big(s_{56} Y_7 + c_{56} Z_7'\big) + S_3 s_{23}\big(s_{56} X_7 + c_{56} X_7'\big), \\
B_{2,2} = {} & \bar{X}_5 s_{23} K\big[K_{11} c_{56}\big(-X_5 \bar{X}_7' + Y_5 Z_7' + Z_5 Y_7\big) + K_{21}\big(X_5 Z_7' + Y_5 \bar{X}_7'\big) + K_{22} Y_7\big] \\
& + \bar{Y}_5 s_{23} K\big[K_{12} c_{56}\big(X_5 \bar{X}_7' - Y_5 Z_7' - Z_5 Y_7\big) + K_{21} Y_7 - K_{22}\big(X_5 Z_7' + Y_5 \bar{X}_7'\big)\big] \\
& - a_{23} c_{12}\big(s_{56} X_7 + c_{56} X_7'\big) - S_3 s_{23}\big(s_{56} Y_7 + c_{56} Z_7'\big), \\
B_{2,3} = {} & a_{23} s_{12}(s_{56} Z_7 + c_{56} \bar{Y}_7), \\
D_{2,1} = {} & \bar{X}_5 s_{12} c_{23} K\big[K_{11} c_{56}\big(X_5 \bar{X}_7' - Y_5 Z_7' - Z_5 Y_7\big) - K_{21}\big(X_5 Z_7' + Y_5 \bar{X}_7'\big) - K_{22} Y_7\big] \\
& + \bar{Y}_5 s_{12} c_{23} K\big[K_{12} c_{56}\big(-X_5 \bar{X}_7' + Y_5 Z_7' + Z_5 Y_7\big) - K_{21} Y_7 + K_{22}\big(X_5 Z_7' + Y_5 \bar{X}_7'\big)\big] \\
& + a_{12}\big(s_{56} X_7 + c_{56} X_7'\big) + S_2 s_{12}\big(s_{56} Y_7 + c_{56} Z_7'\big) + S_3 s_{12} c_{23}\big(s_{56} Y_7 + c_{56} Z_7'\big), \\
D_{2,2} = {} & \bar{X}_5 s_{12} c_{23} K\big[K_{11} c_{56}\big(X_5 c_7 - Y_5 X_7' - Z_5 X_7\big) - K_{21}\big(X_5 X_7' + Y_5 c_7\big) - K_{22} X_7\big] \\
& + \bar{Y}_5 s_{12} c_{23} K\big[K_{12} c_{56}\big(-X_5 c_7 + Y_5 X_7' + Z_5 X_7\big) - K_{21} X_7 + K_{22}\big(X_5 X_7' + Y_5 c_7\big)\big] \\
& - a_{12}\big(s_{56} Y_7 + c_{56} Z_7'\big) + S_2 s_{12}\big(s_{56} X_7 + c_{56} X_7'\big) + S_3 s_{12} c_{23}\big(s_{56} X_7 + c_{56} X_7'\big), \\
D_{2,3} = {} & \bar{X}_5 K\big[K_{11} c_{56}(c_{34} + c_{12} c_{23}(X_5 \bar{X}_7 - Y_5 \bar{Y}_7 - Z_5 Z_7)) - K_{21} c_{12} c_{23}(X_5 \bar{Y}_7 + Y_5 \bar{X}_7) \\
& + K_{22}(c_{34} Z_5 - c_{12} c_{23} \bar{Z}_7)\big] + \bar{Y}_5 K\big[K_{12} c_{56}(-c_{34} + c_{12} c_{23}(-X_5 \bar{X}_7 + Y_5 \bar{Y}_7 \\
& + Z_5 \bar{Z}_7)) + K_{21}(c_{34} Z_5 - c_{12} c_{23} \bar{Z}_7) + K_{22} c_{12} c_{23}(X_5 \bar{Y}_7 + Y_5 \bar{X}_7)\big] \\
& + S_1(s_{56} \bar{Z}_7 + c_{56} \bar{Y}_7) + S_2 c_{12}(s_{56} \bar{Z}_7 + c_{56} \bar{Y}_7) + S_3 c_{12} c_{23}(s_{56} \bar{Z}_7 + c_{56} \bar{Y}_7) \\
& - S_4 s_{45} c_5 + S_6 s_{56} + S_7(s_{56} c_{67} - c_{56} s_{67}) + a_{45} s_5 + a_{71}(s_{56} X_7 + c_{56} X_7'), \\
E_{2,1} = {} & \bar{X}_5 c_{12} s_{23} K\big[K_{11} c_{56}\big(X_5 Z_7' + Y_5 \bar{X}_7'\big) - K_{12} c_{56} Y_7 + K_{21}\big(X_5 \bar{X}_7' - Y_5 Z_7' + Z_5 Y_7\big)\big] \\
& + \bar{Y}_5 c_{12} s_{23} K\big[-K_{11} c_{56} Y_7 - K_{12} c_{56}\big(X_5 Z_7' + Y_5 \bar{X}_7'\big) \\
& + K_{22}\big(-X_5 \bar{X}_7' + Y_5 Z_7' - Z_5 Y_7\big)\big] - c_{56}\big(a_{23} c_7 + S_3 c_{12} s_{23} \bar{X}_7'\big), \\
E_{2,2} = {} & \bar{X}_5 c_{12} s_{23} K\big[K_{11} c_{56}\big(Y_5 c_7 + X_5 X_7'\big) - K_{12} c_{56} X_7 + K_{21}\big(X_5 c_7 - Y_5 X_7' + Z_5 X_7\big)\big] \\
& + \bar{Y}_5 c_{12} s_{23} K\big[-K_{11} c_{56} X_7 - K_{12} c_{56}\big(X_5 X_7' + Y_5 c_7\big) \\
& + K_{22}\big(-X_5 c_7 + Y_5 X_7' - Z_5 X_7\big)\big] + c_{56}\big(a_{23} \bar{X}_7' - S_3 c_{12} s_{23} c_7\big), \\
E_{2,3} = {} & \bar{X}_5 s_{12} s_{23} K\big[-K_{11} c_{56}(X_5 \bar{Y}_7 + Y_5 \bar{X}_7) + K_{12} c_{56} Z_7 \\
& + K_{21}(-X_5 \bar{X}_7 + Y_5 \bar{Y}_7 - Z_5 Z_7)\big] + \bar{Y}_5 s_{12} s_{23} K\big[K_{11} c_{56} Z_7 \\
& + K_{12} c_{56}(X_5 \bar{Y}_7 + Y_5 \bar{X}_7) + K_{22}(X_5 \bar{X}_7 - Y_5 \bar{Y}_7 + Z_5 Z_7)\big] + S_3 c_{56} \bar{X}_7, \\
F_{2,1} = {} & \bar{X}_5 s_{23} K\big[K_{11} c_{56}\big(X_5 X_7' + Y_5 c_7\big) - K_{12} c_{56} X_7 + K_{21}\big(X_5 c_7 - Y_5 X_7' + Z_5 X_7\big)\big] \\
& + \bar{Y}_5 s_{23} K\big[-K_{11} c_{56} X_7 - K_{12} c_{56}\big(X_5 X_7' + Y_5 c_7\big) + K_{22}\big(-X_5 c_7 + Y_5 X_7' - Z_5 X_7\big)\big] \\
& + c_{56}\big(a_{23} c_{12} \bar{X}_7' - S_3 s_{23} c_7\big), \\
F_{2,2} = {} & \bar{X}_5 s_{23} K\big[-K_{11} c_{56}\big(X_5 Z_7' + Y_5 \bar{X}_7'\big) + K_{12} c_{56} Y_7 + K_{21}\big(-X_5 \bar{X}_7' + Y_5 Z_7' - Z_5 Y_7\big)\big] \\
& + \bar{Y}_5 s_{23} K\big[K_{11} c_{56} Y_7 + K_{12} c_{56}\big(X_5 Z_7' + Y_5 \bar{X}_7\big) + K_{22}\big(X_5 \bar{X}_7' - Y_5 Z_7' + Z_5 Y_7\big)\big] \\
& + c_{56}\big(a_{23} c_{12} c_7 + S_3 s_{23} \bar{X}_7'\big), \\
F_{2,3} = {} & -a_{23} s_{12} c_{56} \bar{X}_7, \\
G_{2,1} = {} & \bar{X}_5 s_{12} c_{23} K\big[K_{11} c_{56}\big(X_5 Z_7' + Y_5 \bar{X}_7'\big) - K_{12} c_{56} Y_7 + K_{21}\big(X_5 \bar{X}_7' - Y_5 Z_7' + Z_5 Y_7\big)\big]
\end{aligned}$$

$$
\begin{aligned}
&+\bar{Y}_5 s_{12}c_{23}K\big[-K_{11}c_{56}Y_7 - K_{12}c_{56}\big(X_5Z_7' + Y_5\bar{X}_7'\big) + K_{22}\big(-X_5\bar{X}_7'\\
&+Y_5Z_7' - Z_5Y_7\big)\big] - c_{56}\big(a_{12}c_7 + S_2s_{12}\bar{X}_7' + S_3s_{12}c_{23}\bar{X}_7'\big),\\
G_{2,2} = {}&\bar{X}_5 s_{12}c_{23}K\big[K_{11}c_{56}\big(X_5X_7' + Y_5c_7\big) - K_{12}c_{56}X_7 + K_{21}\big(X_5c_7 - Y_5X_7' + Z_5X_7\big)\big]\\
&+\bar{Y}_5 s_{12}c_{23}K\big[-K_{11}c_{56}X_7 - K_{12}c_{56}\big(X_5X_7' + Y_5c_7\big) + K_{22}\big(-X_5c_7\\
&+Y_5X_7' - Z_5X_7\big)\big] + c_{56}\big(a_{12}\bar{X}_7' - S_2s_{12}c_7 - S_3s_{12}c_{23}c_7\big),\\
G_{2,3} = {}&\bar{X}_5K\big[K_{11}c_{12}c_{23}c_{56}(X_5\bar{Y}_7 + Y_5\bar{X}_7) + K_{12}c_{56}(c_{34}Z_5 - c_{12}c_{23}Z_7)\\
&+K_{21}(-c_{34} + c_{12}c_{23}(X_5\bar{X}_7 - Y_5\bar{Y}_7 + Z_5Z_7))\big] + \bar{Y}_5K\big[K_{11}c_{56}(c_{34}Z_5 - c_{12}c_{23}Z_7)\\
&+K_{12}c_{12}c_{23}c_{56}(-X_5\bar{Y}_7 - Y_5\bar{X}_7) + K_{22}(c_{34} + c_{12}c_{23}(-X_5\bar{X}_7 + Y_5\bar{Y}_7 - Z_5Z_7))\big]\\
&-c_{56}(J_1 + a_{67} + a_{71}c_7 + S_1\bar{X}_7 + S_2c_{12}\bar{X}_7 + S_3c_{12}c_{23}\bar{X}_7), \qquad (9.203)\\
A_{3,1} = {}&\bar{X}_5 c_{12}s_{23}K\big[K_{31}\big(X_5Z_7' + Y_5\bar{X}_7'\big) + K_{32}Y_7 + K_{41}\big(-X_5\bar{X}_7' + Y_5Z_7' + Z_5Y_7\big)\big]\\
&+Y_5 c_{12}s_{23}K\big[K_{31}Y_7 - K_{32}\big(X_5Z_7' + Y_5\bar{X}_7'\big) + K_{42}\big(X_5\bar{X}_7' - Y_5Z_7' - Z_5Y_7\big)\big]\\
&+s_5\big[s_{23}\big(a_{12}s_{12}Y_7 - a_{67}c_{12}Z_7' + a_{71}c_{12}Z_7 - S_1c_{12}X_7 - S_2X_7 - S_7c_{12}s_{67}\bar{X}_7'\big)\\
&-a_{23}c_{12}c_{23}Y_7\big]/\big(s_{56}c_5\big) + c_{12}s_{23}c_{45}\big[a_{67}\bar{X}_7' + S_6Y_7 - S_7s_{71}\big]/\big(s_{45}c_5\big)\\
&-c_{34}\big[a_{23}X_7 + S_3c_{12}s_{23}Y_7\big]/\big(s_{45}c_{56}c_5\big) + S_1c_{12}s_{23}c_7 + S_2s_{23}c_7 - S_6c_{12}s_{23}Z_7'\\
&+S_7c_{12}s_{23}c_{71}c_7 - a_{12}s_{12}s_{23}\bar{X}_7' + a_{23}c_{12}c_{23}\bar{X}_7' - a_{71}c_{12}s_{23}\bar{X}_7,\\
A_{3,2} = {}&\bar{X}_5 c_{12}s_{23}K\big[K_{31}\big(X_5X_7' + Y_5c_7\big) + K_{32}X_7 + K_{41}\big(-X_5c_7 + Y_5X_7' + Z_5X_7\big)\big]\\
&+\bar{Y}_5 c_{12}s_{23}K\big[K_{31}X_7 - K_{32}\big(X_5X_7' + Y_5c_7\big) + K_{42}\big(X_5c_7 - Y_5X_7' - Z_5X_7\big)\big]\\
&+s_5\big[a_{12}s_{12}s_{23}X_7 - a_{23}c_{12}c_{23}X_7 - a_{67}c_{12}s_{23}X_7' + S_1c_{12}s_{23}Y_7 + S_2s_{23}Y_7\\
&-S_7c_{12}s_{23}s_{67}c_7\big]/\big(s_{56}c_5\big) + c_{12}s_{23}\big[a_{67}c_{45}c_7 + a_{71}c_{45} + S_6c_{45}X_7\big]/\big(s_{45}c_5\big)\\
&+c_{34}\big[a_{23}Y_7 - S_3c_{12}s_{23}X_7\big]/\big(s_{45}c_{56}c_5\big) - S_1c_{12}s_{23}\bar{X}_7' - S_2s_{23}\bar{X}_7' - S_6c_{12}s_{23}X_7'\\
&-S_7c_{12}s_{23}s_7 - a_{12}s_{12}s_{23}c_7 + a_{23}c_{12}c_{23}c_7,\\
A_{3,3} = {}&\bar{X}_5 s_{12}s_{23}K\big[-K_{31}(X_5\bar{Y}_7 + Y_5\bar{X}_7) - K_{32}Z_7 + K_{41}(X_5\bar{X}_7 - Y_5\bar{Y}_7 - Z_5Z_7)\big]\\
&+\bar{Y}_5 s_{12}s_{23}K\big[-K_{31}Z_7 + K_{32}(X_5\bar{Y}_7 + Y_5\bar{X}_7) + K_{42}(-X_5\bar{X}_7 + Y_5\bar{Y}_7 + Z_5Z_7)\big]\\
&+s_5\big[a_{12}c_{12}s_{23}Z_7 + a_{23}s_{12}c_{23}Z_7 + a_{67}s_{12}s_{23}\bar{Y}_7 + a_{71}s_{12}s_{23}Y_7\\
&+S_7s_{12}s_{23}s_{67}\bar{X}_7\big]/(s_{56}c_5) - s_{12}s_{23}c_{45}\big[a_{67}\bar{X}_7 + S_1 + S_6Z_7 + S_7c_{71}\big]/(s_{45}c_5)\\
&+[S_3s_{12}s_{23}c_{34}Z_7]/(s_{45}c_{56}c_5) + S_6s_{12}s_{23}\bar{Y}_7 - S_7s_{12}s_{23}s_{71}c_7 - a_{12}c_{12}s_{23}\bar{X}_7\\
&-a_{23}s_{12}c_{23}\bar{X}_7 - a_{71}s_{12}s_{23}\bar{X}_7',\\
B_{3,1} = {}&\bar{X}_5 s_{23}K\big[K_{31}\big(X_5X_7' + Y_5c_7\big) + K_{32}X_7 + K_{41}\big(-X_5c_7 + Y_5X_7' + Z_5X_7\big)\big]\\
&+\bar{Y}_5 s_{23}K\big[K_{31}X_7 - K_{32}\big(X_5X_7' + Y_5c_7\big) + K_{42}\big(X_5c_7 - Y_5X_7' - Z_5X_7\big)\big]\\
&+s_5\big[S_1s_{23}Y_7 + S_2c_{12}s_{23}Y_7 - S_7s_{23}s_{67}c_7 - a_{23}c_{23}X_7 - a_{67}s_{23}X_7'\big]/\big(s_{56}c_5\big)\\
&+s_{23}c_{45}\big[S_6X_7 + a_{67}c_7 + a_{71}\big]/\big(s_{45}c_5\big) + c_{34}\big[-S_3s_{23}X_7 + a_{23}c_{12}Y_7\big]/\big(s_{45}c_{56}c_5\big)\\
&-S_1s_{23}\bar{X}_7' - S_2c_{12}s_{23}\bar{X}_7' - S_6s_{23}X_7' - S_7s_{23}s_7 + a_{23}c_{23}c_7,\\
B_{3,2} = {}&\bar{X}_5 s_{23}K\big[-K_{31}\big(X_5Z_7' + Y_5\bar{X}_7'\big) - K_{32}Y_7 + K_{41}\big(X_5\bar{X}_7' - Y_5Z_7' - Z_5Y_7\big)\big]\\
&+\bar{Y}_5 s_{23}K\big[-K_{31}Y_7 + K_{32}\big(X_5Z_7' + Y_5\bar{X}_7'\big) + K_{42}\big(-X_5\bar{X}_7' + Y_5Z_7' + Z_5Y_7\big)\big]\\
&+s_5\big[S_1s_{23}X_7 + S_2c_{12}s_{23}X_7 + S_7s_{23}c_{71}X_7 + a_{23}c_{23}Y_7 + a_{67}s_{23}Z_7'\\
&-a_{71}s_{23}Z_7\big]/\big(s_{56}c_5\big) + s_{23}c_{45}\big[-S_6Y_7 + S_7s_{71} - a_{67}\bar{X}_7'\big]/\big(s_{45}c_5\big)
\end{aligned}
$$

$$
\begin{aligned}
&+c_{34}\left[S_3s_{23}Y_7+a_{23}c_{12}X_7\right]/\left(s_{45}c_{56}c_5\right)-S_1s_{23}c_7-S_2c_{12}s_{23}c_7+S_6s_{23}Z'_7\\
&-S_7s_{23}c_{71}c_7-a_{23}c_{23}\bar{X}'_7+a_{71}s_{23}\bar{X}_7,\\
B_{3,3}={}&[a_{12}s_{23}c_{45}]/(s_{45}c_5)-[S_2s_{12}s_{23}s_5Z_7]/(s_{56}c_5)\\
&-[a_{23}s_{12}c_{34}Z_7]/(s_{45}c_{56}c_5)+S_2s_{12}s_{23}\bar{X}_7,\\
D_{3,1}={}&\bar{X}_5s_{12}c_{23}K\left[K_{31}\left(X_5Z'_7+Y_5\bar{X}'_7\right)+K_{32}Y_7+K_{41}\left(-X_5\bar{X}'_7+Y_5Z'_7+Z_5Y_7\right)\right]\\
&+\bar{Y}_5s_{12}c_{23}K\left[K_{31}Y_7-K_{32}\left(X_5Z'_7+Y_5\bar{X}'_7\right)+K_{42}\left(X_5X'_7-Y_5Z'_7-Z_5Y_7\right)\right]\\
&+s_5\left[-S_1s_{12}c_{23}X_7-S_7s_{12}c_{23}s_{67}\bar{X}'_7-a_{12}c_{12}c_{23}Y_7+a_{23}s_{12}s_{23}Y_7-a_{67}s_{12}c_{23}Z'_7\right.\\
&\left.+a_{71}s_{12}c_{23}Z_7\right]/\left(s_{56}c_5\right)+s_{12}c_{23}c_{45}\left[S_6Y_7-S_7s_{71}+a_{67}\bar{X}'_7\right]/\left(s_{45}c_5\right)\\
&-c_{34}\left[S_2s_{12}Y_7+S_3s_{12}c_{23}Y_7+a_{12}X_7\right]/\left(s_{45}c_{56}c_5\right)+S_1s_{12}c_{23}c_7-S_6s_{12}c_{23}Z'_7\\
&+S_7s_{12}c_{23}c_{71}c_7+a_{12}c_{12}c_{23}\bar{X}'_7-a_{23}s_{12}s_{23}\bar{X}'_7-a_{71}s_{12}c_{23}\bar{X}_7,\\
D_{3,2}={}&\bar{X}_5s_{12}c_{23}K\left[K_{31}\left(X_5X'_7+Y_5c_7\right)+K_{32}X_7+K_{41}\left(-X_5c_7+Y_5X'_7+Z_5X_7\right)\right]\\
&+\bar{Y}_5s_{12}c_{23}K\left[K_{31}X_7-K_{32}\left(X_5X'_7+Y_5c_7\right)+K_{42}\left(X_5c_7-Y_5X'_7-Z_5X_7\right)\right]\\
&+s_5\left[S_1s_{12}c_{23}Y_7-S_7s_{12}c_{23}s_{67}c_7-a_{12}c_{12}c_{23}X_7+a_{23}s_{12}s_{23}X_7\right.\\
&\left.-a_{67}s_{12}c_{23}X'_7\right]/\left(s_{56}c_5\right)+s_{12}c_{23}c_{45}\left[S_6X_7+a_{67}c_7+a_{71}\right]/\left(s_{45}c_5\right)\\
&+c_{34}\left[-S_2s_{12}X_7-S_3s_{12}c_{23}X_7+a_{12}Y_7\right]/\left(s_{45}c_{56}c_5\right)-S_1s_{12}c_{23}\bar{X}'_7-S_6s_{12}c_{23}X'_7\\
&-S_7s_{12}c_{23}s_7+a_{12}c_{12}c_{23}c_7-a_{23}s_{12}s_{23}c_7,\\
D_{3,3}={}&\bar{X}_5K\left[K_{31}c_{12}c_{23}(X_5\bar{Y}_7+Y_5\bar{X}_7)+K_{32}(-c_{34}Z_5+c_{12}c_{23}Z_7)\right.\\
&\left.+K_{41}(-c_{34}+c_{12}c_{23}(-X_5\bar{X}_7+Y_5\bar{Y}_7+Z_5Z_7))\right]\\
&+\bar{Y}_5K\left[K_{31}(-c_{34}Z_5+c_{12}c_{23}Z_7)-K_{32}c_{12}c_{23}(X_5\bar{Y}_7+Y_5\bar{X}_7)\right.\\
&\left.+K_{42}(c_{34}+c_{12}c_{23}(X_5\bar{X}_7-Y_5\bar{Y}_7-Z_5Z_7))\right]+s_5\left[-S_7c_{12}c_{23}s_{67}\bar{X}_7\right.\\
&+a_{12}s_{12}c_{23}Z_7+a_{23}c_{12}s_{23}Z_7-a_{34}s_{34}c_{45}c_{56}-a_{45}c_{34}s_{45}c_{56}\\
&\left.-a_{67}c_{12}c_{23}\bar{Y}_7-a_{71}c_{12}c_{23}Y_7\right]/(s_{56}c_5)+c_{45}\left[S_1c_{12}c_{23}+S_2c_{23}+S_3+S_4c_{34}\right.\\
&\left.+S_6c_{12}c_{23}Z_7+S_7c_{12}c_{23}c_{71}+a_{67}c_{12}c_{23}\bar{X}_7\right]/(s_{45}c_5)+c_{34}\left[-S_1Z_7-S_2c_{12}Z_7\right.\\
&\left.-S_3c_{12}c_{23}Z_7-S_4Z_5-S_6-S_7c_{67}-a_{45}\bar{X}_5-a_{71}X_7\right]/(s_{45}c_{56}c_5)-S_6c_{12}c_{23}\bar{Y}_7\\
&+S_7c_{12}c_{23}s_{71}c_7-a_{12}s_{12}c_{23}\bar{X}_7-a_{23}c_{12}s_{23}\bar{X}_7+a_{71}c_{12}c_{23}\bar{X}'_7-a_{56}c_{34}c_{56}X_5/s_{56},\\
E_{3,1}={}&\bar{X}_5c_{12}s_{23}K\left[K_{31}\left(-X_5\bar{X}'_7+Y_5Z'_7-Z_5Y_7\right)-K_{41}\left(X_5Z'_7+Y_5\bar{X}'_7\right)+K_{42}Y_7\right]\\
&+\bar{Y}_5c_{12}s_{23}K\left[K_{32}\left(X_5\bar{X}'_7-Y_5Z'_7+Z_5Y_7\right)+K_{41}Y_7+K_{42}\left(X_5Z'_7+Y_5\bar{X}'_7\right)\right]\\
&+c_{56}\left[S_1c_{12}s_{23}X_7+S_2s_{23}X_7+S_7c_{12}s_{23}s_{67}\bar{X}'_7-a_{12}s_{12}s_{23}Y_7+a_{23}c_{12}c_{23}Y_7\right.\\
&\left.+a_{67}c_{12}s_{23}Z'_7-a_{71}c_{12}s_{23}Z_7\right]/s_{56}+S_1c_{12}s_{23}X'_7+S_2s_{23}X'_7+S_6c_{12}s_{23}\bar{X}'_7\\
&+S_7c_{12}s_{23}c_{67}\bar{X}'_7-a_{12}s_{12}s_{23}Z'_7+a_{23}c_{12}c_{23}Z'_7-a_{67}c_{12}s_{23}Y_7-a_{71}c_{12}s_{23}\bar{Y}_7,\\
E_{3,2}={}&\bar{X}_5c_{12}s_{23}K\left[K_{31}\left(-X_5c_7+Y_5X'_7-Z_5X_7\right)-K_{41}\left(X_5X'_7+Y_5c_7\right)+K_{42}X_7\right]\\
&+\bar{Y}_5c_{12}s_{23}K\left[K_{32}\left(X_5c_7-Y_5X'_7+Z_5X_7\right)+K_{41}X_7+K_{42}\left(X_5X'_7+Y_5c_7\right)\right]\\
&+c_{56}\left[-S_1c_{12}s_{23}Y_7-S_2s_{23}Y_7+S_7c_{12}s_{23}s_{67}c_7-a_{12}s_{12}s_{23}X_7+a_{23}c_{12}c_{23}X_7\right.\\
&\left.+a_{67}c_{12}s_{23}X'_7\right]/s_{56}-S_1c_{12}s_{23}Z'_7-S_2s_{23}Z'_7+S_6c_{12}s_{23}c_7+S_7c_{12}s_{23}c_{67}c_7\\
&-a_{12}s_{12}s_{23}X'_7+a_{23}c_{12}c_{23}X'_7-a_{67}c_{12}s_{23}X_7,\\
E_{3,3}={}&\bar{X}_5s_{12}s_{23}K\left[K_{31}(X_5\bar{X}_7-Y_5\bar{Y}_7+Z_5Z_7)+K_{41}(X_5\bar{Y}_7+Y_5\bar{X}_7)-K_{42}Z_7\right]
\end{aligned}
$$

$$
\begin{aligned}
&+ \bar{Y}_5 s_{12}s_{23}K[K_{32}(-X_5\bar{X}_7 + Y_5\bar{Y}_7 - Z_5Z_7) - K_{41}Z_7 - K_{42}(X_5\bar{Y}_7 + Y_5\bar{X}_7)] \\
&+ c_{56}[-S_7 s_{12}s_{23}s_{67}\bar{X}_7 - a_{12}c_{12}s_{23}Z_7 - a_{23}s_{12}c_{23}Z_7 - a_{67}s_{12}s_{23}\bar{Y}_7 \\
&- a_{71}s_{12}s_{23}Y_7]/s_{56} - S_6 s_{12}s_{23}\bar{X}_7 - S_7 s_{12}s_{23}c_{67}\bar{X}_7 - a_{12}c_{12}s_{23}\bar{Y}_7 \\
&- a_{23}s_{12}c_{23}\bar{Y}_7 + a_{67}s_{12}s_{23}Z_7 - a_{71}s_{12}s_{23}Z'_7, \\
F_{3,1} = {}& \bar{X}_5 s_{23}K[K_{31}(-X_5c_7 + Y_5X'_7 - Z_5X_7) - K_{41}(X_5X'_7 + Y_5c_7) + K_{42}X_7] \\
&+ \bar{Y}_5 s_{23}K[K_{32}(X_5c_7 - Y_5X'_7 + Z_5X_7) + K_{41}X_7 + K_{42}(X_5X'_7 + Y_5c_7)] \\
&+ c_{56}[-S_1s_{23}Y_7 - S_2c_{12}s_{23}Y_7 + S_7s_{23}s_{67}c_7 + a_{23}c_{23}X_7 + a_{67}s_{23}X'_7]/s_{56} \\
&- S_1s_{23}Z'_7 - S_2c_{12}s_{23}Z'_7 + S_6s_{23}c_7 + S_7s_{23}c_{67}c_7 + a_{23}c_{23}X'_7 - a_{67}s_{23}X_7, \\
F_{3,2} = {}& \bar{X}_5 s_{23}K[K_{31}(X_5\bar{X}'_7 - Y_5Z'_7 + Z_5Y_7) + K_{41}(X_5Z'_7 + Y_5\bar{X}'_7) - K_{42}Y_7] \\
&+ \bar{Y}_5 s_{23}K[K_{32}(-X_5\bar{X}'_7 + Y_5Z'_7 - Z_5Y_7) - K_{41}Y_7 - K_{42}(X_5Z'_7 + Y_5\bar{X}'_7)] \\
&+ c_{56}[-S_1s_{23}X_7 - S_2c_{12}s_{23}X_7 - S_7s_{23}s_{67}\bar{X}'_7 - a_{23}c_{23}Y_7 - a_{67}s_{23}Z'_7 \\
&+ a_{71}s_{23}Z_7]/s_{56} - S_1s_{23}X'_7 - S_2c_{12}s_{23}X'_7 - S_6s_{23}\bar{X}'_7 - S_7s_{23}c_{67}\bar{X}'_7 - a_{23}c_{23}Z'_7 \\
&+ a_{67}s_{23}Y_7 + a_{71}s_{23}\bar{Y}_7, \\
F_{3,3} = {}& S_2s_{12}s_{23}\bar{Y}_7 + S_2s_{12}s_{23}c_{56}Z_7/s_{56}, \\
G_{3,1} = {}& \bar{X}_5 s_{12}c_{23}K[K_{31}(-X_5\bar{X}'_7 + Y_5Z'_7 - Z_5Y_7) - K_{41}(X_5Z'_7 + Y_5\bar{X}'_7) + K_{42}Y_7] \\
&+ \bar{Y}_5 s_{12}c_{23}K[K_{32}(X_5\bar{X}'_7 - Y_5Z'_7 + Z_5Y_7) + K_{41}Y_7 + K_{42}(X_5Z'_7 + Y_5\bar{X}'_7)] \\
&+ c_{56}[S_1s_{12}c_{23}X_7 + S_7s_{12}c_{23}s_{67}\bar{X}'_7 + a_{12}c_{12}c_{23}Y_7 - a_{23}s_{12}s_{23}Y_7 + a_{67}s_{12}c_{23}Z'_7 \\
&- a_{71}s_{12}c_{23}Z_7]/s_{56} + S_1s_{12}c_{23}X'_7 + S_6s_{12}c_{23}\bar{X}'_7 + S_7s_{12}c_{23}c_{67}\bar{X}'_7 + a_{12}c_{12}c_{23}Z'_7 \\
&- a_{23}s_{12}s_{23}Z'_7 - a_{67}s_{12}c_{23}Y_7 - a_{71}s_{12}c_{23}\bar{Y}_7, \\
G_{3,2} = {}& \bar{X}_5 s_{12}c_{23}K[K_{31}(-X_5c_7 + Y_5X'_7 - Z_5X_7) - K_{41}(X_5X'_7 + Y_5c_7) + K_{42}X_7] \\
&+ \bar{Y}_5 s_{12}c_{23}K[K_{32}(X_5c_7 - Y_5X'_7 + Z_5X_7) + K_{41}X_7 + K_{42}(X_5X'_7 + Y_5c_7)] \\
&+ c_{56}[-S_1s_{12}c_{23}Y_7 + S_7s_{12}c_{23}s_{67}c_7 + a_{12}c_{12}c_{23}X_7 - a_{23}s_{12}s_{23}X_7 \\
&+ a_{67}s_{12}c_{23}X'_7]/s_{56} - S_1s_{12}c_{23}Z'_7 + S_6s_{12}c_{23}c_7 + S_7s_{12}c_{23}c_{67}c_7 + a_{12}c_{12}c_{23}X'_7 \\
&- a_{23}s_{12}s_{23}X'_7 - a_{67}s_{12}c_{23}X_7, \\
G_{3,3} = {}& \bar{X}_5K[K_{31}(c_{34} + c_{12}c_{23}(-X_5\bar{X}_7 + Y_5\bar{Y}_7 - Z_5Z_7)) - K_{41}c_{12}c_{23}(X_5\bar{Y}_7 + Y_5\bar{X}_7) \\
&+ K_{42}(-c_{34}Z_5 + c_{12}c_{23}Z_7)] + \bar{Y}_5K[K_{32}(-c_{34} + c_{12}c_{23}(X_5\bar{X}_7 - Y_5\bar{Y}_7 + Z_5Z_7)) \\
&+ K_{41}(-c_{34}Z_5 + c_{12}c_{23}Z_7) + K_{42}c_{12}c_{23}(X_5\bar{Y}_7 + Y_5\bar{X}_7)] \\
&+ [c_{56}(S_7c_{12}c_{23}s_{67}\bar{X}_7 - a_{12}s_{12}c_{23}Z_7 - a_{23}c_{12}s_{23}Z_7 + a_{67}c_{12}c_{23}\bar{Y}_7 + a_{71}c_{12}c_{23}Y_7) \\
&+ a_{34}s_{34}c_{45} + a_{45}c_{34}s_{45} + a_{56}c_{34}s_{45}c_5]/s_{56} + S_6c_{12}c_{23}\bar{X}_7 + S_7c_{12}c_{23}c_{67}\bar{X}_7 \\
&- a_{12}s_{12}c_{23}\bar{Y}_7 - a_{23}c_{12}s_{23}\bar{Y}_7 - a_{67}c_{12}c_{23}Z_7 + a_{71}c_{12}c_{23}Z'_7, \qquad (9.204) \\
A_{4,1} = {}& \bar{X}_5c_{12}s_{23}K[K_{31}(-X_5\bar{X}'_7 + Y_5Z'_7 + Z_5Y_7) - K_{41}(X_5Z'_7 + Y_5\bar{X}'_7) - K_{42}Y_7] \\
&+ \bar{Y}_5c_{12}s_{23}K[K_{32}(X_5\bar{X}'_7 - Y_5Z'_7 - Z_5Y_7) - K_{41}Y_7 + K_{42}(X_5Z'_7 + Y_5\bar{X}'_7) \\
&+ c_{56}[-S_1c_{12}c_{23}X_7 - S_2s_{23}X_7 - S_7c_{12}s_{23}s_{67}\bar{X}'_7 + a_{12}s_{12}s_{23}Y_7 - a_{23}c_{12}c_{23}Y_7 \\
&- a_{67}c_{12}s_{23}Z'_7 + a_{71}c_{12}s_{23}Z_7]/s_{56} + S_1c_{12}s_{23}X'_7 + S_2s_{23}X'_7 + S_6c_{12}s_{23}\bar{X}'_7 \\
&+ S_7c_{12}s_{23}c_{67}\bar{X}'_7 - a_{12}s_{12}s_{23}Z'_7 + a_{23}c_{12}c_{23}Z'_7 - a_{67}c_{12}s_{23}Y_7 - a_{71}c_{12}s_{23}\bar{Y}_7, \\
A_{4,2} = {}& \bar{X}_5c_{12}s_{23}K[K_{31}(-X_5c_7 + Y_5X'_7 + Z_5X_7) - K_{41}(X_5X'_7 + Y_5c_7) - K_{42}X_7]
\end{aligned}
$$

$$
\begin{aligned}
&+ \bar{Y}_5 c_{12} s_{23} K\left[K_{32}(X_5 c_7 - Y_5 X'_7 - Z_5 X_7) - K_{41} X_7 + K_{42}(X_5 X'_7 + Y_5 c_7)\right] \\
&+ c_{56}\left[S_1 c_{12} s_{23} Y_7 + S_2 s_{23} Y_7 - S_7 c_{12} s_{23} s_{67} c_7 + a_{12} s_{12} s_{23} X_7 - a_{23} c_{12} c_{23} X_7 \right.\\
&\left. - a_{67} c_{12} s_{23} X'_7\right]/s_{56} - S_1 c_{12} s_{23} Z'_7 - S_2 s_{23} Z'_7 + S_6 c_{12} s_{23} c_7 + S_7 c_{12} s_{23} c_{67} c_7 \\
&- a_{12} s_{12} s_{23} X'_7 + a_{23} c_{12} c_{23} X'_7 - a_{67} c_{12} s_{23} X_7, \\
A_{4,3} = {} & \bar{X}_5 s_{12} s_{23} K\left[K_{31}(X_5 \bar{X}_7 - Y_5 \bar{Y}_7 - Z_5 Z_7) + K_{41}(X_5 \bar{Y}_7 + Y_5 \bar{X}_7) + K_{42} Z_7\right] \\
&+ \bar{Y}_5 s_{12} s_{23} K\left[K_{32}(-X_5 \bar{X}_7 + Y_5 \bar{Y}_7 + Z_5 Z_7) + K_{41} Z_7 - K_{42}(X_5 \bar{Y}_7 + Y_5 \bar{X}_7)\right] \\
&+ c_{56}\left[S_7 s_{12} s_{23} s_{67} \bar{X}_7 + a_{12} c_{12} s_{23} Z_7 + a_{23} s_{12} c_{23} Z_7 + a_{67} s_{12} s_{23} \bar{Y}_7 \right.\\
&\left. + a_{71} s_{12} s_{23} Y_7\right]/s_{56} - S_6 s_{12} s_{23} \bar{X}_7 - S_7 s_{12} s_{23} c_{67} \bar{X}_7 - a_{12} c_{12} s_{23} \bar{Y}_7 - a_{23} s_{12} c_{23} \bar{Y}_7 \\
&+ a_{67} s_{12} s_{23} Z_7 - a_{71} s_{12} s_{23} Z'_7, \\
B_{4,1} = {} & \bar{X}_5 s_{23} K\left[K_{31}(-X_5 c_7 + Y_5 X'_7 + Z_5 X_7) - K_{41}(X_5 X'_7 + Y_5 c_7) - K_{42} X_7\right] \\
&+ \bar{Y}_5 s_{23} K\left[K_{32}(X_5 c_7 - Y_5 X'_7 - Z_5 X_7) - K_{41} X_7 + K_{42}(X_5 X'_7 + Y_5 c_7)\right] \\
&+ c_{56}\left[S_1 s_{23} Y_7 + S_2 c_{12} s_{23} Y_7 - S_7 s_{23} s_{67} c_7 - a_{23} c_{23} X_7 - a_{67} s_{23} X'_7\right]/s_{56} - S_1 s_{23} Z'_7 \\
&- S_2 c_{12} s_{23} Z'_7 + S_6 s_{23} c_7 + S_7 s_{23} c_{67} c_7 + a_{23} c_{23} X'_7 - a_{67} s_{23} X_7, \\
B_{4,2} = {} & \bar{X}_5 s_{23} K\left[K_{31}(X_5 \bar{X}'_7 - Y_5 Z'_7 - Z_5 Y_7) + K_{41}(X_5 Z'_7 + Y_5 \bar{X}'_7) + K_{42} Y_7\right] \\
&+ \bar{Y}_5 s_{23} K\left[K_{32}(-X_5 \bar{X}'_7 + Y_5 Z'_7 + Z_5 Y_7) + K_{41} Y_7 - K_{42}(X_5 Z'_7 + Y_5 \bar{X}'_7)\right] \\
&+ c_{56}\left[S_1 s_{23} X_7 + S_2 c_{12} s_{23} X_7 + S_7 s_{23} s_{67} \bar{X}'_7 + a_{23} c_{23} Y_7 + a_{67} s_{23} Z'_7 \right.\\
&\left. - a_{71} s_{23} Z_7\right]/s_{56} - S_1 s_{23} X'_7 - S_2 c_{12} s_{23} X'_7 - S_6 s_{23} \bar{X}'_7 - S_7 s_{23} c_{67} \bar{X}'_7 - a_{23} c_{23} Z'_7 \\
&+ a_{67} s_{23} Y_7 + a_{71} s_{23} \bar{Y}_7, \\
B_{4,3} = {} & S_2 s_{12} s_{23} \bar{Y}_7 - S_2 s_{12} s_{23} c_{56} Z_7/s_{56}, \\
D_{4,1} = {} & \bar{X}_5 s_{12} c_{23} K\left[K_{31}(-X_5 \bar{X}'_7 + Y_5 Z'_7 + Z_5 Y_7) - K_{41}(X_5 Z'_7 + Y_5 \bar{X}'_7) - K_{42} Y_7\right] \\
&+ \bar{Y}_5 s_{12} c_{23} K\left[K_{32}(X_5 \bar{X}'_7 - Y_5 Z'_7 - Z_5 Y_7) - K_{41} Y_7 + K_{42}(X_5 Z'_7 + Y_5 \bar{X}'_7)\right] \\
&+ c_{56}\left[-S_1 s_{12} c_{23} X_7 - S_7 s_{12} c_{23} s_{67} \bar{X}'_7 - a_{12} c_{12} c_{23} Y_7 + a_{23} s_{12} s_{23} Y_7 - a_{67} s_{12} c_{23} Z'_7 \right.\\
&\left. + a_{71} s_{12} c_{23} Z_7\right]/s_{56} + S_1 s_{12} c_{23} X'_7 + S_6 s_{12} c_{23} \bar{X}'_7 + S_7 s_{12} c_{23} c_{67} \bar{X}'_7 + a_{12} c_{12} c_{23} Z'_7 \\
&- a_{23} s_{12} s_{23} Z'_7 - a_{67} s_{12} c_{23} Y_7 - a_{71} s_{12} c_{23} \bar{Y}_7, \\
D_{4,2} = {} & \bar{X}_5 s_{12} c_{23} K\left[K_{31}(-X_5 c_7 + Y_5 X'_7 + Z_5 X_7) - K_{41}(X_5 X'_7 + Y_5 c_7) - K_{42} X_7\right] \\
&+ \bar{Y}_5 s_{12} c_{23} K\left[K_{32}(X_5 c_7 - Y_5 X'_7 - Z_5 X_7) - K_{41} X_7 + K_{42}(X_5 X'_7 + Y_5 c_7)\right] \\
&+ c_{56}\left[S_1 s_{12} c_{23} Y_7 - S_7 s_{12} c_{23} s_{67} c_7 - a_{12} c_{12} c_{23} X_7 + a_{23} s_{12} s_{23} X_7 \right.\\
&\left. - a_{67} s_{12} c_{23} X'_7\right]/s_{56} - S_1 s_{12} c_{23} Z'_7 + S_6 s_{12} c_{23} c_7 + S_7 s_{12} c_{23} c_{67} c_7 + a_{12} c_{12} c_{23} X'_7 \\
&- a_{23} s_{12} s_{23} X'_7 - a_{67} s_{12} c_{23} X_7, \\
D_{4,3} = {} & \bar{X}_5 K\left[K_{31}(-c_{34} + c_{12} c_{23}(-X_5 \bar{X}_7 + Y_5 \bar{Y}_7 + Z_5 Z_7)) - K_{41} c_{12} c_{23}(X_5 \bar{Y}_7 + Y_5 \bar{X}_7) \right.\\
&\left. + K_{42}(c_{34} Z_5 - c_{12} c_{23} Z_7)\right] + \bar{Y}_5 K\left[K_{32}(c_{34} + c_{12} c_{23}(X_5 \bar{X}_7 - Y_5 \bar{Y}_7 - Z_5 Z_7)) \right.\\
&\left. + K_{41}(c_{34} Z_5 - c_{12} c_{23} Z_7) + K_{42} c_{12} c_{23}(X_5 \bar{Y}_7 + Y_5 \bar{X}_7)\right] + \left[c_{56}(-S_7 c_{12} c_{23} s_{67} \bar{X}_7 \right.\\
&+ a_{12} s_{12} c_{23} Z_7 + a_{23} c_{12} s_{23} Z_7 - a_{67} c_{12} c_{23} \bar{Y}_7 - a_{71} c_{12} c_{23} Y_7) - a_{34} s_{34} c_{45} \\
&\left. - a_{45} c_{34} s_{45} - a_{56} c_{34} s_{45} c_5\right]/s_{56} + S_6 c_{12} c_{23} \bar{X}_7 + S_7 c_{12} c_{23} c_{67} \bar{X}_7 - a_{12} s_{12} c_{23} \bar{Y}_7 \\
&- a_{23} c_{12} s_{23} \bar{Y}_7 - a_{67} c_{12} c_{23} Z_7 + a_{71} c_{12} c_{23} Z'_7,
\end{aligned}
$$

$$
\begin{aligned}
E_{4,1} = {} & \bar{X}_5 c_{12} s_{23} K[-K_{31}(X_5 Z'_7 + Y_5 \bar{X}'_7) + K_{32} Y_7 + K_{41}(X_5 \bar{X}'_7 - Y_5 Z'_7 + Z_5 Y_7)] \\
& + \bar{Y}_5 c_{12} s_{23} K[K_{31} Y_7 + K_{32}(X_5 Z'_7 + Y_5 \bar{X}'_7) + K_{42}(-X_5 \bar{X}'_7 + Y_5 Z'_7 - Z_5 Y_7)] \\
& + s_5[-S_1 c_{12} s_{23} X_7 - S_2 s_{23} X_7 - S_7 c_{12} s_{23} s_{67} \bar{X}'_7 + a_{12} s_{12} s_{23} Y_7 - a_{23} c_{12} c_{23} Y_7 \\
& - a_{67} c_{12} s_{23} Z'_7 + a_{71} c_{12} s_{23} Z_7]/(s_{56} c_5) + c_{12} s_{23} c_{45}[S_6 Y_7 \\
& - S_7 s_{71} + a_{67} \bar{X}'_7]/(s_{45} c_5) - c_{34}[S_3 c_{12} s_{23} Y_7 + a_{23} X_7]/(s_{45} c_{56} c_5) - S_1 c_{12} s_{23} c_7 \\
& - S_2 s_{23} c_7 + S_6 c_{12} s_{23} Z'_7 - S_7 c_{12} s_{23} c_{71} c_7 + a_{12} s_{12} s_{23} \bar{X}'_7 \\
& - a_{23} c_{12} c_{23} \bar{X}'_7 + a_{71} c_{12} s_{23} \bar{X}_7, \\
E_{4,2} = {} & \bar{X}_5 c_{12} s_{23} K[-K_{31}(X_5 X'_7 + Y_5 c_7) + K_{32} X_7 + K_{41}(X_5 c_7 - Y_5 X'_7 + Z_5 X_7)] \\
& + \bar{Y}_5 c_{12} s_{23} K[K_{31} X_7 + K_{32}(X_5 X'_7 + Y_5 c_7) + K_{42}(-X_5 c_7 + Y_5 X'_7 - Z_5 X_7)] \\
& + s_5[S_1 c_{12} s_{23} Y_7 + S_2 s_{23} Y_7 - S_7 c_{12} s_{23} s_{67} c_7 + a_{12} s_{12} s_{23} X_7 - a_{23} c_{12} c_{23} X_7 \\
& - a_{67} c_{12} s_{23} X'_7]/(s_{56} c_5) + c_{12} s_{23} c_{45}[S_6 X_7 + a_{67} c_7 + a_{71}]/(s_{45} c_5) \\
& + c_{34}[-S_3 c_{12} s_{23} X_7 + a_{23} Y_7]/(s_{45} c_{56} c_5) + S_1 c_{12} s_{23} \bar{X}'_7 + S_2 s_{23} \bar{X}'_7 \\
& + S_6 c_{12} s_{23} X'_7 + S_7 c_{12} s_{23} s_7 + a_{12} s_{12} s_{23} c_7 - a_{23} c_{12} c_{23} c_7, \\
E_{4,3} = {} & \bar{X}_5 s_{12} s_{23} K[K_{31}(X_5 \bar{Y}_7 + Y_5 \bar{X}_7) - K_{32} Z_7 + K_{41}(-X_5 \bar{X}_7 + Y_5 \bar{Y}_7 - Z_5 Z_7)] \\
& + \bar{Y}_5 s_{12} s_{23} K[-K_{31} Z_7 - K_{32}(X_5 \bar{Y}_7 + Y_5 \bar{X}_7) + K_{42}(X_5 \bar{X}_7 - Y_5 \bar{Y}_7 + Z_5 Z_7)] \\
& + s_5[S_7 s_{12} s_{23} s_{67} \bar{X}_7 + a_{12} c_{12} s_{23} Z_7 + a_{23} s_{12} c_{23} Z_7 + a_{67} s_{12} s_{23} \bar{Y}_7 \\
& + a_{71} s_{12} s_{23} Y_7]/(s_{56} c_5) - s_{12} s_{23} c_{45}[S_1 + S_6 Z_7 + S_7 c_{71} + a_{67} \bar{X}_7]/(s_{45} c_5) \\
& + [S_3 s_{12} s_{23} c_{34} Z_7]/(s_{45} c_{56} c_5) - S_6 s_{12} s_{23} \bar{Y}_7 + S_7 s_{12} s_{23} s_{71} c_7 \\
& + a_{12} c_{12} s_{23} \bar{X}_7 + a_{23} s_{12} c_{23} \bar{X}_7 + a_{71} s_{12} s_{23} \bar{X}'_7, \\
F_{4,1} = {} & \bar{X}_5 s_{23} K[-K_{31}(X_5 X'_7 + Y_5 c_7) + K_{32} X_7 + K_{41}(X_5 c_7 - Y_5 X'_7 + Z_5 X_7)] \\
& + \bar{Y}_5 s_{23} K[K_{31} X_7 + K_{32}(X_5 X'_7 + Y_5 c_7) + K_{42}(-X_5 c_7 + Y_5 X'_7 - Z_5 X_7)] \\
& + s_5[S_1 s_{23} Y_7 + S_2 c_{12} s_{23} Y_7 - S_7 s_{23} s_{67} c_7 - a_{23} c_{23} X_7 - a_{67} s_{23} X'_7]/(s_{56} c_5) \\
& + s_{23} c_{45}[S_6 X_7 + a_{67} c_7 + a_{71}]/(s_{45} c_5) + c_{34}[-S_3 s_{23} X_7 + a_{23} c_{12} Y_7]/(s_{45} c_{56} c_5) \\
& + S_1 s_{23} \bar{X}'_7 + S_2 c_{12} c_{23} \bar{X}'_7 + S_6 s_{23} X'_7 + S_7 s_{23} s_7 - a_{23} c_{23} c_7, \\
F_{4,2} = {} & \bar{X}_5 s_{23} K[K_{31}(X_5 Z'_7 + Y_5 \bar{X}'_7) - K_{32} Y_7 + K_{41}(-X_5 \bar{X}'_7 + Y_5 Z'_7 - Z_5 Y_7)] \\
& + \bar{Y}_5 s_{23} K[-K_{31} Y_7 - K_{32}(X_5 Z'_7 + Y_5 \bar{X}'_7) + K_{42}(X_5 \bar{X}'_7 - Y_5 Z'_7 + Z_5 Y_7)] \\
& + s_5[S_1 s_{23} X_7 + S_2 c_{12} s_{23} X_7 + S_7 s_{23} s_{67} \bar{X}'_7 + a_{23} c_{23} Y_7 + a_{67} s_{23} Z'_7 \\
& - a_{71} s_{23} Z_7]/(s_{56} c_5) + s_{23} c_{45}[-S_6 Y_7 + S_7 s_{71} - a_{67} \bar{X}'_7]/(s_{45} c_5) \\
& + c_{34}[S_3 s_{23} Y_7 + a_{23} c_{12} X_7]/(s_{45} c_{56} c_5) + S_1 s_{23} c_7 + S_2 c_{12} s_{23} c_7 - S_6 s_{23} Z'_7 \\
& + S_7 s_{23} c_{71} c_7 + a_{23} c_{23} \bar{X}'_7 - a_{71} s_{23} \bar{X}_7, \\
F_{4,3} = {} & -S_2 s_{12} s_{23}[\bar{X}_7 + s_5 Z_7/(s_{56} c_5)] + a_{12} s_{23} c_{45}/(s_{45} c_5) - a_{23} s_{12} c_{34} Z_7/(s_{45} c_{56} c_5), \\
G_{4,1} = {} & \bar{X}_5 s_{12} c_{23} K[-K_{31}(X_5 Z'_7 + Y_5 \bar{X}'_7) + K_{32} Y_7 + K_{41}(X_5 \bar{X}'_7 - Y_5 Z'_7 + Z_5 Y_7)] \\
& + \bar{Y}_5 s_{12} c_{23} K[K_{31} Y_7 + K_{32}(X_5 Z'_7 + Y_5 \bar{X}'_7) + K_{42}(-X_5 \bar{X}'_7 + Y_5 Z'_7 - Z_5 Y_7)] \\
& + s_5[-S_1 s_{12} c_{23} X_7 - S_7 s_{12} c_{23} s_{67} \bar{X}'_7 - a_{12} c_{12} c_{23} Y_7 + a_{23} s_{12} s_{23} Y_7 - a_{67} s_{12} c_{23} Z'_7 \\
& + a_{71} s_{12} c_{23} Z_7]/(s_{56} c_5) + s_{12} c_{23} c_{45}[S_6 Y_7 - S_7 s_{71} + a_{67} \bar{X}'_7]/(s_{45} c_5)
\end{aligned}
$$

$$
\begin{aligned}
&-c_{34}\left[S_2s_{12}Y_7 + S_3s_{12}c_{23}Y_7 + a_{12}X_7\right]/\left(s_{45}c_{56}c_5\right) - S_1s_{12}c_{23}c_7 \\
&+ S_6s_{12}c_{23}Z'_7 - S_7s_{12}c_{23}c_{71}c_7 - a_{12}c_{12}c_{23}\bar{X}'_7 + a_{23}s_{12}s_{23}\bar{X}'_7 + a_{71}s_{12}c_{23}\bar{X}_7, \\
G_{4,2} =\ &\bar{X}_5s_{12}c_{23}K\left[-K_{31}\left(X_5X'_7 + Y_5c_7\right) + K_{32}X_7 + K_{41}\left(X_5c_7 - Y_5X'_7 + Z_5X_7\right)\right] \\
&+ \bar{Y}_5s_{12}c_{23}K\left[K_{31}X_7 + K_{32}\left(X_5X'_7 + Y_5c_7\right) + K_{42}\left(-X_5c_7 + Y_5X'_7 - Z_5X_7\right)\right] \\
&+ s_5\left[S_1s_{12}c_{23}Y_7 - S_7s_{12}c_{23}s_{67}c_7 - a_{12}c_{12}c_{23}X_7 + a_{23}s_{12}s_{23}X_7\right. \\
&\left.- a_{67}s_{12}c_{23}X'_7\right]/\left(s_{56}c_5\right) + s_{12}c_{23}c_{45}\left[S_6X_7 + a_{67}c_7 + a_{71}\right]/\left(s_{45}c_5\right) \\
&+ c_{34}\left[-S_2s_{12}X_7 - S_3s_{12}c_{23}X_7 + a_{12}Y_7\right]/\left(s_{45}c_{56}c_5\right) + S_1s_{12}c_{23}\bar{X}'_7 + S_6s_{12}c_{23}X'_7 \\
&+ S_7s_{12}c_{23}s_7 - a_{12}c_{12}c_{23}c_7 + a_{23}s_{12}s_{23}c_7, \\
G_{4,3} =\ &\bar{X}_5K\left[-K_{31}c_{12}c_{23}(X_5\bar{Y}_7 + Y_5\bar{X}_7) + K_{32}(-c_{34}Z_5 + c_{12}c_{23}Z_7)\right. \\
&\left.+ K_{41}(-c_{34} + c_{12}c_{23}(X_5\bar{X}_7 - Y_5\bar{Y}_7 + Z_5Z_7))\right] + \bar{Y}_5K\left[K_{31}(-c_{34}Z_5 + c_{12}c_{23}Z_7)\right. \\
&\left.+ K_{32}c_{12}c_{23}(X_5\bar{Y}_7 + Y_5\bar{X}_7) + K_{42}(c_{34} + c_{12}c_{23}(-X_5\bar{X}_7 + Y_5\bar{Y}_7 - Z_5Z_7))\right] \\
&+ s_5\left[-S_7c_{12}c_{23}s_{67}\bar{X}_7 + a_{12}s_{12}c_{23}Z_7 + a_{23}c_{12}s_{23}Z_7 - a_{34}s_{34}c_{45}c_{56} - a_{45}c_{34}s_{45}c_{56}\right. \\
&\left.- a_{67}c_{12}c_{23}\bar{Y}_7 - a_{71}c_{12}c_{23}Y_7\right]/(s_{56}c_5) + c_{45}\left[S_1c_{12}c_{23} + S_2c_{23} + S_3 + S_4c_{34}\right. \\
&\left.+ S_6c_{12}c_{23}Z_7 + S_7c_{12}c_{23}c_{71} + a_{67}c_{12}c_{23}\bar{X}_7\right]/(s_{45}c_5) - c_{34}\left[a_{45}\bar{X}_5 + S_1Z_7\right. \\
&\left.+ S_2c_{12}Z_7 + S_3c_{12}c_{23}Z_7 + S_4Z_5 + S_6 + S_7c_{67} + a_{71}X_7\right]/(s_{45}c_{56}c_5) \\
&+ S_6c_{12}c_{23}\bar{Y}_7 - S_7c_{12}c_{23}s_{71}c_7 + a_{12}s_{12}c_{23}\bar{X}_7 + a_{23}c_{12}s_{23}\bar{X}_7 \\
&- a_{71}c_{12}c_{23}\bar{X}'_7 - a_{56}c_{34}c_{56}X_5/s_{56}.
\end{aligned}
\tag{9.205}
$$

The following definitions were used in the coefficients:

$$X'_7 = c_{67}s_7, \tag{9.206}$$

$$\bar{X}'_7 = c_{71}s_7, \tag{9.207}$$

$$Z'_7 = s_{67}s_{71} - c_{67}c_{71}c_7. \tag{9.208}$$

At this point there are four equations of the form of Eq. (9.198) whose coefficients are listed in Eq. (9.199). These coefficients have been expanded in terms of the given mechanism parameters in Eqs. (9.202) through (9.205). The solution can proceed in a manner identical to that for the 5R-C mechanism in the previous section with the exception that the four equations are linear in the tan-half-angle of θ_6 rather than the tan-half-angle of θ_3.

The four equations of Eq. set (9.198) are next modified by substituting the tan-half-angle expressions for the sines and cosines of θ_1 and θ_2. The equations may be written as follows after multiplying each by the product $(1 + x_1^2)(1 + x_2^2)$:

$$\left(a_ix_2^2 + b_ix_2 + d_i\right)x_6 + \left(e_ix_2^2 + f_ix_2 + g_i\right) = 0, \qquad i = 1 \ldots 4, \tag{9.209}$$

where

$$
\begin{aligned}
a_i &= a_{i,1}x_1^2 + a_{i,2}x_1 + a_{i,3}, \\
&\ \vdots \\
g_i &= g_{i,1}x_1^2 + g_{i,2}x_1 + g_{i,3}.
\end{aligned}
\tag{9.210}
$$

The coefficients $a_{i,1}$ through $g_{i,3}$ are defined in terms of $A_{i,1}$ through $G_{i,3}$ in Eq. set (9.77). The input/output equation for this mechanism is then obtained as described in Section 9.1.

An 8×8 determinant is expanded to yield a sixteenth-degree polynomial in the tan-half-angle of the output angle, θ_1. Equation (9.81) is this input/output equation.

9.4.2 Determination of θ_2 and θ_6

Section 9.3.2 describes how to determine the corresponding values for the tan-half-angle of the angles θ_2 and θ_3 from the four equations of Eq. set (9.75) for each calculated value of the tan-half-angle of the output angle θ_1. The procedure for solving for the tan-half-angle of θ_2 and θ_6 for this mechanism is identical with the exception that x_3 in Eq. set (9.75) is replaced by x_6 in Eq. set (9.209). Following the solution method outlined in Section 9.3.2, expressions for the tan-half-angle for the corresponding values of θ_2 and θ_6 may be written as

$$x_2 = \frac{-|abeg||adgb| + |adgf||abed|}{|abef||adgb| - |adge||abed|}, \tag{9.211}$$

$$x_6 = \frac{-|abef||adgf| + |adge||abeg|}{|abef||adgb| - |adge||abed|}. \tag{9.212}$$

9.4.3 Determination of θ_3 and θ_4

Corresponding values for θ_3 and θ_4 may be obtained from the following fundamental sine and sine–cosine laws for a spherical heptagon:

$$X_{56712} = s_{34}s_3, \tag{9.213}$$

$$Y_{56712} = s_{34}c_3, \tag{9.214}$$

$$X_{21765} = s_{34}s_4, \tag{9.215}$$

$$Y_{21765} = s_{34}c_4. \tag{9.216}$$

9.4.4 Determination of S_5

The last parameter to be determined is the offset distance S_5. This may be determined by projecting the vector loop equation onto any direction, resulting in one equation in one unknown. Projecting onto the vector S_5 gives

$$S_1Z_{76} + S_2Z_{34} + S_3Z_4 + S_4c_{45} + S_5 + S_6c_{56} + S_7Z_6 + a_{12}U_{1765} + a_{23}U_{345} + a_{34}s_{45}s_4 + a_{67}U_{65} + a_{71}U_{765} = 0. \tag{9.217}$$

This equation can readily be solved for the parameter S_5.

9.4.5 Numerical example

Table 9.3 shows data that were used as input for a numerical example. The calculated values for the sixteen configurations are listed in Table 9.4.

Table 9.3. *RRPRRRR mechanism parameters.*

Link length, cm.	Twist angle, deg.	Joint offset, cm.	Joint angle, deg.
$a_{12} = 2.9$	$\alpha_{12} = 90$	$S_1 = 12.7$	$\theta_1 =$ variable
$a_{23} = 13.2$	$\alpha_{23} = 90$	$S_2 = 1.1$	$\theta_2 =$ variable
$a_{34} = 1.1$	$\alpha_{34} = 90$	$S_3 = 3.1$	$\theta_3 =$ variable
$a_{45} = 16.7$	$\alpha_{45} = 90$	$S_4 = 6.8$	$\theta_4 =$ variable
$a_{56} = 12.7$	$\alpha_{56} = 89$	$S_5 =$ variable	$\theta_5 = 252$
$a_{67} = 19.4$	$\alpha_{67} = 90$	$S_6 = 11.8$	$\theta_6 =$ variable
$a_{71} = 6.0$	$\alpha_{71} = 90$	$S_7 = 6.9$	$\theta_7 = 83$ (input)

Table 9.4. *Calculated configurations for the RRPRRRR spatial mechanism.*

Solution	θ_1, deg.	θ_2, deg.	θ_3, deg.	θ_4, deg.	θ_6, deg.	S_5, cm.
A	−104.45	−15.64	−83.75	−169.89	−108.56	26.57
B	−92.69	−17.54	71.40	−8.92	75.10	−26.82
C	−81.74	−147.90	68.43	9.80	−60.75	32.15
D	−77.36	−151.93	−79.48	166.21	122.33	−33.78
E	−22.26	−106.57	18.84	73.16	−72.93	20.52
F	−12.82	119.26	−9.09	−56.65	−83.58	46.60
G	16.21	−107.69	−167.05	95.41	105.76	−22.59
H	22.23	121.09	173.63	−131.49	98.41	−45.17
I	69.34	−172.85	127.94	168.99	86.65	−34.41
J	74.42	−175.21	−93.24	7.64	−97.05	34.26
K	106.03	−36.04	−99.73	−16.90	131.19	−28.89
L	134.51	−52.30	138.26	−139.97	−58.47	24.60
M	159.46	107.73	−168.75	74.84	−80.65	45.47
N	−164.56	115.68	9.06	104.77	101.29	−47.82
O	−163.88	−73.99	−20.16	−88.53	113.33	−19.20
P	143.01	−58.22	144.98	−130.15	−61.19	23.42

9.5 Summary

It has been shown that it is possible to solve group 3 spatial mechanisms by obtaining four equations of the form

$$(a_i x_j^2 + b_i x_j + d_i)x_k + (e_i x_j^2 + f_i x_j + g_i) = 0, \qquad i = 1 \ldots 4, \tag{9.218}$$

where the coefficients a_i through g_i are quadratic in the tan-half-angle of the output angle. Eliminating x_j and x_k from the set of equations results in a sixteenth-degree input/output

Table 9.5. *RRPRRRR mechanism parameters.*

Link length, cm.	Twist angle, deg.	Joint offset, cm.	Joint angle, deg.
$a_{12} = 9.8$	$\alpha_{12} = 291$	$S_1 = 5.7$	$\theta_1 =$ variable
$a_{23} = 2.9$	$\alpha_{23} = 263$	$S_2 = 0.8$	$\theta_2 =$ variable
$a_{34} = 2.1$	$\alpha_{34} = 147$	$S_3 = 3.6$	$\theta_3 =$ variable
$a_{45} = 4.6$	$\alpha_{45} = 184$	$S_4 = 9.4$	$\theta_4 =$ variable
$a_{56} = 4.5$	$\alpha_{56} = 268$	$S_5 =$ variable	$\theta_5 = 97$
$a_{67} = 3.3$	$\alpha_{67} = 73$	$S_6 = 4.4$	$\theta_6 =$ variable
$a_{71} = 1.4$	$\alpha_{71} = 153$	$S_7 = 0.8$	$\theta_7 = 279$ (input)

equation that can be expressed as

$$\begin{aligned} &|aebd||defg| + |aebf||bdfg| - |aebg||bdeg| \\ &\quad - |aedf||adfg| + |aedg||adeg| + |aefg||abdg| = 0, \end{aligned} \tag{9.219}$$

where the determinant notation $|xyzw|$ is defined by

$$|xyzw| = \begin{vmatrix} x_1 & y_1 & z_1 & w_1 \\ x_2 & y_2 & z_2 & w_2 \\ x_3 & y_3 & z_3 & w_3 \\ x_4 & y_4 & z_4 & w_4 \end{vmatrix}. \tag{9.220}$$

Examples of six-link 5R-C and seven-link 6R-P mechanisms have been presented in this chapter. Alternate inversions of these mechanisms, that is, the location of the C or P joint changes in the serial chain, can be solved in a manner similar to that described in the examples. It should be noted also that several inversions may be solved by simply changing (or cycling) the number of the joints in the example problems so that the example mechanism matches the case to be analyzed.

The symbolic expansion of the coefficients of the four equations can be a tedious process (see Eqs. (9.71) through (9.74) and (9.202) through (9.205)). C language computer code that numerically evaluates these coefficients can, however, be obtained from the authors. Once these coefficients are evaluated for a specific mechanism, the remainder of the analysis is straightforward.

9.6 Problems

1. The mechanism parameters of a group 3 RRPRRRR spatial mechanism are given in Table 9.5. Using the available computer code that expands the coefficients of Eqs. (9.202) through (9.205), write a computer program to determine all the real solutions for the variable parameters θ_1, θ_2, θ_3, θ_4, θ_6, and S_5.

10

Group 4 spatial mechanisms

10.1 Introduction

The solution of the group 4 general 7R spatial mechanism (with seven joint axes that are arbitrarily skew) was described by Ferdinand Freudenstein as the "Mount Everest of kinematic problems." This complicated analysis is presented solely for reference purposes and could be omitted by the vast majority of readers. The derivation given contains much more detail than that presented by Lee and Liang (1988). The intention is to assist any researcher who wishes to develop a computer program for the 7R mechanism analysis. Further, an in-depth study may well lead to a simpler derivation.

It will be shown in this chapter that the input/output equation for the general 7R mechanism can be obtained from four equations of the form

$$\left(a_i x_j^2 + b_i x_j + d_i\right) x_k + \left(e_i x_j^2 + f_i x_j + g_i\right) = 0, \qquad i = 1 \ldots 4, \tag{10.1}$$

where the coefficients a_i through g_i are quadratic in the tan-half-angle of the output parameter. Eliminating the variables x_j and x_k from this set of equations will yield a sixteenth-degree input/output equation in the tan-half-angle of the output angle.

Because the format of (10.1) is identical to that of (9.2), the generation of the input/output equation will be identical to that developed for the group 3 mechanisms once the quadratic coefficients a_i through g_i are obtained. Further, the solution for the parameters x_j and x_k will be the same as presented in Section 9.3.2.

The majority of industrial manipulators in use today consist of an end effector link free to move in space connected serially by six revolute joints to ground. The end effector link, together with the five intermediate links plus ground comprise seven links. Thus, when the close-the-loop process is performed as part of the reverse analysis procedure, a group 4 7R spatial mechanism results where one angle, θ_7, is known. It will be seen that obtaining the coefficients a_i through g_i in Eq. set (10.1) is a lengthy undertaking. The numerical evaluation of these coefficients and the subsequent solution of a sixteenth-degree input/output equation can require a significant computational time. Precise real-time control of industrial manipulators requires that the reverse-analysis calculations be performed as rapidly as possible.

At the outset it appears that a manipulator design that can be modeled by a group 1 spatial mechanism would be preferred to a manipulator that is modeled by a group 4 spatial mechanism, simply because the reverse-analysis procedure is easier. However, the vast majority of industrial manipulators do in fact incorporate six revolute joints with "specialized" geometry. For example, certain link lengths may be set equal to zero, which allows pairs of successive joint axes to intersect. This produces the mechanical design of wrist and shoulder-type joints. It is also common to set to zero or π radians certain twist angles for which adjacent joint axes are parallel. All such special geometry greatly simplifies the reverse analysis and avoids any problems associated with the actuation of slider displacements, which is necessary for a manipulator design modeled by a group 1 mechanism.

This chapter will first present the solution of the general 7R spatial mechanism with arbitrary dimensions. This complicated analysis is presented for reference purposes. Following this, six 7R mechanisms with special geometries will be presented to show how the reverse-analysis procedure can be greatly simplified for such special cases.

10.2 General 7R group 4 spatial mechanism

Shown in Figure 10.1 is a planar representation of the 7R spatial mechanism. It is assumed that all the constant mechanism parameters are known together with the input angle, θ_7. The objective is to obtain corresponding values for the remaining unknown joint displacements. In particular, the problem statement is as follows:

given: $\alpha_{12}, \alpha_{23}, \alpha_{34}, \alpha_{45}, \alpha_{56}, \alpha_{67}, \alpha_{71}$,
$a_{12}, a_{23}, a_{34}, a_{45}, a_{56}, a_{67}, a_{71}$,
$S_1, S_2, S_3, S_4, S_5, S_6, S_7$, and
θ_7 (input angle),

find: $\theta_1, \theta_2, \theta_3, \theta_4, \theta_5$, and θ_6.

The angle θ_1 is the output angle because it is attached to the frame a_{71}, and it will be solved for first. The solution will proceed by first obtaining two pairs of equations that are linear in the tan-half-angle of θ_6 and that also contain the variables $\theta_1, \theta_2, \theta_4$, and θ_5. Once these are obtained, it will be shown how to eliminate the angles θ_4 and θ_5 from these equations in order to obtain four equations of the form of Eq. (10.1).

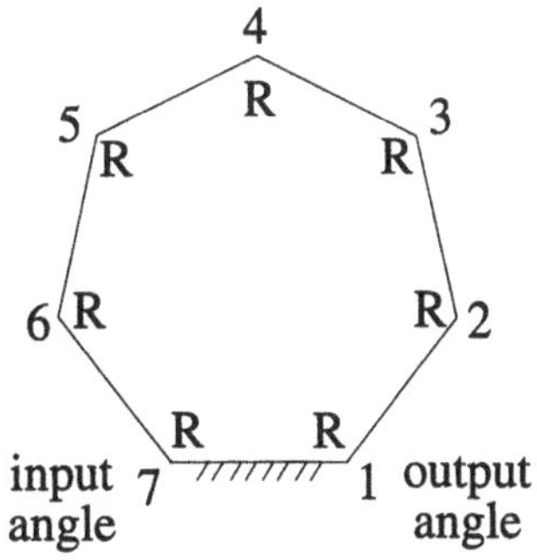

Figure 10.1. Planar representation of 7R spatial mechanism.

10.2.1 Derivation of the first pair of equations

The vector loop equation for the 7R spatial mechanism may be written as

$$\mathbf{R} = \mathbf{R}^{6,23} + \mathbf{R}^{3,56} = \mathbf{0}, \tag{10.2}$$

where

$$\mathbf{R}^{6,23} = S_6\mathbf{S}_6 + a_{67}\mathbf{a}_{67} + S_7\mathbf{S}_7 + a_{71}\mathbf{a}_{71} + S_1\mathbf{S}_1 + a_{12}\mathbf{a}_{12} + S_2\mathbf{S}_2 + a_{23}\mathbf{a}_{23}, \tag{10.3}$$

$$\mathbf{R}^{3,56} = S_3\mathbf{S}_3 + a_{34}\mathbf{a}_{34} + S_4\mathbf{S}_4 + a_{45}\mathbf{a}_{45} + S_5\mathbf{S}_5 + a_{56}\mathbf{a}_{56}. \tag{10.4}$$

In general, the notation $\mathbf{R}^{i,jk}$ will represent the sum of the terms of the vector loop equation beginning with $S_i\mathbf{S}_i$ and ending with $a_{jk}\mathbf{a}_{jk}$.

Projecting Eq. (10.2) onto the direction of the vector $\mathbf{a}_{56}$ and then onto the direction $(\mathbf{S}_6 \times \mathbf{a}_{56})$ gives

$$\mathbf{R}^{6,23} \cdot \mathbf{a}_{56} = -\mathbf{R}^{3,56} \cdot \mathbf{a}_{56}, \tag{10.5}$$

$$\mathbf{R}^{6,23} \cdot (\mathbf{S}_6 \times \mathbf{a}_{56}) = -\mathbf{R}^{3,56} \cdot (\mathbf{S}_6 \times \mathbf{a}_{56}). \tag{10.6}$$

The scalar products on the right side of these equations will be evaluated by using set 10 of the sets of direction cosines for a spherical heptagon. This yields

$$J_1 = -\mathbf{R}^{3,56} \cdot \mathbf{a}_{56} = -(S_3X_{45} + a_{34}W_{45} + S_4X_5 + a_{45}c_5 + a_{56}), \tag{10.7}$$

$$J_2 = -\mathbf{R}^{3,56} \cdot (\mathbf{S}_6 \times \mathbf{a}_{56}) = S_3Y_{45} - a_{34}U^*_{456} + S_4Y_5 + a_{45}s_5c_{56} - S_5s_{56}. \tag{10.8}$$

The scalar products on the left side of Eq. (10.5) will be evaluated by using set 5 from the direction cosine table, and the left side of Eq. (10.6) will be evaluated with set 10. This yields

$$\mathbf{R}^{6,23} \cdot \mathbf{a}_{56} = a_{67}c_6 + S_7\bar{X}_6 + a_{71}W_{76} + S_1X_{76} + a_{12}W_{176} + S_2X_{176} + a_{23}W_{2176}, \tag{10.9}$$

$$\mathbf{R}^{6,23} \cdot (\mathbf{S}_6 \times \mathbf{a}_{56}) = a_{67}s_6 - S_7Y_{12345} + a_{71}U^*_{123456} - S_1Y_{2345} + a_{12}U^*_{23456} - S_2Y_{345} + a_{23}U^*_{3456}. \tag{10.10}$$

Fundamental and subsidiary spherical and polar sine–cosine laws may be used to substitute for the coefficients of S_7, a_{71}, S_1, a_{12}, S_2, and a_{23} in Eq. (10.10) to give

$$\mathbf{R}^{6,23} \cdot (\mathbf{S}_6 \times \mathbf{a}_{56}) = a_{67}s_6 - S_7s_{67}c_6 - a_{71}V_{76} + S_1X^*_{76} - a_{12}V_{176} + S_2X^*_{176} - a_{23}V_{2176}. \tag{10.11}$$

Expanding Eqs. (10.9) and (10.11) and regrouping terms using the expansions

$$\begin{aligned}
\bar{X}_6 &= s_{67}s_6, & X_{76} &= \bar{X}_7c_6 - \bar{Y}_7s_6, & X_{176} &= X_{17}c_6 - Y_{17}s_6,\\
X^*_{76} &= \bar{X}_7s_6 + \bar{Y}_7c_6, & X^*_{176} &= X_{17}s_6 + Y_{17}c_6, & &\\
W_{76} &= c_6c_7 - s_6s_7c_{67}, & W_{176} &= s_6U^*_{176} + c_6W_{17}, & W_{2176} &= s_6U^*_{2176} + c_6W_{217},\\
V_{76} &= -(s_6c_7 + c_6s_7c_{67}), & V_{176} &= c_6U^*_{176} - s_6W_{17}, & V_{2176} &= c_6U^*_{2176} - s_6W_{217},
\end{aligned} \tag{10.12}$$

yields

$$\mathbf{R}^{6,23} \cdot \mathbf{a}_{56} = H_1 c_6 - H_2 s_6, \tag{10.13}$$

$$\mathbf{R}^{6,23} \cdot (\mathbf{S}_6 \times \mathbf{a}_{56}) = H_2 c_6 + H_1 s_6, \tag{10.14}$$

where

$$H_1 = a_{67} + a_{71} c_7 + S_1 \bar{X}_7 + a_{12} W_{17} + S_2 X_{17} + a_{23} W_{217}, \tag{10.15}$$

$$H_2 = -S_7 s_{67} + a_{71} c_{67} s_7 + S_1 \bar{Y}_7 - a_{12} U^*_{176} + S_2 Y_{17} - a_{23} U^*_{2176}. \tag{10.16}$$

Substituting Eqs. (10.7), (10.8), (10.13), and (10.14) into Eqs. (10.5) and (10.6) and rearranging gives

$$H_1 c_6 - H_2 s_6 - J_1 = 0, \tag{10.17}$$

$$H_2 c_6 + H_1 s_6 - J_2 = 0. \tag{10.18}$$

Adding Eq. (10.17) to x_6 times Eq. (10.18) yields

$$H_1 (c_6 + s_6 x_6) + H_2 (c_6 x_6 - s_6) - J_2 x_6 - J_1 = 0. \tag{10.19}$$

Subtracting x_6 times Eq. (10.17) from Eq. (10.18) gives

$$H_1 (s_6 - c_6 x_6) + H_2 (c_6 + s_6 x_6) + J_1 x_6 - J_2 = 0. \tag{10.20}$$

The following trigonometric identities were introduced in Eqs. (9.108) and (9.109):

$$s_6 - c_6 x_6 = x_6, \tag{10.21}$$

$$c_6 + s_6 x_6 = 1. \tag{10.22}$$

Equations (10.19) and (10.20) can be simplified by using these identities and rearranged to give

$$(H_2 + J_2) x_6 - (H_1 - J_1) = 0, \tag{10.23}$$

$$(H_1 + J_1) x_6 + (H_2 - J_2) = 0. \tag{10.24}$$

These two equations are linear in the tan-half-angle of θ_6 and contain the variable parameters $\theta_1, \theta_2, \theta_4$, and θ_5. It will be necessary to eliminate the angles θ_4 and θ_5 from these equations so that the result may be expressed in the format of Eq. (10.1). Two additional equations will first be generated before the angles θ_4 and θ_5 are eliminated.

10.2.2 Some important vector expressions

Prior to developing the next pair of equations that will be linear in the tan-half-angle of θ_6, four new expressions will be obtained. These expressions will be used later.

Using set 6 of the table of direction cosines for a spherical heptagon, it is apparent that

$$\mathbf{R}^{6,23} \cdot \mathbf{a}_{67} = H_1, \tag{10.25}$$

$$\mathbf{R}^{6,23} \cdot (\mathbf{S}_6 \times \mathbf{a}_{67}) = H_2, \tag{10.26}$$

where H_1 and H_2 are given by Eqs. (10.15) and (10.16). Substituting for H_1 and H_2 in Eqs. (10.13) and (10.14) gives

$$\mathbf{R}^{6,23} \cdot \mathbf{a}_{56} = [\mathbf{R}^{6,23} \cdot \mathbf{a}_{67}]c_6 - [\mathbf{R}^{6,23} \cdot (\mathbf{S}_6 \times \mathbf{a}_{67})]s_6, \tag{10.27}$$

$$\mathbf{R}^{6,23} \cdot (\mathbf{S}_6 \times \mathbf{a}_{56}) = [\mathbf{R}^{6,23} \cdot (\mathbf{S}_6 \times \mathbf{a}_{67})]c_6 + [\mathbf{R}^{6,23} \cdot \mathbf{a}_{67}]s_6. \tag{10.28}$$

Also from set 6 of the direction cosine table, the scalar product of $\mathbf{S}_3$ and $\mathbf{a}_{56}$ may be written as

$$\mathbf{S}_3 \cdot \mathbf{a}_{56} = X_{217}c_6 - Y_{217}s_6. \tag{10.29}$$

This equation may also be written as

$$\mathbf{S}_3 \cdot \mathbf{a}_{56} = [\mathbf{S}_3 \cdot \mathbf{a}_{67}]c_6 - [\mathbf{S}_3 \cdot (\mathbf{S}_6 \times \mathbf{a}_{67})]s_6. \tag{10.30}$$

Lastly, set 6 of the direction cosine table may be used to evaluate the scalar product of $\mathbf{S}_3$ with the vector $(\mathbf{S}_6 \times \mathbf{a}_{56})$ as

$$\mathbf{S}_3 \cdot (\mathbf{S}_6 \times \mathbf{a}_{56}) = \begin{vmatrix} X_{217} & Y_{217} & Z_{217} \\ 0 & 0 & 1 \\ c_6 & -s_6 & 0 \end{vmatrix}. \tag{10.31}$$

Expanding this determinant gives

$$\mathbf{S}_3 \cdot (\mathbf{S}_6 \times \mathbf{a}_{56}) = X_{217}s_6 + Y_{217}c_6. \tag{10.32}$$

This may also be written as

$$\mathbf{S}_3 \cdot (\mathbf{S}_6 \times \mathbf{a}_{56}) = [\mathbf{S}_3 \cdot \mathbf{a}_{67}]s_6 + [\mathbf{S}_3 \cdot (\mathbf{S}_6 \times \mathbf{a}_{67})]c_6. \tag{10.33}$$

The four expressions listed in Eqs. (10.27), (10.28), (10.30), and (10.33) will be used in the next section to derive a further pair of equations that are linear in x_6 and that also contain the variable joint angles θ_1, θ_2, θ_4, and θ_5.

10.2.3 Derivation of the second pair of equations

Using Eq. (10.2), we can form the following pair of equations:

$$\frac{1}{2}(\mathbf{R}^{6,23} \cdot \mathbf{R}^{6,23})(\mathbf{S}_3 \cdot \mathbf{a}_{56}) - (\mathbf{R}^{6,23} \cdot \mathbf{S}_3)(\mathbf{R}^{6,23} \cdot \mathbf{a}_{56})$$
$$= \frac{1}{2}(\mathbf{R}^{3,56} \cdot \mathbf{R}^{3,56})(\mathbf{S}_3 \cdot \mathbf{a}_{56}) - (\mathbf{R}^{3,56} \cdot \mathbf{S}_3)(\mathbf{R}^{3,56} \cdot \mathbf{a}_{56}) \tag{10.34}$$

and

$$\frac{1}{2}(\mathbf{R}^{6,23} \cdot \mathbf{R}^{6,23})(\mathbf{S}_3 \cdot \mathbf{S}_6 \times \mathbf{a}_{56}) - (\mathbf{R}^{6,23} \cdot \mathbf{S}_3)(\mathbf{R}^{6,23} \cdot \mathbf{S}_6 \times \mathbf{a}_{56})$$
$$= \frac{1}{2}(\mathbf{R}^{3,56} \cdot \mathbf{R}^{3,56})(\mathbf{S}_3 \cdot \mathbf{S}_6 \times \mathbf{a}_{56}) - (\mathbf{R}^{3,56} \cdot \mathbf{S}_3)(\mathbf{R}^{3,56} \cdot \mathbf{S}_6 \times \mathbf{a}_{56}). \tag{10.35}$$

Substituting Eqs. (10.27) and (10.30) into Eq. (10.34) gives

$$\frac{1}{2}(\mathbf{R}^{6,23}\cdot\mathbf{R}^{6,23})[(\mathbf{S}_3\cdot\mathbf{a}_{67})c_6-(\mathbf{S}_3\cdot\mathbf{S}_6\times\mathbf{a}_{67})s_6]-(\mathbf{R}^{6,23}\cdot\mathbf{S}_3)[(\mathbf{R}^{6,23}\cdot\mathbf{a}_{67})c_6 \\ -(\mathbf{R}^{6,23}\cdot\mathbf{S}_6\times\mathbf{a}_{67})s_6]=\frac{1}{2}(\mathbf{R}^{3,56}\cdot\mathbf{R}^{3,56})(\mathbf{S}_3\cdot\mathbf{a}_{56})-(\mathbf{R}^{3,56}\cdot\mathbf{S}_3)(\mathbf{R}^{3,56}\cdot\mathbf{a}_{56}). \tag{10.36}$$

Regrouping terms gives

$$H_3c_6-H_4s_6-J_3=0, \tag{10.37}$$

where

$$H_3=\frac{1}{2}(\mathbf{R}^{6,23}\cdot\mathbf{R}^{6,23})(\mathbf{S}_3\cdot\mathbf{a}_{67})-(\mathbf{R}^{6,23}\cdot\mathbf{S}_3)(\mathbf{R}^{6,23}\cdot\mathbf{a}_{67}), \tag{10.38}$$

$$H_4=\frac{1}{2}(\mathbf{R}^{6,23}\cdot\mathbf{R}^{6,23})(\mathbf{S}_3\cdot\mathbf{S}_6\times\mathbf{a}_{67})-(\mathbf{R}^{6,23}\cdot\mathbf{S}_3)(\mathbf{R}^{6,23}\cdot\mathbf{S}_6\times\mathbf{a}_{67}), \tag{10.39}$$

$$J_3=\frac{1}{2}(\mathbf{R}^{3,56}\cdot\mathbf{R}^{3,56})(\mathbf{S}_3\cdot\mathbf{a}_{56})-(\mathbf{R}^{3,56}\cdot\mathbf{S}_3)(\mathbf{R}^{3,56}\cdot\mathbf{a}_{56}). \tag{10.40}$$

Similarly, substituting Eqs. (10.28) and (10.33) into Eq. (10.35) yields

$$\frac{1}{2}(\mathbf{R}^{6,23}\cdot\mathbf{R}^{6,23})[(\mathbf{S}_3\cdot\mathbf{a}_{67})s_6+(\mathbf{S}_3\cdot\mathbf{S}_6\times\mathbf{a}_{67})c_6]-(\mathbf{R}^{6,23}\cdot\mathbf{S}_3)[(\mathbf{R}^{6,23}\cdot\mathbf{S}_6\times\mathbf{a}_{67})c_6 \\ +(\mathbf{R}^{6,23}\cdot\mathbf{a}_{67})s_6]=\frac{1}{2}(\mathbf{R}^{3,56}\cdot\mathbf{R}^{3,56})(\mathbf{S}_3\cdot\mathbf{S}_6\times\mathbf{a}_{56})-(\mathbf{R}^{3,56}\cdot\mathbf{S}_3)(\mathbf{R}^{3,56}\cdot\mathbf{S}_6\times\mathbf{a}_{56}). \tag{10.41}$$

Rearranging terms gives

$$H_3s_6+H_4c_6-J_4=0, \tag{10.42}$$

where

$$J_4=\frac{1}{2}(\mathbf{R}^{3,56}\cdot\mathbf{R}^{3,56})(\mathbf{S}_3\cdot\mathbf{S}_6\times\mathbf{a}_{56})-(\mathbf{R}^{3,56}\cdot\mathbf{S}_3)(\mathbf{R}^{3,56}\cdot\mathbf{S}_6\times\mathbf{a}_{56}). \tag{10.43}$$

By analogy with Eqs. (10.17) and (10.18), Eqs. (10.37) and (10.42) yield the following pair of equations:

$$(H_4+J_4)x_6-(H_3-J_3)=0, \tag{10.44}$$

$$(H_3+J_3)x_6+(H_4-J_4)=0, \tag{10.45}$$

which are analogous to Eqs. (10.23) and (10.24).

10.2.4 Expansion of terms H_3, H_4, J_3, and J_4

In order to avoid massive tedious expansions, it is necessary to expand vectors such as $\mathbf{R}^{3,56}$ and $\mathbf{R}^{6,23}$ and to regroup terms. Further, in order to avoid repeating expansions unnecessarily, it is preferable to expand $\mathbf{R}^{3,56}$ in a sequence and to use these results to expand the larger vector $\mathbf{R}^{6,23}$ using as far as possible appropriate exchanges of subscripts and superscripts.

Firstly, substituting

$$\mathbf{R}^{3,56} = \mathbf{R}^{3,45} + \mathbf{R}^{5,56}, \tag{10.46}$$

where

$$\mathbf{R}^{3,45} = S_3\mathbf{S}_3 + a_{34}\mathbf{a}_{34} + S_4\mathbf{S}_4 + a_{45}\mathbf{a}_{45} \tag{10.47}$$

and

$$\mathbf{R}^{5,56} = S_5\mathbf{S}_5 + a_{56}\mathbf{a}_{56}, \tag{10.48}$$

into Eq. (10.40) gives

$$\begin{aligned} J_3 = {} & \frac{1}{2}(\mathbf{R}^{3,45}\cdot\mathbf{R}^{3,45})(\mathbf{S}_3\cdot\mathbf{a}_{56}) + (\mathbf{R}^{3,45}\cdot\mathbf{R}^{5,56})(\mathbf{S}_3\cdot\mathbf{a}_{56}) + \frac{1}{2}(\mathbf{R}^{5,56}\cdot\mathbf{R}^{5,56})(\mathbf{S}_3\cdot\mathbf{a}_{56}) \\ & -(\mathbf{R}^{3,56}\cdot\mathbf{S}_3)(\mathbf{R}^{5,56}\cdot\mathbf{a}_{56}) - (\mathbf{R}^{3,45}\cdot\mathbf{S}_3)(\mathbf{R}^{3,45}\cdot\mathbf{a}_{56}) - (\mathbf{R}^{5,56}\cdot\mathbf{S}_3)(\mathbf{R}^{3,45}\cdot\mathbf{a}_{56}). \end{aligned} \tag{10.49}$$

The expansion of J_3 continues by writing*

$$(\mathbf{R}^{3,45}\times\mathbf{S}_3)\cdot(\mathbf{R}^{5,56}\times\mathbf{a}_{56}) = (\mathbf{R}^{3,45}\cdot\mathbf{R}^{5,56})(\mathbf{S}_3\cdot\mathbf{a}_{56}) - (\mathbf{R}^{3,45}\cdot\mathbf{a}_{56})(\mathbf{R}^{5,56}\cdot\mathbf{S}_3). \tag{10.50}$$

Substituting this expression into Eq. (10.49) and rearranging gives

$$\begin{aligned} J_3 = {} & \frac{1}{2}(\mathbf{R}^{3,45}\cdot\mathbf{R}^{3,45})(\mathbf{S}_3\cdot\mathbf{a}_{56}) - (\mathbf{R}^{3,45}\cdot\mathbf{S}_3)(\mathbf{R}^{3,45}\cdot\mathbf{a}_{56}) + \frac{1}{2}(\mathbf{R}^{5,56}\cdot\mathbf{R}^{5,56})(\mathbf{S}_3\cdot\mathbf{a}_{56}) \\ & - (\mathbf{R}^{3,56}\cdot\mathbf{S}_3)(\mathbf{R}^{5,56}\cdot\mathbf{a}_{56}) + (\mathbf{R}^{3,45}\times\mathbf{S}_3)\cdot(\mathbf{R}^{5,56}\times\mathbf{a}_{56}). \end{aligned} \tag{10.51}$$

The term $\mathbf{R}^{3,45}$ is now written as

$$\mathbf{R}^{3,45} = \mathbf{R}^{3,34} + \mathbf{R}^{4,45}, \tag{10.52}$$

where

$$\mathbf{R}^{3,34} = S_3\mathbf{S}_3 + a_{34}\mathbf{a}_{34}, \tag{10.53}$$

$$\mathbf{R}^{4,45} = S_4\mathbf{S}_4 + a_{45}\mathbf{a}_{45}. \tag{10.54}$$

The first two terms of Eq. (10.51) are of the same format as the terms on the right side of Eq. (10.40), which was expanded to give Eq. (10.51). Thus, these two terms can be expanded by substituting the superscripts (3,34), (4,45), and (3,45) for the superscripts (3,45), (5,56), and (3,56) everywhere in Eq. (10.51) and substituting the result in for the

* This is simply the expression for the scalar product of four vectors **a**, **b**, **c**, and **d** and $(\mathbf{a}\times\mathbf{b})\cdot(\mathbf{c}\times\mathbf{d}) = (\mathbf{a}\cdot\mathbf{c})(\mathbf{b}\cdot\mathbf{d}) - (\mathbf{a}\cdot\mathbf{d})(\mathbf{b}\cdot\mathbf{c})$.

first two terms of Eq. (10.51). This gives

$$J_3 = \frac{1}{2}(\mathbf{R}^{3,34} \cdot \mathbf{R}^{3,34})(\mathbf{S}_3 \cdot \mathbf{a}_{56}) - (\mathbf{R}^{3,34} \cdot \mathbf{S}_3)(\mathbf{R}^{3,34} \cdot \mathbf{a}_{56}) + \frac{1}{2}(\mathbf{R}^{4,45} \cdot \mathbf{R}^{4,45})(\mathbf{S}_3 \cdot \mathbf{a}_{56}) - (\mathbf{R}^{3,45} \cdot \mathbf{S}_3)(\mathbf{R}^{4,45} \cdot \mathbf{a}_{56}) + (\mathbf{R}^{3,34} \times \mathbf{S}_3) \cdot (\mathbf{R}^{4,45} \times \mathbf{a}_{56}) + \frac{1}{2}(\mathbf{R}^{5,56} \cdot \mathbf{R}^{5,56}) \times (\mathbf{S}_3 \cdot \mathbf{a}_{56}) - (\mathbf{R}^{3,56} \cdot \mathbf{S}_3)(\mathbf{R}^{5,56} \cdot \mathbf{a}_{56}) + (\mathbf{R}^{3,45} \times \mathbf{S}_3) \cdot (\mathbf{R}^{5,56} \times \mathbf{a}_{56}). \tag{10.55}$$

This equation may be rearranged in the form

$$J_3 = \frac{1}{2}(\mathbf{R}^{3,34} \cdot \mathbf{R}^{3,34} + \mathbf{R}^{4,45} \cdot \mathbf{R}^{4,45} + \mathbf{R}^{5,56} \cdot \mathbf{R}^{5,56})(\mathbf{S}_3 \cdot \mathbf{a}_{56}) - (\mathbf{R}^{3,34} \cdot \mathbf{S}_3)(\mathbf{R}^{3,34} \cdot \mathbf{a}_{56}) - (\mathbf{R}^{3,45} \cdot \mathbf{S}_3)(\mathbf{R}^{4,45} \cdot \mathbf{a}_{56}) - (\mathbf{R}^{3,56} \cdot \mathbf{S}_3)(\mathbf{R}^{5,56} \cdot \mathbf{a}_{56}) + (\mathbf{R}^{3,34} \times \mathbf{S}_3) \cdot (\mathbf{R}^{4,45} \times \mathbf{a}_{56}) + (\mathbf{R}^{3,45} \times \mathbf{S}_3) \cdot (\mathbf{R}^{5,56} \times \mathbf{a}_{56}). \tag{10.56}$$

The expression for J_4 can be obtained directly from Eq. (10.56) by substituting $(\mathbf{S}_6 \times \mathbf{a}_{56})$ everywhere for $\mathbf{a}_{56}$. Thus,

$$J_4 = \frac{1}{2}(\mathbf{R}^{3,34} \cdot \mathbf{R}^{3,34} + \mathbf{R}^{4,45} \cdot \mathbf{R}^{4,45} + \mathbf{R}^{5,56} \cdot \mathbf{R}^{5,56})(\mathbf{S}_3 \cdot \mathbf{S}_6 \times \mathbf{a}_{56}) - (\mathbf{R}^{3,34} \cdot \mathbf{S}_3)(\mathbf{R}^{3,34} \cdot \mathbf{S}_6 \times \mathbf{a}_{56}) - (\mathbf{R}^{3,45} \cdot \mathbf{S}_3)(\mathbf{R}^{4,45} \cdot \mathbf{S}_6 \times \mathbf{a}_{56}) - (\mathbf{R}^{3,56} \cdot \mathbf{S}_3)(\mathbf{R}^{5,56} \cdot \mathbf{S}_6 \times \mathbf{a}_{56}) + (\mathbf{R}^{3,34} \times \mathbf{S}_3) \cdot (\mathbf{R}^{4,45} \times (\mathbf{S}_6 \times \mathbf{a}_{56})) + (\mathbf{R}^{3,45} \times \mathbf{S}_3) \cdot (\mathbf{R}^{5,56} \times (\mathbf{S}_6 \times \mathbf{a}_{56})). \tag{10.57}$$

The terms H_3 and H_4 will now be expanded. Firstly, it is observed that the form of the term H_3 is very similar to that of J_3. (This is based upon a comparison of Eqs. (10.38) and (10.40).) Thus, it is possible to write the term H_3 by introducing the expression

$$\mathbf{R}^{6,23} = \mathbf{R}^{6,12} + \mathbf{R}^{2,23}, \tag{10.58}$$

where

$$\mathbf{R}^{6,12} = S_6\mathbf{S}_6 + a_{67}\mathbf{a}_{67} + S_7\mathbf{S}_7 + a_{71}\mathbf{a}_{71} + S_1\mathbf{S}_1 + a_{12}\mathbf{a}_{12} \tag{10.59}$$

and

$$\mathbf{R}^{2,23} = S_2\mathbf{S}_2 + a_{23}\mathbf{a}_{23}, \tag{10.60}$$

and then substituting the superscripts (6,12), (2,23), and (6,23) for the superscripts (3,45), (5,56), and (3,56) and the vector $\mathbf{a}_{67}$ for $\mathbf{a}_{56}$ in Eq. (10.51). This yields

$$H_3 = \frac{1}{2}(\mathbf{R}^{6,12} \cdot \mathbf{R}^{6,12})(\mathbf{S}_3 \cdot \mathbf{a}_{67}) - (\mathbf{R}^{6,12} \cdot \mathbf{S}_3)(\mathbf{R}^{6,12} \cdot \mathbf{a}_{67}) + \frac{1}{2}(\mathbf{R}^{2,23} \cdot \mathbf{R}^{2,23})(\mathbf{S}_3 \cdot \mathbf{a}_{67}) - (\mathbf{R}^{6,23} \cdot \mathbf{S}_3)(\mathbf{R}^{2,23} \cdot \mathbf{a}_{67}) + (\mathbf{R}^{6,12} \times \mathbf{S}_3) \cdot (\mathbf{R}^{2,23} \times \mathbf{a}_{67}). \tag{10.61}$$

The last term of this equation may be expanded as

$$(\mathbf{R}^{6,12} \times \mathbf{S}_3) \cdot (\mathbf{R}^{2,23} \times \mathbf{a}_{67}) = (\mathbf{R}^{6,12} \cdot \mathbf{R}^{2,23})(\mathbf{S}_3 \cdot \mathbf{a}_{67}) - (\mathbf{R}^{2,23} \cdot \mathbf{S}_3)(\mathbf{R}^{6,12} \cdot \mathbf{a}_{67}). \tag{10.62}$$

An additional expression may now be written in the form

$$(\mathbf{R}^{6,12} \times \mathbf{a}_{67}) \cdot (\mathbf{R}^{2,23} \times \mathbf{S}_3) = (\mathbf{R}^{6,12} \cdot \mathbf{R}^{2,23})(\mathbf{S}_3 \cdot \mathbf{a}_{67}) - (\mathbf{R}^{2,23} \cdot \mathbf{a}_{67})(\mathbf{R}^{6,12} \cdot \mathbf{S}_3). \tag{10.63}$$

The first term on the right side of Eq. (10.63) appears in Eq. (10.62). Upon substitution, Eq. (10.62) may be written as

$$(\mathbf{R}^{6,12} \times \mathbf{S}_3) \cdot (\mathbf{R}^{2,23} \times \mathbf{a}_{67}) = (\mathbf{R}^{6,12} \times \mathbf{a}_{67}) \cdot (\mathbf{R}^{2,23} \times \mathbf{S}_3) + (\mathbf{R}^{2,23} \cdot \mathbf{a}_{67})(\mathbf{R}^{6,12} \cdot \mathbf{S}_3) - (\mathbf{R}^{2,23} \cdot \mathbf{S}_3)(\mathbf{R}^{6,12} \cdot \mathbf{a}_{67}). \tag{10.64}$$

Substituting this result into Eq. (10.61) yields

$$\begin{aligned} \mathrm{H}_3 = {} & \frac{1}{2}(\mathbf{R}^{6,12} \cdot \mathbf{R}^{6,12})(\mathbf{S}_3 \cdot \mathbf{a}_{67}) - (\mathbf{R}^{6,12} \cdot \mathbf{S}_3)(\mathbf{R}^{6,12} \cdot \mathbf{a}_{67}) + \frac{1}{2}(\mathbf{R}^{2,23} \cdot \mathbf{R}^{2,23})(\mathbf{S}_3 \cdot \mathbf{a}_{67}) \\ & - (\mathbf{R}^{6,23} \cdot \mathbf{S}_3)(\mathbf{R}^{2,23} \cdot \mathbf{a}_{67}) + (\mathbf{R}^{6,12} \times \mathbf{a}_{67}) \cdot (\mathbf{R}^{2,23} \times \mathbf{S}_3) \\ & + (\mathbf{R}^{2,23} \cdot \mathbf{a}_{67})(\mathbf{R}^{6,12} \cdot \mathbf{S}_3) - (\mathbf{R}^{2,23} \cdot \mathbf{S}_3)(\mathbf{R}^{6,12} \cdot \mathbf{a}_{67}). \end{aligned} \tag{10.65}$$

Now, the terms $-(\mathbf{R}^{6,23} \cdot \mathbf{S}_3)(\mathbf{R}^{2,23} \cdot \mathbf{a}_{67}) + (\mathbf{R}^{6,12} \cdot \mathbf{S}_3)(\mathbf{R}^{2,23} \cdot \mathbf{a}_{67})$ are equivalent to $-[(\mathbf{R}^{6,23} - \mathbf{R}^{6,12}) \cdot \mathbf{S}_3](\mathbf{R}^{2,23} \cdot \mathbf{a}_{67})$. This term may be written as $-(\mathbf{R}^{2,23} \cdot \mathbf{S}_3)(\mathbf{R}^{2,23} \cdot \mathbf{a}_{67})$, and hence

$$\begin{aligned} \mathrm{H}_3 = {} & \frac{1}{2}(\mathbf{R}^{6,12} \cdot \mathbf{R}^{6,12})(\mathbf{S}_3 \cdot \mathbf{a}_{67}) - (\mathbf{R}^{6,12} \cdot \mathbf{S}_3)(\mathbf{R}^{6,12} \cdot \mathbf{a}_{67}) + \frac{1}{2}(\mathbf{R}^{2,23} \cdot \mathbf{R}^{2,23})(\mathbf{S}_3 \cdot \mathbf{a}_{67}) \\ & - (\mathbf{R}^{2,23} \cdot \mathbf{S}_3)(\mathbf{R}^{2,23} \cdot \mathbf{a}_{67}) + (\mathbf{R}^{6,12} \times \mathbf{a}_{67}) \cdot (\mathbf{R}^{2,23} \times \mathbf{S}_3) - (\mathbf{R}^{2,23} \cdot \mathbf{S}_3)(\mathbf{R}^{6,12} \cdot \mathbf{a}_{67}). \end{aligned} \tag{10.66}$$

Now,

$$\mathbf{R}^{6,12} = \mathbf{R}^{6,71} + \mathbf{R}^{1,12}, \tag{10.67}$$

where

$$\mathbf{R}^{6,71} = S_6\mathbf{S}_6 + a_{67}\mathbf{a}_{67} + S_7\mathbf{S}_7 + a_{71}\mathbf{a}_{71} \tag{10.68}$$

and

$$\mathbf{R}^{1,12} = S_1\mathbf{S}_1 + a_{12}\mathbf{a}_{12}. \tag{10.69}$$

The first two terms of Eq. (10.66) are of the same format as the terms on the right side of Eq. (10.38). Thus, these two terms can be expanded by substituting the superscripts (6,71), (1,12), and (6,12) for the superscripts (6,12), (2,23), and (6,23) in Eq. (10.66).

This gives

$$\begin{aligned}\mathrm{H}_3 = {} & \frac{1}{2}(\mathbf{R}^{6,71}\cdot\mathbf{R}^{6,71})(\mathrm{S}_3\cdot\mathbf{a}_{67}) - (\mathbf{R}^{6,71}\cdot\mathrm{S}_3)(\mathbf{R}^{6,71}\cdot\mathbf{a}_{67}) + \frac{1}{2}(\mathbf{R}^{1,12}\cdot\mathbf{R}^{1,12})(\mathrm{S}_3\cdot\mathbf{a}_{67}) \\ & - (\mathbf{R}^{1,12}\cdot\mathrm{S}_3)(\mathbf{R}^{1,12}\cdot\mathbf{a}_{67}) + (\mathbf{R}^{6,71}\times\mathbf{a}_{67})\cdot(\mathbf{R}^{1,12}\times\mathrm{S}_3) - (\mathbf{R}^{1,12}\cdot\mathrm{S}_3)(\mathbf{R}^{6,71}\cdot\mathbf{a}_{67}) \\ & + \frac{1}{2}(\mathbf{R}^{2,23}\cdot\mathbf{R}^{2,23})(\mathrm{S}_3\cdot\mathbf{a}_{67}) - (\mathbf{R}^{2,23}\cdot\mathrm{S}_3)(\mathbf{R}^{2,23}\cdot\mathbf{a}_{67}) + (\mathbf{R}^{6,12}\times\mathbf{a}_{67}) \\ & \cdot(\mathbf{R}^{2,23}\times\mathrm{S}_3) - (\mathbf{R}^{2,23}\cdot\mathrm{S}_3)(\mathbf{R}^{6,12}\cdot\mathbf{a}_{67}). \end{aligned} \tag{10.70}$$

Finally, the term $\mathbf{R}^{6,71}$ is defined as

$$\mathbf{R}^{6,71} = \mathbf{R}^{6,67} + \mathbf{R}^{7,71}, \tag{10.71}$$

where

$$\mathbf{R}^{6,67} = \mathrm{S}_6\mathrm{S}_6 + \mathrm{a}_{67}\mathbf{a}_{67} \tag{10.72}$$

and

$$\mathbf{R}^{7,71} = \mathrm{S}_7\mathrm{S}_7 + \mathrm{a}_{71}\mathbf{a}_{71}. \tag{10.73}$$

The first two terms of Eq. (10.70) are of the same form as the terms on the right side of Eq. (10.38). Thus, these two terms can be expanded by substituting the superscripts (6,67), (7,71), and (6,71) for the superscripts (6,12), (2,23), and (6,23) in Eq. (10.66). This gives

$$\begin{aligned}\mathrm{H}_3 = {} & \frac{1}{2}(\mathbf{R}^{6,67}\cdot\mathbf{R}^{6,67})(\mathrm{S}_3\cdot\mathbf{a}_{67}) - (\mathbf{R}^{6,67}\cdot\mathrm{S}_3)(\mathbf{R}^{6,67}\cdot\mathbf{a}_{67}) + \frac{1}{2}(\mathbf{R}^{7,71}\cdot\mathbf{R}^{7,71})(\mathrm{S}_3\cdot\mathbf{a}_{67}) \\ & - (\mathbf{R}^{7,71}\cdot\mathrm{S}_3)(\mathbf{R}^{7,71}\cdot\mathbf{a}_{67}) + (\mathbf{R}^{6,67}\times\mathbf{a}_{67})\cdot(\mathbf{R}^{7,71}\times\mathrm{S}_3) - (\mathbf{R}^{7,71}\cdot\mathrm{S}_3)(\mathbf{R}^{6,67}\cdot\mathbf{a}_{67}) \\ & + \frac{1}{2}(\mathbf{R}^{1,12}\cdot\mathbf{R}^{1,12})(\mathrm{S}_3\cdot\mathbf{a}_{67}) - (\mathbf{R}^{1,12}\cdot\mathrm{S}_3)(\mathbf{R}^{1,12}\cdot\mathbf{a}_{67}) + (\mathbf{R}^{6,71}\times\mathbf{a}_{67})\cdot(\mathbf{R}^{1,12}\times\mathrm{S}_3) \\ & - (\mathbf{R}^{1,12}\cdot\mathrm{S}_3)(\mathbf{R}^{6,71}\cdot\mathbf{a}_{67}) + \frac{1}{2}(\mathbf{R}^{2,23}\cdot\mathbf{R}^{2,23})(\mathrm{S}_3\cdot\mathbf{a}_{67}) - (\mathbf{R}^{2,23}\cdot\mathrm{S}_3)(\mathbf{R}^{2,23}\cdot\mathbf{a}_{67}) \\ & + (\mathbf{R}^{6,12}\times\mathbf{a}_{67})\cdot(\mathbf{R}^{2,23}\times\mathrm{S}_3) - (\mathbf{R}^{2,23}\cdot\mathrm{S}_3)(\mathbf{R}^{6,12}\cdot\mathbf{a}_{67}). \end{aligned} \tag{10.74}$$

This equation can be rearranged as

$$\begin{aligned}\mathrm{H}_3 = {} & \frac{1}{2}(\mathbf{R}^{6,67}\cdot\mathbf{R}^{6,67} + \mathbf{R}^{7,71}\cdot\mathbf{R}^{7.71} + \mathbf{R}^{1,12}\cdot\mathbf{R}^{1,12} + \mathbf{R}^{2,23}\cdot\mathbf{R}^{2,23})(\mathrm{S}_3\cdot\mathbf{a}_{67}) \\ & - (\mathbf{R}^{6,67}\cdot\mathrm{S}_3)(\mathbf{R}^{6,67}\cdot\mathbf{a}_{67}) - (\mathbf{R}^{7,71}\cdot\mathrm{S}_3)(\mathbf{R}^{7,71}\cdot\mathbf{a}_{67}) \\ & + (\mathbf{R}^{6,67}\times\mathbf{a}_{67})\cdot(\mathbf{R}^{7,71}\times\mathrm{S}_3) - (\mathbf{R}^{7,71}\cdot\mathrm{S}_3)(\mathbf{R}^{6,67}\cdot\mathbf{a}_{67}) \\ & - (\mathbf{R}^{1,12}\cdot\mathrm{S}_3)(\mathbf{R}^{1,12}\cdot\mathbf{a}_{67}) + (\mathbf{R}^{6,71}\times\mathbf{a}_{67})\cdot(\mathbf{R}^{1,12}\times\mathrm{S}_3) \\ & - (\mathbf{R}^{1,12}\cdot\mathrm{S}_3)(\mathbf{R}^{6,71}\cdot\mathbf{a}_{67}) - (\mathbf{R}^{2,23}\cdot\mathrm{S}_3)(\mathbf{R}^{2,23}\cdot\mathbf{a}_{67}) \\ & + (\mathbf{R}^{6,12}\times\mathbf{a}_{67})\cdot(\mathbf{R}^{2,23}\times\mathrm{S}_3) - (\mathbf{R}^{2,23}\cdot\mathrm{S}_3)(\mathbf{R}^{6,12}\cdot\mathbf{a}_{67}). \end{aligned} \tag{10.75}$$

The expression for H_4 can be obtained directly from Eq. (10.75) by substituting ($\mathrm{S}_6\ \times$

$\mathbf{a}_{67}$) for $\mathbf{a}_{67}$. Thus,

$$\begin{aligned}
H_4 = {} & \frac{1}{2}(\mathbf{R}^{6,67}\cdot\mathbf{R}^{6,67}+\mathbf{R}^{7,71}\cdot\mathbf{R}^{7.71}+\mathbf{R}^{1,12}\cdot\mathbf{R}^{1,12}+\mathbf{R}^{2,23}\cdot\mathbf{R}^{2,23})(\mathbf{S}_3\cdot\mathbf{S}_6\times\mathbf{a}_{67}) \\
& -(\mathbf{R}^{6,67}\cdot\mathbf{S}_3)(\mathbf{R}^{6,67}\cdot\mathbf{S}_6\times\mathbf{a}_{67})-(\mathbf{R}^{7,71}\cdot\mathbf{S}_3)(\mathbf{R}^{7,71}\cdot\mathbf{S}_6\times\mathbf{a}_{67}) \\
& +(\mathbf{R}^{6,67}\times(\mathbf{S}_6\times\mathbf{a}_{67}))\cdot(\mathbf{R}^{7,71}\times\mathbf{S}_3)-(\mathbf{R}^{7,71}\cdot\mathbf{S}_3)(\mathbf{R}^{6,67}\cdot\mathbf{S}_6\times\mathbf{a}_{67}) \\
& -(\mathbf{R}^{1,12}\cdot\mathbf{S}_3)(\mathbf{R}^{1,12}\cdot\mathbf{S}_6\times\mathbf{a}_{67})+(\mathbf{R}^{6,71}\times(\mathbf{S}_6\times\mathbf{a}_{67}))\cdot(\mathbf{R}^{1,12}\times\mathbf{S}_3) \\
& -(\mathbf{R}^{1,12}\cdot\mathbf{S}_3)(\mathbf{R}^{6,71}\cdot\mathbf{S}_6\times\mathbf{a}_{67})-(\mathbf{R}^{2,23}\cdot\mathbf{S}_3)(\mathbf{R}^{2,23}\cdot\mathbf{S}_6\times\mathbf{a}_{67}) \\
& +(\mathbf{R}^{6,12}\times(\mathbf{S}_6\times\mathbf{a}_{67}))\cdot(\mathbf{R}^{2,23}\times\mathbf{S}_3)-(\mathbf{R}^{2,23}\cdot\mathbf{S}_3)(\mathbf{R}^{6,12}\cdot\mathbf{S}_6\times\mathbf{a}_{67}).
\end{aligned} \tag{10.76}$$

Equations (10.75), (10.76), (10.56), and (10.57) provide expressions for the terms H_3, H_4, J_3, and J_4 in terms of scalar products of various vectors. It is next necessary to expand these scalar products in terms of the constant mechanism parameters and the variable joint angles. This will be accomplished in the next sections.

10.2.5 Detailed expansion of J_3 and J_4

The following expressions that are contained in J_3 must be expanded:

$$\frac{1}{2}(\mathbf{R}^{3,34}\cdot\mathbf{R}^{3,34}+\mathbf{R}^{4,45}\cdot\mathbf{R}^{4,45}+\mathbf{R}^{5,56}\cdot\mathbf{R}^{5,56}),\qquad \mathbf{S}_3\cdot\mathbf{a}_{56},$$

$$\mathbf{R}^{3,56}\cdot\mathbf{S}_3,\qquad \mathbf{R}^{3,45}\cdot\mathbf{S}_3,\qquad \mathbf{R}^{3,34}\cdot\mathbf{S}_3,$$

$$\mathbf{R}^{5,56}\cdot\mathbf{a}_{56},\qquad \mathbf{R}^{4,45}\cdot\mathbf{a}_{56},\qquad \mathbf{R}^{3,34}\cdot\mathbf{a}_{56},$$

$$(\mathbf{R}^{3,34}\times\mathbf{S}_3)\cdot(\mathbf{R}^{4,45}\times\mathbf{a}_{56}),\ (\mathbf{R}^{3,45}\times\mathbf{S}_3)\cdot(\mathbf{R}^{5,56}\times\mathbf{a}_{56}).$$

Additionally, the following terms in J_4 must also be expanded:

$$\mathbf{S}_3\cdot\mathbf{S}_6\times\mathbf{a}_{56},$$

$$\mathbf{R}^{5,56}\cdot\mathbf{S}_6\times\mathbf{a}_{56},\qquad \mathbf{R}^{4,45}\cdot\mathbf{S}_6\times\mathbf{a}_{56},\qquad \mathbf{R}^{3,34}\cdot\mathbf{S}_6\times\mathbf{a}_{56},$$

$$(\mathbf{R}^{3,34}\times\mathbf{S}_3)\cdot(\mathbf{R}^{4,45}\times(\mathbf{S}_6\times\mathbf{a}_{56})),\qquad (\mathbf{R}^{3,45}\times\mathbf{S}_3)\cdot(\mathbf{R}^{5,56}\times(\mathbf{S}_6\times\mathbf{a}_{56})).$$

Each of these terms will now be expanded individually, with the results substituted back into Eqs. (10.56) and (10.57).

(i) $\frac{1}{2}(\mathbf{R}^{3,34}\cdot\mathbf{R}^{3,34}+\mathbf{R}^{4,45}\cdot\mathbf{R}^{4,45}+\mathbf{R}^{5,56}\cdot\mathbf{R}^{5,56})$

These simple scalar products can be readily evaluated using the sets of direction cosines listed in the appendix. The result can be written as

$$\frac{1}{2}(\mathbf{R}^{3,34}\cdot\mathbf{R}^{3,34}+\mathbf{R}^{4,45}\cdot\mathbf{R}^{4,45}+\mathbf{R}^{5,56}\cdot\mathbf{R}^{5,56})=K_1, \tag{10.77}$$

where

$$K_1=\frac{1}{2}\left(S_3^2+a_{34}^2+S_4^2+a_{45}^2+S_5^2+a_{56}^2\right). \tag{10.78}$$

(ii) $\mathbf{S}_3 \cdot \mathbf{a}_{56}$

From set 10 of the table of direction cosines

$$\mathbf{S}_3 \cdot \mathbf{a}_{56} = X_{45}. \tag{10.79}$$

(iii) $\mathbf{R}^{3,56} \cdot \mathbf{S}_3$

This term may be written as

$$\mathbf{R}^{3,56} \cdot \mathbf{S}_3 = (S_3\mathbf{S}_3 + a_{34}\mathbf{a}_{34} + S_4\mathbf{S}_4 + a_{45}\mathbf{a}_{45} + S_5\mathbf{S}_5 + a_{56}\mathbf{a}_{56}) \cdot \mathbf{S}_3. \tag{10.80}$$

Evaluating the scalar products using sets from the table of direction cosines gives

$$\mathbf{R}^{3,56} \cdot \mathbf{S}_3 = J_5, \tag{10.81}$$

where

$$J_5 = S_3 + S_4c_{34} + a_{45}X_4 + S_5Z_4 + a_{56}X_{45}. \tag{10.82}$$

(iv) $\mathbf{R}^{3,45} \cdot \mathbf{S}_3$

The individual scalar products evaluated in (iii) can be used here to obtain

$$\mathbf{R}^{3,45} \cdot \mathbf{S}_3 = S_3 + S_4c_{34} + a_{45}X_4. \tag{10.83}$$

(v) $\mathbf{R}^{3,34} \cdot \mathbf{S}_3$

The individual scalar products evaluated in (iii) can be used here to obtain

$$\mathbf{R}^{3,34} \cdot \mathbf{S}_3 = S_3. \tag{10.84}$$

(vi) $\mathbf{R}^{5,56} \cdot \mathbf{a}_{56}$

This term may be written as

$$\mathbf{R}^{5,56} \cdot \mathbf{a}_{56} = (S_5\mathbf{S}_5 + a_{56}\mathbf{a}_{56}) \cdot \mathbf{a}_{56}. \tag{10.85}$$

Evaluating the scalar products gives

$$\mathbf{R}^{5,56} \cdot \mathbf{a}_{56} = a_{56}. \tag{10.86}$$

(vii) $\mathbf{R}^{4,45} \cdot \mathbf{a}_{56}$

This term may be written as

$$\mathbf{R}^{4,45} \cdot \mathbf{a}_{56} = (S_4\mathbf{S}_4 + a_{45}\mathbf{a}_{45}) \cdot \mathbf{a}_{56}. \tag{10.87}$$

Evaluating the scalar products gives

$$\mathbf{R}^{4,45} \cdot \mathbf{a}_{56} = S_4X_5 + a_{45}c_5. \tag{10.88}$$

(viii) $\mathbf{R}^{3,34} \cdot \mathbf{a}_{56}$

This term may be written as

$$\mathbf{R}^{3,34} \cdot \mathbf{a}_{56} = (S_3\mathbf{S}_3 + a_{34}\mathbf{a}_{34}) \cdot \mathbf{a}_{56}. \tag{10.89}$$

Evaluating the scalar products gives

$$\mathbf{R}^{3,34} \cdot \mathbf{a}_{56} = S_3X_{45} + a_{34}W_{45}. \tag{10.90}$$

(ix) $(\mathbf{R}^{3,34} \times \mathbf{S}_3) \cdot (\mathbf{R}^{4,45} \times \mathbf{a}_{56})$

Expanding the terms $\mathbf{R}^{3,34}$ and $\mathbf{R}^{4,45}$ and recognizing that $\mathbf{S}_3 \times \mathbf{S}_3 = \mathbf{0}$ gives

$$(\mathbf{R}^{3,34} \times \mathbf{S}_3) \cdot (\mathbf{R}^{4,45} \times \mathbf{a}_{56}) = a_{34}(\mathbf{a}_{34} \times \mathbf{S}_3) \cdot [(S_4\mathbf{S}_4 + a_{45}\mathbf{a}_{45}) \times \mathbf{a}_{56}]. \tag{10.91}$$

All the cross products must be performed in terms of the same coordinate system so that each of the resulting terms of the final scalar product, that is, $(\mathbf{a}_{34} \times \mathbf{S}_3)$ and $[(S_4\mathbf{S}_4 + a_{45}\mathbf{a}_{45}) \times \mathbf{a}_{56}]$, will be evaluated in the same coordinate system. Using set 4 from the table of direction cosines with the vector $\mathbf{S}_3$ given by $[s_{34}s_4, s_{34}c_4, c_{34}]^T$, the cross product terms are evaluated as

$$\mathbf{a}_{34} \times \mathbf{S}_3 = \begin{vmatrix} \mathbf{i} & \mathbf{j} & \mathbf{k} \\ c_4 & -s_4 & 0 \\ s_{34}s_4 & s_{34}c_4 & c_{34} \end{vmatrix} = \begin{bmatrix} -c_{34}s_4 \\ -c_{34}c_4 \\ s_{34} \end{bmatrix}, \tag{10.92}$$

$$\mathbf{S}_4 \times \mathbf{a}_{56} = \begin{vmatrix} \mathbf{i} & \mathbf{j} & \mathbf{k} \\ 0 & 0 & 1 \\ c_5 & c_{45}s_5 & s_{45}s_5 \end{vmatrix} = \begin{bmatrix} -c_{45}s_5 \\ c_5 \\ 0 \end{bmatrix}, \tag{10.93}$$

$$\mathbf{a}_{45} \times \mathbf{a}_{56} = \begin{vmatrix} \mathbf{i} & \mathbf{j} & \mathbf{k} \\ 1 & 0 & 0 \\ c_5 & c_{45}s_5 & s_{45}s_5 \end{vmatrix} = \begin{bmatrix} 0 \\ -s_{45}s_5 \\ c_{45}s_5 \end{bmatrix}. \tag{10.94}$$

Substituting these results into Eq. (10.91) gives

$$(\mathbf{R}^{3,34} \times \mathbf{S}_3) \cdot (\mathbf{R}^{4,45} \times \mathbf{a}_{56}) = a_{34} \begin{bmatrix} -c_{34}s_4 \\ -c_{34}c_4 \\ s_{34} \end{bmatrix} \cdot \begin{bmatrix} -S_4c_{45}s_5 \\ S_4c_5 - a_{45}s_{45}s_5 \\ a_{45}c_{45}s_5 \end{bmatrix}. \tag{10.95}$$

Evaluating the scalar product gives

$$(\mathbf{R}^{3,34} \times \mathbf{S}_3) \cdot (\mathbf{R}^{4,45} \times \mathbf{a}_{56}) = a_{34}S_4c_{34}c_{45}s_4s_5 - (a_{34}c_{34}c_4)(S_4c_5 - a_{45}s_{45}s_5) + a_{34}a_{45}s_{34}c_{45}s_5. \tag{10.96}$$

This equation may be regrouped as

$$(\mathbf{R}^{3,34} \times \mathbf{S}_3) \cdot (\mathbf{R}^{4,45} \times \mathbf{a}_{56}) = -a_{34}(S_4c_{34}W_{45} + a_{45}s_5\bar{Y}_4). \tag{10.97}$$

(x) $(\mathbf{R}^{3,45} \times \mathbf{S}_3) \cdot (\mathbf{R}^{5,56} \times \mathbf{a}_{56})$

The vectors $\mathbf{R}^{3,45}$ and $\mathbf{R}^{5,56}$ are expanded to give

$$\begin{aligned}&(\mathbf{R}^{3,45} \times \mathbf{S}_3) \cdot (\mathbf{R}^{5,56} \times \mathbf{a}_{56})\\ &\quad = [(S_3\mathbf{S}_3 + a_{34}\mathbf{a}_{34} + S_4\mathbf{S}_4 + a_{45}\mathbf{a}_{45}) \times \mathbf{S}_3] \cdot [(S_5\mathbf{S}_5 + a_{56}\mathbf{a}_{56}) \times \mathbf{a}_{56}].\end{aligned} \tag{10.98}$$

The individual cross products will all be evaluated using set 11 of the table of direction cosines. This gives

$$\mathbf{a}_{34} \times \mathbf{S}_3 = \begin{vmatrix} \mathbf{i} & \mathbf{j} & \mathbf{k} \\ c_4 & -c_{45}s_4 & s_{45}s_4 \\ X_4 & -Y_4 & Z_4 \end{vmatrix} = \begin{bmatrix} s_4(s_{45}Y_4 - c_{45}Z_4) \\ s_{45}s_4X_4 - c_4Z_4 \\ c_{45}s_4X_4 - c_4Y_4 \end{bmatrix}, \tag{10.99}$$

$$\mathbf{S}_4 \times \mathbf{S}_3 = \begin{vmatrix} \mathbf{i} & \mathbf{j} & \mathbf{k} \\ 0 & s_{45} & c_{45} \\ X_4 & -Y_4 & Z_4 \end{vmatrix} = \begin{bmatrix} s_{45}Z_4 + c_{45}Y_4 \\ c_{45}X_4 \\ -s_{45}X_4 \end{bmatrix}, \tag{10.100}$$

$$\mathbf{a}_{45} \times \mathbf{S}_3 = \begin{vmatrix} \mathbf{i} & \mathbf{j} & \mathbf{k} \\ 1 & 0 & 0 \\ X_4 & -Y_4 & Z_4 \end{vmatrix} = \begin{bmatrix} 0 \\ -Z_4 \\ -Y_4 \end{bmatrix}, \tag{10.101}$$

$$\mathbf{S}_5 \times \mathbf{a}_{56} = \begin{bmatrix} \mathbf{i} & \mathbf{j} & \mathbf{k} \\ 0 & 0 & 1 \\ c_5 & s_5 & 0 \end{bmatrix} = \begin{bmatrix} -s_5 \\ c_5 \\ 0 \end{bmatrix}. \tag{10.102}$$

These results may be used to evaluate the terms $(\mathbf{R}^{3,45} \times \mathbf{S}_3)$ and $(\mathbf{R}^{5,56} \times \mathbf{a}_{56})$ as follows:

$$(\mathbf{R}^{3,45} \times \mathbf{S}_3) = \begin{bmatrix} a_{34}s_4(s_{45}Y_4 - c_{45}Z_4) + S_4(s_{45}Z_4 + c_{45}Y_4) \\ a_{34}(s_{45}s_4X_4 - c_4Z_4) + S_4c_{45}X_4 - a_{45}Z_4 \\ a_{34}(c_{45}s_4X_4 - c_4Y_4) - S_4s_{45}X_4 - a_{45}Y_4 \end{bmatrix}, \tag{10.103}$$

$$(\mathbf{R}^{5,56} \times \mathbf{a}_{56}) = \begin{bmatrix} -S_5c_5 \\ S_5c_5 \\ 0 \end{bmatrix}. \tag{10.104}$$

The term $(s_{45}Y_4 - c_{45}Z_4)$ may be expanded, and

$$s_{45}Y_4 - c_{45}Z_4 = -s_{45}(s_{45}c_{34} + c_{45}s_{34}c_4) - c_{45}(c_{45}c_{34} - s_{45}s_{34}c_4). \tag{10.105}$$

Regrouping and substituting $s_{45}^2 + c_{45}^2 = 1$ yields

$$s_{45}Y_4 - c_{45}Z_4 = -c_{34}. \tag{10.106}$$

Analogously, the term $(s_{45}Z_4 + c_{45}Y_4)$ may be expanded, and

$$s_{45}Z_4 + c_{45}Y_4 = s_{45}(c_{45}c_{34} - s_{45}s_{34}c_4) - c_{45}(s_{45}c_{34} + c_{45}s_{34}c_4). \tag{10.107}$$

Regrouping gives

$$s_{45}Z_4 + c_{45}Y_4 = -s_{34}c_4. \tag{10.108}$$

The term $(s_{45}s_4X_4 - c_4Z_4)$ may be expanded, and

$$s_{45}s_4X_4 - c_4Z_4 = s_{45}s_4(s_{34}s_4) - c_4(c_{45}c_{34} - s_{45}s_{34}c_4), \tag{10.109}$$

which reduces to

$$s_{45}s_4X_4 - c_4Z_4 = s_{34}s_{45} - c_{34}c_{45}c_4. \tag{10.110}$$

Lastly, the term $(c_{45}s_4X_4 - c_4Y_4)$ may be expanded, and

$$c_{45}s_4X_4 - c_4Y_4 = c_{45}s_4(s_{34}s_4) + c_4(s_{45}c_{34} + c_{45}s_{34}c_4), \tag{10.111}$$

which reduces to

$$c_{45}s_4X_4 - c_4Y_4 = -\bar{Y}_4. \tag{10.112}$$

Substituting the results of Eqs. (10.106), (10.108), (10.110), and (10.112) into Eq. (10.103) gives

$$(\mathbf{R}^{3,45} \times \mathbf{S}_3) = \begin{bmatrix} -a_{34}s_4c_{34} - S_4s_{34}c_4 \\ a_{34}(s_{34}s_{45} - c_{34}c_{45}c_4) + S_4c_{45}X_4 - a_{45}Z_4 \\ -a_{34}\bar{Y}_4 - S_4s_{45}X_4 - a_{45}Y_4 \end{bmatrix}. \tag{10.113}$$

Equation (10.113) can be written in the abbreviated form

$$(\mathbf{R}^{3,45} \times \mathbf{S}_3) = \begin{bmatrix} -X_{04} \\ Y_{04} \\ -Z_{04} \end{bmatrix}, \tag{10.114}$$

where

$$X_{04} = a_{34}c_{34}s_4 + S_4s_{34}c_4, \tag{10.115}$$

$$Y_{04} = S_4c_{45}X_4 + a_{34}(s_{45}s_{34} - c_{45}c_{34}c_4) - a_{45}Z_4, \tag{10.116}$$

$$Z_{04} = S_4s_{45}X_4 + a_{34}\bar{Y}_4 + a_{45}Y_4. \tag{10.117}$$

The scalar product of Eqs. (10.114) and (10.104) may now be written as

$$(\mathbf{R}^{3,45} \times \mathbf{S}_3) \cdot (\mathbf{R}^{5,56} \times \mathbf{a}_{56}) = S_5(s_5X_{04} + c_5Y_{04}). \tag{10.118}$$

(xi) $\mathbf{S}_3 \cdot \mathbf{S}_6 \times \mathbf{a}_{56}$

This term may be expanded by using set 5 from the table of direction cosines as follows:

$$\mathbf{S}_3 \cdot \mathbf{S}_6 \times \mathbf{a}_{56} = \begin{vmatrix} X_{2176} & Y_{2176} & Z_{2176} \\ 0 & -s_{56} & c_{56} \\ 1 & 0 & 0 \end{vmatrix}. \tag{10.119}$$

Expanding this determinant gives

$$\mathbf{S}_3 \cdot \mathbf{S}_6 \times \mathbf{a}_{56} = c_{56} Y_{2176} + s_{56} Z_{2176}. \tag{10.120}$$

Substituting the definitions of the terms Y_{2176} and Z_{2176} gives

$$\mathbf{S}_3 \cdot \mathbf{S}_6 \times \mathbf{a}_{56} = c_{56}\left(c_{56} X^*_{2176} - s_{56} Z_{217}\right) + s_{56}\left(s_{56} X^*_{2176} + c_{56} Z_{217}\right). \tag{10.121}$$

This equation may now be written as

$$\mathbf{S}_3 \cdot \mathbf{S}_6 \times \mathbf{a}_{56} = X^*_{2176} = -Y_{45}, \tag{10.122}$$

where a subsidiary sine–cosine law was used to substitute for X^*_{2176}.

(xii) $\mathbf{R}^{5,56} \cdot \mathbf{S}_6 \times \mathbf{a}_{56}$

The vector $\mathbf{R}^{5,56}$ is expanded to give

$$\mathbf{R}^{5,56} \cdot \mathbf{S}_6 \times \mathbf{a}_{56} = (S_5 \mathbf{S}_5 + a_{56} \mathbf{a}_{56}) \cdot \mathbf{S}_6 \times \mathbf{a}_{56}. \tag{10.123}$$

Recognizing that $\mathbf{a}_{56} \cdot (\mathbf{S}_6 \times \mathbf{a}_{56}) = 0$, the equation may be written as

$$\mathbf{R}^{5,56} \cdot \mathbf{S}_6 \times \mathbf{a}_{56} = S_5 \mathbf{S}_5 \cdot \mathbf{S}_6 \times \mathbf{a}_{56}. \tag{10.124}$$

Expanding this scalar triple product by using set 10 from the table of direction cosines yields

$$\mathbf{S}_5 \cdot \mathbf{S}_6 \times \mathbf{a}_{56} = \begin{vmatrix} 0 & s_{56} & c_{56} \\ 0 & 0 & 1 \\ 1 & 0 & 0 \end{vmatrix}, \tag{10.125}$$

and therefore

$$\mathbf{R}^{5,56} \cdot \mathbf{S}_6 \times \mathbf{a}_{56} = S_5 s_{56}. \tag{10.126}$$

(xiii) $\mathbf{R}^{4,45} \cdot \mathbf{S}_6 \times \mathbf{a}_{56}$

Expanding the vector $\mathbf{R}^{4,45}$,

$$\mathbf{R}^{4,45} \cdot \mathbf{S}_6 \times \mathbf{a}_{56} = (S_4 \mathbf{S}_4 + a_{45} \mathbf{a}_{45}) \cdot \mathbf{S}_6 \times \mathbf{a}_{56}. \tag{10.127}$$

The scalar triple products $\mathbf{S}_4 \cdot \mathbf{S}_6 \times \mathbf{a}_{56}$ and $\mathbf{a}_{45} \cdot \mathbf{S}_6 \times \mathbf{a}_{56}$ can be evaluated using set 10 from the table of direction cosines, and

$$\mathbf{S}_4 \cdot \mathbf{S}_6 \times \mathbf{a}_{56} = \begin{vmatrix} X_5 & -Y_5 & Z_5 \\ 0 & 0 & 1 \\ 1 & 0 & 0 \end{vmatrix} = -Y_5, \tag{10.128}$$

$$\mathbf{a}_{45} \cdot \mathbf{S}_6 \times \mathbf{a}_{56} = \begin{vmatrix} c_5 & -s_5 c_{56} & s_5 s_{56} \\ 0 & 0 & 1 \\ 1 & 0 & 0 \end{vmatrix} = -c_{56} s_5. \tag{10.129}$$

Substituting these results into Eq. (10.127) gives

$$\mathbf{R}^{4,45} \cdot \mathbf{S}_6 \times \mathbf{a}_{56} = -(S_4 Y_5 + a_{45} c_{56} s_5). \tag{10.130}$$

(xiv) $\mathbf{R}^{3,34} \cdot \mathbf{S}_6 \times \mathbf{a}_{56}$

Expanding the vector $\mathbf{R}^{3,34}$,

$$\mathbf{R}^{3,34} \cdot \mathbf{S}_6 \times \mathbf{a}_{56} = (S_3\mathbf{S}_3 + a_{34}\mathbf{a}_{34}) \cdot \mathbf{S}_6 \times \mathbf{a}_{56}. \tag{10.131}$$

The scalar triple products $\mathbf{S}_3 \cdot \mathbf{S}_6 \times \mathbf{a}_{56}$ and $\mathbf{a}_{34} \cdot \mathbf{S}_6 \times \mathbf{a}_{56}$ can be evaluated using set 10 from the table of direction cosines, and

$$\mathbf{S}_3 \cdot \mathbf{S}_6 \times \mathbf{a}_{56} = \begin{vmatrix} X_{45} & -Y_{45} & Z_{45} \\ 0 & 0 & 1 \\ 1 & 0 & 0 \end{vmatrix} = -Y_{45}, \tag{10.132}$$

$$\mathbf{a}_{34} \cdot \mathbf{S}_6 \times \mathbf{a}_{56} = \begin{vmatrix} W_{45} & U^*_{456} & U_{456} \\ 0 & 0 & 1 \\ 1 & 0 & 0 \end{vmatrix} = U^*_{456}. \tag{10.133}$$

Substituting these results into Eq. (10.131) gives

$$\mathbf{R}^{3,34} \cdot \mathbf{S}_6 \times \mathbf{a}_{56} = -S_3 Y_{45} + a_{34} U^*_{456}. \tag{10.134}$$

(xv) $(\mathbf{R}^{3,34} \times \mathbf{S}_3) \cdot (\mathbf{R}^{4,45} \times (\mathbf{S}_6 \times \mathbf{a}_{56}))$

Set 4 from the table of direction cosines will be used to evaluate all the scalar and vector products. First, the vector product $\mathbf{S}_6 \times \mathbf{a}_{56}$ is evaluated as

$$\mathbf{S}_6 \times \mathbf{a}_{56} = \begin{vmatrix} \mathbf{i} & \mathbf{j} & \mathbf{k} \\ \bar{X}_5 & \bar{Y}_5 & \bar{Z}_5 \\ c_5 & c_{45}s_5 & s_{45}s_5 \end{vmatrix} = \begin{bmatrix} s_5(s_{45}\bar{Y}_5 - c_{45}\bar{Z}_5) \\ c_5\bar{Z}_5 - s_{45}s_5\bar{X}_5 \\ c_{45}s_5\bar{X}_5 - c_5\bar{Y}_5 \end{bmatrix}. \tag{10.135}$$

This result will be expanded using the definitions for $\bar{X}_5$, $\bar{Y}_5$, and $\bar{Z}_5$. The term $(s_{45}\bar{Y}_5 - c_{45}\bar{Z}_5)$ may thus be written as

$$s_{45}\bar{Y}_5 - c_{45}\bar{Z}_5 = -s_{45}(s_{45}c_{56} + c_{45}s_{56}c_5) - c_{45}(c_{45}c_{56} - s_{45}s_{56}c_6), \tag{10.136}$$

which reduces to

$$s_{45}\bar{Y}_5 - c_{45}\bar{Z}_5 = -c_{56}. \tag{10.137}$$

The term $(c_5\bar{Z}_5 - s_{45}s_5\bar{X}_5)$ may be written as

$$c_5\bar{Z}_5 - s_{45}s_5\bar{X}_5 = c_5(c_{45}c_{56} - s_{45}s_{56}c_5) - s_{45}s_5(s_{56}s_5), \tag{10.138}$$

which reduces to

$$c_5\bar{Z}_5 - s_{45}s_5\bar{X}_5 = -(s_{45}s_{56} - c_{45}c_{56}c_5) = -Z'_5. \tag{10.139}$$

The term $(c_{45}s_5\bar{X}_5 - c_5\bar{Y}_5)$ may be written as

$$c_{45}s_5\bar{X}_5 - c_5\bar{Y}_5 = c_{45}s_5(s_{56}s_5) + c_5(s_{45}c_{56} + c_{45}s_{56}c_5), \tag{10.140}$$

which reduces to

$$c_{45}s_5\bar{X}_5 - c_5\bar{Y}_5 = s_{56}c_{45} + c_{56}s_{45}c_5 = -Y_5. \tag{10.141}$$

Substituting Eqs. (10.137), (10.139), and (10.141) into Eq. (10.135) and recognizing that $c_{56}s_5$ is the definition of the term $\bar{X}'_5$ gives

$$\mathbf{S}_6 \times \mathbf{a}_{56} = \begin{bmatrix} -\bar{X}'_5 \\ -Z'_5 \\ -Y_5 \end{bmatrix}. \tag{10.142}$$

The vector triple product $(\mathbf{R}^{4,45} \times (\mathbf{S}_6 \times \mathbf{a}_{56}))$ will next be evaluated by first expanding the vector $\mathbf{R}^{4,45}$. This gives

$$\mathbf{R}^{4,45} \times (\mathbf{S}_6 \times \mathbf{a}_{56}) = (S_4\mathbf{S}_4 + a_{45}\mathbf{a}_{45}) \times (\mathbf{S}_6 \times \mathbf{a}_{56}). \tag{10.143}$$

The vector products $\mathbf{S}_4 \times (\mathbf{S}_6 \times \mathbf{a}_{56})$ and $\mathbf{a}_{45} \times (\mathbf{S}_6 \times \mathbf{a}_{56})$ are evaluated by using set 4 from the table of direction cosines as follows:

$$\mathbf{S}_4 \times (\mathbf{S}_6 \times \mathbf{a}_{56}) = \begin{vmatrix} \mathbf{i} & \mathbf{j} & \mathbf{k} \\ 0 & 0 & 1 \\ -\bar{X}'_5 & -Z'_5 & -Y_5 \end{vmatrix} = \begin{bmatrix} Z'_5 \\ -\bar{X}'_5 \\ 0 \end{bmatrix}, \tag{10.144}$$

$$\mathbf{a}_{45} \times (\mathbf{S}_6 \times \mathbf{a}_{56}) = \begin{vmatrix} \mathbf{i} & \mathbf{j} & \mathbf{k} \\ 1 & 0 & 0 \\ -\bar{X}'_5 & -Z'_5 & -Y_5 \end{vmatrix} = \begin{bmatrix} 0 \\ Y_5 \\ -Z'_5 \end{bmatrix}. \tag{10.145}$$

Substituting Eqs. (10.144) and (10.145) into Eq. (10.143) gives

$$\mathbf{R}^{4,45} \times (\mathbf{S}_6 \times \mathbf{a}_{56}) = \begin{bmatrix} S_4 Z'_5 \\ -S_4\bar{X}'_5 + a_{45}Y_5 \\ -a_{45}Z'_5 \end{bmatrix}. \tag{10.146}$$

The cross product $(\mathbf{R}^{3,34} \times \mathbf{S}_3)$ is evaluated by expanding the vector $\mathbf{R}^{3,34}$. This gives

$$\mathbf{R}^{3,34} \times \mathbf{S}_3 = (S_3\mathbf{S}_3 + a_{34}\mathbf{a}_{34}) \times \mathbf{S}_3. \tag{10.147}$$

Now, $\mathbf{S}_3 \times \mathbf{S}_3 = \mathbf{0}$ and the vector product $\mathbf{a}_{34} \times \mathbf{S}_3$ is evaluated using set 4 from the table of direction cosines as

$$\mathbf{a}_{34} \times \mathbf{S}_3 = \begin{vmatrix} \mathbf{i} & \mathbf{j} & \mathbf{k} \\ c_4 & -s_4 & 0 \\ X_{21765} & Y_{21765} & Z_{21765} \end{vmatrix} = \begin{vmatrix} \mathbf{i} & \mathbf{j} & \mathbf{k} \\ c_4 & -s_4 & 0 \\ s_{34}s_4 & s_{34}c_4 & c_{34} \end{vmatrix} = \begin{bmatrix} -c_{34}s_4 \\ -c_{34}c_4 \\ s_{34} \end{bmatrix}. \tag{10.148}$$

Note that fundamental sine, sine–cosine, and cosine laws for a spherical heptagon were used in this equation to substitute for the terms X_{21765}, Y_{21765}, and Z_{21765}. Substituting

Eq. (10.148) into Eq. (10.147) gives

$$\mathbf{R}^{3,34} \times \mathbf{S}_3 = \begin{bmatrix} -a_{34}c_{34}s_4 \\ -a_{34}c_{34}c_4 \\ a_{34}s_{34} \end{bmatrix}. \tag{10.149}$$

Forming the scalar product of Eq. (10.146) with Eq. (10.149) gives the result

$$(\mathbf{R}^{3,34} \times \mathbf{S}_3) \cdot (\mathbf{R}^{4,45} \times (\mathbf{S}_6 \times \mathbf{a}_{56})) \\ = -a_{34}\left[S_4(Z_5' c_{34}s_4 - \bar{X}_5' c_{34}c_4) + a_{45}(Y_5 c_{34}c_4 + Z_5' s_{34})\right]. \tag{10.150}$$

The terms Z_5' and $\bar{X}_5'$ may be substituted into the expression $(Z_5' c_{34}s_4 - \bar{X}_5' c_{34}c_4)$ to yield

$$Z_5' c_{34}s_4 - \bar{X}_5' c_{34}c_4 = (s_{56}s_{45} - c_{56}c_{45}c_5)c_{34}s_4 - (c_{56}s_5)c_{34}c_4. \tag{10.151}$$

Regrouping the right side of Eq. (10.151) gives

$$Z_5' c_{34}s_4 - \bar{X}_5' c_{34}c_4 = c_{34}[s_{56}(s_{45}s_4) - c_{56}(s_5c_4 + c_5s_4c_{45})] = c_{34}U_{456}^*. \tag{10.152}$$

Substituting this result into Eq. (10.150) yields

$$(\mathbf{R}^{3,34} \times \mathbf{S}_3) \cdot (\mathbf{R}^{4,45} \times (\mathbf{S}_6 \times \mathbf{a}_{56})) = -a_{34}\left[S_4(c_{34}U_{456}^*) + a_{45}(Y_5 c_{34}c_4 + Z_5' s_{34})\right]. \tag{10.153}$$

(xvi) $(\mathbf{R}^{3,45} \times \mathbf{S}_3) \cdot (\mathbf{R}^{5,56} \times (\mathbf{S}_6 \times \mathbf{a}_{56}))$

Set 11 from the table of direction cosines will be used to evaluate all the scalar and vector products in this expression. Firstly, the vector product $\mathbf{S}_6 \times \mathbf{a}_{56}$ is evaluated as

$$\mathbf{S}_6 \times \mathbf{a}_{56} = \begin{vmatrix} \mathbf{i} & \mathbf{j} & \mathbf{k} \\ X_{71234} & -Y_{71234} & Z_{71234} \\ c_5 & s_5 & 0 \end{vmatrix} = \begin{vmatrix} \mathbf{i} & \mathbf{j} & \mathbf{k} \\ s_{56}s_5 & -s_{56}c_5 & c_{56} \\ c_5 & s_5 & 0 \end{vmatrix} = \begin{bmatrix} -c_{56}s_5 \\ c_{56}c_5 \\ s_{56} \end{bmatrix}. \tag{10.154}$$

Note that fundamental sine, sine-cosine, and cosine laws for a spherical heptagon were used in this equation to substitute for the terms X_{71234}, Y_{71234}, and Z_{71234}.

The expression $(\mathbf{R}^{5,56} \times (\mathbf{S}_6 \times \mathbf{a}_{56}))$ is next evaluated by expanding the term $\mathbf{R}^{5,56}$. This gives

$$\mathbf{R}^{5,56} \times (\mathbf{S}_6 \times \mathbf{a}_{56}) = (S_5\mathbf{S}_5 + a_{56}\mathbf{a}_{56}) \times (\mathbf{S}_6 \times \mathbf{a}_{56}). \tag{10.155}$$

The vector product $(\mathbf{S}_5 \times (\mathbf{S}_6 \times \mathbf{a}_{56}))$ is evaluated first, and

$$\mathbf{S}_5 \times (\mathbf{S}_6 \times \mathbf{a}_{56}) = \begin{vmatrix} \mathbf{i} & \mathbf{j} & \mathbf{k} \\ 0 & 0 & 1 \\ -c_{56}s_5 & c_{56}c_5 & s_{56} \end{vmatrix} = \begin{bmatrix} -c_{56}c_5 \\ -c_{56}s_5 \\ 0 \end{bmatrix}. \tag{10.156}$$

The vector product $(\mathbf{a}_{56} \times (\mathbf{S}_6 \times \mathbf{a}_{56}))$ is evaluated, and

$$\mathbf{a}_{56} \times (\mathbf{S}_6 \times \mathbf{a}_{56}) = \begin{vmatrix} \mathbf{i} & \mathbf{j} & \mathbf{k} \\ c_5 & s_5 & 0 \\ -c_{56}s_5 & c_{56}c_5 & s_{56} \end{vmatrix} = \begin{bmatrix} s_{56}s_5 \\ -s_{56}c_5 \\ c_{56} \end{bmatrix}. \tag{10.157}$$

Substituting Eqs. (10.156) and (10.157) into Eq. (10.155) gives

$$\mathbf{R}^{5,56} \times (\mathbf{S}_6 \times \mathbf{a}_{56}) = \begin{bmatrix} -S_5c_{56}c_5 + a_{56}s_{56}s_5 \\ -S_5c_{56}s_5 - a_{56}s_{56}c_5 \\ a_{56}c_{56} \end{bmatrix}. \tag{10.158}$$

The factor $(\mathbf{R}^{3,45} \times \mathbf{S}_3)$ was previously expanded using set 11 from the table of direction cosines in Eq. (10.117). Using this result, the scalar product $(\mathbf{R}^{3,45} \times \mathbf{S}_3) \cdot (\mathbf{R}^{5,56} \times (\mathbf{S}_6 \times \mathbf{a}_{56}))$ may now be written as

$$\begin{aligned}(\mathbf{R}^{3,45} \times \mathbf{S}_3) \cdot (\mathbf{R}^{5,56} \times (\mathbf{S}_6 \times \mathbf{a}_{56})) = {} & X_{04}(S_5c_{56}c_5 - a_{56}s_{56}s_5) \\ & - Y_{04}(S_5c_{56}s_5 + a_{56}s_{56}c_5) - Z_{04}(a_{56}c_{56}).\end{aligned} \tag{10.159}$$

Now that the sixteen terms in Eqs. (10.56) and (10.57) have been expanded, the terms J_3 and J_4 may be written in the abbreviated forms

$$\begin{aligned}J_3 = {} & K_1X_{45} - S_3(S_3X_{45} + a_{34}W_{45}) - (S_3 + S_4c_{34} + a_{45}X_4)(S_4X_5 + a_{45}c_5) \\ & - J_5a_{56} - a_{34}(S_4c_{34}W_{45} + a_{45}s_5\bar{Y}_4) + S_5(s_5X_{04} + c_5Y_{04}),\end{aligned} \tag{10.160}$$

$$\begin{aligned}J_4 = {} & -K_1Y_{45} + S_3(S_3Y_{45} - a_{34}U^*_{456}) + (S_3 + S_4c_{34} + a_{45}X_4)(S_4Y_5 + a_{45}c_{56}s_5) \\ & - J_5S_5s_{56} - a_{34}\left[S_4(c_{34}U^*_{456}) + a_{45}(Y_5c_{34}c_4 + Z'_5s_{34})\right] + X_{04}(S_5c_{56}c_5 - a_{56}s_{56}s_5) \\ & - Y_{04}(S_5c_{56}s_5 + a_{56}s_{56}c_5) - Z_{04}(a_{56}c_{56}),\end{aligned} \tag{10.161}$$

where K_1 and J_5 were defined in Eqs. (10.78) and (10.82).

10.2.6 Detailed expansion of H_3 and H_4

The following terms that are contained in H_3 (see Eq. (10.75)) must be expanded:

$$\frac{1}{2}(\mathbf{R}^{6,67} \cdot \mathbf{R}^{6,67} + \mathbf{R}^{7,71} \cdot \mathbf{R}^{7,71} + \mathbf{R}^{1,12} \cdot \mathbf{R}^{1,12} + \mathbf{R}^{2,23} \cdot \mathbf{R}^{2,23}), \quad \mathbf{S}_3 \cdot \mathbf{a}_{67},$$

$$\mathbf{R}^{6,67} \cdot \mathbf{S}_3, \quad \mathbf{R}^{7,71} \cdot \mathbf{S}_3, \quad \mathbf{R}^{1,12} \cdot \mathbf{S}_3,$$

$$\mathbf{R}^{2,23} \cdot \mathbf{S}_3, \quad \mathbf{R}^{2,23} \cdot \mathbf{a}_{67}, \quad \mathbf{R}^{1,12} \cdot \mathbf{a}_{67},$$

$$\mathbf{R}^{7,71} \cdot \mathbf{a}_{67}, \quad \mathbf{R}^{6,67} \cdot \mathbf{a}_{67}, \quad \mathbf{R}^{6,71} \cdot \mathbf{a}_{67},$$

$$\mathbf{R}^{6,12} \cdot \mathbf{a}_{67}, \quad (\mathbf{R}^{6,67} \times \mathbf{a}_{67}) \cdot (\mathbf{R}^{7,71} \times \mathbf{S}_3), \; (\mathbf{R}^{6,71} \times \mathbf{a}_{67}) \cdot (\mathbf{R}^{1,12} \times \mathbf{S}_3),$$

$$(\mathbf{R}^{6,12} \times \mathbf{a}_{67}) \cdot (\mathbf{R}^{2,23} \times \mathbf{S}_3).$$

Additionally, the following terms in H_4 (see Eq. (10.76)) must also be expanded:

$$\mathbf{S}_3 \cdot \mathbf{S}_6 \times \mathbf{a}_{67}, \qquad \mathbf{R}^{2,23} \cdot \mathbf{S}_6 \times \mathbf{a}_{67}, \qquad \mathbf{R}^{1,12} \cdot \mathbf{S}_6 \times \mathbf{a}_{67},$$

$$\mathbf{R}^{7,71} \cdot \mathbf{S}_6 \times \mathbf{a}_{67}, \qquad \mathbf{R}^{6,67} \cdot \mathbf{S}_6 \times \mathbf{a}_{67}, \qquad \mathbf{R}^{6,71} \cdot \mathbf{S}_6 \times \mathbf{a}_{67},$$

$$\mathbf{R}^{6,12} \cdot \mathbf{S}_6 \times \mathbf{a}_{67}, \qquad (\mathbf{R}^{6,67} \times (\mathbf{S}_6 \times \mathbf{a}_{67})) \cdot (\mathbf{R}^{7,71} \times \mathbf{S}_3),$$

$$(\mathbf{R}^{6,71} \times (\mathbf{S}_6 \times \mathbf{a}_{67})) \cdot (\mathbf{R}^{1,12} \times \mathbf{S}_3), \qquad (\mathbf{R}^{6,12} \times (\mathbf{S}_6 \times \mathbf{a}_{67})) \cdot (\mathbf{R}^{2,23} \times \mathbf{S}_3).$$

Each of these terms will now be expanded individually, and the results substituted back into Eqs. (10.75) and (10.76) to yield H_3 and H_4.

(i) $\frac{1}{2}(\mathbf{R}^{6,67} \cdot \mathbf{R}^{6,67} + \mathbf{R}^{7,71} \cdot \mathbf{R}^{7,71} + \mathbf{R}^{1,12} \cdot \mathbf{R}^{1,12} + \mathbf{R}^{2,23} \cdot \mathbf{R}^{2,23})$

These simple scalar products are readily evaluated as

$$\frac{1}{2}(\mathbf{R}^{6,67} \cdot \mathbf{R}^{6,67} + \mathbf{R}^{7,71} \cdot \mathbf{R}^{7,71} + \mathbf{R}^{1,12} \cdot \mathbf{R}^{1,12} + \mathbf{R}^{2,23} \cdot \mathbf{R}^{2,23}) = K_2, \tag{10.162}$$

where

$$K_2 = \frac{1}{2}(S_6^2 + a_{67}^2 + S_7^2 + a_{71}^2 + S_1^2 + a_{12}^2 + S_2^2 + a_{23}^2). \tag{10.163}$$

(ii) $\mathbf{S}_3 \cdot \mathbf{a}_{67}$

Set 6 from the table of direction cosines can be used to evaluate this term as

$$\mathbf{S}_3 \cdot \mathbf{a}_{67} = X_{217}. \tag{10.164}$$

(iii) $\mathbf{R}^{6,67} \cdot \mathbf{S}_3$

This term may be written as

$$\mathbf{R}^{6,67} \cdot \mathbf{S}_3 = (S_6\mathbf{S}_6 + a_{67}\mathbf{a}_{67}) \cdot \mathbf{S}_3. \tag{10.165}$$

The scalar products in this equation may be evaluated using set 6 of the table of direction cosines to give

$$\mathbf{R}^{6,67} \cdot \mathbf{S}_3 = S_6 Z_{217} + a_{67} X_{217}. \tag{10.166}$$

(iv) $\mathbf{R}^{7,71} \cdot \mathbf{S}_3$

This term may be written as

$$\mathbf{R}^{7,71} \cdot \mathbf{S}_3 = (S_7\mathbf{S}_7 + a_{71}\mathbf{a}_{71}) \cdot \mathbf{S}_3. \tag{10.167}$$

The scalar products in this equation may be evaluated using set 7 of the table of direction cosines to give

$$\mathbf{R}^{7,71} \cdot \mathbf{S}_3 = S_7 Z_{21} + a_{71} X_{21}. \tag{10.168}$$

(v) $\mathbf{R}^{1,12} \cdot \mathbf{S}_3$

Expanding the vector $\mathbf{R}^{1,12}$,

$$\mathbf{R}^{1,12} \cdot \mathbf{S}_3 = (S_1\mathbf{S}_1 + a_{12}\mathbf{a}_{12}) \cdot \mathbf{S}_3. \tag{10.169}$$

The scalar products in this equation may be evaluated using set 1 of the table of direction cosines to give

$$\mathbf{R}^{1,12} \cdot \mathbf{S}_3 = S_1\bar{Z}_2 + a_{12}\bar{X}_2. \tag{10.170}$$

(vi) $\mathbf{R}^{2,23} \cdot \mathbf{S}_3$

Expanding the vector $\mathbf{R}^{2,23}$,

$$\mathbf{R}^{2,23} \cdot \mathbf{S}_3 = (S_2\mathbf{S}_2 + a_{23}\mathbf{a}_{23}) \cdot \mathbf{S}_3. \tag{10.171}$$

The scalar products in this equation may be evaluated using set 2 of the table of direction cosines to give

$$\mathbf{R}^{2,23} \cdot \mathbf{S}_3 = S_2c_{23}. \tag{10.172}$$

(vii) $\mathbf{R}^{2,23} \cdot \mathbf{a}_{67}$

Expanding the vector $\mathbf{R}^{2,23}$,

$$\mathbf{R}^{2,23} \cdot \mathbf{a}_{67} = (S_2\mathbf{S}_2 + a_{23}\mathbf{a}_{23}) \cdot \mathbf{a}_{67}. \tag{10.173}$$

Evaluating the scalar products using set 14 and then set 15 gives

$$\mathbf{R}^{2,23} \cdot \mathbf{a}_{67} = S_2U_{712} + a_{23}W_{712}. \tag{10.174}$$

(viii) $\mathbf{R}^{1,12} \cdot \mathbf{a}_{67}$

Expanding the vector $\mathbf{R}^{1,12}$,

$$\mathbf{R}^{1,12} \cdot \mathbf{a}_{67} = (S_1\mathbf{S}_1 + a_{12}\mathbf{a}_{12}) \cdot \mathbf{a}_{67}. \tag{10.175}$$

Evaluating the scalar products using set 8 and then set 14 gives

$$\mathbf{R}^{1,12} \cdot \mathbf{a}_{67} = S_1U_{71} + a_{12}W_{71}. \tag{10.176}$$

(ix) $\mathbf{R}^{7,71} \cdot \mathbf{a}_{67}$

Expanding the vector $\mathbf{R}^{7,71}$,

$$\mathbf{R}^{7,71} \cdot \mathbf{a}_{67} = (S_7\mathbf{S}_7 + a_{71}\mathbf{a}_{71}) \cdot \mathbf{a}_{67}. \tag{10.177}$$

Evaluating the scalar products using set 7 gives

$$\mathbf{R}^{7,71} \cdot \mathbf{a}_{67} = a_{71}c_7. \tag{10.178}$$

(x) $\mathbf{R}^{6,67} \cdot \mathbf{a}_{67}$

Expanding the vector $\mathbf{R}^{6,67}$,

$$\mathbf{R}^{6,67} \cdot \mathbf{a}_{67} = (S_6\mathbf{S}_6 + a_{67}\mathbf{a}_{67}) \cdot \mathbf{a}_{67}. \tag{10.179}$$

Evaluating the scalar products using set 6 gives

$$\mathbf{R}^{6,67} \cdot \mathbf{a}_{67} = a_{67}. \tag{10.180}$$

(xi) $\mathbf{R}^{6,71} \cdot \mathbf{a}_{67}$

Expanding the vector $\mathbf{R}^{6,71}$,

$$\mathbf{R}^{6,71} \cdot \mathbf{a}_{67} = (\mathbf{R}^{6,67} + S_7\mathbf{S}_7 + a_{71}\mathbf{a}_{71}) \cdot \mathbf{a}_{67}. \tag{10.181}$$

Using Eq. (10.180) and evaluating the last two scalar products using set 6 gives

$$\mathbf{R}^{6,71} \cdot \mathbf{a}_{67} = a_{67} + a_{71}c_7. \tag{10.182}$$

(xii) $\mathbf{R}^{6,12} \cdot \mathbf{a}_{67}$

Expanding the vector $\mathbf{R}^{6,12}$,

$$\mathbf{R}^{6,12} \cdot \mathbf{a}_{67} = (\mathbf{R}^{6,71} + S_1\mathbf{S}_1 + a_{12}\mathbf{a}_{12}) \cdot \mathbf{a}_{67}. \tag{10.183}$$

Using Eq. (10.182) and evaluating the last two scalar products using set 6 gives

$$\mathbf{R}^{6,12} \cdot \mathbf{a}_{67} = a_{67} + a_{71}c_7 + S_1\bar{X}_7 + a_{12}W_{17}. \tag{10.184}$$

(xiii) $(\mathbf{R}^{6,67} \times \mathbf{a}_{67}) \cdot (\mathbf{R}^{7,71} \times \mathbf{S}_3)$

The vector and scalar products in this term are evaluated using set 7 from the table of direction cosines. Expanding the vector $\mathbf{R}^{6,67}$ gives

$$\mathbf{R}^{6,67} \times \mathbf{a}_{67} = (S_6\mathbf{S}_6 + a_{67}\mathbf{a}_{67}) \times \mathbf{a}_{67}. \tag{10.185}$$

The vector product of $\mathbf{S}_6$ and $\mathbf{a}_{67}$ may be evaluated as

$$\mathbf{S}_6 \times \mathbf{a}_{67} = \begin{vmatrix} \mathbf{i} & \mathbf{j} & \mathbf{k} \\ X_{54321} & Y_{54321} & Z_{54321} \\ c_7 & -s_7 & 0 \end{vmatrix} = \begin{vmatrix} \mathbf{i} & \mathbf{j} & \mathbf{k} \\ s_{67}s_7 & s_{67}c_7 & c_{67} \\ c_7 & -s_7 & 0 \end{vmatrix} = \begin{bmatrix} c_{67}s_7 \\ c_{67}c_7 \\ -s_{67} \end{bmatrix}, \tag{10.186}$$

where fundamental sine, sine–cosine, and cosine laws for a spherical heptagon were used to simplify the direction cosines of $\mathbf{S}_6$. Because $\mathbf{a}_{67} \times \mathbf{a}_{67} = \mathbf{0}$, Eq. (10.185) may be written as

$$\mathbf{R}^{6,67} \times \mathbf{a}_{67} = \begin{bmatrix} S_6c_{67}s_7 \\ S_6c_{67}c_7 \\ -S_6s_{67} \end{bmatrix}. \tag{10.187}$$

The factor $(\mathbf{R}^{7,71} \times \mathbf{S}_3)$ may be expanded as

$$\mathbf{R}^{7,71} \times \mathbf{S}_3 = (S_7\mathbf{S}_7 + a_{71}\mathbf{a}_{71}) \times \mathbf{S}_3. \tag{10.188}$$

The vector product of $\mathbf{S}_7$ and $\mathbf{S}_3$ is written as

$$\mathbf{S}_7 \times \mathbf{S}_3 = \begin{vmatrix} \mathbf{i} & \mathbf{j} & \mathbf{k} \\ 0 & 0 & 1 \\ X_{21} & Y_{21} & Z_{21} \end{vmatrix} = \begin{bmatrix} -Y_{21} \\ X_{21} \\ 0 \end{bmatrix}. \tag{10.189}$$

Further,

$$\mathbf{a}_{71} \times \mathbf{S}_3 = \begin{vmatrix} \mathbf{i} & \mathbf{j} & \mathbf{k} \\ 1 & 0 & 0 \\ X_{21} & Y_{21} & Z_{21} \end{vmatrix} = \begin{bmatrix} 0 \\ -Z_{21} \\ Y_{21} \end{bmatrix}. \tag{10.190}$$

Equation (10.188) may now be written as

$$\mathbf{R}^{7,71} \times \mathbf{S}_3 = \begin{bmatrix} -S_7 Y_{21} \\ S_7 X_{21} - a_{71} Z_{21} \\ a_{71} Y_{21} \end{bmatrix}. \tag{10.191}$$

Finally, the scalar product of Eqs. (10.187) and (10.191) yields

$$(\mathbf{R}^{6,67} \times \mathbf{a}_{67}) \cdot (\mathbf{R}^{7,71} \times \mathbf{S}_3) = S_6 S_7 c_{67} X_{217} + S_6 a_{71}(-s_{67} Y_{21} - c_{67} c_7 Z_{21}). \tag{10.192}$$

(xiv) $(\mathbf{R}^{6,71} \times \mathbf{a}_{67}) \cdot (\mathbf{R}^{1,12} \times \mathbf{S}_3)$

The vector and scalar products in this expression are evaluated using set 1 from the table of direction cosines. The first vector product may be written as

$$\mathbf{R}^{6,71} \times \mathbf{a}_{67} = (S_6 \mathbf{S}_6 + a_{67} \mathbf{a}_{67} + S_7 \mathbf{S}_7 + a_{71} \mathbf{a}_{71}) \times \mathbf{a}_{67}. \tag{10.193}$$

The vector product of $\mathbf{S}_6$ and $\mathbf{a}_{67}$ may be written as

$$\mathbf{S}_6 \times \mathbf{a}_{67} = \begin{vmatrix} \mathbf{i} & \mathbf{j} & \mathbf{k} \\ X_{5432} & Y_{5432} & Z_{5432} \\ W_{65432} & -U^*_{654321} & U_{654321} \end{vmatrix} = \begin{vmatrix} \mathbf{i} & \mathbf{j} & \mathbf{k} \\ X_{71} & -X^*_{71} & Z_7 \\ W_{71} & V_{71} & U_{71} \end{vmatrix}$$
$$= \begin{bmatrix} -U_{71} X^*_{71} - V_{71} Z_7 \\ W_{71} Z^*_7 - X_{71} U_{71} \\ V_{71} X_{71} + W_{71} X^*_{71} \end{bmatrix}. \tag{10.194}$$

Here, subsidiary spherical and polar sine, sine–cosine, and cosine laws have been used to simplify the direction cosines for $\mathbf{S}_6$ and $\mathbf{a}_{67}$. Expanding the right side of Eq. (10.194), regrouping terms, and using the identities $s_1^2 + c_1^2 = 1$, $s_7^2 + c_7^2 = 1$, and $s_{71}^2 + c_{71}^2 = 1$ gives

$$\mathbf{S}_6 \times \mathbf{a}_{67} = \begin{bmatrix} -U^*_{176} \\ -c_1 Z'_7 - s_1 X'_7 \\ \bar{Y}_7 \end{bmatrix}, \tag{10.195}$$

where the terms X'_7 and Z'_7 were introduced in the previous chapter, and

$$X'_7 = c_{67} s_7, \tag{10.196}$$

$$Z'_7 = s_{71} s_{67} - c_{71} c_{67} c_7. \tag{10.197}$$

Now, $\mathbf{a}_{67} \times \mathbf{a}_{67} = \mathbf{0}$, and the vector product of $\mathbf{S}_7$ and $\mathbf{a}_{67}$ can be written as

$$\mathbf{S}_7 \times \mathbf{a}_{67} = \begin{vmatrix} \mathbf{i} & \mathbf{j} & \mathbf{k} \\ X_{65432} & Y_{65432} & Z_{65432} \\ W_{71} & V_{71} & U_{71} \end{vmatrix} = \begin{vmatrix} \mathbf{i} & \mathbf{j} & \mathbf{k} \\ s_{71}s_1 & s_{71}c_1 & c_{71} \\ W_{71} & V_{71} & U_{71} \end{vmatrix}$$

$$= \begin{bmatrix} U_{71}s_{71}c_1 - V_{71}c_{71} \\ -U_{71}s_{71}s_1 + W_{71}c_{71} \\ V_{71}s_{71}s_1 - W_{71}s_{71}c_1 \end{bmatrix}, \tag{10.198}$$

where fundamental sine, sine–cosine, and cosine laws were used to simplify the direction cosines for $\mathbf{S}_7$. The last element of this expression may be expanded as

$$V_{71}s_{71}s_1 - W_{71}s_{71}c_1 = -(s_1c_7 + c_1s_7c_{71})s_{71}s_1 - (c_1c_7 - s_1s_7c_{71})s_{71}c_1. \tag{10.199}$$

Regrouping this expression and substituting $s_1^2 + c_1^2 = 1$ yields

$$V_{71}s_{71}s_1 - W_{71}s_{71}c_1 = -s_{71}c_7. \tag{10.200}$$

Substituting Eq. (10.200) into Eq. (10.198) gives

$$\mathbf{S}_7 \times \mathbf{a}_{67} = \begin{bmatrix} U_{71}s_{71}c_1 - V_{71}c_{71} \\ -U_{71}s_{71}s_1 + W_{71}c_{71} \\ -s_{71}c_7 \end{bmatrix}. \tag{10.201}$$

Similarly, the first two elements of Eq. (10.198) can be expanded and regrouped to yield

$$\mathbf{S}_7 \times \mathbf{a}_{67} = \begin{bmatrix} -V_{17} \\ -s_7s_1 + c_7c_1c_{71} \\ -s_{71}c_7 \end{bmatrix}. \tag{10.202}$$

The vector product of $\mathbf{a}_{71}$ and $\mathbf{a}_{67}$ may be written as

$$\mathbf{a}_{71} \times \mathbf{a}_{67} = \begin{vmatrix} \mathbf{i} & \mathbf{j} & \mathbf{k} \\ c_1 & -s_1 & 0 \\ W_{71} & V_{71} & U_{71} \end{vmatrix} = \begin{bmatrix} -U_{71}s_1 \\ -U_{71}c_1 \\ V_{71}c_1 + W_{71}s_1 \end{bmatrix}. \tag{10.203}$$

The last element of this expression may be written as

$$V_{71}c_1 + W_{71}s_1 = -(s_1c_7 + c_1s_7c_{71})c_1 + (c_1c_7 - s_1s_7c_{71})s_1. \tag{10.204}$$

Regrouping this expression and substituting $s_1^2 + c_1^2 = 1$ yields

$$V_{71}c_1 + W_{71}s_1 = -c_{71}s_7. \tag{10.205}$$

Substituting this result into Eq. (10.203) gives

$$\mathbf{a}_{71} \times \mathbf{a}_{67} = \begin{bmatrix} -U_{71}s_1 \\ -U_{71}c_1 \\ -c_{71}s_7 \end{bmatrix}. \tag{10.206}$$

Substituting Eqs. (10.195), (10.202), and (10.206) into Eq. (10.193) gives

$$\mathbf{R}^{6,71} \times \mathbf{a}_{67} = \begin{bmatrix} -S_6 U^*_{176} - S_7 V_{17} - a_{71} U_{71} s_1 \\ S_6(-c_1 Z'_7 - s_1 X'_7) + S_7(-s_7 s_1 + c_7 c_1 c_{71}) - a_{71} U_{71} c_1 \\ S_6 \bar{Y}_7 - S_7 s_{71} c_7 - a_{71} c_{71} s_7 \end{bmatrix}. \quad (10.207)$$

Expanding the vector $\mathbf{R}^{1,12}$,

$$\mathbf{R}^{1,12} \times \mathbf{S}_3 = (S_1 \mathbf{S}_1 + a_{12} \mathbf{a}_{12}) \times \mathbf{S}_3. \quad (10.208)$$

The vector product $\mathbf{S}_1 \times \mathbf{S}_3$ may be expressed as

$$\mathbf{S}_1 \times \mathbf{S}_3 = \begin{vmatrix} \mathbf{i} & \mathbf{j} & \mathbf{k} \\ 0 & 0 & 1 \\ \bar{X}_2 & \bar{Y}_2 & \bar{Z}_2 \end{vmatrix} = \begin{bmatrix} -\bar{Y}_2 \\ \bar{X}_2 \\ 0 \end{bmatrix}. \quad (10.209)$$

The vector product $\mathbf{a}_{12} \times \mathbf{S}_3$ may be expressed as

$$\mathbf{a}_{12} \times \mathbf{S}_3 = \begin{vmatrix} \mathbf{i} & \mathbf{j} & \mathbf{k} \\ 1 & 0 & 0 \\ \bar{X}_2 & \bar{Y}_2 & \bar{Z}_2 \end{vmatrix} = \begin{bmatrix} 0 \\ -\bar{Z}_2 \\ \bar{Y}_2 \end{bmatrix}. \quad (10.210)$$

Substituting the results of Eqs. (10.209) and (10.210) into Eq. (10.208) yields

$$\mathbf{R}^{1,12} \times \mathbf{S}_3 = \begin{bmatrix} -S_1 \bar{Y}_2 \\ S_1 \bar{X}_2 - a_{12} \bar{Z}_2 \\ a_{12} \bar{Y}_2 \end{bmatrix}. \quad (10.211)$$

Evaluating the scalar product of the vectors $(\mathbf{R}^{6,71} \times \mathbf{a}_{67})$ and $(\mathbf{R}^{1,12} \times \mathbf{S}_3)$ gives

$$\begin{aligned} &(\mathbf{R}^{6,71} \times \mathbf{a}_{67}) \cdot (\mathbf{R}^{1,12} \times \mathbf{S}_3) \\ &\quad = \left[-S_6 U^*_{176} - S_7 V_{17} - a_{71} U_{71} s_1\right](-S_1 \bar{Y}_2) + \left[S_6(-c_1 Z'_7 - s_1 X'_7)\right. \\ &\qquad + S_7(-s_7 s_1 + c_7 c_1 c_{71}) - a_{71} U_{71} c_1\left.\right](S_1 \bar{X}_2 - a_{12} \bar{Z}_2) \\ &\qquad + [S_6 \bar{Y}_7 - S_7 s_{71} c_7 - a_{71} c_{71} s_7](a_{12} \bar{Y}_2). \end{aligned} \quad (10.212)$$

(xv) $(\mathbf{R}^{6,12} \times \mathbf{a}_{67}) \cdot (\mathbf{R}^{2,23} \times \mathbf{S}_3)$

Set 1 of the table of direction cosines will be used to evaluate all the scalar and vector products of this term. Expanding the vector $\mathbf{R}^{6,12}$,

$$\mathbf{R}^{6,12} \times \mathbf{a}_{67} = (\mathbf{R}^{6,71} + S_1 \mathbf{S}_1 + a_{12} \mathbf{a}_{12}) \times \mathbf{a}_{67}, \quad (10.213)$$

where the vector product of $\mathbf{R}^{6,71}$ and $\mathbf{a}_{67}$ has been evaluated using set 1 of the table of direction cosines in Eq. (10.207). The vector product of $\mathbf{S}_1$ and $\mathbf{a}_{67}$ may be written as

$$\mathbf{S}_1 \times \mathbf{a}_{67} = \begin{vmatrix} \mathbf{i} & \mathbf{j} & \mathbf{k} \\ 0 & 0 & 1 \\ W_{71} & V_{71} & U_{71} \end{vmatrix} = \begin{bmatrix} -V_{71} \\ W_{71} \\ 0 \end{bmatrix}, \quad (10.214)$$

where the substitutions $W_{71} = W_{65432}$, $V_{71} = -U^*_{654321}$, and $U_{71} = U_{654321}$ have been made.

The vector product of $\mathbf{a}_{12}$ and $\mathbf{a}_{67}$ may be written as

$$\mathbf{a}_{12} \times \mathbf{a}_{67} = \begin{vmatrix} \mathbf{i} & \mathbf{j} & \mathbf{k} \\ 1 & 0 & 0 \\ W_{71} & V_{71} & U_{71} \end{vmatrix} = \begin{bmatrix} 0 \\ -U_{71} \\ V_{71} \end{bmatrix}. \tag{10.215}$$

Substituting the results of Eqs. (10.207), (10.214), and (10.215) into Eq. (10.213) gives

$$\mathbf{R}^{6,12} \times \mathbf{a}_{67} = \begin{bmatrix} -S_6U^*_{176} - S_7V_{17} - a_{71}U_{71}s_1 - S_1V_{71} \\ S_6(-c_1Z'_7 - s_1X'_7) + S_7(-s_7s_1 + c_7c_1c_{71}) - a_{71}U_{71}c_1 + S_1W_{71} - a_{12}U_{71} \\ S_6\bar{Y}_7 - S_7s_{71}c_7 - a_{71}c_{71}s_7 + a_{12}V_{71} \end{bmatrix}. \tag{10.216}$$

Now,

$$\mathbf{R}^{2,23} \times \mathbf{S}_3 = (S_2\mathbf{S}_2 + a_{23}\mathbf{a}_{23}) \times \mathbf{S}_3. \tag{10.217}$$

The vector product $\mathbf{S}_2 \times \mathbf{S}_3$ is written as

$$\mathbf{S}_2 \times \mathbf{S}_3 = \begin{vmatrix} \mathbf{i} & \mathbf{j} & \mathbf{k} \\ 0 & -s_{12} & c_{12} \\ \bar{X}_2 & \bar{Y}_2 & \bar{Z}_2 \end{vmatrix} = \begin{bmatrix} -s_{12}\bar{Z}_2 - c_{12}\bar{Y}_2 \\ c_{12}\bar{X}_2 \\ s_{12}\bar{X}_2 \end{bmatrix}. \tag{10.218}$$

The first element of this vector can be simplified by introducing the definitions of $\bar{Y}_2$ and $\bar{Z}_2$ and regrouping terms to give

$$\mathbf{S}_2 \times \mathbf{S}_3 = \begin{bmatrix} s_{23}c_2 \\ c_{12}\bar{X}_2 \\ s_{12}\bar{X}_2 \end{bmatrix}. \tag{10.219}$$

The vector product $\mathbf{a}_{23} \times \mathbf{S}_3$ may be written as

$$\mathbf{a}_{23} \times \mathbf{S}_3 = \begin{vmatrix} \mathbf{i} & \mathbf{j} & \mathbf{k} \\ c_2 & s_2c_{12} & s_2s_{12} \\ \bar{X}_2 & \bar{Y}_2 & \bar{Z}_2 \end{vmatrix} = \begin{bmatrix} s_2c_{12}\bar{Z}_2 - s_2s_{12}\bar{Y}_2 \\ s_2s_{12}\bar{X}_2 - c_2\bar{Z}_2 \\ c_2\bar{Y}_2 - s_2c_{12}\bar{X}_2 \end{bmatrix}. \tag{10.220}$$

This expression can be simplified by substituting the definitions of $\bar{X}_2$, $\bar{Y}_2$, and $\bar{Z}_2$, substituting $s_2^2 + c_2^2 = 1$ and $s_{12}^2 + c_{12}^2 = 1$, and then regrouping terms to yield

$$\mathbf{a}_{23} \times \mathbf{S}_3 = \begin{bmatrix} \bar{X}'_2 \\ Z'_2 \\ Y_2 \end{bmatrix}, \tag{10.221}$$

where $\bar{X}'_2 = c_{23}s_2$ and $Z'_2 = s_{12}s_{23} - c_{12}c_{23}c_2$.

Substituting Eqs. (10.219) and (10.221) into Eq. (10.217) yields

$$\mathbf{R}^{2,23} \times \mathbf{S}_3 = \begin{bmatrix} S_2s_{23}c_2 + a_{23}\bar{X}'_2 \\ S_2c_{12}\bar{X}_2 + a_{23}Z'_2 \\ S_2s_{12}\bar{X}_2 + a_{23}Y_2 \end{bmatrix}. \tag{10.222}$$

Evaluating the scalar product of the vectors $(\mathbf{R}^{6,12} \times \mathbf{a}_{67})$ and $(\mathbf{R}^{2,23} \times \mathbf{S}_3)$ gives

$$\begin{aligned}
&(\mathbf{R}^{6,12} \times \mathbf{a}_{67}) \cdot (\mathbf{R}^{2,23} \times \mathbf{S}_3) \\
&\quad = \left[-S_6 U^*_{176} - S_7 V_{17} - a_{71} U_{71} s_1 - S_1 V_{71}\right](S_2 s_{23} c_2 + a_{23} \bar{X}'_2) + \left[S_6(-c_1 Z'_7 - s_1 X'_7)\right. \\
&\qquad + S_7(-s_7 s_1 + c_7 c_1 c_{71}) - a_{71} U_{71} c_1 + S_1 W_{71} - a_{12} U_{71}\left.\right](S_2 c_{12} \bar{X}_2 + a_{23} Z'_2) \\
&\qquad + [S_6 \bar{Y}_7 - S_7 s_{71} c_7 - a_{71} c_{71} s_7 + a_{12} V_{71}](S_2 s_{12} \bar{X}_2 + a_{23} Y_2).
\end{aligned} \tag{10.223}$$

(xvi) $\mathbf{S}_3 \cdot \mathbf{S}_6 \times \mathbf{a}_{67}$

Set 6 of the table of direction cosines will be used to evaluate this term as

$$\mathbf{S}_3 \cdot \mathbf{S}_6 \times \mathbf{a}_{67} = \begin{vmatrix} X_{217} & Y_{217} & Z_{217} \\ 0 & 0 & 1 \\ 1 & 0 & 0 \end{vmatrix} = Y_{217}. \tag{10.224}$$

(xvii) $\mathbf{R}^{2,23} \cdot \mathbf{S}_6 \times \mathbf{a}_{67}$

Now,

$$\mathbf{R}^{2,23} \cdot \mathbf{S}_6 \times \mathbf{a}_{67} = (S_2 \mathbf{S}_2 + a_{23} \mathbf{a}_{23}) \cdot (\mathbf{S}_6 \times \mathbf{a}_{67}). \tag{10.225}$$

The vector product $\mathbf{S}_6 \times \mathbf{a}_{67}$ may be evaluated using set 6 from the table of direction cosines as

$$\mathbf{S}_6 \times \mathbf{a}_{67} = \begin{vmatrix} \mathbf{i} & \mathbf{j} & \mathbf{k} \\ 0 & 0 & 1 \\ 1 & 0 & 0 \end{vmatrix} = \begin{bmatrix} 0 \\ 1 \\ 0 \end{bmatrix}. \tag{10.226}$$

Forming the scalar product of $\mathbf{S}_2$ and $\mathbf{a}_{23}$ with $(\mathbf{S}_6 \times \mathbf{a}_{67})$, Eq. (10.225) can be expressed in the form

$$\mathbf{R}^{2,23} \cdot \mathbf{S}_6 \times \mathbf{a}_{67} = S_2 Y_{17} - a_{23} U^*_{2176}. \tag{10.227}$$

(xviii) $\mathbf{R}^{1,12} \cdot \mathbf{S}_6 \times \mathbf{a}_{67}$

Now,

$$\mathbf{R}^{1,12} \cdot \mathbf{S}_6 \times \mathbf{a}_{67} = (S_1 \mathbf{S}_1 + a_{12} \mathbf{a}_{12}) \cdot (\mathbf{S}_6 \times \mathbf{a}_{67}). \tag{10.228}$$

Using set 6 of the table of direction cosines and Eq. (10.226),

$$\mathbf{R}^{1,12} \cdot \mathbf{S}_6 \times \mathbf{a}_{67} = S_1 \bar{Y}_7 - a_{12} U^*_{176}. \tag{10.229}$$

(xix) $\mathbf{R}^{7,71} \cdot \mathbf{S}_6 \times \mathbf{a}_{67}$

Now,

$$\mathbf{R}^{7,71} \cdot \mathbf{S}_6 \times \mathbf{a}_{67} = (S_7 \mathbf{S}_7 + a_{71} \mathbf{a}_{71}) \cdot (\mathbf{S}_6 \times \mathbf{a}_{67}). \tag{10.230}$$

Using set 6 of the table of direction cosines and Eq. (10.226),

$$\mathbf{R}^{7,71} \cdot \mathbf{S}_6 \times \mathbf{a}_{67} = -S_7 s_{67} + a_{71} c_{67} s_7. \tag{10.231}$$

(xx) $\mathbf{R}^{6,67} \cdot \mathbf{S}_6 \times \mathbf{a}_{67}$

Now,

$$\mathbf{R}^{6,67} \cdot \mathbf{S}_6 \times \mathbf{a}_{67} = (\mathrm{S}_6\mathbf{S}_6 + \mathrm{a}_{67}\mathbf{a}_{67}) \cdot (\mathbf{S}_6 \times \mathbf{a}_{67}), \tag{10.232}$$

and clearly

$$\mathbf{R}^{6,67} \cdot \mathbf{S}_6 \times \mathbf{a}_{67} = 0 \tag{10.233}$$

because $\mathbf{S}_6 \cdot \mathbf{S}_6 \times \mathbf{a}_{67} = 0$ and $\mathbf{a}_{67} \cdot \mathbf{S}_6 \times \mathbf{a}_{67} = 0$.

(xxi) $\mathbf{R}^{6,71} \cdot \mathbf{S}_6 \times \mathbf{a}_{67}$

Now,

$$\mathbf{R}^{6,71} \cdot \mathbf{S}_6 \times \mathbf{a}_{67} = \mathbf{R}^{6,67} \cdot \mathbf{S}_6 \times \mathbf{a}_{67} + \mathbf{R}^{7,71} \cdot \mathbf{S}_6 \times \mathbf{a}_{67}. \tag{10.234}$$

From Eqs. (10.233) and (10.231),

$$\mathbf{R}^{6,71} \cdot \mathbf{S}_6 \times \mathbf{a}_{67} = -\mathrm{S}_7 s_{67} + a_{71} c_{67} s_7. \tag{10.235}$$

(xxii) $\mathbf{R}^{6,12} \cdot \mathbf{S}_6 \times \mathbf{a}_{67}$

Now,

$$\mathbf{R}^{6,12} \cdot \mathbf{S}_6 \times \mathbf{a}_{67} = \mathbf{R}^{6,71} \cdot \mathbf{S}_6 \times \mathbf{a}_{67} + \mathbf{R}^{1,12} \cdot \mathbf{S}_6 \times \mathbf{a}_{67}. \tag{10.236}$$

From Eqs. (10.235) and (10.229),

$$\mathbf{R}^{6,12} \cdot \mathbf{S}_6 \times \mathbf{a}_{67} = -\mathrm{S}_7 s_{67} + a_{71} c_{67} s_7 + \mathrm{S}_1 \bar{\mathrm{Y}}_7 - a_{12} \mathrm{U}^*_{176}. \tag{10.237}$$

(xxiii) $(\mathbf{R}^{6,67} \times (\mathbf{S}_6 \times \mathbf{a}_{67})) \cdot (\mathbf{R}^{7,71} \times \mathbf{S}_3)$

Set 7 of the table of direction cosines will be used to evaluate the scalar and vector products in this expression. The first term in the expression may be written as

$$\mathbf{R}^{6,67} \times (\mathbf{S}_6 \times \mathbf{a}_{67}) = (\mathrm{S}_6\mathbf{S}_6 + \mathrm{a}_{67}\mathbf{a}_{67}) \times (\mathbf{S}_6 \times \mathbf{a}_{67}). \tag{10.238}$$

The vector product $\mathbf{S}_6 \times \mathbf{a}_{67}$ can be evaluated as

$$\mathbf{S}_6 \times \mathbf{a}_{67} = \begin{vmatrix} \mathbf{i} & \mathbf{j} & \mathbf{k} \\ X_{54321} & Y_{54321} & Z_{54321} \\ c_7 & -s_7 & 0 \end{vmatrix} = \begin{vmatrix} \mathbf{i} & \mathbf{j} & \mathbf{k} \\ s_{67}s_7 & s_{67}c_7 & c_{67} \\ c_7 & -s_7 & 0 \end{vmatrix} = \begin{bmatrix} c_{67}s_7 \\ c_{67}c_7 \\ -s_{67} \end{bmatrix}, \tag{10.239}$$

where the fundamental sine, sine–cosine, and cosine laws $X_{54321} = s_{67}s_7$, $Y_{54321} = s_{67}c_7$, and $Z_{54321} = c_{67}$ were used to simplify the direction cosines of $\mathbf{S}_6$.

The vector product $\mathbf{S}_6 \times (\mathbf{S}_6 \times \mathbf{a}_{67})$ may now be written as

$$\mathbf{S}_6 \times (\mathbf{S}_6 \times \mathbf{a}_{67}) = \begin{vmatrix} \mathbf{i} & \mathbf{j} & \mathbf{k} \\ s_{67}s_7 & s_{67}c_7 & c_{67} \\ c_{67}s_7 & c_{67}c_7 & -s_{67} \end{vmatrix} = \begin{bmatrix} -c_7 \\ s_7 \\ 0 \end{bmatrix}. \tag{10.240}$$

The vector product of $\mathbf{a}_{67}$ and $(\mathbf{S}_6 \times \mathbf{a}_{67})$ may now be written as

$$\mathbf{a}_{67} \times (\mathbf{S}_6 \times \mathbf{a}_{67}) = \begin{vmatrix} \mathbf{i} & \mathbf{j} & \mathbf{k} \\ c_7 & -s_7 & 0 \\ c_{67}s_7 & c_{67}c_7 & -s_{67} \end{vmatrix} = \begin{bmatrix} s_{67}s_7 \\ s_{67}c_7 \\ c_{67} \end{bmatrix}. \tag{10.241}$$

Substituting Eqs. (10.240) and (10.241) into Eq. (10.238) yields

$$\mathbf{R}^{6,67} \times (\mathbf{S}_6 \times \mathbf{a}_{67}) = \begin{bmatrix} -S_6c_7 + a_{67}s_{67}s_7 \\ S_6s_7 + a_{67}s_{67}c_7 \\ a_{67}c_{67} \end{bmatrix}. \tag{10.242}$$

The factor $(\mathbf{R}^{7,71} \times \mathbf{S}_3)$ has previously been evaluated using set 7 of the table of direction cosines. The result is stated in Eq. (10.191). Forming the scalar product of $(\mathbf{R}^{6,67} \times (\mathbf{S}_6 \times \mathbf{a}_{67}))$ and $(\mathbf{R}^{7,71} \times \mathbf{S}_3)$ yields the result

$$\begin{aligned}(\mathbf{R}^{6,67} &\times (\mathbf{S}_6 \times \mathbf{a}_{67})) \cdot (\mathbf{R}^{7,71} \times \mathbf{S}_3) \\ &= (-S_7Y_{21})(-S_6c_7 + a_{67}s_{67}s_7) + (S_7X_{21} - a_{71}Z_{21})(S_6s_7 + a_{67}s_{67}c_7) \\ &\quad + (a_{71}Y_{21})(a_{67}c_{67}).\end{aligned} \tag{10.243}$$

(xxiv) $(\mathbf{R}^{6,71} \times (\mathbf{S}_6 \times \mathbf{a}_{67})) \cdot (\mathbf{R}^{1,12} \times \mathbf{S}_3)$

Set 7 of the table of direction cosines will be used to evaluate all the vector and scalar products in this expression. Expanding the vector $\mathbf{R}^{6,71}$,

$$\mathbf{R}^{6,71} \times (\mathbf{S}_6 \times \mathbf{a}_{67}) = (\mathbf{R}^{6,67} + S_7\mathbf{S}_7 + a_{71}\mathbf{a}_{71}) \times (\mathbf{S}_6 \times \mathbf{a}_{67}). \tag{10.244}$$

The vector product $\mathbf{S}_6 \times \mathbf{a}_{67}$ was previously calculated using set 7, and the results are presented in Eq. (10.239). Also, the vector product $\mathbf{R}^{6,67} \times (\mathbf{S}_6 \times \mathbf{a}_{67})$ is listed in Eq. (10.242). The vector product $\mathbf{S}_7 \times (\mathbf{S}_6 \times \mathbf{a}_{67})$ may be written as

$$\mathbf{S}_7 \times (\mathbf{S}_6 \times \mathbf{a}_{67}) = \begin{vmatrix} \mathbf{i} & \mathbf{j} & \mathbf{k} \\ 0 & 0 & 1 \\ c_{67}s_7 & c_{67}c_7 & -s_{67} \end{vmatrix} = \begin{bmatrix} -c_{67}c_7 \\ c_{67}s_7 \\ 0 \end{bmatrix}. \tag{10.245}$$

The vector product $\mathbf{a}_{71} \times (\mathbf{S}_6 \times \mathbf{a}_{67})$ may be written as

$$\mathbf{a}_{71} \times (\mathbf{S}_6 \times \mathbf{a}_{67}) = \begin{vmatrix} \mathbf{i} & \mathbf{j} & \mathbf{k} \\ 1 & 0 & 0 \\ c_{67}s_7 & c_{67}c_7 & -s_{67} \end{vmatrix} = \begin{bmatrix} 0 \\ s_{67} \\ c_{67}c_7 \end{bmatrix}. \tag{10.246}$$

Substituting the results of Eqs. (10.242), (10.245), and (10.246) into Eq. (10.244) gives

$$\mathbf{R}^{6,71} \times (\mathbf{S}_6 \times \mathbf{a}_{67}) = \begin{bmatrix} -S_6c_7 + a_{67}s_{67}s_7 - S_7c_{67}c_7 \\ S_6s_7 + a_{67}s_{67}c_7 + S_7c_{67}s_7 + a_{71}s_{67} \\ a_{67}c_{67} + a_{71}c_{67}c_7 \end{bmatrix}. \tag{10.247}$$

Now,

$$\mathbf{R}^{1,12} \times \mathbf{S}_3 = (S_1\mathbf{S}_1 + a_{12}\mathbf{a}_{12}) \times \mathbf{S}_3. \tag{10.248}$$

The vector product $\mathbf{S}_1 \times \mathbf{S}_3$ may be written as

$$\mathbf{S}_1 \times \mathbf{S}_3 = \begin{vmatrix} \mathbf{i} & \mathbf{j} & \mathbf{k} \\ 0 & -s_{71} & c_{71} \\ X_{21} & Y_{21} & Z_{21} \end{vmatrix} = \begin{bmatrix} -s_{71}Z_{21} - c_{71}Y_{21} \\ c_{71}X_{21} \\ s_{71}X_{21} \end{bmatrix}. \tag{10.249}$$

Expanding Y_{21} and Z_{21} and substituting $s_{71}^2 + c_{71}^2 = 1$ in the first component of $\mathbf{S}_1 \times \mathbf{S}_3$ gives

$$\mathbf{S}_1 \times \mathbf{S}_3 = \begin{bmatrix} -X_{21}^* \\ c_{71}X_{21} \\ s_{71}X_{21} \end{bmatrix}. \tag{10.250}$$

The vector product $\mathbf{a}_{12} \times \mathbf{S}_3$ may be written as

$$\mathbf{a}_{12} \times \mathbf{S}_3 = \begin{vmatrix} \mathbf{i} & \mathbf{j} & \mathbf{k} \\ c_1 & s_1c_{71} & U_{17} \\ X_{21} & Y_{21} & Z_{21} \end{vmatrix} = \begin{bmatrix} s_1c_{71}Z_{21} - U_{17}Y_{21} \\ U_{17}X_{21} - c_1Z_{21} \\ c_1Y_{21} - s_1c_{71}X_{21} \end{bmatrix}. \tag{10.251}$$

Expanding the terms X_{21}, Y_{21}, Z_{21}, and U_{17}, and substituting $s_1^2 + c_1^2 = 1$ and $s_{71}^2 + c_{71}^2 = 1$ gives

$$\mathbf{a}_{12} \times \mathbf{S}_3 = \begin{bmatrix} s_1Z_2 \\ -\bar{Y}_2s_{71} - c_{71}c_1\bar{Z}_2 \\ c_{71}\bar{Y}_2 - s_{71}c_1\bar{Z}_2 \end{bmatrix}. \tag{10.252}$$

Substituting Eqs. (10.250) and (10.252) into Eq. (10.248) yields

$$\mathbf{R}^{1,12} \times \mathbf{S}_3 = \begin{bmatrix} -S_1X_{21}^* + a_{12}s_1Z_2 \\ S_1c_{71}X_{21} - a_{12}(\bar{Y}_2s_{71} + c_{71}c_1\bar{Z}_2) \\ S_1s_{71}X_{21} + a_{12}(c_{71}\bar{Y}_2 - s_{71}c_1\bar{Z}_2) \end{bmatrix}. \tag{10.252}$$

Evaluating a scalar product of Eqs. (10.252) and (10.247) yields

$$\begin{aligned}(\mathbf{R}^{6,71} \times (\mathbf{S}_6 \times \mathbf{a}_{67})) \cdot (\mathbf{R}^{1,12} \times \mathbf{S}_3) &= [-S_1X_{21}^* + a_{12}s_1Z_2](-S_6c_7 + a_{67}s_{67}s_7 - S_7c_{67}c_7) \\ &+ [S_1c_{71}X_{21} - a_{12}(\bar{Y}_2s_{71} + c_{71}c_1\bar{Z}_2)](S_6s_7 + a_{67}s_{67}c_7 + S_7c_{67}s_7 + a_{71}s_{67}) \\ &+ [S_1s_{71}X_{21} + a_{12}(c_{71}\bar{Y}_2 - s_{71}c_1\bar{Z}_2)](a_{67}c_{67} + a_{71}c_{67}c_7).\end{aligned} \tag{10.253}$$

(xxv) $(\mathbf{R}^{6,12} \times (\mathbf{S}_6 \times \mathbf{a}_{67})) \cdot (\mathbf{R}^{2,23} \times \mathbf{S}_3)$

Set 1 of the table of direction cosines will be used to evaluate all the vector and scalar products in this expression. The first factor may be written as

$$\mathbf{R}^{6,12} \times (\mathbf{S}_6 \times \mathbf{a}_{67}) = (S_6\mathbf{S}_6 + a_{67}\mathbf{a}_{67} + S_7\mathbf{S}_7 + a_{71}\mathbf{a}_{71} + S_1\mathbf{S}_1 + a_{12}\mathbf{a}_{12}) \times (\mathbf{S}_6 \times \mathbf{a}_{67}). \tag{10.254}$$

The vector product $\mathbf{S}_6 \times \mathbf{a}_{67}$ was previously calculated using set 1, and the results are presented in Eq. (10.195). The vector product $\mathbf{S}_6 \times (\mathbf{S}_6 \times \mathbf{a}_{67})$ may be written as

$$\mathbf{S}_6 \times (\mathbf{S}_6 \times \mathbf{a}_{67}) = \begin{vmatrix} \mathbf{i} & \mathbf{j} & \mathbf{k} \\ X_{71} & -X^*_{71} & Z_7 \\ -U^*_{176} & -c_1 Z'_7 - s_1 X'_7 & \bar{Y}_7 \end{vmatrix} = \begin{bmatrix} -X^*_{71}\bar{Y}_7 + (c_1 Z'_7 + s_1 X'_7) Z_7 \\ -X_{71}\bar{Y}_7 - Z_7 U^*_{176} \\ X_{71}(-c_1 Z'_7 - s_1 X'_7) - X^*_{71} U^*_{176} \end{bmatrix}, \quad (10.255)$$

where the subsidiary sine, sine–cosine, and cosine laws $X_{5432} = X_{71}$, $Y_{5432} = -X^*_{71}$, and $Z_{5432} = Z_7$ were used to simplify the direction cosines of vector $\mathbf{S}_6$. Expanding and regrouping terms and introducing the trigonometric identities $s^2_{67} + c^2_{67} = 1$, $s^2_{71} + c^2_{71} = 1$, $s^2_7 + c^2_7 = 1$, and $s^2_1 + c^2_1 = 1$ yields

$$\mathbf{S}_6 \times (\mathbf{S}_6 \times \mathbf{a}_{67}) = \begin{bmatrix} -W_{71} \\ -V_{71} \\ -U_{71} \end{bmatrix}. \quad (10.256)$$

The vector product $\mathbf{a}_{67} \times (\mathbf{S}_6 \times \mathbf{a}_{67})$ may be written as

$$\mathbf{a}_{67} \times (\mathbf{S}_6 \times \mathbf{a}_{67}) = \begin{vmatrix} \mathbf{i} & \mathbf{j} & \mathbf{k} \\ W_{71} & V_{71} & U_{71} \\ -U^*_{176} & -c_1 Z'_7 - s_1 X'_7 & \bar{Y}_7 \end{vmatrix} = \begin{bmatrix} V_{71}\bar{Y}_7 + U_{71}(c_1 Z'_7 + s_1 X'_7) \\ -W_{71}\bar{Y}_7 - U_{71} U^*_{176} \\ W_{71}(-c_1 Z'_7 - s_1 X'_7) + V_{71} U^*_{176} \end{bmatrix}, \quad (10.257)$$

where the subsidiary polar sine, sine–cosine, and cosine laws $U_{71} = U_{654321}$, $V_{71} = -U^*_{654321}$, $W_{71} = W_{65432}$ were used to simplify the direction cosines of vector $\mathbf{a}_{67}$. Expanding and regrouping terms and introducing the trigonometric identities $s^2_{71} + c^2_{71} = 1$, $s^2_7 + c^2_7 = 1$, and $s^2_1 + c^2_1 = 1$ yields

$$\mathbf{a}_{67} \times (\mathbf{S}_6 \times \mathbf{a}_{67}) = \begin{bmatrix} X_{71} \\ -X^*_{71} \\ Z_7 \end{bmatrix}. \quad (10.258)$$

The vector product $\mathbf{S}_7 \times (\mathbf{S}_6 \times \mathbf{a}_{67})$ may be written as

$$\mathbf{S}_7 \times (\mathbf{S}_6 \times \mathbf{a}_{67}) = \begin{vmatrix} \mathbf{i} & \mathbf{j} & \mathbf{k} \\ s_{71}s_1 & s_{71}c_1 & c_{71} \\ -U^*_{176} & -c_1 Z'_7 - s_1 X'_7 & \bar{Y}_7 \end{vmatrix} = \begin{bmatrix} s_{71}c_1\bar{Y}_7 + c_{71}(c_1 Z'_7 + s_1 X'_7) \\ -s_{71}s_1\bar{Y}_7 - c_{71} U^*_{176} \\ s_{71}s_1(-c_1 Z'_7 - s_1 X'_7) + s_{71}c_1 U^*_{176} \end{bmatrix}, \quad (10.259)$$

where the fundamental sine, sine–cosine, and cosine laws $X_{65432} = s_{71}s_1$, $Y_{65432} = s_{71}c_1$, and $Z_{65432} = c_{71}$ were used to simplify the direction cosines of vector $\mathbf{S}_7$. Expanding and regrouping terms and introducing the trigonometric identities $s^2_{71} + c^2_{71} = 1$ and $s^2_1 + c^2_1 = 1$

yields

$$\mathbf{S}_7 \times (\mathbf{S}_6 \times \mathbf{a}_{67}) = \begin{bmatrix} -c_{67}W_{71} \\ -c_{67}V_{71} \\ -c_{67}U_{71} \end{bmatrix}. \tag{10.260}$$

The vector product $\mathbf{a}_{71} \times (\mathbf{S}_6 \times \mathbf{a}_{67})$ may be written as

$$\mathbf{a}_{71} \times (\mathbf{S}_6 \times \mathbf{a}_{67}) = \begin{vmatrix} \mathbf{i} & \mathbf{j} & \mathbf{k} \\ c_1 & -s_1 & 0 \\ -U^*_{176} & -c_1Z'_7 - s_1X'_7 & \bar{Y}_7 \end{vmatrix} = \begin{bmatrix} -s_1\bar{Y}_7 \\ -c_1\bar{Y}_7 \\ c_1(-c_1Z'_7 - s_1X'_7) - s_1U^*_{176} \end{bmatrix}. \tag{10.261}$$

Expanding, regrouping, and substituting $s_1^2 + c_1^2 = 1$ in the last element of this vector gives

$$\mathbf{a}_{71} \times (\mathbf{S}_6 \times \mathbf{a}_{67}) = \begin{bmatrix} -s_1\bar{Y}_7 \\ -c_1\bar{Y}_7 \\ -Z'_7 \end{bmatrix}. \tag{10.262}$$

The vector product $\mathbf{S}_1 \times (\mathbf{S}_6 \times \mathbf{a}_{67})$ may be written as

$$\mathbf{S}_1 \times (\mathbf{S}_6 \times \mathbf{a}_{67}) = \begin{vmatrix} \mathbf{i} & \mathbf{j} & \mathbf{k} \\ 0 & 0 & 1 \\ -U^*_{176} & -c_1Z'_7 - s_1X'_7 & \bar{Y}_7 \end{vmatrix} = \begin{bmatrix} c_1Z'_7 + s_1X'_7 \\ -U^*_{176} \\ 0 \end{bmatrix}. \tag{10.263}$$

The vector product $\mathbf{a}_{12} \times (\mathbf{S}_6 \times \mathbf{a}_{67})$ may be written as

$$\mathbf{a}_{12} \times (\mathbf{S}_6 \times \mathbf{a}_{67}) = \begin{vmatrix} \mathbf{i} & \mathbf{j} & \mathbf{k} \\ 1 & 0 & 0 \\ -U^*_{176} & -c_1Z'_7 - s_1X'_7 & \bar{Y}_7 \end{vmatrix} = \begin{bmatrix} 0 \\ -\bar{Y}_7 \\ -c_1Z'_7 - s_1X'_7 \end{bmatrix}. \tag{10.264}$$

The factor $(\mathbf{R}^{6,12} \times (\mathbf{S}_6 \times \mathbf{a}_{67}))$ may now be expressed by substituting the results of Eqs. (10.256), (10.258), (10.260), (10.262), (10.263), and (10.264) into Eq. (10.254) to give

$$\mathbf{R}^{6,12} \times (\mathbf{S}_6 \times \mathbf{a}_{67}) = \begin{bmatrix} -S_6W_{71} + a_{67}X_{71} - S_7c_{67}W_{71} - a_{71}s_1\bar{Y}_7 + S_1(c_1Z'_7 + s_1X'_7) \\ -S_6V_{71} - a_{67}X^*_{71} - S_7c_{67}V_{71} - a_{71}c_1\bar{Y}_7 - S_1U^*_{176} - a_{12}\bar{Y}_7 \\ -S_6U_{71} + a_{67}Z_7 - S_7c_{67}U_{71} - a_{71}Z'_7 - a_{12}(c_1Z'_7 + s_1X'_7) \end{bmatrix}. \tag{10.265}$$

The term $(\mathbf{R}^{2,23} \times \mathbf{S}_3)$ was previously expanded in terms of set 1 of the table of direction cosines. The results are expressed in Eq. (10.222). The scalar product $(\mathbf{R}^{6,12} \times (\mathbf{S}_6 \times \mathbf{a}_{67})) \cdot (\mathbf{R}^{2,23} \times \mathbf{S}_3)$ may now be written as

$$\begin{aligned} &(\mathbf{R}^{6,12} \times (\mathbf{S}_6 \times \mathbf{a}_{67})) \cdot (\mathbf{R}^{2,23} \times \mathbf{S}_3) \\ &\quad = [-S_6W_{71} + a_{67}X_{71} - S_7c_{67}W_{71} - a_{71}s_1\bar{Y}_7 + S_1(c_1Z'_7 + s_1X'_7)][S_2s_{23}c_2 + a_{23}\bar{X}'_2] \\ &\quad + [-S_6V_{71} - a_{67}X^*_{71} - S_7c_{67}V_{71} - a_{71}c_1\bar{Y}_7 - S_1U^*_{176} - a_{12}\bar{Y}_7][S_2c_{12}\bar{X}_2 + a_{23}Z'_2] \\ &\quad + [-S_6U_{71} + a_{67}Z_7 - S_7c_{67}U_{71} - a_{71}Z'_7 - a_{12}(c_1Z'_7 + s_1X'_7)][S_2s_{12}\bar{X}_2 + a_{23}Y_2]. \end{aligned} \tag{10.266}$$

All twenty five-terms listed at the beginning of this section have now been expanded. The results of these expansions will now be substituted into the expressions for the terms H_3 and H_4 as defined in Eqs. (10.75) and (10.76), which gives

$$\begin{aligned}
H_3 = {} & K_2X_{217} - (S_6Z_{217} + a_{67}X_{217})a_{67} - (S_7Z_{21} + a_{71}X_{21})a_{71}c_7 \\
& + [S_6S_7c_{67}X_{217} - S_6a_{71}(s_{67}Y_{21} + c_{67}c_7Z_{21})] - (S_7Z_{21} + a_{71}X_{21})a_{67} \\
& - (S_1Z_2 + a_{12}\bar{X}_2)(S_1U_{71} + a_{12}W_{71}) + \big[S_6U^*_{176} + S_7V_{17} \\
& + a_{71}U_{71}s_1\big](S_1\bar{Y}_2) + \big[S_6\big(-c_1Z'_7 - s_1X'_7\big) + S_7(-s_7s_1 + c_7c_1c_{71}) \\
& - a_{71}U_{71}c_1\big](S_1\bar{X}_2 - a_{12}\bar{Z}_2) + [S_6\bar{Y}_7 - S_7s_{71}c_7 - a_{71}c_{71}s_7](a_{12}\bar{Y}_2) \\
& - (S_1Z_2 + a_{12}\bar{X}_2)(a_{67} + a_{71}c_7) - S_2c_{23}(S_2U_{712} + a_{23}W_{712}) + \big[-S_6U^*_{176} \\
& - S_7V_{17} - a_{71}U_{71}s_1 - S_1V_{71}\big] \times \big(S_2s_{23}c_2 + a_{23}\bar{X}'_2\big) + \big[S_6\big(-c_1Z'_7 \\
& - s_1X'_7\big) + S_7(-s_7s_1 + c_7c_1c_{71}) - a_{71}U_{71}c_1 + S_1W_{71} - a_{12}U_{71}\big]\big(S_2c_{12}\bar{X}_2 \\
& + a_{23}Z'_2\big) + [s_6\bar{Y}_7 - S_7s_{71}c_7 - a_{71}c_{71}s_7 + a_{12}V_{71}](S_2s_{12}\bar{X}_2 + a_{23}Y_2) \\
& - S_2c_{23}(a_{67} + a_{71}c_7 + S_1\bar{X}_7 + a_{12}W_{17}),
\end{aligned} \tag{10.267}$$

$$\begin{aligned}
H_4 = {} & K_2Y_{217} - (S_7Z_{12} + a_{71}X_{21})(-S_7s_{67} + a_{71}c_{67}s_7) + (-S_7Y_{21})(-S_6c_7 \\
& + a_{67}s_{67}s_7) + (S_7X_{21} - a_{71}Z_{21})(S_6s_7 + a_{67}s_{67}c_7) + (a_{71}Y_{21})(a_{67}c_{67}) \\
& - (S_1Z_2 + a_{12}\bar{X}_2)\big(S_1\bar{Y}_7 - a_{12}U^*_{176}\big) + \big[-S_1X^*_{21} + a_{12}s_1Z_2\big](-S_6c_7 + a_{67}s_{67}s_7 \\
& - S_7c_{67}c_7) + \big[S_1c_{71}X_{21} - a_{12}(\bar{Y}_2s_{71} + c_{71}c_1\bar{Z}_2)\big](S_6s_7 + a_{67}s_{67}c_7 \\
& + S_7c_{67}s_7 + a_{71}s_{67}) + \big[S_1s_{71}X_{21} + a_{12}(c_{71}\bar{Y}_2 - s_{71}c_1\bar{Z}_2)\big](a_{67}c_{67} + a_{71}c_{67}c_7) \\
& - (S_1Z_2 + a_{12}\bar{X}_2)(-S_7s_{67} + a_{71}c_{67}s_7) - S_2c_{23}\big(S_2Y_{17} - a_{23}U^*_{2176}\big) + \big[-S_6W_{71} \\
& + a_{67}X_{71} - S_7c_{67}W_{71} - a_{71}s_1\bar{Y}_7 + S_1\big(c_1Z'_7 + s_1X'_7\big)\big]\big[S_2s_{23}c_2 + a_{23}\bar{X}'_2\big] \\
& + \big[-S_6V_{71} - a_{67}X^*_{71} - S_7c_{67}V_{71} - a_{71}c_1\bar{Y}_7 - S_1U^*_{176} - a_{12}\bar{Y}_7\big]\big[S_2c_{12}\bar{X}_2 + a_{23}Z'_2\big] \\
& + \big[-S_6U_{71} + a_{67}Z_7 - S_7c_{67}U_{71} - a_{71}Z'_7 - a_{12}\big(c_1Z'_7 + s_1X'_7\big)\big]\big[S_2s_{12}\bar{X}_2 + a_{23}Y_2\big] \\
& - S_2c_{23}\big(-S_7s_{67} + a_{71}c_{67}s_7 + S_1\bar{Y}_7 - a_{12}U^*_{176}\big).
\end{aligned} \tag{10.268}$$

10.2.7 Regrouping of the terms H_3, H_4, J_3, and J_4

The terms H_3, H_4, J_3, and J_4 are defined respectively by Eqs. (10.267), (10.268), (10.160), and (10.161). These equations can be regrouped as follows:

$$H_3 = (L_1c_1 + L_2s_1 + L_3)c_2 + (L_4c_1 + L_5s_1 + L_6)s_2 + L_8s_1 + L_9) = 0, \tag{10.269}$$

$$\begin{aligned}
H_4 = {} & (L_{10}c_1 + L_{11}s_1 + L_{12})c_2 + (L_{13}c_1 + L_{14}s_1 + L_{15})s_2 \\
& + (L_{16}c_1 + L_{17}s_1 + L_{18}) = 0,
\end{aligned} \tag{10.270}$$

$$J_3 = (L_{19}c_5 + L_{20}s_5 + L_{21})c_4 + (L_{22}c_5 + L_{23}s_5 + L_{24})s_4 + (L_{25}c_5 + L_{26}s_5 + L_{27}) = 0, \tag{10.271}$$

$$\begin{aligned}
J_4 = {} & (L_{28}c_5 + L_{29}s_5 + L_{30})c_4 + (L_{31}c_5 + L_{32}s_5 + L_{33})s_4 \\
& + (L_{34}c_5 + L_{35}s_5 + L_{36}) = 0,
\end{aligned} \tag{10.272}$$

where

$$\begin{aligned}
L_1 = {} & S_1\left(S_2 s_{23}\bar{X}'_7 + S_6 c_{12} s_{23} X'_7 + S_7 c_{12} s_{23} s_7 + a_{12} s_{12} s_{23} c_7 - a_{23} c_{12} c_{23} c_7\right) \\
& + S_2\left(S_6 s_{23} X'_7 + S_7 s_{23} s_7 - a_{23} c_{23} c_7\right) + S_6\left(S_7 c_{12} s_{23} c_{71} X'_7 - a_{12} s_{12} s_{23} Z'_7\right. \\
& \left. + a_{23} c_{12} c_{23} Z'_7 - a_{67} c_{12} s_{23} Y_7 - a_{71} c_{12} s_{23} \bar{Y}_7\right) + S_7(a_{12} s_{12} s_{23} c_{71} c_7 \\
& - a_{23} c_{12} c_{23} c_{71} c_7 + a_{67} c_{12} s_{23} s_{71} + a_{71} c_{12} s_{23} s_{71} c_7) + a_{12}\left(a_{23} s_{12} c_{23} \bar{X}'_7\right. \\
& \left. - a_{71} s_{12} s_{23} \bar{X}_7\right) + a_{23} a_{71} c_{12} c_{23} \bar{X}_7 + \left(K_2 - a_{67}^2\right) c_{12} s_{23} \bar{X}'_7, \\
L_2 = {} & S_1\left(S_2 s_{23} c_7 - S_6 c_{12} s_{23} Z'_7 + S_7 c_{12} s_{23} c_{71} c_7 - a_{12} s_{12} s_{23} \bar{X}'_7\right. \\
& \left. + a_{23} c_{12} c_{23} \bar{X}'_7 - a_{71} c_{12} s_{23} \bar{X}_7\right) + S_2\left(-S_6 s_{23} Z'_7 + S_7 s_{23} c_{71} c_7\right. \\
& \left. + a_{23} c_{23} \bar{X}'_7 - a_{71} s_{23} \bar{X}_7\right) + S_6\left(S_7 c_{12} s_{23} c_{67} c_7 - a_{12} s_{12} s_{23} X'_7\right. \\
& \left. + a_{23} c_{12} c_{23} X'_7 - a_{67} c_{12} s_{23} X_7\right) + S_7(-a_{12} s_{12} s_{23} s_7 + a_{23} c_{12} c_{23} s_7) \\
& + a_{12} a_{23} s_{12} c_{23} c_7 - a_{67} a_{71} c_{12} s_{23} + \left(K_2 - a_{67}^2 - a_{71}^2\right) c_{12} s_{23} c_7, \\
L_3 = {} & S_1(a_{67} s_{12} s_{23} + a_{71} s_{12} s_{23} c_7) + S_6\left(-S_7 s_{12} s_{23} c_{67} \bar{X}_7 - a_{12} c_{12} s_{23} \bar{Y}_7 - a_{23} s_{12} c_{23} \bar{Y}_7\right. \\
& \left. + a_{67} s_{12} s_{23} Z_7 - a_{71} s_{12} s_{23} Z'_7\right) + S_7(a_{12} c_{12} s_{23} s_{71} c_7 + a_{23} s_{12} c_{23} s_{71} c_7 + a_{67} s_{12} s_{23} c_{71} \\
& + a_{71} s_{12} s_{23} c_{71} c_7) + a_{12}\left(a_{23} c_{12} c_{23} \bar{X}_7 + a_{71} c_{12} s_{23} \bar{X}'_7\right) + a_{23} a_{71} s_{12} c_{23} \bar{X}'_7 \\
& + \left(-K_2 + S_1^2 + a_{67}^2\right) s_{12} s_{23} \bar{X}_7, \\
L_4 = {} & S_1\left(S_2 c_{12} s_{23} c_7 - S_6 s_{23} Z'_7 + S_7 s_{23} c_{71} c_7 + a_{23} c_{23} \bar{X}'_7 - a_{71} s_{23} \bar{X}_7\right) + S_2\left(-S_6 c_{12} s_{23} Z'_7\right. \\
& \left. + S_7 c_{12} s_{23} c_{71} c_7 - a_{12} s_{12} s_{23} \bar{X}'_7 + a_{23} c_{12} c_{23} \bar{X}'_7 - a_{71} c_{12} s_{23} \bar{X}_7\right) + S_6\left(S_7 s_{23} c_{67} c_7\right. \\
& \left. + a_{23} c_{23} X'_7 - a_{67} s_{23} X_7\right) + S_7 a_{23} c_{23} s_7 - a_{67} a_{71} s_{23} + \left(K_2 - a_{12}^2 - a_{67}^2 - a_{71}^2\right) s_{23} c_7, \\
L_5 = {} & S_1\left(-S_2 c_{12} s_{23} \bar{X}'_7 - S_6 s_{23} X'_7 - S_7 s_{23} s_7 + a_{23} c_{23} c_7\right) + S_2\left(-S_6 c_{12} s_{23} X'_7\right. \\
& \left. - S_7 c_{12} s_{23} s_7 - a_{12} s_{12} s_{23} c_7 + a_{23} c_{12} c_{23} c_7\right) + S_6\left(-S_7 s_{23} c_{71} X'_7 - a_{23} c_{23} Z'_7\right. \\
& \left. + a_{67} s_{23} Y_7 + a_{71} s_{23} \bar{Y}_7\right) + S_7(a_{23} c_{23} c_{71} c_7 - a_{67} s_{23} s_{71} - a_{71} s_{23} s_{71} c_7) - a_{23} a_{71} c_{23} \bar{X}_7 \\
& + \left(-K_2 + a_{12}^2 + a_{67}^2\right) s_{23} \bar{X}'_7, \\
L_6 = {} & -S_1 a_{12} s_{23} \bar{X}_7 + S_2\left(S_6 s_{12} s_{23} \bar{Y}_7 - S_7 s_{12} s_{23} s_{71} c_7 - a_{12} c_{12} s_{23} \bar{X}_7\right. \\
& \left. - a_{23} s_{12} c_{23} \bar{X}_7 - a_{71} s_{12} s_{23} \bar{X}'_7\right) + a_{12}(-a_{67} s_{23} - a_{71} s_{23} c_7), \\
L_7 = {} & S_1\left(S_6 s_{12} c_{23} X'_7 + S_7 s_{12} c_{23} s_7 - a_{12} c_{12} c_{23} c_7 + a_{23} s_{12} s_{23} c_7\right) - S_2 a_{12} c_{23} c_7 \\
& + S_6\left(S_7 s_{12} c_{23} c_{71} X'_7 + a_{12} c_{12} c_{23} Z'_7 - a_{23} s_{12} s_{23} Z'_7 - a_{67} s_{12} c_{23} Y_7 - a_{71} s_{12} c_{23} \bar{Y}_7\right) \\
& + S_7(-a_{12} c_{12} c_{23} c_{71} c_7 + a_{23} s_{12} s_{23} c_{71} c_7 + a_{67} s_{12} c_{23} s_{71} + a_{71} s_{12} c_{23} s_{71} c_7) \\
& + a_{12}(a_{23} c_{12} s_{23} \bar{X}'_7 + a_{71} c_{12} c_{23} \bar{X}_7) - a_{23} a_{71} s_{12} s_{23} \bar{X}_7 + \left(K_2 - S_2^2 - a_{67}^2\right) s_{12} c_{23} \bar{X}'_7, \\
L_8 = {} & S_1\left(-S_6 s_{12} c_{23} Z'_7 + S_7 s_{12} c_{23} c_{71} c_7 + a_{12} c_{12} c_{23} \bar{X}'_7 - a_{23} s_{12} s_{23} \bar{X}'_7 - a_{71} s_{12} c_{23} \bar{X}_7\right) \\
& + S_2 a_{12} c_{23} \bar{X}'_7 + S_6\left(S_7 s_{12} c_{23} c_{67} c_7 + a_{12} c_{12} c_{23} X'_7 - a_{23} s_{12} s_{23} X'_7 - a_{67} s_{12} c_{23} X_7\right) \\
& + S_7(a_{12} c_{12} c_{23} s_7 - a_{23} s_{12} s_{23} s_7) + a_{12} a_{23} c_{12} s_{23} c_7 - a_{67} a_{71} s_{12} c_{23} \\
& + \left(K_2 - S_2^2 - a_{67}^2 - a_{71}^2\right) s_{12} c_{23} c_7, \\
L_9 = {} & S_1(-S_2 c_{23} \bar{X}_7 - a_{67} c_{12} c_{23} - a_{71} c_{12} c_{23} c_7) + S_2(-a_{67} c_{23} - a_{71} c_{23} c_7) \\
& + S_6\left(S_7 c_{12} c_{23} c_{67} \bar{X}_7 - a_{12} s_{12} c_{23} \bar{Y}_7 - a_{23} c_{12} s_{23} \bar{Y}_7 - a_{67} c_{12} c_{23} Z_7 + a_{71} c_{12} c_{23} Z'_7\right) \\
& + S_7(a_{12} s_{12} c_{23} s_{71} c_7 + a_{23} c_{12} s_{23} s_{71} c_7 - a_{67} c_{12} c_{23} c_{71} - a_{71} c_{12} c_{23} c_{71} c_7)
\end{aligned}$$

$$+a_{12}(-a_{23}s_{12}s_{23}\bar{X}_7 + a_{71}s_{12}c_{23}\bar{X}'_7) + a_{23}a_{71}c_{12}s_{23}\bar{X}'_7$$
$$+ (K_2 - S_1^2 - S_2^2 - a_{67}^2)c_{12}c_{23}\bar{X}_7, \tag{10.273}$$

$$\begin{aligned}
L_{10} = {} & S_1(S_2s_{23}Z'_7 - S_6c_{12}s_{23}c_7 - S_7c_{12}s_{23}c_{67}c_7 + a_{12}s_{12}s_{23}X'_7 - a_{23}c_{12}c_{23}X'_7 \\
& + a_{67}c_{12}s_{23}X_7) + S_2(-S_6s_{23}c_7 - S_7s_{23}c_{67}c_7 - a_{23}c_{23}X'_7 + a_{67}s_{23}X_7) \\
& + S_6(-S_7c_{12}s_{23}c_{71}c_7 + a_{12}s_{12}s_{23}\bar{X}'_7 - a_{23}c_{12}c_{23}\bar{X}'_7 + a_{71}c_{12}s_{23}\bar{X}_7) \\
& + S_7(a_{12}s_{12}s_{23}c_{71}X'_7 - a_{23}c_{12}c_{23}c_{71}X'_7 + a_{67}c_{12}s_{23}c_{71}X_7 + a_{71}c_{12}s_{23}c_{67}\bar{X}_7) \\
& + a_{12}(a_{23}s_{12}c_{23}Z'_7 - a_{67}s_{12}s_{23}Y_7 - a_{71}s_{12}s_{23}\bar{Y}_7) + a_{23}(a_{67}c_{12}c_{23}Y_7 \\
& + a_{71}c_{12}c_{23}\bar{Y}_7) - a_{67}a_{71}c_{12}s_{23}Z_7 + K_2c_{12}s_{23}Z'_7 - S_7^2c_{12}s_{23}s_{67}s_{71}, \\
L_{11} = {} & S_1(S_2s_{23}X'_7 + S_6c_{12}s_{23}\bar{X}'_7 + S_7c_{12}s_{23}c_{71}X'_7 - a_{12}s_{12}s_{23}Z'_7 + a_{23}c_{12}c_{23}Z'_7 \\
& - a_{67}c_{12}s_{23}Y_7 - a_{71}c_{12}s_{23}\bar{Y}_7) + S_2(S_6s_{23}\bar{X}'_7 + S_7s_{23}c_{71}X'_7 + a_{23}c_{23}Z'_7 - a_{67}s_{23}Y_7 \\
& - a_{71}s_{23}\bar{Y}_7) + S_6(S_7c_{12}s_{23}s_7 + a_{12}s_{12}s_{23}c_7 - a_{23}c_{12}c_{23}c_7) + S_7(a_{12}s_{12}s_{23}c_{67}c_7 \\
& - a_{23}c_{12}c_{23}c_{67}c_7 + a_{67}c_{12}s_{23}s_{67}c_7 + a_{71}c_{12}s_{23}s_{67}) + a_{12}(a_{23}s_{12}c_{23}X'_7 \\
& - a_{67}s_{12}s_{23}X_7) + a_{23}a_{67}c_{12}c_{23}X_7 + (K_2 - a_{71}^2)c_{12}s_{23}X'_7, \\
L_{12} = {} & S_1(-S_7s_{12}s_{23}s_{67} + a_{71}s_{12}s_{23}X'_7) + S_6(S_7s_{12}s_{23}s_{71}c_7 + a_{12}c_{12}s_{23}\bar{X}_7 + a_{23}s_{12}c_{23}\bar{X}_7 \\
& + a_{71}s_{12}s_{23}\bar{X}'_7) + S_7(a_{12}c_{12}s_{23}c_{67}\bar{X}_7 + a_{23}s_{12}c_{23}c_{67}\bar{X}_7 - a_{67}s_{12}s_{23}s_{71}X_7 \\
& + a_{71}s_{12}s_{23}c_{71}X'_7) + a_{12}(a_{23}c_{12}c_{23}\bar{Y}_7 - a_{67}c_{12}s_{23}Z_7 + a_{71}c_{12}s_{23}Z'_7) \\
& + a_{23}(-a_{67}s_{12}c_{23}Z_7 + a_{71}s_{12}c_{23}Z'_7) - a_{67}a_{71}s_{12}s_{23}Y_7 \\
& + (-K_2 + S_1^2)s_{12}s_{23}\bar{Y}_7 - S_7^2s_{12}s_{23}s_{67}c_{71}, \\
L_{13} = {} & S_1(S_2c_{12}s_{23}X'_7 + S_6s_{23}\bar{X}'_7 + S_7s_{23}c_{71}X'_7 + a_{23}c_{23}Z'_7 - a_{67}s_{23}Y_7 - a_{71}s_{23}\bar{Y}_7) \\
& + S_2(S_6c_{12}s_{23}\bar{X}'_7 + S_7c_{12}s_{23}c_{71}X'_7 - a_{12}s_{12}s_{23}Z'_7 + a_{23}c_{12}c_{23}Z'_7 - a_{67}c_{12}s_{23}Y_7 \\
& - a_{71}c_{12}s_{23}\bar{Y}_7) + S_6(S_7s_{23}s_7 - a_{23}c_{23}c_7) + S_7(-a_{23}c_{23}c_{67}c_7 + a_{67}s_{23}s_{67}c_7 \\
& + a_{71}s_{23}s_{67}) + a_{23}a_{67}c_{23}X_7 + (K_2 - a_{12}^2 - a_{71}^2)s_{23}X'_7, \\
L_{14} = {} & S_1(-S_2c_{12}s_{23}Z'_7 + S_6s_{23}c_7 + S_7s_{23}c_{67}c_7 + a_{23}c_{23}X'_7 - a_{67}s_{23}X_7) \\
& + S_2(S_6c_{12}s_{23}c_7 + S_7c_{12}s_{23}c_{67}c_7 - a_{12}s_{12}s_{23}X'_7 + a_{23}c_{12}c_{23}X'_7 - a_{67}c_{12}s_{23}X_7) \\
& + S_6(S_7s_{23}c_{71}c_7 + a_{23}c_{23}\bar{X}'_7 - a_{71}s_{23}\bar{X}_7) + S_7(a_{23}c_{23}c_{71}X'_7 - a_{67}s_{23}c_{71}X_7 \\
& - a_{71}s_{23}c_{67}\bar{X}_7) + a_{23}(-a_{67}c_{23}Y_7 - a_{71}c_{23}\bar{Y}_7) + a_{67}a_{71}s_{23}Z_7 \\
& + (-K_2 + a_{12}^2)s_{23}Z'_7 + S_7^2s_{23}s_{67}s_{71}, \\
L_{15} = {} & -S_1a_{12}s_{23}\bar{Y}_7 + S_2(-S_6s_{12}s_{23}\bar{X}_7 - S_7s_{12}s_{23}c_{67}\bar{X}_7 - a_{12}c_{12}s_{23}\bar{Y}_7 - a_{23}s_{12}c_{23}\bar{Y}_7 \\
& + a_{67}s_{12}s_{23}Z_7 - a_{71}s_{12}s_{23}Z'_7) + S_7a_{12}s_{23}s_{67} - a_{12}a_{71}s_{23}X'_7, \\
L_{16} = {} & S_1(-S_6s_{12}c_{23}c_7 - S_7s_{12}c_{23}c_{67}c_7 - a_{12}c_{12}c_{23}X'_7 + a_{23}s_{12}s_{23}X'_7 + a_{67}s_{12}c_{23}X_7) \\
& - S_2a_{12}c_{23}X'_7 + S_6(-S_7s_{12}c_{23}c_{71}c_7 - a_{12}c_{12}c_{23}\bar{X}'_7 + a_{23}s_{12}s_{23}\bar{X}'_7 + a_{71}s_{12}c_{23}\bar{X}_7) \\
& + S_7(-a_{12}c_{12}c_{23}c_{71}X'_7 + a_{23}s_{12}s_{23}c_{71}X'_7 + a_{67}s_{12}c_{23}c_{71}X_7 + a_{71}s_{12}c_{23}c_{67}\bar{X}_7) \\
& + a_{12}(a_{23}c_{12}s_{23}Z'_7 + a_{67}c_{12}c_{23}Y_7 + a_{71}c_{12}c_{23}\bar{Y}_7) + a_{23}(-a_{67}s_{12}s_{23}Y_7 \\
& - a_{71}s_{12}s_{23}\bar{Y}_7) - a_{67}a_{71}s_{12}c_{23}Z_7 + (K_2 - S_2^2)s_{12}c_{23}Z'_7 - S_7^2s_{12}c_{23}s_{67}s_{71},
\end{aligned}$$

$$\begin{aligned}
L_{17} = {} & S_1\big(S_6 s_{12} c_{23} \bar{X}'_7 + S_7 s_{12} c_{23} c_{71} X'_7 + a_{12} c_{12} c_{23} Z'_7 - a_{23} s_{12} s_{23} Z'_7 - a_{67} s_{12} c_{23} Y_7 \\
& - a_{71} s_{12} c_{23} \bar{Y}_7\big) + S_2 a_{12} c_{23} Z'_7 + S_6(S_7 s_{12} c_{23} s_7 - a_{12} c_{12} c_{23} c_7 + a_{23} s_{12} s_{23} c_7) \\
& + S_7(-a_{12} c_{12} c_{23} c_{67} c_7 + a_{23} s_{12} s_{23} c_{67} c_7 + a_{67} s_{12} c_{23} s_{67} c_7 + a_{71} s_{12} c_{23} s_{67}) \\
& + a_{12}\big(a_{23} c_{12} s_{23} X'_7 + a_{67} c_{12} c_{23} X_7\big) - a_{23} a_{67} s_{12} s_{23} X_7 + \big(K_2 - S_2^2 - a_{71}^2\big) s_{12} c_{23} X'_7, \\
L_{18} = {} & S_1\big(-S_2 c_{23} \bar{Y}_7 + S_7 c_{12} c_{23} s_{67} - a_{71} c_{12} c_{23} X'_7\big) + S_2\big(S_7 c_{23} s_{67} - a_{71} c_{23} X'_7\big) \\
& + S_6\big(-S_7 c_{12} c_{23} s_{71} c_7 + a_{12} s_{12} c_{23} \bar{X}_7 + a_{23} c_{12} s_{23} \bar{X}_7 - a_{71} c_{12} c_{23} \bar{X}'_7\big) \\
& + S_7\big(a_{12} s_{12} c_{23} c_{67} \bar{X}_7 + a_{23} c_{12} s_{23} c_{67} \bar{X}_7 + a_{67} c_{12} c_{23} s_{71} X_7 - a_{71} c_{12} c_{23} c_{71} X'_7\big) \\
& + a_{12}\big(-a_{23} s_{12} s_{23} \bar{Y}_7 - a_{67} s_{12} c_{23} Z_7 + a_{71} s_{12} c_{23} Z'_7\big) + a_{23}\big(-a_{67} c_{12} s_{23} Z_7 \\
& + a_{71} c_{12} s_{23} Z'_7\big) + a_{67} a_{71} c_{12} c_{23} Y_7 + \big(K_2 - S_1^2 - S_2^2\big) c_{12} c_{23} \bar{Y}_7 + S_7^2 c_{12} c_{23} s_{67} c_{71},
\end{aligned} \tag{10.274}$$

$$\begin{aligned}
L_{19} &= -S_3 a_{34} - S_4 a_{34} c_{34} + S_5(-a_{34} c_{34} c_{45} + a_{45} s_{34} s_{45}), \\
L_{20} &= S_4 S_5 s_{34} + a_{34} a_{45} c_{34} s_{45} + \big(K_1 - S_3^2 - a_{56}^2\big) s_{34} c_{45}, \\
L_{21} &= S_5 a_{56} s_{34} s_{45}, \\
L_{22} &= S_4 S_5 s_{34} c_{45} + \big(K_1 - S_3^2 - a_{45}^2 - a_{56}^2\big) s_{34}, \\
L_{23} &= S_3 a_{34} c_{45} + S_4(a_{34} c_{34} c_{45} - a_{45} s_{34} s_{45}) + S_5 a_{34} c_{34}, \\
L_{24} &= -a_{45} a_{56} s_{34}, \\
L_{25} &= -S_3 a_{45} - S_4 a_{45} c_{34} + S_5(a_{34} s_{34} s_{45} - a_{45} c_{34} c_{45}), \\
L_{26} &= -S_3 S_4 s_{45} + a_{34} a_{45} s_{34} c_{45} + \big(K_1 - S_3^2 - S_4^2 - a_{56}^2\big) c_{34} s_{45}, \\
L_{27} &= -S_3 a_{56} - S_4 a_{56} c_{34} - S_5 a_{56} c_{34} c_{45},
\end{aligned} \tag{10.275}$$

$$\begin{aligned}
L_{28} = {} & S_4 S_5 s_{34} c_{56} + a_{34}(a_{45} c_{34} s_{45} c_{56} + a_{56} c_{34} c_{45} s_{56}) - a_{45} a_{56} s_{34} s_{45} s_{56} \\
& + \big(K_1 - S_3^2\big) s_{34} c_{45} c_{56}, \\
L_{29} = {} & S_3 a_{34} c_{56} + S_4(a_{34} c_{34} c_{56} - a_{56} s_{34} s_{56}) + S_5(a_{34} c_{34} c_{45} c_{56} \\
& - a_{45} s_{34} s_{45} c_{56} - a_{56} s_{34} c_{45} s_{56}), \\
L_{30} = {} & a_{34}(a_{45} c_{34} c_{45} s_{56} + a_{56} c_{34} s_{45} c_{56}) + a_{45} a_{56} s_{34} c_{45} c_{56} \\
& + (-K_1 + S_3^2 + S_5^2) s_{34} s_{45} s_{56}, \\
L_{31} = {} & S_3 a_{34} c_{45} c_{56} + S_4(a_{34} c_{34} c_{45} c_{56} - a_{45} s_{34} s_{45} c_{56} - a_{56} s_{34} c_{45} s_{56}) \\
& + S_5(a_{34} c_{34} c_{56} - a_{56} s_{34} s_{56}), \\
L_{32} = {} & -S_4 S_5 s_{34} c_{45} c_{56} - a_{34} a_{56} c_{34} s_{56} + \big(-K_1 + S_3^2 + a_{45}^2\big) s_{34} c_{56}, \\
L_{33} = {} & -S_3 a_{34} s_{45} s_{56} + S_4(-a_{34} c_{34} s_{45} s_{56} - a_{45} s_{34} c_{45} s_{56} - a_{56} s_{34} s_{45} c_{56}) - S_5 a_{45} s_{34} s_{56}, \\
L_{34} = {} & -S_3 S_4 s_{45} c_{56} + a_{34}(a_{45} s_{34} c_{45} c_{56} - a_{56} s_{34} s_{45} s_{56}) + a_{45} a_{56} c_{34} c_{45} s_{56} \\
& + \big(K_1 - S_3^2 - S_4^2\big) c_{34} s_{45} c_{56}, \\
L_{35} = {} & S_3 a_{45} c_{56} + S_4 a_{45} c_{34} c_{56} + S_5(-a_{34} s_{34} s_{45} c_{56} + a_{45} c_{34} c_{45} c_{56} - a_{56} c_{34} s_{45} s_{56}), \\
L_{36} = {} & S_3(-S_4 c_{45} s_{56} - S_5 s_{56}) - S_4 S_5 c_{34} s_{56} + a_{34}(-a_{45} s_{34} s_{45} s_{56} + a_{56} s_{34} c_{45} c_{56}) \\
& + a_{45} a_{56} c_{34} s_{45} c_{56} + \big(K_1 - S_3^2 - S_4^2 - S_5^2\big) c_{34} c_{45} s_{56}.
\end{aligned} \tag{10.276}$$

10.2.8 Grouping of the four equations

Equations (10.23), (10.24), (10.44), and (10.45) are linear in the tan-half-angle of θ_6 and linear in the sines and cosines of the angles θ_1 and θ_2. They are also linear in the sines and cosines of the angles θ_4 and θ_5. It is necessary to eliminate these last two angles without increasing the degree of the sines and cosines of θ_1 and θ_2 in order to obtain four equations of the form of Eq. (10.1).

It is possible to express Eqs. (10.23), (10.24), (10.44), and (10.45) in the form

$$\begin{aligned}&(M_i c_2 + N_i s_2 + O_i)x_6 + (P_i c_2 + Q_i s_2 + R_i)\\&\quad = \left(M'_i c_4 + N'_i s_4 + O'_i\right)x_6 + \left(P'_i c_4 + Q'_i s_4 + R'_i\right),\end{aligned} \tag{10.277}$$

where $i = 1 \ldots 4$ and

$$\begin{aligned}M_i &= M_{i,1}c_1 + M_{i,2}s_1 + M_{i,3},\\N_i &= N_{i,1}c_1 + N_{i,2}s_1 + N_{i,3},\\O_i &= O_{i,1}c_1 + O_{i,2}s_1 + O_{i,3},\\P_i &= P_{i,1}c_1 + P_{i,2}s_1 + P_{i,3},\\Q_i &= Q_{i,1}c_1 + Q_{i,2}s_1 + Q_{i,3},\\R_i &= R_{i,1}c_1 + R_{i,2}s_1 + R_{i,3}\end{aligned} \tag{10.278}$$

and

$$\begin{aligned}M'_i &= M'_{i,1}c_5 + M'_{i,2}s_5 + M'_{i,3},\\N'_i &= N'_{i,1}c_5 + N'_{i,2}s_5 + N'_{i,3},\\O'_i &= O'_{i,1}c_5 + O'_{i,2}s_5,\\P'_i &= P'_{i,1}c_5 + P'_{i,2}s_5 + P'_{i,3},\\Q'_i &= Q'_{i,1}c_5 + Q'_{i,2}s_5 + Q'_{i,3},\\R'_i &= R'_{i,1}c_5 + R'_{i,2}s_5.\end{aligned} \tag{10.279}$$

Note that all of the coefficients, that is, $M_{i,1}$ through $R'_{i,2}$, can be numerically evaluated in terms of the given constant mechanism parameters.

The expansions of the coefficients are given as follows:

$$\begin{aligned}&M_{1,1} = a_{23}X'_7, && M_{1,2} = -a_{23}Z'_7, && M_{1,3} = 0,\\&N_{1,1} = -a_{23}c_{12}Z'_7, && N_{1,2} = -a_{23}c_{12}X'_7, && N_{1,3} = a_{23}s_{12}\bar{Y}_7,\end{aligned}$$

$$\begin{aligned}O_{1,1} &= S_2 s_{12} Z'_7 + a_{12}X'_7,\\O_{1,2} &= S_2 s_{12} X'_7 - a_{12}Z'_7,\\O_{1,3} &= S_1\bar{Y}_7 + S_2 c_{12}\bar{Y}_7 - S_3 c_{34}c_{45}s_{56} - S_4 c_{45}s_{56} - S_5 s_{56} - S_7 s_{67} + a_{71}X'_7,\end{aligned}$$

$$\begin{aligned}&P_{1,1} = -a_{23}c_7, && P_{1,2} = a_{23}\bar{X}'_7, && P_{1,3} = 0,\\&Q_{1,1} = a_{23}c_{12}\bar{X}'_7, && Q_{1,2} = a_{23}c_{12}c_7, && Q_{1,3} = -a_{23}s_{12}\bar{X}_7,\end{aligned}$$

$$\begin{aligned}R_{1,1} &= -S_2 s_{12}\bar{X}'_7 - a_{12}c_7,\\R_{1,2} &= -S_2 s_{12}c_7 + a_{12}\bar{X}'_7,\end{aligned}$$

$$
\begin{aligned}
&R_{1,3} = -S_1\bar{X}_7 - S_2c_{12}\bar{X}_7 - a_{56} - a_{67} - a_{71}c_7,\\
&M'_{1,1} = S_3s_{34}c_{45}c_{56}, \quad M'_{1,2} = -a_{34}c_{56}, \quad M'_{1,3} = -S_3s_{34}s_{45}s_{56},\\
&N'_{1,1} = -a_{34}c_{45}c_{56}, \quad N'_{1,2} = -S_3s_{34}c_{56}, \quad N'_{1,3} = a_{34}s_{45}s_{56},\\
&O'_{1,1} = S_3c_{34}s_{45}c_{56} + S_4s_{45}c_{56} \quad O'_{1,2} = -a_{45}c_{56},\\
&P'_{1,1} = a_{34}, \quad P'_{1,2} = S_3s_{34}c_{45}, \quad P'_{1,3} = 0,\\
&Q'_{1,1} = S_3s_{34}, \quad Q'_{1,2} = -a_{34}c_{45}, \quad Q'_{1,3} = 0,\\
&R'_{1,1} = a_{45}, \quad R'_{1,2} = S_3c_{34}s_{45} + S_4s_{45},
\end{aligned}
\tag{10.280}
$$

$$
\begin{aligned}
&M_{2,1} = a_{23}c_7, \quad M_{2,2} = -a_{23}\bar{X}'_7, \quad M_{2,3} = 0,\\
&N_{2,1} = -a_{23}c_{12}\bar{X}'_7, \quad N_{2,2} = -a_{23}c_{12}c_7, \quad N_{2,3} = a_{23}s_{12}\bar{X}_7,\\
&O_{2,1} = S_2s_{12}\bar{X}'_7 + a_{12}c_7,\\
&O_{2,2} = S_2s_{12}c_7 - a_{12}\bar{X}'_7,\\
&O_{2,3} = S_1\bar{X}_7 + S_2c_{12}\bar{X}_7 - a_{56} + a_{67} + a_{71}c_7,\\
&P_{2,1} = a_{23}X'_7, \quad P_{2,2} = -a_{23}Z'_7, \quad P_{2,3} = 0,\\
&Q_{2,1} = -a_{23}c_{12}Z'_7, \quad Q_{2,2} = -a_{23}c_{12}X'_7, \quad Q_{2,3} = a_{23}s_{12}\bar{Y}_7,\\
&R_{2,1} = S_2s_{12}Z'_7 + a_{12}X'_7,\\
&R_{2,2} = S_2s_{12}X'_7 - a_{12}Z'_7,\\
&R_{2,3} = S_1\bar{Y}_7 + S_2c_{12}\bar{Y}_7 + S_3c_{34}c_{45}s_{56} + S_4c_{45}s_{56} + S_5s_{56} - S_7s_{67} + a_{71}X'_7,\\
&M'_{2,1} = a_{34}, \quad M'_{2,2} = S_3s_{34}c_{45}, \quad M'_{2,3} = 0,\\
&N'_{2,1} = S_3s_{34}, \quad N'_{2,2} = -a_{34}c_{45}, \quad N'_{2,3} = 0,\\
&O'_{2,1} = a_{45}, \quad O'_{2,2} = S_3c_{34}s_{45} + S_4s_{45},\\
&P'_{2,1} = -S_3s_{34}c_{45}c_{56}, \quad P'_{2,2} = a_{34}c_{56}, \quad P'_{2,3} = S_3s_{34}s_{45}s_{56},\\
&Q'_{2,1} = a_{34}c_{45}c_{56}, \quad Q'_{2,2} = S_3s_{34}c_{56}, \quad Q'_{2,3} = -a_{34}s_{45}s_{56},\\
&R'_{2,1} = -S_3c_{34}s_{45}c_{56} - S_4s_{45}c_{56},\\
&R'_{2,2} = a_{45}c_{56},
\end{aligned}
\tag{10.281}
$$

$$
\begin{aligned}
&M_{3,1} = L_{10}, \quad M_{3,2} = L_{11}, \quad M_{3,3} = L_{12},\\
&N_{3,1} = L_{13}, \quad N_{3,2} = L_{14}, \quad N_{3,3} = L_{15},\\
&O_{3,1} = L_{16}, \quad O_{3,2} = L_{17}, \quad O_{3,3} = L_{18} + L_{36},\\
&P_{3,1} = -L_1, \quad P_{3,2} = -L_2, \quad P_{3,3} = -L_3,\\
&Q_{3,1} = -L_4, \quad Q_{3,2} = -L_5, \quad Q_{3,3} = -L_6,\\
&R_{3,1} = -L_7, \quad R_{3,2} = -L_8, \quad R_{3,3} = -L_9 + L_{27},\\
&M'_{3,1} = -L_{28}, \quad M'_{3,2} = -L_{29}, \quad M'_{3,3} = -L_{30},\\
&N'_{3,1} = -L_{31}, \quad N'_{3,2} = -L_{32}, \quad N'_{3,3} = -L_{33},\\
&O'_{3,1} = -L_{34}, \quad O'_{3,2} = -L_{35},\\
&P'_{3,1} = -L_{19}, \quad P'_{3,2} = -L_{20}, \quad P'_{3,3} = -L_{21},\\
&Q'_{3,1} = -L_{22}, \quad Q'_{3,2} = -L_{23}, \quad Q'_{3,3} = -L_{24},\\
&R'_{3,1} = -L_{25}, \quad R'_{3,3} = -L_{26},
\end{aligned}
\tag{10.282}
$$

$$
\begin{aligned}
&M_{4,1} = L_1, \quad M_{4,2} = L_2, \quad M_{4,3} = L_3,\\
&N_{4,1} = L_4, \quad N_{4,2} = L_5, \quad N_{4,3} = L_6,\\
&O_{4,1} = L_7, \quad O_{4,2} = L_8, \quad O_{4,3} = L_9 + L_{27},\\
&P_{4,1} = L_{10}, \quad P_{4,2} = L_{11}, \quad P_{4,3} = L_{12},\\
&Q_{4,1} = L_{13}, \quad Q_{4,2} = L_{14}, \quad Q_{4,3} = L_{15},
\end{aligned}
$$

$$
\begin{array}{lll}
R_{4,1} = L_{16}, & R_{4,2} = L_{17}, & R_{4,3} = L_{18} - L_{36}, \\
M'_{4,1} = -L_{19}, & M'_{4,2} = -L_{20}, & M'_{4,3} = -L_{21}, \\
N'_{4,1} = -L_{22}, & N'_{4,2} = -L_{23}, & N'_{4,3} = -L_{24}, \\
O'_{4,1} = -L_{25}, & O'_{4,2} = -L_{26}, & \\
P'_{4,1} = L_{28}, & P'_{4,2} = L_{29}, & P'_{4,3} = L_{30}, \\
Q'_{4,1} = L_{31}, & Q'_{4,2} = L_{32}, & Q'_{4,3} = L_{33}, \\
R'_{4,1} = L_{34}, & R'_{4,2} = L_{35}. &
\end{array}
\tag{10.283}
$$

The four equations represented by Eq. (10.277) may be written in matrix format as follows:

$$
\mathbf{T}_1\mathbf{a} = \mathbf{T}_2\mathbf{b}, \tag{10.284}
$$

where

$$
\mathbf{T}_1 =
\begin{bmatrix}
M_{1,1} & M_{1,2} & M_{1,3} & N_{1,1} & N_{1,2} & N_{1,3} & O_{1,1} & O_{1,2} & O_{1,3} & P_{1,1} & P_{1,2} & P_{1,3} & Q_{1,1} & Q_{1,2} & Q_{1,3} & R_{1,1} & R_{1,2} & R_{1,3} \\
M_{2,1} & M_{2,2} & M_{2,3} & N_{2,1} & N_{2,2} & N_{2,3} & O_{2,1} & O_{2,2} & O_{2,3} & P_{2,1} & P_{2,2} & P_{2,3} & Q_{2,1} & Q_{2,2} & Q_{2,3} & R_{2,1} & R_{2,2} & R_{2,3} \\
M_{3,1} & M_{3,2} & M_{3,3} & N_{3,1} & N_{3,2} & N_{3,3} & O_{3,1} & O_{3,2} & O_{3,3} & P_{3,1} & P_{3,2} & P_{3,3} & Q_{3,1} & Q_{3,2} & Q_{3,3} & R_{3,1} & R_{3,2} & R_{3,3} \\
M_{4,1} & M_{4,2} & M_{4,3} & N_{4,1} & N_{4,2} & N_{4,3} & O_{4,1} & O_{4,2} & O_{4,3} & P_{4,1} & P_{4,2} & P_{4,3} & Q_{4,1} & Q_{4,2} & Q_{4,3} & R_{4,1} & R_{4,2} & R_{4,3}
\end{bmatrix},
\tag{10.285}
$$

$$
\mathbf{a} = [c_1c_2x_6,\ s_1c_2x_6,\ c_2x_6,\ c_1s_2x_6,\ s_1s_2x_6,\ s_2x_6,\ c_1x_6,\ s_1x_6,\ x_6,\ c_1c_2,\ s_1c_2,\ c_2,\ c_1s_2,\ s_1s_2,\ s_2,\ c_1,\ s_1,\ 1]^T, \tag{10.286}
$$

$$
\mathbf{T}_2 =
\begin{bmatrix}
M'_{1,1} & M'_{1,2} & M'_{1,3} & N'_{1,1} & N'_{1,2} & N'_{1,3} & O'_{1,1} & O'_{1,2} & P'_{1,1} & P'_{1,2} & P'_{1,3} & Q'_{1,1} & Q'_{1,2} & Q'_{1,3} & R'_{1,1} & R'_{1,2} \\
M'_{2,1} & M'_{2,2} & M'_{2,3} & N'_{2,1} & N'_{2,2} & N'_{2,1} & O'_{2,1} & O'_{2,2} & P'_{2,1} & P'_{2,2} & P'_{2,3} & Q'_{2,1} & Q'_{2,2} & Q'_{2,3} & R'_{2,1} & R'_{2,2} \\
M'_{3,1} & M'_{3,2} & M'_{3,3} & N'_{3,1} & N'_{3,2} & N'_{3,3} & O'_{3,1} & O'_{3,2} & P'_{3,1} & P'_{3,2} & P'_{3,3} & Q'_{3,1} & Q'_{3,2} & Q'_{3,3} & R'_{3,1} & R'_{3,2} \\
M'_{4,1} & M'_{4,2} & M'_{4,3} & N'_{4,1} & N'_{4,2} & N'_{4,3} & O'_{4,1} & O'_{4,2} & P'_{4,1} & P'_{4,2} & P'_{4,3} & Q'_{4,1} & Q'_{4,2} & Q'_{4,3} & R'_{4,1} & R'_{4,2}
\end{bmatrix},
\tag{10.287}
$$

$$
\mathbf{b} = [c_4c_5x_6,\ c_4s_5x_6,\ c_4x_6,\ s_4c_5x_6,\ s_4s_5x_6,\ s_4x_6,\ c_5x_6,\ s_5x_6,\ c_4c_5,\ c_4s_5,\ c_4,\ s_4c_5,\ s_4s_5,\ s_4,\ c_5,\ s_5]^T. \tag{10.288}
$$

Note that all the elements of matrices $\mathbf{T}_1$ and $\mathbf{T}_2$ are known in terms of the given mechanism parameters.

It is necessary to express vector **b** in terms of the constant mechanism parameters, the output angle θ_1, the angle θ_2, and the tan-half-angle of θ_6. Once this is accomplished, tan-half-angle substitutions will be made for the sines and cosines of the variable joint parameters, and the matrix equation, Eq. (10.284), can be regrouped to represent four equations of the form of Eq. (10.1). Such an expression for **b** will be obtained in the next section.

10.2.9 Elimination of θ_4 and θ_5 to obtain input–output equation

An expression for **b** in Eq. (10.284) will be obtained by generating sixteen additional equations that are (i) linear in the sines and cosine of θ_4 and θ_5, (ii) linear in the variable x_6, and (iii) linear in the sines and cosines of the variables θ_1 and θ_2. The solution of these sixteen linear equations will then be substituted into Eq. (10.284), the result of which will, upon regrouping, be four equations of the form of Eq. (10.1). The sixteen equations are generated as follows:

(i) Half-Angle Law

The following subsidiary half-angle law may be written for a spherical heptagon:

$$x_6(X_{217} + X_{45}) + (Y_{217} + Y_{45}) = 0. \tag{10.289}$$

This equation is regrouped as

$$X_{45}x_6 + Y_{45} = -X_{217}x_6 - Y_{217}. \tag{10.290}$$

Expanding X_{45} and Y_{45} gives

$$(X_4c_5 - Y_4s_5)x_6 + c_{56}(X_4s_5 + Y_4c_5) - s_{56}Z_4 = -X_{217}x_6 - Y_{217}. \tag{10.291}$$

Expanding X_4, Y_4, and Z_4 and regrouping terms gives

$$\begin{aligned}&(s_{34}c_{45})c_4s_5x_6 + (s_{34})s_4c_5x_6 + (c_{34}s_{45})s_5x_6 + (-s_{34}c_{45}c_{56})c_4c_5 + (s_{34}s_{45}s_{56})c_4\\ &\quad +(s_{34}c_{56})s_4s_5 + (-c_{34}s_{45}c_{56})c_5 = -X_{217}x_6 - Y_{217} + c_{34}c_{45}s_{56}.\end{aligned} \tag{10.292}$$

(ii) Half-Angle Law

The following subsidiary half-angle law may be written for a spherical heptagon:

$$x_6(Y_{217} - Y_{45}) - (X_{217} - X_{45}) = 0. \tag{10.293}$$

This equation is regrouped as

$$-Y_{45}x_6 + X_{45} = -Y_{217}x_6 + X_{217}. \tag{10.294}$$

Expanding X_{45} and Y_{45} gives

$$-[c_{56}(X_4s_5 + Y_4c_5) - s_{56}Z_4]x_6 + (X_4c_5 - Y_4s_5) = -Y_{217}x_6 + X_{217}. \tag{10.295}$$

Expanding X_4, Y_4, and Z_4 and regrouping terms gives

$$\begin{aligned}&(s_{34}c_{45}c_{56})c_4c_5x_6 + (-s_{34}s_{45}s_{56})c_4x_6 + (-s_{34}c_{56})s_4s_5x_6 + (c_{34}s_{45}c_{56})c_5x_6\\ &\quad + (s_{34}c_{45})c_4s_5 + (s_{34})s_4c_5 + (c_{34}s_{45})s_5 = (-Y_{217} - c_{34}c_{45}s_{56})x_6 + X_{217}.\end{aligned} \tag{10.296}$$

(iii) and (iv) Secondary Half-Angle Law

The following subsidiary sine and sine–cosine laws may be written for a spherical heptagon:

$$X_{2176} = X_{45}, \tag{10.297}$$

$$X^*_{2176} = -Y_{45}. \tag{10.298}$$

Substituting dual angles into these equations gives

$$X_{02176} = X_{045}, \tag{10.299}$$

$$X^*_{02176} = -Y_{045}. \tag{10.300}$$

Equations (10.299) and (10.300) may be expanded as

$$-S_6(X_{217}s_6 + Y_{217}c_6) + X_{0217}c_6 - Y_{0217}s_6 = X_{045}, \quad (10.301)$$

$$S_6(X_{217}c_6 - Y_{217}s_6) + X_{0217}s_6 + Y_{0217}c_6 = -Y_{045}. \quad (10.302)$$

The following trigonometric identities were introduced in Eqs. (9.108) and (9.109):

$$s_6 - c_6x_6 = x_6, \quad (10.303)$$

$$c_6 + s_6x_6 = 1. \quad (10.304)$$

Adding Eq. (10.301) to x_6 times Eq. (10.302) and using the above trigonometric identities yields

$$-S_6(X_{217}x_6 + Y_{217}) + X_{0217} - Y_{0217}x_6 = X_{045} - Y_{045}x_6. \quad (10.305)$$

Subtracting Eq. (10.301) times x_6 from Eq. (10.302) and using the trigonometric identities gives

$$S_6(X_{217} - Y_{217}x_6) + X_{0217}x_6 + Y_{0217} = -(X_{045}x_6 + Y_{045}). \quad (10.306)$$

Equations (10.305) and (10.306) may be rearranged as

$$-Y_{045}x_6 + X_{045} = (-Y_{0217} - S_6X_{217})x_6 + (X_{0217} - S_6Y_{217}), \quad (10.307)$$

$$X_{045}x_6 + Y_{045} = (-X_{0217} + S_6Y_{217})x_6 - (Y_{0217} + S_6X_{217}). \quad (10.308)$$

The terms X_{045} and Y_{045} may be expanded as

$$X_{045} = -S_5X_{45}^* + X_{04}c_5 - Y_{04}s_5, \quad (10.309)$$

$$Y_{045} = S_5c_{56}X_{45} - a_{56}Z_{45} + c_{56}(X_{04}s_5 + Y_{04}c_5) - s_{56}Z_{04}, \quad (10.310)$$

where

$$X_{04} = S_4s_{34}c_4 + a_{34}c_{34}s_4, \quad (10.311)$$

$$Y_{04} = S_4c_{45}X_4 - a_{45}Z_4 + a_{34}(s_{45}s_{34} - c_{45}c_{34}c_4), \quad (10.312)$$

$$Z_{04} = S_4s_{45}X_4 + a_{34}\bar{Y}_4 + a_{45}Y_4. \quad (10.313)$$

Substituting these expressions into Eqs. (10.307) and (10.308) gives the third and fourth of the sixteen equations as

$$\begin{aligned} &K_3c_4c_5x_6 + K_4c_4s_5x_6 + K_5c_4x_6 + K_6s_4c_5x_6 + K_7s_4s_5x_6 + K_8s_4x_6 + K_9c_5x_6 \\ &\quad + K_{10}s_5x_6 + K_{11}c_4c_5 + K_{12}c_4s_5 + K_{13}s_4c_5 + K_{14}s_4s_5 + K_{15}c_5 \\ &\quad + K_{16}s_5 = (-Y_{0217} - S_6X_{217} + K_{17})x_6 + (X_{0217} - S_6Y_{217}), \quad (10.314) \\ &K_{11}c_4c_5x_6 + K_{12}c_4s_5x_6 + K_{13}s_4c_5x_6 + K_{14}s_4s_5x_6 + K_{15}c_5x_6 + K_{16}s_5x_6 \\ &\quad - K_3c_4c_5 - K_4c_4s_5 - K_5c_4 - K_6s_4c_5 - K_7s_4s_5 - K_8s_4 - K_9c_5 \\ &\quad - K_{10}s_5 = (-X_{0217} + S_6Y_{217})x_6 - (Y_{0217} + S_6X_{217} + K_{17}), \quad (10.315) \end{aligned}$$

where

$$
\begin{aligned}
K_3 &= (a_{34}c_{34}c_{45}c_{56} - a_{45}s_{34}s_{45}c_{56} - a_{56}s_{34}c_{45}s_{56}),\\
K_4 &= (-S_4 s_{34}c_{56} - S_5 s_{34}c_{45}c_{56}),\\
K_5 &= (-a_{34}c_{34}s_{45}s_{56} - a_{45}s_{34}c_{45}s_{56} - a_{56}s_{34}s_{45}c_{56}),\\
K_6 &= (-S_4 s_{34}c_{45}c_{56} - S_5 s_{34}c_{56}),\\
K_7 &= (-a_{34}c_{34}c_{56} + a_{56}s_{34}s_{56}),\\
K_8 &= (S_4 s_{34}s_{45}s_{56}),\\
K_9 &= (-a_{34}s_{34}s_{45}c_{56} + a_{45}c_{34}c_{45}c_{56} - a_{56}c_{34}s_{45}s_{56}),\\
K_{10} &= (-S_5 c_{34}s_{45}c_{56}),\\
K_{11} &= (S_4 s_{34} + S_5 s_{34}c_{45}),\\
K_{12} &= (a_{34}c_{34}c_{45} - a_{45}s_{34}s_{45}),\\
K_{13} &= (a_{34}c_{34}),\\
K_{14} &= (-S_4 s_{34}c_{45} - S_5 s_{34}),\\
K_{15} &= (S_5 c_{34}s_{45}),\\
K_{16} &= (-a_{34}s_{34}s_{45} + a_{45}c_{34}c_{45}),\\
K_{17} &= (a_{34}s_{34}c_{45}s_{56} + a_{45}c_{34}s_{45}s_{56} - a_{56}c_{34}c_{45}c_{56}).
\end{aligned}
\tag{10.316}
$$

(v) Subsidiary Cosine Law

A subsidiary cosine law for a spherical heptagon is written as

$$Z_{45} = Z_{217}. \tag{10.317}$$

Expanding Z_{45} gives

$$s_{56}(X_4 s_5 + Y_4 c_5) + c_{56}Z_4 = Z_{217}. \tag{10.318}$$

Expanding X_4, Y_4, and Z_4 and rearranging terms gives

$$(-s_{34}c_{45}s_{56})c_4c_5 + (-s_{34}s_{45}c_{56})c_4 + (s_{34}s_{56})s_4s_5 + (-c_{34}s_{45}s_{56})c_5 = Z_{217} - c_{34}c_{45}c_{56}. \tag{10.319}$$

(vi) Secondary Cosine Law

Dual angles may be substituted into Eq. (10.317) to give

$$Z_{045} = Z_{0217}. \tag{10.320}$$

The term Z_{045} may be expanded as

$$Z_{045} = a_{56}Y_{45} + S_5 s_{56}X_{45} + a_{45}(-s_{56}c_5Z_4 + c_{56}Y_4) + S_4 s_{34}X_{54} + a_{34}Y_{54}. \tag{10.321}$$

Substituting this expression into Eq. (10.320) and then expanding the terms X_{45}, Y_{45}, X_{54}, Y_{54}, X_4, Y_4, Z_4, $\bar{X}_4$, and $\bar{Y}_4$ and rearranging gives

$$
\begin{aligned}
&K_{18}c_4c_5 + K_{19}c_4s_5 + K_{20}c_4 + K_{21}s_4c_5\\
&\quad + K_{22}s_4s_5 + K_{23}s_4 + K_{24}c_5 + K_{25}s_5 = Z_{0217} + K_{26},
\end{aligned}
\tag{10.322}
$$

where

$$\begin{aligned}
K_{18} &= (-a_{34}c_{34}c_{45}s_{56} + a_{45}s_{34}s_{45}s_{56} - a_{56}s_{34}c_{45}c_{56}),\\
K_{19} &= (S_4 s_{34}s_{56} + S_5 s_{34}c_{45}s_{56}),\\
K_{20} &= (-a_{34}c_{34}s_{45}c_{56} - a_{45}s_{34}c_{45}c_{56} + a_{56}s_{34}s_{45}s_{56}),\\
K_{21} &= (S_4 s_{34}c_{45}s_{56} + S_5 s_{34}s_{56}),\\
K_{22} &= (a_{34}c_{34}s_{56} + a_{56}s_{34}c_{56}),\\
K_{23} &= (S_4 s_{34}s_{45}c_{56}),\\
K_{24} &= (a_{34}s_{34}s_{45}s_{56} - a_{45}c_{34}c_{45}s_{56} - a_{56}c_{34}s_{45}c_{56}),\\
K_{25} &= (S_5 c_{34}s_{45}s_{56}),\\
K_{26} &= (a_{34}s_{34}c_{45}c_{56} + a_{45}c_{34}s_{45}c_{56} + a_{56}c_{34}c_{45}s_{56}).
\end{aligned} \tag{10.323}$$

(vii) Projection of Vector Loop Equation

The vector loop equation can be written as the sum of the vectors $\mathbf{R}^{6,23}$ and $\mathbf{R}^{3,56}$. Projecting these terms onto the vector $\mathbf{S}_3$ yields

$$\mathbf{R}^{6,23} \cdot \mathbf{S}_3 = -\mathbf{R}^{3,56} \cdot \mathbf{S}_3. \tag{10.324}$$

The left side of this equation may be written as

$$\mathbf{R}^{6,23} \cdot \mathbf{S}_3 = (S_6\mathbf{S}_6 + a_{67}\mathbf{a}_{67} + S_7\mathbf{S}_7 + a_{71}\mathbf{a}_{71} + S_1\mathbf{S}_1 + a_{12}\mathbf{a}_{12} + S_2\mathbf{S}_2 + a_{23}\mathbf{a}_{23}) \cdot \mathbf{S}_3. \tag{10.325}$$

Evaluating the scalar products gives

$$\mathbf{R}^{6,23} \cdot \mathbf{S}_3 = H_5, \tag{10.326}$$

where

$$H_5 = S_6 Z_{712} + a_{67}U_{7123} + S_7 Z_{12} + a_{71}U_{123} + S_1 Z_2 + a_{12}U_{23} + S_2 c_{23}. \tag{10.327}$$

The right side of Eq. (10.324) may be written as

$$-\mathbf{R}^{3,56} \cdot \mathbf{S}_3 = -(S_3\mathbf{S}_3 + a_{34}\mathbf{a}_{34} + S_4\mathbf{S}_4 + a_{45}\mathbf{a}_{45} + S_5\mathbf{S}_5 + a_{56}\mathbf{a}_{56}) \cdot \mathbf{S}_3. \tag{10.328}$$

Evaluating the scalar products gives

$$-\mathbf{R}^{3,56} \cdot \mathbf{S}_3 = -(S_3 + S_4 c_{34} + a_{45}U_{43} + S_5 Z_4 + a_{56}U_{543}). \tag{10.329}$$

Substituting Eqs. (10.326) and (10.329) into Eq. (10.324) and then expanding U_{43}, Z_4, and U_{543} and rearranging gives

$$\begin{aligned}
&(-a_{56}s_{34}c_{45})c_4 s_5 + (S_5 s_{34}s_{45})c_4 + (-a_{56}s_{34})s_4 c_5\\
&\quad + (-a_{45}s_{34})s_4 + (-a_{56}c_{34}s_{45})s_5 = H_5 + S_3 + S_4 c_{34} + S_5 c_{34}c_{45}.
\end{aligned} \tag{10.330}$$

(viii) Projection of Vector Loop Equation

The vector loop equation can be written as the sum of the vectors $\mathbf{R}^{6,23}$ and $\mathbf{R}^{3,56}$. Projecting these terms onto the vector $\mathbf{S}_6$ yields

$$\mathbf{R}^{6,23} \cdot \mathbf{S}_6 = -\mathbf{R}^{3,56} \cdot \mathbf{S}_6. \tag{10.331}$$

The left side of this equation may be written as

$$\mathbf{R}^{6,23} \cdot \mathbf{S}_6 = (S_6\mathbf{S}_6 + a_{67}\mathbf{a}_{67} + S_7\mathbf{S}_7 + a_{71}\mathbf{a}_{71} + S_1\mathbf{S}_1 + a_{12}\mathbf{a}_{12} + S_2\mathbf{S}_2 + a_{23}\mathbf{a}_{23}) \cdot \mathbf{S}_6. \tag{10.332}$$

Evaluating the scalar products gives

$$\mathbf{R}^{6,23} \cdot \mathbf{S}_6 = H_6, \tag{10.333}$$

where

$$H_6 = S_6 + S_7c_{67} + a_{71}U_{76} + S_1Z_7 + a_{12}U_{176} + S_2Z_{17} + a_{23}U_{2176}. \tag{10.334}$$

The right side of Eq. (10.331) may be written as

$$-\mathbf{R}^{3,56} \cdot \mathbf{S}_6 = -(S_3\mathbf{S}_3 + a_{34}\mathbf{a}_{34} + S_4\mathbf{S}_4 + a_{45}\mathbf{a}_{45} + S_5\mathbf{S}_5 + a_{56}\mathbf{a}_{56}) \cdot \mathbf{S}_6. \tag{10.335}$$

Evaluating the scalar products gives

$$-\mathbf{R}^{3,56} \cdot \mathbf{S}_6 = -(S_3Z_{45} + a_{34}U_{456} + S_4Z_5 + a_{45}U_{56} + S_5c_{56}). \tag{10.336}$$

Substituting Eqs. (10.333) and (10.336) into Eq. (10.331) and then expanding Z_{45}, U_{456}, Z_5, and U_{56} and rearranging gives

$$\begin{aligned}&(S_3s_{34}c_{45}s_{56})c_4c_5 + (-a_{34}s_{56})c_4s_5 + (S_3s_{34}s_{45}c_{56})c_4 + (-a_{34}c_{45}s_{56})s_4c_5\\&\quad + (-S_3s_{34}s_{56})s_4s_5 + (-a_{34}s_{45}c_{56})s_4 + K_{27}c_5 + (-a_{45}s_{56})s_5\\&\quad = H_6 + S_3c_{34}c_{45}c_{56} + S_4c_{45}c_{56} + S_5c_{56},\end{aligned} \tag{10.337}$$

where

$$K_{27} = (S_4s_{45}s_{56} + S_3c_{34}s_{45}s_{56}). \tag{10.338}$$

(ix) Self-Scalar Product

The vector loop equation can be written as the sum of the vectors $\mathbf{R}^{6,23}$ and $\mathbf{R}^{3,56}$. Thus, it can be written that

$$\mathbf{R}^{6,23} = -\mathbf{R}^{3,56}. \tag{10.339}$$

Taking the self-scalar product of each side of this equation and dividing by two gives

$$\frac{1}{2}(\mathbf{R}^{6,23} \cdot \mathbf{R}^{6,23}) = \frac{1}{2}(\mathbf{R}^{3,56} \cdot \mathbf{R}^{3,56}). \tag{10.340}$$

The left side of this equation can be evaluated as

$$\frac{1}{2}(\mathbf{R}^{6,23} \cdot \mathbf{R}^{6,23}) = H_7, \tag{10.341}$$

where

$$H_7 = K_2 + S_6(S_7c_{67} + a_{71}X_7 + S_1Z_7 + a_{12}X_{71} + S_2Z_{71} + a_{23}X_{712}) + a_{67}(a_{71}c_7 + S_1\bar{X}_7 + a_{12}W_{17} + S_2X_{17} + a_{23}W_{217}) + S_7(S_1c_{71} + a_{12}X_1 + S_2Z_1 + a_{23}X_{12}) + a_{71}(a_{12}c_1 + S_2\bar{X}_1 + a_{23}W_{21}) + S_1(S_2c_{12} + a_{23}X_2) + a_{12}(a_{23}c_2) \quad (10.342)$$

and where K_2 has been previously defined in Eq. (10.163).

The right side of Eq. (10.340) can be expanded as

$$\frac{1}{2}(\mathbf{R}^{3,56} \cdot \mathbf{R}^{3,56}) = K_1 + S_3(S_4c_{34} + a_{45}X_4 + S_5Z_4 + a_{56}X_{45}) + a_{34}(a_{45}c_4 + S_5\bar{X}_4 + a_{56}W_{54}) + sS_4(S_5c_{45} + a_{56}X_5) + a_{45}(a_{56}c_5), \quad (10.343)$$

where K_1 is defined in Eq. (10.78).

Substituting Eqs. (10.341) and (10.343) into Eq. (10.340) and then substituting the definitions for the terms that contain θ_4 and θ_5 and regrouping gives

$$(a_{34}a_{56})c_4c_5 + (S_3a_{56}s_{34}c_{45})c_4s_5 + K_{28}c_4 + (S_3a_{56}s_{34})s_4c_5 + (-a_{34}a_{56}c_{45})s_4s_5 + K_{29}s_4 + (a_{45}a_{56})c_5 + K_{30}s_5 = H_7 - S_3S_4c_{34} - S_3S_5c_{34}c_{45} - S_4S_5c_{45} - K_1, \quad (10.344)$$

where

$$\begin{aligned} K_{28} &= (a_{34}a_{45} - S_3S_5s_{34}s_{45}), \\ K_{29} &= (a_{34}S_5s_{45} + S_3a_{45}s_{34}), \\ K_{30} &= (S_3a_{56}c_{34}s_{45} + S_4a_{56}s_{45}). \end{aligned} \quad (10.345)$$

(x) Projection of Self-Scalar Product

Because $\mathbf{R}^{6,23} = -\mathbf{R}^{3,56}$, it is possible to construct the following expression:

$$\frac{1}{2}(\mathbf{R}^{6,23} \cdot \mathbf{R}^{6,23})(\mathbf{S}_6 \cdot \mathbf{S}_3) - (\mathbf{R}^{6,23} \cdot \mathbf{S}_6)(\mathbf{R}^{6,23} \cdot \mathbf{S}_3) = \frac{1}{2}(\mathbf{R}^{3,56} \cdot \mathbf{R}^{3,56})(\mathbf{S}_6 \cdot \mathbf{S}_3) - (\mathbf{R}^{3,56} \cdot \mathbf{S}_6)(\mathbf{R}^{3,56} \cdot \mathbf{S}_3). \quad (10.346)$$

The left side of this equation will be designated as H_8, and the right side will be named J_8. Thus, it may be written that

$$H_8 = J_8, \quad (10.347)$$

where

$$H_8 = \frac{1}{2}(\mathbf{R}^{6,23} \cdot \mathbf{R}^{6,23})(\mathbf{S}_6 \cdot \mathbf{S}_3) - (\mathbf{R}^{6,23} \cdot \mathbf{S}_6)(\mathbf{R}^{6,23} \cdot \mathbf{S}_3) \quad (10.348)$$

and

$$J_8 = \frac{1}{2}(\mathbf{R}^{3,56} \cdot \mathbf{R}^{3,56})(\mathbf{S}_6 \cdot \mathbf{S}_3) - (\mathbf{R}^{3,56} \cdot \mathbf{S}_6)(\mathbf{R}^{3,56} \cdot \mathbf{S}_3). \quad (10.349)$$

The format of H_8 is the same as that of H_3 written in Eq. (10.38), with the term $\mathbf{a}_{67}$ now replaced by $\mathbf{S}_6$. Thus, an expression for H_8 may be obtained by rewriting Eq. (10.75), replacing each occurrence of $\mathbf{a}_{67}$ with $\mathbf{S}_6$. This yields

$$\begin{aligned}H_8 = {} & \frac{1}{2}(\mathbf{R}^{6,67}\cdot\mathbf{R}^{6,67}+\mathbf{R}^{7,71}\cdot\mathbf{R}^{7,71}+\mathbf{R}^{1,12}\cdot\mathbf{R}^{1,12}+\mathbf{R}^{2,23}\cdot\mathbf{R}^{2,23})(\mathbf{S}_3\cdot\mathbf{S}_6)\\ & -(\mathbf{R}^{6,67}\cdot\mathbf{S}_3)(\mathbf{R}^{6,67}\cdot\mathbf{S}_6)-(\mathbf{R}^{7,71}\cdot\mathbf{S}_3)(\mathbf{R}^{7,71}\cdot\mathbf{S}_6)\\ & +(\mathbf{R}^{6,67}\times\mathbf{S}_6)\cdot(\mathbf{R}^{7,71}\times\mathbf{S}_3)-(\mathbf{R}^{7,71}\cdot\mathbf{S}_3)(\mathbf{R}^{6,67}\cdot\mathbf{S}_6)\\ & -(\mathbf{R}^{1,12}\cdot\mathbf{S}_3)(\mathbf{R}^{1,12}\cdot\mathbf{S}_6)+(\mathbf{R}^{6,71}\times\mathbf{S}_6)\cdot(\mathbf{R}^{1,12}\times\mathbf{S}_3)\\ & -(\mathbf{R}^{1,12}\cdot\mathbf{S}_3)(\mathbf{R}^{6,71}\cdot\mathbf{S}_6)-(\mathbf{R}^{2,23}\cdot\mathbf{S}_3)(\mathbf{R}^{2,23}\cdot\mathbf{S}_6)\\ & +(\mathbf{R}^{6,12}\times\mathbf{S}_6)\cdot(\mathbf{R}^{2,23}\times\mathbf{S}_3)-(\mathbf{R}^{2,23}\cdot\mathbf{S}_3)(\mathbf{R}^{6,12}\cdot\mathbf{S}_6).\end{aligned}\tag{10.350}$$

The format of J_8 as listed in Eq. (10.349) is similar to the format of J_3 listed in Eq. (10.40) with the vector $\mathbf{S}_6$ replacing $\mathbf{a}_{56}$. An expression for J_8 may now be written based on Eq. (10.56) as follows:

$$\begin{aligned}J_8 = {} & \frac{1}{2}(\mathbf{R}^{3,34}\cdot\mathbf{R}^{3,34}+\mathbf{R}^{4,45}\cdot\mathbf{R}^{4,45}+\mathbf{R}^{5,56}\cdot\mathbf{R}^{5,56})(\mathbf{S}_3\cdot\mathbf{S}_6)-(\mathbf{R}^{3,34}\cdot\mathbf{S}_3)(\mathbf{R}^{3,34}\cdot\mathbf{S}_6)\\ & -(\mathbf{R}^{3,45}\cdot\mathbf{S}_3)(\mathbf{R}^{4,45}\cdot\mathbf{S}_6)-(\mathbf{R}^{3,56}\cdot\mathbf{S}_3)(\mathbf{R}^{5,56}\cdot\mathbf{S}_6)+(\mathbf{R}^{3,34}\times\mathbf{S}_3)\cdot(\mathbf{R}^{4,45}\times\mathbf{S}_6)\\ & +(\mathbf{R}^{3,45}\times\mathbf{S}_3)\cdot(\mathbf{R}^{5,56}\times\mathbf{S}_6).\end{aligned}\tag{10.351}$$

The following terms that are contained in H_8 and J_8 must now be expanded:

$$\begin{array}{llll}\mathbf{S}_3\cdot\mathbf{S}_6, & \mathbf{R}^{6,67}\cdot\mathbf{S}_6, & \mathbf{R}^{7,71}\cdot\mathbf{S}_6, & \mathbf{R}^{1,12}\cdot\mathbf{S}_6,\\ \mathbf{R}^{2,23}\cdot\mathbf{S}_6, & \mathbf{R}^{6,71}\cdot\mathbf{S}_6, & \mathbf{R}^{6,12}\cdot\mathbf{S}_6, & \mathbf{R}^{3,34}\cdot\mathbf{S}_6,\\ \mathbf{R}^{4,45}\cdot\mathbf{S}_6, & \mathbf{R}^{5,56}\cdot\mathbf{S}_6, & & \end{array}$$

$$\begin{array}{l}(\mathbf{R}^{6,67}\times\mathbf{S}_6)\cdot(\mathbf{R}^{7,71}\times\mathbf{S}_3),\\ (\mathbf{R}^{6,71}\times\mathbf{S}_6)\cdot(\mathbf{R}^{1,12}\times\mathbf{S}_3),\\ (\mathbf{R}^{6,12}\times\mathbf{S}_6)\cdot(\mathbf{R}^{2,23}\times\mathbf{S}_3),\\ (\mathbf{R}^{3,34}\times\mathbf{S}_3)\cdot(\mathbf{R}^{4,45}\times\mathbf{S}_6),\\ (\mathbf{R}^{3,45}\times\mathbf{S}_3)\cdot(\mathbf{R}^{5,56}\times\mathbf{S}_6).\end{array}$$

(a) $\mathbf{S}_3\cdot\mathbf{S}_6$

This term can be evaluated using set 13 of the table of direction cosines to give

$$\mathbf{S}_3\cdot\mathbf{S}_6 = Z_{712}.\tag{10.352}$$

It may also be evaluated from set 3 as

$$\mathbf{S}_3\cdot\mathbf{S}_6 = Z_{54}.\tag{10.353}$$

The results of Eq. (10.352) will be substituted into H_8, and Eq. (10.353) will be substituted into J_8.

(b) $\mathbf{R}^{6,67} \cdot \mathbf{S}_6$

This expression may be written as

$$\mathbf{R}^{6,67} \cdot \mathbf{S}_6 = (S_6\mathbf{S}_6 + a_{67}\mathbf{a}_{67}) \cdot \mathbf{S}_6. \tag{10.354}$$

Evaluating the scalar products gives

$$\mathbf{R}^{6,67} \cdot \mathbf{S}_6 = S_6. \tag{10.355}$$

(c) $\mathbf{R}^{7,71} \cdot \mathbf{S}_6$

This expression may be written as

$$\mathbf{R}^{7,71} \cdot \mathbf{S}_6 = (S_7\mathbf{S}_7 + a_{71}\mathbf{a}_{71}) \cdot \mathbf{S}_6. \tag{10.356}$$

Evaluating the scalar products gives

$$\mathbf{R}^{7,71} \cdot \mathbf{S}_6 = S_7 c_{67} + a_{71} X_7. \tag{10.357}$$

(d) $\mathbf{R}^{1,12} \cdot \mathbf{S}_6$

This expression may be written as

$$\mathbf{R}^{1,12} \cdot \mathbf{S}_6 = (S_1\mathbf{S}_1 + a_{12}\mathbf{a}_{12}) \cdot \mathbf{S}_6. \tag{10.358}$$

Evaluating the scalar products gives

$$\mathbf{R}^{1,12} \cdot \mathbf{S}_6 = S_7 Z_7 + a_{12} X_{71}. \tag{10.359}$$

(e) $\mathbf{R}^{2,23} \cdot \mathbf{S}_6$

This expression may be written as

$$\mathbf{R}^{2,23} \cdot \mathbf{S}_6 = (S_2\mathbf{S}_2 + a_{23}\mathbf{a}_{23}) \cdot \mathbf{S}_6. \tag{10.360}$$

Evaluating the scalar products gives

$$\mathbf{R}^{2,23} \cdot \mathbf{S}_6 = S_2 Z_{71} + a_{23} X_{712}. \tag{10.361}$$

(f) $\mathbf{R}^{6,71} \cdot \mathbf{S}_6$

This expression may be written as

$$\mathbf{R}^{6,71} \cdot \mathbf{S}_6 = (\mathbf{R}^{6,67} + \mathbf{R}^{7,71}) \cdot \mathbf{S}_6. \tag{10.362}$$

Using the results of Eqs. (10.355) and (10.357) gives

$$\mathbf{R}^{6,71} \cdot \mathbf{S}_6 = S_6 + S_7 c_{67} + a_{71} X_7. \tag{10.363}$$

(g) $\mathbf{R}^{6,12} \cdot \mathbf{S}_6$

This expression may be written as

$$\mathbf{R}^{6,12} \cdot \mathbf{S}_6 = (\mathbf{R}^{6,71} + \mathbf{R}^{1,12}) \cdot \mathbf{S}_6. \tag{10.364}$$

Using the results of Eqs. (10.363) and (10.359) gives

$$\mathbf{R}^{6,12} \cdot \mathbf{S}_6 = S_6 + S_7 c_{67} + a_{71} X_7 + S_1 Z_7 + a_{12} X_{71}. \tag{10.365}$$

(h) $\mathbf{R}^{3,34} \cdot \mathbf{S}_6$

This expression may be written as

$$\mathbf{R}^{3,34} \cdot \mathbf{S}_6 = (S_3 \mathbf{S}_3 + a_{34} \mathbf{a}_{34}) \cdot \mathbf{S}_6. \tag{10.366}$$

Evaluating the scalar products gives

$$\mathbf{R}^{3,34} \cdot \mathbf{S}_6 = S_3 Z_{54} + a_{34} X_{54}. \tag{10.367}$$

(i) $\mathbf{R}^{4,45} \cdot \mathbf{S}_6$

This expression may be written as

$$\mathbf{R}^{4,45} \cdot \mathbf{S}_6 = (S_4 \mathbf{S}_4 + a_{45} \mathbf{a}_{45}) \cdot \mathbf{S}_6. \tag{10.368}$$

Evaluating the scalar products gives

$$\mathbf{R}^{4,45} \cdot \mathbf{S}_6 = S_4 Z_5 + a_{45} \bar{X}_5. \tag{10.369}$$

(j) $\mathbf{R}^{5,56} \cdot \mathbf{S}_6$

This expression may be written as

$$\mathbf{R}^{5,56} \cdot \mathbf{S}_6 = (S_5 \mathbf{S}_5 + a_{56} \mathbf{a}_{56}) \cdot \mathbf{S}_6. \tag{10.370}$$

Evaluating the scalar products gives

$$\mathbf{R}^{5,56} \cdot \mathbf{S}_6 = S_5 c_{56}. \tag{10.371}$$

(k) $(\mathbf{R}^{6,67} \times \mathbf{S}_6) \cdot (\mathbf{R}^{7,71} \times \mathbf{S}_3)$

The scalar and vector products in this expression will all be evaluated using set 7 from the table of direction cosines. The first term of this expression may be written as

$$\mathbf{R}^{6,67} \times \mathbf{S}_6 = (S_6 \mathbf{S}_6 + a_{67} \mathbf{a}_{67}) \times \mathbf{S}_6. \tag{10.372}$$

The vector product of $\mathbf{S}_6$ with itself is zero. The vector product $\mathbf{a}_{67} \times \mathbf{S}_6 = -\mathbf{S}_6 \times \mathbf{a}_{67}$

(see Eq. (10.186)). Thus Eq. (10.372) may be written as

$$\mathbf{R}^{6,67} \times \mathbf{S}_6 = \begin{bmatrix} -a_{67}c_{67}s_7 \\ -a_{67}c_{67}c_7 \\ a_{67}s_{67} \end{bmatrix}. \tag{10.373}$$

The vector product of $\mathbf{R}^{7,71}$ and $\mathbf{S}_3$ was previously evaluated using set 7, and the results are presented in Eq. (10.191). The scalar product of Eqs. (10.373) and (10.191) yields the result

$$(\mathbf{R}^{6,67} \times \mathbf{S}_6) \cdot (\mathbf{R}^{7,71} \times \mathbf{S}_3) = -a_{67}S_7c_{67}X_{217} + a_{67}a_{71}(s_{67}Y_{21} + c_{67}c_7Z_{21}). \tag{10.374}$$

(l) $(\mathbf{R}^{6,71} \times \mathbf{S}_6) \cdot (\mathbf{R}^{1,12} \times \mathbf{S}_3)$

Set 1 from the table of direction cosines will be used to evaluate all the scalar and vector products in this expression. The first term in this expression may be written as

$$\mathbf{R}^{6,71} \times \mathbf{S}_6 = (S_6\mathbf{S}_6 + a_{67}\mathbf{a}_{67} + S_7\mathbf{S}_7 + a_{71}\mathbf{a}_{71}) \times \mathbf{S}_6. \tag{10.375}$$

Now, $\mathbf{S}_6 \times \mathbf{S}_6 = \mathbf{0}$. The vector product $\mathbf{S}_6 \times \mathbf{a}_{67}$ was evaluated in Eq. (10.195). Thus,

$$\mathbf{a}_{67} \times \mathbf{S}_6 = \begin{bmatrix} U^*_{176} \\ c_1Z'_7 + s_1X'_7 \\ -\bar{Y}_7 \end{bmatrix}. \tag{10.376}$$

The vector product $\mathbf{S}_7 \times \mathbf{S}_6$ may be written as

$$\mathbf{S}_7 \times \mathbf{S}_6 = \begin{vmatrix} \mathbf{i} & \mathbf{j} & \mathbf{k} \\ s_{71}s_1 & s_{71}c_1 & c_{71} \\ X_{71} & -X^*_{71} & Z_7 \end{vmatrix} = \begin{bmatrix} s_{71}c_1Z_7 + c_{71}X^*_{71} \\ c_{71}X_{71} - s_{71}s_1Z_7 \\ -s_{71}s_1X^*_{71} - s_{71}c_1X_{71} \end{bmatrix}, \tag{10.377}$$

where fundamental and subsidiary sine, sine–cosine, and cosine laws were used to simplify the direction cosines of $\mathbf{S}_6$ and $\mathbf{S}_7$. Expanding and regrouping the elements of this vector and recognizing that $s_1^2 + c_1^2 = 1$ gives

$$\mathbf{S}_7 \times \mathbf{S}_6 = \begin{bmatrix} -s_{67}W_{71} \\ -s_{67}V_{71} \\ -s_{67}U_{71} \end{bmatrix}. \tag{10.378}$$

The vector product $\mathbf{a}_{71} \times \mathbf{S}_6$ may be written as

$$\mathbf{a}_{71} \times \mathbf{S}_6 = \begin{vmatrix} \mathbf{i} & \mathbf{j} & \mathbf{k} \\ c_1 & -s_1 & 0 \\ X_{71} & -X^*_{71} & Z_7 \end{vmatrix} = \begin{bmatrix} -s_1Z_7 \\ -c_1Z_7 \\ -c_1X^*_{71} + s_1X_{71} \end{bmatrix}. \tag{10.379}$$

Expanding the last term of this vector gives

$$\mathbf{a}_{71} \times \mathbf{S}_6 = \begin{bmatrix} -s_1 Z_7 \\ -c_1 Z_7 \\ -Y_7 \end{bmatrix}. \tag{10.380}$$

Substituting the results of Eqs. (10.376), (10.378), and (10.380) into Eq. (10.375) gives

$$\mathbf{R}^{6,71} \times \mathbf{S}_6 = \begin{bmatrix} a_{67} U^*_{176} - S_7 s_{67} W_{71} - a_{71} s_1 Z_7 \\ a_{67}(c_1 Z'_7 + s_1 X'_7) - S_7 s_{67} V_{71} - a_{71} c_1 Z_7 \\ -a_{67} \bar{Y}_7 - S_7 s_{67} U_{71} - a_{71} Y_7 \end{bmatrix}. \tag{10.381}$$

The factor $(\mathbf{R}^{1,12} \times \mathbf{S}_3)$ has been previously evaluated in terms of set 1 from the table of direction cosines. The result is presented in Eq. (10.211). Performing a scalar product of Eqs. (10.381) and (10.211) gives the result

$$\begin{aligned}(\mathbf{R}^{6,71} \times \mathbf{S}_6) \cdot (\mathbf{R}^{1,12} \times \mathbf{S}_3) = \\ \left[a_{67} U^*_{176} - S_7 s_{67} W_{71} - a_{71} s_1 Z_7\right](-S_1 \bar{Y}_2) + \left[a_{67}(c_1 Z'_7 + s_1 X'_7) - S_7 s_{67} V_{71} \right. \\ \left. - a_{71} c_1 Z_7\right](S_1 \bar{X}_2 - a_{12} \bar{Z}_2) + [-a_{67} \bar{Y}_7 - S_7 s_{67} U_{71} - a_{71} Y_7](a_{12} \bar{Y}_2).\end{aligned} \tag{10.382}$$

(m) $(\mathbf{R}^{6,12} \times \mathbf{S}_6) \cdot (\mathbf{R}^{2,23} \times \mathbf{S}_3)$

Set 1 from the table of direction cosines will be used to evaluate all the scalar and vector products in this expression. The left factor may be written as

$$(\mathbf{R}^{6,12} \times \mathbf{S}_6) = (\mathbf{R}^{6,71} + S_1 \mathbf{S}_1 + a_{12} \mathbf{a}_{12}) \times \mathbf{S}_6. \tag{10.383}$$

The vector product of $\mathbf{R}^{6,71}$ and $\mathbf{S}_6$ has been evaluated in set 1 (see Eq. (10.381)). The vector product of $\mathbf{S}_1$ and $\mathbf{S}_6$ may be written as

$$\mathbf{S}_1 \times \mathbf{S}_6 = \begin{vmatrix} \mathbf{i} & \mathbf{j} & \mathbf{k} \\ 0 & 0 & 1 \\ X_{71} & -X^*_{71} & Z_7 \end{vmatrix} = \begin{bmatrix} X^*_{71} \\ X_{71} \\ 0 \end{bmatrix}. \tag{10.384}$$

The vector product of $\mathbf{a}_{12}$ and $\mathbf{S}_6$ may be written as

$$\mathbf{a}_{12} \times \mathbf{S}_6 = \begin{vmatrix} \mathbf{i} & \mathbf{j} & \mathbf{k} \\ 1 & 0 & 0 \\ X_{71} & -X^*_{71} & Z_7 \end{vmatrix} = \begin{bmatrix} 0 \\ -Z_7 \\ -X^*_{71} \end{bmatrix}. \tag{10.385}$$

Substituting Eqs. (10.381), (10.384), and (10.385) into Eq. (10.383) gives

$$\mathbf{R}^{6,12} \times \mathbf{S}_6 = \begin{bmatrix} a_{67} U^*_{176} - S_7 s_{67} W_{71} - a_{71} s_1 Z_7 + S_1 X^*_7 \\ a_{67}(c_1 Z'_7 + s_1 X'_7) - S_7 s_{67} V_{71} - a_{71} c_1 Z_7 + S_1 X_{71} - a_{12} Z_7 \\ -a_{67} \bar{Y}_7 - S_7 s_{67} U_{71} - a_{71} Y_7 - a_{12} X^*_{71} \end{bmatrix}. \tag{10.386}$$

The factor $(\mathbf{R}^{2,23} \times \mathbf{S}_3)$ was previously evaluated using set 1 in Eq. (10.222). The scalar product of Eqs. (10.386) and (10.222) gives the result

$$(\mathbf{R}^{6,12} \times \mathbf{S}_6) \cdot (\mathbf{R}^{2,23} \times \mathbf{S}_3) = \\ [a_{67}U^*_{176} - S_7 s_{67} W_{71} - a_{71} s_1 Z_7 + S_1 X^*_{71}](S_2 s_{23} c_2 + a_{23} \bar{X}'_2) + [a_{67}(c_1 Z'_7 + s_1 X'_7) \\ - S_7 s_{67} V_{71} - a_{71} c_1 Z_7 + S_1 X_{71} - a_{12} Z_7](S_2 c_{12} \bar{X}_2 + a_{23} Z'_2) + [-a_{67} \bar{Y}_7 - S_7 s_{67} U_{71} \\ - a_{71} Y_7 - a_{12} X^*_{71}](S_2 s_{12} \bar{X}_2 + a_{23} Y_2). \tag{10.387}$$

(n) $(\mathbf{R}^{3,34} \times \mathbf{S}_3) \cdot (\mathbf{R}^{4,45} \times \mathbf{S}_6)$

Set 4 from the table of direction cosines will be used to evaluate all the scalar and vector products in this term. The first term may be written as

$$\mathbf{R}^{3,34} \times \mathbf{S}_3 = (S_3\mathbf{S}_3 + a_{34}\mathbf{a}_{34}) \times \mathbf{S}_3. \tag{10.388}$$

Now, $\mathbf{S}_3 \times \mathbf{S}_3 = \mathbf{0}$. The vector product $\mathbf{a}_{34} \times \mathbf{S}_3$ was previously determined in Eq. (10.92). Eq. (10.388) can thus be written as

$$\mathbf{R}^{3,34} \times \mathbf{S}_3 = \begin{bmatrix} -a_{34}c_{34}s_4 \\ -a_{34}c_{34}c_4 \\ a_{34}s_{34} \end{bmatrix}. \tag{10.389}$$

The term $(\mathbf{R}^{4,45} \times \mathbf{S}_6)$ may be written as

$$\mathbf{R}^{4,45} \times \mathbf{S}_6 = (S_4\mathbf{S}_4 + a_{45}\mathbf{a}_{45}) \times \mathbf{S}_6. \tag{10.390}$$

The vector product $\mathbf{S}_4 \times \mathbf{S}_6$ may be written as

$$\mathbf{S}_4 \times \mathbf{S}_6 = \begin{vmatrix} \mathbf{i} & \mathbf{j} & \mathbf{k} \\ 0 & 0 & 1 \\ \bar{X}_5 & \bar{Y}_5 & \bar{Z}_5 \end{vmatrix} = \begin{bmatrix} -\bar{Y}_5 \\ \bar{X}_5 \\ 0 \end{bmatrix}. \tag{10.391}$$

The vector product $\mathbf{a}_{45} \times \mathbf{S}_6$ may be written as

$$\mathbf{a}_{45} \times \mathbf{S}_6 = \begin{vmatrix} \mathbf{i} & \mathbf{j} & \mathbf{k} \\ 1 & 0 & 0 \\ \bar{X}_5 & \bar{Y}_5 & \bar{Z}_5 \end{vmatrix} = \begin{bmatrix} 0 \\ -\bar{Z}_5 \\ \bar{Y}_5 \end{bmatrix}. \tag{10.392}$$

Substituting Eqs. (10.391) and (10.392) into Eq. (10.390) gives

$$\mathbf{R}^{4,45} \times \mathbf{S}_6 = \begin{bmatrix} -S_4\bar{Y}_5 \\ S_4\bar{X}_5 - a_{45}\bar{Z}_5 \\ a_{45}\bar{Y}_5 \end{bmatrix}. \tag{10.393}$$

Forming the scalar product of Eqs. (10.389) and (10.393) gives the result

$$(\mathbf{R}^{3,34} \times \mathbf{S}_3) \cdot (\mathbf{R}^{4,45} \times \mathbf{S}_6) = (-a_{34}c_{34}s_4)(-S_4\bar{Y}_5) + (-a_{34}c_{34}c_4)(S_4\bar{X}_5 - a_{45}\bar{Z}_5) \\ + (a_{34}s_{34})(a_{45}\bar{Y}_5). \tag{10.394}$$

(o) $(\mathbf{R}^{3,45} \times \mathbf{S}_3) \cdot (\mathbf{R}^{5,56} \times \mathbf{S}_6)$

Set 11 of the table of direction cosines will be used to evaluate all scalar and vector products in this expression. The first term has previously been expanded, and the results are presented in Eq. (10.117). The second term may be written as

$$\mathbf{R}^{5,56} \times \mathbf{S}_6 = (S_5\mathbf{S}_5 + a_{56}\mathbf{a}_{56}) \times \mathbf{S}_6. \tag{10.395}$$

The vector product $\mathbf{S}_5 \times \mathbf{S}_6$ may be written as

$$\mathbf{S}_5 \times \mathbf{S}_6 = \begin{vmatrix} \mathbf{i} & \mathbf{j} & \mathbf{k} \\ 0 & 0 & 1 \\ s_{56}s_5 & -s_{56}c_5 & c_{56} \end{vmatrix} = \begin{bmatrix} s_{56}c_5 \\ s_{56}s_5 \\ 0 \end{bmatrix}, \tag{10.396}$$

where fundamental sine, sine–cosine, and cosine laws were used to simplify the direction cosines of the vector $\mathbf{S}_6$.

The vector product $\mathbf{a}_{56} \times \mathbf{S}_6$ may be written as

$$\mathbf{a}_{56} \times \mathbf{S}_6 = \begin{vmatrix} \mathbf{i} & \mathbf{j} & \mathbf{k} \\ c_5 & s_5 & 0 \\ s_{56}s_5 & -s_{56}c_5 & c_{56} \end{vmatrix} = \begin{bmatrix} c_{56}s_5 \\ -c_{56}c_5 \\ -s_{56} \end{bmatrix}. \tag{10.397}$$

Substituting Eqs. (10.396) and (10.397) into Eq. (10.395) gives

$$\mathbf{R}^{5,56} \times \mathbf{S}_6 = \begin{bmatrix} S_5s_{56}c_5 + a_{56}c_{56}s_5 \\ S_5s_{56}s_5 - a_{56}c_{56}c_5 \\ -a_{56}s_{56} \end{bmatrix}. \tag{10.398}$$

Evaluating the scalar product of Eqs. (10.117) and (10.398) gives the result

$$(\mathbf{R}^{3,45} \times \mathbf{S}_3) \cdot (\mathbf{R}^{5,56} \times \mathbf{S}_6) = \\ -X_{04}[S_5s_{56}c_5 + a_{56}c_{56}s_5] + Y_{04}[S_5s_{56}s_5 - a_{56}c_{56}c_5] + Z_{04}[a_{56}s_{56}]. \tag{10.399}$$

The terms H_8 and J_8 may now be expanded by using the previous results to yield

$$\begin{aligned} H_8 = {} & K_2Z_{712} - S_6(S_6Z_{712} + a_{67}X_{217}) - (S_7Z_{21} + a_{71}X_{21})(S_7c_{67} + a_{71}X_7) \\ & + \left[-a_{67}S_7c_{67}X_{217} + a_{67}a_{71}(s_{67}Y_{21} + c_{67}c_7Z_{21})\right] - S_6(S_7Z_{21} + a_{71}X_{21}) \\ & - (S_1Z_2 + a_{12}\bar{X}_2)(S_1Z_7 + a_{12}X_{71}) + \left[a_{67}U^*_{176} - S_7s_{67}W_{71} - a_{71}s_1Z_7\right](-S_1\bar{Y}_2) \\ & + \left[a_{67}(c_1Z'_7 + s_1X'_7) - S_7s_{67}V_{71} - a_{71}c_1Z_7\right](S_1\bar{X}_2 - a_{12}\bar{Z}_2) + \left[-a_{67}\bar{Y}_7\right. \\ & \left. - S_7s_{67}U_{71} - a_{71}Y_7\right](a_{12}\bar{Y}_2) - (S_1Z_2 + a_{12}\bar{X}_2)(S_6 + S_7c_{67} + a_{71}X_7) \\ & - S_2c_{23}(S_2Z_{71} + a_{23}X_{712}) + \left[a_{67}U^*_{176} - S_7s_{67}W_{71} - a_{71}s_1Z_7 + S_1X^*_{71}\right] \\ & \times (S_2s_{23}c_2 + a_{23}\bar{X}'_2) + \left[a_{67}(c_1Z'_7 + s_1X'_7) - S_7s_{67}V_{71} - a_{71}c_1Z_7 + S_1X_{71}\right. \\ & \left. - a_{12}Z_7\right](S_2c_{12}\bar{X}_2 + a_{23}Z'_2) + \left[-a_{67}\bar{Y}_7 - S_7s_{67}U_{71} - a_{71}Y_7 - a_{12}X^*_{71}\right] \\ & \times (S_2s_{12}\bar{X}_2 + a_{23}Y_2) - S_2c_{23}(S_6 + S_7c_{67} + a_{71}X_7 + S_1Z_7 + a_{12}X_{71}), \end{aligned} \tag{10.400}$$

$$
\begin{aligned}
J_8 = {} & K_1 Z_{54} - S_3(S_3 Z_{54} + a_{34} X_{54}) - (S_3 + S_4 c_{34} + a_{45} X_4)(S_4 Z_5 + a_{45} \bar{X}_5) - J_5 S_5 c_{56} \\
& + (-a_{34} c_{34} s_4)(-S_4 \bar{Y}_5) + (-a_{34} c_{34} c_4)(S_4 \bar{X}_5 - a_{45} \bar{Z}_5) + (a_{34} s_{34})(a_{45} \bar{Y}_5) \\
& - X_{04}[S_5 s_{56} c_5 + a_{56} c_{56} s_5] + Y_{04}[S_5 s_{56} s_5 - a_{56} c_{56} c_5] + Z_{04}[a_{56} s_{56}],
\end{aligned}
\tag{10.401}
$$

where K_1 is defined in Eq. (10.78), K_2 is defined in Eq. (10.163), and J_5 is defined in Eq. (10.82).

Equation (10.347) may be factored into the format

$$
\begin{aligned}
& K_{31} c_4 c_5 + K_{32} c_4 s_5 + K_{33} c_4 + K_{34} s_4 c_5 + K_{35} s_4 s_5 + K_{36} s_4 + K_{37} c_5 \\
& \quad + K_{38} s_5 = H_8 - K_{39},
\end{aligned}
\tag{10.402}
$$

where

$$
\begin{aligned}
K_{31} = {} & s_{56}\left[(s_{34} c_{45}/2)\left(S_3^2 - S_4^2 - S_5^2 - a_{34}^2 - a_{45}^2 - a_{56}^2\right) - a_{34} a_{45} c_{34} s_{45} - S_4 S_5 s_{34}\right] \\
& + c_{56} a_{56}[a_{34} c_{34} c_{45} - a_{45} s_{34} s_{45}], \\
K_{32} = {} & s_{56}\left[-a_{34} S_4 c_{34} - a_{34} S_5 c_{34} c_{45} - a_{34} S_3 + a_{45} S_5 s_{34} s_{45}\right] + c_{56} a_{56} s_{34}[-S_4 - S_5 c_{45}], \\
K_{33} = {} & s_{56} a_{56}[-a_{34} c_{34} s_{45} - a_{45} s_{34} c_{45}] + c_{56}\left[(s_{34} s_{45}/2)\right. \\
& \left. \times \left(S_3^2 - S_4^2 + S_5^2 - a_{34}^2 - a_{45}^2 - a_{56}^2\right) + a_{34} a_{45} c_{34} c_{45}\right], \\
K_{34} = {} & s_{56}\left[-a_{34} S_4 c_{34} c_{45} + a_{45} S_4 s_{34} s_{45} - a_{34} S_3 c_{45} - a_{34} S_5 c_{34}\right] + c_{56} a_{56} s_{34}\left[-S_5 - S_4 c_{45}\right], \\
K_{35} = {} & s_{56} s_{34}\left[\left(-S_3^2 + S_4^2 + S_5^2 + a_{34}^2 - a_{45}^2 + a_{56}^2\right)/2 + S_4 S_5 c_{45}\right] - c_{56} a_{34} a_{56} c_{34}, \\
K_{36} = {} & s_{56}\left[a_{56} S_4 s_{34} s_{45}\right] + c_{56}\left[-a_{34} S_3 s_{45} - a_{45} S_5 s_{34} - a_{45} S_4 s_{34} c_{45} - a_{34} S_4 c_{34} s_{45}\right], \\
K_{37} = {} & s_{56}\left[(c_{34} s_{45}/2)\left(S_3^2 + S_4^2 - S_5^2 - a_{34}^2 - a_{45}^2 - a_{56}^2\right) - a_{34} a_{45} s_{34} c_{45} + S_3 S_4 s_{45}\right] \\
& + c_{56} a_{56}[-a_{34} s_{34} s_{45} + a_{45} c_{34} c_{45}], \\
K_{38} = {} & s_{56}\left[-a_{45} S_4 c_{34} - a_{45} S_5 c_{34} c_{45} + a_{34} S_5 s_{34} s_{45} - a_{45} S_3\right] - c_{56}\left[a_{56} S_5 c_{34} s_{45}\right], \\
K_{39} = {} & s_{56} a_{56}[-a_{34} s_{34} c_{45} - a_{45} c_{34} s_{45}] + c_{56}\left[(c_{34} c_{45}/2)\left(-S_3^2 - S_4^2 - S_5^2 + a_{34}^2\right.\right. \\
& \left.\left. + a_{45}^2 + a_{56}^2\right) - S_3 S_5 - a_{34} a_{45} s_{34} s_{45} - S_3 S_4 c_{45} - S_4 S_5 c_{34}\right].
\end{aligned}
\tag{10.403}
$$

(xi) through (xvi)

The fifth through tenth equations that have been generated do not contain the tan-half-angle of θ_6. Therefore, the final six equations can be obtained by multiplying these equations by the term x_6.

The sixteen equations may be written in matrix form as

$$
\mathbf{A}\mathbf{b} = \mathbf{c},
\tag{10.404}
$$

where **b** was previously defined in Eq. (10.280) but is repeated here along with **A** and **c** as

$$
\mathbf{A} = \left[\begin{array}{cccccccccccccccc}
0 & s_{34}c_{45} & 0 & s_{34} & 0 & 0 & 0 & c_{34}s_{45} & -s_{34}c_{45}c_{56} & 0 & s_{34}s_{45}s_{56} & 0 & s_{34}c_{56} & 0 & -c_{34}s_{45}c_{56} & 0 \\
s_{34}c_{45}c_{56} & 0 & -s_{34}s_{45}s_{56} & 0 & -s_{34}c_{56} & 0 & c_{34}s_{45}c_{56} & 0 & 0 & s_{34}c_{45} & 0 & s_{34} & 0 & 0 & 0 & c_{34}s_{45} \\
K_3 & K_4 & K_5 & K_6 & K_7 & K_8 & K_9 & K_{10} & K_{11} & K_{12} & 0 & K_{13} & K_{14} & 0 & K_{15} & K_{16} \\
K_{11} & K_{12} & 0 & K_{13} & K_{14} & 0 & K_{15} & K_{16} & -K_3 & -K_4 & -K_5 & -K_6 & -K_7 & -K_8 & -K_9 & -K_{10} \\
0 & 0 & 0 & 0 & 0 & 0 & 0 & 0 & -s_{34}c_{45}s_{56} & 0 & -s_{34}s_{45}c_{56} & 0 & s_{34}s_{56} & 0 & -c_{34}s_{45}s_{56} & 0 \\
0 & 0 & 0 & 0 & 0 & 0 & 0 & 0 & K_{18} & K_{19} & K_{20} & K_{21} & K_{22} & K_{23} & K_{24} & K_{25} \\
0 & 0 & 0 & 0 & 0 & 0 & 0 & 0 & 0 & -a_{56}s_{34}c_{45} & S_5s_{34}s_{45} & -a_{56}s_{34} & 0 & -a_{45}s_{34} & 0 & -a_{56}c_{34}s_{45} \\
0 & 0 & 0 & 0 & 0 & 0 & 0 & 0 & S_3s_{34}c_{45}s_{56} & -a_{34}s_{56} & S_3s_{34}s_{45}c_{56} & -a_{34}c_{45}s_{56} & -S_3s_{34}s_{56} & -a_{34}s_{45}c_{56} & K_{27} & -a_{45}s_{56} \\
0 & 0 & 0 & 0 & 0 & 0 & 0 & 0 & a_{34}a_{56} & S_3a_{56}s_{34}c_{45} & K_{28} & S_3a_{56}s_{34} & -a_{34}a_{56}c_{45} & K_{29} & a_{45}a_{56} & K_{30} \\
0 & 0 & 0 & 0 & 0 & 0 & 0 & 0 & K_{31} & K_{32} & K_{33} & K_{34} & K_{35} & K_{36} & K_{37} & K_{38} \\
-s_{34}c_{45}s_{56} & 0 & -s_{34}s_{45}c_{56} & 0 & s_{34}s_{56} & 0 & -c_{34}s_{45}s_{56} & 0 & 0 & 0 & 0 & 0 & 0 & 0 & 0 & 0 \\
K_{18} & K_{19} & K_{20} & K_{21} & K_{22} & K_{23} & K_{24} & K_{25} & 0 & 0 & 0 & 0 & 0 & 0 & 0 & 0 \\
0 & -a_{56}s_{34}c_{45} & S_5s_{34}s_{45} & -a_{56}s_{34} & 0 & -a_{45}s_{34} & 0 & -a_{56}c_{34}s_{45} & 0 & 0 & 0 & 0 & 0 & 0 & 0 & 0 \\
S_3s_{34}c_{45}s_{56} & -a_{34}s_{56} & S_3s_{34}s_{45}c_{56} & -a_{34}c_{45}s_{56} & -S_3s_{34}s_{56} & -a_{34}s_{45}c_{56} & K_{27} & -a_{45}s_{56} & 0 & 0 & 0 & 0 & 0 & 0 & 0 & 0 \\
a_{34}a_{56} & S_3a_{56}s_{34}c_{45} & K_{28} & S_3a_{56}s_{34} & -a_{34}a_{56}c_{45} & K_{29} & a_{45}a_{56} & K_{30} & 0 & 0 & 0 & 0 & 0 & 0 & 0 & 0 \\
K_{31} & K_{32} & K_{33} & K_{34} & K_{35} & K_{36} & K_{37} & K_{38} & 0 & 0 & 0 & 0 & 0 & 0 & 0 & 0
\end{array}\right], \tag{10.405}
$$

$$\mathbf{b} = \begin{bmatrix} c_4c_5x_6 \\ c_4s_5x_6 \\ c_4x_6 \\ s_4c_5x_6 \\ s_4s_5x_6 \\ s_4x_6 \\ c_5x_6 \\ s_5x_6 \\ c_4c_5 \\ c_4s_5 \\ c_4 \\ s_4c_5 \\ s_4s_5 \\ s_4 \\ c_5 \\ s_5 \end{bmatrix}, \tag{10.406}$$

$$\mathbf{c} = \begin{bmatrix} [-X_{217}]x_6 - Y_{217} + c_{34}c_{45}s_{56} \\ [-Y_{217} - c_{34}c_{45}s_{56}]x_6 + X_{217} \\ [-Y_{0217} - S_6X_{217} + K_{17}]x_6 + X_{0217} - S_6Y_{217} \\ [-X_{0217} + S_6Y_{217}]x_6 - Y_{0217} - S_6X_{217} - K_{17} \\ Z_{217} - c_{34}c_{45}c_{56} \\ Z_{0217} + K_{26} \\ H_5 + S_3 + S_4c_{34} + S_5c_{34}c_{45} \\ H_6 + S_3c_{34}c_{45}c_{56} + S_4c_{45}c_{56} + S_5c_{56} \\ H_7 - S_3S_4c_{34} - S_3S_5c_{34}c_{45} - S_4S_5c_{45} - K_1 \\ H_8 - K_{39} \\ [Z_{217} - c_{34}c_{45}c_{56}]x_6 \\ [Z_{0217} + K_{26}]x_6 \\ [H_5 + S_3 + S_4c_{34} + S_5c_{34}c_{45}]x_6 \\ [H_6 + S_3c_{34}c_{45}c_{56} + S_4c_{45}c_{56} + S_5c_{56}]x_6 \\ [H_7 - S_3S_4c_{34} - S_3S_5c_{34}c_{45} - S_4S_5c_{45} - K_1]x_6 \\ [H_8 - K_{39}]x_6 \end{bmatrix}. \tag{10.407}$$

Note that the matrix **A** is completely defined in terms of the constant mechanism parameters, whereas the elements of **c** are linear in the sines and cosines of θ_1 and θ_2 and the tan-half-angle of θ_6.

The vector **c** may now be factored into the format

$$\mathbf{c} = \mathbf{T}_3\mathbf{a}, \tag{10.408}$$

where **a** was previously defined in Eq.(10.278) and $\mathbf{T}_3$ is defined as

$$\mathbf{T}_3 = \begin{bmatrix} M''_{1,1} & M''_{1,2} & M''_{1,3} & N''_{1,1} & N''_{1,2} & N''_{1,3} & O''_{1,1} & O''_{1,2} & O''_{1,3} & P''_{1,1} & P''_{1,2} & P''_{1,3} & Q''_{1,1} & Q''_{1,2} & Q''_{1,3} & R''_{1,1} & R''_{1,2} & R''_{1,3} \\ M''_{2,1} & M''_{2,2} & M''_{2,3} & N''_{2,1} & N''_{2,2} & N''_{2,3} & O''_{2,1} & O''_{2,2} & O''_{2,3} & P''_{2,1} & P''_{2,2} & P''_{2,3} & Q''_{2,1} & Q''_{2,2} & Q''_{2,3} & R''_{2,1} & R''_{2,2} & R''_{2,3} \\ \vdots & & & & & & & & & & & & & & & & & \\ M''_{16,1} & M''_{16,2} & M''_{16,3} & N''_{16,1} & N''_{16,2} & N''_{16,3} & O''_{16,1} & O''_{16,2} & O''_{16,3} & P''_{16,1} & P''_{16,2} & P''_{16,3} & Q''_{16,1} & Q''_{16,2} & Q''_{16,3} & R''_{16,1} & R''_{16,2} & R''_{16,3} \end{bmatrix}. \tag{10.409}$$

The terms in the matrix $\mathbf{T}_3$, which are all defined in terms of the constant mechanism parameters, are written as

$$\begin{aligned}
&M''_{1,1} = -c_{71}c_{12}s_{23}s_7, && M''_{1,2} = -c_{12}s_{23}c_7, && M''_{1,3} = s_{71}s_{12}s_{23}s_7,\\
&N''_{1,1} = -s_{23}c_7, && N''_{1,2} = c_{71}s_{23}s_7, && N''_{1,3} = 0,\\
&O''_{1,1} = -c_{71}s_{12}c_{23}s_7, && O''_{1,2} = -s_{12}c_{23}c_7, && O''_{1,3} = -s_{71}c_{12}c_{23}s_7,\\
&P''_{1,1} = -c_{12}s_{23}Z'_7, && P''_{1,2} = -c_{12}s_{23}X'_7, && P''_{1,3} = s_{12}s_{23}\bar{Y}_7,\\
&Q''_{1,1} = -s_{23}X'_7, && Q''_{1,2} = s_{23}Z'_7, && Q''_{1,3} = 0,\\
&R''_{1,1} = -s_{12}c_{23}Z'_7, && R''_{1,2} = -s_{12}c_{23}X'_7, && R''_{1,3} = -c_{12}c_{23}\bar{Y}_7 + c_{34}c_{45}s_{56},
\end{aligned} \tag{10.410}$$

$$\begin{aligned}
&M''_{2,1} = P''_{1,1}, && M''_{2,2} = P''_{1,2}, && M''_{2,3} = P''_{1,3},\\
&N''_{2,1} = Q''_{1,1}, && N''_{2,2} = Q''_{1,2}, && N''_{2,3} = 0,\\
&O''_{2,1} = R''_{1,1}, && O''_{2,2} = R''_{1,2}, && O''_{2,3} = -c_{12}c_{23}\bar{Y}_7 - c_{34}c_{45}s_{56},\\
&P''_{2,1} = -M''_{1,1}, && P''_{2,2} = -M''_{1,2}, && P''_{2,3} = -M''_{1,3},\\
&Q''_{2,1} = -N''_{1,1}, && Q''_{2,2} = -N''_{1,2}, && Q''_{2,3} = 0,\\
&R''_{2,1} = -O''_{1,1}, && R''_{2,2} = -O''_{1,2}, && R''_{2,3} = -O''_{1,3},
\end{aligned} \tag{10.411}$$

$$\begin{aligned}
M''_{3,1} &= -S_1c_{12}s_{23}X'_7 - S_2s_{23}X'_7 - S_6c_{12}s_{23}\bar{X}'_7 - S_7c_{71}c_{12}s_{23}X'_7 + a_{12}s_{12}s_{23}Z'_7\\
&\quad - a_{23}c_{12}c_{23}Z'_7 + a_{67}c_{12}s_{23}Y_7 + a_{71}c_{12}s_{23}\bar{Y}_7,\\
M''_{3,2} &= S_1c_{12}s_{23}Z'_7 + S_2s_{23}Z'_7 - S_6c_{12}s_{23}c_7 - S_7c_{12}s_{23}c_{67}c_7 + a_{12}s_{12}s_{23}X'_7\\
&\quad - a_{23}c_{12}c_{23}X'_7 + a_{67}c_{12}s_{23}X_7,\\
M''_{3,3} &= S_6s_{12}s_{23}\bar{X}_7 + S_7c_{67}s_{12}s_{23}\bar{X}_7 + a_{12}c_{12}s_{23}\bar{Y}_7\\
&\quad + a_{23}s_{12}c_{23}\bar{Y}_7 - a_{67}s_{12}s_{23}Z_7 + a_{71}s_{12}s_{23}Z'_7,\\
N''_{3,1} &= S_1s_{23}Z'_7 + S_2c_{12}s_{23}Z'_7 - S_6s_{23}c_7 - S_7s_{23}c_{67}c_7 - a_{23}c_{23}X'_7 + a_{67}s_{23}X_7,\\
N''_{3,2} &= S_1s_{23}X'_7 + S_2c_{12}s_{23}X'_7 + S_6s_{23}\bar{X}'_7 + S_7s_{23}c_{71}X'_7 + a_{23}c_{23}Z'_7\\
&\quad - a_{67}s_{23}Y_7 - a_{71}s_{23}\bar{Y}_7,\\
N''_{3,3} &= -S_2s_{12}s_{23}\bar{Y}_7,\\
O''_{3,1} &= -S_1s_{12}c_{23}X'_7 - S_6s_{12}c_{23}\bar{X}'_7 - S_7s_{12}c_{23}c_{71}X'_7 - a_{12}c_{12}c_{23}Z'_7 + a_{23}s_{12}s_{23}Z'_7\\
&\quad + a_{67}s_{12}c_{23}Y_7 + a_{71}s_{12}c_{23}\bar{Y}_7,\\
O''_{3,2} &= S_1s_{12}c_{23}Z'_7 - S_6s_{12}c_{23}c_7 - S_7s_{12}c_{23}c_{67}c_7 - a_{12}c_{12}c_{23}X'_7\\
&\quad + a_{23}s_{12}s_{23}X'_7 + a_{67}s_{12}c_{23}X_7,\\
O''_{3,3} &= -S_6c_{12}c_{23}\bar{X}_7 - S_7c_{12}c_{23}c_{67}\bar{X}_7 + a_{12}s_{12}c_{23}\bar{Y}_7 + a_{23}c_{12}s_{23}\bar{Y}_7 + a_{67}c_{12}c_{23}Z_7\\
&\quad - a_{71}c_{12}c_{23}Z'_7 + a_{34}s_{34}c_{45}s_{56} + a_{45}c_{34}s_{45}s_{56} - a_{56}c_{34}c_{45}c_{56},\\
P''_{3,1} &= S_1c_{12}s_{23}c_7 + S_2s_{23}c_7 - S_6c_{12}s_{23}Z'_7 + S_7c_{12}s_{23}c_{71}c_7 - a_{12}s_{12}s_{23}\bar{X}'_7\\
&\quad + a_{23}c_{12}c_{23}\bar{X}'_7 - a_{71}c_{12}s_{23}\bar{X}_7,\\
P''_{3,2} &= -S_1c_{12}s_{23}\bar{X}'_7 - S_2s_{23}\bar{X}'_7 - S_6c_{12}s_{23}X'_7 - S_7c_{12}s_{23}s_7 - a_{12}s_{12}s_{23}c_7 + a_{23}c_{12}c_{23}c_7,\\
P''_{3,3} &= S_6s_{12}s_{23}\bar{Y}_7 - S_7s_{12}s_{23}s_{71}c_7 - a_{12}c_{12}s_{23}\bar{X}_7 - a_{23}s_{12}c_{23}\bar{X}_7 - a_{71}s_{12}s_{23}\bar{X}'_7,
\end{aligned}$$

$$\begin{aligned}
Q''_{3,1} &= -S_1 s_{23}\bar{X}'_7 - S_2 c_{12} s_{23}\bar{X}'_7 - S_6 s_{23} X'_7 - S_7 s_{23} s_7 + a_{23} c_{23} c_7,\\
Q''_{3,2} &= -S_1 s_{23} c_7 - S_2 c_{12} s_{23} c_7 + S_6 s_{23} Z'_7 - S_7 s_{23} c_{71} c_7 - a_{23} c_{23}\bar{X}'_7 + a_{71} s_{23}\bar{X}_7,\\
Q''_{3,3} &= S_2 s_{12} s_{23}\bar{X}_7,\\
R''_{3,1} &= S_1 s_{12} c_{23} c_7 - S_6 s_{12} c_{23} Z'_7 + S_7 s_{12} c_{23} c_{71} c_7 + a_{12} c_{12} c_{23}\bar{X}'_7 - a_{23} s_{12} s_{23}\bar{X}'_7\\
&\quad -a_{71} s_{12} c_{23}\bar{X}_7,\\
R''_{3,2} &= -S_1 s_{12} c_{23}\bar{X}'_7 - S_6 s_{12} c_{23} X'_7 - S_7 s_{12} c_{23} s_7 + a_{12} c_{12} c_{23} c_7 - a_{23} s_{12} s_{23} c_7,\\
R''_{3,3} &= -S_6 c_{12} c_{23}\bar{Y}_7 + S_7 c_{12} c_{23} s_{71} c_7 - a_{12} s_{12} c_{23}\bar{X}_7 - a_{23} c_{12} s_{23}\bar{X}_7 + a_{71} c_{12} c_{23}\bar{X}'_7,
\end{aligned} \tag{10.412}$$

$$\begin{aligned}
&M''_{4,1} = -P''_{3,1}, \quad M''_{4,2} = -P''_{3,2}, \quad M''_{4,3} = -P''_{3,3},\\
&N''_{4,1} = -Q''_{3,1}, \quad N''_{4,2} = -Q''_{3,2}, \quad N''_{4,3} = -Q''_{3,3},\\
&O''_{4,1} = -R''_{3,1}, \quad O''_{4,2} = -R''_{3,2}, \quad O''_{4,3} = -R''_{3,3},\\
&P''_{4,1} = M''_{3,1}, \quad P''_{4,2} = M''_{3,2}, \quad P''_{4,3} = M''_{3,3},\\
&Q''_{4,1} = N''_{3,1}, \quad Q''_{4,2} = N''_{3,2}, \quad Q''_{4,3} = N''_{3,3},\\
&R''_{4,1} = O''_{3,1}, \quad R''_{4,2} = O''_{3,2}, \quad R''_{4,3} = O''_{3,3} - 2K_{17},
\end{aligned} \tag{10.413}$$

$$\begin{aligned}
&M''_{5,1} = M''_{5,2} = M''_{5,3} = 0, \quad N''_{5,1} = N''_{5,2} = N''_{5,3} = 0, \quad O''_{5,1} = O''_{5,2} = O''_{5,3} = 0,\\
&P''_{5,1} = c_{12} s_{23} Y_7, \quad P''_{5,2} = c_{12} s_{23} X_7, \quad P''_{5,3} = -s_{12} s_{23} Z_7,\\
&Q''_{5,1} = s_{23} X_7, \quad Q''_{5,2} = -s_{23} Y_7, \quad Q''_{5,3} = 0,\\
&R''_{5,1} = s_{12} c_{23} Y_7, \quad R''_{5,2} = s_{12} c_{23} X_7, \quad R''_{5,3} = c_{12} c_{23} Z_7 - c_{34} c_{45} c_{56},
\end{aligned} \tag{10.414}$$

$$\begin{aligned}
M''_{6,1} &= M''_{6,2} = M''_{6,3} = 0, \quad N''_{6,1} = N''_{6,2} = N''_{6,3} = 0, O''_{6,1} = O''_{6,2} = O''_{6,3} = 0,\\
P''_{6,1} &= S_1 c_{12} s_{23} X_7 + S_2 s_{23} X_7 + S_7 c_{12} s_{23} c_{71} X_7 - a_{12} s_{12} s_{23} Y_7 + a_{23} c_{12} c_{23} Y_7\\
&\quad + a_{67} c_{12} s_{23} Z'_7 - a_{71} c_{12} s_{23} Z_7,\\
P''_{6,2} &= -S_1 c_{12} s_{23} Y_7 - S_2 s_{23} Y_7 + S_7 c_{12} s_{23} s_{67} c_7 - a_{12} s_{12} s_{23} X_7 + a_{23} c_{12} c_{23} X_7\\
&\quad + a_{67} c_{12} s_{23} X'_7,\\
P''_{6,3} &= -S_7 s_{12} s_{23} s_{71} X_7 - a_{12} c_{12} s_{23} Z_7 - a_{23} s_{12} c_{23} Z_7 - a_{67} s_{12} s_{23}\bar{Y}_7 - a_{71} s_{12} s_{23} Y_7,\\
Q''_{6,1} &= -S_1 s_{23} Y_7 - S_2 c_{12} s_{23} Y_7 + S_7 s_{23} s_{67} c_7 + a_{23} c_{23} X_7 + a_{67} s_{23} X'_7,\\
Q''_{6,2} &= -S_1 s_{23} X_7 - S_2 c_{12} s_{23} X_7 - S_7 s_{23} c_{71} X_7 - a_{23} c_{23} Y_7 - a_{67} s_{23} Z'_7 + a_{71} s_{23} Z_7,\\
Q''_{6,3} &= S_2 s_{12} s_{23} Z_7,\\
R''_{6,1} &= S_1 s_{12} c_{23} X_7 + S_7 s_{12} c_{23} c_{71} X_7 + a_{12} c_{12} c_{23} Y_7 - a_{23} s_{12} s_{23} Y_7 + a_{67} s_{12} c_{23} Z'_7\\
&\quad - a_{71} s_{12} c_{23} Z_7,\\
R''_{6,2} &= -S_1 s_{12} c_{23} Y_7 + S_7 s_{12} c_{23} s_{67} c_7 + a_{12} c_{12} c_{23} X_7 - a_{23} s_{12} s_{23} X_7 + a_{67} s_{12} c_{23} X'_7,\\
R''_{6,3} &= S_7 c_{12} c_{23} s_{71} X_7 - a_{12} s_{12} c_{23} Z_7 - a_{23} c_{12} s_{23} Z_7 + a_{67} c_{12} c_{23}\bar{Y}_7 + a_{71} c_{12} c_{23} Y_7 + K_{26},
\end{aligned} \tag{10.415}$$

$$\begin{aligned}
M''_{7,1} &= M''_{7,2} = M''_{7,3} = 0, N''_{7,1} = N''_{7,2} = N''_{7,3} = 0, O''_{7,1} = O''_{7,2} = O''_{7,3} = 0,\\
P''_{7,1} &= S_6 c_{12} s_{23} Y_7 - S_7 c_{12} s_{23} s_{71} + a_{67} c_{12} s_{23}\bar{X}'_7,
\end{aligned}$$

$$
\begin{aligned}
P''_{7,2} &= S_6c_{12}s_{23}X_7 + a_{67}c_{12}s_{23}c_7 + a_{71}c_{12}s_{23}, \\
P''_{7,3} &= -S_1s_{12}s_{23} - S_6s_{12}s_{23}Z_7 - S_7s_{12}s_{23}c_{71} - a_{67}s_{12}s_{23}\bar{X}_7, \\
Q''_{7,1} &= S_6s_{23}X_7 + a_{67}s_{23}c_7 + a_{71}s_{23}, \\
Q''_{7,2} &= -S_6s_{23}Y_7 + S_7s_{23}s_{71} - a_{67}s_{23}\bar{X}'_7, \\
Q''_{7,3} &= a_{12}s_{23}, \\
R''_{7,1} &= S_6s_{12}c_{23}Y_7 - S_7s_{12}c_{23}s_{71} + a_{67}s_{12}c_{23}\bar{X}'_7, \\
R''_{7,2} &= S_6s_{12}c_{23}X_7 + a_{67}s_{12}c_{23}c_7 + a_{71}s_{12}c_{23}, \\
R''_{7,3} &= S_1c_{12}c_{23} + S_2c_{23} + S_3 + S_4c_{34} + S_5c_{34}c_{45} + S_6c_{12}c_{23}Z_7 + S_7c_{12}c_{23}c_{71} \\
&\quad + a_{67}c_{12}c_{23}\bar{X}_7,
\end{aligned}
\tag{10.416}
$$

$$
\begin{aligned}
M''_{8,1} &= M''_{8,2} & N''_{8,1} &= N''_{8,2} & O''_{8,1} &= O''_{8,2} \\
&= M''_{8,3} = 0, & &= N''_{8,3} = 0, & &= O''_{8,3} = 0, \\
P''_{8,1} &= a_{23}X_7, & P''_{8,2} &= -a_{23}Y_7, & P''_{8,3} &= 0, \\
Q''_{8,1} &= -a_{23}c_{12}Y_7, & Q''_{8,2} &= -a_{23}c_{12}X_7, & Q''_{8,3} &= a_{23}s_{12}Z_7, \\
R''_{8,1} &= S_2s_{12}Y_7 + a_{12}X_7, \\
R''_{8,2} &= S_2s_{12}X_7 - a_{12}Y_7, \\
R''_{8,3} &= S_1Z_7 + S_2c_{12}Z_7 + S_3c_{34}c_{45}c_{56} + S_4c_{45}c_{56} + S_5c_{56} + S_6 + S_7c_{67} + a_{71}X_7,
\end{aligned}
\tag{10.417}
$$

$$
\begin{aligned}
M''_{9,1} &= M''_{9,2} = M''_{9,3} = 0,\ N''_{9,1} = N''_{9,2} = N''_{9,3} = 0,\ O''_{9,1} = O''_{9,2} = O''_{9,3} = 0, \\
P''_{9,1} &= a_{23}(S_6X_7 + a_{67}c_7 + a_{71}), \\
P''_{9,2} &= a_{23}\big(-S_6Y_7 + S_7s_{71} - a_{67}\bar{X}'_7\big), \\
P''_{9,3} &= a_{12}a_{23}, \\
Q''_{9,1} &= a_{23}\big(-S_6c_{12}Y_7 + S_7c_{12}s_{71} - a_{67}c_{12}\bar{X}'_7\big), \\
Q''_{9,2} &= a_{23}(-S_6c_{12}X_7 - a_{67}c_{12}c_7 - a_{71}c_{12}), \\
Q''_{9,3} &= a_{23}(S_1s_{12} + S_6s_{12}Z_7 + S_7s_{12}c_{71} + a_{67}s_{12}\bar{X}_7), \\
R''_{9,1} &= S_6(S_2s_{12}Y_7 + a_{12}X_7) + S_2\big(-S_7s_{12}s_{71} + a_{67}s_{12}\bar{X}'_7\big) + a_{12}(a_{67}c_7 + a_{71}), \\
R''_{9,2} &= S_6(S_2s_{12}X_7 - a_{12}Y_7) + S_2(a_{67}s_{12}c_7 + a_{71}s_{12}) + a_{12}\big(S_7s_{71} - a_{67}\bar{X}'_7\big), \\
R''_{9,3} &= S_6(S_1Z_7 + S_2c_{12}Z_7 + S_7c_{67} + a_{71}X_7) + S_2(S_1c_{12} + S_7c_{12}c_{71} \\
&\quad + a_{67}c_{12}\bar{X}_7) + S_1(S_7c_{71} + a_{67}\bar{X}_7) + S_3(-S_4c_{34} - S_5c_{34}c_{45}) \\
&\quad + S_3(-S_4c_{34} - S_5c_{34}c_{45}) - S_4S_5c_{45} + a_{67}a_{71}c_7 - K_1 + K_2,
\end{aligned}
\tag{10.418}
$$

$$
\begin{aligned}
M''_{10,1} &= M''_{10,2} = M''_{10,3} = 0, \\
N''_{10,1} &= N''_{10,2} = N''_{10,3} = 0, \\
O''_{10,1} &= O''_{10,2} = O''_{10,3} = 0,
\end{aligned}
$$

$$
\begin{aligned}
P''_{10,1} &= S_1\big(S_2s_{23}Y_7 - S_7c_{12}s_{23}s_{67}c_7 + a_{12}s_{12}s_{23}X_7 - a_{23}c_{12}c_{23}X_7 - a_{67}c_{12}s_{23}X'_7 \\
&\quad + S_2\big(-S_7s_{23}s_{67}c_7 - a_{23}c_{23}X_7 - a_{67}s_{23}X'_7\big) + S_6\big(S_7c_{12}s_{23}s_{71} - a_{67}c_{12}s_{23}\bar{X}'_7\big) \\
&\quad - S_6^2c_{12}s_{23}Y_7 + S_7(a_{12}s_{12}s_{23}c_{71}X_7 - a_{23}c_{12}c_{23}c_{71}X_7 - a_{67}c_{12}s_{23}c_{71}X'_7
\end{aligned}
$$

$$
\begin{aligned}
&+ a_{71}c_{12}s_{23}s_{71}X_7) + S_7^2 c_{12}s_{23}c_{67}s_{71} + a_{12}\big(a_{23}s_{12}c_{23}Y_7 + a_{67}s_{12}s_{23}Z_7' \\
&- a_{71}s_{12}s_{23}Z_7\big) + a_{23}\big(-a_{67}c_{12}c_{23}Z_7' + a_{71}c_{12}c_{23}Z_7\big) + a_{67}a_{71}c_{12}s_{23}\bar{Y}_7 \\
&+ K_2 c_{12}s_{23}Y_7,
\end{aligned}
$$

$$
\begin{aligned}
P''_{10,2} = {} & S_1\big(S_2 s_{23}X_7 + S_7 c_{12}s_{23}c_{71}X_7 - a_{12}s_{12}s_{23}Y_7 + a_{23}c_{12}c_{23}Y_7 + a_{67}c_{12}s_{23}Z_7' \\
&- a_{71}c_{12}s_{23}Z_7\big) + S_2\big(S_7 s_{23}c_{71}X_7 + a_{23}c_{23}Y_7 + a_{67}s_{23}Z_7' - a_{71}s_{23}Z_7\big) \\
&+ S_6(-a_{67}c_{12}s_{23}c_7 - a_{71}c_{12}s_{23}) - S_6^2 c_{12}s_{23}X_7 + S_7(a_{12}s_{12}s_{23}s_{67}c_7 \\
&- a_{23}c_{12}c_{23}s_{67}c_7 - a_{67}c_{12}s_{23}c_{67}c_7 - a_{71}c_{12}s_{23}c_{67}) + a_{12}(a_{23}s_{12}c_{23}X_7 \\
&+ a_{67}s_{12}s_{23}X_7') - a_{23}a_{67}c_{12}c_{23}X_7' - a_{71}^2 c_{12}s_{23}X_7 + K_2 c_{12}s_{23}X_7,
\end{aligned}
$$

$$
\begin{aligned}
P''_{10,3} = {} & S_1^2 s_{12}s_{23}Z_7 + S_1(S_6 s_{12}s_{23} + S_7 s_{12}s_{23}c_{67} + a_{71}s_{12}s_{23}X_7) + S_6^2 s_{12}s_{23}Z_7 \\
&+ S_6(S_7 s_{12}s_{23}c_{71} + a_{67}s_{12}s_{23}\bar{X}_7) + S_7^2 s_{12}s_{23}c_{67}c_{71} + S_7(a_{12}c_{12}s_{23}s_{71}X_7 \\
&+ a_{23}s_{12}c_{23}s_{71}X_7 + a_{67}s_{12}s_{23}c_{67}\bar{X}_7 + a_{71}s_{12}s_{23}c_{71}X_7) + a_{12}(a_{23}c_{12}c_{23}Z_7 \\
&+ a_{67}c_{12}s_{23}\bar{Y}_7 + a_{71}c_{12}s_{23}Y_7) + a_{23}(a_{67}s_{12}c_{23}\bar{Y}_7 + a_{71}s_{12}c_{23}Y_7) \\
&+ a_{67}a_{71}s_{12}s_{23}Z_7' - K_2 s_{12}s_{23}Z_7,
\end{aligned}
$$

$$
\begin{aligned}
Q''_{10,1} = {} & S_1\big(S_2 c_{12}s_{23}X_7 + S_7 s_{23}c_{71}X_7 + a_{23}c_{23}Y_7 + a_{67}s_{23}Z_7' - a_{71}s_{23}Z_7\big) \\
&+ S_2\big(S_7 c_{12}s_{23}c_{71}X_7 - a_{12}s_{12}s_{23}Y_7 + a_{23}c_{12}c_{23}Y_7 + a_{67}c_{12}s_{23}Z_7' \\
&- a_{71}c_{12}s_{23}Z_7\big) - S_6^2 s_{23}X_7 + S_6(-a_{67}s_{23}c_7 - a_{71}s_{23}) + S_7(-a_{23}c_{23}s_{67}c_7 \\
&- a_{67}s_{23}c_{67}c_7 - a_{71}s_{23}c_{67}) - a_{12}^2 s_{23}X_7 - a_{23}a_{67}c_{23}X_7' - a_{71}^2 s_{23}X_7 + K_2 s_{23}X_7,
\end{aligned}
$$

$$
\begin{aligned}
Q''_{10,2} = {} & S_1\big(-S_2 c_{12}s_{23}Y_7 + S_7 s_{23}s_{67}c_7 + a_{23}c_{23}X_7 + a_{67}s_{23}X_7'\big) + S_2\big(S_7 c_{12}s_{23}s_{67}c_7 \\
&- a_{12}s_{12}s_{23}X_7 + a_{23}c_{12}c_{23}X_7 + a_{67}c_{12}s_{23}X_7'\big) + S_6^2 s_{23}Y_7 + S_6\big(-S_7 s_{23}s_{71} \\
&+ a_{67}s_{23}\bar{X}_7'\big) - S_7^2 s_{23}c_{67}s_{71} + S_7\big(a_{23}c_{23}c_{71}X_7 + a_{67}s_{23}c_{71}X_7' - a_{71}s_{23}s_{71}X_7\big) \\
&+ a_{12}^2 s_{23}Y_7 + a_{23}\big(a_{67}c_{23}Z_7' - a_{71}c_{23}Z_7\big) - a_{67}a_{71}s_{23}\bar{Y}_7 - K_2 s_{23}Y_7,
\end{aligned}
$$

$$
\begin{aligned}
Q''_{10,3} = {} & -S_1 a_{12}s_{23}Z_7 + S_2(-S_7 s_{12}s_{23}s_{71}X_7 - a_{12}c_{12}s_{23}Z_7 - a_{23}s_{12}c_{23}Z_7 - a_{67}s_{12}s_{23}\bar{Y}_7 \\
&- a_{71}s_{12}s_{23}Y_7) - S_6 a_{12}s_{23} - S_7 a_{12}s_{23}c_{67} - a_{12}a_{71}s_{23}X_7,
\end{aligned}
$$

$$
\begin{aligned}
R''_{10,1} = {} & S_1\big(-S_7 s_{12}c_{23}s_{67}c_7 - a_{12}c_{12}c_{23}X_7 + a_{23}s_{12}s_{23}X_7 - a_{67}s_{12}c_{23}X_7'\big) - S_2^2 s_{12}c_{23}Y_7 \\
&- S_2 a_{12}c_{23}X_7 - S_6^2 s_{12}c_{23}Y_7 + S_6\big(S_7 s_{12}c_{23}s_{71} - a_{67}s_{12}c_{23}\bar{X}_7'\big) + S_7^2 s_{12}c_{23}c_{67}s_{71} \\
&+ S_7\big(-a_{12}c_{12}c_{23}c_{71}X_7 + a_{23}s_{12}s_{23}c_{71}X_7 - a_{67}s_{12}c_{23}c_{71}X_7' + a_{71}s_{12}c_{23}s_{71}X_7\big) \\
&+ a_{12}\big(a_{23}c_{12}s_{23}Y_7 - a_{67}c_{12}c_{23}Z_7' + a_{71}c_{12}c_{23}Z_7\big) + a_{23}\big(a_{67}s_{12}s_{23}Z_7' \\
&- a_{71}s_{12}s_{23}Z_7\big) + a_{67}a_{71}s_{12}c_{23}\bar{Y}_7 + K_2 s_{12}c_{23}Y_7,
\end{aligned}
$$

$$
\begin{aligned}
R''_{10,2} = {} & S_1\big(S_7 s_{12}c_{23}c_{71}X_7 + a_{12}c_{12}c_{23}Y_7 - a_{23}s_{12}s_{23}Y_7 + a_{67}s_{12}c_{23}Z_7' - a_{71}s_{12}c_{23}Z_7\big) \\
&- S_2^2 s_{12}c_{23}X_7 + S_2 a_{12}c_{23}Y_7 - S_6^2 s_{12}c_{23}X_7 + S_6(-a_{67}s_{12}c_{23}c_7 - a_{71}s_{12}c_{23}) \\
&+ S_7(-a_{12}c_{12}c_{23}s_{67}c_7 + a_{23}s_{12}s_{23}s_{67}c_7 - a_{67}s_{12}c_{23}c_{67}c_7 - a_{71}s_{12}c_{23}c_{67}) \\
&+ a_{12}\big(a_{23}c_{12}s_{23}X_7 - a_{67}c_{12}c_{23}X_7'\big) + a_{23}a_{67}s_{12}s_{23}c_{67}s_7 - a_{71}^2 s_{12}c_{23}X_7 \\
&+ K_2 s_{12}c_{23}X_7,
\end{aligned}
$$

$$
\begin{aligned}
R''_{10,3} = {} & -S_1^2 c_{12}c_{23}Z_7 + S_1(-S_2 c_{23}Z_7 - S_6 c_{12}c_{23} - S_7 c_{12}c_{23}c_{67} - a_{71}c_{12}c_{23}X_7) \\
&- S_2^2 c_{12}c_{23}Z_7 + S_2(-S_6 c_{23} - S_7 c_{23}c_{67} - a_{71}c_{23}X_7) - S_6^2 c_{12}c_{23}Z_7
\end{aligned}
$$

$$+S_6(-S_7c_{12}c_{23}c_{71}-a_{67}c_{12}c_{23}\bar{X}_7)-S_7^2c_{12}c_{23}c_{67}c_{71}+S_7(a_{12}s_{12}c_{23}s_{71}X_7$$
$$+a_{23}c_{12}s_{23}s_{71}X_7-a_{67}c_{12}c_{23}c_{67}\bar{X}_7-a_{71}c_{12}c_{23}c_{71}X_7)+a_{12}(-a_{23}s_{12}s_{23}Z_7$$
$$+a_{67}s_{12}c_{23}\bar{Y}_7+a_{71}s_{12}c_{23}Y_7)+a_{23}(a_{67}c_{12}s_{23}\bar{Y}_7+a_{71}c_{12}s_{23}Y_7)$$
$$-a_{67}a_{71}c_{12}c_{23}Z_7'+K_2c_{12}c_{23}Z_7-K_{39}, \quad (10.419)$$

$$M''_{i,j}=P''_{(i-6),j},\quad N''_{i,j}=Q''_{(i-6),j},\quad O''_{i,j}=R''_{(i-6),j},$$
$$P''_{i,j}=0,\quad Q''_{i,j}=0,\quad R''_{i,j}=0, \quad (10.420)$$
$$i=11\ldots16,\quad j=1\ldots3.$$

The vector **b** in Eq. (10.404) may be solved for by inverting the matrix **A** to yield

$$\mathbf{b}=\mathbf{A}^{-1}\mathbf{c}. \quad (10.421)$$

Using Eq. (10.408) to substitute for the vector **c** gives

$$\mathbf{b}=\mathbf{A}^{-1}\mathbf{T}_3\mathbf{a}. \quad (10.422)$$

Substituting this result into Eq. (10.284) yields

$$\mathbf{T}_1\mathbf{a}=\mathbf{T}_2\mathbf{A}^{-1}\mathbf{T}_3\mathbf{a}. \quad (10.423)$$

This equation can be rearranged as

$$\mathbf{Ta}=\mathbf{0}, \quad (10.424)$$

where

$$\mathbf{T}=[\mathbf{T}_1-\mathbf{T}_2\mathbf{A}^{-1}\mathbf{T}_3]. \quad (10.425)$$

The matrix **T** will have four rows and eighteen columns. All elements of this 4×18 matrix are expressed in terms of the constant mechanism parameters and the input angle, θ_7.

Equation (10.424) represents four equations of the form

$$[(t_{i,1}c_1+t_{i,2}s_1+t_{i,3})c_2+(t_{i,4}c_1+t_{i,5}s_1+t_{i,6})s_2$$
$$+(t_{i,7}c_1+t_{i,8}s_1+t_{i,9})]x_6+[(t_{i,10}c_1+t_{i,11}s_1+t_{i,12})c_2$$
$$+(t_{i,13}c_1+t_{i,14}s_1+t_{i,15})s_2+(t_{i,16}c_1+t_{i,17}s_1+t_{i,18})]=0,\quad i=1\ldots4, \quad (10.426)$$

where $t_{i,j}$ represents the element from the ith row and j^{th} column of the matrix **T**. The four equations of Eq. set (10.426) may be modified by substituting the tan-half-angle expressions for the sines and cosines of θ_1 and θ_2. The equations may be written as follows after multiplying each by the product $(1+x_1^2)(1+x_2^2)$:

$$\big[(a_{i,1}x_1^2+a_{i,2}x_1+a_{i,3})x_2^2+(b_{i,4}x_1^2+b_{i,5}x_1+b_{i,6})x_2+(d_{i,7}x_1^2+d_{i,8}x_1+d_{i,9})\big]x_6$$
$$+\big[(e_{i,10}x_1^2+e_{i,11}x_1+e_{i,12})x_2^2+(f_{i,13}x_1^2+f_{i,14}x_1+f_{i,15})x_2$$
$$+(g_{i,16}x_1^2+g_{i,17}x_1+g_{i,18})\big]=0,\quad i=1\ldots4, \quad (10.427)$$

where

$$
\begin{aligned}
a_{i,1} &= t_{i,9} - t_{i,3} - t_{i,7} + t_{i,1},\\
a_{i,2} &= 2(t_{i,8} - t_{i,2}),\\
a_{i,3} &= t_{i,9} - t_{i,3} + t_{i,7} - t_{i,1},\\
b_{i,4} &= 2(t_{i,6} - t_{i,4}),\\
b_{i,5} &= 4t_{i,5},\\
b_{i,6} &= 2(t_{i,6} + t_{i,4}),\\
d_{i,7} &= t_{i,9} + t_{i,3} - t_{i,7} - t_{i,1},\\
d_{i,8} &= 2(t_{i,8} + t_{i,2}),\\
d_{i,9} &= t_{i,9} + t_{i,3} + t_{i,7} + t_{i,1},\\
e_{i,10} &= t_{i,18} - t_{i,12} - t_{i,16} + t_{i,10},\\
e_{i,11} &= 2(t_{i,17} - t_{i,11}),\\
e_{i,12} &= t_{i,18} - t_{i,12} + t_{i,16} - t_{i,10},\\
f_{i,13} &= 2(t_{i,15} - t_{i,13}),\\
f_{i,14} &= 4t_{i,14},\\
f_{i,15} &= 2(t_{i,15} + t_{i,13}),\\
g_{i,16} &= t_{i,18} + t_{i,12} - t_{i,16} - t_{i,10},\\
g_{i,17} &= 2(t_{i,17} + t_{i,11}),\\
g_{i,18} &= t_{i,18} + t_{i,12} + t_{i,16} + t_{i,10}.
\end{aligned}
\tag{10.428}
$$

Equation (10.427) represents four equations that are of the form of Eq. (10.1). The input/output equation for the mechanism can be obtained from these equations as described in Section 9.1. An 8×8 determinant is expanded to yield a sixteenth-degree polynomial in the tan-half-angle of the output angle, θ_1.

10.2.10 Determination of θ_2 and θ_6

Section 9.3.2 describes how to determine the corresponding values for the tan-half-angle of the angles θ_2 and θ_3 from the four equations of Eq. set (9.75) for each calculated value of the tan-half-angle of the output angle, θ_1. The procedure for solving for the tan-half-angle of θ_2 and θ_6 for this mechanism is identical with the exception that x_3 in Eq. set (9.75) is replaced by x_6 in Eq. set (10.427). Following the solution method outlined in Section 9.3.2, expressions for the tan-half-angle for the corresponding values of θ_2 and θ_6 may be written as

$$x_2 = \frac{-|abeg||adgb| + |adgf||abed|}{|abef||adgb| - |adge||abed|}, \tag{10.429}$$

$$x_6 = \frac{-|abef||adgf| + |adge||abeg|}{|abef||adgb| - |adge||abed|}. \tag{10.430}$$

10.2.11 Determination of θ_4 and θ_5

At this point of the analysis, corresponding values for the angles θ_1, θ_2, and θ_6 have been determined. Thus, for each of these solution sets it is possible to determine numerical values for the components of vector **a** as defined in Eq. (10.286). The vector **a** may then be substituted into Eq. (10.422) to solve for the vector **b**. The sine and cosine of θ_4 and θ_5 are now known, as they are the eleventh, fourteenth, fifteenth, and sixteenth components of vector **b**. Unique corresponding values for θ_4 and θ_5 can now be computed because the sine and cosine of these angles are known.

10.2.12 Determination of θ_3

Fundamental sine and sine–cosine laws for a spherical heptagon may be written as

$$X_{56712} = s_{34}s_3, \tag{10.431}$$

$$Y_{56712} = s_{34}c_3. \tag{10.432}$$

The sine and cosine of the corresponding value of θ_3 may be obtained by evaluating the left-hand sides of these equations using the previously calculated solution set of angles.

10.2.13 Numerical example

The analysis of the 7R group 4 spatial mechanism has been completed, and it was shown that a maximum of sixteen solution configurations exist. Table 10.1 shows data that were used as input for a numerical example. The calculated values for the sixteen configurations are listed in Table 10.2. Figure 10.2 shows the sixteen configurations of the mechanism.

10.3 RRRSR spatial mechanism

A significant simplification in the solution of group 4 spatial mechanisms occurs when special geometric conditions exist in the mechanism. The first case to be considered is the RRRSR spatial mechanism that is shown in Figure 10.3 with link a_{71} fixed to ground.

Table 10.1. *7R mechanism parameters.*

Link length, cm.	Twist angle, deg.	Joint offset, cm.	Joint angle, deg.
$a_{12} = 8.7$	$\alpha_{12} = 90$	$S_1 = 3.0$	θ_1 = variable
$a_{23} = 2.5$	$\alpha_{23} = 90$	$S_2 = 1.3$	θ_2 = variable
$a_{34} = 9$	$\alpha_{34} = 90$	$S_3 = 0.7$	θ_3 = variable
$a_{45} = 0.1$	$\alpha_{45} = 90$	$S_4 = 3.4$	θ_4 = variable
$a_{56} = 8.2$	$\alpha_{56} = 90$	$S_5 = 0$	θ_5 = variable
$a_{67} = 7.1$	$\alpha_{67} = 90$	$S_6 = 4.7$	θ_6 = variable
$a_{71} = 3.7$	$\alpha_{71} = 90$	$S_7 = 1.8$	$\theta_7 = 278$ (input)

Table 10.2. *Calculated configurations for the 7R spatial mechanism.*

Solution	θ_1, deg.	θ_2, deg.	θ_3, deg.	θ_4, deg.	θ_5, deg.	θ_6, deg.
A	−97.56	105.97	14.43	−132.72	−6.80	30.15
B	−90.68	−166.91	7.78	138.09	−11.94	−144.51
C	−63.36	22.39	−151.65	72.99	−177.58	177.11
D	−62.07	−93.71	179.09	−71.14	−151.34	77.06
E	−31.38	−44.61	155.20	−134.21	−104.87	105.70
F	1.35	179.15	73.01	−30.02	162.09	118.78
G	5.37	164.37	−30.70	154.33	−67.30	−102.08
H	34.22	−23.83	−122.82	28.51	106.28	−113.47
I	117.34	−70.97	−167.15	78.65	−23.09	−77.78
J	−147.03	157.60	−58.82	−11.98	−115.06	107.32
K	114.05	13.31	−162.78	−71.40	−13.94	−148.40
L	−167.58	165.47	51.26	166.27	27.04	−99.17
M	146.06	−41.07	176.07	30.20	−59.81	−97.72
N	−134.71	−29.64	−102.22	−164.93	59.33	118.96
O	78.45	119.63	3.05	72.83	−165.38	−43.64
P	62.08	−161.89	29.35	−59.19	−179.21	165.13

The ball and socket joint, which is designated by the letter S, is modeled in the figure by three cointersecting revolute joint axes, $\mathbf{S}_2$, $\mathbf{S}_3$, and $\mathbf{S}_4$. The special geometric values for this case are

$$a_{23} = a_{34} = S_3 = 0. \tag{10.433}$$

In this analysis the angle θ_7 will be the known input value. The angle θ_1 will be solved for first and is the output angle. Specifically, the problem can be stated as

given:

$\alpha_{12}, \alpha_{23}, \alpha_{34}, \alpha_{45}, \alpha_{56}, \alpha_{67}, \alpha_{71},$
$a_{12}, a_{45}, a_{56}, a_{67}, a_{71},$
$S_1, S_2, S_4, S_5, S_6, S_7,$
$a_{23} = a_{34} = S_3 = 0$, and
$\theta_7,$

find: $\theta_1, \theta_2, \theta_3, \theta_4, \theta_5,$ and θ_6.

It will be shown that a maximum of eight solution configurations exist for this mechanism.

10.3.1 Determination of input–output equation

It will be shown that the input–soutput equation can be obtained from two equations that contain the output angle, θ_1, and an extra angle, θ_5. Eliminating the angle θ_5 from the pair of equations will result in a fourth-degree input–output equation in the tan-half-angle of θ_1.

The first equation is obtained from a projection of the vector loop equation onto the direction of the vector $\mathbf{S}_6$. Using set 10 from the table of direction cosines, this projection

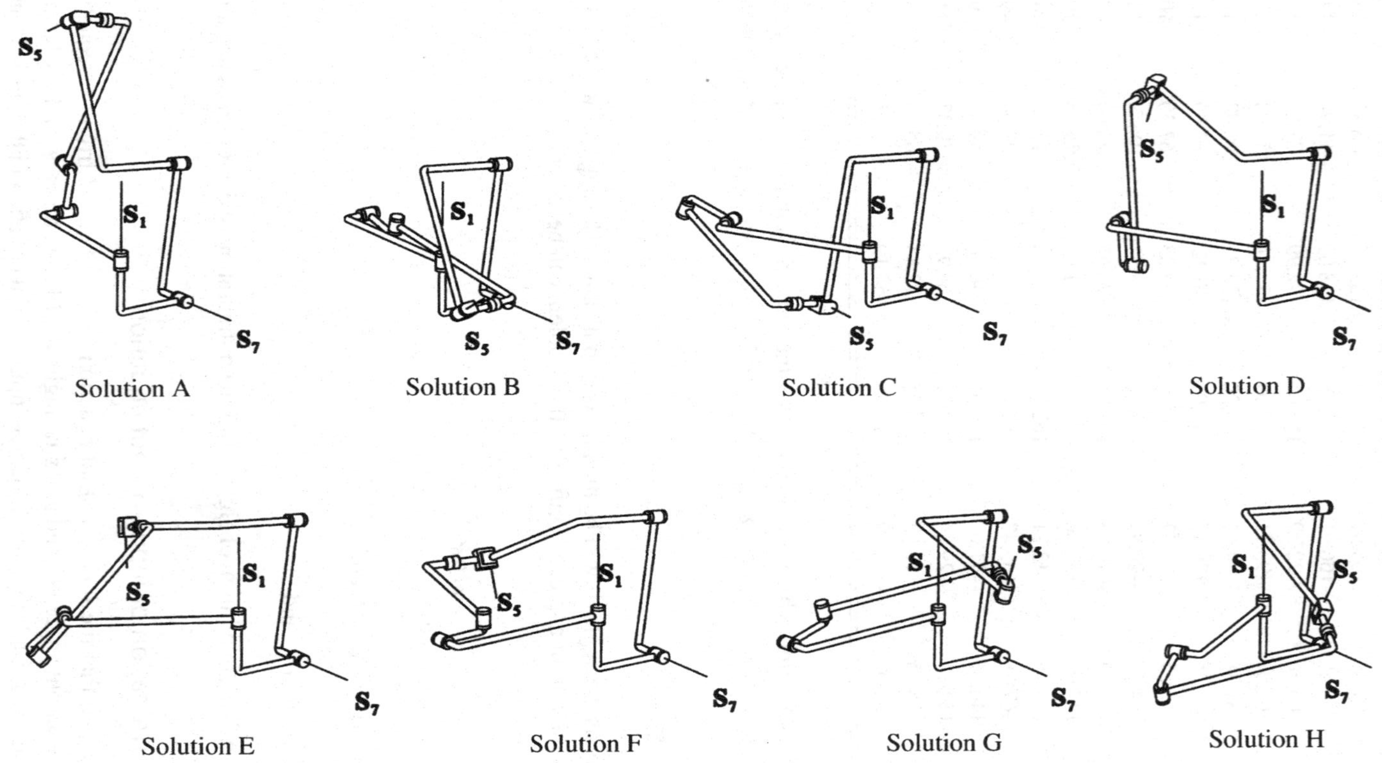

Figure 10.2. Sixteen configurations of the 7R spatial mechanism.

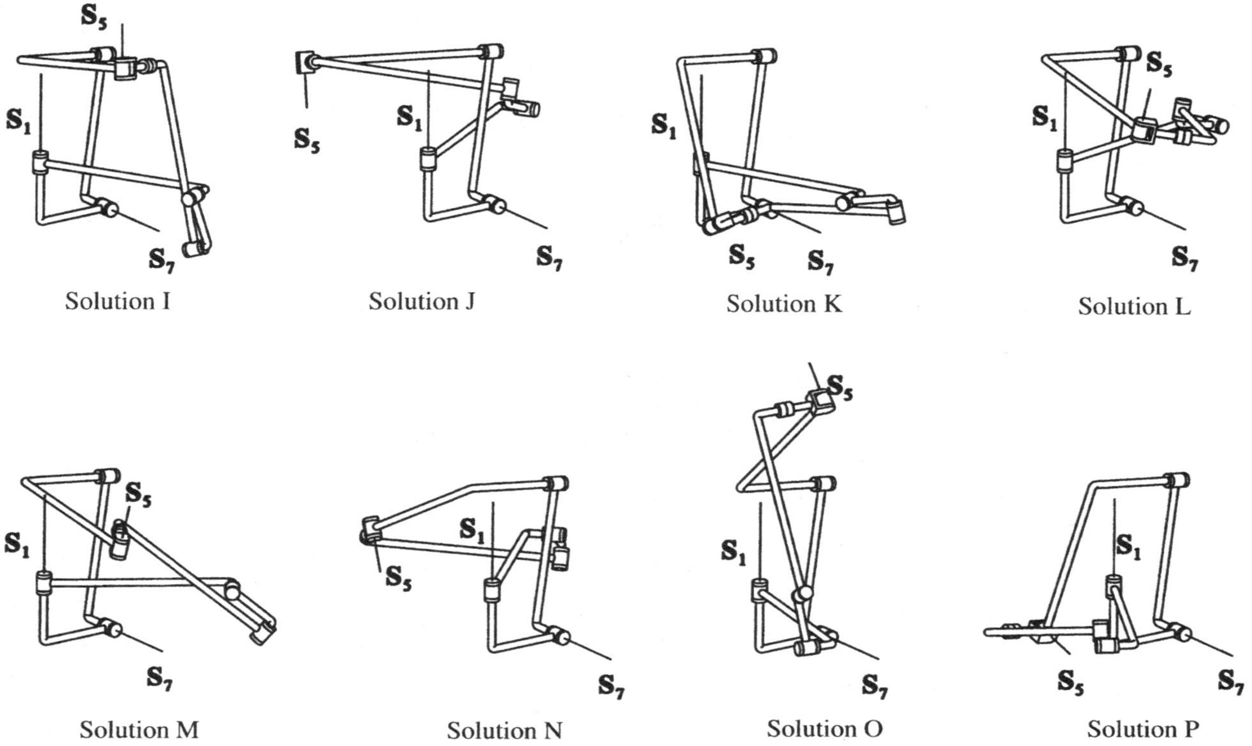

Figure 10.2. (*cont.*)

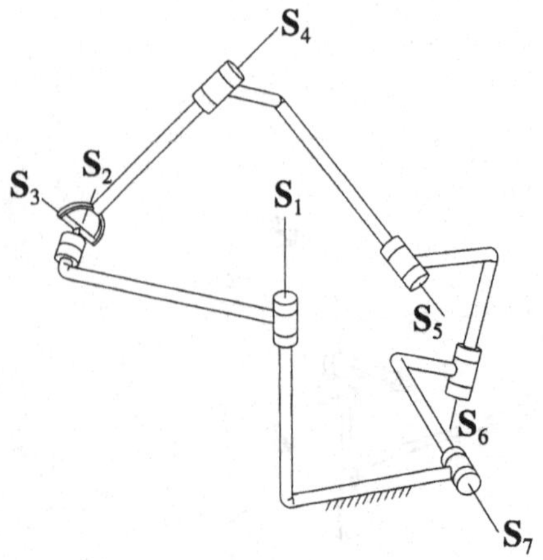

Figure 10.3. RRRSR spatial mechanism.

may be written as

$$S_1Z_{2345} + S_2Z_{345} + S_4Z_5 + S_5c_{56} + S_6 + S_7Z_{12345} + a_{12}U_{23456}$$
$$+ a_{45}U_{56} + a_{71}U_{123456} = 0. \tag{10.434}$$

Subsidiary and fundamental cosine laws for a spherical heptagon may be used to substitute for the terms Z_{2345}, Z_{345}, and Z_{12345}. Also, subsidiary and fundamental polar sine laws for a spherical heptagon may be used to substitute for the terms U_{23456} and U_{123456} to give

$$S_1Z_7 + S_2Z_{71} + S_4Z_5 + S_5c_{56} + S_6 + S_7c_{67} + a_{12}U_{176} + a_{45}U_{56} + a_{71}U_{76} = 0. \tag{10.435}$$

Terms that contain the angle θ_5 are transferred to the right side of the equation to give

$$S_1Z_7 + S_2Z_{71} + S_5c_{56} + S_6 + S_7c_{67} + a_{12}U_{176} + a_{71}U_{76} = -S_4Z_5 - a_{45}U_{56}. \tag{10.436}$$

Expanding Z_5 and U_{56} and then regrouping terms gives

$$S_1Z_7 + S_2Z_{71} + S_4c_{45}c_{56} + S_5c_{56} + S_6 + S_7c_{67} + a_{12}U_{176} + a_{71}U_{76}$$
$$= (S_4s_{45}s_{56})c_5 - (a_{45}s_{56})s_5. \tag{10.437}$$

The left side of Eq. (10.437) contains θ_1 as the only unknown parameter, whereas the right side has θ_5 as the only unknown.

The second equation is obtained by taking a self-scalar product of the vector loop equation. The vector loop equation may be written as

$$S_6\mathbf{S}_6 + a_{67}\mathbf{a}_{67} + S_7\mathbf{S}_7 + a_{71}\mathbf{a}_{71} + S_1\mathbf{S}_1 + a_{12}\mathbf{a}_{12} + S_2\mathbf{S}_2$$
$$= -(S_4\mathbf{S}_4 + a_{45}\mathbf{a}_{45} + S_5\mathbf{S}_5 + a_{56}\mathbf{a}_{56}). \tag{10.438}$$

The self-scalar product of both sides of Eq. (10.438) can be expressed in the form

$$\begin{aligned}L &+ S_6\mathbf{S}_6\cdot(S_7\mathbf{S}_7 + a_{71}\mathbf{a}_{71} + S_1\mathbf{S}_1 + a_{12}\mathbf{a}_{12} + S_2\mathbf{S}_2) + a_{67}\mathbf{a}_{67}\cdot(a_{71}\mathbf{a}_{71} + S_1\mathbf{S}_1 + a_{12}\mathbf{a}_{12}\\ &+ S_2\mathbf{S}_2) + S_7\mathbf{S}_7\cdot(S_1\mathbf{S}_1 + a_{12}\mathbf{a}_{12} + S_2\mathbf{S}_2) + a_{71}\mathbf{a}_{71}\cdot(a_{12}\mathbf{a}_{12} + S_2\mathbf{S}_2)\\ &+ S_1\mathbf{S}_1\cdot(S_2\mathbf{S}_2) = S_4\mathbf{S}_4\cdot(S_5\mathbf{S}_5 + a_{56}\mathbf{a}_{56}) + a_{45}\mathbf{a}_{45}\cdot(a_{56}\mathbf{a}_{56}),\end{aligned} \tag{10.439}$$

where L is defined as

$$L = (S_6^2 + a_{67}^2 + S_7^2 + a_{71}^2 + S_1^2 + a_{12}^2 + S_2^2 - S_4^2 - a_{45}^2 - S_5^2 - a_{56}^2)/2. \tag{10.440}$$

The scalar products of mutually perpendicular vectors, such as, for example, $\mathbf{S}_6$ and $\mathbf{a}_{67}$, equal zero, and all these terms have been deleted from Eq. (10.439). Evaluating the scalar products using the sets of direction cosines listed in the appendix yields

$$\begin{aligned}L &+ S_6(S_7c_{67} + a_{71}X_7 + S_1Z_7 + a_{12}X_{71} + S_2Z_{71}) + a_{67}(a_{71}c_7 + S_1\bar{X}_7 + a_{12}W_{17}\\ &+ S_2X_{17}) + S_7(S_1c_{71} + a_{12}X_1 + S_2Z_1) + a_{71}(a_{12}c_1 + S_2\bar{X}_1)\\ &+ S_1(S_2c_{12}) = S_4(S_5c_{45} + a_{56}X_5) + a_{45}a_{56}c_5.\end{aligned} \tag{10.441}$$

Expanding the definition of the term X_5 and rearranging this equation gives

$$\begin{aligned}L &+ S_6(S_7c_{67} + a_{71}X_7 + S_1Z_7 + a_{12}X_{71} + S_2Z_{71}) + a_{67}(a_{71}c_7 + S_1\bar{X}_7\\ &+ a_{12}W_{17} + S_2X_{17}) + S_7(S_1c_{71} + a_{12}X_1 + S_2Z_1) + a_{71}(a_{12}c_1 + S_2\bar{X}_1)\\ &+ S_1(S_2c_{12}) - S_4S_5c_{45} = a_{56}[(S_4s_{45})s_5 + (a_{45})c_5].\end{aligned} \tag{10.442}$$

Equations (10.437) and (10.442) may be written respectively as

$$Q_{71} = s_{56}(S_4s_{45}c_5 - a_{45}s_5), \tag{10.443}$$

$$R_{71} = a_{56}(S_4s_{45}s_5 + a_{45}c_5), \tag{10.444}$$

where

$$Q_{71} = S_1Z_7 + S_2Z_{71} + S_4c_{45}c_{56} + S_5c_{56} + S_6 + S_7c_{67} + a_{12}U_{176} + a_{71}U_{76}, \tag{10.445}$$

$$\begin{aligned}R_{71} &= L + S_6(S_7c_{67} + a_{71}X_7 + S_1Z_7 + a_{12}X_{71} + S_2Z_{71}) + a_{67}(a_{71}c_7\\ &\quad + S_1\bar{X}_7 + a_{12}W_{17} + S_2X_{17}) + S_7(S_1c_{71} + a_{12}X_1 + S_2Z_1)\\ &\quad + a_{71}(a_{12}c_1 + S_2\bar{X}_1) + S_1(S_2c_{12}) - S_4S_5c_{45}.\end{aligned} \tag{10.446}$$

The terms Q_{71} and R_{71} contain the angle θ_1 as the only unknown parameter and can thus be expressed in the form

$$Q_{71} = K_1c_1 + K_2s_1 + K_3, \tag{10.447}$$

$$R_{71} = K_4c_1 + K_5s_1 + K_6, \tag{10.448}$$

where

$$
\begin{aligned}
K_1 &= S_2 s_{12} Y_7 + a_{12} X_7, \\
K_2 &= S_2 s_{12} X_7 - a_{12} Y_7, \\
K_3 &= S_1 Z_7 + S_2 c_{12} Z_7 + S_4 c_{45} c_{56} + S_5 c_{56} + S_6 + S_7 c_{67} + a_{71} X_7, \\
K_4 &= S_2(S_6 s_{12} Y_7 - S_7 s_{12} s_{71} + a_{67} s_{12} c_{71} s_7) + S_6 a_{12} X_7 + a_{12}(a_{67} c_7 + a_{71}), \\
K_5 &= S_2(S_6 s_{12} X_7 + a_{67} s_{12} c_7 + a_{71} s_{12}) + a_{12}(-S_6 Y_7 + S_7 s_{71} - a_{67} c_{71} s_7), \\
K_6 &= L + S_1(S_2 c_{12} + S_6 Z_7 + S_7 c_{71} + a_{67} \bar{X}_7) + S_2(S_6 c_{12} Z_7 + S_7 c_{12} c_{71} + a_{67} c_{12} \bar{X}_7) \\
&\quad - S_4 S_5 c_{45} + S_6(S_7 c_{67} + a_{71} X_7) + a_{67} a_{71} c_7.
\end{aligned} \tag{10.449}
$$

Subtracting $a_{56}x_5$ times Eq. (10.443) from s_{56} times Eq. (10.444), where x_5 is the tan-half-angle of θ_5, gives

$$s_{56}R_{71} - a_{56}x_5Q_{71} = a_{56}s_{56}[S_4s_{45}(s_5 - x_5c_5) + a_{45}(c_5 + x_5s_5)]. \tag{10.450}$$

Using the trigonometric identities listed in Eqs. (9.108) and (9.109) gives

$$s_{56}R_{71} - a_{56}x_5Q_{71} = a_{56}s_{56}(S_4s_{45}x_5 + a_{45}). \tag{10.451}$$

Regrouping this equation gives

$$(-a_{56}Q_{71} - S_4a_{56}s_{45}s_{56})x_5 + (s_{56}R_{71} - a_{45}a_{56}s_{56}) = 0. \tag{10.452}$$

Adding a_{56} times Eq. (10.443) to $s_{56}x_5$ times Eq. (10.444) gives

$$a_{56}Q_{71} + s_{56}x_5R_{71} = a_{56}s_{56}[S_4s_{45}(s_5x_5 + c_5) + a_{45}(c_5x_5 - s_5)]. \tag{10.453}$$

Simplifying this equation by substituting the results of the trigonometric identities and rearranging yields

$$(s_{56}R_{71} + a_{45}a_{56}s_{56})x_5 + (a_{56}Q_{71} - S_4a_{56}s_{45}s_{56}) = 0. \tag{10.454}$$

Equations (10.452) and (10.454) can be factored into the form

$$(A_ic_1 + B_is_1 + D_i)x_5 + (E_ic_1 + F_is_1 + G_i) = 0, \quad i = 1, 2, \tag{10.455}$$

where the coefficients A_i through G_i are defined as

$$
\begin{aligned}
&A_1 = -a_{56}K_1, && B_1 = -a_{56}K_2, && D_1 = -a_{56}K_3 - S_4a_{56}s_{45}s_{56}, \\
&E_1 = s_{56}K_4, && F_1 = s_{56}K_5, && G_1 = s_{56}K_6 - a_{45}a_{56}s_{56}, \\
&A_2 = s_{56}K_4, && B_2 = s_{56}K_5, && D_2 = s_{56}K_6 + a_{45}a_{56}s_{56}, \\
&E_2 = a_{56}K_1, && F_2 = a_{56}K_2, && G_2 = a_{56}K_3 - S_4a_{56}s_{45}s_{56}.
\end{aligned} \tag{10.456}
$$

Substituting the tan-half-angle identities for the sine and cosine of θ_1 in Eq. (10.455) and then multiplying throughout by $(1 + x_1^2)$ and regrouping gives

$$\left(a_ix_1^2 + b_ix_1 + d_i\right)x_5 + \left(e_ix_1^2 + f_ix_1 + g_i\right) = 0, \quad i = 1, 2, \tag{10.457}$$

where

$$a_i = D_i - A_i, \qquad b_i = 2B_i, \qquad d_i = D_i + A_i,$$
$$e_i = G_i - E_i, \qquad f_i = 2F_i, \qquad g_i = G_i + E_i. \tag{10.458}$$

Equation (10.457) represents two equations that are linear in the variable x_5. A common solution for x_5 will exist for these two equations only if they are linearly dependent (see discussion in Section 8.2). Thus, for a common solution of x_5 to exist, the coefficients of Eq. set (10.457) must satisfy the condition

$$\begin{vmatrix} (a_1x_1^2 + b_1x_1 + d_1) & (e_1x_1^2 + f_1x_1 + g_1) \\ (a_2x_1^2 + b_2x_1 + d_2) & (e_2x_1^2 + f_2x_1 + g_2) \end{vmatrix} = 0. \tag{10.459}$$

Expanding this determinant will yield the following fourth-degree input–output equation:

$$(a_1e_2 - a_2e_1)x_1^4 + (a_1f_2 - a_2f_1 + b_1e_2 - b_2e_1)x_1^3 + (a_1g_2 - a_2g_1 + b_1f_2 - b_2f_1 + d_1e_2 - d_2e_1)x_1^2 + (b_1g_2 - b_2g_1 + d_1f_2 - d_2f_1)x_1 + (d_1g_2 - d_2g_1) = 0. \tag{10.460}$$

A corresponding value for θ_1 can be obtained for each value of x_1 from

$$\theta_1 = 2\tan^{-1}(x_1). \tag{10.461}$$

10.3.2 Determination of θ_5

The value of the $x_5 = \tan(\theta_5/2)$ that corresponds to each calculated value of θ_1 can be found from either of the linear equations of Eq. set (10.457). Thus, x_5 may be calculated from

$$x_5 = \frac{-(e_1x_1^2 + f_1x_1 + g_1)}{a_1x_1^2 + b_1x_1 + d_1} \tag{10.462}$$

or

$$x_5 = \frac{-(e_2x_1^2 + f_2x_1 + g_2)}{a_2x_1^2 + b_2x_1 + d_2}. \tag{10.463}$$

The value of $\theta_5 = 2\tan^{-1}(x_5)$ (see Eq. (10.461)).

10.3.3 Determination of θ_6

The corresponding value of θ_6 may be obtained from two projections of the vector loop equation. Using set 6 from the table of direction cosines, the vector loop equation can be projected onto the direction of the vectors $\mathbf{a}_{67}$ and $(\mathbf{S}_6 \times \mathbf{a}_{67})$ to yield

$$S_1\bar{X}_7 + S_2X_{17} + S_4X_{3217} + S_5X_{43217} + a_{12}W_{17} + a_{45}W_{43217} + a_{56}c_6 + a_{67} + a_{71}c_7 = 0, \tag{10.464}$$

$$S_1\bar{Y}_7 + S_2Y_{17} + S_4Y_{3217} + S_5Y_{43217} - S_7s_{67} - a_{12}U^*_{176} - a_{45}U^*_{432176} - a_{56}s_6 + a_{71}c_{67}s_7 = 0. \tag{10.465}$$

Substituting fundamental and subsidiary spherical sine and sine–cosine laws for the terms X_{3217}, X_{43217}, Y_{3217}, and Y_{43217} and subsidiary polar sine–cosine and cosine laws for the terms W_{43217} and U^*_{432176} gives

$$S_1\bar{X}_7 + S_2X_{17} + S_4X_{56} + S_5X_6 + a_{12}W_{17} + a_{45}W_{56} + a_{56}c_6 + a_{67} + a_{71}c_7 = 0, \tag{10.466}$$

$$S_1\bar{Y}_7 + S_2Y_{17} - S_4X^*_{56} + S_5s_{56}c_6 - S_7s_{67} - a_{12}U^*_{176} + a_{45}V_{56} - a_{56}s_6 + a_{71}c_{67}s_7 = 0. \tag{10.467}$$

Each of these equations contains the sine and cosine of θ_6 as its only unknown parameter and can be expressed in the form

$$[S_4X_5 + a_{45}c_5 + a_{56}]c_6 + [-S_4Y_5 + S_5s_{56} - a_{45}c_{56}s_5]s_6 + [S_1\bar{X}_7 + S_2X_{17} + a_{12}W_{17} + a_{67} + a_{71}c_7] = 0, \tag{10.468}$$

$$[-S_4Y_5 + S_5s_{56} - a_{45}c_{56}s_5]c_6 + [-S_4X_5 - a_{45}c_5 - a_{56}]s_6 + \left[S_1\bar{Y}_7 + S_2Y_{17} - S_7s_{67} - a_{12}U^*_{176} + a_{71}c_{67}s_7\right] = 0. \tag{10.469}$$

The expressions in brackets can be numerically evaluated, as they are defined in terms of the constant mechanism parameters, the input angle, and the previously calculated joint parameters. Equations (10.468) and (10.469) thus represent two linear equations in the two unknowns, s_6 and c_6. Solving for s_6 and c_6 will yield the unique corresponding value for the angle θ_6.

10.3.4 Determination of θ_3

The following spherical cosine law may be written for a spatial heptagon:

$$Z_{1765} = Z_3. \tag{10.470}$$

Expanding the definition of the term Z_3 and solving for c_3 gives

$$c_3 = \frac{c_{23}c_{34} - Z_{1765}}{s_{23}s_{34}}. \tag{10.471}$$

Because it is not possible to solve for a unique value for the sine of θ_3, two values of θ_3 will exist for each set of angles $[\theta_1, \theta_5, \theta_6]$. Thus, a total of eight solution configurations will exist for the RRRSR spatial mechanism.

10.3.5 Determination of θ_2

The following sine and sine–cosine laws may be written for a spherical heptagon:

$$X_{5671} = X_{32}, \tag{10.472}$$

$$Y_{5671} = -X^*_{32}. \tag{10.473}$$

Expanding X_{32} and X^*_{32} and rearranging gives

$$(\bar{X}_3)c_2 + (-\bar{Y}_3)s_2 + (-X_{5671}) = 0, \tag{10.474}$$

$$(\bar{Y})_3c_2 + (\bar{X}_3)s_2 + (Y_{5671}) = 0. \tag{10.475}$$

All the terms in parentheses are defined in terms of the constant mechanism parameters and the previously calculated joint angles. Thus, these two equations represent two linear equations in the unknowns s_2 and c_2. A unique corresponding value for θ_2 can be determined from the sine and cosine values.

10.3.6 Determination of θ_4

The following fundamental sine and sine–cosine laws may be written for a spherical heptagon:

$$X_{67123} = s_{45}s_4, \tag{10.476}$$

$$Y_{67123} = s_{45}c_4. \tag{10.477}$$

Upon solving for the sine and cosine of θ_4, a unique corresponding value for this angle can be obtained.

10.3.7 Numerical example

It has been shown that a total of eight solution configurations exist for the RRRSR spatial mechanism. Although this is a group 4 spatial mechanism, the special geometry of having $a_{23} = a_{34} = S_3 = 0$ greatly simplifies the analysis. A fourth-degree input/output equation was obtained to solve for the angle θ_1. Unique corresponding values were then determined for the angles θ_5 and θ_6. Two values for the angles θ_2, θ_3, and θ_4 were obtained for each of the four sets of angles $[\theta_1, \theta_5, \theta_6]$, thus giving a total of eight solutions.

The mechanism dimensions of a numerical example are listed in Table 10.3. The resulting eight solution configurations are listed in Table 10.4 and are drawn in Figure 10.4. It is apparent from the figure that there are four classes of solutions. For example, solution B is the same as solution A except that θ_4 has been rotated an additional 180 degrees,

Table 10.3. *RRRSR mechanism parameters.*

Link length, cm.	Twist angle, deg.	Joint offset, cm.	Joint angle, deg.
$a_{12} = 9.9$	$\alpha_{12} = 60$	$S_1 = 8.5$	$\theta_1 =$ variable
$a_{23} = 0$	$\alpha_{23} = 90$	$S_2 = 2.0$	$\theta_2 =$ variable
$a_{34} = 0$	$\alpha_{34} = 90$	$S_3 = 0$	$\theta_3 =$ variable
$a_{45} = 7.6$	$\alpha_{45} = 90$	$S_4 = 8.3$	$\theta_4 =$ variable
$a_{56} = 8.0$	$\alpha_{56} = 60$	$S_5 = 8.6$	$\theta_5 =$ variable
$a_{67} = 3.2$	$\alpha_{67} = 75$	$S_6 = 4.9$	$\theta_6 =$ variable
$a_{71} = 7.7$	$\alpha_{71} = 90$	$S_7 = 8.6$	$\theta_7 = 21$ (input)

Table 10.4. *Calculated configurations for the RRRSR spatial mechanism.*

Solution	θ_1, deg.	θ_2, deg.	θ_3, deg.	θ_4, deg.	θ_5, deg.	θ_6, deg.
A	−114.34	−144.83	106.86	−12.35	−124.13	−53.55
B	−114.34	35.17	−106.86	167.65	−124.13	−53.55
C	−38.38	148.30	134.67	−52.23	10.29	−141.63
D	−38.38	−31.70	−134.67	127.77	10.29	−141.63
E	134.78	80.24	63.10	−121.61	11.94	−82.62
F	134.78	−99.76	−63.10	58.39	11.94	−82.62
G	−148.58	−79.59	112.94	−75.77	−124.61	−14.16
H	−148.58	100.41	−112.94	104.23	−124.61	−14.16

causing the vector $\mathbf{S}_3$ to point in the opposite direction. The angle θ_3 is the negative of its value for solution A, and θ_2 is advanced by 180 degrees.

10.4 RRSRR spatial mechanism

The RRSRR spatial mechanism is similar to the previous case in that the ball and socket joint can be modeled by three intersecting revolute joint axes $\mathbf{S}_3$, $\mathbf{S}_4$, and $\mathbf{S}_5$ as shown in Figure 10.5. The special geometric values for this case are

$$a_{34} = a_{45} = S_4 = 0. \tag{10.478}$$

In this analysis, the angle θ_7 will be the known input value. The angle θ_1 will be solved for first and will be referred to as the output angle. Specifically, the problem can be stated as

given: $\alpha_{12}, \alpha_{23}, \alpha_{34}, \alpha_{45}, \alpha_{56}, \alpha_{67}, \alpha_{71}$,
$a_{12}, a_{23}, a_{56}, a_{67}, a_{71}$,
$S_1, S_2, S_3, S_5, S_6, S_7$,
$a_{34} = a_{45} = S_4 = 0$,
and θ_7,

find: $\theta_1, \theta_2, \theta_3, \theta_4, \theta_5$, and θ_6.

A maximum of eight solution configurations exist for this mechanism also.

10.4.1 Determination of input/output equation

It will be shown that the input/output equation can be obtained from two equations that contain the input angle θ_7, the output angle θ_1, and the extra angle θ_2. Elimination of θ_2 from this pair of equations will result in a fourth-degree input/output equation in the tan-half-angle of θ_1.

The first equation is obtained from a projection of the vector loop equation onto the direction of the vector $\mathbf{S}_6$. Using set 6 of the table of direction cosines, this projection may be written as

$$S_1\bar{Z}_7 + S_2Z_{17} + S_3Z_{217} + S_5Z_{43217} + S_6 + S_7c_{67} + a_{12}U_{176} + a_{23}U_{2176} + a_{71}U_{76} = 0. \tag{10.479}$$

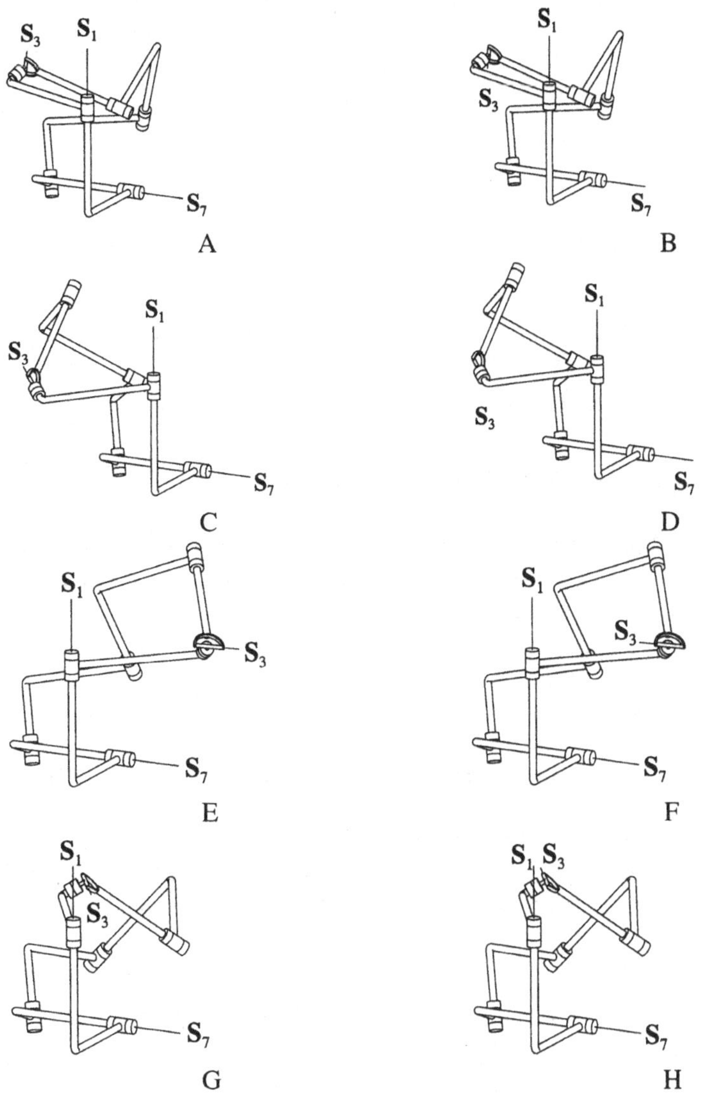

Figure 10.4. Eight configurations of the RRRSR spatial mechanism.

The fundamental cosine law $Z_{43217} = c_{56}$ is substituted to give

$$S_1\bar{Z}_7 + S_2Z_{17} + S_3Z_{217} + S_5c_{56} + S_6 + S_7c_{67} + a_{12}U_{176} + a_{23}U_{2176} + a_{71}U_{76} = 0. \tag{10.480}$$

The second equation is obtained by writing the vector loop equation as

$$a_{67}\mathbf{a}_{67} + S_7\mathbf{S}_7 + a_{71}\mathbf{a}_{71} + S_1\mathbf{S}_1 + a_{12}\mathbf{a}_{12} + S_2\mathbf{S}_2 + a_{23}\mathbf{a}_{23} + S_3\mathbf{S}_3$$
$$= -(S_5\mathbf{S}_5 + a_{56}\mathbf{a}_{56} + S_6\mathbf{S}_6) \tag{10.481}$$

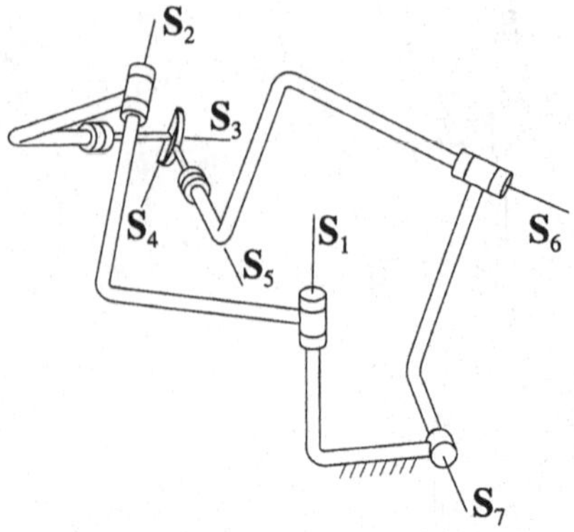

Figure 10.5. RRSRR spatial mechanism.

and then taking the self-scalar product, which can be written as

$$\begin{aligned} L &+ a_{67}\mathbf{a}_{67} \cdot (a_{71}\mathbf{a}_{71} + S_1\mathbf{S}_1 + a_{12}\mathbf{a}_{12} + S_2\mathbf{S}_2 + a_{23}\mathbf{a}_{23} + S_3\mathbf{S}_3) + S_7\mathbf{S}_7 \cdot (S_1\mathbf{S}_1 \\ &+ a_{12}\mathbf{a}_{12} + S_2\mathbf{S}_2 + a_{23}\mathbf{a}_{23} + S_3\mathbf{S}_3) + a_{71}\mathbf{a}_{71} \cdot (a_{12}\mathbf{a}_{12} + S_2\mathbf{S}_2 + a_{23}\mathbf{a}_{23} + S_3\mathbf{S}_3) \\ &+ S_1\mathbf{S}_1 \cdot (S_2\mathbf{S}_2 + a_{23}\mathbf{a}_{23} + S_3\mathbf{S}_3) + a_{12}\mathbf{a}_{12} \cdot (a_{23}\mathbf{a}_{23} + S_3\mathbf{S}_3) \\ &+ S_2\mathbf{S}_2 \cdot S_3\mathbf{S}_3 = S_5\mathbf{S}_5 \cdot S_6\mathbf{S}_6, \end{aligned} \tag{10.482}$$

where L is defined as

$$L = \left(a_{67}^2 + S_7^2 + a_{71}^2 + S_1^2 + a_{12}^2 + S_2^2 + a_{23}^2 + S_3^2 - S_5^2 - a_{56}^2 - S_6^2\right)/2. \tag{10.483}$$

The scalar product of mutually perpendicular vectors, such as, for example, $\mathbf{a}_{67}$ and $\mathbf{S}_7$, equals zero, and all these terms have been deleted from Eq. (10.482). Evaluating the scalar products using the sets of direction cosines listed in the appendix yields

$$\begin{aligned} L &+ a_{67}(a_{71}c_7 + S_1\bar{X}_7 + a_{12}W_{17} + S_2X_{17} + a_{23}W_{217} + S_3X_{217}) + S_7(S_1c_{71} + a_{12}X_1 \\ &+ S_2Z_1 + a_{23}X_{12} + S_3Z_{12}) + a_{71}(a_{12}c_1 + S_2\bar{X}_1 + a_{23}W_{21} + S_3X_{21}) + S_1(S_2c_{12} \\ &+ a_{23}X_2 + S_3Z_2) + a_{12}(a_{23}c_2 + S_3\bar{X}_2) + S_2S_3c_{23} = S_5S_6c_{56}. \end{aligned} \tag{10.484}$$

Equations (10.482) and (10.484) may be factored into the form

$$(A_ic_1 + B_is_1 + D_i)c_2 + (E_ic_1 + F_is_1 + G_i)s_2 + (H_ic_1 + I_is_1 + J_i) = 0, \tag{10.485}$$

where $i = 1, 2$ and

$$\begin{aligned} A_1 &= S_3c_{12}s_{23}Y_7 + a_{23}X_7, \\ B_1 &= S_3c_{12}s_{23}X_7 - a_{23}Y_7, \\ D_1 &= -S_3s_{12}s_{23}Z_7, \\ E_1 &= S_3s_{23}X_7 - a_{23}c_{12}Y_7, \\ F_1 &= -S_3s_{23}Y_7 - a_{23}c_{12}X_7, \\ G_1 &= a_{23}s_{12}Z_7, \\ H_1 &= S_2s_{12}Y_7 + S_3s_{12}c_{23}Y_7 + a_{12}X_7, \end{aligned}$$

$$\begin{aligned}
I_1 &= S_2 s_{12} X_7 + S_3 s_{12} c_{23} X_7 - a_{12} Y_7, \\
J_1 &= S_1 Z_7 + S_2 c_{12} Z_7 + S_3 c_{12} c_{23} Z_7 + S_5 c_{56} + S_6 + S_7 c_{67} \\
&\quad + a_{71} U_{76},
\end{aligned} \tag{10.486}$$

$$\begin{aligned}
A_2 &= S_3(-S_7 c_{12} s_{23} s_{71} + a_{67} c_{12} s_{23} \bar{X}'_7) + a_{23}(a_{67} c_7 + a_{71}), \\
B_2 &= S_3(a_{67} c_{12} s_{23} c_7 + a_{71} c_{12} s_{23}) + a_{23}(S_7 s_{71} - a_{67} \bar{X}'_7), \\
D_2 &= S_3(-S_1 s_{12} s_{23} - S_7 s_{12} s_{23} c_{71} - a_{67} s_{12} s_{23} \bar{X}_7) + a_{12} a_{23}, \\
E_2 &= S_3(a_{67} s_{23} c_7 + a_{71} s_{23}) + a_{23}(S_7 c_{12} s_{71} - a_{67} c_{12} \bar{X}'_7), \\
F_2 &= S_3(S_7 s_{23} s_{71} - a_{67} s_{23} \bar{X}'_7) + a_{23}(-a_{67} c_{12} c_7 - a_{71} c_{12}), \\
G_2 &= S_3 a_{12} s_{23} + a_{23}(S_1 s_{12} + S_7 s_{12} c_{71} + a_{67} s_{12} \bar{X}_7), \\
H_2 &= S_3(-S_7 s_{12} c_{23} s_{71} + a_{67} s_{12} c_{23} \bar{X}'_7) + S_2(-S_7 s_{12} s_{71} + a_{67} s_{12} \bar{X}'_7) + a_{12}(a_{67} c_7 + a_{71}), \\
I_2 &= S_3(a_{67} s_{12} c_{23} c_7 + a_{71} s_{12} c_{23}) + S_2(a_{67} s_{12} c_7 + a_{71} s_{12}) + a_{12}(S_7 s_{71} - a_{67} \bar{X}'_7), \\
J_2 &= S_3(S_1 c_{12} c_{23} + S_2 c_{23} + S_7 c_{12} c_{23} c_{71} + a_{67} c_{12} c_{23} \bar{X}_7) + S_2(S_1 c_{12} + S_7 c_{12} c_{71} \\
&\quad + a_{67} c_{12} \bar{X}_7) + S_1(S_7 c_{71} + a_{67} \bar{X}_7) - S_5 S_6 c_{56} + a_{67} a_{71} c_7 + L
\end{aligned} \tag{10.487}$$

and where $\bar{X}'_7 = c_{71} s_7$.

Substituting the tan-half-angle identities for the sine and cosine of θ_2 in Eq. set (10.485) and then multiplying throughout by $(1 + x_2^2)$ gives the two quadratic equations

$$L_i x_2^2 + M_i x_2 + N_i = 0, \tag{10.488}$$

where $i = 1, 2$ and

$$\begin{aligned}
L_i &= (H_i - A_i)c_1 + (I_i - B_i)s_1 + (J_i - D_i), \\
M_i &= (2E_i)c_1 + (2F_i)s_1 + (2G_i), \\
N_i &= (H_i + A_i)c_1 + (I_i + B_i)S_i + (J_i + D_i).
\end{aligned} \tag{10.489}$$

The solution of two equations of the type represented by (10.488) was presented in Section 8.2. According to Bezout's solution method, which is described in Section 8.2.2, the coefficients L_i, M_i, and N_i must satisfy the following condition in order for the two equations of Eq. set (10.488) to have a common solution for x_2:

$$\begin{vmatrix} L_1 & M_1 \\ L_2 & M_2 \end{vmatrix} \begin{vmatrix} M_1 & N_1 \\ M_2 & N_2 \end{vmatrix} - \begin{vmatrix} L_1 & N_1 \\ L_2 & N_2 \end{vmatrix}^2 = 0. \tag{10.490}$$

Typically, the tan-half-angle of the output angle, θ_1, is substituted into Eq. set (10.488) so that the coefficients L_i, M_i, and N_i can be expressed as second-degree polynomials in x_1. Expansion of Eq. (10.490) would then result in an eighth-degree input/output equation. In this case, however, the three determinants in Eq. (10.490) will be expanded in terms of the sine and cosine of θ_1, and it will be shown that each determinant expansion will reduce to an expression that is linear in terms of the sine and cosine of θ_1. Once these linear expressions are obtained for each determinant, the tan-half-angle substitution will be made for θ_1 and the resulting input/output equation will be of degree four.

The first determinant may be written as

$$\begin{vmatrix} L_1 & M_1 \\ L_2 & M_2 \end{vmatrix} = \begin{vmatrix} (H_1 - A_1)c_1 + (I_1 - B_1)s_1 + (J_1 - D_1) & (2E_1)c_1 + (2F_1)s_1 + (2G_1) \\ (H_2 - A_2)c_1 + (I_2 - B_2)s_1 + (J_2 - D_2) & (2E_2)c_1 + (2F_2)s_1 + (2G_2) \end{vmatrix}. \tag{10.491}$$

Expanding the determinant gives

$$\begin{aligned} \begin{vmatrix} L_1 & M_1 \\ L_2 & M_2 \end{vmatrix} &= 2c_1^2[E_2(H_1 - A_1) - E_1(H_2 - A_2)] + 2s_1^2[F_2(I_1 - B_1) - F_1(I_2 - B_2)] \\ &\quad + 2s_1c_1[F_2(H_1 - A_1) + E_2(I_1 - B_1) - F_1(H_2 - A_2) - E_1(I_2 - B_2)] \\ &\quad + 2c_1[E_2(J_1 - D_1) - E_1(J_2 - D_2) + G_2(H_1 - A_1) - G_1(H_2 - A_2)] \\ &\quad + 2s_1[F_2(J_1 - D_1) - F_1(J_2 - D_2) + G_2(I_1 - B_1) - G_1(I_2 - B_2)] \\ &\quad + 2[G_2(J_1 - D_1) - G_1(J_2 - D_2)]. \end{aligned} \tag{10.492}$$

Substituting for the coefficients A_1 through J_2, regrouping, and using the trigonometric identities $s_1^2 + c_1^2 = 1$ and $s_7^2 + c_7^2 = 1$ gives

$$\begin{vmatrix} L_1 & M_1 \\ L_2 & M_2 \end{vmatrix} = P_1c_1 + Q_1s_1 + R_1, \tag{10.493}$$

where

$$\begin{aligned} P_1 &= 2[E_2(J_1 - D_1) - E_1(J_2 - D_2) + G_2(H_1 - A_1) - G_1(H_2 - A_2)], \\ Q_1 &= 2[F_2(J_1 - D_1) - F_1(J_2 - D_2) + G_2(I_1 - B_1) - G_1(I_2 - B_2)], \\ R_1 &= 2[G_2(J_1 - D_1) - G_1(J_2 - D_2)] + 2S_3^2[(S_7s_{23}s_{71}X_7 + a_{67}s_{23}\bar{Y}_7 + a_{71}s_{23}Y_7) \\ &\quad \times (s_{12}c_{23} - c_{12}s_{23})] + 2S_3S_2[S_7s_{12}s_{23}s_{71}X_7 + a_{67}s_{12}s_{23}\bar{Y}_7 + a_{71}s_{12}s_{23}Y_7] \\ &\quad + 2a_{23}^2[-S_7c_{12}s_{71}X_7 - a_{67}c_{12}\bar{Y}_7 - a_{71}c_{12}Y_7] \\ &\quad + 2a_{23}a_{12}[S_7c_{12}s_{71}X_7 + a_{67}c_{12}\bar{Y}_7 + a_{71}c_{12}Y_7]. \end{aligned} \tag{10.494}$$

The second determinant in Eq. (10.490) may be written as

$$\begin{vmatrix} M_1 & N_1 \\ M_2 & N_2 \end{vmatrix} = \begin{vmatrix} (2E_1)c_1 + (2F_1)s_1 + (2G_1) & (H_1 + A_1)c_1 + (I_1 + B_1)s_1 + (J_1 + D_1) \\ (2E_2)c_1 + (2F_2)s_1 + (2G_2) & (H_2 + A_2)c_1 + (I_2 + B_2)s_1 + (J_2 + D_2) \end{vmatrix}. \tag{10.495}$$

Expanding the determinant gives

$$\begin{aligned} \begin{vmatrix} M_1 & N_1 \\ M_2 & N_2 \end{vmatrix} &= 2c_1^2[E_1(H_2 + A_2) - E_2(H_1 + A_1)] + 2s_1^2[F_1(I_2 + B_2) - F_2(I_1 + B_1)] \\ &\quad + 2s_1c_1[F_1(H_2 + A_2) + E_1(I_2 + B_2) - F_2(H_1 + A_1) - E_2(I_1 + B_1)] \\ &\quad + 2c_1[E_1(J_2 + D_2) - E_2(J_1 + D_1) + G_1(H_2 + A_2) - G_2(H_1 + A_1)] \\ &\quad + 2s_1[F_1(J_2 + D_2) - F_2(J_1 + D_1) + G_1(I_2 + B_2) - G_2(I_1 + B_1)] \\ &\quad + 2[G_1(J_2 + D_2) - G_2(J_1 + D_1)]. \end{aligned} \tag{10.496}$$

Substituting for the coefficients A_1 through J_2, regrouping, and using the trigonometric identity $s_1^2 + c_1^2 = 1$ and $s_7^2 + c_7^2 = 1$ gives

$$\begin{vmatrix} M_1 & N_1 \\ M_2 & N_2 \end{vmatrix} = P_2 c_1 + Q_2 s_1 + R_2, \tag{10.497}$$

where

$$\begin{aligned} P_2 &= 2[E_1(J_2 + D_2) - E_2(J_1 + D_1) + G_1(H_2 + A_2) - G_2(H_1 + A_1)], \\ Q_2 &= 2[F_1(J_2 + D_2) - F_2(J_1 + D_1) + G_1(I_2 + B_2) - G_2(I_1 + B_1)], \\ R_2 &= 2[G_1(J_2 + D_2) - G_2(J_1 + D_1)] \\ &\quad + 2S_3^2[(-S_7 s_{23} s_{71} X_7 - a_{67} s_{23} \bar{Y}_7 - a_{71} s_{23} Y_7) \times (s_{12} c_{23} + c_{12} s_{23})] \\ &\quad + 2S_3 S_2[-S_7 s_{12} s_{23} s_{71} X_7 - a_{67} s_{12} s_{23} \bar{Y}_7 - a_{71} s_{12} s_{23} Y_7] + 2a_{23}^2[-S_7 c_{12} s_{71} X_7 \\ &\quad - a_{67} c_{12} \bar{Y}_7 - a_{71} c_{12} Y_7] + 2a_{23} a_{12}[-S_7 c_{12} s_{71} X_7 - a_{67} c_{12} \bar{Y}_7 - a_{71} c_{12} Y_7]. \end{aligned} \tag{10.498}$$

The third determinant in Eq. (10.490) may be written as

$$\begin{vmatrix} L_1 & N_1 \\ L_2 & N_2 \end{vmatrix} = \begin{vmatrix} (H_1 - A_1)c_1 + (I_1 - B_1)s_1 + (J_1 - D_1) & (H_1 + A_1)c_1 + (I_1 + B_1)s_1 + (J_1 + D_1) \\ (H_2 - A_2)c_1 + (I_2 - B_2)s_1 + (J_2 - D_2) & (H_2 + A_2)c_1 + (I_2 + B_2)s_1 + (J_2 + D_2) \end{vmatrix}. \tag{10.499}$$

Expanding the determinant gives

$$\begin{aligned} \begin{vmatrix} L_1 & N_1 \\ L_2 & N_2 \end{vmatrix} &= c_1^2[(H_1 - A_1)(H_2 + A_2) - (H_2 - A_2)(H_1 + A_1)] + s_1^2[(I_1 - B_1) \\ &\quad \times (I_2 + B_2) - (I_2 - B_2)(I_1 + B_1)] + s_1 c_1[(I_1 - B_1)(H_2 + A_2) \\ &\quad + (H_1 - A_1)(I_2 + B_2) - (I_2 - B_2)(H_1 + A_1) - (H_2 - A_2)(I_1 + B_1)] \\ &\quad + c_1[(H_1 - A_1)(J_2 + D_2) - (H_2 - A_2)(J_1 + D_1) + (J_1 - D_1) \\ &\quad \times (H_2 + A_2) - (J_2 - D_2)(H_1 + A_1)] + s_1[(I_1 - B_1)(J_2 + D_2) \\ &\quad - (I_2 - B_2)(J_1 + D_1) + (J_1 - D_1)(I_2 + B_2) - (J_2 - D_2)(I_1 + B_1)] \\ &\quad + [(J_1 - D_1)(J_2 + D_2) - (J_2 - D_2)(J_1 + D_1)]. \end{aligned} \tag{10.500}$$

Substituting for the coefficients A_1 through J_2, regrouping, and using the trigonometric identities $s_1^2 + c_1^2 = 1$ and $s_7^2 + c_7^2 = 1$ gives

$$\begin{vmatrix} L_1 & N_1 \\ L_2 & N_2 \end{vmatrix} = P_3 c_1 + Q_3 s_1 + R_3, \tag{10.501}$$

where

$$
\begin{aligned}
P_3 = {} & (H_1 - A_1)(J_2 + D_2) - (H_2 - A_2)(J_1 + D_1) + (J_1 - D_1)(H_2 + A_2) \\
& - (J_2 - D_2)(H_1 + A_1), \\
Q_3 = {} & (I_1 - B_1)(J_2 + D_2) - (I_2 - B_2)(J_1 + D_1) + (J_1 - D_1)(I_2 + B_2) \\
& - (J_2 - D_2)(I_1 + B_1), \\
R_3 = {} & (J_1 - D_1)(J_2 + D_2) - (J_2 - D_2)(J_1 + D_1) + 2S_3 a_{23}[S_7 s_{12} c_{23} s_{71} X_7 \\
& + a_{67} s_{12} c_{23} \bar{Y}_7 + a_{71} s_{12} c_{23} Y_7] + 2S_3 a_{12}[-S_7 c_{12} s_{23} s_{71} X_7 - a_{67} c_{12} s_{23} \bar{Y}_7 \\
& - a_{71} c_{12} s_{23} Y_7] + 2S_2 a_{23}[S_7 s_{12} s_{71} X_7 + a_{67} s_{12} \bar{Y}_7 + a_{71} s_{12} Y_7].
\end{aligned} \tag{10.502}
$$

The tan-half-angle identities for the sine and cosine of θ_1 can be substituted into Eqs. (10.493), (10.497), and (10.501) to give

$$
\begin{vmatrix} L_1 & M_1 \\ L_2 & M_2 \end{vmatrix} = \frac{p_1 x_1^2 + q_1 x_1 + r_1}{1 + x_1^2}, \tag{10.503}
$$

$$
\begin{vmatrix} M_1 & N_1 \\ M_2 & N_2 \end{vmatrix} = \frac{p_2 x_1^2 + q_2 x_1 + r_2}{1 + x_1^2}, \tag{10.504}
$$

$$
\begin{vmatrix} L_1 & N_1 \\ L_2 & N_2 \end{vmatrix} = \frac{p_3 x_1^2 + q_3 x_1 + r_3}{1 + x_1^2}, \tag{10.505}
$$

where

$$
p_i = R_i - P_i, \quad q_i = 2Q_i, \quad r_i = R_i + P_i, \quad i = 1 \ldots 3. \tag{10.506}
$$

Substituting Eqs. (10.503) through (10.505) into Eq. (10.490) and multiplying by $(1+x_1^2)^2$ gives the fourth-degree input/output equation

$$
(p_1 x_1^2 + q_1 x_1 + r_1)(p_2 x_1^2 + q_2 x_2 + r_2) - (p_3 x_1^2 + q_3 x_1 + r_3)^2 = 0. \tag{10.507}
$$

Corresponding values of θ_1 for each value of x_1 that satisfies Eq. (10.507) can be obtained from

$$
\theta_1 = 2 \tan^{-1}(x_1). \tag{10.508}
$$

10.4.2 Determination of θ_2

Corresponding values of θ_2 will be obtained for each value of θ_1. As explained in Section 8.2.2, the corresponding value for x_2 may be calculated from either

$$
x_2 = \frac{-\begin{vmatrix} M_1 & N_1 \\ M_2 & N_2 \end{vmatrix}}{\begin{vmatrix} L_1 & N_2 \\ L_2 & N_2 \end{vmatrix}} \tag{10.509}
$$

or

$$x_2 = \frac{-\begin{vmatrix} L_1 & N_1 \\ L_2 & N_2 \end{vmatrix}}{\begin{vmatrix} L_1 & M_1 \\ L_2 & M_2 \end{vmatrix}}. \tag{10.510}$$

The angle θ_2 is obtained from x_2 as follows:

$$\theta_2 = 2\tan^{-1}(x_2). \tag{10.511}$$

10.4.3 Determination of θ_6

A unique corresponding value of θ_6 can be obtained by projecting the vector loop equation on the directions of vectors $\mathbf{a}_{67}$ and $\mathbf{S}_6 \times \mathbf{a}_{67}$. Using set 6 from the table of direction cosines gives the two equations

$$S_1\bar{X}_7 + S_2X_{17} + S_3X_{217} + S_5X_{43217} + a_{12}W_{17} + a_{23}W_{217} + a_{56}c_6 + a_{67} + a_{71}c_7 = 0, \tag{10.512}$$

$$S_1\bar{Y}_7 + S_2Y_{17} + S_3Y_{217} + S_5Y_{43217} - S_7s_{67} - a_{12}U^*_{176} - a_{23}U^*_{2176} - a_{56}s_6 + a_{71}c_{67}s_7 = 0. \tag{10.513}$$

Upon substituting the fundamental sine law $X_{43217} = s_{56}s_6$ and sine–cosine law $Y_{43217} = s_{56}c_6$ for a spherical heptagon, these two equations may be expressed in the form

$$K_1c_6 + K_2s_6 + K_3 = 0, \tag{10.514}$$

$$K_4c_6 + K_5s_6 + K_6 = 0, \tag{10.515}$$

where

$$\begin{aligned}
K_1 &= a_{56}, \\
K_2 &= S_5s_{56}, \\
K_3 &= S_1\bar{X}_7 + S_2X_{17} + S_3X_{217} + a_{12}W_{17} + a_{23}W_{217} + a_{67} + a_{71}c_7, \\
K_4 &= S_5s_{56}, \\
K_5 &= -a_{56}, \\
K_6 &= S_1\bar{Y}_7 + S_2Y_{17} + S_3Y_{217} - S_7s_{67} - a_{12}U^*_{176} - a_{23}U^*_{2176} + a_{71}c_{67}s_7.
\end{aligned} \tag{10.516}$$

Equations (10.514) and (10.515) are two linear equations in the variables s_6 and c_6. Thus, unique values for these parameters, and a unique value for θ_6, can be determined for each set of solutions for the angles θ_1 and θ_2.

10.4.4 Determination of θ_4

The following spherical cosine law may be written for a spherical heptagon:

$$Z_{2176} = Z_4. \tag{10.517}$$

Expanding the definition of the term Z_4 and solving for c_4 gives

$$c_4 = \frac{c_{34}c_{45} - Z_{2176}}{s_{34}s_{45}}. \tag{10.518}$$

Two distinct values of θ_4 exist that satisfy this equation for each set of angles $[\theta_1, \theta_2, \theta_6]$. Thus, a total of eight solution configurations will exist for the RRSRR spatial mechanism.

10.4.5 Determination of θ_5

The following sine and sine–cosine laws may be written for a spherical heptagon:

$$X_{2176} = X_{45}, \tag{10.519}$$

$$Y_{2176} = -X^*_{45}. \tag{10.520}$$

Expanding the definitions of X_{45} and X^*_{45} and rearranging gives

$$(X_4)c_5 - (Y_4)s_5 + (-X_{2176}) = 0, \tag{10.521}$$

$$(Y_4)c_5 + (X_4)s_5 + (Y_{2176}) = 0. \tag{10.522}$$

All the terms in parentheses are defined in terms of the constant mechanism parameters and the previously calculated joint angles. Thus, these two equations represent two linear equations in the unknowns s_5 and c_5. A unique corresponding value for θ_5 can be determined from the sine and cosine values.

10.4.6 Determination of θ_3

The following fundamental sine and sine–cosine laws may be written for a spherical heptagon:

$$X_{17654} = s_{23}s_3, \tag{10.523}$$

$$Y_{17654} = s_{23}c_3. \tag{10.524}$$

A unique corresponding value for θ_3 can be found by solving these equations for s_3 and c_3.

10.4.7 Numerical example

It has been shown that a total of eight solution configurations exist for the RRSRR spatial mechanism. Although this is a group 4 spatial mechanism, the special geometry of having a_{34}, a_{45}, and S_4 equal zero greatly simplifies the analysis. A fourth-degree input/output equation was obtained to solve for the angle θ_1. Unique corresponding values were then determined for the angles θ_2 and θ_6. Pairs of values for the angles θ_3, θ_4, and θ_5 were obtained for each of the four sets of angles $[\theta_1, \theta_2, \theta_6]$, thus giving a total of eight solutions.

The mechanism dimensions of a numerical example are listed in Table 10.5. The resulting eight solution configurations are listed in Table 10.6 and are drawn in Figure 10.6. It is apparent from the figure that there are four classes of solutions. For example, solution B is the same as solution A except that θ_5 has been rotated an additional 180 degrees,

Table 10.5. *RRSRR mechanism parameters.*

Link length, cm.	Twist angle, deg.	Joint offset, cm.	Joint angle, deg.
$a_{12} = 8.3$	$\alpha_{12} = 60$	$S_1 = 5.8$	$\theta_1 =$ variable
$a_{23} = 6.8$	$\alpha_{23} = 60$	$S_2 = 8.4$	$\theta_2 =$ variable
$a_{34} = 0$	$\alpha_{34} = 90$	$S_3 = 9.0$	$\theta_3 =$ variable
$a_{45} = 0$	$\alpha_{45} = 90$	$S_4 = 0$	$\theta_4 =$ variable
$a_{56} = 6.8$	$\alpha_{56} = 60$	$S_5 = 9.0$	$\theta_5 =$ variable
$a_{67} = 8.3$	$\alpha_{67} = 60$	$S_6 = 8.4$	$\theta_6 =$ variable
$a_{71} = 5.0$	$\alpha_{71} = 90$	$S_7 = 5.8$	$\theta_7 = 289$ (input)

Table 10.6. *Calculated configurations for the RRSRR spatial mechanism.*

Solution	θ_1, deg.	θ_2, deg.	θ_3, deg.	θ_4, deg.	θ_5, deg.	θ_6, deg.
A	−73.02	83.66	−161.30	65.57	−168.92	92.85
B	−73.02	83.66	18.70	−65.57	11.08	92.85
C	−23.85	155.71	84.16	49.81	−154.55	19.38
D	−23.85	155.71	−95.84	−49.81	25.45	19.38
E	−22.81	−37.85	−137.80	59.30	89.73	−175.37
F	−22.81	−37.85	42.20	−59.30	−90.27	−175.37
G	−6.47	164.44	53.06	53.78	−164.45	12.20
H	−6.47	164.44	−126.94	−53.78	15.55	12.20

causing the vector $\mathbf{S}_4$ to point in the opposite direction. The angle θ_4 is the negative of its value for solution A, and θ_3 is advanced by 180 degrees.

10.5 RSTR spatial mechanism

The ball and socket joint of the RSTR spatial mechanism can be modeled by three intersecting revolute joints, and the Hooke joint can be modeled by two intersecting revolute joints as shown in Figure 10.7. The special geometric values for this case are

$$a_{23} = a_{45} = a_{56} = S_5 = 0. \tag{10.525}$$

In this analysis, the angle θ_7 will be the known input value. The angle θ_1 will be solved for first and will be referred to as the output angle. Specifically, the problem can be stated as

given: $\alpha_{12}, \alpha_{23}, \alpha_{34}, \alpha_{45}, \alpha_{56}, \alpha_{67}, \alpha_{71}$,
$a_{12}, a_{34}, a_{67}, a_{71}$,
$S_1, S_2, S_3, S_4, S_6, S_7$,
$a_{23} = a_{45} = a_{56} = S_5 = 0$, and
θ_7,

find: $\theta_1, \theta_2, \theta_3, \theta_4, \theta_5$, and θ_6 .

It will be shown that a maximum of eight solution configurations exist for this mechanism.

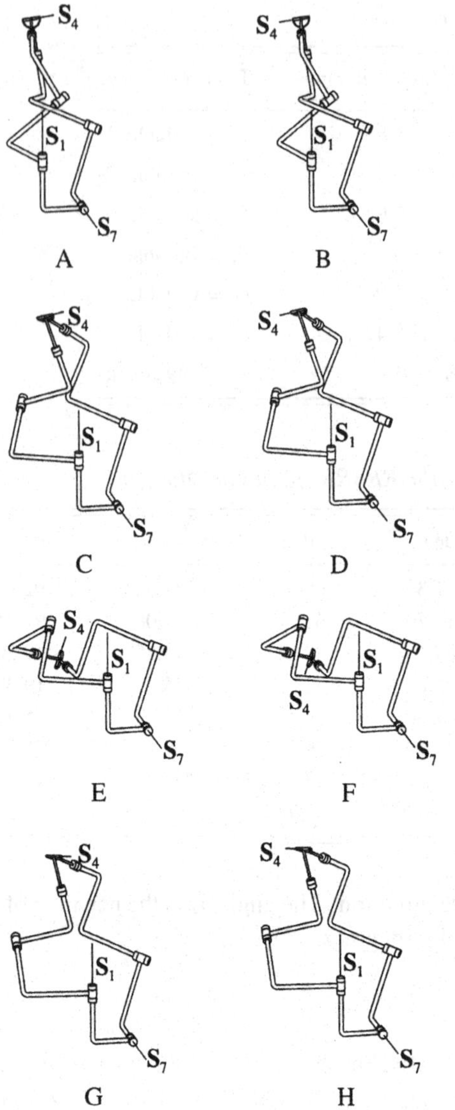

Figure 10.6. Eight configurations of the RRSRR spatial mechanism.

10.5.1 Determination of input/output equation

The vector loop equation for the RSTR mechanism may be written as

$$S_6\mathbf{S}_6 + a_{67}\mathbf{a}_{67} + S_7\mathbf{S}_7 + a_{71}\mathbf{a}_{71} + S_1\mathbf{S}_1 + a_{12}\mathbf{a}_{12} + S_2\mathbf{S}_2 = -S_3\mathbf{S}_3 - a_{34}\mathbf{a}_{34} - S_4\mathbf{S}_4. \tag{10.526}$$

Evaluating the self-scalar product of each side of this equation yields

$$\begin{aligned} L &+ S_6\mathbf{S}_6 \cdot (S_7\mathbf{S}_7 + a_{71}\mathbf{a}_{71} + S_1\mathbf{S}_1 + a_{12}\mathbf{a}_{12} + S_2\mathbf{S}_2) + a_{67}\mathbf{a}_{67} \cdot (a_{71}\mathbf{a}_{71} + S_1\mathbf{S}_1 \\ &+ a_{12}\mathbf{a}_{12} + S_2\mathbf{S}_2) + S_7\mathbf{S}_7 \cdot (S_1\mathbf{S}_1 + a_{12}\mathbf{a}_{12} + S_2\mathbf{S}_2) + a_{71}\mathbf{a}_{71} \cdot (a_{12}\mathbf{a}_{12} + S_2\mathbf{S}_2) \\ &+ S_1\mathbf{S}_1 \cdot (S_2\mathbf{S}_2) = S_3\mathbf{S}_3 \cdot S_4\mathbf{S}_4, \end{aligned} \tag{10.527}$$

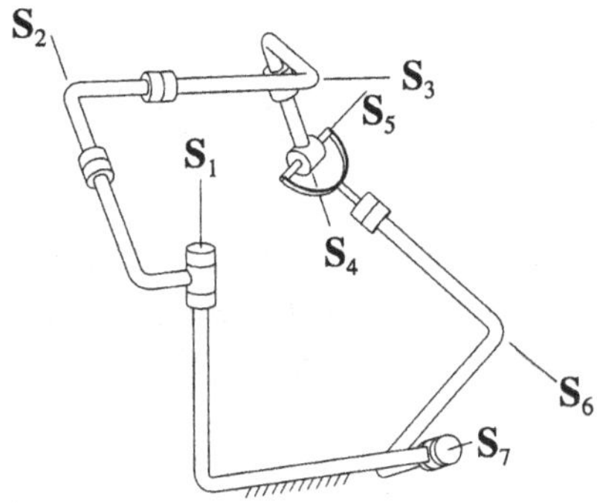

Figure 10.7. RSTR spatial mechanism.

where L is defined as

$$L = (S_6^2 + a_{67}^2 + S_7^2 + a_{71}^2 + S_1^2 + a_{12}^2 + S_2^2 - S_3^2 - a_{34}^2 - S_4^2)/2. \tag{10.528}$$

The scalar products of mutually perpendicular vectors, such as, for example, $\mathbf{S}_6$ and $\mathbf{a}_{67}$, equal zero, and all these terms have been omitted from Eq. (10.527). Evaluating the scalar products using the sets of direction cosines listed in the appendix yields

$$\begin{aligned} &L + S_6(S_7c_{67} + a_{71}X_7 + S_1Z_7 + a_{12}X_{71} + S_2Z_{71}) \\ &\quad + a_{67}(a_{71}c_7 + S_1\bar{X}_7 + a_{12}W_{17} + S_2X_{17}) + S_7(S_1c_{71} + a_{12}X_1 + S_2Z_1) \\ &\quad + a_{71}(a_{12}c_1 + S_2\bar{X}_1) + S_1S_2c_{12} = S_3S_4c_{34}. \end{aligned} \tag{10.529}$$

Expanding all terms that contain the parameter θ_1 and rearranging the equation gives

$$\begin{aligned} &[S_2(S_6s_{12}Y_7 - S_7s_{12}s_{71} + a_{67}s_{12}\bar{X}_7') + a_{12}(S_6X_7 + a_{67}c_7 + a_{71})]c_1 \\ &\quad + [S_2(S_6s_{12}X_7 + a_{67}s_{12}c_7 + a_{71}s_{12}) + a_{12}(-S_6Y_7 + S_7s_{71} - a_{67}\bar{X}_7')]s_1 \\ &\quad + [S_2(S_1c_{12} + S_6c_{12}Z_7 + S_7c_{12}c_{71} + a_{67}c_{12}\bar{X}_7) + S_6(S_1Z_7 + S_7c_{67} + a_{71}X_7) \\ &\quad + S_1(S_7c_{71} + a_{67}\bar{X}_7) - S_3S_4c_{34} + a_{67}a_{71}c_7 + L] = 0. \end{aligned} \tag{10.530}$$

All the expressions in brackets are defined in terms of the given constant mechanism parameters and the input angle. Thus, Eq. (10.530) is the input/ output equation for this mechanism, and it can be solved for two values of θ_1 via the technique described in Section 6.7.2(c).

10.5.2 Determination of θ_3

Projecting the vector loop equation onto the direction of the vector $\mathbf{S}_2$ using set 14 from the table of direction cosines gives

$$\begin{aligned} &S_1c_{12} + S_2 + S_3Z_{45671} + a_{34}U_{456712} + S_4Z_{5671} + S_6Z_{71} + a_{67}U_{712} \\ &\quad + S_7Z_1 + a_{71}U_{12} = 0. \end{aligned} \tag{10.531}$$

Substituting the fundamental and subsidiary cosine laws $Z_{45671} = c_{23}$ and $Z_{5671} = Z_3$ and the fundamental polar sine law $U_{456712} = s_{23}s_3$ yields

$$S_1c_{12} + S_2 + S_3c_{23} + a_{34}s_{23}s_3 + S_4Z_3 + S_6Z_{71} + a_{67}U_{712} + S_7Z_1 + a_{71}U_{12} = 0. \tag{10.532}$$

Expanding Z_3 and regrouping gives

$$[-S_4s_{23}s_{34}]c_3 + [a_{34}s_{23}]s_3 + [S_1c_{12} + S_2 + S_3c_{23} + S_4c_{23}c_{34} + S_6Z_{71} + a_{67}U_{712} + S_7Z_1 + a_{71}U_{12}] = 0. \tag{10.533}$$

All expressions in brackets are defined in terms of the given mechanism parameters, the input angle θ_7, and the output angle θ_1. Thus, for each of the two previously calculated output angles, two corresponding values of θ_3 can be determined from Eq. (10.533).

10.5.3 Determination of θ_2

Projecting the vector loop equation onto the direction of the vector $\mathbf{a}_{12}$ and $\mathbf{S}_2 \times \mathbf{a}_{12}$ using set 14 from the table of direction cosines gives

$$a_{12} + S_3X_{45671} + a_{34}W_{45671} + S_4X_{5671} + S_6X_{71} + a_{67}W_{71} + S_7X_1 + a_{71}c_1 = 0, \tag{10.534}$$

$$S_1s_{12} - S_3Y_{45671} + a_{34}U^*_{456712} - S_4Y_{5671} - S_6Y_{71} + a_{67}U^*_{712} - S_7Y_1 - a_{71}c_{12}s_1 = 0. \tag{10.535}$$

Substituting the fundamental and subsidiary sine and sine–cosine laws $X_{45671} = s_{23}s_2$, $X_{5671} = X_{32}$, $Y_{45671} = s_{23}c_2$, and $Y_{5671} = -X^*_{32}$ and the subsidiary polar sine–cosine and cosine laws $U^*_{456712} = -V_{32}$ and $W_{45671} = W_{32}$ gives

$$a_{12} + S_3s_{23}s_2 + a_{34}W_{32} + S_4X_{32} + S_6X_{71} + a_{67}W_{71} + S_7X_1 + a_{71}c_1 = 0, \tag{10.536}$$

$$S_1s_{12} - S_3s_{23}c_2 - a_{34}V_{32} + S_4X^*_{32} - S_6Y_{71} + a_{67}U^*_{712} - S_7Y_1 - a_{71}c_{12}s_1 = 0. \tag{10.537}$$

Expanding the terms W_{32}, X_{32}, V_{32}, and X^*_{32} and regrouping yields

$$Ac_2 + Bs_2 = D_1, \tag{10.538}$$

$$-Bc_2 + As_2 = D_2, \tag{10.539}$$

where

$$\begin{aligned} A &= S_4\bar{X}_3 + a_{34}c_3, \\ B &= S_3s_{23} - S_4\bar{Y}_3 - a_{34}X'_3, \\ D_1 &= -(S_6X_{71} + S_7X_1 + a_{12} + a_{67}W_{71} + a_{71}c_1), \\ D_2 &= -S_1s_{12} + S_6Y_{71} + S_7Y_1 - a_{67}U^*_{712} + a_{71}\bar{X}'_1. \end{aligned} \tag{10.540}$$

Adding B times Eq. (10.538) to A times Eq. (10.539) and solving for S_2 yields

$$s_2 = \frac{BD_1 + AD_2}{A^2 + B^2}. \tag{10.541}$$

Subtracting B times Eq. (10.539) from A times Eq. (10.538) and solving for c_2 yields

$$c_2 = \frac{AD_1 - BD_2}{A^2 + B^2}. \tag{10.542}$$

The coefficients A, B, D_1, and D_2 can be evaluated numerically for each of the four sets of values of $[\theta_7, \theta_1, \theta_3]$. A unique corresponding value for θ_2 for each set can be determined from the calculated values of the sine and cosine of θ_2 of Eqs. (10.541) and (10.542).

10.5.4 Determination of θ_5

The following subsidiary cosine law may be written for a spherical heptagon:

$$Z_{7123} = \bar{Z}_5. \tag{10.543}$$

Expanding $\bar{Z}_5$ and solving for c_5 gives

$$c_5 = \frac{c_{45}c_{56} - Z_{7123}}{s_{45}s_{56}}. \tag{10.544}$$

Thus, the cosine of θ_5 can be evaluated for each of the four sets of values of $[\theta_7, \theta_1, \theta_2, \theta_3]$. Therefore, two values of θ_5 correspond to each of the previous four solution sets, and a total of eight sets of values of $[\theta_7, \theta_1, \theta_2, \theta_3, \theta_5]$ exist.

10.5.5 Determination of θ_4

The following subsidiary sine and sine–cosine laws may be written for a spherical heptagon:

$$X_{7123} = X_{54}, \tag{10.545}$$

$$Y_{7123} = -X^*_{54}. \tag{10.546}$$

Expanding X_{54} and X^*_{54} and rearranging yields

$$\bar{X}_5c_4 - \bar{Y}_5s_4 - X_{7123} = 0, \tag{10.547}$$

$$\bar{X}_5s_4 + \bar{Y}_5c_4 + Y_{7123} = 0. \tag{10.548}$$

These two equations may be solved for c_4 and s_4 for each of the solution sets $[\theta_7, \theta_1, \theta_2, \theta_3, \theta_5]$, and a unique corresponding value for θ_4 is thus determined.

10.5.6 Determination of θ_6

The following fundamental sine and sine–cosine laws may be written for a spherical heptagon:

$$X_{43217} = s_{56}s_6, \tag{10.549}$$

$$Y_{43217} = s_{56}c_6. \tag{10.550}$$

Each of the eight solution sets $[\theta_7, \theta_1, \theta_2, \theta_3, \theta_4, \theta_5]$ are substituted into these equations to yield corresponding values for the sine and cosine of θ_6. A unique corresponding value of θ_6 is thus determined.

Table 10.7. *RSTR mechanism parameters.*

Link length, cm.	Twist angle, deg.	Joint offset, cm.	Joint angle, deg.
$a_{12} = 2.3$	$\alpha_{12} = 60$	$S_1 = 7.6$	θ_1 = variable
$a_{23} = 0$	$\alpha_{23} = 90$	$S_2 = 8.7$	θ_2 = variable
$a_{34} = 4.6$	$\alpha_{34} = 65$	$S_3 = 9.6$	θ_3 = variable
$a_{45} = 0$	$\alpha_{45} = 90$	$S_4 = 4.8$	θ_4 = variable
$a_{56} = 0$	$\alpha_{56} = 90$	$S_5 = 0$	θ_5 = variable
$a_{67} = 6.8$	$\alpha_{67} = 75$	$S_6 = 9.1$	θ_6 = variable
$a_{71} = 9.1$	$\alpha_{71} = 65$	$S_7 = 6.2$	$\theta_7 = 322$ (input)

Table 10.8. *Calculated configurations for the RSTR spatial mechanism.*

Solution	θ_1, deg.	θ_2, deg.	θ_3, deg.	θ_4, deg.	θ_5, deg.	θ_6, deg.
A	256.38	93.99	264.91	−107.78	125.36	−102.02
B	256.38	93.99	264.91	72.22	−125.36	77.98
C	256.38	49.62	1.90	157.35	77.59	−45.16
D	256.38	49.62	1.90	−22.65	−77.59	134.84
E	−2.34	−28.13	206.68	42.14	77.26	−145.46
F	−2.34	−28.13	206.68	−137.86	−77.26	34.54
G	−2.34	−83.21	60.12	147.50	137.49	−66.79
H	−2.34	−83.21	60.12	−32.50	−137.49	113.21

10.5.7 Numerical example

It has been shown that a total of eight solution configurations exist for the RSTR spatial mechanism. Although this is a group 4 spatial mechanism, the special geometry of having a_{23}, a_{45}, a_{56}, and S_5 equal zero greatly simplifies the analysis. An input/output equation was obtained that was linear in the sines and cosines of the angle θ_1. Two corresponding values for the angle θ_3 were next determined, followed by unique corresponding values for the angle θ_2. For each of the four solution sets of θ_1, θ_2, and θ_3, two corresponding values were computed for the angle θ_5. Lastly, for each of the eight solution sets of $\theta_1, \theta_2, \theta_3$, and θ_5, unique corresponding values of θ_4 and θ_6 were computed.

The mechanism dimensions of a numerical example are listed in Table 10.7. The resulting eight solution configurations are listed in Table 10.8 and are drawn in Figure 10.8. It is apparent from the figure that there are four classes of solutions. For example, solution B is the same as solution A except that θ_6 has been rotated by an additional 180 degrees, causing the vector $\mathbf{S}_5$ to point in the opposite direction. The angle θ_5 is the negative of its value for solution A, and θ_4 is advanced by 180 degrees.

10.6 RTTT spatial mechanism, case 1: $\alpha_{45} = 0$

An RTTT spatial mechanism is shown in Figure 10.9 (see also Lin (1987)). In this mechanism, the first and second, third and fourth, and fifth and sixth joint axes intersect and the second through sixth joint offset distances equal zero. For the case to be analyzed

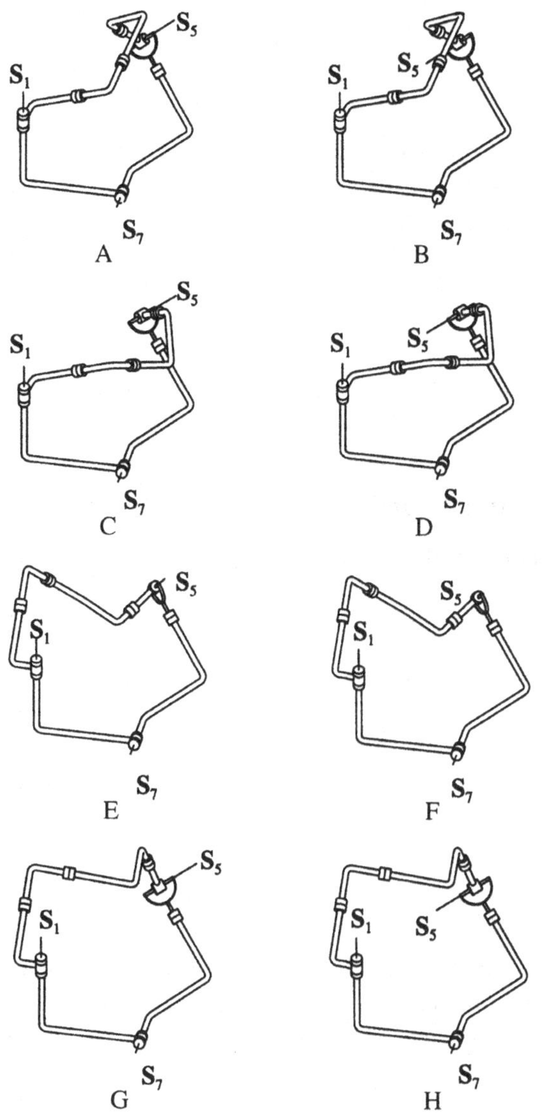

Figure 10.8. Eight configurations of the RSTR spatial mechanism.

here, the fourth and fifth joint axes will be assumed to be parallel. Also, the joint angles of the Hooke joints, that is, α_{12}, α_{34}, and α_{56}, will be set equal to 90 degrees. Specifically, the problem can be stated as

given: $\alpha_{23}, \alpha_{67}, \alpha_{71}$,
$a_{23}, a_{45}, a_{67}, a_{71}$,
S_1, S_7,
$\alpha_{12} = \alpha_{34} = \alpha_{56} = \pi/2$,
$\alpha_{45} = 0$,
$a_{12} = a_{34} = a_{56} = S_2 = S_3 = S_4 = S_5 = S_6 = 0$, and
θ_7 (input angle),

find: $\theta_1, \theta_2, \theta_3, \theta_4, \theta_5$, and θ_6 .

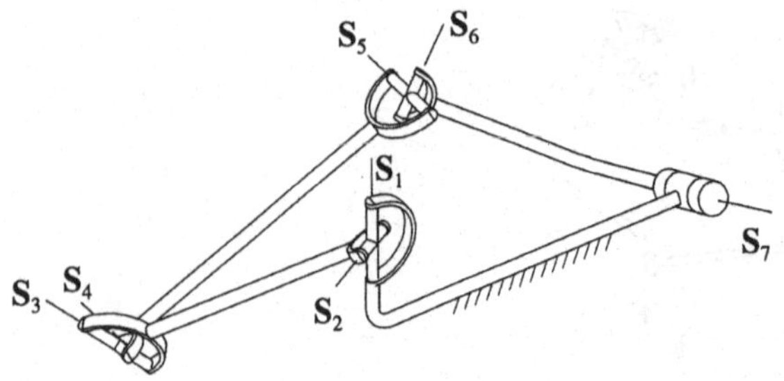

Figure 10.9. RTTT spatial mechanism.

It will be shown that a maximum of sixteen solution configurations exist for this mechanism.

10.6.1 Determination of input/output equation

The vector loop equation for the RTTT mechanism may be written as

$$a_{67}\mathbf{a}_{67} + S_7\mathbf{S}_7 + a_{71}\mathbf{a}_{71} + S_1\mathbf{S}_1 + a_{23}\mathbf{a}_{23} = -a_{45}\mathbf{a}_{45}. \tag{10.551}$$

Taking the self-scalar product of this equation gives

$$\begin{aligned} L &+ a_{67}\mathbf{a}_{67} \cdot (a_{71}\mathbf{a}_{71} + S_1\mathbf{S}_1 + a_{23}\mathbf{a}_{23}) + S_7\mathbf{S}_7 \cdot (S_1\mathbf{S}_1 + a_{23}\mathbf{a}_{23}) + a_{71}\mathbf{a}_{71} \cdot a_{23}\mathbf{a}_{23} \\ &+ S_1\mathbf{S}_1 \cdot a_{23}\mathbf{a}_{23} = 0, \end{aligned} \tag{10.552}$$

where

$$L = \left(a_{67}^2 + S_7^2 + a_{71}^2 + S_1^2 + a_{23}^2 - a_{45}^2\right)/2. \tag{10.553}$$

The scalar products of mutually perpendicular vectors, such as, for example, $\mathbf{a}_{67}$ and $\mathbf{S}_7$, equal zero, and all these terms have been omitted from Eq. (10.552). Evaluating the scalar products using the sets of direction cosines listed in the appendix yields

$$L + a_{67}(a_{71}c_7 + S_1\bar{X}_7 + a_{23}W_{712}) + S_7(S_1c_{71} + a_{23}X_{12}) + a_{23}(a_{71}W_{12} + S_1X_2) = 0. \tag{10.554}$$

Expanding all terms that contain the parameter θ_2 and rearranging yields

$$A_1c_2 + B_1s_2 + D_1 = 0, \tag{10.555}$$

where

$$\begin{aligned} A_1 &= a_{23}(S_7X_1 + a_{67}W_{71} + a_{71}c_1), \\ B_1 &= a_{23}\left(S_1s_{12} - S_7Y_1 + a_{67}U^*_{712} - a_{71}c_{12}s_1\right), \\ D_1 &= L + S_1(a_{67}\bar{X}_7 + S_7c_{71}) + a_{67}a_{71}c_7. \end{aligned} \tag{10.556}$$

Expanding the terms that contain the parameter θ_1 and substituting the value of $\pi/2$ for

α_{12} and regrouping gives

$$\begin{aligned} A_1 &= F_2c_1 + F_3s_1, \\ B_1 &= a_{23}(S_1 + S_7c_{71} + a_{67}\bar{X}_7), \\ D_1 &= F_1, \end{aligned} \tag{10.557}$$

where

$$\begin{aligned} F_1 &= L + S_1(a_{67}\bar{X}_7 + S_7c_{71}) + a_{67}a_{71}c_7, \\ F_2 &= a_{23}(a_{67}c_7 + a_{71}), \\ F_3 &= a_{23}(S_7s_{71} - a_{67}c_{71}s_7). \end{aligned} \tag{10.558}$$

A second equation that contains the output angle θ_1, and the angle θ_2 can be obtained from the secondary cosine law

$$Z_{0712} = Z_{045}. \tag{10.559}$$

The right side of this equation may be expanded as

$$Z_{045} = a_{56}Y_{45} + S_5s_{56}X_{45} + a_{45}(-s_{56}c_5Z_4 + c_{56}Y_4) + a_{34}Y_{54} + S_4s_{34}X_{54}. \tag{10.560}$$

Substituting $S_4 = S_5 = a_{34} = a_{56} = 0$ reduces this equation to

$$Z_{045} = a_{45}(-s_{56}c_5Z_4 + c_{56}Y_4). \tag{10.561}$$

Substituting the value of α_{56} into this equation and the value of α_{34} and α_{45} into Z_4 yields

$$Z_{045} = 0. \tag{10.562}$$

The left side of Eq. (10.559) may be expanded as

$$\begin{aligned} Z_{0712} &= a_{23}Y_{712} + S_2s_{23}X_{712} + a_{12}(c_{23}Y_{71} - s_{23}c_2Z_{71}) + S_1(-X^*_{71}\bar{X}_2 - X_{71}\bar{Y}_2) \\ &\quad + a_{67}Y_{217} + S_7s_{67}X_{217} + a_{71}(c_{67}Y_{21} - s_{67}c_7Z_{21}). \end{aligned} \tag{10.563}$$

Substituting $a_{12} = S_2 = 0$ into Eq. (10.563) and equating it to zero reduces this equation to

$$a_{23}Y_{712} + S_1(-X^*_{71}\bar{X}_2 - X_{71}\bar{Y}_2) + a_{67}Y_{217} + S_7s_{67}X_{217} + a_{71}(c_{67}Y_{21} - s_{67}c_7Z_{21}) = 0. \tag{10.564}$$

Expanding all terms that contain the parameter θ_2 and rearranging yields

$$A_2c_2 + B_2s_2 + D_2 = 0, \tag{10.565}$$

where

$$\begin{aligned} A_2 &= S_1c_{12}s_{23}X_{71} - S_7s_{23}s_{67}U^*_{712} + a_{23}c_{23}Y_{71} + a_{67}s_{23}(-s_{12}\bar{Y}_7 + c_{12}X'_7\bar{X}'_1 \\ &\quad + c_{12}c_1Z'_7) + a_{71}s_{23}(-s_{67}c_7Y_1 + c_{67}Z'_1), \\ B_2 &= -S_1s_{23}X^*_{71} + S_7s_{23}s_{67}W_{71} + a_{23}c_{23}X_{71} - a_{67}s_{23}U^*_{176} + a_{71}s_{23}s_1Z_7, \\ D_2 &= S_1s_{12}c_{23}X_{71} + S_7c_{23}s_{67}X_{17} - a_{23}s_{23}Z_{71} + a_{67}c_{23}Y_{17} \\ &\quad + a_{71}c_{23}(-s_{67}c_7Z_1 + c_{67}\bar{Y}_1). \end{aligned} \tag{10.566}$$

Substituting $s_{12} = 1$ and $c_{12} = 0$ into these coefficients yields

$$\begin{aligned}
A_2 &= -S_7 s_{23} s_{67} U^*_{712} + a_{23} c_{23} Y_{71} - a_{67} s_{23} \bar{Y}_7 + a_{71} s_{23} (-s_{67} c_7 Y_1 + c_{67} Z'_1), \\
B_2 &= -S_1 s_{23} X^*_{71} + S_7 s_{23} s_{67} W_{71} + a_{23} c_{23} X_{71} - a_{67} s_{23} U^*_{176} + a_{71} s_{23} s_1 Z_7, \\
D_2 &= S_1 c_{23} X_{71} + S_7 c_{23} s_{67} X_{17} - a_{23} s_{23} Z_{71} + a_{67} c_{23} Y_{17} + a_{71} c_{23} (-s_{67} c_7 Z_1 + c_{67} \bar{Y}_1).
\end{aligned} \tag{10.567}$$

Expanding the terms that contain the parameter θ_1 and substituting the value of $\pi/2$ for α_{12} and regrouping gives

$$\begin{aligned}
A_2 &= G_1, \\
B_2 &= G_2 c_1 + G_3 s_1, \\
D_2 &= G_4 c_1 + G_5 s_1,
\end{aligned} \tag{10.568}$$

where

$$\begin{aligned}
G_1 &= -S_7 s_{23} s_{67} \bar{X}_7 - a_{23} c_{23} Z_7 - a_{67} s_{23} \bar{Y}_7 - a_{71} s_{23} Y_7, \\
G_2 &= -S_1 s_{23} Y_7 + S_7 s_{23} s_{67} c_7 + a_{23} c_{23} X_7 + a_{67} s_{23} X'_7, \\
G_3 &= -S_1 s_{23} X_7 - S_7 s_{23} s_{67} \bar{X}'_7 - a_{23} c_{23} Y_7 - a_{67} s_{23} Z'_7 + a_{71} s_{23} Z_7, \\
G_4 &= S_1 c_{23} X_7 + S_7 c_{23} s_{67} \bar{X}'_7 - a_{23} s_{23} Y_7 + a_{67} c_{23} Z'_7 - a_{71} c_{23} Z_7, \\
G_5 &= -S_1 c_{23} Y_7 + S_7 c_{23} s_{67} c_7 - a_{23} s_{23} X_7 + a_{67} c_{23} X'_7.
\end{aligned} \tag{10.569}$$

Equations (10.555) and (10.565) represent two equations in the two variables θ_1 and θ_2. Substituting the tan-half-angle relations $c_2 = (1 - x_2^2)/(1 + x_2^2)$ and $s_2 = 2x_2/(1 + x_2^2)$, where $x_2 = \tan(\theta_2/2)$, into these equations, multiplying throughout by $(1 + x_2^2)$, and regrouping gives

$$a_i x_2^2 + b_i x_2 + d_i = 0, \tag{10.570}$$

where $i = 1, 2$ and

$$\begin{aligned}
a_1 &= -F_2 c_1 - F_3 s_1 + F_1, & b_1 &= 2B_1, \\
d_1 &= F_2 c_1 + F_3 s_1 + F_1, & a_2 &= G_4 c_1 + G_5 s_1 - G_1, \\
b_2 &= 2G_2 c_1 + 2G_3 s_1, & d_2 &= G_4 c_1 + G_5 s_1 + G_1.
\end{aligned} \tag{10.571}$$

The variable x_2 can be eliminated from Eq. set (10.570) as discussed in Section 8.2. According to Bezout's solution method, the coefficients a_i, b_i, and d_i must satisfy the following condition in order for the two equations of Eq. set (10.570) to have a common solution for x_2:

$$\begin{vmatrix} a_1 & b_1 \\ a_2 & b_2 \end{vmatrix} \begin{vmatrix} b_1 & d_1 \\ b_2 & d_2 \end{vmatrix} - \begin{vmatrix} a_1 & d_1 \\ a_2 & d_2 \end{vmatrix}^2 = 0. \tag{10.572}$$

The notation $|jk|$ will be used to represent a determinant $\begin{vmatrix} j_1 & k_1 \\ j_2 & k_2 \end{vmatrix}$. The determinants $|ab|$, $|bd|$, and $|ad|$ can be expanded using Eq. (10.571) as

$$\begin{aligned}
|ab| &= \sigma^{(+)}, \\
|bd| &= \sigma^{(-)}, \\
|ad| &= (-2F_2 G_4) c_1^2 + (-2F_3 G_5) s_1^2 + (-2F_2 G_5 - 2F_3 G_4) s_1 c_1 + 2F_1 G_1,
\end{aligned} \tag{10.573}$$

where

$$\sigma^{(\pm)} = (-2F_2G_2)c_1^2 + (-2F_3G_3)s_1^2 + (-2F_3G_2 - 2F_2G_3)s_1c_1 \pm (2F_1G_2 - 2B_1G_4)c_1 \pm (2F_1G_3 - 2B_1G_5)s_1 + 2B_1G_1. \tag{10.574}$$

Substituting the tan-half-angle identities for s_1 and c_1 into Eq. (10.573) and then regrouping gives

$$|ab| = \frac{2}{(1+x_1^2)^2}\big[(B_1G_1 - F_1G_2 + B_1G_4 - F_2G_2)x_1^4 + 2(F_2G_3 + F_3G_2 - B_1G_5 + F_1G_3)x_1^3 + 2(B_1G_1 - 2F_3G_3 + F_2G_2)x_1^2 + 2(-F_3G_2 - F_2G_3 - B_1G_5 + F_1G_3)x_1 + (B_1G_1 - F_2G_2 + F_1G_2 - B_1G_4)\big], \tag{10.575}$$

$$|bd| = \frac{2}{(1+x_1^2)^2}\big[(B_1G_1 + F_1G_2 - B_1G_4 - F_2G_2)x_1^4 + 2(F_2G_3 + F_3G_2 + B_1G_5 - F_1G_3)x_1^3 + 2(B_1G_1 - 2F_3G_3 + F_2G_2)x_1^2 + 2(-F_3G_2 - F_2G_3 + B_1G_5 - F_1G_3)x_1 + (B_1G_1 - F_2G_2 - F_1G_2 + B_1G_4)\big], \tag{10.576}$$

$$|ad| = \frac{2}{(1+x_1^2)^2}\big[(F_1G_1 - F_2G_4)x_1^4 + 2(F_2G_5 + F_3G_4)x_1^3 + 2(F_1G_1 + F_2G_4 - 2F_3G_5)x_1^2 + 2(-F_2G_5 - F_3G_4)x_1 + (F_1G_1 - F_2G_4)\big]. \tag{10.577}$$

Substituting Eqs. (10.575) through (10.577) into Eq. (10.572) and then dividing throughout by $4/(1+x_1^2)^4$ and regrouping gives the following skew reciprocal polynomial†:

$$m_0x_1^8 + m_1x_1^7 + m_2x_1^6 + m_3x_1^5 + m_4x_1^4 - m_3x_1^3 + m_2x_1^2 - m_1x_1 + m_0 = 0, \tag{10.578}$$

where

$$\begin{aligned}
m_0 &= B_1^2G_1^2 + F_2^2G_2^2 - F_1^2G_2^2 - F_1^2G_1^2 - F_2^2G_4^2 - B_1^2G_4^2 + 2F_1F_2G_1G_4 - 2B_1F_2G_1G_2 \\
&\quad + 2F_1B_1G_2G_4, \\
m_1 &= 4\big[-F_2F_3G_2^2 + B_1^2G_4G_5 + F_2^2G_4G_5 + F_2F_3G_4^2 - F_2^2G_2G_3 + F_1^2G_2G_3 + B_1F_2G_1G_3 \\
&\quad - B_1F_1G_2G_5 + B_1F_3G_1G_2 - F_1F_3G_1G_4 - F_1F_2G_1G_5 - B_1F_1G_3G_4\big], \\
m_2 &= 4\big[F_1^2(-G_1^2 - G_3^2) + F_2^2(-G_2^2 + G_3^2 + G_4^2 - G_5^2) + F_3^2(G_2^2 - G_4^2) + B_1^2(G_1^2 - G_5^2) \\
&\quad + 2B_1(F_1G_3G_5 - F_3G_1G_3) + 2F_3(2F_2G_2G_3 + F_1G_1G_5 - 2F_2G_4G_5)\big], \\
m_3 &= 4\big[F_1^2G_2G_3 + 3F_2^2(G_2G_3 - G_4G_5) + 4F_3^2(G_4G_5 - G_2G_3) + B_1^2G_4G_5 \\
&\quad + F_2F_3(3G_2^2 - 4G_3^2 - 3G_4^2 + 4G_5^2) + F_1(-B_1G_2G_5 - B_1G_3G_4 \\
&\quad - F_3G_1G_4) + G_1(B_1F_2G_3 - F_1F_2G_5 + B_1F_3G_2)\big], \\
m_4 &= 2\big[B_1^2(3G_1^2 + G_4^2 - 4G_5^2) + F_1^2(-3G_1^2 + G_2^2 - 4G_3^2) \\
&\quad + F_2^2(3G_2^2 - 4G_3^2 - 3G_4^2 + 4G_5^2) + F_3^2(-4G_2^2 + 8G_3^2 + 4G_4^2 - 8G_5^2) \\
&\quad + F_1(8F_3G_1G_5 + 8B_1G_3G_5 - 2B_1G_2G_4 - 2F_2G_1G_4) \\
&\quad + F_3(-8B_1G_1G_3 + 16F_2G_4G_5 - 16F_2G_2G_3) + 2F_2B_1G_1G_2\big].
\end{aligned} \tag{10.579}$$

† This kind of polynomial was defined as skew reciprocal in Lin (1987) to distinguish it from an eighth-degree polynomial $\sum_{i=0}^{8} m_ix^i = 0$ where $m_i = m_{8-i}$ $(i = 0\ldots3)$, which was defined as reciprocal by Todhunter (1988).

Dividing all terms in Eq. (10.578) by x_1^4 gives

$$m_0\left(x_1^4+\frac{1}{x_1^4}\right)+m_1\left(x_1^3-\frac{1}{x_1^3}\right)+m_2\left(x_1^2+\frac{1}{x_1^2}\right)+m_3\left(x_1-\frac{1}{x_1}\right)+m_4=0. \tag{10.580}$$

This equation can be simplified by using the trigonometric identity

$$t_1=\frac{2x_1}{1-x_1^2}, \tag{10.581}$$

where $t_1=\tan\theta_1$. Rearranging Eq. (10.581) gives

$$x_1-\frac{1}{x_1}=-\frac{2}{t_1}. \tag{10.582}$$

Squaring both sides of Eq. (10.582) gives

$$x_1^2+\frac{1}{x_1^2}-2=\frac{4}{t_1^2}. \tag{10.583}$$

Rearranging this equation gives

$$x_1^2+\frac{1}{x_1^2}=\frac{4}{t_1^2}+2. \tag{10.584}$$

Equating the product of the left sides of Eqs. (10.582) and (10.584) with the product of the right sides of these equations gives

$$\left(x_1-\frac{1}{x_1}\right)\left(x_1^2+\frac{1}{x_1^2}\right)=\left(-\frac{2}{t_1}\right)\left(\frac{4}{t_1^2}+2\right). \tag{10.585}$$

Expanding this equation gives

$$x_1^3-\frac{1}{x_1^3}=-\frac{8}{t_1^3}-\frac{4}{t_1}+x_1-\frac{1}{x_1}. \tag{10.586}$$

Substituting Eq. (10.582) into the right side of Eq. (10.587) gives

$$x_1^3-\frac{1}{x_1^3}=-\frac{8}{t_1^3}-\frac{6}{t_1}. \tag{10.587}$$

Squaring both sides of Eq. (10.584) and rearranging gives

$$x_1^4+\frac{1}{x_1^4}=\frac{16}{t_1^4}+\frac{16}{t_1^2}+2. \tag{10.588}$$

Substituting Eqs. (10.582), (10.584), (10.587), and (10.588) into Eq. (10.580) gives

$$m_0\left(\frac{16}{t_1^4}+\frac{16}{t_1^2}+2\right)+m_1\left(-\frac{8}{t_1^3}-\frac{6}{t_1}\right)+m_2\left(\frac{4}{t_1^2}+2\right)+m_3\left(-\frac{2}{t_1}\right)+m_4=0. \tag{10.589}$$

Multiplying throughout by t_1^4 gives

$$m_0(16 + 16t_1^2 + 2t_1^4) + m_1(-8t_1 - 6t_1^3) + m_2(4t_1^2 + 2t_1^4) + m_3(-2t_1^3) + m_4t_1^4 = 0. \qquad (10.590)$$

Rearranging this equation yields

$$(2m_0 + 2m_2 + m_4)t_1^4 + (-6m_1 - 2m_3)t_1^3 + (16m_0 + 4m_2)t_1^2 + (-8m_1)t_1 + (16m_0) = 0. \qquad (10.591)$$

Equation (10.591) is a fourth-order input/output equation in terms of the tangent of the output angle. Four values of t_1 may be determined from this equation. Two unique values of θ_1 (which differ by 180 degrees) correspond to each value of t_1, and thus a total of eight distinct values of θ_1 exist.

10.6.2 Determination of θ_2

Corresponding values of θ_2 can be obtained for each of the eight values of θ_1. As explained in Section 8.2.2, the corresponding value for x_2 may be calculated from either

$$x_2 = \frac{-\begin{vmatrix} b_1 & d_1 \\ b_2 & d_2 \end{vmatrix}}{\begin{vmatrix} a_1 & d_1 \\ a_2 & d_2 \end{vmatrix}} \qquad (10.592)$$

or

$$x_2 = \frac{-\begin{vmatrix} a_1 & d_1 \\ a_2 & d_2 \end{vmatrix}}{\begin{vmatrix} a_1 & b_1 \\ a_2 & b_2 \end{vmatrix}}. \qquad (10.593)$$

The angle θ_2 is obtained from x_2 as follows:

$$\theta_2 = 2\tan^{-1}(x_2). \qquad (10.594)$$

10.6.3 Determination of θ_4

Using the direction cosines listed in the appendix for a spherical heptagon to project the vector loop equation onto the direction of the vector S_3 yields

$$a_{67}X_{217} + S_7Z_{21} + a_{71}X_{21} + S_1Z_2 + a_{45}X_4 = 0. \qquad (10.595)$$

Expanding the definition of the term X_4 and solving for the sine of θ_4 gives

$$s_4 = -(a_{67}X_{217} + S_7Z_{21} + a_{71}X_{21} + S_1Z_2)/a_{45}. \qquad (10.596)$$

Two values of θ_4 exist that will satisfy this equation for each of the eight sets of values of θ_1 and θ_2. Thus, a total of sixteen solution sets of the angles $(\theta_1, \theta_2, \theta_4)$ exist.

10.6.4 Determination of θ_3

Using set 13 of the table of direction cosines to project the vector loop equation onto the directions of the vectors $\mathbf{a}_{23}$ and $(\mathbf{S}_3 \times \mathbf{a}_{23})$ gives

$$a_{67}W_{712} + S_7X_{12} + a_{71}W_{12} + S_1X_2 + a_{23} + a_{45}W_{56712} = 0, \tag{10.597}$$

$$a_{67}U^*_{7123} - S_7Y_{12} + a_{71}U^*_{123} - S_1Y_2 + a_{45}U^*_{567123} = 0. \tag{10.598}$$

Substituting the subsidiary polar sine–cosine and cosine laws $U^*_{567123} = -V_{43}$ and $W_{56712} = W_{43}$ gives

$$a_{67}W_{712} + S_7X_{12} + a_{71}W_{12} + S_1X_2 + a_{23} + a_{45}W_{43} = 0, \tag{10.599}$$

$$a_{67}U^*_{7123} - S_7Y_{12} + a_{71}U^*_{123} - S_1Y_2 - a_{45}V_{43} = 0. \tag{10.600}$$

Expanding the definitions of V_{43} and W_{43}, substituting $s_{34} = 1$ and $c_{34} = 0$, and regrouping gives

$$c_3 = -(a_{67}W_{712} + S_7X_{12} + a_{71}W_{12} + S_1X_2 + a_{23})/(a_{45}c_4), \tag{10.601}$$

$$s_3 = -\left(a_{67}U^*_{7123} - S_7Y_{12} + a_{71}U^*_{123} - S_1Y_2\right)/(a_{45}c_4). \tag{10.602}$$

Thus, for each of the sixteen sets of solutions of $(\theta_1, \theta_2, \theta_4)$ a unique corresponding value of θ_3 can be determined.

10.6.5 Determination of θ_5

The following fundamental sine and sine–cosine laws may be written for a spherical heptagon:

$$X_{71234} = s_{56}s_5, \tag{10.603}$$

$$Y_{71234} = s_{56}c_5. \tag{10.604}$$

Substituting $\alpha_{56} = \pi/2$ reduces these equations to

$$s_5 = X_{71234}, \tag{10.605}$$

$$c_5 = Y_{71234}. \tag{10.606}$$

Thus, a unique corresponding value of θ_5 can be computed for each of the sixteen solution sets of $(\theta_1, \theta_2, \theta_3, \theta_4)$.

10.6.6 Determination of θ_6

The following fundamental sine and sine–cosine laws may be written for a spherical heptagon:

$$X_{43217} = s_{56}s_6, \tag{10.607}$$

$$Y_{43217} = s_{56}c_6. \tag{10.608}$$

Table 10.9. *RTTT mechanism parameters, case 1.*

Link length, cm.	Twist angle, deg.	Joint offset, cm.	Joint angle, deg.
$a_{12} = 0$	$\alpha_{12} = 90$	$S_1 = 2.3$	θ_1 = variable
$a_{23} = 8.3$	$\alpha_{23} = 60$	$S_2 = 0$	θ_2 = variable
$a_{34} = 0$	$\alpha_{34} = 90$	$S_3 = 0$	θ_3 = variable
$a_{45} = 8.6$	$\alpha_{45} = 0$	$S_4 = 0$	θ_4 = variable
$a_{56} = 0$	$\alpha_{56} = 90$	$S_5 = 0$	θ_5 = variable
$a_{67} = 4.9$	$\alpha_{67} = 75$	$S_6 = 0$	θ_6 = variable
$a_{71} = 8.6$	$\alpha_{71} = 65$	$S_7 = 4.6$	$\theta_7 = 217$ (input)

Substituting $\alpha_{56} = \pi/2$ reduces these equations to

$$s_6 = X_{43217}, \tag{10.609}$$

$$c_6 = Y_{43217}. \tag{10.610}$$

Thus, a unique corresponding value of θ_6 can be computed for each of the sixteen solution sets of $(\theta_1, \theta_2, \theta_3, \theta_4)$.

10.6.7 Numerical example

It has been shown that a total of sixteen solution configurations exist for the RTTT spatial mechanism. Although this is a group 4 spatial mechanism, the special geometry greatly simplifies the analysis. An input/output equation was obtained which was fourth-degree linear in the tangent of the angle θ_1. Eight distinct values of θ_1 correspond to these four values of $\tan(\theta_1)$. A unique corresponding value for the angle θ_2 was next determined for each value of θ_1. Two values of θ_4 were next calculated for each of the eight sets of values (θ_1, θ_2). Unique corresponding values for θ_3, θ_5, and θ_6 were next determined for each of the sixteen sets of values of $(\theta_1, \theta_2, \theta_4)$.

The mechanism dimensions of a numerical example are listed in Table 10.9. The resulting sixteen solution configurations are listed in Table 10.10 and are drawn in Figure 10.10. It is apparent in the figure that there are four classes of solutions. Each class has four cases, that is, two configurations for the Hooke joint and two configurations for the ball and socket joint.

10.7 RTTT spatial mechanism, case 2: $\alpha_{23} = \alpha_{45} = 90$ deg.

An RTTT spatial mechanism is shown in Figure 10.11 (see also Lin (1987)). In this mechanism, the first and second, third and fourth, and fifth and sixth joint axes intersect and the second through sixth joint offset distances equal zero. For the case to be analyzed here, the second and third joint axes and the fourth and fifth joint axes will be assumed to be perpendicular. Also, the joint angles of the Hooke joints, that is, α_{12}, α_{34}, and α_{56}, will be set equal to 90 degrees. Specifically, the problem can be stated as

Table 10.10. *Calculated configurations for the RTTT spatial mechanism, case 1.*

Solution	θ_1, deg.	θ_2, deg.	θ_3, deg.	θ_4, deg.	θ_5, deg.	θ_6, deg.
A	−61.29	−38.51	131.93	20.94	59.64	−1.99
B	−61.29	−38.51	−48.07	159.06	120.36	178.01
C	118.71	−141.49	−131.93	−20.94	120.36	178.01
D	118.71	−141.49	48.07	−159.06	59.64	−1.99
E	−20.09	−159.31	131.42	19.37	177.83	55.41
F	−20.09	−159.31	−48.59	160.63	2.17	−124.59
G	159.91	−20.69	−131.42	−19.37	2.17	−124.59
H	159.91	−20.69	48.59	−160.63	177.83	55.41
I	65.16	−101.69	−137.46	32.12	67.94	−115.43
J	65.16	−101.69	42.54	147.88	112.06	64.58
K	−114.84	−78.31	137.46	−32.12	112.06	64.58
L	−114.84	−78.31	−42.54	−147.88	67.94	−115.43
M	55.32	126.17	153.56	−45.82	−39.39	−31.36
N	55.32	126.17	−26.44	−134.18	−140.61	148.64
O	−124.68	53.83	−153.56	45.82	−140.61	148.64
P	−124.68	53.83	26.44	134.18	−39.39	−31.36

given: α_{67}, α_{71},
$a_{23}, a_{45}, a_{67}, a_{71}$,
S_1, S_7,
$\alpha_{12} = \alpha_{23} = \alpha_{34} = \alpha_{45} = \alpha_{56} = \pi/2$,
$a_{12} = a_{34} = a_{56} = S_2 = S_3 = S_4 = S_5 = S_6 = 0$, and
θ_7 (input angle),
find: $\theta_1, \theta_2, \theta_3, \theta_4, \theta_5$, and θ_6 .

It will be shown that a maximum of sixteen solution configurations exist for this mechanism.

10.7.1 Determination of input/output equation

The input/output equation is obtained from a pair of equations of the form

$$a_i x_2^2 + b_i x_2 + d_i = 0, \tag{10.611}$$

where $i = 1, 2$ and the coefficients a_i, b_i, d_i are functions of the output parameter θ_1. The first equation is the same self-scalar product of the vector loop equation as was derived in Section 10.6.1. This equation was expanded, and the coefficients a_1, b_1, and d_1 are defined in Eq. (10.571). The derivation of the second equation is more complicated, and it is derived from three equations that contain the unknown joint variables $\theta_1, \theta_2, \theta_3$, and θ_4.

(i) Projection of the Vector Loop Equation Along ($\mathbf{S}_1 \times \mathbf{a}_{12}$)

The vector loop equation for the RTTT mechanism may be written as

$$a_{67}\mathbf{a}_{67} + S_7\mathbf{S}_7 + a_{71}\mathbf{a}_{71} + S_1\mathbf{S}_1 + a_{23}\mathbf{a}_{23} + a_{45}\mathbf{a}_{45} = \mathbf{0}. \tag{10.612}$$

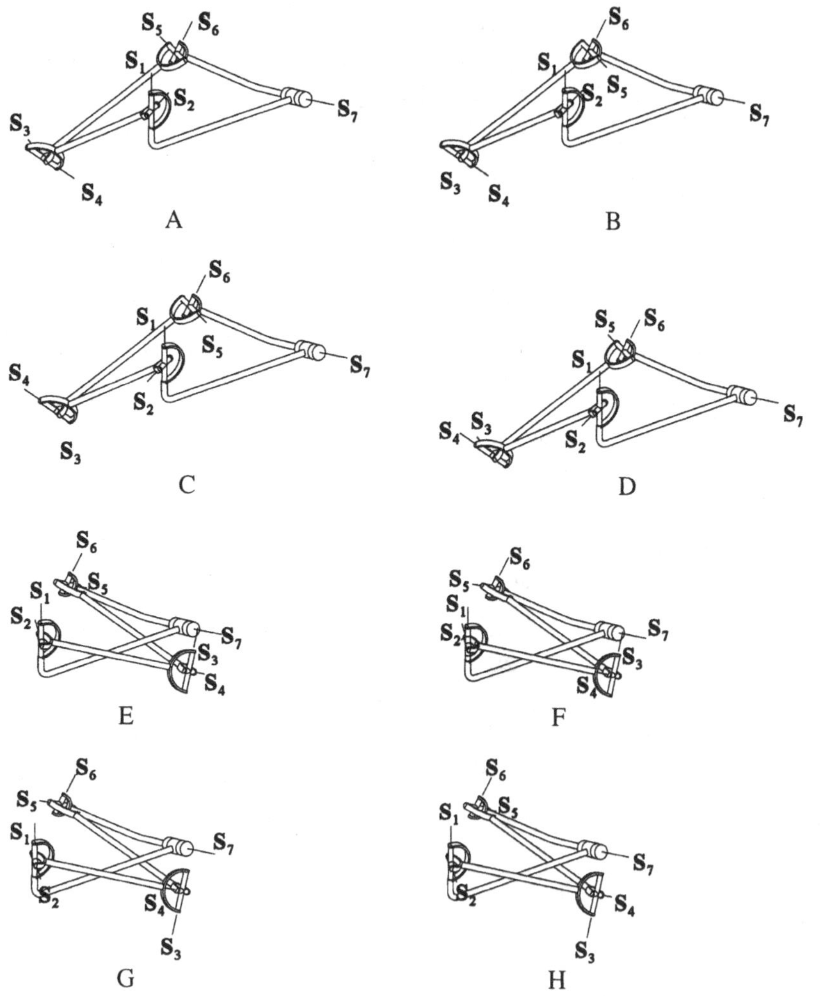

Figure 10.10. Sixteen configurations of the RTTT spatial mechanism.

Projecting this equation onto the direction of $(\mathbf{S}_1 \times \mathbf{a}_{12})$ by using set 1 of the table of direction cosines gives

$$-a_{67}U^*_{654321} + S_7Y_{65432} - a_{71}s_1 + a_{23}c_{12}s_2 - a_{45}U^*_{4321} = 0. \tag{10.613}$$

The spherical and polar sine–cosine laws $Y_{65432} = s_{71}c_1$ and $U^*_{654321} = -V_{71}$ may be substituted into this equation to give

$$a_{67}V_{71} + S_7s_{71}c_1 - a_{71}s_1 + a_{23}c_{12}s_2 - a_{45}U^*_{4321} = 0. \tag{10.614}$$

Expanding U^*_{4321} and substituting the value of $\pi/2$ for the twist angles α_{12} through α_{56} reduces this equation to

$$a_{67}V_{71} + S_7s_{71}c_1 - a_{71}s_1 - a_{45}s_3c_4 = 0. \tag{10.615}$$

This equation may be rewritten as

$$P_1 = a_{45}s_3c_4, \tag{10.616}$$

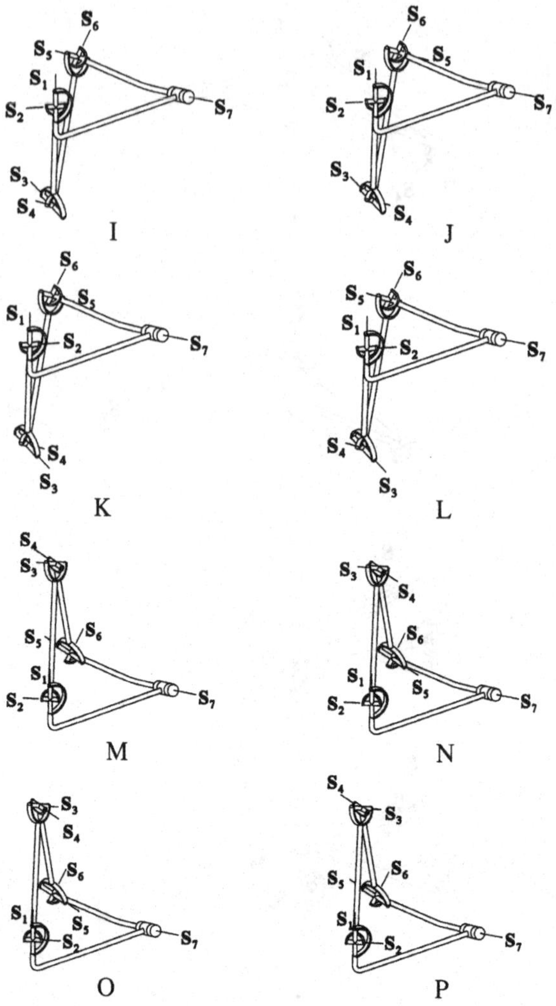

Figure 10.10. (*cont.*)

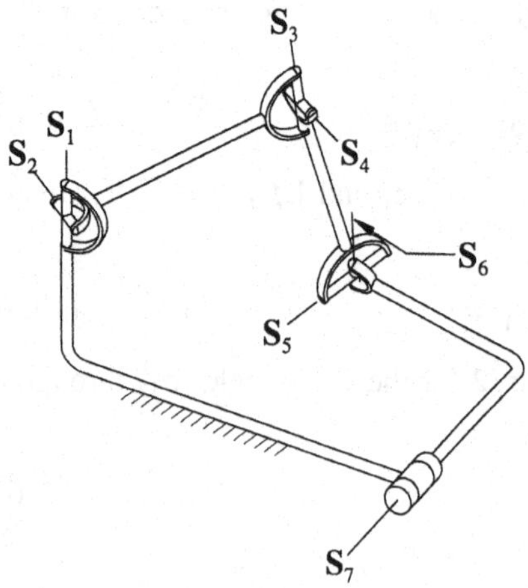

Figure 10.11. RTTT spatial mechanism.

where

$$P_1 = a_{67}V_{71} + S_7 s_{71} c_1 - a_{71} s_1. \tag{10.617}$$

The term P_1 may be factored as

$$P_1 = P_{1a}c_1 + P_{1b}s_1 + P_{1c}, \tag{10.618}$$

where

$$P_{1a} = S_7 s_{71} - a_{67}c_{71}s_7, \qquad P_{1b} = -a_{67}c_7 - a_{71}, \qquad P_{1c} = 0. \tag{10.619}$$

(ii) Secondary Cosine Law

A secondary cosine law for a spatial heptagon may be written as

$$Z_{07123} = \bar{Z}_{05}. \tag{10.620}$$

The term $\bar{Z}_{05}$ may be written as

$$\bar{Z}_{05} = -S_5 s_{45} s_{56} S_5 + a_{45}\bar{Y}_5 + a_{56}\bar{Y}_5. \tag{10.621}$$

Substituting $S_5 = a_{56} = 0$ and expanding $\bar{Y}_5$ and substituting values for α_{45} and α_{56} gives

$$\bar{Z}_{05} = 0. \tag{10.622}$$

The term Z_{07123} may be expanded as

$$\begin{aligned} Z_{07123} = {} & a_{34}Y_{7123} + S_3 s_{34} X_{7123} + a_{23}(c_{34}Y_{712} - s_{34}c_3 Z_{712}) + S_2(-\bar{X}_3 X^*_{712} - \bar{Y}_3 X_{712}) \\ & + a_{12}(Z_{71}X^*_{32} + Y_{71}\bar{Z}_3) + S_1(-X_7 X^*_{321} - Y_7 X_{321}) + a_{67}Y_{3217} + S_7 s_{67} X_{3217} \\ & + a_{71}(c_{67}Y_{321} - s_{67}c_7 Z_{321}). \end{aligned} \tag{10.623}$$

Substituting $S_2 = S_3 = a_{12} = a_{34} = 0$ reduces this equation to

$$\begin{aligned} Z_{07123} = {} & a_{23}(c_{34}Y_{712} - s_{34}c_3 Z_{712}) + S_1(-X_7 X^*_{321} - Y_7 X_{321}) + a_{67}Y_{3217} + S_7 s_{67} X_{3217} \\ & + a_{71}(c_{67}Y_{321} - s_{67}c_7 Z_{321}). \end{aligned} \tag{10.624}$$

Equating this equation to zero, expanding all terms that contain θ_2 and θ_3, and regrouping gives

$$Q_1 c_3 + Q_2 s_3 = 0, \tag{10.625}$$

where

$$Q_1 = P_2 c_2 + P_3 s_2 + P_4, \tag{10.626}$$

$$Q_2 = P_5 c_2 + P_6 s_2 \tag{10.627}$$

and where

$$
\begin{aligned}
P_2 &= -a_{23}Y_{71},\\
P_3 &= -a_{23}X_{71},\\
P_4 &= -S_1X_{71} + S_7s_{67}V_{71} + a_{67}(s_{67}Z_1 + c_{67}(-s_1s_7 + c_1c_7c_{71}))\\
&\quad + a_{71}(s_{67}c_7Z_1 + c_{67}c_{71}c_1),\\
P_5 &= -S_1X^*_{71} + S_7s_{67}W_{71} + a_{67}(-s_{67}X_1 - c_{67}V_{17}) + a_{71}(c_{67}c_{71}s_1 - s_{67}c_7X_1),\\
P_6 &= S_7s_{67}U_{71} + a_{67}(s_{67}Y_1 - c_{67}s_{71}c_7) + a_{71}(s_{67}c_7Y_1 - c_{67}s_{71}).
\end{aligned}
\tag{10.628}
$$

The terms P_2 through P_6 may be factored into the form

$$P_i = P_{ia}c_1 + P_{ib}s_1 + P_{ic}, \tag{10.629}$$

where $i = 2\ldots6$ and

$$
\begin{aligned}
P_{2a} &= 0,\\
P_{2b} &= 0,\\
P_{2c} &= a_{23}Z_7,\\
P_{3a} &= -a_{23}X_7,\\
P_{3b} &= a_{23}Y_7,\\
P_{3c} &= 0,\\
P_{4a} &= -S_1X_7 - S_7c_{71}X_7 - a_{67}Z'_7 + a_{71}Z_7,\\
P_{4b} &= S_1Y_7 - S_7s_{67}c_7 - a_{67}c_{67}s_7,\\
P_{4c} &= 0,\\
P_{5a} &= -P_{4b},\\
P_{5b} &= P_{4a},\\
P_{5c} &= 0,\\
P_{6a} &= 0,\\
P_{6b} &= 0,\\
P_{6c} &= S_7s_{71}X_7 + a_{67}\bar{Y}_7 + a_{71}Y_7.
\end{aligned}
\tag{10.630}
$$

(iii) Self-Scalar Product of the Vector Loop Equation

The vector loop equation for the mechanism may be written as

$$a_{67}\mathbf{a}_{67} + S_7\mathbf{S}_7 + a_{71}\mathbf{a}_{71} + S_1\mathbf{S}_1 = -a_{23}\mathbf{a}_{23} - a_{45}\mathbf{a}_{45}. \tag{10.631}$$

Taking the self-scalar product of each side of this equation gives

$$K + a_{67}\mathbf{a}_{67} \cdot (a_{71}\mathbf{a}_{71} + S_1\mathbf{S}_1) + S_7\mathbf{S}_7 \cdot S_1\mathbf{S}_1 = a_{23}\mathbf{a}_{23} \cdot a_{45}\mathbf{a}_{45}, \tag{10.632}$$

where

$$K = (a_{67}^2 + S_7^2 + a_{71}^2 + S_1^2 - a_{23}^2 - a_{45}^2)/2. \tag{10.633}$$

The scalar products of mutually perpendicular vectors, such as, for example, $\mathbf{a}_{67}$ and $\mathbf{S}_7$, equal zero, and all these terms have been omitted from Eq. (10.632). Evaluating the scalar products using the sets of direction cosines listed in the appendix yields

$$K + a_{67}(a_{71}c_7 + S_1\bar{X}_7) + S_1S_7c_{71} = a_{23}a_{45}W_{43}. \tag{10.634}$$

Expanding the definition of W_{43} and substituting $c_{34} = 0$ gives the result

$$P_7 = a_{23}a_{45}c_3c_4, \tag{10.635}$$

where

$$P_7 = K + a_{67}(a_{71}c_7 + S_1\bar{X}_7) + S_1S_7c_{71}. \tag{10.636}$$

The second equation that contains θ_1 and θ_2 as the only unknown parameters will next be obtained by manipulation of Eqs. (10.616), (10.625), and (10.636). Multiplying Eq. (10.616) by $a_{23}c_3$ and Eq. (10.635) by $-s_3$ and adding gives

$$P_1a_{23}c_3 - P_7s_3 = 0. \tag{10.637}$$

Equations (10.625) and (10.637) are linear homogeneous equations in the variables s_3 and c_3. A solution will exist only if the equations are linearly dependent (note that the trivial solution of $s_3 = c_3 = 0$ is not physically possible). Because the two equations must be linearly dependent, it may be written that

$$a_{23}P_1Q_2 + Q_1P_7 = 0. \tag{10.638}$$

Expanding Q_1 and Q_2 using Eqs. (10.626) and (10.627) and regrouping gives

$$(a_{23}P_1P_5 + P_7P_2)c_2 + (a_{23}P_1P_6 + P_7P_3)s_2 + P_7P_4 = 0. \tag{10.639}$$

Substituting the expressions for P_1 through P_6, that is, Eqs. (10.618) and (10.629), and regrouping gives

$$(H_1c_1^2 + H_2s_1^2 + H_3s_1c_1 + H_4)c_2 + (H_5c_1 + H_6s_1)s_2 + (H_7c_1 + H_8s_1) = 0, \tag{10.640}$$

where

$$\begin{aligned}
H_1 &= -a_{23}P_{1a}P_{4b},\\
H_2 &= a_{23}P_{1b}P_{4a},\\
H_3 &= a_{23}(P_{1a}P_{4a} - P_{1b}P_{4b}),\\
H_4 &= P_{2c}P_7,\\
H_5 &= a_{23}P_{1a}P_{6c} + P_{3a}P_7,\\
H_6 &= a_{23}P_{1b}P_{6c} + P_{3b}P_7,\\
H_7 &= P_{4a}P_7,\\
H_8 &= P_{4b}P_7.
\end{aligned} \tag{10.641}$$

Substituting the tan-half-angle identities $s_2 = 2x_2/(1 + x_2^2)$ and $c_2 = (1 - x_2^2)/(1 + x_2^2)$ into Eq. (10.640), multiplying throughout by $(1 + x_2^2)$, and regrouping gives

$$a_2x_2^2 + b_2x_2 + d_2 = 0, \tag{10.642}$$

where

$$\begin{aligned}
a_2 &= (H_7c_1 + H_8s_1) - \left(H_1c_1^2 + H_2s_1^2 + H_3s_1c_1 + H_4\right),\\
b_2 &= 2(H_5c_1 + H_6s_1),\\
d_2 &= (H_7c_1 + H_8s_1) + \left(H_1c_1^2 + H_2s_1^2 + H_3s_1c_1 + H_4\right).
\end{aligned} \tag{10.643}$$

Equation (10.643) will be paired with the first equation from the previous section, that is, Eq. (10.570), whose coefficients a_1, b_1, and d_1 are defined in Eq. (10.571). The variable x_2 can be eliminated from Eqs. (10.570) and (10.642) as discussed in Section 8.2. According to Bezout's method, the coefficients a_i through d_i must satisfy the following condition in order for the two equations to have a common solution for x_2:

$$|ab||bd| - |ad|^2 = 0, \tag{10.644}$$

where the notation $|jk|$ is used to represent the determinant $\begin{vmatrix} j_1 & k_1 \\ j_2 & k_2 \end{vmatrix}$. The determinants $|ab|$, $|bd|$, and $|ad|$ are expanded as

$$\begin{aligned}
|ab| &= \delta^{(+)},\\
|bd| &= \delta^{(-)},\\
|ad| &= 2(F_1H_1 - F_2H_7)c_1^2 + 2(F_1H_2 - F_3H_8)s_1^2\\
&\quad + 2(F_1H_3 - F_2H_8 - F_3H_7)s_1c_1 + 2(F_1H_4),
\end{aligned} \tag{10.645}$$

where

$$\begin{aligned}
\delta^{(\pm)} &= 2(B_1H_1 - F_2H_5)c_1^2 + 2(B_1H_2 - F_3H_6)s_1^2 + 2(B_1H_3 - F_2H_6 - F_3H_5)s_1c_1\\
&\quad \pm 2(-B_1H_7 + F_1H_5)c_1 \pm 2(-B_1H_8 + F_1H_6)s_1 + 2B_1H_4.
\end{aligned} \tag{10.646}$$

Substituting the tan-half-angle identities for s_1 and c_1 into the determinants of Eq. (10.645) and then regrouping gives

$$\begin{aligned}
|ab| &= \frac{2}{\left(1 + x_1^2\right)^2}\{[B_1(H_1 + H_4 + H_7) - H_5(F_1 + F_2)]x_1^4 + [-2B_1(H_3 + H_8)\\
&\quad + 2H_6(F_1 + F_2) + 2F_3H_5]x_1^3 + [2B_1(-H_1 + 2H_2 + H_4) + 2F_2H_5\\
&\quad - 4F_3H_6]x_1^2 + [2B_1(H_3 - H_8) + 2H_6(F_1 - F_2) - 2F_3H_5]x_1\\
&\quad + [B_1(H_1 + H_4 - H_7) + H_5(F_1 - F_2)]\},
\end{aligned} \tag{10.647}$$

$$\begin{aligned}
|bd| &= \frac{2}{\left(1 + x_1^2\right)^2}\{[B_1(H_1 + H_4 - H_7) + H_5(F_1 - F_2)]x_1^4 + [2B_1(-H_3 + H_8)\\
&\quad + 2H_6(-F_1 + F_2) + 2F_3H_5]x_1^3 + [2B_1(-H_1 + 2H_2 + H_4) + 2F_2H_5\\
&\quad - 4F_3H_6]x_1^2 + [2B_1(H_3 + H_8) - 2H_6(F_1 + F_2) - 2F_3H_5]x_1\\
&\quad + [B_1(H_1 + H_4 + H_7) - H_5(F_1 + F_2)]\},
\end{aligned} \tag{10.648}$$

$$\begin{aligned}
|ad| &= \frac{2}{\left(1 + x_1^2\right)^2}\{[F_1(H_1 + H_4) - F_2H_7]x_1^4 + 2[-F_1H_3 + F_2H_8 + F_3H_7]x_1^3\\
&\quad + [2F_1(-H_1 + H_2 + H_4) + 2F_2H_7 - 4F_3H_8]x_1^2\\
&\quad + 2[F_1H_3 - F_2H_8 - F_3H_7]x_1 + [F_1(H_1 + H_4) - F_2H_7]\}.
\end{aligned} \tag{10.649}$$

Substituting Eqs. (10.647) through (10.649) into Eq. (10.644) and then dividing throughout by $4/(1+x_1^2)^4$ and regrouping gives the following skew reciprocal polynomial:

$$m_0x_1^8 + m_1x_1^7 + m_2x_1^6 + m_3x_1^5 + m_4x_1^4 - m_3x_1^3 + m_2x_1^2 - m_1x_1 + m_0 = 0, \quad (10.650)$$

where

$$\begin{aligned}
m_0 &= B_1^2(H_1^2 + H_4^2 - H_7^2 + 2H_1H_4) + 2B_1H_5(F_1H_7 - F_2H_4 - F_2H_1) + F_1^2(-H_1^2 - H_4^2 \\
&\quad - H_5^2 - 2H_1H_4) + 2F_1F_2H_7(H_1 + H_4) + F_2^2(H_5^2 - H_7^2), \\
m_1 &= 4B_1^2(-H_1H_3 - H_3H_4 + H_7H_8) + 4B_1[H_5(-F_1H_8 + F_3H_1 + F_2H_3 + F_3H_4) \\
&\quad + H_6(-F_1H_7 + F_2H_1 + F_2H_4)] + 4F_1^2(H_1H_3 + H_3H_4 + H_5H_6) \\
&\quad + 4F_1[F_2(-H_1H_8 - H_3H_7 - H_4H_8) + F_3H_7(-H_1 - H_4)] \\
&\quad + 4F_2^2(H_7H_8 - H_5H_6) + 4F_2F_3(H_7^2 - H_5^2), \\
m_2 &= 4F_1^2(H_1^2 - H_3^2 - H_4^2 - H_6^2 - 2H_1H_2 - 2H_2H_4) + 8F_1[F_2(-H_1H_7 + H_2H_7 \\
&\quad + H_3H_8) + F_3(H_1H_8 + H_3H_7 + H_4H_8) + B_1H_6H_8] + 4B_1^2(-H_1^2 + H_3^2 + H_4^2 \\
&\quad - H_8^2 + 2H_1H_2 + 2H_2H_4) + 8B_1[F_2(-H_2H_5 - H_3H_6 + H_1H_5) \\
&\quad + F_3(-H_1H_6 - H_3H_5 - H_4H_6)] + 4F_2^2(-H_5^2 + H_6^2 + H_7^2 - H_8^2) \\
&\quad + 16F_2F_3(H_5H_6 - H_7H_8) + 4F_3^2(H_5^2 - H_7^2), \\
m_3 &= 4F_1^2(H_3H_4 + H_5H_6 - 3H_1H_3 + 4H_2H_3) + 4F_1[F_2(3H_3H_7 + 3H_1H_8 - 4H_2H_8 \\
&\quad - H_4H_8) - B_1(H_5H_8 + H_6H_7) + F_3(3H_1H_7 - 4H_3H_8 - 4H_2H_7 - H_4H_7)] \\
&\quad + 4B_1^2(-H_3H_4 + H_7H_8 + 3H_1H_3 - 4H_2H_3) + 4B_1[F_2(4H_2H_6 - 3H_3H_5 \\
&\quad + H_4H_6 - 3H_1H_6) + F_3(H_4H_5 + 4H_3H_6 - 3H_1H_5 + 4H_2H_5)] + 12F_2^2(H_5H_6 \\
&\quad - H_7H_8) + 16F_3^2(H_7H_8 - H_5H_6) + 4F_2F_3(-4H_6^2 - 3H_7^2 + 4H_8^2 + 3H_5^2), \\
m_4 &= 2F_1^2(-3H_1^2 - 3H_4^2 - 8H_2^2 + H_5^2 + 4H_3^2 - 4H_6^2 + 8H_1H_2 - 8H_2H_4 + 2H_1H_4) \\
&\quad + 4F_1[F_2(-H_4H_7 - 4H_2H_7 + 3H_1H_7 - 4H_3H_8) + B_1(4H_6H_8 - H_5H_7) \\
&\quad + F_3(-4H_3H_7 - 4H_1H_8 + 8H_2H_8 + 4H_4H_8)] + 2B_1^2(3H_4^2 - 4H_8^2 - 4H_3^2 \\
&\quad + 8H_2^2 + 3H_1^2 + H_7^2 + 8H_2H_4 - 8H_1H_2 - 2H_1H_4) + 4B_1[F_2(4H_3H_6 - 3H_1H_5 \\
&\quad + H_4H_5 + 4H_2H_5) + F_3(-4H_4H_6 + 4H_3H_5 + 4H_1H_6 - 8H_2H_6)] \\
&\quad + 2F_2^2(-3H_7^2 + 3H_5^2 - 4H_6^2 + 4H_8^2) + 32F_2F_3(H_7H_8 - H_5H_6) \\
&\quad + 8F_3^2(H_7^2 - 2H_8^2 + 2H_6^2 - H_5^2).
\end{aligned} \quad (10.651)$$

It was shown in Section 10.6.1 how a skew reciprocal polynomial of the form of Eq. (10.650) may be written in the form

$$\begin{aligned}
&(2m_0 + 2m_2 + m_4)t_1^4 + (-6m_1 - 2m_3)t_1^3 + (16m_0 + 4m_2)t_1^2 + (-8m_1)t_1 \\
&\quad + (16m_0) = 0,
\end{aligned} \quad (10.652)$$

where t_1 is equal to $\tan(\theta_1)$. Equation (10.652) is a fourth-order input/output equation in terms of the tangent of the output angle. Four values of t_1 may be determined from this equation. Two unique values of θ_1 (which differ by 180 degrees) correspond to each value of t_1, and thus a total of eight distinct values of θ_1 exist.

Table 10.11. *RTTT mechanism parameters, case 2.*

Link length, cm.	Twist angle, deg.	Joint offset, cm.	Joint angle, deg.
$a_{12} = 0$	$\alpha_{12} = 90$	$S_1 = 5$	θ_1 = variable
$a_{23} = 7$	$\alpha_{23} = 90$	$S_2 = 0$	θ_2 = variable
$a_{34} = 0$	$\alpha_{34} = 90$	$S_3 = 0$	θ_3 = variable
$a_{45} = 7$	$\alpha_{45} = 90$	$S_4 = 0$	θ_4 = variable
$a_{56} = 0$	$\alpha_{56} = 90$	$S_5 = 0$	θ_5 = variable
$a_{67} = 5$	$\alpha_{67} = 55$	$S_6 = 0$	θ_6 = variable
$a_{71} = 10$	$\alpha_{71} = 75$	$S_7 = 7$	θ_7 = 190 (input)

10.7.2 Determination of remaining joint angles

The joint parameters θ_2 through θ_6 can be determined using the same equations listed in Sections 10.6.2 through 10.6.6. The current values of $\alpha_{23} = \alpha_{45} = 90$ degrees are now substituted into the definitions of the terms in these equations.

10.7.3 Numerical example

It has been shown that a total of sixteen solution configurations exist for the RTTT spatial mechanism. Although this is a group 4 spatial mechanism, the special geometry greatly simplifies the analysis. An input/output equation was obtained that was fourth-degree linear in the tangent of the angle θ_1. Eight distinct values of θ_1 correspond to these four values of $\tan(\theta_1)$. A unique corresponding value for the angle θ_2 was next determined for each value of θ_1. Two values of θ_4 were next calculated for each of the eight sets of values (θ_1, θ_2). Unique corresponding values for θ_3, θ_5, and θ_6 were next determined for each of the sixteen sets of values of $(\theta_1, \theta_2, \theta_4)$.

The mechanism dimensions of a numerical example are listed in Table 10.11. The resulting sixteen solution configurations are listed in Table 10.12 and are drawn in Figure 10.12. It is apparent in the figure that there are two classes of solutions. Each class has eight cases, that is, two configurations for each of the three Hooke joints of the mechanism.

10.8 RRR-R-RRR spatial mechanism

An RRR-R-RRR spatial mechanism is shown in Figure 10.13. The notation R-R-R is used to indicate that the third, fourth, and fifth joint axes are parallel. Specifically, the problem can be stated as

given: $\alpha_{12}, \alpha_{23}, \alpha_{56}, \alpha_{67}, \alpha_{71}$,
$a_{12}, a_{23}, a_{34}, a_{45}, a_{56}, a_{67}, a_{71}$,
$S_1, S_2, S_3, S_4, S_5, S_6, S_7, \alpha_{34} = \alpha_{45} = 0$, and
θ_7
(input angle),

find: $\theta_1, \theta_2, \theta_3, \theta_4, \theta_5$, and θ_6.

It will be shown that a maximum of eight solution configurations exist for this mechanism.

Table 10.12. *Calculated configurations for the RTTT spatial mechanism, case 2.*

Solution	θ_1, deg.	θ_2, deg.	θ_3, deg.	θ_4, deg.	θ_5, deg.	θ_6, deg.
A	38.65	175.86	69.05	−69.53	−95.35	156.26
B	38.65	175.86	−110.95	−110.47	−84.65	−23.74
C	−141.35	4.14	−69.05	69.53	−84.65	−23.74
D	−141.35	4.14	110.95	110.47	−95.35	156.26
E	53.29	173.27	6.94	−82.77	−107.95	−156.79
F	53.29	173.27	−173.06	−97.23	−72.05	23.21
G	−126.71	6.73	−6.94	82.77	−72.05	23.21
H	−126.71	6.73	173.06	97.23	−107.95	−156.79
I	78.52	−105.69	−76.28	58.20	−176.79	127.65
J	78.52	−105.69	103.72	121.80	−3.21	−52.35
K	−101.48	−74.31	76.28	−58.20	−3.21	−52.35
L	−101.48	−74.31	−103.72	−121.80	−176.79	127.65
M	−88.14	−70.90	80.63	−39.88	−2.41	−47.18
N	−88.14	−70.90	−99.37	−140.12	−177.59	132.82
O	91.86	−109.10	−80.63	39.88	−177.59	132.82
P	91.86	−109.10	99.37	140.12	−2.41	−47.18

10.8.1 Determination of θ_6

The angle θ_6 will be solved for first for this mechanism. It will be shown that a fourth-degree polynomial in the tan-half-angle of θ_6 is obtained from two equations that contain the unknown joint parameters θ_6 and θ_2.

The first equation is obtained from a projection of the vector loop equation onto the direction of $\mathbf{S}_5$. The vector loop equation may be written as

$$S_1\mathbf{S}_1 + a_{12}\mathbf{a}_{12} + S_2\mathbf{S}_2 + a_{23}\mathbf{a}_{23} + S_3\mathbf{S}_3 + a_{34}\mathbf{a}_{34} + S_4\mathbf{S}_4 + a_{45}\mathbf{a}_{45} + S_5\mathbf{S}_5 + a_{56}\mathbf{a}_{56} + S_6\mathbf{S}_6 + a_{67}\mathbf{a}_{67} + S_7\mathbf{S}_7 + a_{71}\mathbf{a}_{71} = \mathbf{0}. \tag{10.653}$$

Using set 5 from the table of direction cosines for a spherical heptagon, the projection of the vector loop equation may be written as

$$S_1Z_{76} + a_{12}U_{1765} + S_2Z_{176} + a_{23}U_{21765} + S_3Z_{2176} + a_{34}U_{321765} + S_4Z_{32176} + S_5 + S_6c_{56} + a_{67}U_{65} + S_7\bar{Z}_6 + a_{71}U_{765} = 0. \tag{10.654}$$

The spherical cosine laws and polar sine laws $Z_{176} = Z_{34}$, $Z_{2176} = Z_4$, $Z_{32176} = c_{45}$, $U_{1765} = U_{2345}$, $U_{21765} = U_{345}$, and $U_{321765} = U_{45}$ are substituted into this equation to give

$$S_1Z_{76} + a_{12}U_{2345} + S_2Z_{34} + a_{23}U_{345} + S_3Z_4 + a_{34}U_{45} + S_4c_{45} + S_5 + S_6c_{56} + a_{67}U_{65} + S_7\bar{Z}_6 + a_{71}U_{765} = 0. \tag{10.655}$$

Expanding U_{2345}, Z_{34}, U_{345}, Z_4, and U_{45} and substituting $\alpha_{34} = \alpha_{45} = 0$ simplifies this

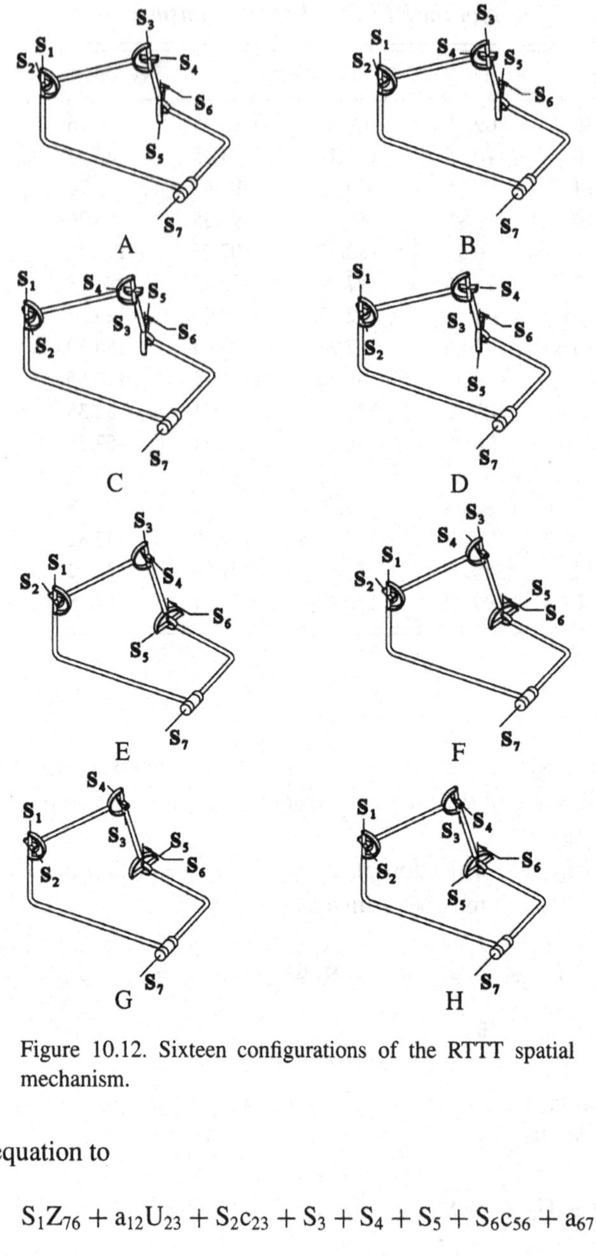

Figure 10.12. Sixteen configurations of the RTTT spatial mechanism.

equation to

$$S_1Z_{76} + a_{12}U_{23} + S_2c_{23} + S_3 + S_4 + S_5 + S_6c_{56} + a_{67}U_{65} + S_7\bar{Z}_6 + a_{71}U_{765} = 0. \tag{10.656}$$

Expanding Z_{76}, U_{23}, U_{65}, $\bar{Z}_6$, and U_{765} and regrouping gives

$$As_2 + B = 0, \tag{10.657}$$

where

$$\begin{aligned} A &= a_{12}s_{23}, \\ B &= B_1c_6 + B_2s_6 + B_3 \end{aligned} \tag{10.658}$$

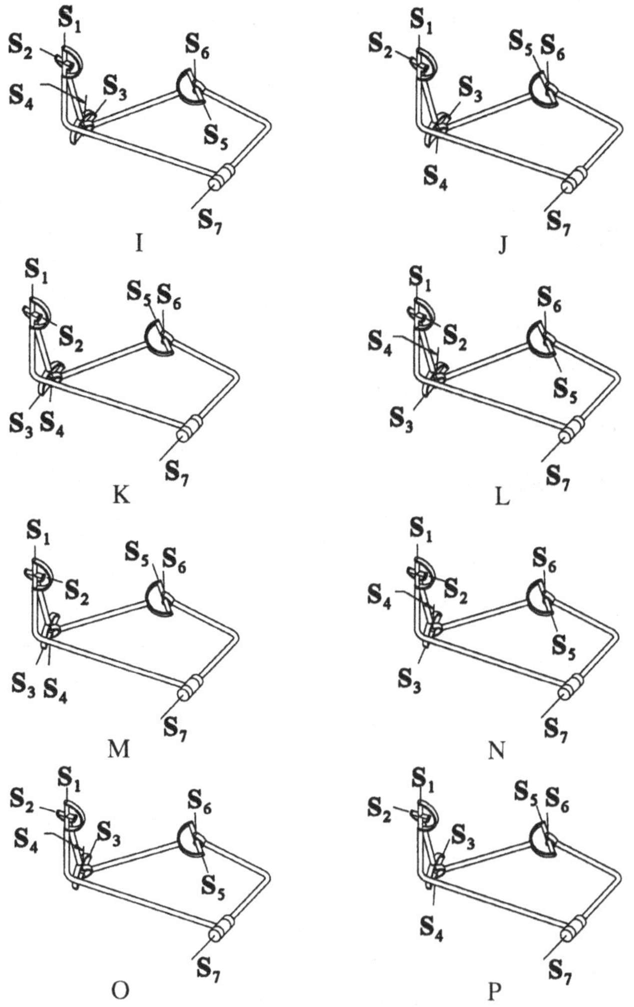

Figure 10.12. *(cont.)*

and where

$$B_1 = S_1 s_{56} \bar{Y}_7 - S_7 s_{56} s_{67} + a_{71} s_{56} c_{67} s_7,$$
$$B_2 = S_1 s_{56} \bar{X}_7 + a_{67} s_{56} + a_{71} s_{56} c_7,$$
$$B_3 = S_1 c_{56} \bar{Z}_7 + S_2 c_{23} + S_3 + S_4 + S_5 + S_6 c_{56} + S_7 c_{56} c_{67} + a_{71} c_{56} X_7. \quad (10.659)$$

The second equation is obtained from the subsidiary cosine law

$$Z_{76} = Z_{234}. \quad (10.660)$$

Expanding Z_{234} and substituting $\alpha_{34} = \alpha_{45} = 0$ gives

$$Z_{76} = Z_2. \quad (10.661)$$

Expanding Z_{76} and Z_2 and regrouping gives

$$Dc_2 + E = 0, \quad (10.662)$$

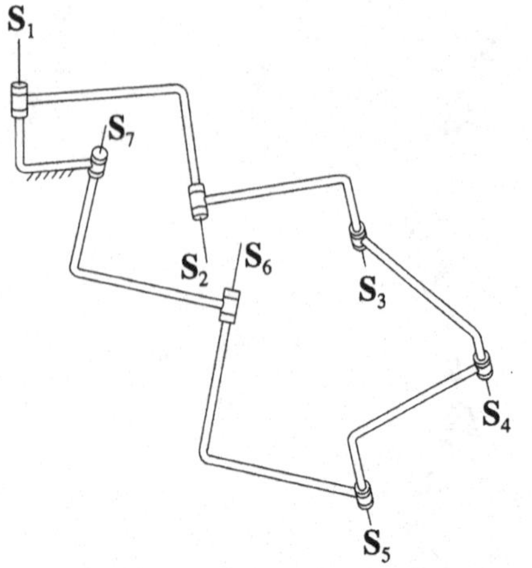

Figure 10.13. RRR-R-RRR spatial mechanism.

where

$$\begin{aligned} D &= s_{12}s_{23}, \\ E &= E_1c_6 + E_2s_6 + E_3 \end{aligned} \tag{10.663}$$

and where

$$\begin{aligned} E_1 &= s_{56}\bar{Y}_7, \\ E_2 &= s_{56}\bar{X}_7, \\ E_3 &= c_{56}\bar{Z}_7 - c_{12}c_{23}. \end{aligned} \tag{10.664}$$

The parameter θ_2 may be eliminated from Eqs. (10.657) and (10.662) by subtracting $a_{12}x_2$ times Eq. (10.662) from s_{12} times Eq. (10.657) to give

$$s_{12}As_2 - a_{12}x_2Dc_2 + s_{12}B - a_{12}x_2E = 0. \tag{10.665}$$

Because $s_{12}A = a_{12}D$, this equation may be written as

$$s_{12}A(s_2 - x_2c_2) + s_{12}B - a_{12}x_2E = 0. \tag{10.666}$$

Because $(s_2 - x_2c_2) = x_2$ (see Eq. (9.108)),

$$(s_{12}A - a_{12}E)x_2 + (s_{12}B) = 0. \tag{10.667}$$

Adding $s_{12}x_2$ times Eq. (10.657) to a_{12} times Eq. (10.662) gives

$$s_{12}x_2As_2 + a_{12}Dc_2 + s_{12}x_2B + a_{12}E = 0. \tag{10.668}$$

Again, because $s_{12}A = a_{12}D$, this equation may be written as

$$s_{12}A(x_2s_2 + c_2) + s_{12}x_2B + a_{12}E = 0. \tag{10.669}$$

Because $(x_2s_2 + c_2) = 1$ (see Eq. (9.109)),

$$(s_{12}B)x_2 + (s_{12}A + a_{12}E) = 0. \tag{10.670}$$

Equations (10.667) and (10.670) are linear in the variable x_2. In order for there to be a common value of x_2 that satisfies the two equations, the equations must be linearly dependent. As such, the coefficients of these two equations must satisfy the following expression:

$$(s_{12}A - a_{12}E)(s_{12}A + a_{12}E) - (s_{12}B)^2 = 0. \tag{10.671}$$

Multiplying the first terms gives

$$s_{12}^2A^2 - a_{12}^2E^2 - s_{12}^2B^2 = 0. \tag{10.672}$$

Substituting Eqs. (10.658) and (10.663) for B and E and regrouping gives

$$F_1c_6^2 + F_2s_6^2 + F_3s_6c_6 + F_4c_6 + F_5s_6 + F_6 = 0, \tag{10.673}$$

where

$$\begin{aligned}
F_1 &= -a_{12}^2E_1^2 - s_{12}^2B_1^2,\\
F_2 &= -a_{12}^2E_2^2 - s_{12}^2B_2^2,\\
F_3 &= -2\left(a_{12}^2E_1E_2 + s_{12}^2B_1B_2\right),\\
F_4 &= -2\left(a_{12}^2E_1E3 + s_{12}^2B_1B_3\right),\\
F_5 &= -2\left(a_{12}^2E_2E_3 + s_{12}^2B_2B_3\right),\\
F_6 &= -a_{12}^2E_3^2 + s_{12}^2\left(A^2 - B_3^2\right).
\end{aligned} \tag{10.674}$$

Substituting the trigonometric identities $s_6 = (2x_6)/(1 + x_6^2)$ and $c_6 = (1 - x_6^2)/(1 + x_6^2)$ for the sine and cosine of θ_6, multiplying throughout by $(1 + x_6^2)^2$, and regrouping gives

$$f_4x_6^4 + f_3x_6^3 + f_2x_6^2 + f_1x_6 + f_0 = 0, \tag{10.675}$$

where

$$\begin{aligned}
f_4 &= F_1 - F_4 + F_6,\\
f_3 &= -2(F_3 - F_5),\\
f_2 &= 2(-F_1 + 2F_2 + F_6),\\
f_1 &= 2(F_3 + F_5),\\
f_0 &= F_1 + F_4 + F_6.
\end{aligned} \tag{10.676}$$

Equation (10.675) is a fourth-order equation that can be solved for x_6. Four values of x_6 and thereby four distinct values of θ_6 can be obtained from this equation. It is interesting to note that this equation does not contain the link lengths a_{34} and a_{45}, the perpendicular distances between the parallel axes.

10.8.2 Determination of θ_2

A corresponding value of θ_2 may be obtained for each value of θ_6 from either Eq. (10.667) or Eq. (10.670). Numerical values for B and E can be obtained for each calculated value of θ_6. The tan-half-angle of θ_2 may then be obtained from either

$$x_2 = \frac{-s_{12}B}{s_{12}A - a_{12}E} \tag{10.677}$$

or

$$x_2 = \frac{-(s_{12}A + a_{12}E)}{s_{12}B}. \tag{10.678}$$

The angle θ_2 is obtained from x_2 via the equation

$$\theta_2 = 2\tan^{-1}(x_2). \tag{10.679}$$

10.8.3 Determination of θ_1

The following sine and sine–cosine laws may be written for a spherical heptagon:

$$X_{6712} = X_{43}, \tag{10.680}$$

$$Y_{6712} = -X^*_{43}. \tag{10.681}$$

Expanding the right sides of this pair of equations and substituting $\alpha_{34} = \alpha_{45} = 0$ yields

$$X_{6712} = 0, \tag{10.682}$$

$$Y_{6712} = 0. \tag{10.683}$$

Expanding the definitions of the terms X_{6712} and Y_{6712} and regrouping gives

$$[X_{67}c_2 - c_{12}Y_{67}s_2]c_1 + [-Y_{67}c_2 - c_{12}X_{67}s_2]s_1 + [s_{12}s_2Z_{67}] = 0, \tag{10.684}$$

$$[c_{23}(X_{67}s_2 + c_{12}Y_{67}c_2) - s_{12}s_{23}Y_{67}]c_1 + [c_{23}(-Y_{67}s_2 + c_{12}X_{67}c_2) - s_{12}s_{23}X_{67}]s_1 \\ + [-s_{12}c_{23}Z_{67}c_2 - c_{12}s_{23}Z_{67}] = 0. \tag{10.685}$$

All the terms in brackets may be numerically evaluated for each set of solution values of (θ_6, θ_2). Equations (10.684) and (10.685) thus represent two linear equations in the variables c_1 and s_1. The solution of these two linear equations will yield the unique corresponding value for the sine and cosine of θ_1, and thus a unique value for the angle θ_1.

10.8.4 Determination of θ_4

Projecting the vector loop equation onto the direction of $\mathbf{a}_{56}$ and $(\mathbf{S}_5 \times \mathbf{a}_{56})$ using set 5 of the table of direction cosines yields

$$S_1X_{76} + a_{12}W_{176} + S_2X_{176} + a_{23}W_{2176} + S_3X_{2176} + a_{34}W_{32176} + S_4X_{32176} \\ + a_{45}c_5 + a_{56} + a_{67}c_6 + S_7\bar{X}_6 + a_{71}W_{76} = 0, \tag{10.686}$$

$$S_1Y_{76} - a_{12}U^*_{1765} + S_2Y_{176} - a_{23}U^*_{21765} + S_3Y_{2176} - a_{34}U^*_{321765} + S_4Y_{32176} \\ - a_{45}s_5 - S_6s_{56} + a_{67}c_{56}s_6 + S_7\bar{Y}_6 - a_{71}U^*_{765} = 0. \tag{10.687}$$

Substituting the spherical and polar sine and sine–cosine laws $X_{32176} = s_{45}s_5$, $W_{32176} = W_{45}$, $Y_{32176} = s_{45}c_5$, and $U^*_{321765} = -V_{45}$ gives

$$S_1X_{76} + a_{12}W_{176} + S_2X_{176} + a_{23}W_{2176} + S_3X_{2176} + a_{34}W_{45} + S_4s_{45}s_5 + a_{45}c_5 + a_{56} + a_{67}c_6 + S_7\bar{X}_6 + a_{71}W_{76} = 0, \quad (10.688)$$

$$S_1Y_{76} - a_{12}U^*_{1765} + S_2Y_{176} - a_{23}U^*_{21765} + S_3Y_{2176} + a_{34}V_{45} + S_4s_{45}c_5 - a_{45}s_5 - S_6s_{56} + a_{67}c_{56}s_6 + S_7\bar{Y}_6 - a_{71}U^*_{765} = 0. \quad (10.689)$$

Expanding V_{45} and W_{45} and substituting $\alpha_{34} = \alpha_{45} = 0$ yields

$$S_1X_{76} + a_{12}W_{176} + S_2X_{176} + a_{23}W_{2176} + S_3X_{2176} + a_{34}(c_4c_5 - s_4s_5) + a_{45}c_5 + a_{56} + a_{67}c_6 + S_7\bar{X}_6 + a_{71}W_{76} = 0, \quad (10.690)$$

$$S_1Y_{76} - a_{12}U^*_{1765} + S_2Y_{176} - a_{23}U^*_{21765} + S_3Y_{2176} - a_{34}(s_5c_4 + c_5s_4) - a_{45}s_5 - S_6s_{56} + a_{67}c_{56}s_6 + S_7\bar{Y}_6 - a_{71}U^*_{765} = 0. \quad (10.691)$$

Introducing the notation $s_{4+5} = \sin(\theta_4 + \theta_5)$ and $c_{4+5} = \cos(\theta_4 + \theta_5)$ and recognizing that $\sin(\theta_4 + \theta_5) = s_4c_5 + c_4s_5$ and $\cos(\theta_4 + \theta_5) = c_4c_5 - s_4s_5$ gives

$$S_1X_{76} + a_{12}W_{176} + S_2X_{176} + a_{23}W_{2176} + S_3X_{2176} + a_{34}c_{4+5} + a_{45}c_5 + a_{56} + a_{67}c_6 + S_7\bar{X}_6 + a_{71}W_{76} = 0, \quad (10.692)$$

$$S_1Y_{76} - a_{12}U^*_{1765} + S_2Y_{176} - a_{23}U^*_{21765} + S_3Y_{2176} - a_{34}s_{4+5} - a_{45}s_5 - S_6s_{56} + a_{67}c_{56}s_6 + S_7\bar{Y}_6 - a_{71}U^*_{765} = 0. \quad (10.693)$$

These equations may be rearranged as

$$a_{34}c_{4+5} + a_{45}c_5 = P_{2176}, \quad (10.694)$$

$$a_{34}s_{4+5} + a_{45}s_5 = Q_{2176}, \quad (10.695)$$

where

$$P_{2176} = -S_1X_{76} - a_{12}W_{176} - S_2X_{176} - a_{23}W_{2176} - S_3X_{2176} - a_{56} - a_{67}c_6 - S_7\bar{X}_6 - a_{71}W_{76}, \quad (10.696)$$

$$Q_{2176} = S_1Y_{76} - a_{12}U^*_{1765} + S_2Y_{176} - a_{23}U^*_{21765} + S_3Y_{2176} - S_6s_{56} + a_{67}c_{56}s_6 + S_7\bar{Y}_6 - a_{71}U^*_{765}. \quad (10.697)$$

The terms P_{2176} and Q_{2176} can be numerically determined for each of the four solution sets of $(\theta_6, \theta_2, \theta_1)$.

Squaring and adding Eqs. (10.694) and (10.695) yields

$$a_{34}^2 + a_{45}^2 + 2a_{34}a_{45}(c_{4+5}c_5 + s_{4+5}s_5) = P_{2176}^2 + Q_{2176}^2. \quad (10.698)$$

Because the cosine of the difference of two angles α and β may be written as $\cos(\alpha - \beta) =$

$c_\alpha c_\beta + s_\alpha s_\beta$, this equation may be written as

$$a_{34}^2 + a_{45}^2 + 2a_{34}a_{45}\cos((\theta_4 + \theta_5) - \theta_5) = P_{1762}^2 + Q_{1762}^2, \quad (10.699)$$

which thus reduces to

$$a_{34}^2 + a_{45}^2 + 2a_{34}a_{45}c_4 = P_{2176}^2 + Q_{2176}^2. \quad (10.700)$$

Solving this equation for c_4 gives

$$c_4 = \frac{P_{2176}^2 + Q_{2176}^2 - a_{34}^2 - a_{45}^2}{2a_{34}a_{45}}. \quad (10.701)$$

Two distinct values of θ_4 will satisfy Eq. (10.701) for each of the four sets of solutions $(\theta_6, \theta_2, \theta_1)$. Thus, a total of eight solution configurations exist for this mechanism.

10.8.5 Determination of θ_5

Equations (10.694) and (10.695) may be expanded as

$$a_{34}(c_4c_5 - s_4s_5) + a_{45}c_5 = P_{2176}, \quad (10.702)$$

$$a_{34}(s_4c_5 + c_4s_5) + a_{45}s_5 = Q_{2176}. \quad (10.703)$$

Regrouping these equations gives

$$(a_{34}c_4 + a_{45})c_5 - (a_{34}s_4)s_5 = P_{2176}, \quad (10.704)$$

$$(a_{34}s_4)c_5 + (a_{34}c_4 + a_{45})s_5 = Q_{2176}. \quad (10.705)$$

These two equations that are linear in the variables s_5 and c_5 can be used to determine unique corresponding values of θ_5 for each of the eight solution sets $(\theta_6, \theta_1, \theta_2, \theta_4)$.

10.8.6 Determination of θ_3

The following fundamental sine and sine–cosine laws may be written for a spherical heptagon:

$$X_{17654} = s_{23}s_3, \quad (10.706)$$

$$Y_{17654} = s_{23}c_3. \quad (10.707)$$

Numerical values can be determined for X_{17654} and Y_{17654} for each of the eight solution sets $(\theta_6, \theta_1, \theta_2, \theta_4, \theta_5)$. The calculated values of s_3 and c_3 yield the unique corresponding value for θ_3.

10.8.7 Numerical example

It has been shown that a total of eight solution configurations exist for the RRR-R-RRR spatial mechanism. Although this is a group 4 spatial mechanism, the special geometry of having three parallel joint axes greatly simplifies the analysis. Several industrial manipulators incorporate this geometry for this reason.

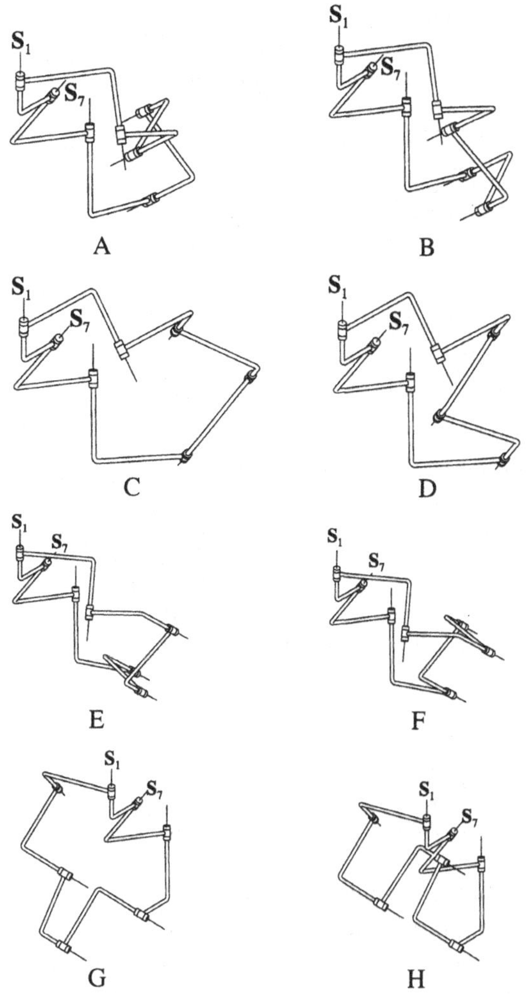

Figure 10.14. Eight configurations of the RRR-R-RRR spatial mechanism.

The mechanism dimensions of a numerical example are listed in Table 10.13. The resulting eight solution configurations are listed in Table 10.14 and are drawn in Figure 10.14. It is apparent in the figure that there are four classes of solutions. Each class has two configurations for the three parallel joint axes.

10.9 Summary

Robot manipulators that contain six revolute joints are very common. This is because rotary electric and hydraulic actuators are readily available. The reverse analysis of a general six-revolute robot, however, will require the solution of a general group 4 spatial mechanism once the close-the-loop step is accomplished. The solution of this mechanism is very complicated, and a computer program is available from the authors. The

Table 10.13. *RRR-R-RRR mechanism parameters.*

Link length, cm.	Twist angle, deg.	Joint offset, cm.	Joint angle, deg.
$a_{12} = 10.9$	$\alpha_{12} = 139$	$S_1 = 4.6$	θ_1 = variable
$a_{23} = 9.9$	$\alpha_{23} = 80$	$S_2 = 7.5$	θ_2 = variable
$a_{34} = 10.6$	$\alpha_{34} = 0$	$S_3 = 7.5$	θ_3 = variable
$a_{45} = 9.8$	$\alpha_{45} = 0$	$S_4 = 4.0$	θ_4 = variable
$a_{56} = 11.2$	$\alpha_{56} = 243$	$S_5 = 7.5$	θ_5 = variable
$a_{67} = 10.5$	$\alpha_{67} = 307$	$S_6 = 9.6$	θ_6 = variable
$a_{71} = 5.2$	$\alpha_{71} = 34$	$S_7 = 11.8$	$\theta_7 = 20$ (input)

Table 10.14. *Calculated configurations for the RRR-R-RRR spatial mechanism.*

Solution	θ_1, deg.	θ_2, deg.	θ_3, deg.	θ_4, deg.	θ_5, deg.	θ_6, deg.
A	147.64	−37.31	−36.68	93.13	115.41	−20.73
B	147.64	−37.31	51.71	−93.13	−146.72	−20.73
C	177.69	14.00	50.49	107.98	44.33	14.44
D	177.69	14.00	152.29	−107.98	158.49	14.44
E	127.24	32.29	131.03	80.95	−24.74	71.30
F	127.24	32.29	−151.86	−80.95	60.03	71.30
G	−69.32	−75.75	36.85	143.93	−65.07	140.08
H	−69.32	−75.75	167.04	−143.93	92.59	140.08

reverse-analysis computations require approximately 0.2 seconds when run on a Unix workstation with a MIPS 4400 CPU operating at a speed of 150 MHZ.

To simplify the reverse-analysis procedure, specialized geometries are typically incorporated into robot manipulators that contain six revolute joints. Most common are the cases where three axes intersect at a point or three axes are parallel. Sections 10.3 through 10.8 show how the reverse analysis is simplified for these cases, and Chapter 11 will present specific examples of industrial manipulators.

10.10 Problems

1. An RTTT spatial mechanism has the following dimensions (angles are in degrees, lengths are in cm):

$\alpha_{12} = 90$ $\alpha_{23} = 60$ $\alpha_{34} = 90$ $\alpha_{45} = 0$
$\alpha_{56} = 90$ $\alpha_{67} = 75$ $\alpha_{71} = 65$
$a_{12} = 0$ $a_{23} = 5.8$ $a_{34} = 0$ $a_{45} = 8.2$
$a_{56} = 0$ $a_{67} = 4.5$ $a_{71} = 6.6$
$S_1 = 5.6$ $S_2 = S_3 = S_4 = S_5 = S_6 = 0$ $S_7 = 2.5$
$\theta_7 = 235$

Determine the sets of values for the angles θ_1 through θ_6.

2. An RSTR spatial mechanism has the following dimensions (angles are in degrees, lengths are in cm):

$\alpha_{12} = 60$	$\alpha_{23} = 90$	$\alpha_{34} = 65$	$\alpha_{45} = 90$
$\alpha_{56} = 90$	$\alpha_{67} = 75$	$\alpha_{71} = 65$	
$a_{12} = 5.1$	$a_{23} = 0$	$a_{34} = 4.3$	$a_{45} = 0$
$a_{56} = 0$	$a_{67} = 2.8$	$a_{71} = 8.0$	
$S_1 = 9.3$	$S_2 = 9.7$	$S_3 = 8.8$	$S_4 = 9.4$
$S_5 = 0$	$S_6 = 3.9$	$S_7 = 6.2$	
$\theta_7 = 18$			

Determine the sets of values for the angles θ_1 through θ_6.

3. The directions of the fourth, fifth, and sixth joint axes of a 7R spatial mechanism are parallel. The value of θ_7 is given. Explain how to solve for the remaining joint parameters, θ_1 through θ_6.

11

Case studies

11.1 Introduction

The majority of industrial robots contain six revolute joints. Closing the loop will yield a one-degree-of-freedom 7R spatial mechanism with the angle θ_7 known. The general 7R spatial mechanism is a group 4 mechanism that is computationally difficult to solve. It will be shown, however, that most industrial robots have special geometries that greatly simplify the reverse analysis process. Examples of special geometries are joint axes that are parallel or that intersect. Three common industrial robots and two other manipulators will be analyzed in this chapter.

11.2 Puma industrial robot

The Puma 560 robot is shown in Figure 11.1. A kinematic model of this robot with joint axis and link vectors labeled is shown in Figure 11.2. The constant mechanism parameters are listed in Table 11.1.

The parameter S_6 is a free choice that must be made in order to specify the location of the origin of the sixth coordinate system. One input to the reverse-analysis problem is the location of the tool point measured in terms of the sixth coordinate system. This cannot be specified if the physical location of the origin of the sixth coordinate system is not known. A value of S_6 equal to four inches is selected to locate the origin of the sixth coordinate system at the center of the robot's tool mounting plate.

The reverse-analysis problem statement is as follows:

given: S_6 and the direction of $\mathbf{a}_{67}$ relative to $\mathbf{S}_6$ in order to establish the sixth coordinate system,
$^6\mathbf{P}_{\text{tool}}$: the location of the tool point in the sixth coordinate system,
$^F\mathbf{P}_{\text{tool}}$: the desired location of the tool point in the fixed coordinate system, and
$^F\mathbf{S}_6$, $^F\mathbf{a}_{67}$: the desired orientation of the robot end effector,

find: $\phi_1, \theta_2, \theta_3, \theta_4, \theta_5, \theta_6$: the joint angle parameters that will position and orient the end effector as desired.

The solution to this problem proceeds as described in Chapter 5. From the given information, Eq. (5.3) can be used to determine the position of the origin of the sixth

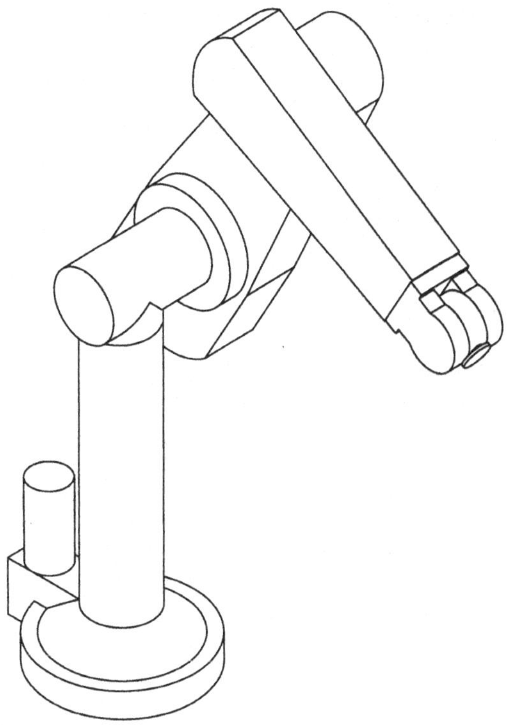

Figure 11.1. Puma robot.

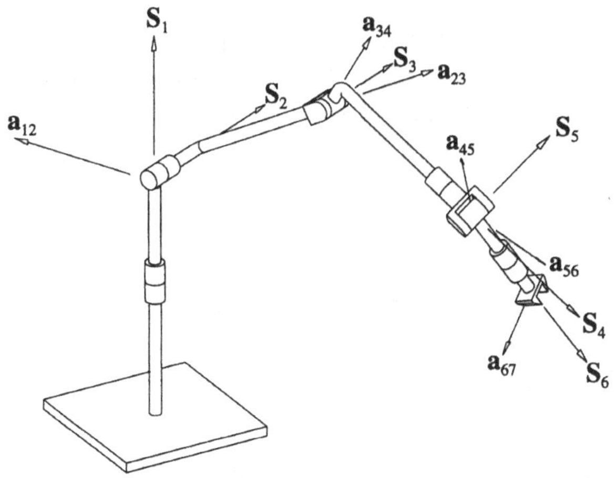

Figure 11.2. Kinematic model of Puma robot.

coordinate system measured in the fixed coordinate system. A hypothetical closure link is then created to form a closed-loop spatial mechanism. The link length a_{67} and the twist angle α_{67} were arbitrarily selected as zero and ninety degrees respectively. With these two choices, the direction of the vector S_7 is known in terms of the fixed coordinate system. Further, the hypothetical seventh joint axis is known to pass through the origin of the sixth coordinate system.

Table 11.1. *Mechanism parameters for Puma robot.*

Link length, in.	Twist angle, deg.	Joint offset, in.	Joint angle, deg.
$a_{12} = 0$	$\alpha_{12} = 90$		$\phi_1 =$ variable
$a_{23} = 17$	$\alpha_{23} = 0$	$S_2 = 5.9$	$\theta_2 =$ variable
$a_{34} = 0.8$	$\alpha_{34} = 270$	$S_3 = 0$	$\theta_3 =$ variable
$a_{45} = 0$	$\alpha_{45} = 90$	$S_4 = 17$	$\theta_4 =$ variable
$a_{56} = 0$	$\alpha_{56} = 90$	$S_5 = 0$	$\theta_5 =$ variable
			$\theta_6 =$ variable

Table 11.2. *Mechanism parameters for closed-loop Puma mechanism.*

Link length, in.	Twist angle, deg.	Joint offset, in.	Joint angle, deg.
$a_{12} = 0$	$\alpha_{12} = 90$	$S_1 =$ C.L.	$\phi_1 =$ variable
$a_{23} = 17$	$\alpha_{23} = 0$	$S_2 = 5.9$	$\theta_2 =$ variable
$a_{34} = 0.8$	$\alpha_{34} = 270$	$S_3 = 0$	$\theta_3 =$ variable
$a_{45} = 0$	$\alpha_{45} = 90$	$S_4 = 17$	$\theta_4 =$ variable
$a_{56} = 0$	$\alpha_{56} = 90$	$S_5 = 0$	$\theta_5 =$ variable
$a_{67} = 0^*$	$\alpha_{67} = 90^*$	$S_6 = 4^*$	$\theta_6 =$ variable
$a_{71} =$ C.L.	$\alpha_{71} =$ C.L.	$S_7 =$ C.L.	$\theta_7 =$ C.L.

* = User-selected value
C.L. = Calculated during the close-the-loop procedure

Chapter 5 shows how the six close-the-loop parameters (S_7, S_1, a_{71}, θ_7, α_{71}, and γ_1) can now be determined. The analysis will proceed assuming that values for these six parameters have been calculated. Table 11.2 shows the mechanism parameters for the newly formed closed-loop spatial mechanism.

11.2.1 Solution for θ_1 and ϕ_1

The vector loop equation for the closed-loop Puma mechanism is as follows:

$$S_1\mathbf{S}_1 + S_2\mathbf{S}_2 + a_{23}\mathbf{a}_{23} + a_{34}\mathbf{a}_{34} + S_4\mathbf{S}_4 + S_6\mathbf{S}_6 + S_7\mathbf{S}_7 + a_{71}\mathbf{a}_{71} = \mathbf{0}. \tag{11.1}$$

Expressing the vectors in terms of set 14 of the table of direction cosines for the spatial heptagon yields

$$S_1\begin{pmatrix}0\\-s_{12}\\c_{12}\end{pmatrix} + S_2\begin{pmatrix}0\\0\\1\end{pmatrix} + a_{23}\begin{pmatrix}c_2\\-s_2\\0\end{pmatrix} + a_{34}\begin{pmatrix}W_{45671}\\-U^*_{456712}\\U_{456712}\end{pmatrix} + S_4\begin{pmatrix}X_{5671}\\Y_{5671}\\Z_{5671}\end{pmatrix}$$

$$+ S_6\begin{pmatrix}X_{71}\\Y_{71}\\Z_{71}\end{pmatrix} + S_7\begin{pmatrix}X_1\\Y_1\\Z_1\end{pmatrix} + a_{71}\begin{pmatrix}c_1\\s_1c_{12}\\U_{12}\end{pmatrix} = \begin{pmatrix}0\\0\\0\end{pmatrix}. \tag{11.2}$$

Subsidiary spatial and polar sine, sine–cosine, and cosine laws can be used to simply the terms for the vectors $\mathbf{a}_{34}$ and $\mathbf{S}_4$. Thus, Eq. (11.2) can be written as

$$S_1\begin{pmatrix}0\\-s_{12}\\c_{12}\end{pmatrix}+S_2\begin{pmatrix}0\\0\\1\end{pmatrix}+a_{23}\begin{pmatrix}c_2\\-s_2\\0\end{pmatrix}+a_{34}\begin{pmatrix}W_{32}\\V_{32}\\U_{32}\end{pmatrix}+S_4\begin{pmatrix}X_{32}\\-X^*_{32}\\\bar{Z}_3\end{pmatrix}$$
$$+S_6\begin{pmatrix}X_{71}\\Y_{71}\\Z_{71}\end{pmatrix}+S_7\begin{pmatrix}X_1\\Y_1\\Z_1\end{pmatrix}+a_{71}\begin{pmatrix}c_1\\s_1c_{12}\\U_{12}\end{pmatrix}=\begin{pmatrix}0\\0\\0\end{pmatrix}. \quad (11.3)$$

The Z component equation (which is equivalent to projecting the vector loop equation onto the $\mathbf{S}_2$ axis) is

$$S_1c_{12}+S_2+a_{34}U_{32}+S_4\bar{Z}_3+S_6Z_{71}+S_7Z_1+a_{71}U_{12}=0. \quad (11.4)$$

Now, $U_{32}=s_3s_{23}$, and because $\alpha_{23}=0$, then $U_{32}=0$. Similarly, $\bar{Z}_3=c_{34}c_{23}-s_{34}s_{23}c_3$, and because $\alpha_{23}=0$ and $\alpha_{34}=270°$, $\bar{Z}_3=0$. Further, $c_{12}=0$ because $\alpha_{12}=90°$. Thus, Eq. (11.4) reduces to

$$S_2+S_6Z_{71}+S_7Z_1+a_{71}U_{12}=0. \quad (11.5)$$

This equation contains θ_1 as its only unknown. Expanding Z_{71}, Z_1, and U_{12} yields

$$S_2+S_6[s_{12}(X_7s_1+Y_7c_1)+c_{12}Z_7]+S_7[c_{12}c_{71}-s_{12}s_{71}c_1]+a_{71}[s_1s_{12}]=0. \quad (11.6)$$

The equation can be simplified by substituting $s_{12}=1$ and $c_{12}=0$. Thus, Eq. (11.6) may be written as

$$S_2+S_6[X_7s_1+Y_7c_1]+S_7[-s_{71}c_1]+a_{71}[s_1]=0. \quad (11.7)$$

Grouping the s_1 and c_1 terms yields

$$[S_6Y_7-S_7s_{71}]c_1+[S_6X_7+a_{71}]s_1+[S_2]=0. \quad (11.8)$$

The terms within the brackets in Eq. (11.8) can be calculated from the known values. Thus, Eq. (11.8) represents an equation of the form $Ac_1+Bs_1+D=0$, where A, B, and D are constants. It was shown in Section 6.7.2 how this type of equation can be solved to yield two values of θ_1, that is, θ_{1a} and θ_{1b}. Figure 11.3 shows the solution tree for the mechanism thus far. The two associated values for the angle ϕ_1 can be calculated as $(\theta_{1a}-\gamma_1)$ and $(\theta_{1b}-\gamma_1)$.

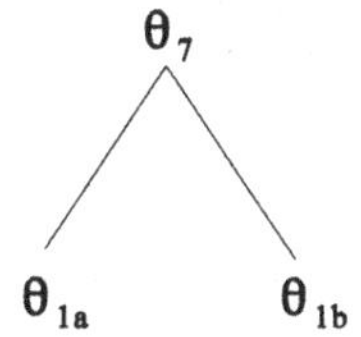

Figure 11.3. Puma solution tree.

11.2.2 Solution for θ_3

Substituting for $s_{12} = 1$ and $c_{12} = 0$ into the X and Y components of Eq. (11.3) yields

$$a_{23}c_2 + a_{34}W_{32} + S_4X_{32} + S_6X_{71} + S_7X_1 + a_{71}c_1 = 0, \tag{11.9}$$

$$-S_1 - a_{23}s_2 + a_{34}V_{32} - S_4X^*_{32} + S_6Y_{71} + S_7Y_1 = 0. \tag{11.10}$$

All terms that do not contain the unknown variables θ_2 and θ_3 will be moved to the right-hand side of Eqs. (11.9) and (11.10) to give

$$a_{23}c_2 + a_{34}W_{32} + S_4X_{32} = A, \tag{11.11}$$

$$-a_{23}s_2 + a_{34}V_{32} - S_4X^*_{32} = B, \tag{11.12}$$

where

$$A = -S_6X_{71} - S_7X_1 - a_{71}c_1, \tag{11.13}$$

$$B = S_1 - S_6Y_{71} - S_7Y_1. \tag{11.14}$$

Expanding W_{32}, V_{32}, X_{32}, and X^*_{32} gives

$$a_{23}c_2 + a_{34}[c_2c_3 - s_2s_3c_{23}] + S_4[(s_{34}s_3)c_2 + (s_{23}c_{34} + c_{23}s_{34}c_3)s_2] = A, \tag{11.15}$$

$$-a_{23}s_2 - a_{34}[s_2c_3 + c_2s_3c_{23}] - S_4[(s_{34}s_3)s_2 - (s_{23}c_{34} + c_{23}s_{34}c_3)c_2] = B. \tag{11.16}$$

Substituting for $\alpha_{23} = 0°$ and $\alpha_{34} = 270°$ gives

$$a_{23}c_2 + a_{34}[c_2c_3 - s_2s_3] + S_4[-s_3c_2 - c_3s_2] = A, \tag{11.17}$$

$$-a_{23}s_2 - a_{34}[s_2c_3 + c_2s_3] - S_4[-s_3s_2 + c_3c_2] = B. \tag{11.18}$$

These equations can be written as

$$a_{23}c_2 + a_{34}c_{2+3} - S_4s_{2+3} = A, \tag{11.19}$$

$$-a_{23}s_2 - a_{34}s_{2+3} - S_4c_{2+3} = B, \tag{11.20}$$

where s_{2+3} and c_{2+3} represent the sine and cosine of $(\theta_2 + \theta_3)$ respectively.

Equations (11.19) and (11.20) represent two equations in the two unknowns θ_2 and θ_3. These variables will be solved for by adding the squares of Eq. (11.19) and Eq. (11.20). Squaring the equations gives

$$a_{23}^2c_2^2 + a_{34}^2c_{2+3}^2 + S_4^2s_{2+3}^2 + 2a_{23}a_{34}c_2c_{2+3} - 2a_{23}S_4c_2s_{2+3} - 2a_{34}S_4c_{2+3}s_{2+3} = A^2, \tag{11.21}$$

$$a_{23}^2s_2^2 + a_{34}^2s_{2+3}^2 + S_4^2c_{2+3}^2 + 2a_{23}a_{34}s_2s_{2+3} + 2a_{23}S_4s_2c_{2+3} + 2a_{34}S_4s_{2+3}c_{2+3} = B^2 \tag{11.22}$$

and adding yields

$$a_{23}^2 + a_{34}^2 + S_4^2 + 2a_{23}a_{34}[c_2c_{2+3} + s_2s_{2+3}] + 2a_{23}S_4[s_2c_{2+3} - c_2s_{2+3}] = A^2 + B^2. \tag{11.23}$$

Now, $[c_2c_{2+3} + s_2s_{2+3}] = \cos[(\theta_2 + \theta_3) - \theta_2] = c_3$ and $[s_2c_{2+3} - c_2s_{2+3}] = \sin[\theta_2 - (\theta_2 + \theta_3)] = \sin(-\theta_3) = -s_3$, which gives

$$a_{23}^2 + a_{34}^2 + S_4^2 + 2a_{23}a_{34}c_3 - 2a_{23}S_4s_3 = A^2 + B^2. \tag{11.24}$$

Regrouping this equation gives

$$c_3[2a_{23}a_{34}] + s_3[-2a_{23}S_4] + \left[a_{23}^2 + a_{34}^2 + S_4^2 - A^2 - B^2\right] = 0. \tag{11.25}$$

This equation contains only θ_3 as an unknown. Values for A and B will first be obtained for $\theta_1 = \theta_{1a}$. Two corresponding values for θ_3 will then be found by solving Eq. (11.25). Next, values for A and B will be obtained for $\theta_1 = \theta_{1b}$. Two additional corresponding values for θ_3 will then be found. The current solution tree is shown in Figure 11.4

11.2.3 Solution for θ_2

Equations (11.17) and (11.18) can be used to solve for the angle θ_2. Regrouping these equations gives

$$c_2[a_{23} + a_{34}c_3 - S_4s_3] + s_2[-a_{34}s_3 - S_4c_3] = A, \tag{11.26}$$

$$s_2[-a_{23} - a_{34}c_3 + S_4s_3] + c_2[-a_{34}s_3 - S_4c_3] = B. \tag{11.27}$$

Equations (11.26) and (11.27) represent two equations in the two unknowns c_2 and s_2. Thus, a unique corresponding value of θ_2 can be found for each θ_1, θ_3 pair. In other words, θ_{1a} and θ_{3a} will be substituted into the equations to yield θ_{2a}. Similarly, θ_{1a} and θ_{3b} will be substituted into the equations to yield θ_{2b}. The current solution tree is shown in Figure 11.5.

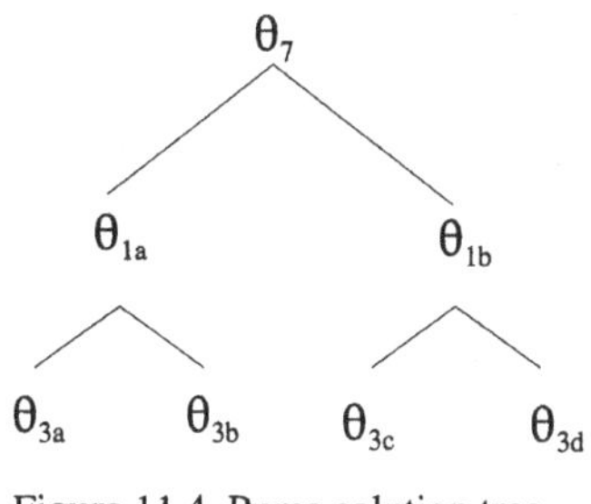

Figure 11.4. Puma solution tree.

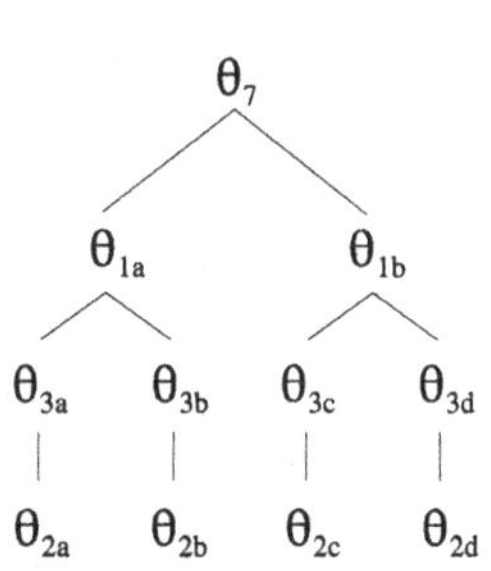

Figure 11.5. Puma solution tree.

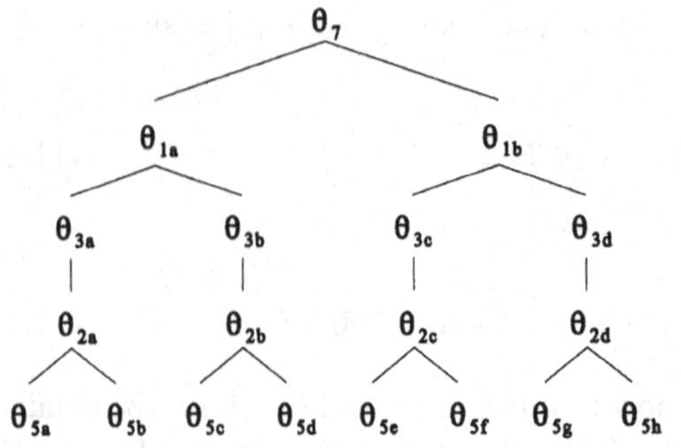

Figure 11.6. Puma solution tree.

11.2.4 Solution for θ_5

The angle θ_5 can be readily determined from the spherical equation

$$Z_{7123} = \bar{Z}_5. \tag{11.28}$$

Expanding the right-hand side of this equation and substituting for α_{45} and α_{56} yields

$$c_5 = -Z_{7123}. \tag{11.29}$$

Thus, for each combination of θ_7, θ_1, θ_3, and θ_2, a value for c_5 and thereby two values for θ_5 can be determined. The solution tree for this analysis is now shown in Figure 11.6.

11.2.5 Solution for θ_4

Corresponding values for θ_4 can be obtained from the following two subsidiary spherical equations:

$$X_{54} = X_{7123}, \tag{11.30}$$

$$X^*_{54} = -Y_{7123}. \tag{11.31}$$

Expanding the left-hand sides of these equations yields

$$\bar{X}_5 c_4 - \bar{Y}_5 s_4 = X_{7123}, \tag{11.32}$$

$$\bar{X}_5 s_4 + \bar{Y}_5 c_4 = -Y_{7123}. \tag{11.33}$$

Expanding $\bar{X}_5$ and $\bar{Y}_5$ gives

$$(s_{56}s_5)c_4 + (s_{45}c_{56} + c_{45}s_{56}c_5)s_4 = X_{7123}, \tag{11.34}$$

$$(s_{56}s_5)s_4 - (s_{45}c_{56} + c_{45}s_{56}c_5)c_4 = -Y_{7123}. \tag{11.35}$$

Inserting the values $\alpha_{45} = \alpha_{56} = 90°$ yields

$$s_5 c_4 = X_{7123}, \tag{11.36}$$

$$s_5 s_4 = -Y_{7123}. \tag{11.37}$$

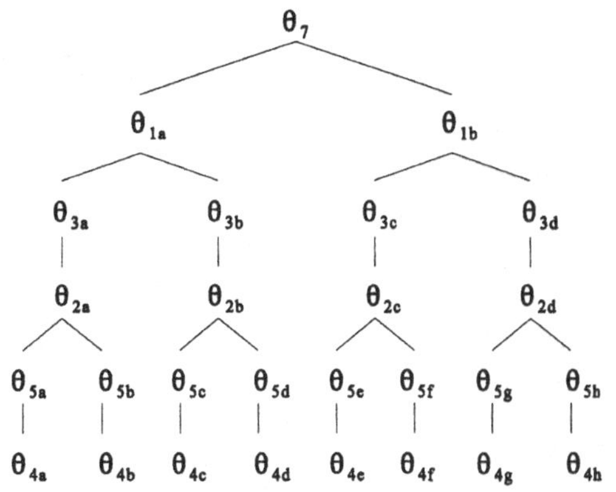

Figure 11.7. Puma solution tree.

Solving for s_4 and c_4 gives

$$c_4 = \frac{X_{7123}}{s_5}, \tag{11.38}$$

$$s_4 = -\frac{Y_{7123}}{s_5}. \tag{11.39}$$

Substituting the previously calculated values for the eight sets of corresponding values for $\theta_7, \theta_1, \theta_2, \theta_3$, and θ_5 will yield a corresponding value for θ_4. The current solution tree is shown in Figure 11.7. It should be noted that the solution for θ_4 becomes indeterminate if θ_5 equals 0 or 180 degrees. This special case will be discussed in Section 11.5.

11.2.6 Solution for θ_6

The angle θ_6 is the last remaining joint angle to be determined. It will be calculated from the following two fundamental spherical sine and sine–cosine laws:

$$X_{43217} = s_{56}s_6, \tag{11.40}$$

$$Y_{43217} = s_{56}c_6. \tag{11.41}$$

Because $\alpha_{56} = 90°$, these equations reduce to

$$s_6 = X_{43217}, \tag{11.42}$$

$$c_6 = Y_{43217}. \tag{11.43}$$

Thus, a corresponding value for θ_6 can be found for each of the eight sets of values for $\theta_7, \theta_1, \theta_3, \theta_2, \theta_5$, and θ_4. The final solution tree for the Puma robot is shown in Figure 11.8. The solution of the reverse analysis for the Puma robot is complete.

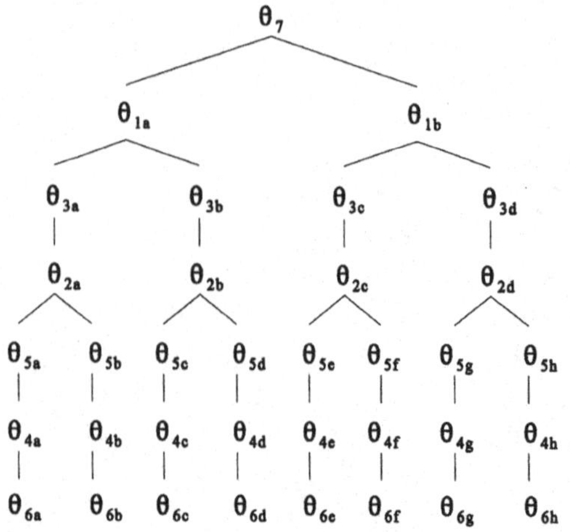

Figure 11.8. Final Puma solution tree.

11.2.7 Numerical example

As a numerical example, the following information was specified for the Puma manipulator:

$$S_6 = 4\,\text{in.},\quad {}^6\mathbf{P}_{\text{tool}} = \begin{bmatrix} 5 \\ 3 \\ 7 \end{bmatrix}\ \text{in.},\quad {}^F\mathbf{P}_{\text{tool}} = \begin{bmatrix} 24.112 \\ 20.113 \\ 18.167 \end{bmatrix}\ \text{in.},$$

$$^F\mathbf{S}_6 = \begin{bmatrix} 0.079 \\ -0.787 \\ 0.612 \end{bmatrix},\quad {}^F\mathbf{a}_{67} = \begin{bmatrix} 0.997 \\ 0.064 \\ -0.047 \end{bmatrix}.$$

This specified position and orientation is identical to that which was calculated in the numerical example of the forward analysis for the Puma robot in Section 4.2. Thus, one of the solution sets for this reverse-analysis problem must be identical to the input data used previously in the forward-analysis procedure. Table 11.3 shows the results of the reverse-position analysis. Solution set 1 matches the input data used in the forward-analysis example in Section 4.2.

The eight configurations of the Puma robot that position and orient the end effector as specified are shown in Figure 11.9. It is apparent in the figure that four classes of solutions exist, with each class having two configurations for the ball and socket joint.

11.3 GE P60 manipulator

The GE P60 manipulator is shown in Figure 11.10. A kinematic model of the manipulator showing joint axis and link vectors is shown in Figure 11.11. The constant mechanism parameters are listed in Table 11.4.

The value of the offset distance S_6 is a free choice that will define the location of the origin of the sixth coordinate system. A value of 15.24 cm will be used, as this value will

Table 11.3. *Eight solution sets for the Puma robot (angles in degrees).*

Solution	ϕ_1	θ_2	θ_3	θ_4	θ_5	θ_6
A	−135.0	150.0	−60.0	45.0	60.0	−30.0
B	−135.0	150.0	−60.0	−135.0	−60.0	150.0
C	−135.0	177.321	−114.611	38.370	80.585	−49.186
D	−135.0	177.321	−114.611	−141.630	−80.585	130.814
E	66.072	2.679	−60.0	−156.158	75.676	−67.944
F	66.072	2.679	−60.0	23.842	−75.676	112.056
G	66.072	30.0	−114.611	−149.846	51.230	−54.193
H	66.072	30.0	−114.611	30.154	−51.230	125.807

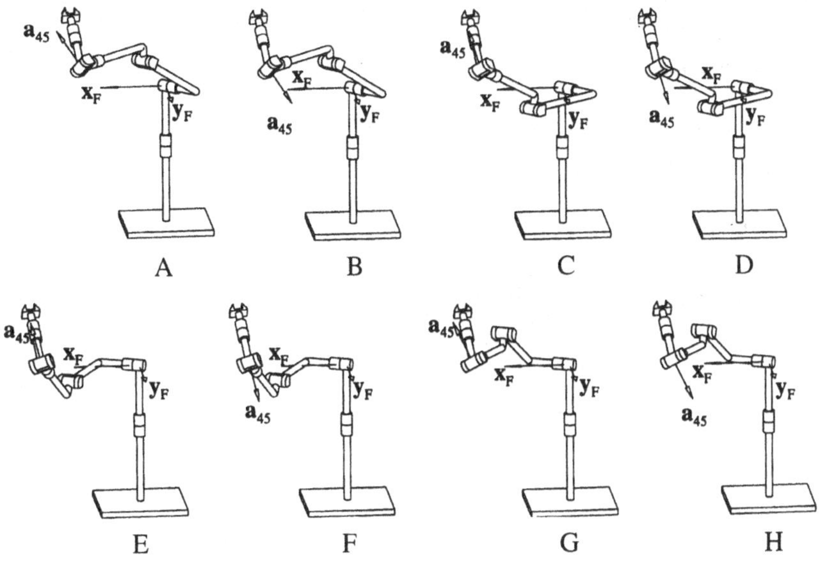

Figure 11.9. Eight solution configurations of the Puma robot.

locate the origin of the sixth coordinate system at the center of the tool mounting plate of the manipulator.

The reverse-analysis problem statement is identical to that for the Puma robot. This problem statement is repeated as

given: S_6 and the direction of $\mathbf{a}_{67}$ relative to $\mathbf{S}_6$ in order to establish the sixth coordinate system,
${}^6\mathbf{P}_{\text{tool}}$: the location of the tool point in the sixth coordinate system,
${}^F\mathbf{P}_{\text{tool}}$: the desired location of the tool point in the fixed coordinate system, and
${}^F\mathbf{S}_6$, ${}^F\mathbf{a}_{67}$: the desired orientation of the robot end effector,

find: $\phi_1, \theta_2, \theta_3, \theta_4, \theta_5, \theta_6$: the joint angle parameters that will position and orient the end effector as desired.

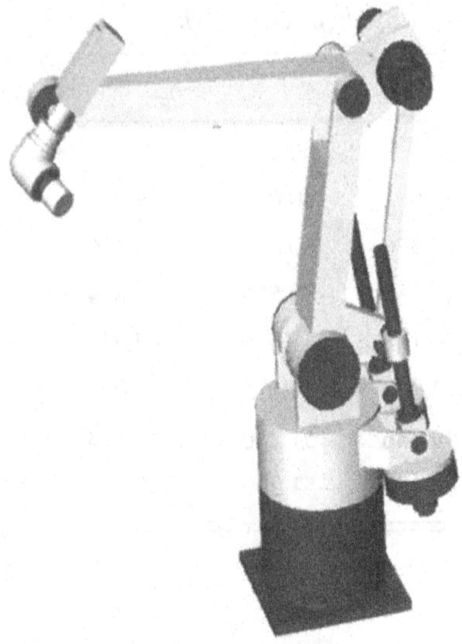

Figure 11.10. GE P60 robot.

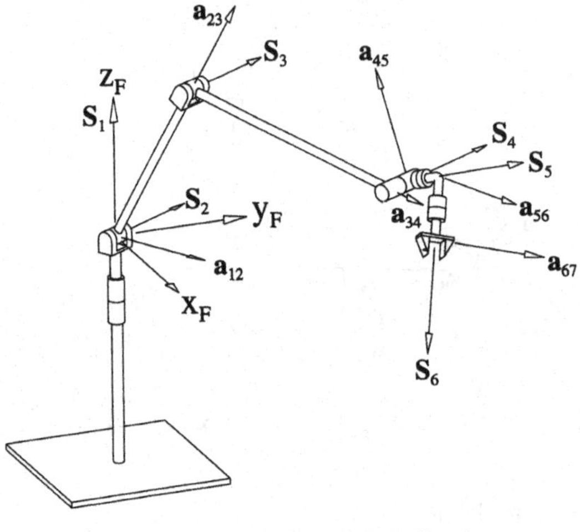

Figure 11.11. Kinematic model of GE P60 robot.

As with the Puma robot, the solution to this problem proceeds as described in Chapter 5. From the given information, Eq. (5.3) can be used to determine the position of the origin of the sixth coordinate system measured in the fixed coordinate system. A hypothetical closure link is then created to form a closed-loop spatial mechanism. The link length a_{67} and the twist angle α_{67} were arbitrarily selected as zero and ninety degrees respectively. With these two choices, the direction of the vector S_7 is known in terms of the fixed coordinate system. Further, the hypothetical seventh joint axis is known to pass through the origin of the sixth coordinate system.

Chapter 5 shows how the six close-the-loop parameters (S_7, S_1, a_{71}, θ_7, α_{71}, and γ_1) can now be determined. The analysis will proceed assuming that values for these six parameters have been calculated. Table 11.5 shows the mechanism parameters for the newly formed closed-loop spatial mechanism.

Table 11.4. *Mechanism parameters for the GE P60 robot.*

Link length, cm.	Twist angle, deg.	Joint offset, cm.	Joint angle, deg.
$a_{12} = 0$	$\alpha_{12} = 270$		ϕ_1 = variable
$a_{23} = 70$	$\alpha_{23} = 0$	$S_2 = 0$	θ_2 = variable
$a_{34} = 90$	$\alpha_{34} = 0$	$S_3 = 0$	θ_3 = variable
$a_{45} = 0$	$\alpha_{45} = 270$	$S_4 = 9.8$	θ_4 = variable
$a_{56} = 0$	$\alpha_{56} = 90$	$S_5 = 14.5$	θ_5 = variable
			θ_6 = variable

Table 11.5. *Mechanism parameters for closed-loop GE P60 mechanism.*

Link length, cm.	Twist angle, deg.	Joint offset, cm.	Joint angle, deg.
$a_{12} = 0$	$\alpha_{12} = 270$	S_1 = C.L.	ϕ_1 = variable
$a_{23} = 70$	$\alpha_{23} = 0$	$S_2 = 0$	θ_2 = variable
$a_{34} = 90$	$\alpha_{34} = 0$	$S_3 = 0$	θ_3 = variable
$a_{45} = 0$	$\alpha_{45} = 270$	$S_4 = 9.8$	θ_4 = variable
$a_{56} = 0$	$\alpha_{56} = 90$	$S_5 = 14.5$	θ_5 = variable
$a_{67} = 0$*	$\alpha_{67} = 90$*	$S_6 = 15.24$*	θ_6 = variable
a_{71} = C.L.	α_{71} = C.L.	S_7 = C.L.	θ_7 = C.L.

* = User-selected value

C.L. = Calculated during the close-the-loop procedure

11.3.1 Solution for θ_1 and ϕ_1

The vector loop equation for the closed-loop mechanism is

$$S_1\mathbf{S}_1 + a_{23}\mathbf{a}_{23} + a_{34}\mathbf{a}_{34} + S_4\mathbf{S}_4 + S_5\mathbf{S}_5 + S_6\mathbf{S}_6 + S_7\mathbf{S}_7 + a_{71}\mathbf{a}_{71} = \mathbf{0}. \tag{11.44}$$

As was the case with the Puma robot, the vector loop equation will be projected on the S_2 direction. This is accomplished as follows:

$$\begin{aligned} &S_1(\mathbf{S}_1 \cdot \mathbf{S}_2) + a_{23}(\mathbf{a}_{23} \cdot \mathbf{S}_2) + a_{34}(\mathbf{a}_{34} \cdot \mathbf{S}_2) + S_4(\mathbf{S}_4 \cdot \mathbf{S}_2) \\ &\quad + S_5(\mathbf{S}_5 \cdot \mathbf{S}_2) + S_6(\mathbf{S}_6 \cdot \mathbf{S}_2) + S_7(\mathbf{S}_7 \cdot \mathbf{S}_2) + a_{71}(\mathbf{a}_{71} \cdot \mathbf{S}_2) = 0. \end{aligned} \tag{11.45}$$

It is apparent from the geometry shown in Figure 11.10 that

$$\mathbf{S}_1 \cdot \mathbf{S}_2 = 0, \tag{11.46}$$

$$\mathbf{a}_{23} \cdot \mathbf{S}_2 = 0, \tag{11.47}$$

$$\mathbf{a}_{34} \cdot \mathbf{S}_2 = 0, \tag{11.48}$$

$$\mathbf{S}_4 \cdot \mathbf{S}_2 = 1, \tag{11.49}$$

$$\mathbf{S}_5 \cdot \mathbf{S}_2 = 0. \tag{11.50}$$

Equation (11.45) thus reduces to

$$S_4 + S_6(\mathbf{S}_6 \cdot \mathbf{S}_2) + S_7(\mathbf{S}_7 \cdot \mathbf{S}_2) + a_{71}(\mathbf{a}_{71} \cdot \mathbf{S}_2) = 0. \tag{11.51}$$

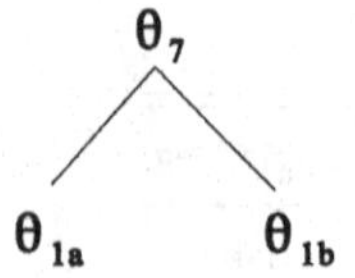

Figure 11.12. GE P60 solution tree.

Set 14 from the direction cosine tables for a spherical heptagon as listed in the appendix is used to evaluate the remaining scalar products. Equation (11.51) can then be written as

$$S_4 + S_6 Z_{71} + S_7 Z_1 + a_{71} U_{12} = 0. \tag{11.52}$$

This equation contains θ_1 as its only unknown. Expanding Z_{71}, Z_1, and U_{12} yields

$$S_4 + S_6[s_{12}(X_7 s_1 + Y_7 c_1) + c_{12} Z_7] + S_7[c_{12} c_{71} - s_{12} s_{71} c_1] + a_{71}[s_1 s_{12}] = 0. \tag{11.53}$$

Because $\alpha_{12} = 270°$, this equation can be simplified as

$$S_4 + S_6[-X_7 s_1 - Y_7 c_1] + S_7[s_{71} c_1] + a_{71}[-s_1] = 0. \tag{11.54}$$

Grouping the s_1 and c_1 terms yields

$$c_1[-S_6 Y_7 + S_7 s_{71}] + s_1[-S_6 X_7 - a_{71}] + [S_4] = 0. \tag{11.55}$$

The terms within the brackets in Eq. (11.55) can be calculated from the known values. Thus, Eq. (11.55) represents an equation of the form $Ac_1 + Bs_1 + D = 0$, where A, B, and D are constants. It was shown in Section 6.7.2 how this type of equation can be solved to yield two values of θ_1, that is, θ_{1a} and θ_{1b}. Figure 11.12 shows the solution tree for the mechanism thus far. The two associated values for the angle ϕ_1 can be calculated as $(\theta_{1a} - \gamma_1)$ and $(\theta_{1b} - \gamma_1)$.

11.3.2 Solution for θ_5

A planar representation of the closed-loop spatial mechanism and its equivalent spherical mechanism is shown in Figure 11.13. Because the vectors $\mathbf{S}_2$, $\mathbf{S}_3$, and $\mathbf{S}_4$ are parallel, the equivalent spherical mechanism will be a spherical pentagon for which the angle between the second and third links will equal $(\theta_2 + \theta_3 + \theta_4)$.

From the equivalent spherical pentagon, the following spherical cosine law can be written:

$$Z_{17} = Z_5. \tag{11.56}$$

It should be noted that this equation can be readily obtained from the spherical cosine law for a spherical heptagon, $Z_{71} = Z_{543}$, with $\alpha_{23} = \alpha_{34} = 0$.

Expanding the right side of Eq. (11.56) yields

$$Z_{17} = c_{56} c_{45} - s_{56} s_{45} c_5. \tag{11.57}$$

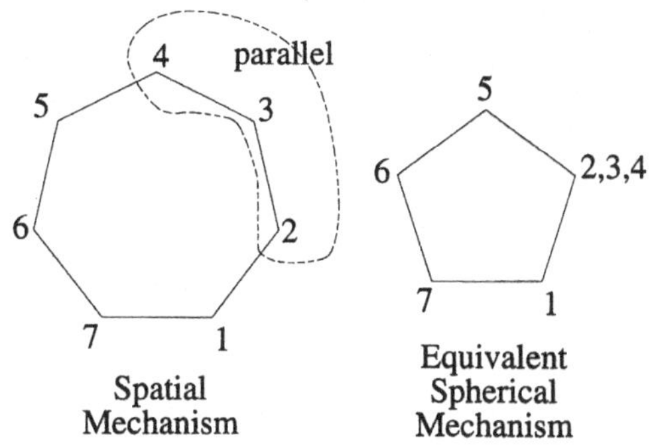

Figure 11.13. Planar representations.

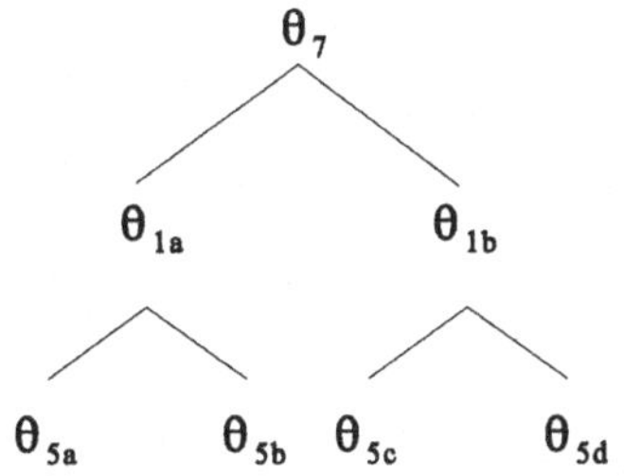

Figure 11.14. GE P60 solution tree.

Upon substituting the values $\alpha_{45} = 270°$ and $\alpha_{56} = 90°$, this equation reduces to

$$Z_{17} = c_5. \tag{11.58}$$

The previously solved-for value of θ_{1a} will be substituted into Eq. (11.58) to yield two corresponding values of θ_5. The process is repeated by using the calculated value for θ_{1b} in the equation to yield a further two other corresponding valued for θ_5. The current solution tree for the problem is shown in Figure 11.14.

11.3.3 Solution for θ_6

The following two equations may be written for the equivalent spherical pentagon shown in Figure 11.12:

$$X_{17} = X_{56}, \tag{11.59}$$

$$Y_{17} = -X^*_{56}. \tag{11.60}$$

Expanding the right sides of these equations yields

$$X_{17} = X_5 c_6 - Y_5 s_6, \tag{11.61}$$

$$Y_{17} = -X_5 s_6 - Y_5 c_6. \tag{11.62}$$

Substituting the definitions for X_5 and Y_5 gives

$$X_{17} = (s_{45} s_5) c_6 + (s_{56} c_{45} + c_{56} s_{45} c_5) s_6, \tag{11.63}$$

$$Y_{17} = -(s_{45} s_5) s_6 + (s_{56} c_{45} + c_{56} s_{45} c_5) c_6. \tag{11.64}$$

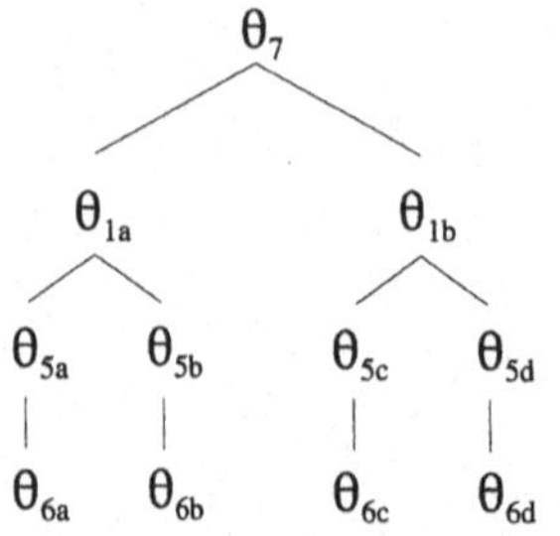

Figure 11.15. GE P60 solution tree.

Substituting values for $\alpha_{45} = 270°$ and $\alpha_{56} = 90°$ yields

$$X_{17} = -s_5c_6, \tag{11.65}$$

$$Y_{17} = s_5s_6. \tag{11.66}$$

Solving for c_6 and s_6 gives

$$c_6 = -\frac{X_{17}}{s_5}, \tag{11.67}$$

$$s_6 = \frac{Y_{17}}{s_5}. \tag{11.68}$$

Substituting the previously calculated values for the four sets of corresponding angles θ_1, θ_7, and θ_5 will yield a corresponding value for θ_6. The current solution tree is shown in Figure 11.15.

It should be noted that the solution for θ_6 will be indeterminate when $\theta_5 = 0°$ or $\theta_5 = 180°$. This special case, which also occurred for the Puma robot, will be discussed in Section 11.5.

11.3.4 Solution for θ_3

At this point of the analysis, all joint angles are known except for θ_2, θ_3, and θ_4. These are the angles for the three joint axes that are parallel. From Figure 11.11 it would appear logical to determine the coordinates of the intersection point of the vectors $\mathbf{S}_4$ and $\mathbf{a}_{34}$ in terms of the first coordinate system and then solve a planar triangle (two sides of which are a_{23} and a_{34}) for the angle θ_3. Once θ_3 is known, corresponding values for θ_2 and θ_4 would be computed.

Following this method, the solution will proceed by obtaining two equations that are projections of the vector loop equation (Eq. 11.44). These two equations are obtained by projecting the loop equation onto $\mathbf{a}_{12}$ (the X axis of the first coordinate system) and a vector perpendicular to $\mathbf{S}_2$ and $\mathbf{a}_{12}$ (the vector $\mathbf{S}_1$, the Z axis of the first coordinate system). Using sets 1 and 14 of the sets of direction cosines for a spherical heptagon as listed in the appendix, the projection of the vector loop equation onto the vector $\mathbf{a}_{12}$ may be written as

$$a_{23}c_2 + a_{34}W_{32} + S_4X_{32} + S_5X_{671} + S_6X_{71} + S_7X_1 + a_{71}c_1 = 0. \tag{11.69}$$

Similarly, set 14 is used to obtain the projection of the vector loop equation onto the vector perpendicular to S_2 and $\mathbf{a}_{12}$ as follows:

$$-S_1 s_{12} - a_{23} s_2 - a_{34} U^*_{456712} + S_4 Y_{5671} + S_5 Y_{671} + S_6 Y_{71} + S_7 Y_1 + a_{71} s_1 c_{12} = 0. \tag{11.70}$$

The subsidiary equations

$$U^*_{456712} = -V_{32} \tag{11.71}$$

and

$$Y_{5671} = -X^*_{32} \tag{11.72}$$

are substituted into Eq. (11.70) to yield

$$-S_1 s_{12} - a_{23} s_2 + a_{34} V_{32} - S_4 X^*_{32} + S_5 Y_{671} + S_6 Y_{71} + S_7 Y_1 + a_{71} s_1 c_{12} = 0. \tag{11.73}$$

The terms X_{32} in Eq. (11.69) and X^*_{32} in Eq. (11.73) equal zero after their definitions are expanded and the constant mechanism dimensions are substituted. Equations (11.69) and (11.73) may now be rearranged to yield respectively

$$a_{23} c_2 + a_{34} W_{32} = -S_5 X_{671} - S_6 X_{71} - S_7 X_1 - a_{71} c_1, \tag{11.74}$$

$$-a_{23} s_2 + a_{34} V_{32} = -S_1 - S_5 Y_{671} - S_6 Y_{71} - S_7 Y_1. \tag{11.75}$$

The right-hand sides of these equations will be denoted by K_1 and K_2, and thus

$$K_1 = -S_5 X_{671} - S_6 X_{71} - S_7 X_1 - a_{71} c_1, \tag{11.76}$$

$$K_2 = -S_1 - S_5 Y_{671} - S_6 Y_{71} - S_7 Y_1. \tag{11.77}$$

Because K_1 and K_2 are defined only in terms of constant mechanism parameters and previously calculated joint angles, these values can be numerically calculated for each of the four solution sets thus far.

Expanding the left sides of Eqs. (11.74) and (11.75) yields

$$a_{23} c_2 + a_{34}(c_2 c_3 - s_2 s_3 c_{23}) = K_1, \tag{11.78}$$

$$-a_{23} s_2 - a_{34}(s_2 c_3 + c_2 s_3 c_{23}) = K_2. \tag{11.79}$$

Because $\alpha_{23} = 0$, these equations reduce to

$$a_{23} c_2 + a_{34}(c_2 c_3 - s_2 s_3) = K_1, \tag{11.80}$$

$$-a_{23} s_2 - a_{34}(s_2 c_3 + c_2 s_3) = K_2 \tag{11.81}$$

or

$$a_{23} c_2 + a_{34} c_{2+3} = K_1, \tag{11.82}$$

$$-a_{23} s_2 - a_{34} s_{2+3} = K_2, \tag{11.83}$$

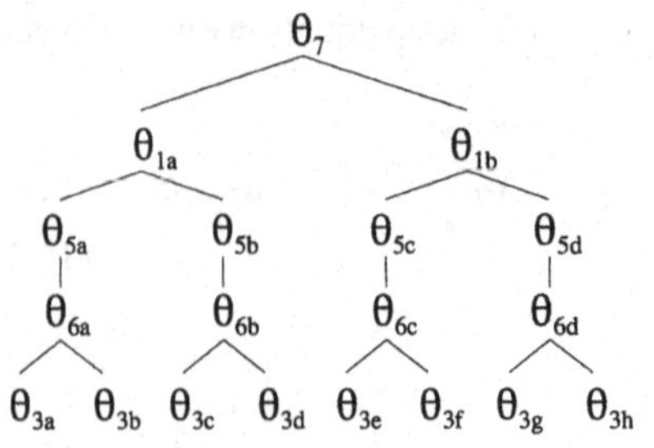

Figure 11.16. GE P60 solution tree.

where the abbreviations s_{2+3} and c_{2+3} have been introduced as

$$s_{2+3} = \sin(\theta_2 + \theta_3), \tag{11.84}$$

$$c_{2+3} = \cos(\theta_2 + \theta_3). \tag{11.85}$$

Squaring and adding Eqs. (11.82) and (11.83) gives

$$a_{23}^2(s_2^2 + c_2^2) + a_{34}^2(s_{2+3}^2 + c_{2+3}^2) + 2a_{23}a_{34}(c_2 c_{2+3} + s_2 s_{2+3}) = K_1^2 + K_2^2. \tag{11.86}$$

Now, $(s_2^2 + c_2^2) = (s_{2+3}^2 + c_{2+3}^2) = 1$. The third term in parentheses equals the cosine of $[\theta_2 - (\theta_2 + \theta_3)]$, which equals the cosine of $[-\theta_3]$. Thus, Eq. (11.86) may be written as

$$a_{23}^2 + a_{34}^2 + 2a_{23}a_{34}\cos(-\theta_3) = K_1^2 + K_2^2. \tag{11.87}$$

Because $\cos(-\theta_3) = \cos\theta_3$,

$$a_{23}^2 + a_{34}^2 + 2a_{23}a_{34}c_3 = K_1^2 + K_2^2. \tag{11.88}$$

Solving for c_3 gives

$$c_3 = \frac{K_1^2 + K_2^2 - a_{23}^2 - a_{34}^2}{2a_{23}a_{34}}. \tag{11.89}$$

Equation (11.89) can be used to determine two values for θ_3 for each of the four sets of values for θ_7, θ_1, and θ_6. The current solution tree for the manipulator is shown in Figure 11.16.

11.3.5 Solution for θ_2

The angle θ_2 can be obtained from Eqs. (11.80) and (11.81). Regrouping these equations gives

$$c_2(a_{23} + a_{34}c_3) + s_2(-a_{34}s_3) = K_1, \tag{11.90}$$

$$c_2(-a_{34}s_3) + s_2(-a_{23} - a_{34}c_3) = K_2. \tag{11.91}$$

These two equations contain two unknowns, that is, c_2 and s_2. Input of sets of the previously calculated joint parameters will yield values for the corresponding sine and cosine of θ_2.

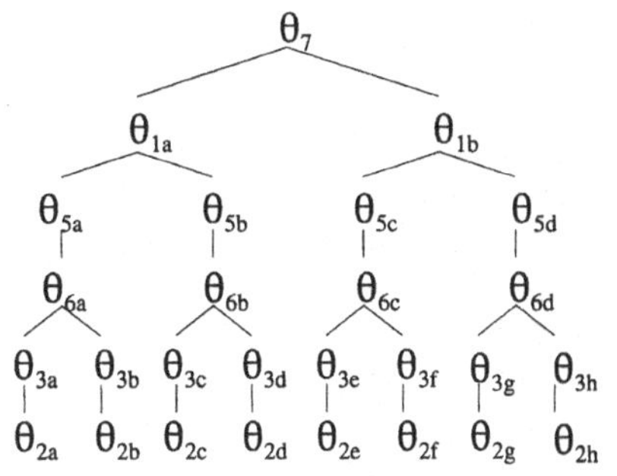

Figure 11.17. GE P60 solution tree.

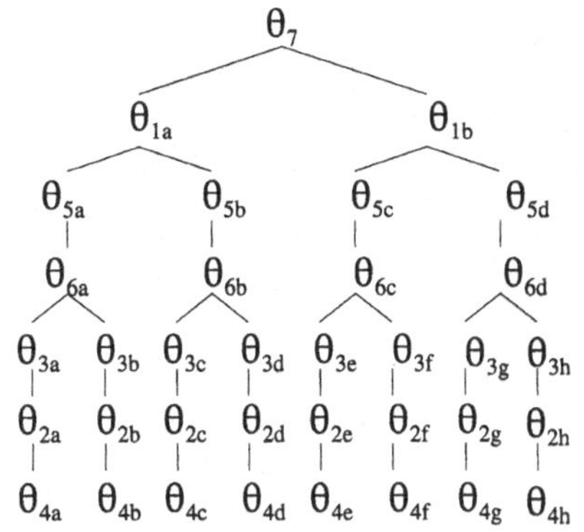

Figure 11.18. Final solution tree for GE P60 robot.

A unique corresponding value for θ_2 is then determined. The solution tree for the problem thus far is shown in Figure 11.17.

11.3.6 Solution for θ_4

The corresponding value for the angle θ_4 will be obtained from the following two fundamental sine and sine–cosine laws:

$$X_{67123} = s_{45}s_4, \tag{11.92}$$

$$Y_{67123} = s_{45}c_4. \tag{11.93}$$

Substituting $\alpha_{45} = 270^\circ$ results in the following solution for the sine and cosine of θ_4:

$$s_4 = -X_{67123}, \tag{11.94}$$

$$c_4 = -Y_{67123}. \tag{11.95}$$

Thus, a unique corresponding value for θ_4 can be determined for each set of previously calculated joint parameters. The final solution tree for the GE P60 robot is shown in Figure 11.18.

Table 11.6. *Eight solution sets for the GE P60 robot (angles in degrees).*

Solution	ϕ_1	θ_2	θ_3	θ_4	θ_5	θ_6
A	−139.443	142.825	73.355	60.073	144.493	−52.334
B	−139.443	−133.183	−73.355	122.790	144.493	−52.334
C	−139.443	123.396	103.879	−131.022	−144.493	127.666
D	−139.443	−114.585	−103.879	−45.284	−144.493	127.666
E	30.730	−63.241	103.385	−149.362	37.692	143.774
F	30.730	58.127	−103.385	−63.960	37.692	143.774
G	30.730	−48.562	73.855	45.490	−37.692	−36.226
H	30.730	36.027	−73.855	108.610	−37.692	−36.226

Table 11.7. *Mechanism parameters for the Cincinnati Milacron T3-776 robot.*

Link length, in.	Twist angle, deg.	Joint offset, in.	Joint angle, deg.
$a_{12} = 0$	$\alpha_{12} = 90$		ϕ_1 = variable
$a_{23} = 44$	$\alpha_{23} = 0$	$S_2 = 0$	θ_2 = variable
$a_{34} = 0$	$\alpha_{34} = 90$	$S_3 = 0$	θ_3 = variable
$a_{45} = 0$	$\alpha_{45} = 61$	$S_4 = 55$	θ_4 = variable
$a_{56} = 0$	$\alpha_{56} = 61$	$S_5 = 0$	θ_5 = variable
			θ_6 = variable

11.3.7 Numerical example

As a numerical example, the following information was specified for the GE P60 manipulator:

$$S_6 = 15.24\,\text{cm.},\quad {}^6\underline{P}_{\text{tool}} = \begin{bmatrix} 2 \\ 3 \\ 5 \end{bmatrix}\text{cm.},\quad {}^F\underline{P}_{\text{tool}} = \begin{bmatrix} 80.0 \\ 80.0 \\ 18.0 \end{bmatrix}\text{cm.},$$

$${}^F\underline{S}_6 = \begin{bmatrix} -0.5774 \\ 0.5774 \\ 0.5774 \end{bmatrix},\quad {}^F\underline{a}_{67} = \begin{bmatrix} 0.4082 \\ 0.8165 \\ -0.4082 \end{bmatrix}.$$

Table 11.6 shows the results of the reverse-position analysis.

11.4 Cincinnati Milacron T3-776 manipulator

The Cincinnati Milacron T3-776 robot is shown in Figure 11.19. A kinematic model of this robot with joint axis and link vectors labeled is shown in Figure 11.20. The constant mechanism parameters are listed in Table 11.7. It is apparent that the geometry of this manipulator is very similar to that of the Puma robot, that is, the second and third joint axes are parallel and the last three joint axes intersect at a point. The detailed solution will therefore be very similar to that for the Puma robot. This solution will be followed by a general discussion of a geometric solution that does not require that the hypothetical closure link be determined.

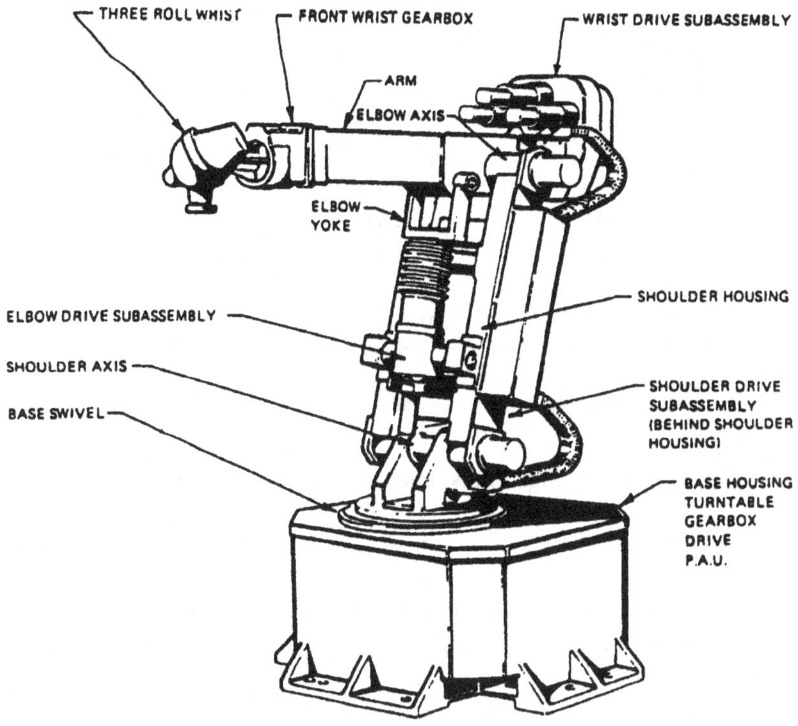

Figure 11.19. Cincinnati Milacron T3-776 robot.

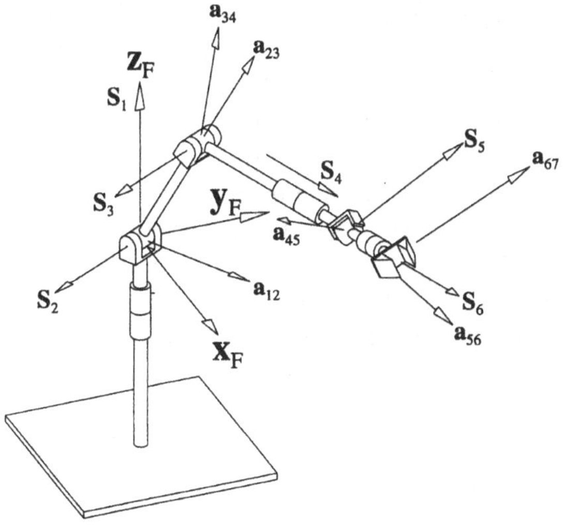

Figure 11.20. Kinematic model of Cincinnati Milacron T3-776 robot.

As before, the parameter S_6 is a free choice that must be made in order to specify the location of the origin of the sixth coordinate system. One input to the reverse-analysis problem is the location of the tool point measured in terms of the sixth coordinate system. This cannot be specified if the physical location of the origin of the sixth coordinate system is not known. A value of S_6 equal to six inches is selected, as this positions the origin of the sixth coordinate system at the center of the robot's tool mounting plate.

Table 11.8. *Mechanism parameters for closed-loop T3-776 mechanism.*

Link length, in.	Twist angle, deg.	Joint offset, in.	Joint angle, deg.
$a_{12} = 0$	$\alpha_{12} = 90$	$S_1 =$ C.L.	$\phi_1 =$ variable
$a_{23} = 44$	$\alpha_{23} = 0$	$S_2 = 0$	$\theta_2 =$ variable
$a_{34} = 0$	$\alpha_{34} = 90$	$S_3 = 0$	$\theta_3 =$ variable
$a_{45} = 0$	$\alpha_{45} = 61$	$S_4 = 55$	$\theta_4 =$ variable
$a_{56} = 0$	$\alpha_{56} = 61$	$S_5 = 0$	$\theta_5 =$ variable
$a_{67} = 0^*$	$\alpha_{67} = 90^*$	$S_6 = 6^*$	$\theta_6 =$ variable
$a_{71} =$ C.L.	$\alpha_{71} =$ C.L.	$S_7 =$ C.L.	$\theta_7 =$ C.L.

* = User-selected value
C.L. = Calculated during the close-the-loop procedure

The reverse analysis problem statement is repeated again as follows:

given: S_6 and the direction of $\mathbf{a}_{67}$ relative to $\mathbf{S}_6$ in order to establish the sixth coordinate system,
$^6\mathbf{P}_{tool}$: the location of the tool point in the sixth coordinate system,
$^F\mathbf{P}_{tool}$: the desired location of the tool point in the fixed coordinate system, and
$^F\mathbf{S}_6$, $^F\mathbf{a}_{67}$: the desired orientation of the robot end effector,

find: $\phi_1, \theta_2, \theta_3, \theta_4, \theta_5, \theta_6$: the joint angle parameters that will position and orient the end effector as desired.

The solution to this problem proceeds as described in Chapter 5. From the given information, Eq. (5.3) can be used to determine the position of the origin of the sixth coordinate system measured in the fixed coordinate system. A hypothetical closure link is then created to form a closed-loop spatial mechanism. The link length a_{67} and the twist angle α_{67} were arbitrarily selected as zero and ninety degrees respectively. With these two choices, the direction of the vector $\mathbf{S}_7$ is known in terms of the fixed coordinate system. Further, the hypothetical seventh joint axis is known to pass through the origin of the sixth coordinate system.

Chapter 5 shows how the six close-the-loop parameters (S_7, S_1, a_{71}, θ_7, α_{71}, and γ_1) can now be determined. The analysis will proceed assuming that values for these six parameters have been calculated. Table 11.8 shows the mechanism parameters for the newly formed closed-loop spatial mechanism.

11.4.1 Solution for θ_1 and ϕ_1

The vector loop equation for the closed-loop Cincinnati Milacron T3-776 mechanism is as follows:

$$S_1\mathbf{S}_1 + a_{23}\mathbf{a}_{23} + S_4\mathbf{S}_4 + S_6\mathbf{S}_6 + S_7\mathbf{S}_7 + a_{71}\mathbf{a}_{71} = \mathbf{0}. \tag{11.96}$$

Expressing the vectors in terms of set 14 of the table of direction cosines for the spatial heptagon yields

$$S_1\begin{pmatrix}0\\-s_{12}\\c_{12}\end{pmatrix}+a_{23}\begin{pmatrix}c_2\\-s_2\\0\end{pmatrix}+S_4\begin{pmatrix}X_{5671}\\Y_{5671}\\Z_{5671}\end{pmatrix}$$
$$+S_6\begin{pmatrix}X_{71}\\Y_{71}\\Z_{71}\end{pmatrix}+S_7\begin{pmatrix}X_1\\Y_1\\Z_1\end{pmatrix}+a_{71}\begin{pmatrix}c_1\\s_1c_{12}\\U_{12}\end{pmatrix}=\begin{pmatrix}0\\0\\0\end{pmatrix}. \tag{11.97}$$

Subsidiary spatial and polar sine, sine–cosine, and cosine laws can be used to simply the terms for the vector $\mathbf{S}_4$. Thus, Eq. (11.97) can be written as

$$S_1\begin{pmatrix}0\\-s_{12}\\c_{12}\end{pmatrix}+a_{23}\begin{pmatrix}c_2\\-s_2\\0\end{pmatrix}+S_4\begin{pmatrix}X_{32}\\-X^*_{32}\\\bar{Z}_3\end{pmatrix}$$
$$+S_6\begin{pmatrix}X_{71}\\Y_{71}\\Z_{71}\end{pmatrix}+S_7\begin{pmatrix}X_1\\Y_1\\Z_1\end{pmatrix}+a_{71}\begin{pmatrix}c_1\\s_1c_{12}\\U_{12}\end{pmatrix}=\begin{pmatrix}0\\0\\0\end{pmatrix}. \tag{11.98}$$

The Z component equation (which is equivalent to projecting the vector loop equation onto the $\mathbf{S}_2$ axis) is

$$S_1c_{12}+S_4\bar{Z}_3+S_6Z_{71}+S_7Z_1+a_{71}U_{12}=0. \tag{11.99}$$

Because $\alpha_{23}=0°$ and $\alpha_{34}=90°$, $\bar{Z}_3=0$, and because $\alpha_{12}=90°$, $c_{12}=0$, Eq. (11.99) thus reduces to

$$S_6Z_{71}+S_7Z_1+a_{71}U_{12}=0. \tag{11.100}$$

This equation contains θ_1 as its only unknown. Expanding Z_{71}, Z_1, and U_{12} yields

$$S_6[s_{12}(X_7s_1+Y_7c_1)+c_{12}Z_7]+S_7[c_{12}c_{71}-s_{12}s_{71}c_1]+a_{71}[s_1s_{12}]=0. \tag{11.101}$$

This equation can be simplified by substituting $s_{12}=1$ and $c_{12}=0$ and may be written as

$$S_6[X_7s_1+Y_7c_1]+S_7[-s_{71}c_1]+a_{71}[s_1]=0. \tag{11.102}$$

Grouping the s_1 and c_1 terms yields

$$[S_6Y_7-S_7s_{71}]c_1+[S_6X_7+a_{71}]s_1=0. \tag{11.103}$$

Upon expanding X_7 and Y_7 with $\alpha_{67}=90°$, Eq. (11.103) may be written as

$$[-S_6c_7c_{71}-S_7s_{71}]c_1+[S_6s_7+a_{71}]s_1=0. \tag{11.104}$$

The terms within the brackets in Eq. (11.104) can be calculated from the known values. Rearranging this equation gives

$$\tan\theta_1=\frac{s_1}{c_1}=\frac{(S_6c_7c_{71}+S_7s_{71})}{(S_6s_7+a_{71})}. \tag{11.105}$$

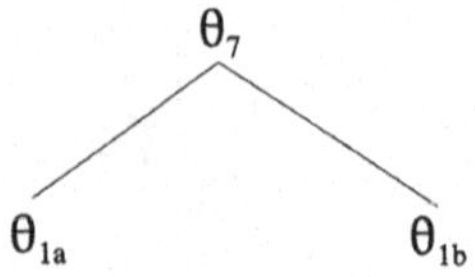

Figure 11.21. T3-776 solution tree.

Two distinct values of θ_1 will satisfy Eq. (11.105). Figure 11.21 shows the solution tree for the mechanism thus far. The two associated values for the angle ϕ_1 can be calculated as $(\theta_{1a} - \gamma_1)$ and $(\theta_{1b} - \gamma_1)$.

11.4.2 Solution for θ_3

Substituting $s_{12} = 1$ and $c_{12} = 0$ into the X and Y components of Eq. (11.98) yields

$$a_{23}c_2 + S_4X_{32} + S_6X_{71} + S_7X_1 + a_{71}c_1 = 0, \tag{11.106}$$

$$-S_1 - a_{23}s_2 - S_4X^*_{32} + S_6Y_{71} + S_7Y_1 = 0. \tag{11.107}$$

All terms that do not contain the unknown variables θ_2 and θ_3 will be transfered to the right-hand side of Eqs. (11.106) and (11.107) to give

$$a_{23}c_2 + S_4X_{32} = A, \tag{11.108}$$

$$a_{23}s_2 + S_4X^*_{32} = B, \tag{11.109}$$

where

$$A = -S_6X_{71} - S_7X_1 - a_{71}c_1, \tag{11.110}$$

$$B = -S_1 + S_6Y_{71} + S_7Y_1. \tag{11.111}$$

Expanding X_{32} and X^*_{32} gives

$$a_{23}c_2 + S_4[(s_{34}s_3)c_2 + (s_{23}c_{34} + c_{23}s_{34}c_3)s_2] = A, \tag{11.112}$$

$$a_{23}s_2 + S_4[(s_{34}s_3)s_2 - (s_{23}c_{34} + c_{23}s_{34}c_3)c_2] = B. \tag{11.113}$$

Substituting for $\alpha_{23} = 0°$ and $\alpha_{34} = 90°$ gives

$$a_{23}c_2 + S_4[s_3c_2 + c_3s_2] = A, \tag{11.114}$$

$$a_{23}s_2 + S_4[s_3s_2 - c_3c_2] = B. \tag{11.115}$$

These equations can be written as

$$a_{23}c_2 + S_4s_{2+3} = A, \tag{11.116}$$

$$a_{23}s_2 - S_4c_{2+3} = B, \tag{11.117}$$

where s_{2+3} and c_{2+3} represent the sine and cosine of $(\theta_2 + \theta_3)$ respectively.

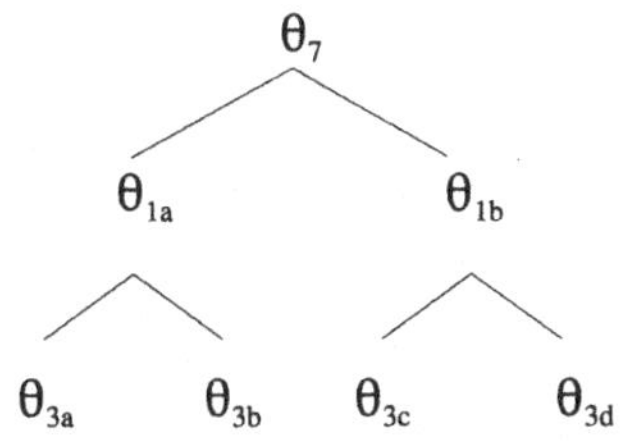

Figure 11.22. T3-776 solution tree.

Equations (11.114) and (11.115) represent two equations in the two unknowns θ_2 and θ_3. These variables will be solved for by adding the squares of Eqs. (11.116) and (11.117). Squaring the equations gives

$$a_{23}^2 c_2^2 + S_4^2 s_{2+3}^2 + 2a_{23}S_4 c_2 s_{2+3} = A^2, \tag{11.118}$$

$$a_{23}^2 s_2^2 + S_4^2 c_{2+3}^2 - 2a_{23}S_4 s_2 c_{2+3} = B^2. \tag{11.119}$$

Adding yields

$$a_{23}^2 + S_4^2 + 2a_{23}S_4[c_2 s_{2+3} - s_2 c_{2+3}] = A^2 + B^2. \tag{11.120}$$

Recognizing that $[c_2 s_{2+3} - s_2 c_{2+3}] = \sin[(\theta_2 + \theta_3) - \theta_2] = \sin(\theta_3) = s_3$ gives

$$a_{23}^2 + S_4^2 + 2a_{23}S_4 s_3 = A^2 + B^2. \tag{11.121}$$

Solving this equation for s_3 gives

$$s_3 = \frac{A^2 + B^2 - a_{23}^2 - S_4^2}{2a_{23}S_4}. \tag{11.122}$$

Two values for θ_3 exist that satisfy this equation. The current solution tree is shown in Figure 11.22.

11.4.3 Solution for θ_2

Equations (11.114) and (11.115) can be used to solve for the angle θ_2. Regrouping these equations gives

$$c_2[a_{23} + S_4 s_3] + s_2[S_4 c_3] = A, \tag{11.123}$$

$$s_2[a_{23} + S_4 s_3] + c_2[-S_4 c_3] = B. \tag{11.124}$$

Equations (11.123) and (11.124) represent two equations in the two unknowns c_2 and s_2. Thus, a unique corresponding value of θ_2 can be found for each θ_1, θ_3 pair. In other words, θ_{1a} and θ_{3a} will be substituted into the equations to yield θ_{2a}. Similarly, θ_{1a} and θ_{3b} will be substituted into the equations to yield θ_{2b}. The current solution tree is shown in Figure 11.23.

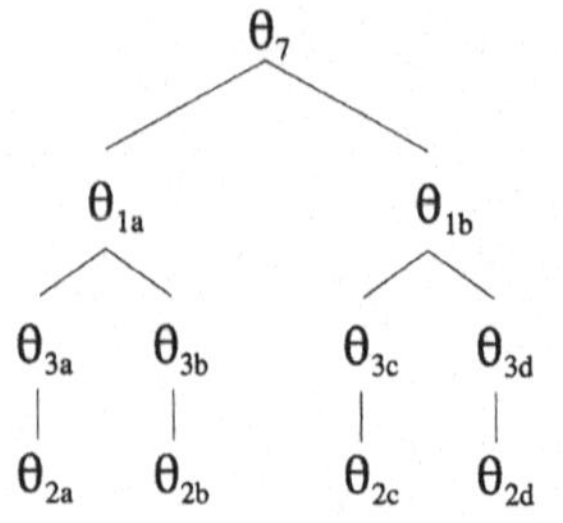

Figure 11.23. T3-776 solution tree.

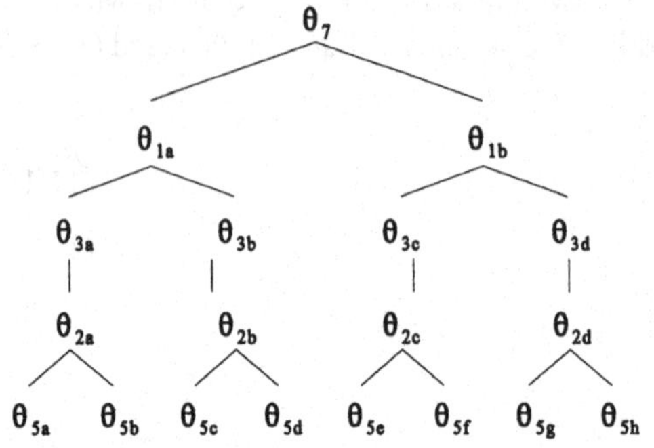

Figure 11.24. T3-776 solution tree.

11.4.4 Solution for θ_5

The angle θ_5 can be readily determined from the spherical equation

$$Z_{7123} = \bar{Z}_5. \tag{11.125}$$

Expanding the right-hand side of this equation yields

$$Z_{7123} = c_{56}c_{45} - s_{56}s_{45}c_5. \tag{11.126}$$

Solving this equation for c_5 yields

$$c_5 = \frac{c_{56}c_{45} - Z_{7123}}{s_{56}s_{45}}. \tag{11.127}$$

Thus, for each combination of θ_7, θ_1, θ_3, and θ_2, a value for c_5 and thereby two values for θ_5 can be determined. The solution tree for this analysis is now shown in Figure 11.24.

11.4.5 Solution for θ_4

Corresponding values for θ_4 can be obtained from the following two subsidiary spherical equations:

$$X_{54} = X_{7123}, \tag{11.128}$$

$$X_{54}^* = -Y_{7123}. \tag{11.129}$$

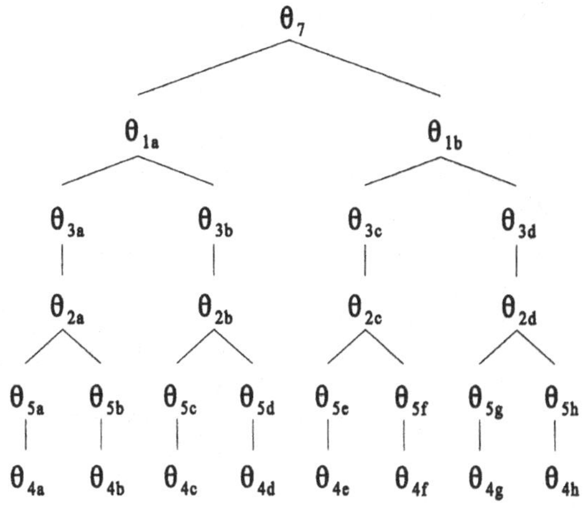

Figure 11.25. T3-776 solution tree.

Expanding the left-hand sides of the above equations yields

$$\bar{X}_5 c_4 - \bar{Y}_5 s_4 = X_{7123}, \tag{11.130}$$

$$\bar{X}_5 s_4 + \bar{Y}_5 c_4 = -Y_{7123}, \tag{11.131}$$

where

$$\bar{X}_5 = s_{56} s_5, \tag{11.132}$$

$$\bar{Y}_5 = -(s_{45} c_{56} + c_{45} s_{56} c_5). \tag{11.133}$$

Equations (11.130) and (11.131) represent two equations in the two unknowns, s_4 and c_4. Substituting the previously calculated values for the eight sets of corresponding values for $\theta_7, \theta_1, \theta_2, \theta_3$, and θ_5 will yield a corresponding value for θ_4. The current solution tree is shown in Figure 11.25.

It should be noted that the solution for θ_4 becomes indeterminate if $\theta_5 = 180°$. In this case, $\bar{X}_5 = \bar{Y}_5 = 0$ ($\bar{Y}_5 = 0$ because $\alpha_{45} = \alpha_{56}$). When this case occurs, the vector $\mathbf{S}_6$ becomes collinear with the vector $\mathbf{S}_4$. This special case will be discussed in Section 11.5.

11.4.6 Solution for θ_6

The angle θ_6 is the last remaining joint angle to be determined. It will be calculated from the following two fundamental spherical sine and sine–cosine laws:

$$X_{43217} = s_{56} s_6, \tag{11.134}$$

$$Y_{43217} = s_{56} c_6. \tag{11.135}$$

Thus, a corresponding value for θ_6 can be found for each of the eight sets of values for $\theta_7, \theta_1, \theta_3, \theta_2, \theta_5$, and θ_4. The final solution tree for the Cincinnati Milacron T3-776 robot is shown in Figure 11.26. The solution of the reverse analysis for the manipulator is complete.

Table 11.9. *Eight solution sets for the T3-776 robot (angles in degrees).*

Solution	ϕ_1	θ_2	θ_3	θ_4	θ_5	θ_6
A	36.945	84.358	−23.095	70.853	127.506	100.818
B	36.945	84.358	−23.095	−136.036	−127.506	−106.070
C	36.945	−47.830	−156.905	163.422	97.464	−6.476
D	36.945	−47.830	−156.905	−62.672	−97.464	127.430
E	−143.055	−132.170	−23.095	−16.578	97.464	−6.476
F	−143.055	−132.170	−23.095	117.328	−97.464	127.430
G	−143.055	95.642	−156.905	−109.147	127.506	100.818
H	−143.055	95.642	−156.905	43.964	−127.506	−106.070

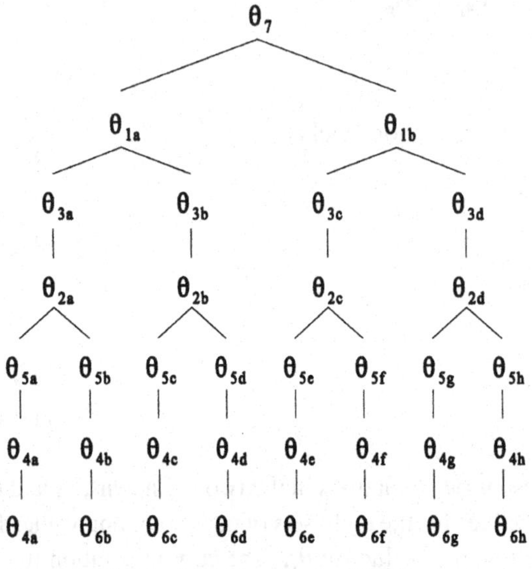

Figure 11.26. Final solution tree for T3-776 robot.

11.4.7 Numerical example

As a numerical example, the following information was specified for the T3-776 manipulator:

$$S_6 = 6\,\text{in.},\quad {}^6\mathbf{P}_{\text{tool}} = \begin{bmatrix}5\\3\\7\end{bmatrix}\ \text{in.},\quad {}^F\mathbf{P}_{\text{tool}} = \begin{bmatrix}55.0\\33.0\\23.0\end{bmatrix}\ \text{in.},$$

$${}^F\mathbf{S}_6 = \begin{bmatrix}1.0\\0.0\\0.0\end{bmatrix},\quad {}^F\mathbf{a}_{67} = \begin{bmatrix}0.0\\0.707\\0.707\end{bmatrix}.$$

Table 11.9 shows the results of the reverse-position analysis.

11.4.8 Geometric solution

It is possible to directly calculate the joint angle values for the Cincinnati Milacron T3-776 manipulator for the reverse-analysis problem statement without performing the close-the-loop step. This direct geometric solution is possible because of the simplicity of the geometry of the manipulator, that is, the first two joint axes intersect, the last three joint axes intersect, the second and third joint axes are parallel, and the vectors $\mathbf{S}_1$, $\mathbf{a}_{12}$, $\mathbf{a}_{23}$, and $\mathbf{S}_4$ are coplanar.

The geometric analysis begins by first obtaining the coordinates of the point at the center of the ball and socket joint in terms of the fixed coordinate system. This is accomplished by first using Eq. (5.3) to obtain the coordinates of the origin of the sixth coordinate system measured with respect to the fixed system. The coordinates of the center of the ball and socket joint (the point of intersection of the vectors $\mathbf{S}_4$, $\mathbf{S}_5$, and $\mathbf{S}_6$), ${}^{\mathrm{F}}\mathbf{P}_{\mathrm{BS}}$, are then determined as

$$ {}^{\mathrm{F}}\mathbf{P}_{\mathrm{BS}} = {}^{\mathrm{F}}\mathbf{P}_{6\,\mathrm{orig}} - S_6\,{}^{\mathrm{F}}\mathbf{S}_6. \tag{11.136} $$

Figure 11.27 shows the kinematic diagram of the T3-776 manipulator with the vector $\mathbf{P}_{\mathrm{BS}}$ drawn. It is apparent in the figure that the vectors $\mathbf{a}_{12}$, $\mathbf{a}_{23}$, $\mathbf{S}_1$, $\mathbf{S}_4$, and $\mathbf{P}_{\mathrm{BS}}$ all lie in the same plane. Because of this, the vector ${}^{\mathrm{F}}\mathbf{a}_{12}$ must equal a unit vector that is parallel or antiparallel to the vector ${}^{\mathrm{F}}\mathbf{P}_{\mathrm{BS}}$ with its Z component subtracted. This can be written as

$$ {}^{\mathrm{F}}\mathbf{a}_{12} = \pm\frac{{}^{\mathrm{F}}\mathbf{P}_{\mathrm{BS}} - ({}^{\mathrm{F}}\mathbf{P}_{\mathrm{BS}}\cdot\mathbf{k})\mathbf{k}}{\left|{}^{\mathrm{F}}\mathbf{P}_{\mathrm{BS}} - ({}^{\mathrm{F}}\mathbf{P}_{\mathrm{BS}}\cdot\mathbf{k})\mathbf{k}\right|}. \tag{11.137} $$

Alternately, if the vector ${}^{\mathrm{F}}\mathbf{P}_{\mathrm{BS}}$ is written as ${}^{\mathrm{F}}P_{\mathrm{BSx}}\mathbf{i} + {}^{\mathrm{F}}P_{\mathrm{BSy}}\mathbf{j} + {}^{\mathrm{F}}P_{\mathrm{BSz}}\mathbf{k}$, Eq. (11.137) can be written as

$$ {}^{\mathrm{F}}\mathbf{a}_{12} = \pm\frac{{}^{\mathrm{F}}P_{\mathrm{BSx}}\mathbf{i} + {}^{\mathrm{F}}P_{\mathrm{BSy}}\mathbf{j}}{\left|{}^{\mathrm{F}}P_{\mathrm{BSx}}\mathbf{i} + {}^{\mathrm{F}}P_{\mathrm{BSy}}\mathbf{j}\right|}. \tag{11.138} $$

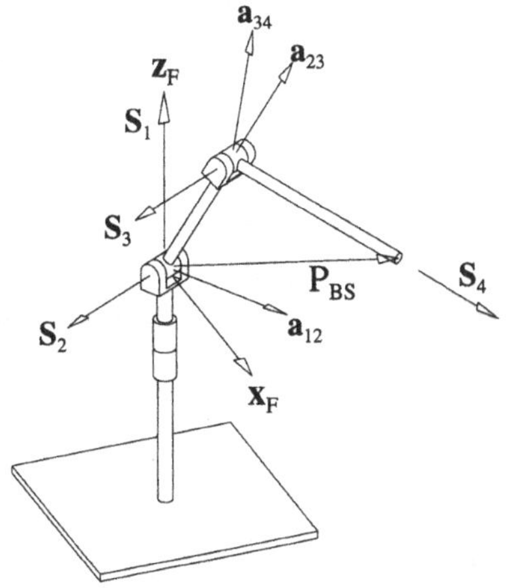

Figure 11.27. Modified kinematic diagram of the T3-776 manipulator.

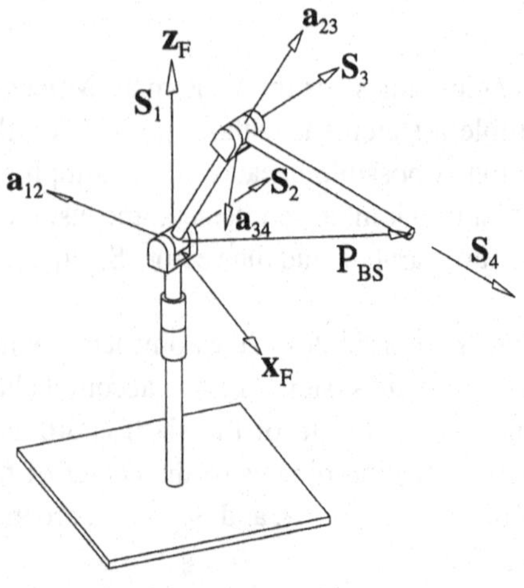

Figure 11.28. Alternate direction for $\mathbf{a}_{12}$.

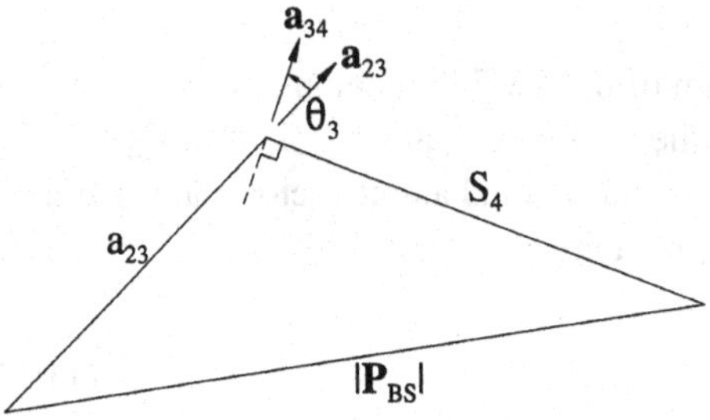

Figure 11.29. Planar triangle.

Figure 11.27 shows the manipulator for the "plus" case of Eq. (11.138), and Figure 11.28 shows the "minus" case. The sine and cosine of ϕ_1 that correspond to each of the two configurations can be calculated from

$$\cos(\phi_1) = {}^{\mathrm{F}}\mathbf{a}_{12} \cdot \mathbf{i} \tag{11.139}$$

and

$$\sin(\phi_1) = \left(\mathbf{i} \times {}^{\mathrm{F}}\mathbf{a}_{12}\right) \cdot \mathbf{k}. \tag{11.140}$$

The two unique values for ϕ_1 are thus determined.

The angle θ_3 will be determined next. Because the vector $\mathbf{P}_{\mathrm{BS}}$ is known in terms of the fixed coordinate system, the magnitude of this vector represents the scalar distance of the center of the ball and socket joint from the origin. A planar triangle is formed whose three sides are all known as shown in Figure 11.29 ($\mathbf{S}_2$ and $\mathbf{S}_3$ point out of the page in the figure).

A cosine law for the planar triangle is written as

$$\left|{}^{\mathrm{F}}\mathbf{P}_{\mathrm{BS}}\right|^2 = a_{23}^2 + S_4^2 - 2a_{23}S_4\cos(\theta_3 + \pi/2). \tag{11.141}$$

Recognizing that $\cos(\theta_3 + \pi/2)$ equals $-s_3$ and then solving for s_3 yields

$$s_3 = \frac{|\mathbf{P}_{\mathrm{BS}}|^2 - a_{23}^2 - S_4^2}{2a_{23}S_4}. \tag{11.142}$$

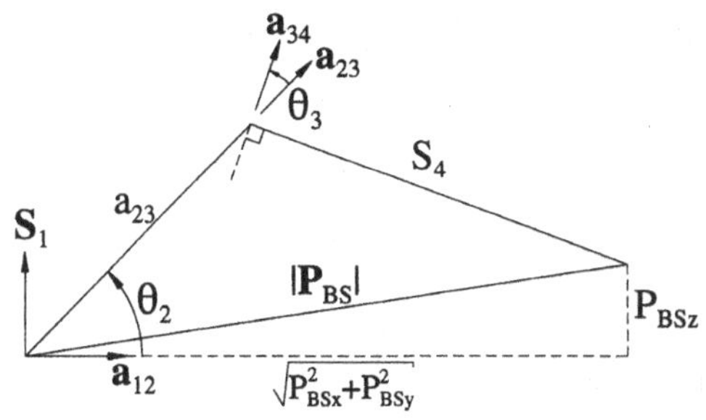

Figure 11.30. Planar triangle.

Two values of θ_3 will satisfy this equation. It is interesting to compare this result with Eq. (11.122). Further, it is interesting to note that the values for θ_3 are not dependent on ϕ_1.

The angle θ_2 can be determined for the case where $\mathbf{a}_{12}$ points towards the ball and socket joint by again considering the planar triangle, which is redrawn in Figure 11.30 with the vectors $\mathbf{S}_2$ and $\mathbf{S}_3$ coming out of the page. Projecting two sides of the triangle onto the vectors $\mathbf{a}_{12}$ and $\mathbf{S}_1$ yields the following two equations:

$$a_{23}c_2 + S_4\cos(\theta_2 + \theta_3 - \pi/2) = \sqrt{P_{BSx}^2 + P_{BSy}^2}, \tag{11.143}$$

$$a_{23}s_2 + S_4\sin(\theta_2 + \theta_3 - \pi/2) = P_{BSz}. \tag{11.144}$$

Recognizing that $\cos(\theta_2 + \theta_3 - \pi/2) = s_{2+3}$ and that $\sin(\theta_2 + \theta_3 - \pi/2) = -c_{2+3}$, these two equations may be written as

$$a_{23}c_2 + S_4 s_{2+3} = \sqrt{P_{BSx}^2 + P_{BSy}^2}, \tag{11.145}$$

$$a_{23}s_2 - S_4 c_{2+3} = P_{BSz}. \tag{11.146}$$

These two equations may be used to determine the unique corresponding value of θ_2 for each of the two previously calculated values of θ_3. It is interesting to compare Eqs. (11.145) and (11.146) with Eqs. (11.116) and (11.117). A similar solution can be obtained by projection of θ_2 for the case where $\mathbf{a}_{12}$ points away from the ball and socket joint.

The remaining angles to be determined are θ_4, θ_5, and θ_6. In Sections 11.4.4 through 11.4.6, the angles were readily found from appropriate spherical sine, sine–cosine, and cosine laws. However, if the close-the-loop step is not performed, then the previous approach is not valid because the angle θ_7 is not known.

The solution can proceed, however, by writing the general transformation equation as

$${}^{F}_{6}\mathbf{T} = {}^{F}_{1}\mathbf{T}\,{}^{1}_{2}\mathbf{T}\,{}^{2}_{3}\mathbf{T}\,{}^{3}_{4}\mathbf{T}\,{}^{4}_{5}\mathbf{T}\,{}^{5}_{6}\mathbf{T}. \tag{11.147}$$

The orientation part of this equation may be written as

$${}^{F}_{6}\mathbf{R} = {}^{F}_{1}\mathbf{R}\,{}^{1}_{2}\mathbf{R}\,{}^{2}_{3}\mathbf{R}\,{}^{3}_{4}\mathbf{R}\,{}^{4}_{5}\mathbf{R}\,{}^{5}_{6}\mathbf{R}. \tag{11.148}$$

The matrix ${}^{F}_{6}\mathbf{R}$ is known because the orientation of the sixth coordinate system relative to the fixed system was given in the reverse-analysis problem statement. The general rotation matrix ${}^{i}_{j}\mathbf{R}$, where $j = i + 1$, was given as the upper left 3×3 matrix in Eq. (3.7) and is repeated here as

$$ {}^{i}_{j}\mathbf{R} = \begin{bmatrix} c_j & -s_j & 0 \\ s_j c_{ij} & c_j c_{ij} & -s_{ij} \\ s_j s_{ij} & c_j s_{ij} & c_{ij} \end{bmatrix} . \tag{11.149} $$

The rotation matrix ${}^{F}_{1}\mathbf{R}$ was given as the upper 3×3 matrix in Eq. (3.9) as

$$ {}^{F}_{1}\mathbf{R} = \begin{bmatrix} \cos(\phi_1) & -\sin(\phi_1) & 0 \\ \sin(\phi_1) & \cos(\phi_1) & 0 \\ 0 & 0 & 1 \end{bmatrix} . \tag{11.150} $$

Because values are known for the angles ϕ_1, θ_2, and θ_3, the rotation matrices ${}^{F}_{1}\mathbf{R}$, ${}^{1}_{2}\mathbf{R}$, and ${}^{2}_{3}\mathbf{R}$ are fully defined. Moving these matrices to the left side of Eq. (11.148) yields

$$ {}^{3}_{2}\mathbf{R}\,{}^{2}_{1}\mathbf{R}\,{}^{1}_{F}\mathbf{R}\,{}^{F}_{6}\mathbf{R} = {}^{3}_{4}\mathbf{R}\,{}^{4}_{5}\mathbf{R}\,{}^{5}_{6}\mathbf{R}. \tag{11.151} $$

All terms on the left-hand side of Eq. (11.151) are known, and the resulting 3×3 matrix can be computed for each of the combination of values for the angles ϕ_1, θ_2, and θ_3. This resulting matrix is equal to ${}^{3}_{6}\mathbf{R}$, and Eq. (11.151) is rewritten as

$$ {}^{3}_{6}\mathbf{R} = {}^{3}_{4}\mathbf{R}\,{}^{4}_{5}\mathbf{R}\,{}^{5}_{6}\mathbf{R}. \tag{11.152} $$

This equation is rearranged by moving the matrix ${}^{3}_{4}\mathbf{R}$ to the left-hand side and then expanding the matrices ${}^{4}_{3}\mathbf{R}$, ${}^{4}_{5}\mathbf{R}$, and ${}^{5}_{6}\mathbf{R}$ and substituting the numerical value for α_{34} to yield

$$ \begin{bmatrix} c_4 & 0 & s_4 \\ -s_4 & 0 & c_4 \\ 0 & -1 & 0 \end{bmatrix} {}^{3}_{6}\mathbf{R} = \begin{bmatrix} c_5 & -s_5 & 0 \\ s_5 c_{45} & c_5 c_{45} & -s_{45} \\ s_5 s_{45} & c_5 s_{45} & c_{45} \end{bmatrix} \begin{bmatrix} c_6 & -s_6 & 0 \\ s_6 c_{56} & c_6 c_{56} & -s_{56} \\ s_6 s_{56} & c_6 s_{56} & c_{56} \end{bmatrix} . \tag{11.153} $$

The matrix ${}^{3}_{6}\mathbf{R}$ is known and will be written as

$$ {}^{3}_{6}\mathbf{R} = \begin{bmatrix} a & b & d \\ e & f & g \\ h & i & j \end{bmatrix} . \tag{11.154} $$

Performing the matrix multiplication on both sides of Eq. (11.154) and equating the third-row, third-column element of the results yields the following equation:

$$ -g = -s_{56} s_{45} c_5 + c_{56} c_{45}. \tag{11.155} $$

This equation can be solved for the cosine of θ_5, which indicates that two values of θ_5 will exist for each combination of ϕ_1, θ_2, and θ_3.

Equating the first-row, third-column elements and the second-row, third-column elements of Eq. (11.153) results in the following two equations:

$$ d c_4 + j s_4 = s_5 s_{56}, \tag{11.156} $$

$$ -d s_4 + j c_4 = -c_{45} s_{56} c_5 - s_{45} c_{56}. \tag{11.157} $$

These two equations can be used to solve for the corresponding values for the sine and cosine of θ_4.

Equating the first-row, first-column elements and the second-row, first-column elements of Eq. (11.153) results in the following two equations:

$$ac_4 + hs_4 = c_5c_6 - s_5s_6c_{56}, \tag{11.158}$$

$$-as_4 + hc_4 = s_5c_6c_{45} + c_{45}c_{56}c_5c_6. \tag{11.159}$$

These two equations can be used to solve for the corresponding values for the sine and cosine of θ_6.

At this point, the geometric solution of the reverse-analysis problem for the Cincinnati Milacron T3-776 manipulator is complete. It was shown that it is possible to perform the reverse-analysis without performing the close-the-loop step. However, it should be noted that the geometric solution was successful in large part because of the simple geometry of the robot being analyzed.

Performing the close-the-loop step allows for the use of the spherical equations derived in Chapter 6. Without these spherical equations, one is left with the task of expanding the transformation matrices in Eq. (11.147) (which is more general than expanding the rotation matrices of Eq. (11.148)) and then rearranging the matrices, performing the matrix multiplications, and looking for corresponding elements that yield appropriate solution equations. Although this approach will work for simple manipulators, it will not be sufficient for more complex cases.

11.5 Special configurations

It was noted for the Puma robot that when $\sin\theta_5 = 0$, the solution for θ_4 is indeterminate (see Section 11.2.5). When this case occurs, the joint axis vectors $\mathbf{S}_4$ and $\mathbf{S}_6$ become collinear. The reverse position analysis can proceed, however, by selecting any arbitrary value for θ_4. The corresponding calculated value for θ_6 will orient the end effector as specified.

A similar situation occurred for the Cincinnati Milacron manipulator. It was noted in Section 11.4.5 that the solution for the angle θ_4 became indeterminate when $\theta_5 = 180°$. For this case, the joint axis vectors $\mathbf{S}_4$ and $\mathbf{S}_6$ again become collinear. The analysis can proceed by selecting an arbitrary value for θ_4 and then solving for the corresponding value of θ_6.

A special configuration of the GE P60 manipulator occurred when $\sin\theta_5 = 0$ (see Section 11.3.3). In this case, four joint axis vectors become parallel. Because only three parallel joint axes are necessary to position and orient an object in the plane perpendicular to the joint axes, the fourth parallel joint axis is redundant. An arbitrary value can be selected for θ_6, and corresponding values for θ_2, θ_3, and θ_4 can be determined.

At each of the special configurations, the joint axes of the manipulator became linearly dependent. In each case, the reverse-position analysis could continue, however, by making an arbitrary selection for the appropriate joint parameter. Examples of when the joint axes of a manipulator become linearly dependent are when four axes intersect at a point, four axes are parallel, or two joint axes are collinear.

It is important to be able to identify all the cases where the joint axes become linearly dependent for a particular manipulator (and to avoid these configurations when possible). When a manipulator is in such a linearly dependent configuration, it will in most cases not be able to move its end effector at the user-commanded velocity. In an attempt to control the velocity of the end effector, some joint velocities may approach infinity. This topic is discussed in detail in a companion book to this text that introduces the screw theory technique and details the forward-and reverse-velocity analyses for a serial robot manipulator.

11.6 Space station remote manipulator system (SSRMS)

The conceptual design of a SSRMS is shown in Figure 11.31. As shown in the figure, this is a seven-axis manipulator in which the first and second joint axes intersect; the third, fourth, and fifth joint axes are parallel; and the sixth and seventh joint axes intersect. The kinematic diagram for the manipulator is shown in Figure 11.32, and the constant

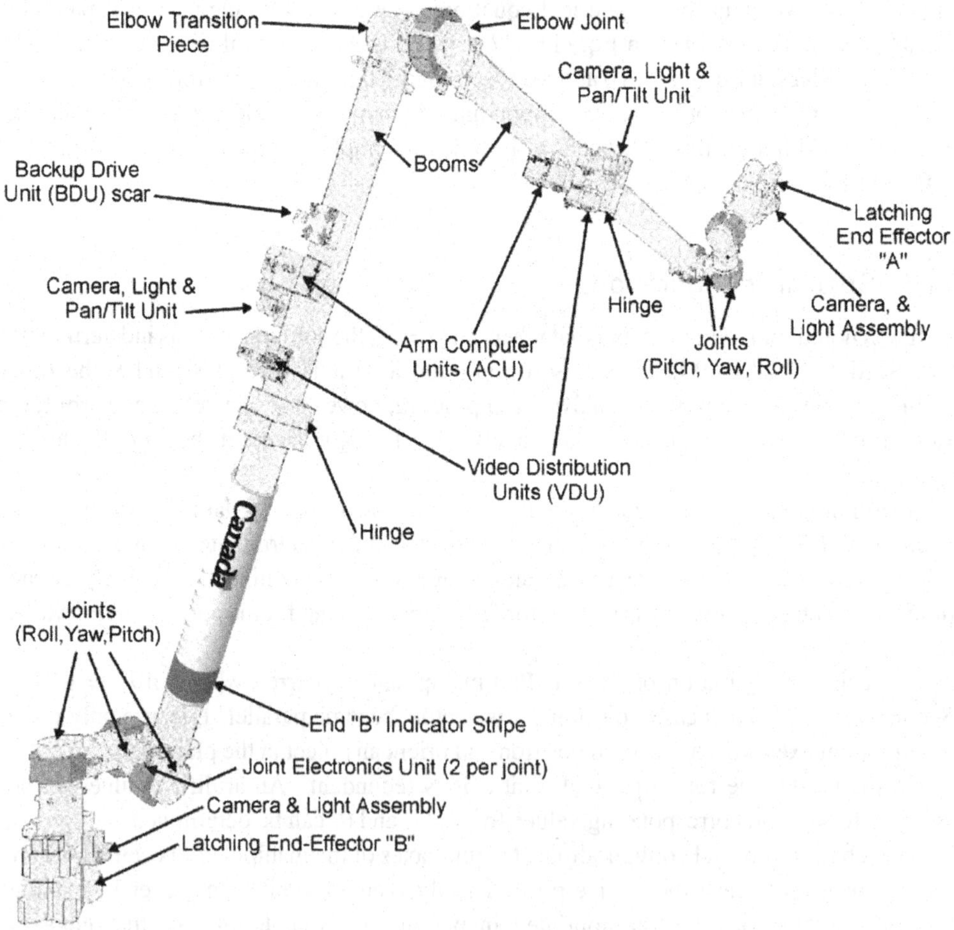

Figure 11.31. Space station remote manipulator system.

Table 11.10. *Mechanism parameters for the SSRMS.*

Link length, mm.	Twist angle, deg.	Joint offset, mm.	Joint angle, deg.
$a_{12} = 0$	$\alpha_{12} = 90$		ϕ_1 = variable
$a_{23} = 380$	$\alpha_{23} = 270$	$S_2 = 635$	θ_2 = user input
$a_{34} = 6850$	$\alpha_{34} = 0$	$S_3 = 504$	θ_3 = variable
$a_{45} = 6850$	$\alpha_{45} = 0$	$S_4 = 504$	θ_4 = variable
$a_{56} = 380$	$\alpha_{56} = 90$	$S_5 = 504$	θ_5 = variable
$a_{67} = 0$	$\alpha_{67} = 90$	$S_6 = 635$	θ_6 = variable
			θ_7 = variable

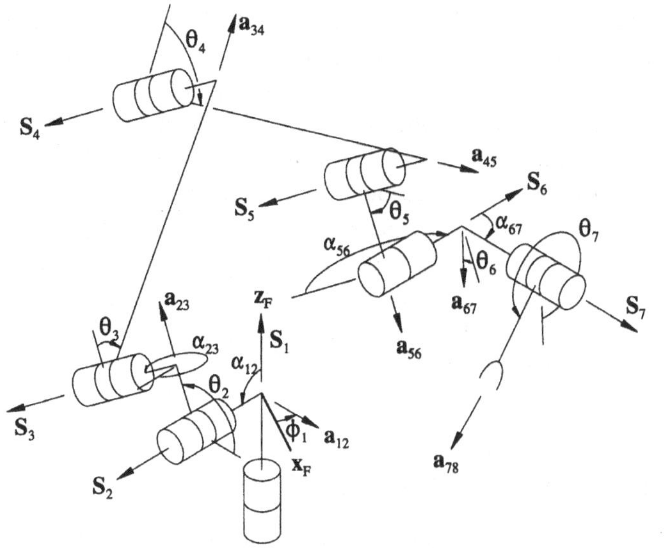

Figure 11.32. SSRMS kinematic diagram.

mechanism parameters are listed in Table 11.10. The manipulator is classified as being redundant because only six joint axes are necessary to position and orient the end effector arbitrarily in space. The reverse kinematic position analysis will proceed, however, by having the user specify one of the joint angle parameters in addition to specifying the desired position and orientation of the end effector (see Crane (1991)).

In the present analysis, the user must specify θ_2 in addition to the desired end effector position and orientation. This strategy offers a distinct advantage in that the parameter θ_2 has a physical meaning for the operator. This angle governs the orientation of the longest links of the manipulator (a_{34} and a_{45} in Figure 11.32) with respect to the XY plane through the base of the robot. The prior specification of θ_2 will enable the user to take better advantage of the redundancy of the system by being able to position the longest links of the manipulator to move over or around obstacles in the work space.

As with the other manipulators, the parameter S_7 is a free choice that must be made in order to specify the location of the origin of the seventh coordinate system. A value of 800 mm will be used for this analysis.

The reverse-analysis problem statement is presented as follows:

given: $\mathbf{S}_7$ and the direction of $\mathbf{a}_{78}$ relative to $\mathbf{S}_7$ in order to establish the seventh coordinate system,
$^7\mathbf{P}_{tool}$: the location of the tool point in the seventh coordinate system,
$^F\mathbf{P}_{tool}$: the desired location of the tool point in the fixed coordinate system,
$^F\mathbf{S}_7$, $^F\mathbf{a}_{78}$: the desired orientation of the robot end effector, and
θ_2: the redundancy parameter,

find: $\phi_1, \theta_3, \theta_4, \theta_5, \theta_6, \theta_7$: the joint angle parameters that will position and orient the end effector as desired.

The solution to this problem proceeds as described in Chapter 5. From the given information, a slightly modified version of Eq. (5.3) can be used to determine the position of the origin of the seventh coordinate system measured in the fixed coordinate system. A hypothetical closure link is then created to form a closed-loop spatial mechanism. The link length a_{78} and the twist angle α_{78} were arbitrarily selected as zero and ninety degrees respectively. With these two choices, the direction of the vector $\mathbf{S}_8$ is known in terms of the fixed coordinate system. Further, the hypothetical eighth joint axis is known to pass through the origin of the seventh coordinate system.

Chapter 5 shows how the six close-the-loop parameters (S_8, S_1, a_{81}, θ_8, α_{81}, and γ_1) can now be determined. The analysis will proceed assuming that values for these six parameters have been calculated. Table 11.11 shows the mechanism parameters for the newly formed closed-loop spatial mechanism.

11.6.1 Development of an equivalent six-degree-of-freedom manipulator

The first step of this reverse kinematic analysis is to reduce the manipulator to an equivalent six-degree-of-freedom device. Shown in Figure 11.33a is a close-up drawing of the first three joints of the manipulator. The value of θ_2 has been specified by the user, and joint 3 becomes, in effect, the second unknown joint angle of the system. For this

Table 11.11. *Mechanism parameters for closed-loop SSRMS mechanism.*

Link length, mm.	Twist angle, deg.	Joint offset, mm.	Joint angle, deg.
$a_{12} = 0$	$\alpha_{12} = 90$	$S_1 = $ C.L.	$\phi_1 = $ variable
$a_{23} = 380$	$\alpha_{23} = 270$	$S_2 = 635$	$\theta_2 = $ user input
$a_{34} = 6850$	$\alpha_{34} = 0$	$S_3 = 504$	$\theta_3 = $ variable
$a_{45} = 6850$	$\alpha_{45} = 0$	$S_4 = 504$	$\theta_4 = $ variable
$a_{56} = 380$	$\alpha_{56} = 90$	$S_5 = 504$	$\theta_5 = $ variable
$a_{67} = 0$	$\alpha_{67} = 90$	$S_6 = 635$	$\theta_6 = $ variable
$a_{78} = 0$*	$\alpha_{78} = 90$*	$S_7 = 800$*	$\theta_7 = $ variable
$a_{81} = $ C.L.	$\alpha_{81} = $ C.L.	$S_8 = $ C.L.	$\theta_8 = $ C.L.

* = User-selected value
C.L. = Calculated during the close-the-loop procedure

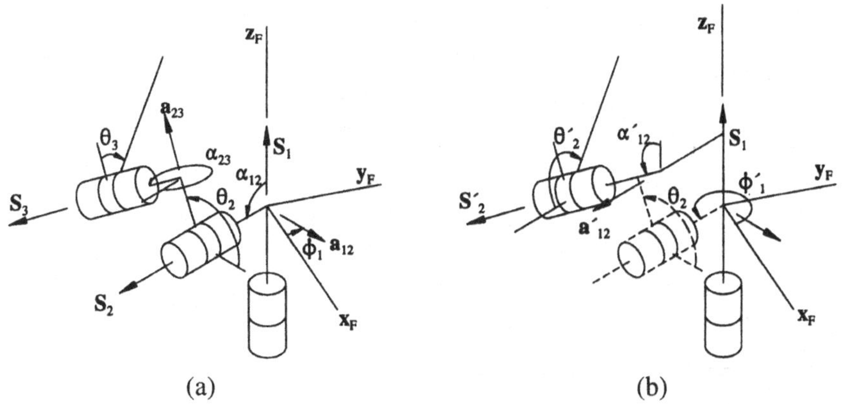

Figure 11.33. Development of equivalent six-axis manipulator.

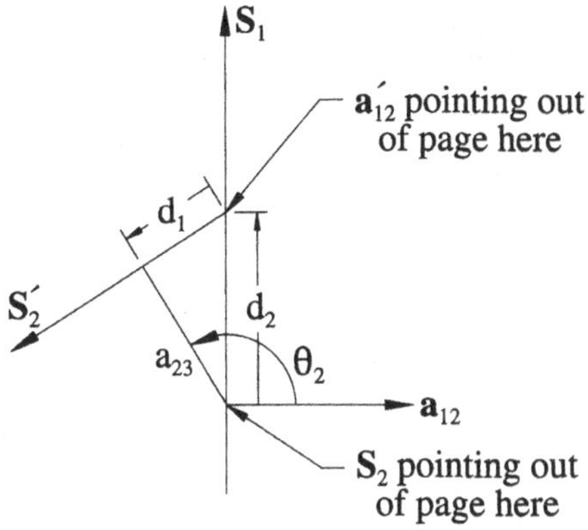

Figure 11.34. Determination of S'_1 and S'_2.

reason, the axis $\mathbf{S}_3$ has been relabeled as $\mathbf{S}'_2$ in Figure 11.33b. It is now necessary to determine

(1) the perpendicular distance between the joint axes $\mathbf{S}_1$ and $\mathbf{S}'_2 (a'_{12})$,
(2) the twist angle between the two axes $\mathbf{S}_1$ and $\mathbf{S}'_2 (\alpha'_{12})$,
(3) the current effective link length S'_2, and
(4) the current effective link length S'_1

as a function of the input parameter θ_2. With knowledge of these four values, an equivalent six-axis manipulator can be modeled.

Figure 11.34 shows a drawing of the first three joints of the manipulator with the axis $\mathbf{S}_2$ coming out of the page. It can be seen in the figure that the lengths d_1 and d_2 are given by

$$d_1 = -a_{23}/\tan\theta_2, \tag{11.160}$$

$$d_2 = a_{23}/\sin\theta_2. \tag{11.161}$$

Table 11.12. *Mechanism parameters for modified closed-loop SSRMS mechanism.*

Link length, mm.	Twist angle, deg.	Joint offset, mm.	Joint angle, deg.
$a_{12} = 635$	$\alpha_{12} = \theta_2^{**}$	$S_1 = C.L + d_2^{**}$	$\phi_1 =$ variable
$a_{23} = 6850$	$\alpha_{23} = 0$	$S_2 = 504 + d_1^{**}$	$\theta_2 =$ variable
$a_{34} = 6850$	$\alpha_{34} = 0$	$S_3 = 504$	$\theta_3 =$ variable
$a_{45} = 380$	$\alpha_{45} = 90$	$S_4 = 504$	$\theta_4 =$ variable
$a_{56} = 0$	$\alpha_{56} = 90$	$S_5 = 635$	$\theta_5 =$ variable
$a_{67} = 0^*$	$\alpha_{67} = 90^*$	$S_6 = 800^*$	$\theta_6 =$ variable
$a_{71} =$ C.L.	$\alpha_{71} =$ C.L.	$S_7 =$ C.L.	$\theta_7 =$ C.L.

* = User-selected value
C.L. = Calculated during the close-the-loop procedure
** = Calculated as a function of the original input angle θ_2

The new effective link lengths S_2' and S_1' can be determined as

$$S_2' = S_3 + d_1, \tag{11.162}$$

$$S_1' = S_1 + d_2. \tag{11.163}$$

From Figure 11.33b, it can be seen that

$$a_{12}' = S_2, \tag{11.164}$$

$$\alpha_{12}' = \theta_2, \tag{11.165}$$

$$\phi_1 = \phi_1' - 270°, \tag{11.166}$$

$$\theta_3 = \theta_2' - 90°. \tag{11.167}$$

By applying Eqs. (11.160) through (11.167), a new equivalent six-axis manipulator is constructed where certain link length and offset values are a function of θ_2. This manipulator is shown in Figure 11.35 without the use of the primed notation used in the equations. Table 11.12 shows the mechanism parameters for the equivalent six-axis manipulator after the close-the-loop procedure has been completed. Also, it should be noted that the values d_1 and d_2 approach infinity when θ_2 is near 0 or 180 degrees. If θ_2 approaches one of these values, the manipulator is in a special configuration because four joint axes are parallel. No solutions to the reverse kinematic problem are determined for this case.

11.6.2 Calculation of $\phi_{1\text{actual}}$

Throughout the remaining sections of the kinematic analysis, the primed notation for the equivalent six-axis manipulator will be discontinued. The joint angle values of the actual SSRMS will be distinguished from the joint angle numbers of the equivalent six-axis manipulator by writing the actual angles with an additional subscript, actual. For example, the fourth joint angle of the actual SSRMS will be written as $\theta_{4\text{actual}}$. The symbol θ_4 will refer to the value of the fourth joint angle of the equivalent six-axis manipulator.

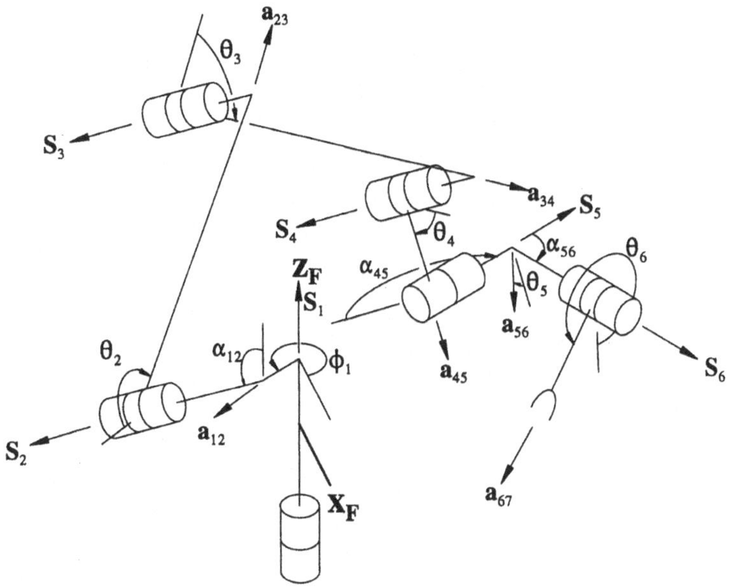

Figure 11.35. Equivalent six-axis manipulator.

The vector loop equation of the equivalent closed-loop spatial mechanism can be written

$$S_1\mathbf{S}_1 + S_2\mathbf{S}_2 + S_3\mathbf{S}_3 + S_4\mathbf{S}_4 + S_5\mathbf{S}_5 + S_6\mathbf{S}_6 + S_7\mathbf{S}_7 + a_{12}\mathbf{a}_{12} + a_{23}\mathbf{a}_{23} + a_{34}\mathbf{a}_{34} + a_{45}\mathbf{a}_{45} + a_{71}\mathbf{a}_{71} = \mathbf{0}. \quad (11.168)$$

This equation can be projected onto the vector $\mathbf{S}_2$ (set 14 of the table of direction cosines for a spherical heptagon in the appendix) to give the following equation:

$$S_6 Z_{71} + S_5 Z_{671} + S_4 Z_{5671} + S_3 Z_{45671} + S_2 + S_1 c_{12} + S_7 Z_1 + a_{45} U_{56712} + a_{34} U_{456712} + a_{71} U_{12} = 0. \quad (11.169)$$

Because the vectors $\mathbf{S}_2$, $\mathbf{S}_3$, and $\mathbf{S}_4$ are parallel, Eq. (11.169) may be rewritten as

$$S_6 Z_{71} + S_5 Z_{671} + (S_2 + S_3 + S_4) + S_1 c_{12} + S_7 Z_1 + a_{45} U_{56712} + a_{34} U_{456712} + a_{71} U_{12} = 0. \quad (11.170)$$

Substitution of the SSRMS mechanism parameters of the equivalent six-axis manipulator reduces Eq. (11.170) to the following expression:

$$c_1[-S_6 s_{12} c_{71} c_7 - S_7 s_{12} s_{71}] + s_1[S_6 s_{12} s_7 + a_{71} s_{12}] + [S_2 + S_3 + S_4 + S_1 c_{12} - S_6 c_{12} s_{71} c_7 + S_7 c_{12} c_{71}] = 0. \quad (11.171)$$

Equation (11.171) is the input/output equation for the closed-loop mechanism. The only unknown in the equation is θ_1. Two values of θ_1 can be determined that satisfy this equation. The solution tree of the manipulator to this point is shown in Figure 11.36.

The angle ϕ_1 can be calculated by subtracting the close-the-loop variable γ_1 from each

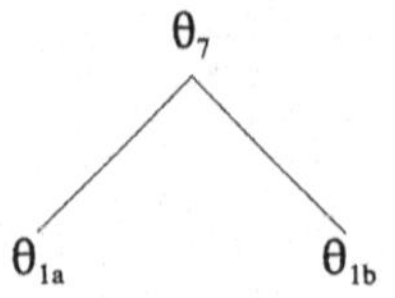

Figure 11.36. SSRMS solution tree.

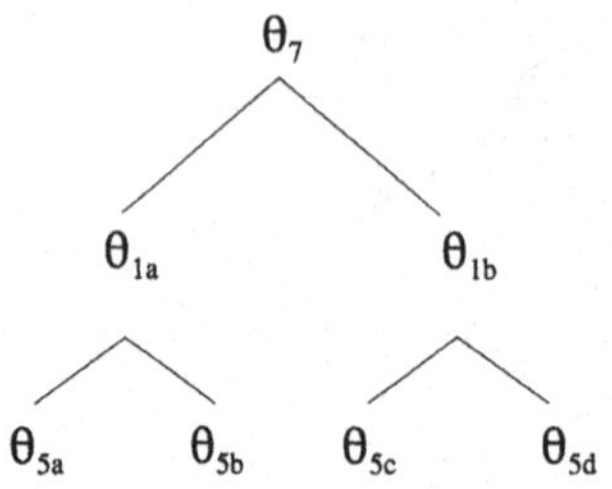

Figure 11.37. SSRMS solution tree.

value of θ_1. The angle $\phi_{1\text{actual}}$, the first joint angle of the actual manipulator, can then be found from Eq. (11.166), where the angles ϕ_1' and ϕ_1 in that equation are respectively the angles ϕ_1 and $\phi_{1\text{actual}}$ referred to here.

11.6.3 Calculation of θ_5 ($\theta_{6\text{actual}}$)

θ_5 can be determined from the following spherical cosine law for a spatial heptagon:

$$Z_{7123} = Z_5. \tag{11.172}$$

This equation reduces to the following result because the vectors $\mathbf{S}_2$, $\mathbf{S}_3$, and $\mathbf{S}_4$ are parallel:

$$Z_{71} = Z_5. \tag{11.173}$$

Substitution of the mechanism parameter values into Eq. (11.173) yields the following expression for θ_5:

$$c_5 = -s_{12}[s_7 s_1 - c_{71} c_7 c_1] + c_{12} s_{71} c_7. \tag{11.174}$$

Two values of θ_5 can be determined that satisfy this equation. The current solution tree for the manipulator is shown in Figure 11.37.

11.6.4 Calculation of θ_6 ($\theta_{7\text{actual}}$)

Two equations that contain θ_6 can be written as follows:

$$X_{17} = X_{56}, \tag{11.175}$$

$$Y_{17} = -X_{56}^*. \tag{11.176}$$

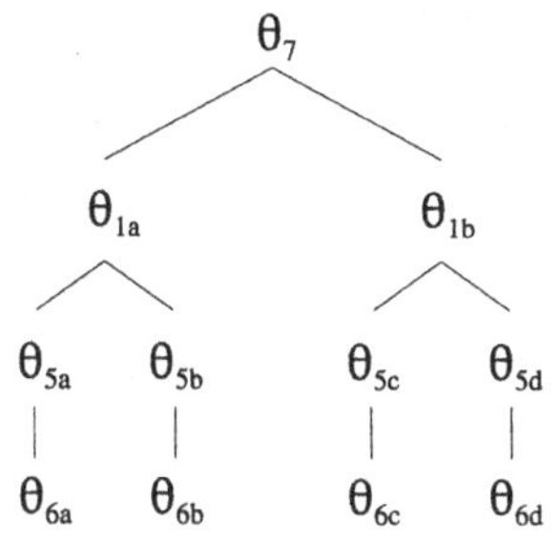

Figure 11.38. SSRMS solution tree.

These equations are spherical sine and sine–cosine laws for a spherical pentagon and can be used in this case because vectors $\mathbf{S}_2$, $\mathbf{S}_3$, and $\mathbf{S}_4$ are parallel.

Substitution of the mechanism parameters into Eqs. (11.175) and (11.176) and solving for the sine and cosine of θ_6 gives

$$c_6 = X_{17}/s_5, \tag{11.177}$$

$$s_6 = -Y_{17}/s_5. \tag{11.178}$$

One value of θ_6 will simultaneously satisfy Eqs. (11.177) and (11.178). The solution tree for the manipulator is shown in Figure 11.38. It must be noted that these equations cannot be solved for θ_6 if θ_5 equals 0 or π. If this occurs, four axes of the manipulator will become parallel and the manipulator is in a special configuration.

11.6.5 Calculation of θ_3 ($\theta_{4\text{actual}}$)

The vector loop equation for the closed-loop mechanism was written as Eq. (11.168). In order to solve for θ_1, this equation was projected onto the $\mathbf{S}_2$ vector. In order to solve for θ_3, Eq. (11.168) will be projected onto the $\mathbf{a}_{12}$ vector and also onto a direction perpendicular to $\mathbf{a}_{12}$ and $\mathbf{S}_2$. These two equations may be written as

$$S_6X_{71} + S_5X_{671} + S_7X_1 + a_{45}W_{5671} + a_{12} + a_{71}c_1 = -a_{34}W_{45671} - a_{23}c_2, \tag{11.179}$$

$$S_6Y_{71} + S_5Y_{671} - S_1s_{12} + S_7Y_1 - a_{45}U^*_{56712} + a_{71}s_1c_{12} = a_{34}U^*_{456712} + a_{23}s_2. \tag{11.180}$$

The left sides of Eqs. (11.179) and (11.180) contain only known mechanism parameters and joint angle values. The right sides of the equations contain θ_2 and θ_3.

Two new terms, P_{5671} and Q_{5671}, will represent the left-hand sides of Eqs (11.179) and (11.180) and are defined as follows:

$$P_{5671}=S_6X_{71} + S_5X_{671} + S_7X_1 + a_{45}W_{5671} + a_{12} + a_{71}c_1, \tag{11.181}$$

$$Q_{5671}=S_6Y_{71} + S_5Y_{671} - S_1s_{12} + S_7Y_1 - a_{45}U^*_{56712} + a_{71}s_1c_{12}. \tag{11.182}$$

The terms on the right-hand side of Eqs. (11.179) and (11.180) may be expanded as

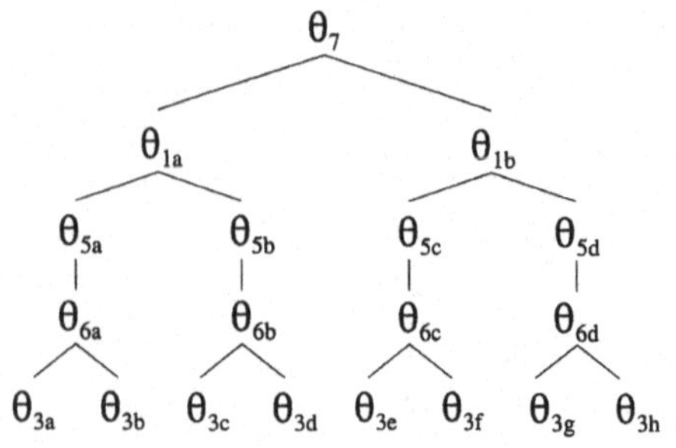

Figure 11.39. SSRMS solution tree.

follows:

$$U^*_{456712} = -V_{32} = s_2c_3 + c_2s_3c_{23}, \tag{11.183}$$

$$W_{45671} = W_{32} = c_2c_3 - s_2s_3c_{23}. \tag{11.184}$$

Substitution of these terms with $\alpha_{23} = 0°$ into Eqs. (11.179) and (11.180) results in the following two equations:

$$P_{5671} = -a_{34}c_{2+3} - a_{23}c_2, \tag{11.185}$$

$$Q_{5671} = a_{34}s_{2+3} + a_{23}s_2, \tag{11.186}$$

where c_{2+3} and s_{2+3} represent the cosine and sine of $(\theta_2 + \theta_3)$.

Squaring and adding Eqs. (11.185) and (11.186) gives

$$P^2_{5671} + Q^2_{5671} = a^2_{34} + a^2_{23} + 2a_{34}a_{23}c_3. \tag{11.187}$$

Equation (11.187) may be solved to yield the value of $\cos\theta_3$. Two values of θ_3 can satisfy this equation, and the current solution tree for the manipulator is shown in Figure 11.39.

11.6.6 Calculation of $\theta_2(\theta_{3actual})$

Equations (11.185) and (11.186) can be used to determine corresponding values for θ_2. Multiplying Eq. (11.185) by $-(a_{34}c_3 + a_{23})$ and Eq. (11.186) by $(a_{34}s_3)$ and summing gives

$$\left[(a_{34}c_3 + a_{23})^2 + (a_{34}s_3)^2\right]c_2 = -(a_{34}c_3 + a_{23})P_{5671} + (a_{34}s_3)Q_{5671}. \tag{11.188}$$

Multiplying Eq. (11.186) by $(a_{34}c_3 + a_{23})$ and Eq. (11.185) by $(a_{34}s_3)$ and summing gives

$$\left[(a_{34}c_3 + a_{23})^2 + (a_{34}s_3)^2\right]s_2 = (a_{34}c_3 + a_{23})Q_{5671} + (a_{34}s_3)P_{5671}. \tag{11.189}$$

Equations (11.188) and (11.189) can be used to calculate a corresponding value for θ_2. Equation (11.167) is then used to determine the angle $\theta_{3actual}$, where the angles θ_2' and θ_3 in that equation are respectively the angles θ_2 and $\theta_{3actual}$ referred to here. The current solution tree for the manipulator is shown in Figure 11.40.

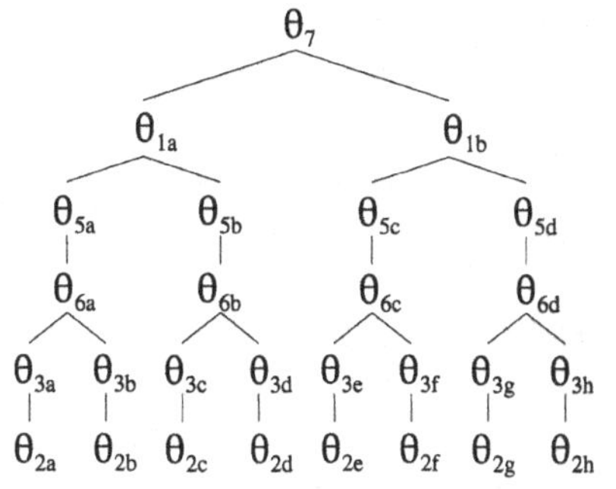

Figure 11.40. SSRMS solution tree.

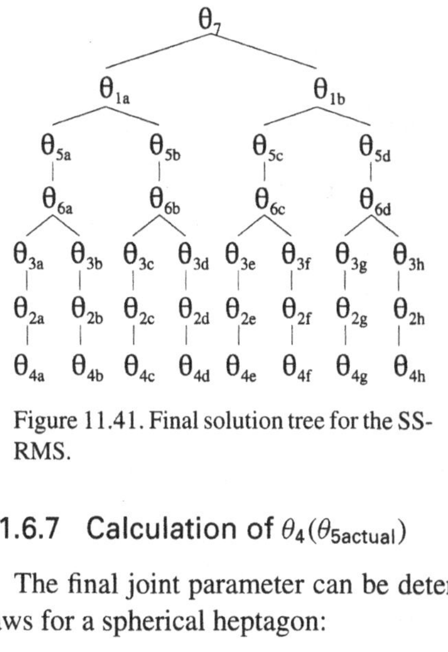

Figure 11.41. Final solution tree for the SS-RMS.

11.6.7 Calculation of $\theta_4(\theta_{5\text{actual}})$

The final joint parameter can be determined from the following sine and sine–cosine laws for a spherical heptagon:

$$X_{67123} = s_{45}s_4, \tag{11.190}$$

$$Y_{67123} = s_{45}c_4. \tag{11.191}$$

Substituting $\alpha_{45} = 90^\circ$ gives

$$s_4 = X_{67123}, \tag{11.192}$$

$$c_4 = Y_{67123}. \tag{11.193}$$

A corresponding value of θ_4 can be found from these two equations. The final solution tree shown in Figure 11.41 indicates that there are eight possible configurations of the manipulator.

The reverse kinematic analysis of the SSRMS is complete. It has been shown that eight solution configurations exist for a given value of θ_2 and a given position and orientation of the manipulator. A similar outcome of eight solution configurations will result if a different angle from θ_2 is given as an input parameter, again assuming that θ_2 is not close to 0 or 180 degrees.

Table 11.13. *Eight solution sets for the SSRMS (angles in degrees).*

Solution	ϕ_1	θ_3	θ_4	θ_5	θ_6	θ_7
A	167.297	−90.076	54.169	−98.092	73.628	−37.030
B	167.297	−35.907	−54.169	−43.923	73.628	−37.030
C	167.297	−103.164	68.847	80.317	−73.628	142.970
D	167.297	−34.317	−68.847	149.164	−73.628	142.970
E	23.873	51.782	39.154	−52.505	28.260	11.479
F	23.873	−269.065	−39.154	−13.352	28.260	11.479
G	23.873	45.626	38.338	134.466	−28.260	−168.521
H	23.873	83.964	−38.338	172.804	−28.260	−168.521

11.6.8 Numerical example

As a numerical example, the following information was specified for the SSRMS:

$$S_7 = 800\,\text{mm.,} \quad {}^7\mathbf{P}_{\text{tool}} = \begin{bmatrix} 100 \\ 50 \\ 60 \end{bmatrix} \text{mm.,} \quad {}^F\mathbf{P}_{\text{tool}} = \begin{bmatrix} -600.0 \\ 12400.0 \\ 3500.0 \end{bmatrix} \text{mm.,}$$

$${}^F\mathbf{S}_7 = \begin{bmatrix} 0.5774 \\ 0.5774 \\ -0.5774 \end{bmatrix}, \quad {}^F\mathbf{a}_{78} = \begin{bmatrix} 0.2673 \\ 0.5345 \\ 0.8018 \end{bmatrix}.$$

The angle θ_2 was chosen to be 30.0 degrees. Table 11.13 shows the results of the reverse-position analysis.

11.7 Modified flight telerobotic servicer (FTS) manipulator system

The original design for the flight telerobotic servicer (FTS) manipulator system consists of a series of links connected by seven revolute joints. The first and second joint axes intersect; the third, fourth, and fifth axes are parallel; and the sixth and seventh joint axes intersect (see Figure 11.42). The configuration of this manipulator is very similar to that of the space station remote manipulator system that was analyzed in the previous section.

A modification was made to the basic FTS manipulator design whereby the first and second joint axes would still intersect when only the third and fourth axes are parallel, and then the fifth and sixth axes would intersect. A drawing of this new configuration is shown in Figure 11.43.

Because the modified FTS manipulator system is a seven-degree-of-freedom device, it will be assumed that the angle θ_7 is specified along with the desired position and orientation of the end effector and that the remaining joint angles ($\phi_1, \theta_2, \theta_3, \theta_4, \theta_5$, and θ_6) must be solved for.

A kinematic model of the modified FTS is shown in Figure 11.44, with joint axis vectors and link vectors labeled. The mechanism parameters are listed in Table 11.14.

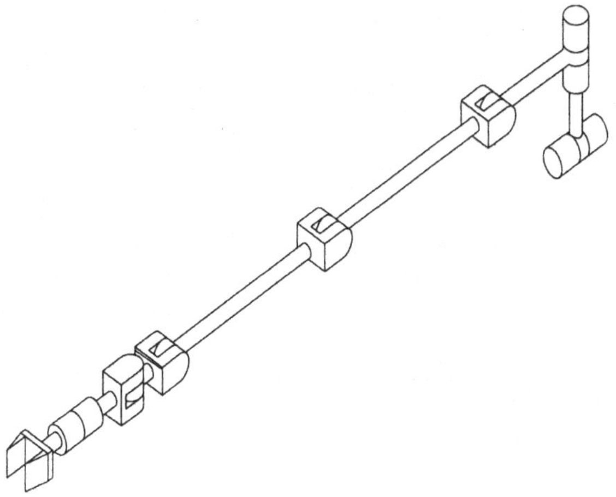

Figure 11.42. Flight telerobotic servicer.

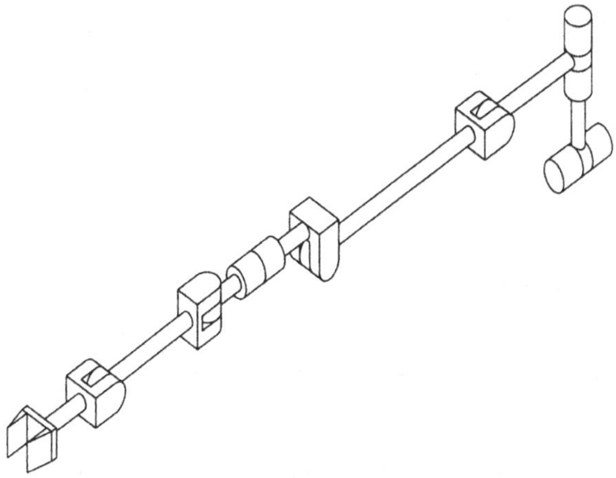

Figure 11.43. Modified flight telerobotic servicer.

As with the other manipulators, the parameter S_7 is a free choice that must be made in order to specify the location of the origin of the seventh coordinate system. A value of zero will be used for this analysis.

The reverse analysis problem statement (see Crane (1992)) is presented as follows:

given: S_7 and the direction of $\mathbf{a}_{78}$ relative to $\mathbf{S}_7$ in order to establish the seventh coordinate system,

$^7\mathbf{P}_{\text{tool}}$: the location of the tool point in the seventh coordinate system,

$^F\mathbf{P}_{\text{tool}}$: the desired location of the tool point in the fixed coordinate system,

$^F\mathbf{S}_7$, $^F\mathbf{a}_{78}$: the desired orientation of the robot end effector, and

θ_7: the redundancy parameter,

find: $\phi_1, \theta_2, \theta_3, \theta_4, \theta_5, \theta_6$: the joint angle parameters that will position and orient the end effector as desired.

Table 11.14. *Mechanism parameters for modified FTS.*

Link length, in.	Twist angle, deg.	Joint offset, in.	Joint angle, deg.
$a_{12} = 0$	$\alpha_{12} = 90$		$\phi_1 = $ variable
$a_{23} = 9$	$\alpha_{23} = 90$	$S_2 = 6.55$	$\theta_2 = $ variable
$a_{34} = 18$	$\alpha_{34} = 0$	$S_3 = 0$	$\theta_3 = $ variable
$a_{45} = 2.62$	$\alpha_{45} = 90$	$S_4 = 0$	$\theta_4 = $ variable
$a_{56} = 0$	$\alpha_{56} = 90$	$S_5 = 18$	$\theta_5 = $ variable
$a_{67} = 4$	$\alpha_{67} = 90$	$S_6 = 0$	$\theta_6 = $ variable
			$\theta_7 = $ user input

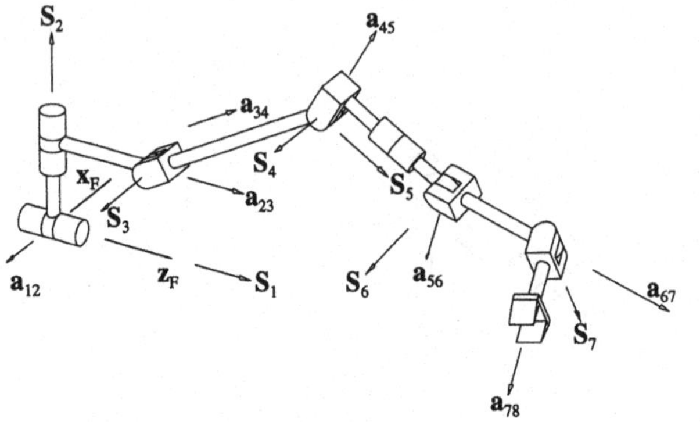

Figure 11.44. Kinematic diagram of modified FTS.

11.7.1 Development of an equivalent six-degree-of-freedom manipulator

The solution to the reverse position problem will proceed as described in Chapter 5. From the given information, a slightly modified version of Eq. (5.3) can be used to determine the position of the origin of the seventh coordinate system measured in the fixed coordinate system, ${}^{F}\mathbf{P}_{7\text{orig}}$, as follows:

$$ {}^{F}\mathbf{P}_{7\text{orig}} = {}^{F}\mathbf{P}_{\text{tool}} - ({}^{7}\mathbf{P}_{\text{tool}} \cdot \mathbf{i})^{F}\mathbf{a}_{78} - ({}^{7}\mathbf{P}_{\text{tool}} \cdot \mathbf{j})^{F}\mathbf{S}_7 \times {}^{F}\mathbf{a}_{78} - ({}^{7}\mathbf{P}_{\text{tool}} \cdot \mathbf{k})^{F}\mathbf{S}_7. \quad (11.194)$$

With these coordinates known, the transformation matrix ${}^{F}_{7}\mathrm{T}$ can be written as

$$ {}^{F}_{7}\mathbf{T} = \begin{bmatrix} {}^{F}\mathbf{a}_{78} & {}^{F}\mathbf{S}_7 \times {}^{F}\mathbf{a}_{78} & {}^{F}\mathbf{S}_7 & {}^{F}\mathbf{P}_{7\text{orig}} \\ 0 & 0 & 0 & 1 \end{bmatrix}. \quad (11.195)$$

The transformation that relates the sixth coordinate system to the fixed can be calculated from

$$ {}^{F}_{6}\mathbf{T} = {}^{6}_{7}\mathbf{T}\,{}^{F}_{7}\mathbf{T}, \quad (11.196)$$

where ${}^6_7\mathbf{T}$ is defined by Eq. (3.7) as

$$ {}^6_7\mathbf{T} = \begin{bmatrix} c_7 & -s_7 & 0 & a_{67} \\ s_7 c_{67} & c_7 c_{67} & -s_{67} & -s_{67} S_7 \\ s_7 s_{67} & c_7 s_{67} & c_{67} & -c_{67} S_7 \\ 0 & 0 & 0 & 1 \end{bmatrix}. \quad (11.197) $$

Because the matrix ${}^F_6\mathbf{T}$ is known, the coordinates of the origin of the sixth coordinate system are known in terms of the fixed coordinate system, that is, ${}^F\mathbf{P}_{6\text{orig}}$. Also, the orientation vectors ${}^F\mathbf{a}_{67}$ and ${}^F\mathbf{S}_6$ are known.

The reverse-analysis problem statement can now be restated as follows:

given: ${}^F\mathbf{P}_{6\text{orig}}$: the location of the origin of the sixth coordinate system measured with respect to the fixed coordinate system and
${}^F\mathbf{S}_6$, ${}^F\mathbf{a}_{67}$: the orientation of the sixth coordinate system,

find: $\phi_1, \theta_2, \theta_3, \theta_4, \theta_5, \theta_6$: the joint angle parameters that will position and orient the sixth coordinate system as desired.

For typical six-axis manipulators such as the Puma, GE P60 , and Cincinnati Milacron T3-776 manipulators, the values of α_{67} and a_{67} are free choices that may be arbitrarily selected. The same condition holds true in this case, and the twist angle α_{67} will be selected as ninety degrees and the link length a_{67} will be chosen as zero.

Chapter 5 shows how the six close-the-loop parameters (S_7, S_1, a_{71}, θ_7, α_{71}, and γ_1) can now be determined. The analysis will proceed assuming that values for these six parameters have been calculated. Table 11.15 shows the mechanism parameters for the newly formed closed-loop spatial mechanism. Note that the calculated value for θ_7 from the close-the-loop procedure will be used in future calculations as opposed to the original user-inputted value of θ_7 for the original manipulator.

Table 11.15. *Mechanism parameters for closed-loop modified FTS mechanism.*

Link length, in.	Twist angle, deg.	Joint offset, in.	Joint angle, deg.
$a_{12} = 0$	$\alpha_{12} = 90$	$S_1 = $ C.L.	$\phi_1 = $ variable
$a_{23} = 9$	$\alpha_{23} = 90$	$S_2 = 6.55$	$\theta_2 = $ variable
$a_{34} = 18$	$\alpha_{34} = 0$	$S_3 = 0$	$\theta_3 = $ variable
$a_{45} = 2.62$	$\alpha_{45} = 90$	$S_4 = 0$	$\theta_4 = $ variable
$a_{56} = 0$	$\alpha_{56} = 90$	$S_5 = 18$	$\theta_5 = $ variable
$a_{67} = 0^*$	$\alpha_{67} = 90^*$	$S_6 = 0$	$\theta_6 = $ variable
$a_{71} = $ C.L.	$\alpha_{71} = $ C.L.	$S_7 = $ C.L.	$\theta_7 = $ C.L.

* = User-selected value
C.L. = Calculated during the close-the-loop procedure

11.7.2 Expansion of required equations

In order to solve for the joint angle parameter ϕ_1, it is necessary to obtain an equation that contains θ_1 as its only unknown. Subtracting the known value γ_1 from each of the values of θ_1 that satisfy this equation will yield the possible values for ϕ_1.

The analysis begins by listing the following seven equations that contain the unknowns $\theta_1, \theta_2, \theta_4$, and θ_6:

(i) *Projection of vector loop equation on the vector* $\mathbf{S}_3$.

The vector loop equation for the closed-loop spatial mechanism can be written as

$$S_1\mathbf{S}_1 + S_2\mathbf{S}_2 + S_5\mathbf{S}_5 + S_7\mathbf{S}_7 + a_{23}\mathbf{a}_{23} + a_{34}\mathbf{a}_{34} + a_{45}\mathbf{a}_{45} + a_{71}\mathbf{a}_{71} = \mathbf{0}. \tag{11.198}$$

Projecting this equation onto the vector $\mathbf{S}_3$ and expanding the scalar products using the direction cosines of a spatial heptagon (set 13 and set 3 of the appendix) gives the equation

$$S_1Z_2 + S_2c_{23} + S_5\bar{Z}_4 + S_7Z_{12} + a_{45}U_{43} + a_{71}U_{123} = 0. \tag{11.199}$$

Substituting the constant mechanism parameters reduces Eq. (11.199) to

$$-S_1c_2 + S_7(s_1s_2s_{71} - c_2c_{71}) + a_{71}s_2c_1 = 0. \tag{11.200}$$

Rearranging this equation yields

$$(a_{71}c_1 + S_7s_1s_{71})s_2 + (-S_1 - S_7c_{71})c_2 = 0. \tag{11.201}$$

(ii) *Secondary sine law.*

A secondary sine law may be written as

$$X_{06712} = X_{043}. \tag{11.202}$$

The right-hand side of Eq. (11.202) can be expanded as follows:

$$X_{043} = -S_3(\bar{X}_4s_3 + \bar{Y}_4c_3) + c_3\bar{X}_{04} - s_3\bar{Y}_{04}, \tag{11.203}$$

where

$$\bar{X}_{04} = S_4(s_{45}c_4) + a_{45}(c_{45}s_4), \tag{11.204}$$

$$\bar{Y}_{04} = S_4(c_{34}s_{45}s_4) + a_{45}(s_{34}s_{45} - c_{34}c_{45}c_4) - a_{34}(c_{34}c_{45} - s_{34}s_{45}c_4). \tag{11.205}$$

Substituting the constant mechanism parameters yields $X_{043} = 0$.

The left side of Eq. (11.202) can be expanded as follows:

$$X_{06712} = S_2(-X_{671}s_2 - Y_{671}c_2) + c_2X_{0671} - s_2Y_{0671}, \tag{11.206}$$

where

$$X_{0671} = S_1(-X_{67}s_1 - Y_{67}c_1) + c_1X_{067} - s_1Y_{067}, \tag{11.207}$$

$$Y_{0671} = S_1c_{12}(X_{67}c_1 - Y_{67}s_1) - a_{12}Z_{671} + c_{12}(X_{067}s_1 + Y_{067}c_1) - s_{12}Z_{067}, \tag{11.208}$$

$$X_{067} = S_7(-X_6s_7 - Y_6c_7) + c_7X_{06} - s_7Y_{06}, \tag{11.209}$$

$$Y_{067} = S_7c_{71}(X_6c_7 - Y_6s_7) - a_{71}Z_{67} + c_{71}(X_{06}s_7 + Y_{06}c_7) - s_{71}Z_{06}, \tag{11.210}$$

$$Z_{067} = S_7s_{71}(X_6c_7 - Y_6s_7) + a_{71}Y_{67} + s_{71}(X_{06}s_7 + Y_{06}c_7) + c_{71}Z_{06}, \tag{11.211}$$

$$X_{06} = S_6s_{56}c_6 + a_{56}c_{56}s_6, \tag{11.212}$$

$$Y_{06} = S_6c_{67}s_{56}s_6 + a_{56}(s_{67}s_{56} - c_{67}c_{56}c_6) - a_{67}Z_6, \tag{11.213}$$

$$Z_{06} = S_6s_{67}s_{56}s_6 + a_{56}\bar{Y}_6 + a_{67}Y_6. \tag{11.214}$$

Substitution of the constant mechanism parameters into Eqs. (11.206) through (11.214) reduces Eq. (11.202) to

$$\begin{aligned} &S_2[c_2(s_{71}s_6s_7 - c_{71}c_6) + s_2(-s_6c_7c_1 + s_1(c_{71}s_6s_7 + s_{71}c_6))] + S_1c_2[c_1(-c_{71}s_6s_7 \\ &\quad - s_{71}c_6) - s_6c_7s_1] + S_7[c_2(-c_1s_6s_7 - s_1c_{71}s_6c_7) + s_2s_{71}s_6c_7] \\ &\quad + a_{71}[c_2s_1(s_{71}s_6s_7 - c_{71}c_6) + s_2(c_{71}s_6s_7 + s_{71}c_6)] = 0. \end{aligned} \tag{11.215}$$

Regrouping this equation gives

$$\begin{aligned} &[S_2(-s_6c_7c_1 + s_1(c_{71}s_6s_7 + s_{71}c_6)) + S_7s_{71}s_6c_7 + a_{71}(c_{71}s_6s_7 + s_{71}c_6)]s_2 \\ &\quad + [S_2(s_{71}s_6s_7 - c_{71}c_6) + S_1(c_1(-c_{71}s_6s_7 - s_{71}c_6) - s_6c_7s_1) \\ &\quad + S_7(-c_1s_6s_7 - s_1c_{71}s_6c_7) + a_{71}s_1(s_{71}s_6s_7 - c_{71}c_6)]c_2 = 0. \end{aligned} \tag{11.216}$$

(iii) *Secondary cosine law.*

A secondary cosine law may be written as

$$Z_{0671} = Z_{043}. \tag{11.217}$$

The right side of Eq. (11.217) may be expanded as follows:

$$\begin{aligned} Z_{043} = {} & a_{23}(c_{23}(\bar{X}_4s_3 + \bar{Y}_4c_3) - s_{23}\bar{Z}_4) \\ & + S_3s_{23}(\bar{X}_4c_3 - \bar{Y}_4s_3) + s_{23}(\bar{X}_{04}s_3 + \bar{Y}_{04}c_3) + c_{23}\bar{Z}_{04}, \end{aligned} \tag{11.218}$$

where

$$\bar{Z}_{04} = S_4(s_{34}s_{45}s_4) - a_{34}(s_{34}c_{45} + c_{34}s_{45}c_4) - a_{45}(c_{34}s_{45} + s_{34}c_{45}c_4) \tag{11.219}$$

and the terms $\bar{X}_{04}$, and $\bar{Y}_{04}$ are defined in Eqs. (11.204) and (11.205). Substitution of the constant mechanism parameters yields $Z_{043} = 0$.

The term Z_{0671} can be expanded as follows:

$$Z_{0671} = S_1 s_{12}(X_{67}c_1 - Y_{67}s_1) + a_{12}Y_{671} + s_{12}(X_{067}s_1 + Y_{067}c_1) + c_{12}Z_{067}, \quad (11.220)$$

where the terms X_{067}, Y_{067}, and Z_{067} are defined in Eqs. (11.209) through (11.211).
Substitution of the constant mechanism parameters into Eq. (11.220) yields

$$S_1[s_6c_7c_1 - s_1(c_{71}s_6s_7 + s_{71}c_6)] + S_7[c_{71}s_6c_7c_1 - s_6s_7s_1]$$
$$+ a_{71}c_1[-s_{71}s_6s_7 + c_{71}c_6] = 0. \quad (11.221)$$

Rearranging this equation yields

$$[S_1(c_7c_1 - s_1c_{71}s_7) + S_7(c_{71}c_7c_1 - s_7s_1) + a_{71}(-c_1s_{71}s_7)]s_6$$
$$+ [-S_1s_1s_{71} + a_{71}c_1c_{71}]c_6 = 0. \quad (11.222)$$

(iv) *Spherical cosine law.*

A spherical cosine law may be written for the closed-loop mechanism as

$$Z_{6712} = \bar{Z}_4. \quad (11.223)$$

The right-hand side of this equation reduces to zero upon substitution of the constant mechanism parameters. Substituting the constant mechanism parameters into the left side of this equation yields

$$(s_6c_7c_1 - (c_{71}s_6s_7 + s_{71}c_6)s_1)s_2 + (-s_{71}s_6s_7 + c_{71}c_6)c_2 = 0. \quad (11.224)$$

(v) *Projection of the vector loop equation on the vector* $\mathbf{S}_5$.

The vector loop equation for the closed-loop mechanism is listed in Eq. (11.198). Projecting this equation onto the vector $\mathbf{S}_5$ and expanding the scalar products using the direction cosines of a spherical heptagon (set 11 and set 5 of the appendix) gives the equation

$$S_1Z_{76} + S_2Z_{176} + S_5 + S_7Z_6 + a_{23}U_{21765} + a_{34}U_{45} + a_{71}U_{765} = 0. \quad (11.225)$$

Substituting the constant mechanism parameters into this equation and transfering the term $a_{34}U_{45}$ to the right-hand side yields

$$S_1[s_{71}s_6s_7 - c_{71}c_6] + S_2[(s_1c_7 + c_{71}c_1s_7)s_6 + s_{71}c_1c_6] + S_5 - S_7c_6 + a_{23}[s_2(-c_6c_{71}$$
$$+ s_6s_7s_{71}) + c_2(-s_1c_6s_{71} - s_1s_6s_7c_{71} + c_1s_6c_7)] + a_{71}s_6c_7 = -a_{34}s_4. \quad (11.226)$$

(vi) *Secondary cosine law.*

A secondary cosine law for the closed-loop mechanism may be written as

$$Z_{06712} = Z_{04}. \quad (11.227)$$

The right-hand side of Eq. (11.227) may be expanded as follows:

$$Z_{04} = S_4(s_{45}s_{34}s_4) - a_{45}(s_{45}c_{34} + c_{45}s_{34}c_4) - a_{34}(c_{45}s_{34} + s_{45}c_{34}c_4). \quad (11.228)$$

Substitution of the constant mechanism parameters reduces Eq. (11.228) to

$$Z_{04} = -a_{45} - a_{34}c_4. \quad (11.229)$$

Expanding the left side of Eq. (11.227) yields

$$Z_{06712} = S_2 s_{23}[X_{671}c_2 - Y_{671}s_2] + a_{23}Y_{6712} + s_{23}(X_{0671}s_2 + Y_{0671}c_2) + c_{23}Z_{0671}. \quad (11.230)$$

The terms X_{0671} and Y_{0671} are defined in Eqs. (11.207) and (11.208), whereas Z_{0671} is defined in Eq. (11.220).

Substitution of the constant mechanism parameters into Eq. (11.230) and equating the result with that of Eq. (11.229) yields

$$\begin{aligned} &a_{23}[-s_6c_7s_1 - c_1(c_{71}s_6s_7 + s_{71}c_6)] + c_2[S_2(s_6c_7c_1 - s_1(c_{71}s_6s_7 + s_{71}c_6)) \\ &\quad - S_7s_{71}s_6c_7 - a_{71}(c_{71}s_6s_7 + s_{71}c_6)] + s_2[S_2(s_{71}s_6s_7 - c_{71}c_6) \\ &\quad + S_1(-s_6c_7s_1 - c_1(c_{71}s_6s_7 + s_{71}c_6)) + S_7(-c_1s_6s_7 - s_1c_{71}s_6c_7) \\ &\quad + a_{71}s_1(s_{71}s_6s_7 - c_{71}c_6)] = -a_{45} - a_{34}c_4. \end{aligned} \quad (11.231)$$

(vii) *Self-scalar product.*

Equation (11.198), the vector loop equation, may be rearranged as

$$S_1\mathbf{S}_1 + S_2\mathbf{S}_2 + S_7\mathbf{S}_7 + a_{23}\mathbf{a}_{23} + a_{71}\mathbf{a}_{71} = -a_{34}\mathbf{a}_{34} - a_{45}\mathbf{a}_{45} - S_5\mathbf{S}_5. \quad (11.232)$$

Projecting the left and right sides of Eq. (11.232) upon themselves, dividing both sides of the equation by two, and expressing the individual scalar products according to the direction cosines listed in the appendix yields

$$\begin{aligned} &K + S_1(S_2c_{12} + S_7c_{71} + a_{23}U_{21}) + S_2(S_7Z_1 + a_{71}U_{12}) \\ &\quad + a_{23}(S_7X_{12} + a_{71}W_{12}) = a_{34}(a_{45}c_4 + S_5U_{45}), \end{aligned} \quad (11.233)$$

where

$$K = \frac{1}{2}(S_1^2 + S_2^2 + S_7^2 + a_{23}^2 + a_{71}^2 - a_{34}^2 - a_{45}^2 - S_5^2). \quad (11.234)$$

Substitution of the constant mechanism parameters reduces Eq. (11.233) to

$$\begin{aligned} &K + S_1(S_7c_{71} + a_{23}s_2) + S_2(-S_7s_{71}c_1 + a_{71}s_1) \\ &\quad + a_{23}(S_7(s_2c_{71} + s_1c_2s_{71}) + a_{71}c_1c_2) = a_{34}(a_{45}c_4 + S_5s_4). \end{aligned} \quad (11.235)$$

11.7.3 Determination of ϕ_1

The analysis proceeds by manipulating Eqs. (11.201), (11.216), (11.224), (11.226), (11.231), and (11.235) until they reduce to one equation that contains the variables θ_1 and θ_6. This new equation, when used together with Eq. (11.222) (which also only contains θ_1 and θ_6), will yield a sixteenth-degree input/output equation in the variable θ_1.

The procedure begins by multiplying Eq. (11.224) by S_2 and adding it to Eq. (11.216) to yield

$$[S_7 s_{71} s_6 c_7 + a_{71}(c_{71} s_6 s_7 + s_{71} c_6)]s_2 + [S_1(c_1(-c_{71} s_6 s_7 - s_{71} c_6) - s_6 c_7 s_1) + S_7(-c_1 s_6 s_7 - s_1 c_{71} s_6 c_7) + a_{71} s_1 (s_{71} s_6 s_7 - c_{71} c_6)]c_2 = 0. \tag{11.236}$$

Adding c_2 times Eq. (11.236) to s_2 times Eq. (11.231) gives

$$a_{71} s_1 (s_6 s_7 s_{71} - c_6 c_{71}) + S_7(-c_1 s_6 s_7 - s_1 s_6 c_7 c_{71}) + S_1(-c_1(s_6 s_7 c_{71} + s_{71} c_6) - s_1 s_6 c_7) + a_{23} s_2 (-s_1 s_6 c_7 + c_1(-s_6 s_7 c_{71} - c_6 s_{71})) + S_2\left[(s_6 s_7 s_{71} - c_6 c_{71})s_2^2 + (c_1 s_6 c_7 - s_1(c_6 s_{71} + s_6 s_7 c_{71}))s_2 c_2\right] = -(a_{45} + a_{34} c_4)s_2. \tag{11.237}$$

Subtracting $S_2 c_2$ times Eq. (11.224) from Eq. (11.237) yields

$$a_{71} s_1 (s_6 s_7 s_{71} - c_6 c_{71}) + S_7(-c_1 s_6 s_7 - s_1 s_6 c_7 c_{71}) + S_1(-c_1(s_6 s_7 c_{71} + s_{71} c_6) - s_1 s_6 c_7) + a_{23} s_2 (-s_1 s_6 c_7 + c_1(-s_6 s_7 c_{71} - c_6 s_{71})) + S_2(s_6 s_7 s_{71} - c_6 c_{71}) = -(a_{45} + a_{34} c_4)s_2. \tag{11.238}$$

The analysis proceeds by multiplying Eq. (11.226) by s_2 and Eq. (11.224) by $-a_{23}c_2$. Summing the results and substituting $s_2^2 + c_2^2 = 1$ gives

$$s_2[S_1(s_{71} s_6 s_7 - c_{71} c_6) + S_2((s_1 c_7 + c_{71} c_1 s_7)s_6 + s_{71} c_1 c_6) + S_5 - S_7 c_6 + a_{71} s_6 c_7] + a_{23}(-c_6 c_{71} + s_6 s_7 s_{71}) = -a_{34} s_4 s_2. \tag{11.239}$$

Subtracting $a_{23}c_2$ times Eq. (11.201) from s_2 times Eq. (11.235) yields

$$s_2[K + S_1 S_7 c_{71} + S_2 a_{71} s_1 - S_2 S_7 s_{71} c_1] + a_{23} S_7 c_{71} + S_1 a_{23} = a_{45}(a_{34} c_4 s_2) + S_5(a_{34} s_4 s_2). \tag{11.240}$$

The terms $(a_{34}c_4s_2)$ and $(a_{34}s_4s_2)$ in Eq. (11.240) may be replaced by direct substitution of Eqs. (11.238) and (11.239) to yield the result

$$s_2[K + S_1 S_7 c_{71} + S_2 a_{71} s_1 - S_2 S_7 s_{71} c_1] + a_{23} S_7 c_{71} + S_1 a_{23} = -a_{45}\{a_{71} s_1 (s_6 s_7 s_{71} - c_6 c_{71}) + S_7(-c_1 s_6 s_7 - s_1 s_6 c_7 c_{71}) + S_1(-c_1(s_6 s_7 c_{71} + s_{71} c_6) - s_1 s_6 c_7) + a_{23} s_2(-s_1 s_6 c_7 + c_1(-s_6 s_7 c_{71} - c_6 s_{71})) + S_2(s_6 s_7 s_{71} - c_6 c_{71}) + a_{45} s_2\} - S_5\{s_2[S_1(s_{71} s_6 s_7 - c_{71} c_6) + S_2((s_1 c_7 + c_{71} c_1 s_7)s_6 + s_{71} c_1 c_6) + S_5 - S_7 c_6 + a_{71} s_6 c_7] + a_{23}(-c_6 c_{71} + s_6 s_7 s_{71})\}. \tag{11.241}$$

This equation may be regrouped and written in the form

$$A_{16}s_2 = -B_{16}, \tag{11.242}$$

where

$$\begin{aligned} A_{16} = {} & K + S_1S_7c_{71} + S_2a_{71}s_1 - S_2S_7s_{71}c_1 + a_{45}a_{23}(-s_1s_6c_7 + c_1(-s_6s_7c_{71} - c_6s_{71})) \\ & + a_{45}^2 + S_5[S_1(s_{71}s_6s_7 - c_{71}c_6) + S_2((s_1c_7 + c_{71}c_1s_7)s_6 + s_{71}c_1c_6) \\ & + S_5 - S_7c_6 + a_{71}s_6c_7] \end{aligned} \tag{11.243}$$

and

$$\begin{aligned} B_{16} = {} & a_{23}S_7c_{71} + S_1a_{23} + a_{45}[a_{71}s_1(s_6s_7s_{71} - c_6c_{71}) + S_7(-c_1s_6s_7 - s_1s_6c_7c_{71}) \\ & + S_1(-c_1(s_6s_7c_{71} + s_{71}c_6) - s_1s_6c_7) + S_2(s_6s_7s_{71} - c_6c_{71})] \\ & + S_5a_{23}(-c_6c_{71} + s_6s_7s_{71}). \end{aligned} \tag{11.244}$$

Equation (11.242) is significant in that it contains only the joint angle parameters θ_1, θ_2, and θ_6. The parameter θ_4 has been eliminated from a manipulation of previous equations.

Subtracting s_2 times Eq. (11.236) from c_2 times Eq. (11.231) yields

$$\begin{aligned} & a_{23}c_2(-s_6c_7s_1 - c_1(c_{71}s_6s_7 + s_{71}c_6)) + S_2c_2^2(s_6c_7c_1 - s_1(c_{71}s_6s_7 + s_{71}c_6)) \\ & + S_2s_2c_2(s_{71}s_6s_7 - c_{71}c_6) - S_7s_{71}s_6c_7 - a_{71}(c_{71}s_6s_7 + s_{71}c_6) = -c_2(a_{45} + a_{34}c_4). \end{aligned} \tag{11.245}$$

Adding S_2s_2 times Eq. (11.224) to Eq. (11.245) and substituting $s_2^2 + c_2^2 = 1$ gives

$$\begin{aligned} & a_{23}c_2(-s_6c_7s_1 - c_1(c_{71}s_6s_7 + s_{71}c_6)) + S_2(s_6c_7c_1 - s_1(c_{71}s_6s_7 + s_{71}c_6)) \\ & - S_7s_{71}s_6c_7 - a_{71}(c_{71}s_6s_7 + s_{71}c_6) = -c_2(a_{45} + a_{34}c_4). \end{aligned} \tag{11.246}$$

Now, Eq. (11.224) may be written as

$$(s_6c_7c_1 - (c_{71}s_6s_7 + s_{71}c_6)s_1)s_2 = (s_{71}s_6s_7 - c_{71}c_6)c_2. \tag{11.247}$$

Multiplying this equation throughout by s_2/c_2 gives

$$(s_6c_7c_1 - (c_{71}s_6s_7 + s_{71}c_6)s_1)s_2^2/c_2 = (s_{71}s_6s_7 - c_{71}c_6)s_2. \tag{11.248}$$

Replacing term s_2^2 by $(1 - c_2^2)$, Eq. (11.248) is then regrouped to give

$$\begin{aligned} (s_6c_7c_1 - (c_{71}s_6s_7 + s_{71}c_6)s_1)/c_2 = {} & (s_{71}s_6s_7 - c_{71}c_6)s_2 \\ & + (s_6c_7c_1 - (c_{71}s_6s_7 + s_{71}c_6)s_1)c_2. \end{aligned} \tag{11.249}$$

Multiplying throughout by c_2 yields

$$\begin{aligned} (s_6c_7c_1 - (c_{71}s_6s_7 + s_{71}c_6)s_1) = {} & (s_{71}s_6s_7 - c_{71}c_6)s_2c_2 \\ & + (s_6c_7c_1 - (c_{71}s_6s_7 + s_{71}c_6)s_1)c_2^2. \end{aligned} \tag{11.250}$$

Upon multiplying Eq. (11.226) by c_2, the right-hand side of Eq. (11.250) may be substituted into the result to give

$$c_2[S_1(s_{71}s_6s_7 - c_{71}c_6) + S_2((s_1c_7 + c_{71}c_1s_7)s_6 + s_{71}c_1c_6) + S_5 - S_7c_6 + a_{71}s_6c_7] \\ + a_{23}[(s_6c_7c_1 - (c_{71}s_6s_7 + s_{71}c_6)s_1)] = -a_{34}s_4c_2. \quad (11.251)$$

Adding $a_{23}s_2$ times Eq. (11.201) to c_2 times Eq. (11.235) gives

$$c_2[K + S_1S_7c_{71} + S_2(-S_7s_{71}c_1 + a_{71}s_1)] + a_{23}(a_{71}c_1 + S_7s_1s_{71}) \\ = a_{45}(a_{34}c_4c_2) + S_5(a_{34}s_4c_2). \quad (11.252)$$

The terms $(a_{34}c_4c_2)$ and $(a_{34}s_4c_2)$ in the previous equation may be directly substituted using the results of Eqs. (11.246) and (11.251). This gives

$$c_2[K + S_1S_7c_{71} + S_2(-S_7s_{71}c_1 + a_{71}s_1)] + a_{23}(a_{71}c_1 + S_7s_1s_{71}) \\ = -a_{45}\{a_{23}c_2(-s_6c_7s_1 - c_1(c_{71}s_6s_7 + s_{71}c_6)) + S_2(s_6c_7c_1 - s_1(c_{71}s_6s_7 + s_{71}c_6)) \\ - S_7s_{71}s_6c_7 - a_{71}(c_{71}s_6s_7 + s_{71}c_6) + a_{45}c_2\} - S_5\{c_2[S_1(s_{71}s_6s_7 - c_{71}c_6) \\ + S_2((s_1c_7 + c_{71}c_1s_7)s_6 + s_{71}c_1c_6) + S_5 - S_7c_6 + a_{71}s_6c_7] \\ + a_{23}[(s_6c_7c_1 - (c_{71}s_6s_7 + s_{71}c_6)s_1)]\}. \quad (11.253)$$

This equation may be regrouped and written as

$$A_{16}c_2 = -D_{16}, \quad (11.254)$$

where A_{16} is defined in Eq. (11.243) and

$$D_{16} = a_{23}(a_{71}c_1 + S_7s_1s_{71}) + a_{45}S_2(s_6c_7c_1 - s_1(c_{71}s_6s_7 + s_{71}c_6)) + a_{45}(-S_7s_{71}s_6c_7 \\ - a_{71}(c_{71}s_6s_7 + s_{71}c_6)) + a_{23}S_5(s_6c_7c_1 - (c_{71}s_6s_7 + s_{71}c_6)s_1). \quad (11.255)$$

Summing the squares of Eqs. (11.242) and (11.254) in order to eliminate θ_2 yields

$$A_{16}^2 = B_{16}^2 + D_{16}^2. \quad (11.256)$$

Equation (11.256) contains only the parameters θ_1 and θ_6, but it must be noted that the equation is not linear in the sines and cosines of these variables.

The terms A_{16}, B_{16}, and D_{16} are expressed in the following form:

$$A_{16} = (N_{A1}c_1 + N_{A2}s_1 + N_{A3})c_6 + (N_{A4}c_1 + N_{A5}s_1 + N_{A6})s_6 \\ + (N_{A7}c_1 + N_{A8}s_1 + N_{A9}), \\ B_{16} = (N_{B1}c_1 + N_{B2}s_1 + N_{B3})c_6 + (N_{B4}c_1 + N_{B5}s_1 + N_{B6})s_6 \\ + (N_{B7}c_1 + N_{B8}s_1 + N_{B9}), \\ D_{16} = (N_{D1}c_1 + N_{D2}s_1 + N_{D3})c_6 + (N_{D4}c_1 + N_{D5}s_1 + N_{D6})s_6 \\ + (N_{D7}c_1 + N_{D8}s_1 + N_{D9}), \quad (11.257)$$

where

$$
\begin{aligned}
&N_{A1} = S_2 S_5 s_{71} - a_{23} a_{45} s_{71}, && N_{A2} = 0,\\
&N_{A3} = -S_1 S_5 c_{71} - S_5 S_7, && N_{A4} = -a_{23} a_{45} c_{71} s_7 + S_2 S_5 c_{71} s_7,\\
&N_{A5} = S_2 S_5 c_7 - a_{23} a_{45} c_7, && N_{A6} = S_1 S_5 s_{71} s_7 + S_5 a_{71} c_7,\\
&N_{A7} = -S_2 S_7 s_{71}, && N_{A8} = S_2 a_{71},\\
&N_{A9} = K + S_5^2 + a_{45}^2 + S_1 S_7 c_{71}, &&\\
&N_{B1} = -S_1 a_{45} s_{71}, && N_{B2} = -a_{45} a_{71} c_{71},\\
&N_{B3} = -S_2 a_{45} c_{71} - S_5 a_{23} c_{71}, && N_{B4} = -S_1 a_{45} c_{71} s_7 - S_7 a_{45} s_7,\\
&N_{B5} = -S_7 a_{45} c_{71} c_7 + a_{45} a_{71} s_{71} s_7 && N_{B6} = S_2 a_{45} s_{71} s_7 + S_5 a_{23} s_{71} s_7,\\
&\qquad - S_1 a_{45} c_7, &&\\
&N_{B7} = 0, && N_{B8} = 0,\\
&N_{B9} = S_7 a_{23} c_{71} + S_1 a_{23}, &&\\
&N_{D1} = 0, && N_{D2} = -S_2 a_{45} s_{71} - S_5 a_{23} s_{71},\\
&N_{D3} = -a_{45} a_{71} s_{71}, && N_{D4} = S_5 a_{23} c_7 + S_2 a_{45} c_7,\\
&N_{D5} = -S_2 a_{45} c_{71} s_7 - a_{23} S_5 c_{71} s_7, && N_{D6} = -S_7 a_{45} s_{71} c_7 - a_{45} a_{71} c_{71} s_7,\\
&N_{D7} = a_{23} a_{71}, && N_{D8} = S_7 a_{23} s_{71},\\
&N_{D9} = 0. &&
\end{aligned}
\tag{11.258}
$$

To convert Eq. (11.256) to a polynomial form, a tan-half-angle substitution is made in Eq. set (11.257) by letting

$$x_1 = \tan\frac{\theta_1}{2} \tag{11.259}$$

and

$$x_6 = \tan\frac{\theta_6}{2}. \tag{11.260}$$

The sines and cosines of θ_1 and θ_6 can then be replaced by the trigonometric identities

$$s_i = \frac{2x_i}{1+x_i^2}, \qquad c_i = \frac{1-x_i^2}{1+x_i^2}, \qquad i = 1, 6. \tag{11.261}$$

Substituting the trigonometric identities into Eq. set (11.257) and then multiplying each equation by $(1+x_1^2)(1+x_6^2)$ results in

$$
\begin{aligned}
A_{16} &= \left(n_{A1}x_1^2 + n_{A2}x_1 + n_{A3}\right)x_6^2 + \left(n_{A4}x_1^2 + n_{A5}x_1 + n_{A6}\right)x_6\\
&\quad + \left(n_{A7}x_1^2 + n_{A8}x_1 + n_{A9}\right),\\
B_{16} &= \left(n_{B1}x_1^2 + n_{B2}x_1 + n_{B3}\right)x_6^2 + \left(n_{B4}x_1^2 + n_{B5}x_1 + n_{B6}\right)x_6\\
&\quad + \left(n_{B7}x_1^2 + n_{B8}x_1 + n_{B9}\right),\\
D_{16} &= \left(n_{D1}x_1^2 + n_{D2}x_1 + n_{D3}\right)x_6^2 + \left(n_{D4}x_1^2 + N_{D5}x_1 + n_{D6}\right)x_6\\
&\quad + \left(n_{D7}x_1^2 + n_{D8}x_1 + n_{D9}\right),
\end{aligned}
\tag{11.262}
$$

where the coefficients in Equation set (11.262) are defined in a similar way to those developed in Section 8.2. These coefficients are defined as follows:

$$
\begin{aligned}
&n_{\zeta 1}=N_{\zeta 1}-N_{\zeta 3}-N_{\zeta 7}+N_{\zeta 9}, && n_{\zeta 2}=2(N_{\zeta 8}-N_{\zeta 2}),\\
&n_{\zeta 3}=-N_{\zeta 1}-N_{\zeta 3}+N_{\zeta 7}+N_{\zeta 9}, && n_{\zeta 4}=2(N_{\zeta 6}-N_{\zeta 4}),\\
&n_{\zeta 5}=4N_{\zeta 5}, && n_{\zeta 6}=2(N_{\zeta 6}+N_{\zeta 4}),\\
&n_{\zeta 7}=-N_{\zeta 1}+N_{\zeta 3}-N_{\zeta 7}+N_{\zeta 9}, && n_{\zeta 8}=2(N_{\zeta 8}+N_{\zeta 2}),\\
&n_{\zeta 9}=N_{\zeta 1}+N_{\zeta 3}+N_{\zeta 7}+N_{\zeta 9},
\end{aligned}
\tag{11.263}
$$

where ζ equals A, B, or D.

Now that the terms A_{16}, B_{16}, and D_{16} have been expressed as nested second-order polynomials in the variables x_1 and x_6, the squares of these terms will be nested fourth-order polynomials in the same variables. The square of the A_{16} term is presented as follows:

$$
\begin{aligned}
A_{16}^2 = {} & \left[n_{A1}^2x_1^4+2n_{A1}n_{A2}x_1^3+(n_{A2}^2+2n_{A1}n_{A3})x_1^2+2n_{A2}n_{A3}x_1+n_{A3}^2\right]x_6^4\\
&+\left[2n_{A1}n_{A4}x_1^4+(2n_{A2}n_{A4}+2n_{A1}n_{A5})x_1^3+(2n_{A2}n_{A5}+2n_{A3}n_{A4}+2n_{A1}n_{A6})x_1^2\right.\\
&\left.+(2n_{A3}n_{A5}+2n_{A2}n_{A6})x_1+2n_{A3}n_{A6}\right]x_6^3+\left[(n_{A4}^2+2n_{A1}n_{A7})x_1^4\right.\\
&+(2n_{A4}n_{A5}+2n_{A1}n_{A8}+2n_{A2}n_{A7})x_1^3+(n_{A5}^2+2n_{A3}n_{A7}+2n_{A1}n_{A9}\\
&+2n_{A2}n_{A8}+2n_{A4}n_{A6})x_1^2+(2n_{A2}n_{A9}+2n_{A5}n_{A6}+2n_{A3}n_{A8})x_1\\
&\left.+2n_{A3}n_{A9}+n_{A6}^2\right]x_6^2+\left[2n_{A4}n_{A7}x_1^4+(2n_{A4}n_{A8}+2n_{A5}n_{A7})x_1^3\right.\\
&\left.+(2n_{A6}n_{A7}+2n_{A4}n_{A9}+2n_{A5}n_{A8})x_1^2+(2n_{A5}n_{A9}+2n_{A6}n_{A8})x_1+2n_{A6}n_{A9}\right]x_6\\
&+\left[n_{A7}^2x_1^4+2n_{A7}n_{A8}x_1^3+(n_{A8}^2+2n_{A7}n_{A9})x_1^2+2n_{A8}n_{A9}x_1+n_{A9}^2\right].
\end{aligned}
\tag{11.264}
$$

Similar expressions can be obtained for B_{16}^2 and D_{16}^2.

Rearranging Eq. (11.256) yields

$$A_{16}^2-B_{16}^2-D_{16}^2=0. \tag{11.265}$$

Because (11.265) is a nested fourth-order equation in the variables x_1 and x_6, it may be written as

$$Ax_6^4+Bx_6^3+Cx_6^2+Dx_6+E=0, \tag{11.266}$$

where

$$
\begin{aligned}
A&=A_4x_1^4+A_3x_1^3+A_2x_1^2+A_1x_1+A_0,\\
B&=B_4x_1^4+B_3x_1^3+B_2x_1^2+B_1x_1+B_0,\\
C&=C_4x_1^4+C_3x_1^3+C_2x_1^2+C_1x_1+C_0,\\
D&=D_4x_1^4+D_3x_1^3+D_2x_1^2+D_1x_1+D_0,\\
E&=E_4x_1^4+E_3x_1^3+E_2x_1^2+E_1x_1+E_0.
\end{aligned}
\tag{11.267}
$$

Equation (11.266) contains only the variables θ_1 and θ_6. This equation will be used in conjunction with Eq. (11.222), which also contains only these variables, in order to eliminate θ_6 from the pair of equations. The result will be an input/output equation that contains θ_1 as its only unknown. The tan-half-angle substitutions listed in Eq. (11.261) are substituted into Eq. (11.222) to yield

$$Fx_6^2 + Gx_6 - F = 0, \tag{11.268}$$

where

$$\begin{aligned} F &= F_2x_1^2 + F_1x_1 + F_0, \\ G &= G_2x_1^2 + G_1x_1 + G_0. \end{aligned} \tag{11.269}$$

The terms F_2 through G_0 are defined as follows:

$$\begin{aligned} F_2 &= a_{71}c_{71}, \qquad F_1 = 2S_1s_{71}, \qquad F_0 = -F_2, \\ G_2 &= -2S_1c_7 + 2a_{71}s_{71}s_7 - 2S_7c_{71}c_7, \\ G_1 &= -4S_1c_{71}s_7 - 4S_7s_7, \\ G_0 &= -G_2. \end{aligned} \tag{11.270}$$

The parameter θ_6 can be eliminated from Eqs. (11.266) and (11.268) by multiplying Eq. (11.266) by 1 and x_6 and Eq. (11.268) by 1, x_6, x_6^2, and x_6^3. In this manner, a total of six equations are created that contain five unknowns (x_6, x_6^2, x_6^3, x_6^4, and x_6^5). These equations can be written in matrix form as

$$\mathbf{Mx} = \mathbf{0}, \tag{11.271}$$

where

$$\mathbf{M} = \begin{bmatrix} 0 & A & B & C & D & E \\ A & B & C & D & E & 0 \\ 0 & 0 & 0 & F & G & -F \\ 0 & 0 & F & G & -F & 0 \\ 0 & F & G & -F & 0 & 0 \\ F & G & -F & 0 & 0 & 0 \end{bmatrix} \quad \text{and} \quad \mathbf{x} = \begin{bmatrix} x_6^5 \\ x_6^4 \\ x_6^3 \\ x_6^2 \\ x_6 \\ 1 \end{bmatrix}. \tag{11.272}$$

The condition that must exist in order for these six equations to have a common set of roots is that the set of equations be linearly dependent. This will occur if the determinant of the matrix **M** equals 0. Expansion of the determinant of **M** yields

$$\begin{aligned} |\mathbf{M}| = 0 = &-(AEG^4) + FG^3(-AD + BE) + F^2G^2(-AC + BD - 4AE - CE) \\ &+ F^3G(-AB + BC - 3AD - CD + 3BE + DE) + F^4(-A^2 + B^2 - 2AC \\ &- C^2 + 2BD + D^2 - 2AE - 2CE - E^2). \end{aligned} \tag{11.273}$$

Because the coefficients A through E are fourth-degree polynomials in the tan-half-angle of θ_1 and the coefficients F and G are second-degree polynomials in the tan-half-angle of θ_1, Eq. (11.273) represents a sixteenth-degree input/output equation. Upon solving this equation for the tan-half angle of θ_1, a maximum of sixteen real solutions can exist. Corresponding values for the parameter ϕ_1 can be found by subtracting the close-the-loop variable γ_1 from each real value of θ_1.

11.7.4 Determination of θ_6

A unique corresponding value of θ_6 for each value of θ_1 is found by first substituting θ_1 into the Eq. sets (11.267) and (11.269). Thus, numerical values are obtained for the coefficients A through G in Eqs. (11.266) and (11.268). Because Eq. (11.268) is quadratic in the tan-half-angle of θ_6, the two values of x_6 that satisfy this equation are solved for. Each of these two values of x_6 are then substituted into Eq. (11.266). In general, only one of the solutions to Eq. (11.268) will also satisfy Eq. (11.266). In this manner, the value of x_6 that simultaneously satisfies both Eq. (11.266) and Eq. (11.268) is easily solved for.

11.7.5 Determination of θ_2

The determination of θ_2 begins by substituting known values of θ_1 and θ_6 into Eqs. (11.243), (11.244), and (11.255) to obtain numerical values for the terms A_{16}, B_{16}, and D_{16}. Equations (11.242) and (11.254) are then used to determine values for the sine and cosine of θ_2 (and thus the unique corresponding value of θ_2) as follows:

$$s_2 = -B_{16}/A_{16}, \tag{11.274}$$

$$c_2 = -D_{16}/A_{16}. \tag{11.275}$$

11.7.6 Determination of θ_5

At this point, sets of values for the angles θ_1, θ_6, and θ_2 have been determined. Corresponding values for θ_5 are found by utilizing the spherical sine and sine–cosine equations

$$X_{2176} = X_{45} \tag{11.276}$$

and

$$Y_{2176} = -X^*_{45}. \tag{11.277}$$

Expanding the terms in these equations and substituting the constant mechanism parameters reduces these two equations to

$$s_5 = c_6(s_2c_1c_7 - (c_{71}s_2s_1 + s_{71}c_2)s_7) + s_6(s_{71}s_1s_2 - c_{71}c_2), \tag{11.278}$$

$$c_5 = -c_1s_2s_7 - (c_{71}s_1s_2 + s_{71}c_2)c_7. \tag{11.279}$$

The corresponding value for θ_5 is readily found from these two equations.

11.7.7 Determination of θ_4

At this point, values for the parameters θ_1, θ_6, θ_2, and θ_5 are known. Corresponding values for θ_4 can be found by using the following two secondary cosine laws:

$$X_{021765} = a_{34}c_{34}s_4 \tag{11.280}$$

and

$$Y_{021765} = a_{34}c_{34}c_4. \tag{11.281}$$

The right-hand sides of these equations have been simplified because s_{34} equals zero.

The following terms in Eq. (11.280) are defined as follows:

$$
\begin{aligned}
X_{021765} &= -S_5(X_{2176}s_5 + Y_{2176}c_5) + c_5X_{02176} - s_5Y_{02176},\\
Y_{021765} &= S_5c_{45}(X_{2176}c_5 - Y_{2176}s_5) - a_{45}[s_{45}(X_{2176}s_5 + Y_{2176}c_5) + c_{45}Z_{2176}]\\
&\quad + c_{45}(X_{02176}s_5 + Y_{02176}c_5) - s_{45}Z_{02176},
\end{aligned}
\tag{11.282}
$$

$$
\begin{aligned}
X_{02176} &= -S_6(X_{217}s_6 + Y_{217}c_6) + c_6X_{0217} - s_6Y_{0217},\\
Y_{02176} &= S_6c_{56}(X_{217}c_6 - Y_{217}s_6) - a_{56}[s_{56}(X_{217}s_6 + Y_{217}c_6) + c_{56}Z_{217}]\\
&\quad + c_{56}(X_{0217}s_6 + Y_{0217}c_6) - s_{56}Z_{0217},\\
Z_{02176} &= S_6s_{56}(X_{217}c_6 - Y_{217}s_6) + a_{56}[c_{56}(X_{217}s_6 + Y_{217}c_6) - s_{56}Z_{217}]\\
&\quad + s_{56}(X_{0217}s_6 + Y_{0217}c_6) + c_{56}Z_{0217},
\end{aligned}
\tag{11.283}
$$

$$
\begin{aligned}
X_{0217} &= -S_7(X_{21}s_7 + Y_{21}c_7) + c_7X_{021} - s_7Y_{021},\\
Y_{0217} &= S_7c_{67}(X_{21}c_7 - Y_{21}s_7) - a_{67}[s_{67}(X_{21}s_7 + Y_{21}c_7) + c_{67}Z_{21}]\\
&\quad + c_{67}(X_{021}s_7 + Y_{021}c_7) - s_{67}Z_{021},\\
Z_{0217} &= S_7s_{67}(X_{21}c_7 - Y_{21}s_7) + a_{67}[c_{67}(X_{21}s_7 + Y_{21}c_7) - s_{67}Z_{21}]\\
&\quad + s_{67}(X_{021}s_7 + Y_{021}c_7) + c_{67}Z_{021},
\end{aligned}
\tag{11.284}
$$

$$
\begin{aligned}
X_{021} &= -S_1(\bar{X}_2s_1 + \bar{Y}_2c_1) + c_1\bar{X}_{02} - s_1\bar{Y}_{02},\\
Y_{021} &= S_1c_{71}(\bar{X}_2c_1 - \bar{Y}_2s_1) - a_{71}[s_{71}(\bar{X}_2s_1 + \bar{Y}_2c_1) + c_{71}\bar{Z}_2]\\
&\quad + c_{71}(\bar{X}_{02}s_1 + \bar{Y}_{02}c_1) - s_{71}\bar{Z}_{02},\\
Z_{021} &= S_1s_{71}(\bar{X}_2c_1 - \bar{Y}_2s_1) + a_{71}[c_{71}(\bar{X}_2s_1 + \bar{Y}_2c_1) - s_{71}\bar{Z}_2]\\
&\quad + s_{71}(\bar{X}_{02}s_1 + \bar{Y}_{02}c_1) + c_{71}\bar{Z}_{02},
\end{aligned}
\tag{11.285}
$$

$$
\begin{aligned}
\bar{X}_{02} &= S_2s_{23}c_2 + a_{23}c_{23}s_2,\\
\bar{Y}_{02} &= S_2c_{12}s_{23}s_2 - a_{12}(c_{12}c_{23} - s_{12}s_{23}c_2) + a_{23}(s_{12}s_{23} - c_{12}c_{23}c_2),\\
\bar{Z}_{02} &= S_2s_{12}s_{23}s_2 - a_{12}(s_{12}c_{23} + c_{12}s_{23}c_2) - a_{23}(c_{12}s_{23} + s_{12}c_{23}c_2).
\end{aligned}
\tag{11.286}
$$

Substitution of the constant mechanism parameters into Eqs. (11.280) and (11.281) using Eq. (11.282) to Eq. (11.286) yields

$$
\begin{aligned}
a_{34}s_4 &= S_1(-c_5c_6c_7s_2s_1 + c_5s_6s_{71}s_2c_1 - c_5c_6s_7c_{71}s_2c_1 + s_5c_7c_{71}s_2c_1 - s_5s_7s_2s_1)\\
&\quad + S_2(-c_5c_6s_7c_{71}c_2s_1 + s_5s_7c_1c_2 + c_5c_6s_7s_{71}s_2 + c_5c_6c_7c_1c_2 + c_5s_6s_{71}c_2s_1\\
&\quad + s_5c_7c_{71}c_2s_1 + c_5s_6c_{71}s_2 - s_5c_7s_{71}s_2) + S_5(-s_5c_6s_2c_1c_7 + s_5c_6s_7c_{71}s_2s_1\\
&\quad + s_5c_6s_7s_{71}c_2 - s_5s_6s_{71}s_2s_1 + s_5s_6c_{71}c_2 + c_5s_2c_1s_7 + c_5c_7c_{71}s_2s_1 + c_5c_7s_{71}c_2)\\
&\quad + S_7(-s_5s_7c_{71}s_2s_1 - s_5s_7s_{71}c_2 - c_5c_6s_2c_1s_7 - c_5c_6c_7c_{71}s_2s_1 - c_5c_6c_7s_{71}c_2\\
&\quad + s_5s_2c_1c_7) + a_{23}(-c_5c_6s_7c_{71}c_1 - s_5s_7s_1 - c_5c_6c_7s_1 + c_5s_6s_{71}c_1 + s_5c_7c_{71}c_1)\\
&\quad + a_{71}(c_5s_6s_{71}c_2 + c_5s_6c_{71}s_2s_1 - s_5c_7s_{71}s_2s_1 + s_5c_7c_{71}c_2\\
&\quad + c_5c_6s_7s_{71}s_2s_1 - c_5c_6s_7c_{71}c_2),
\end{aligned}
\tag{11.287}
$$

$$a_{34}c_4 = S_1(c_6s_{71}s_2c_1 + s_6s_7c_{71}s_2c_1 + s_6c_7s_2s_1) + S_2(s_6s_7c_{71}c_2s_1 - s_6c_7c_1c_2$$
$$+ c_6s_{71}c_2s_1 - s_6s_7s_{71}s_2 + c_6c_{71}s_2) + S_7(s_6s_2c_1s_7 + s_6c_7c_{71}s_2s_1 + s_6c_7s_{71}c_2)$$
$$+ a_{23}(s_6s_7c_{71}c_1 + s_6c_7s_1 + c_6s_{71}c_1) + a_{45}(-s_5c_6s_2c_1c_7 + s_5c_6s_7c_{71}s_2s_1$$
$$+ s_5c_6s_7s_{71}c_2 - s_5s_6s_{71}s_2s_1 + s_5s_6c_{71}c_2 + c_5s_2c_1s_7 + c_5c_7c_{71}s_2s_1$$
$$+ c_5c_7s_{71}c_2) + a_{71}(s_6s_7c_{71}c_2 + c_6c_{71}s_2s_1 + c_6s_{71}c_2 - s_6s_7s_{71}s_2s_1). \quad (11.288)$$

Equations (11.287) and (11.288) can be used to solve for the unique corresponding value of θ_4.

11.7.8 Determination of θ_3

At this point in the analysis, all the joint angle parameters except for θ_3 have been solved for. This remaining parameter can be determined from the following two spherical sine and sine–cosine laws:

$$X_{17654} = s_{23}s_3, \quad (11.289)$$

$$Y_{17654} = s_{23}c_3. \quad (11.290)$$

Substitution of the constant mechanism parameters into the definitions reduces Eqs. (11.289) and (11.290) to the following:

$$s_3 = s_4[c_1(s_6s_7c_{71} + c_6s_{71}) + s_1s_6c_7]$$
$$+ c_4[c_1(c_5c_6s_7c_{71} - c_5s_6s_{71} - s_5c_7c_{71}) + s_1(c_5c_6c_7 + s_5s_7)], \quad (11.291)$$

$$c_3 = s_4[c_1(c_5c_6s_7c_{71} - c_5s_6s_{71} - s_5c_7c_{71}) + s_1(c_5c_6c_7 + s_5s_7)]$$
$$+ c_4[c_1(-s_6s_7c_{71} - c_6s_{71}) - s_1s_6c_7]. \quad (11.292)$$

The corresponding value for θ_3 can be determined directly from the preceding two equations.

At this point the kinematic analysis of the modified flight telerobotic servicer manipulator system is complete. It has been shown that a maximum of sixteen manipulator configurations can position and orient the end effector at some specified position and orientation.

11.7.9 Numerical example

As a numerical example, the following information was specified for the modified flight telerobotic servicer robot:

$$S_7 = 0\,\text{in.}, \quad {}^7\mathbf{P}_{\text{tool}} = \begin{bmatrix} 4.5 \\ 2.2 \\ 1.5 \end{bmatrix}\text{in.}, \quad {}^F\mathbf{P}_{\text{tool}} = \begin{bmatrix} -1 \\ 2 \\ -1 \end{bmatrix}\text{in.},$$

$$^F\mathbf{S}_7 = \begin{bmatrix} -0.0864 \\ 0.7197 \\ 0.6889 \end{bmatrix}, \quad {}^F\mathbf{a}_{78} = \begin{bmatrix} 0.9670 \\ -0.1058 \\ 0.2318 \end{bmatrix}, \quad \theta_7 = 0^\circ.$$

Table 11.16 shows the results of the reverse position analysis.

Table 11.16. *Sixteen solution sets for the modified flight telerobotic servicer robot (angles in degrees).*

Solution	ϕ_1	θ_2	θ_3	θ_4	θ_5	θ_6
A	−104.482	−49.826	72.184	−118.649	175.212	121.049
B	−106.983	135.225	97.660	−139.822	2.462	30.827
C	−75.843	−136.779	−124.485	−79.201	−85.125	−86.359
D	−56.188	−156.038	67.593	−108.187	63.242	92.721
E	−97.802	−68.174	−125.186	−75.985	−156.442	−81.131
F	−109.592	139.633	−121.969	−55.800	−6.261	−106.611
G	−15.375	−165.079	4.587	−84.478	47.445	91.812
H	77.213	−126.224	−172.724	−154.508	8.704	−71.679
I	−131.262	−22.198	69.601	−106.368	−152.935	89.528
J	71.617	42.311	174.740	−166.342	175.644	−116.373
K	73.947	−133.449	−46.450	−42.875	−2.328	164.278
L	−120.974	−28.026	−120.607	−85.177	160.163	−92.269
M	18.375	−164.565	−163.085	−129.971	−38.727	−88.911
N	73.726	46.120	−71.569	−30.520	−177.783	−4.451
O	139.894	−18.261	−165.291	−138.865	124.830	−87.700
P	186.478	−14.673	3.033	−81.689	−138.102	91.380

11.8 Summary

Several examples of the reverse position analysis have been presented in this chapter. The first three examples were of industrial robots comprising six revolute joints. The last two examples were seven-axis manipulators, one of whose joint angles was specified in addition to the desired position and orientation of the end effector.

Each of the manipulators would be classified as a group 4 mechanism once the hypothetical closure link is determined. However, many special conditions exist for these manipulators, such as parallel or intersecting joint axes. These special conditions greatly simplify the solution technique. For the cases of the Puma, GE, and Cincinnati Milacron robots, the eight solutions can be obtained via three separate two-solution equations. Each of these equations can be solved very rapidly via computer. As a result, the overall reverse position analysis can be performed rapidly, that is, in "real time."

The last example of the modified flight telerobotic servicer (FTS) manipulator demonstrates how a relatively simple geometry (two axes intersecting, followed by two parallel axes, followed by two intersecting axes) can in fact be relatively complex to solve. The solution of the modified manipulator is much more complex than that of the original FTS. This complexity will affect the controllability of the manipulator, as more computation time will be required to perform the reverse position analysis. The complexity of the resulting solution should be taken into account during the design phase of the manipulator.

11.9 Problems

1. Write a computer program that performs the reverse analysis for each of the following industrial robots:

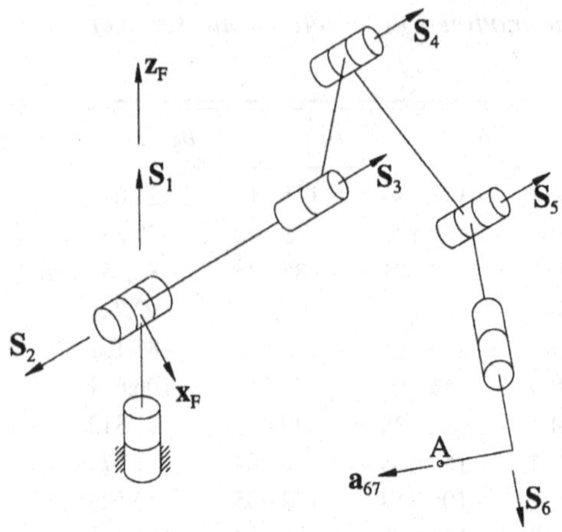

Figure 11.45. Robot manipulator.

(a) Cincinnati Milacron T3-776

(b) GE P60

(c) Puma 560

Check your results by performing a forward analysis of each solution set.

2. A 6R manipulator is shown in Figure 11.45. The following facts are known:

(i) The first and second axes intersect and are perpendicular.

(ii) The second and third axes intersect and are perpendicular.

(iii) The third, fourth, and fifth axes are parallel. The fourth and fifth offset values are zero.

(iv) The fifth and sixth axes intersect and are perpendicular.

(a) Tabulate the mechanism dimensions (link lengths, offsets, and twist angles). Indicate which of these values are equal to zero.

(b) Assume that the coordinates of point A are given together with the direction cosines of $\mathbf{S}_6$ and $\mathbf{a}_{67}$ (all in terms of the fixed coordinate system). List the names of the variables that become known when you close the loop.

(c) Write the vector loop equation for the mechanism.

(d) Obtain an equation that contains only the variables θ_7 and θ_6. Expand the equation as far as necessary in order to show that the only unknowns in the equation are θ_7 and θ_6. How many values of θ_6 will satisfy this equation?

12

Quaternions

12.1 Rigid-body rotations using rotation matrices

In Chapter 2 it was shown how to represent the position and orientation of one coordinate system relative to another. Further, it was shown how to transform the coordinates of a point from one coordinate system to another.

The techniques introduced in Chapter 2 can also be used to define the rotation of a rigid body in space. Any rigid body can be thought of as a collection of points. Suppose that the coordinates of all the points of a body are known in terms of a coordinate system A. The body is then rotated γ degrees about a unit vector $\mathbf{m}$ that passes through the origin of the A coordinate system. The objective is to determine the coordinates of all the points in the rigid body after the rotation is accomplished (see Figure 12.1).

This problem is equivalent to determining the coordinates of all the points in a rigid body in terms of a coordinate system B that is initially coincident with coordinate system A but is then rotated $-\gamma$ about the $\mathbf{m}$ axis vector (see Figure 12.2). This problem was solved in Chapter 2 using rotation matrices.

An alternate solution using quaternions will be introduced in this chapter. Quaternions in many instances may represent a more computationally efficient method of computing rotations of a rigid body compared to the rotation matrix approach. An increase in computational efficiency implies that fewer addition and multiplication operations are required. Quaternions and quaternion algebra will be discussed in the next sections, followed by their application to the rigid-body rotation problem.

12.2 Quaternions

A real quaternion is defined as a set of four real numbers written in a definite order. Two quaternions, q_1 and q_2, may be written as

$$\begin{aligned} q_1 &= (d_1, a_1, b_1, c_1), \\ q_2 &= (d_2, a_2, b_2, c_2). \end{aligned} \tag{12.1}$$

The quaternion q_1 will equal q_2 if and only if $d_1 = d_2$, $a_1 = a_2$, $b_1 = b_2$, and $c_1 = c_2$.

The sum of q_1 and q_2 is defined as

$$q_1 + q_2 = (d_1 + d_2, a_1 + a_2, b_1 + b_2, c_1 + c_2), \tag{12.2}$$

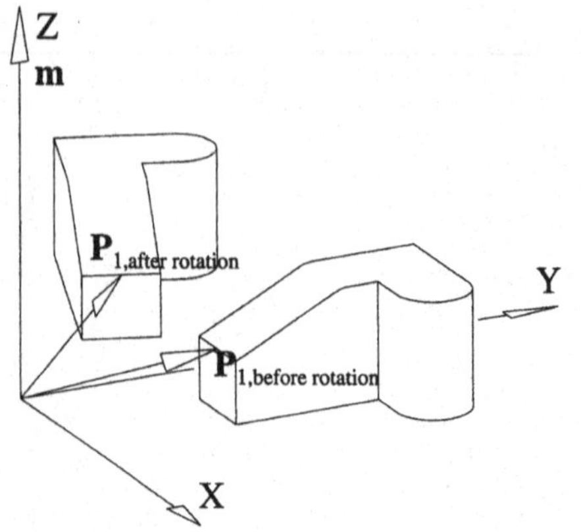

Figure 12.1. Rotation of a rigid body 70 degrees about the Z axis.

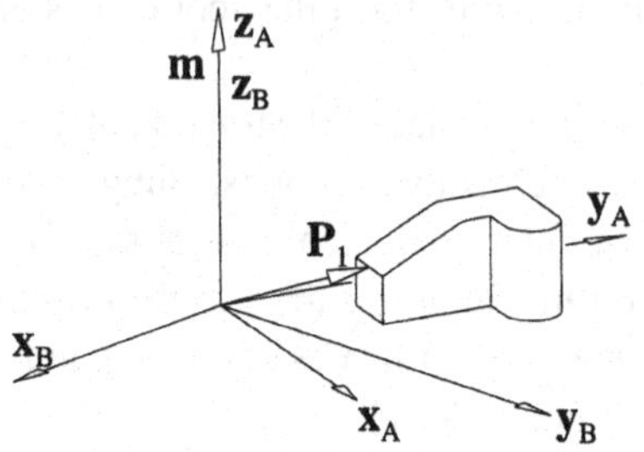

Figure 12.2. Rigid-body rotation represented by rotating coordinate system B −70 degrees about the Z axis.

whereas the difference of the two quaternions is defined as

$$q_1 - q_2 = (d_1 - d_2, a_1 - a_2, b_1 - b_2, c_1 - c_2). \qquad (12.3)$$

A quaternion q_1 that is multiplied by a scalar λ may be written as

$$\lambda q_1 = (\lambda d_1, \lambda a_1, \lambda b_1, \lambda c_1). \qquad (12.4)$$

Multiplying a quaternion by -1 results in

$$-1 q_1 = -q_1 = (-d_1, -a_1, -b_1, -c_1). \qquad (12.5)$$

Lastly, the zero quaternion is defined as (0, 0, 0, 0) and is simply written as 0.

12.3 Quaternion algebra

From the definitions presented in the previous section, it should be apparent that quaternions will observe the following algebraic rules (let p, q, and r be quaternions and λ and

μ be real scalars):

$$\begin{aligned} p + q &= q + p, \\ (p + q) + r &= p + (q + r), \\ \lambda q &= q\lambda, \\ (\lambda\mu)q &= \lambda(\mu q), \\ (\lambda + \mu)q &= \lambda q + \mu q, \\ \lambda(p + q) &= \lambda p + \lambda q. \end{aligned} \tag{12.6}$$

Quaternion multiplication must yet be defined. In order to simplify the resulting expression, the following four quaternion units are defined:

$$\begin{aligned} 1 &= (1, 0, 0, 0), \\ i &= (0, 1, 0, 0), \\ j &= (0, 0, 1, 0), \\ k &= (0, 0, 0, 1). \end{aligned} \tag{12.7}$$

Thus, any quaternion may now be written in the form

$$q = (d, a, b, c) = d1 + ai + bj + ck. \tag{12.8}$$

The product of the two quaternions q_1 and q_2 will now be written as

$$q_1q_2 = (d_1 1 + a_1 i + b_1 j + c_1 k)(d_2 1 + a_2 i + b_2 j + c_2 k). \tag{12.9}$$

Applying the distributive law as in regular algebra yields

$$\begin{aligned} q_1q_2 = {} & d_1d_2(1)(1) + a_1a_2(i)(i) + b_1b_2(j)(j) + c_1c_2(k)(k) + d_1 1(a_2 i + b_2 j + c_2 k) \\ & + a_1 i(d_2 1 + b_2 j + c_2 k) + b_1 j(d_2 1 + a_2 i + c_2 k) + c_1 k(d_2 1 + a_2 i + b_2 j), \end{aligned} \tag{12.10}$$

$$\begin{aligned} q_1q_2 = {} & d_1d_2(1)(1) + a_1a_2(i)(i) + b_1b_2(j)(j) + c_1c_2(k)(k) + d_1a_2 1i + d_1b_2 1j \\ & + d_1c_2 1k + a_1d_2 i1 + a_1b_2 ij + a_1c_2 ik + b_1d_2 j1 + b_1a_2 ji + b_1c_2 jk \\ & + c_1d_2 k1 + c_1a_2 ki + c_1b_2 kj. \end{aligned} \tag{12.11}$$

The individual quaternion products in this equation are defined as follows:

$$\begin{aligned} ij &= k, & ji &= -k, \\ jk &= i, & kj &= -i, \\ ki &= j, & ik &= -j, \\ 1i &= i, & i1 &= i, \\ 1j &= j, & j1 &= j, \\ 1k &= k, & k1 &= k, \\ ii &= -1, & jj &= -1, \qquad kk = -1, \qquad (1)(1) = 1. \end{aligned} \tag{12.12}$$

With these definitions, the quaternion product q_1q_2 may be written as

$$q_1q_2 = d_1d_2 - a_1a_2 - b_1b_2 - c_1c_2 + d_1(a_2i + b_2j + c_2k) + d_2(a_1i + b_1j + c_1k) + \begin{vmatrix} i & j & k \\ a_1 & b_1 & c_1 \\ a_2 & b_2 & c_2 \end{vmatrix}. \tag{12.13}$$

In general, the product $q_1q_2 \neq q_2q_1$. The exception is when the final determinant in Eq. (12.13) vanishes.

A quaternion q, where $q = d + ai + bj + ck$, can be considered as the sum of a scalar, d, and a vector, $v = ai + bj + ck$. The symbols S_q and V_q will be used to represent the scalar and vector parts of quaternion q. Thus

$$\begin{aligned} S_q &= d, \\ V_q &= ai + bj + ck, \\ q &= S_q + V_q. \end{aligned} \tag{12.14}$$

From Eq. (12.13) it is apparent that $q_1q_2 = q_2q_1$ only when one of the vector parts of the quaternions equals zero or when the vector parts of the two quaternions are proportional. Either case causes the determinant to vanish.

Multiplication of two vectors, v_1 and v_2, which are in effect two quaternions with no scalar component, is defined from Eq. (12.13) as

$$v_1v_2 = -a_1a_2 - b_1b_2 - c_1c_2 + \begin{vmatrix} i & j & k \\ a_1 & b_1 & c_1 \\ a_2 & b_2 & c_2 \end{vmatrix}, \tag{12.15}$$

where it is apparent that the scalar part of the result is equal to $-\mathbf{v}_1 \cdot \mathbf{v}_2$ and the vector part is equal to $\mathbf{v}_1 \times \mathbf{v}_2$.

Lastly, it is apparent (and left to the reader to prove) that quaternion multiplication is associative and distributive with respect to addition as follows:

$$\begin{aligned} (pq)r &= p(qr), \\ p(q + r) &= pq + pr, \\ (p + q)r &= pr + qr. \end{aligned} \tag{12.16}$$

12.4 Conjugate and norm of a quaternion

The conjugate of a quaternion $q = d + ai + bj + ck$ will be denoted by K_q, and it is defined as

$$K_q = S_q - V_q = d - ai - bj - ck. \tag{12.17}$$

Because the vector parts of a quaternion and its conjugate differ only in sign, the quaternion product of a quaternion and its conjugate is commutative. That is,

$$qK_q = K_q q = d^2 + a^2 + b^2 + c^2. \tag{12.18}$$

This quaternion product is a scalar and will be defined as the norm of q, that is, N_q. Thus,

$$N_q = qK_q = d^2 + a^2 + b^2 + c^2. \tag{12.19}$$

When $N_q = 1$, then q is referred to as a unit quaternion.

The product of a pair of quaternions $q = q_1 q_2$, where $q_1 = (d_1, a_1, b_1, c_1)$ and $q_2 = (d_2, a_2, b_2, c_2)$, is given by (see Eq. (12.13))

$$q = d_1 d_2 - a_1 a_2 - b_1 b_2 - c_1 c_2 + d_1(a_2 i + b_2 j + c_2 k) + d_2(a_1 i + b_1 j + c_1 k) + \begin{vmatrix} i & j & k \\ a_1 & b_1 & c_1 \\ a_2 & b_2 & c_2 \end{vmatrix}. \tag{12.20}$$

The conjugate of q may be written as

$$K_q = d_1 d_2 - a_1 a_2 - b_1 b_2 - c_1 c_2 - d_1(a_2 i + b_2 j + c_2 k) - d_2(a_1 i + b_1 j + c_1 k) - \begin{vmatrix} i & j & k \\ a_1 & b_1 & c_1 \\ a_2 & b_2 & c_2 \end{vmatrix}. \tag{12.21}$$

Exchanging the last two rows of the determinant and multiplying them by -1 gives

$$K_q = d_1 d_2 - a_1 a_2 - b_1 b_2 - c_1 c_2 + d_1(-a_2 i - b_2 j - c_2 k) + d_2(-a_1 i - b_1 j - c_1 k) + \begin{vmatrix} i & j & k \\ -a_2 & -b_2 & -c_2 \\ -a_1 & -b_1 & -c_1 \end{vmatrix}. \tag{12.22}$$

The right-hand side of Eq. (12.22) is equal to $K_{q_2} K_{q_1}$, and thus

$$K_q = K_{q_2} K_{q_1}. \tag{12.23}$$

Hence, the conjugate of the product of two quaternions is equal to the product of their conjugates taken in reverse order.

The norm of q is equal to the product qK_q. This norm may be written as

$$N_q = qK_q = (q_1 q_2)(K_{q_2} K_{q_1}). \tag{12.24}$$

Regrouping the order of multiplication gives

$$N_q = qK_q = q_1 (q_2 K_{q_2}) K_{q_1}. \tag{12.25}$$

Now, $N_{q_2} = q_2 K_{q_2}$, and thus N_q may be written as

$$N_q = q_1 (N_{q_2}) K_{q_1}. \tag{12.26}$$

Because the norm of a quaternion is a scalar quantity,

$$N_q = N_{q_2}(q_1 K_{q_1}). \tag{12.27}$$

Lastly, $N_{q_1} = q_1 K_{q_1}$, and thus

$$N_q = N_{q_2} N_{q_1}. \tag{12.28}$$

The norm of the product of two quaternions is therefore equal to the product of the individual norms.

12.5 Quaternion division

The definition of a norm of a quaternion was presented in the previous section as

$$N_q = K_q q, \tag{12.29}$$

and therefore

$$\frac{K_q}{N_q} q = 1, \tag{12.30}$$

provided $N_q \neq 0$. Because the vector parts of q and K_q are parallel,

$$q \frac{K_q}{N_q} = 1. \tag{12.31}$$

The term $\frac{K_q}{N_q}$ is defined as the reciprocal of q and is written as

$$q^{-1} = \frac{K_q}{N_q}. \tag{12.32}$$

From Eqs. (12.30) and (12.31) it is apparent that

$$qq^{-1} = q^{-1}q = 1. \tag{12.33}$$

From Eq. (12.28), the norm of (qq^{-1}) must equal $N_q N_{q^{-1}}$, which equals one. Therefore,

$$N_{q^{-1}} = \frac{1}{N_q}. \tag{12.34}$$

To divide a quaternion p by a nonzero quaternion q, it is necessary to solve either the equation

$$r_1 q = p \tag{12.35}$$

or

$$q r_2 = p \tag{12.36}$$

for the result r_1 or r_2. In general, the solutions for the two equations will be different.

Postmultiplying both sides of Eq. (12.35) by q^{-1} gives

$$r_1 = pq^{-1}, \tag{12.37}$$

and premultiplying both sides of Eq. (12.36) by q^{-1} gives

$$r_2 = q^{-1}p. \tag{12.38}$$

Because the two solutions r_1 and r_2 are different, the symmetrical notation p/q cannot be used. Rather, the notation of Eqs. (12.37) or (12.38) will be used. These may be called the left-hand quotient of p divided by q and the right-hand quotient of p divided by q respectively. These two quotients are defined whenever $q \neq 0$.

It is interesting to note that the norm of the two solutions r_1 and r_2 are equal. Taking the norms of both sides of Eqs. (12.37) and (12.38) and using the result of Eq. (12.34) will result in

$$N_{r_1} = N_{r_2} = \frac{N_p}{N_q}. \tag{12.39}$$

It may thus be stated that the norm of either quotient of two quaternions is equal to the quotient of their individual norms.

Lastly, it will be shown how to obtain the inverse of a multiple product of quaternions. Suppose that q is defined as

$$q = q_1 q_2 q_3 \cdots q_n. \tag{12.40}$$

The inverse of q may be written as

$$q^{-1} = (q_1 q_2 q_3 \cdots q_n)^{-1} = \frac{K_q}{N_q}, \tag{12.41}$$

$$q^{-1} = \frac{K_{q_n} \cdots K_{q_3} K_{q_2} K_{q_1}}{N_{q_n} \cdots N_{q_3} N_{q_2} N_{q_1}} = q_n^{-1} \cdots q_3^{-1} q_2^{-1} q_1^{-1}. \tag{12.42}$$

Thus, the inverse of a multiple of quaternions is equal to the product of the inverses of the individual quaternions taken in reverse order.

12.6 Rigid-body rotation

Figure 12.3 shows a vector **r** drawn from a point O that is to be rotated about the unit vector **s** by an angle of 2θ. In this way the vector **r** will be transformed to the vector **r′**. It will be shown in this section that this transformation can be accomplished by the quaternion operator $q(\ )q^{-1}$ and

$$r' = qrq^{-1}, \tag{12.43}$$

where q is defined by the unit quaternion $(\cos\theta + \mathbf{s}\sin\theta)$ and where $V_r = \mathbf{r}$ and $S_r = 0$.

The quaternion r and the unit quaternion q may be written as

$$q = \cos\theta + \sin\theta(s_x i + s_y j + s_z k), \tag{12.44}$$

$$r = r_x i + r_y j + r_z k. \tag{12.45}$$

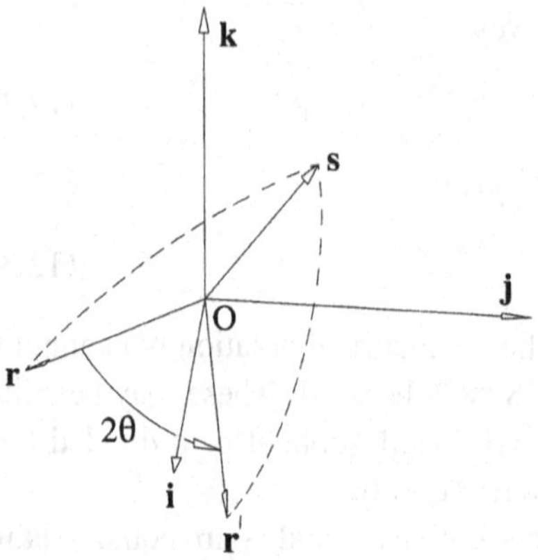

Figure 12.3. Rotation of vector **r** to position **r**′.

The quaternion, r′, is now defined by the expression

$$r' = qrq^{-1}. \tag{12.46}$$

Firstly, let the quaternion t = qr. Therefore,

$$t = qr = -\sin\theta(s_x r_x + s_y r_y + s_z r_z) + \cos\theta(r_x i + r_y j + r_z k) + \sin\theta \begin{vmatrix} i & j & k \\ s_x & s_y & s_z \\ r_x & r_y & r_z \end{vmatrix}, \tag{12.47}$$

$$t = -\sin\theta(s_x r_x + s_y r_y + s_z r_z) + i[r_x\cos\theta + \sin\theta(s_y r_z - s_z r_y)] + j[r_y\cos\theta + \sin\theta(s_z r_x - s_x r_z)] + k[r_z\cos\theta + \sin\theta(s_x r_y - s_y r_x)]. \tag{12.48}$$

The quaternion t will now be written as

$$t = d + ai + bj + ck, \tag{12.49}$$

where

$$\begin{aligned} d &= -\sin\theta(s_x r_x + s_y r_y + s_z r_z), \\ a &= r_x\cos\theta + \sin\theta(s_y r_z - s_z r_y), \\ b &= r_y\cos\theta + \sin\theta(s_z r_x - s_x r_z), \\ c &= r_z\cos\theta + \sin\theta(s_x r_y - s_y r_x). \end{aligned} \tag{12.50}$$

The quaternion $q^{-1} = K_q$ because $N_q = 1$, as q is a unit quaternion. Equation (12.46) may now be written as

$$r' = tK_q = (d + ai + bj + ck)[\cos\theta - \sin\theta(s_x i + s_y j + s_z k)]. \tag{12.51}$$

This may be expanded as

$$r' = d\cos\theta + \sin\theta(as_x + bs_y + cs_z) + d\sin\theta(-s_x i - s_y j - s_z k) + \cos\theta(ai + bj + ck)$$
$$+ \sin\theta \begin{vmatrix} i & j & k \\ a & b & c \\ -s_x & -s_y & -s_z \end{vmatrix}. \quad (12.52)$$

The scalar part of r′ can be written as

$$S_{r'} = d\cos\theta + \sin\theta(as_x + bs_y + cs_z). \quad (12.53)$$

Substituting the expressions from Eq. set (12.50) gives

$$S_{r'} = -\sin\theta\cos\theta(s_x r_x + s_y r_y + s_z r_z) + \sin\theta[s_x(r_x\cos\theta + \sin\theta(s_y r_z - s_z r_y))$$
$$+ s_y(r_y\cos\theta + \sin\theta(s_z r_x - s_x r_z)) + s_z(r_z\cos\theta + \sin\theta(s_x r_y - s_y r_x))], \quad (12.54)$$

which reduces to $S_{r'} = 0$. Thus, it has been shown that r′ is a quaternion with no scalar component.

The norm of the right and left sides of Eq. (12.46) can be written as

$$N_{r'} = N_q N_r N_{q^{-1}}. \quad (12.55)$$

The order of multiplication can be rearranged on the right side of this equation. Further, from Eq. (12.34), the product $N_q N_{q^{-1}}$ will equal one, Eq. (12.55) simplifies to

$$N_{r'} = N_r, \quad (12.56)$$

and the norm of r equals the norm of r′. Because r′ and r have no scalar components, it can be said that the sum of the squares of the vector components of r′ will equal the sum of the squares of the vector components of r. Interpreting r and r′ as vectors, it can be stated that the magnitude of the vector **r**′ will equal the magnitude of the vector **r**. Thus, it is now known that the operator $q(\ )q^{-1}$ transforms a vector to a vector of equal length. This must be the case if the operator represents a pure rotation of a vector, where the vector represents the coordinates of a point in a rigid body.

The task at hand is still to show that the quaternion operator $q(r)q^{-1}$ will rotate the vector **r** about the **s** axis by an angle of 2θ if q equals $(\cos\theta + \mathbf{s}\sin\theta)$. At present all that is known is that the result of the operation will be a vector of the same magnitude as **r**. Equation (12.52) will now be expanded to show that the operation does indeed result in a rotation of the vector about the **s** axis.

Substituting Eq. (12.50) into Eq. (12.52) and regrouping gives

$$r' = i\{\sin^2\theta[s_x(s_x r_x + s_y r_y + s_z r_z) + s_y(s_x r_y - r_x s_y) + s_z(s_x r_z - s_z r_x)]$$
$$+ 2\sin\theta\cos\theta(s_y r_z - r_y s_z) + \cos^2\theta\, r_x\} + j\{\sin^2\theta[s_y(s_x r_x + s_y r_y + s_z r_z)$$
$$+ s_x(s_y r_x - s_x r_y) + s_z(s_y r_z - s_z r_y)] + 2\sin\theta\cos\theta(s_z r_x - s_x r_z) + \cos^2\theta\, r_y\}$$
$$+ k\{\sin^2\theta[s_z(s_x r_x + s_y r_y + s_z r_z) + s_x(s_z r_x - s_x r_z) + s_y(s_z r_y - s_y r_z)]$$
$$+ 2\sin\theta\cos\theta(s_x r_y - s_y r_x) + \cos^2\theta\, r_z\}. \quad (12.57)$$

Rearranging this equation gives

$$\begin{aligned} r' = {} & i\{\sin^2\theta\,[s_x(s_x r_x + 2s_y r_y + 2s_z r_z) + r_x(-s_y^2 - s_z^2)] + 2\sin\theta\cos\theta(s_y r_z \\ & - r_y s_z) + \cos^2\theta r_x\} + j\{\sin^2\theta\,[s_y(2s_x r_x + s_y r_y + 2s_z r_z) + r_y(-s_x^2 - s_z^2)] \\ & + 2\sin\theta\cos\theta(s_z r_x - s_x r_z) + \cos^2\theta r_y\} + k\{\sin^2\theta\,[s_z(2s_x r_x + 2s_y r_y + s_z r_z) \\ & + r_z(-s_x^2 - s_y^2)] + 2\sin\theta\cos\theta(s_x r_y - s_y r_x) + \cos^2\theta r_z\}. \end{aligned} \tag{12.58}$$

The axis vector, **s**, is a unit vector, and therefore the expressions $-s_y^2 - s_z^2 = s_x^2 - 1$, $-s_x^2 - s_z^2 = s_y^2 - 1$, and $-s_x^2 - s_y^2 = s_z^2 - 1$ may be substituted into Eq. (12.58) to give

$$\begin{aligned} r' = {} & i\{\sin^2\theta[s_x(2s_x r_x + 2s_y r_y + 2s_z r_z) - r_x] + 2\sin\theta\cos\theta(s_y r_z - r_y s_z) + \cos^2\theta\, r_x\} \\ & + j\{\sin^2\theta[s_y(2s_x r_x + 2s_y r_y + 2s_z r_z) - r_y] + 2\sin\theta\cos\theta(s_z r_x - s_x r_z) + \cos^2\theta\, r_y\} \\ & + k\{\sin^2\theta[s_z(2s_x r_x + 2s_y r_y + 2s_z r_z) - r_z] + 2\sin\theta\cos\theta(s_x r_y - s_y r_x) + \cos^2\theta\, r_z\}, \end{aligned} \tag{12.59}$$

which is the result of the quaternion operation $q(r)q^{-1}$.

In Section 2.8.1 the rotation matrix was developed for the case where one coordinate system was rotated about an axis by a specified angle relative to the other. Assuming that coordinate systems A and B are initially aligned, and B is then rotated about the axis **s** (which passes through the origin) by the angle 2θ, the rotation matrix ${}^{A}_{B}\mathbf{R}$ is (see Eq. (2.59))

$${}^{A}_{B}\mathbf{R} = \begin{bmatrix} s_x^2(1-\cos 2\theta)+\cos 2\theta & s_x s_y(1-\cos 2\theta)-s_z\sin 2\theta & s_x s_z(1-\cos 2\theta)+s_y\sin 2\theta \\ s_x s_y(1-\cos 2\theta)+s_z\sin 2\theta & s_y^2(1-\cos 2\theta)+\cos 2\theta & s_y s_z(1-\cos 2\theta)-s_x\sin 2\theta \\ s_x s_z(1-\cos 2\theta)-s_y\sin 2\theta & s_y s_z(1-\cos 2\theta)+s_x\sin 2\theta & s_z^2(1-\cos 2\theta)+\cos 2\theta \end{bmatrix}. \tag{12.60}$$

The coordinates of a point that has been rotated by 2θ about the axis **s** can be determined by calculating the coordinates of the point in terms of a coordinate system that has been rotated by an angle of -2θ about the axis **s**. The rotation matrix ${}^{C}_{D}\mathbf{R}$, which relates coordinate systems C and D where they are initially aligned and then D is rotated -2θ about the axis **s**, is given by

$${}^{C}_{D}\mathbf{R} = \begin{bmatrix} s_x^2(1-\cos 2\theta)+\cos 2\theta & s_x s_y(1-\cos 2\theta)+s_z\sin 2\theta & s_x s_z(1-\cos 2\theta)-s_y\sin 2\theta \\ s_x s_y(1-\cos 2\theta)-s_z\sin 2\theta & s_y^2(1-\cos 2\theta)+\cos 2\theta & s_y s_z(1-\cos 2\theta)+s_x\sin 2\theta \\ s_x s_z(1-\cos 2\theta)+s_y\sin 2\theta & s_y s_z(1-\cos 2\theta)-s_x\sin 2\theta & s_z^2(1-\cos 2\theta)+\cos 2\theta \end{bmatrix}. \tag{12.61}$$

Assuming that the coordinates of a point R in the C coordinate system are ${}^{C}\mathbf{P}_R = [r_x, r_y, r_z]^T$, the coordinates of this point in the D system are calculated from

$${}^{D}\mathbf{P}_R = {}^{D}_{C}\mathbf{R}\,{}^{C}\mathbf{P}_R, \tag{12.62}$$

where ${}^{D}_{C}\mathbf{R}$ is the transpose of ${}^{C}_{D}\mathbf{R}$. Expanding Eq. (12.62) gives

$$
{}^{D}\mathbf{P}_{R} = \begin{bmatrix} r_x\left(s_x^2(1-\cos 2\theta)+\cos 2\theta\right) + r_y(s_x s_y(1-\cos 2\theta) - s_z\sin 2\theta) + r_z(s_x s_z(1-\cos 2\theta) + s_y \sin 2\theta) \\ r_x(s_x s_y(1-\cos 2\theta) + s_z\sin 2\theta) + r_y\left(s_y^2(1-\cos 2\theta)+\cos 2\theta\right) + r_z(s_y s_z(1-\cos 2\theta) - s_x\sin 2\theta) \\ r_x(s_x s_z(1-\cos 2\theta) - s_y\sin 2\theta) + r_y(s_y s_z(1-\cos 2\theta) + s_x \sin 2\theta) + r_z\left(s_z^2(1-\cos 2\theta)+\cos 2\theta\right) \end{bmatrix}. \tag{12.63}
$$

Rearranging the terms in Eq. (12.63) gives

$$
{}^{D}\mathbf{P}_{R} = \begin{bmatrix} (1-\cos 2\theta)\left(r_x s_x^2 + r_y s_x s_y + r_z s_x s_z\right) + \cos 2\theta\, r_x + \sin 2\theta(s_y r_z - s_z r_y) \\ (1-\cos 2\theta)\left(r_y s_y^2 + r_x s_x s_y + r_z s_y s_z\right) + \cos 2\theta\, r_y + \sin 2\theta(s_z r_x - s_x r_z) \\ (1-\cos 2\theta)\left(r_z s_z^2 + r_x s_x s_z + r_y s_y s_z\right) + \cos 2\theta\, r_z + \sin 2\theta(s_x r_y - s_y r_x) \end{bmatrix}. \tag{12.64}
$$

Substituting the trigonometric identities $\sin 2\theta = 2\sin\theta\cos\theta$ and $\cos 2\theta = \cos^2\theta - \sin^2\theta$ gives

$$
{}^{D}\mathbf{P}_{R} = \begin{bmatrix} (1-\cos^2\theta+\sin^2\theta)\left(r_x s_x^2 + r_y s_x s_y + r_z s_x s_z\right) + (\cos^2\theta - \sin^2\theta)r_x + 2\sin\theta\cos\theta(s_y r_z - s_z r_y) \\ (1-\cos^2\theta+\sin^2\theta)\left(r_y s_y^2 + r_x s_x s_y + r_z s_y s_z\right) + (\cos^2\theta - \sin^2\theta)r_y + 2\sin\theta\cos\theta(s_z r_x - s_x r_z) \\ (1-\cos^2\theta+\sin^2\theta)\left(r_z s_z^2 + r_x s_x s_z + r_y s_y s_z\right) + (\cos^2\theta - \sin^2\theta)r_z + 2\sin\theta\cos\theta(s_x r_y - s_y r_x) \end{bmatrix}. \tag{12.65}
$$

Substituting $1 - \cos^2\theta = \sin^2\theta$ into Eq. (12.65) and rearranging gives

$$
{}^{D}\mathbf{P}_{R} = \begin{bmatrix} \sin^2\theta[s_x(2r_x s_x + 2r_y s_y + 2r_z s_z) - r_x] + 2\sin\theta\cos\theta(s_y r_z - s_z r_y) + \cos^2\theta r_x \\ \sin^2\theta[s_y(2r_y s_y + 2r_x s_x + 2r_z s_z) - r_y] + 2\sin\theta\cos\theta(s_z r_x - s_x r_z) + \cos^2\theta r_y \\ \sin^2\theta[s_z(2r_z s_z + 2r_x s_x + 2r_y s_y) - r_z] + 2\sin\theta\cos\theta(s_x r_y - s_y r_x) + \cos^2\theta r_z \end{bmatrix}. \tag{12.66}
$$

Comparing Eqs. (12.66) and (12.59), it is apparent that the quaterion operator $q(\ )q^{-1}$, where $q = \cos\theta + \mathbf{s}\sin\theta$, is equivalent to a rotation about the axis **s** by the angle 2θ. Thus, this operator may be used to transform points of a rigid body that undergo rigid-body rotation.

12.7 Example problems

12.7.1 Problem 1

The rigid body shown in Figure 12.4 is rotated about the axis $\mathbf{m}_1 = (3\mathbf{i} + 2\mathbf{j})$ by an angle of sixty degrees. It is then rotated about the axis $\mathbf{m}_2 = (-\mathbf{j} + 2\mathbf{k})$ by an angle of 115 degrees. Both axis $\mathbf{m}_1$ and $\mathbf{m}_2$ are measured with respect to the fixed coordinate system shown in the figure. Determine the coordinates of the point $(5, 2, 3)^T$ after the two rotations have been accomplished.

Figure 12.5 shows the rigid body at its original position and after each of the two rotations. The quaternion operator that models the first rotation is $q_1(\,)q_1^{-1}$, where

$$
q_1 = \cos(30^\circ) + \frac{1}{\sqrt{13}}\sin(30^\circ)(3i + 2j). \tag{12.67}
$$

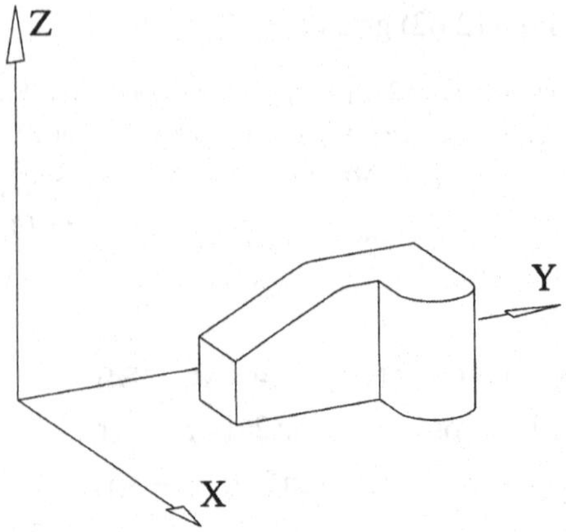

Figure 12.4. Rigid-body rotation.

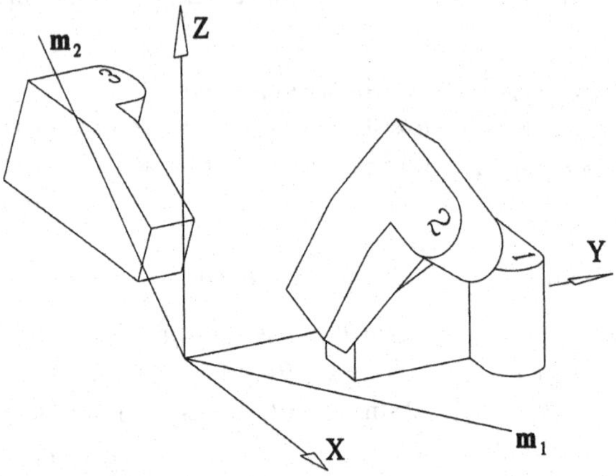

Figure 12.5. Two successive rigid-body rotations.

The operator that models the second rotation is $q_2(\,)q_2^{-1}$, where

$$q_2 = \cos(57.5°) + \frac{1}{\sqrt{5}}\sin(57.5°)(-j + 2k). \tag{12.68}$$

The coordinates of the point $(5, 2, 3)^T$ after the two rotations may be calculated as

$$q_2 q_1 (5i + 2j + 3k) q_1^{-1} q_2^{-1}. \tag{12.69}$$

The solution is $(-3.054, 4.624, 2.701)^T$.

12.7.2 Problem 2

For the previous problem, determine the axis and angle of rotation that will return the rigid body to its original position.

The quaternion operator that transformed a point in the previous problem was $q_2q_1(\,)q_1^{-1}q_2^{-1}$, where q_1 and q_2 were defined in Eqs. (12.67) and (12.68). Performing the multiplication q_2q_1 gives

$$q_2q_1 = 0.5699 + 0.0143i + 0.1362j + 0.8102k. \tag{12.70}$$

This product can be interpreted as representing the net rotation of the body about a single axis. The scalar part of the result equals the cosine of half the net angle of rotation. The magnitude of the vector part of the product will equal the sine of half the net angle of rotation. Letting θ equal the angle of rotation, the cosine and sine of $\theta/2$ may be written as

$$\cos(\theta/2) = 0.5699, \tag{12.71}$$

$$\sin(\theta/2) = 0.8217. \tag{12.72}$$

The angle $\theta/2$ equals 55.255 degrees, and θ equals 110.51 degrees. Equation (12.70) can now be written as

$$q_2q_1 = \cos(55.255^\circ) + \sin(55.255^\circ)[0.0174i + 0.1658j + 0.9860k]. \tag{12.73}$$

By inspection, it is apparent that the net motion is a rotation of 110.51° about the axis $(0.0174i + 0.1658j + 0.9860k)$. The rigid body can be returned to its original position by rotating it -110.51° about the same axis vector.

12.7.3 Problem 3

Show that the quaternion multiplication in example problem 1 transforms the points of the rigid body shown in Figure 12.4 as if they have first been rotated about the $\mathbf{m}_2$ axis by 115 degrees and then about the $\mathbf{m}_1$ axis (measured in terms of a coordinate system that has been modified by the first rotation) by 60 degrees.

The quaternion operator that is being described in this problem is

$$q_2q_1(\,)q_1^{-1}q_2^{-1}, \tag{12.74}$$

where q_1 and q_2 are given in Eqs. (12.67) and (12.68).

The new interpretation for the quaternion operator is shown graphically in Figure 12.6. In (b), the original object has been rotated 115° about the vector $\mathbf{m}_2$. In (c), the modified coordinate system is shown, and in (d) the object has been rotated about the vector $\mathbf{m}_1$, which is defined in terms of the modified coordinate system. The final position and orientation of the rigid body is the same as that shown in Figure 12.5.

The most straightforward means of demonstrating the result will be to use rotation matrices to calculate the coordinates of a general point that has been rotated as given by the problem statement and then comparing the results to the coordinates of the point that have been transformed by the quaternion operator of Eq. (12.74).

In Section 2.8.1, it was shown how to form the rotation matrix that would describe the relationship between two coordinate systems, A and B, that were initially coincident. Coordinate system B was then rotated about the unit vector $\mathbf{m}$ by an angle of θ. The resulting rotation matrix that relates the A and B coordinate systems was shown in Eq. (2.59) to be

$${}^{A}_{B}\mathbf{R} = \begin{bmatrix} m_x^2 v\theta + c\theta & m_x m_y v\theta - m_z s\theta & m_x m_z v\theta + m_y s\theta \\ m_x m_y v\theta + m_z s\theta & m_y^2 v\theta + c\theta & m_y m_z v\theta - m_x s\theta \\ m_x m_z v\theta - m_y s\theta & m_y m_z v\theta + m_x s\theta & m_z^2 v\theta + c\theta \end{bmatrix}, \tag{12.75}$$

where $s\theta$ and $c\theta$ represent the sine and cosine of the angle of rotation and $v\theta = (1 - c\theta)$.

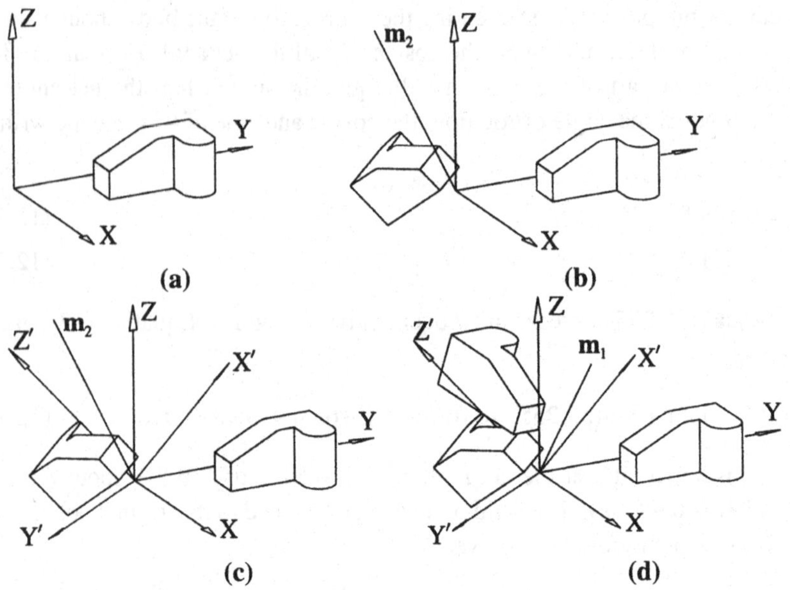

Figure 12.6. Interpretation of successive rotations.

The coordinates of a point that has been rotated about the axis $\mathbf{m}_2$ by an angle of 115 degrees is equivalent to the coordinates of a stationary point as seen in a new coordinate system that has been rotated by -115 degrees about the vector $\mathbf{m}_2$. Thus, the coordinates of a general point that has been rotated 115° about the $\mathbf{m}_2$ axis can be determined from the equation

$$^{\mathrm{B}}\mathbf{P}_1 = {}^{\mathrm{B}}_{\mathrm{A}}\mathbf{R}\,{}^{\mathrm{A}}\mathbf{P}_1, \tag{12.76}$$

where $^{\mathrm{A}}\mathbf{P}_1$ represents the initial coordinates of the point, $^{\mathrm{B}}\mathbf{P}_1$ represents the coordinates of the point after rotation, and ${}^{\mathrm{B}}_{\mathrm{A}}\mathbf{R}$ is evaluated as the transpose of Eq. (12.75)* with θ equal to $-115°$ as

$${}^{\mathrm{B}}_{\mathrm{A}}\mathbf{R} = \begin{bmatrix} -0.4226 & -0.8106 & -0.4053 \\ 0.8106 & -0.1381 & -0.5690 \\ 0.4053 & -0.5690 & 0.7155 \end{bmatrix}. \tag{12.77}$$

Figure 12.7(a) shows the original object, and Figure 12.7(b) shows the object with the original A coordinate system and the modified B coordinate system.

The second transformation is defined as a rotation of sixty degrees about the vector $\mathbf{m}_1$, where this vector is measured in terms of the modified coordinate system. It must be noted that the modified coordinate system is the A coordinate system and that the reference, or stationary, coordinate system is the B coordinate system. Thus, the next axis of rotation is $3\mathbf{i} + 2\mathbf{j}$ as measured in the A coordinate system. This axis of rotation can be calculated in the B coordinate system by transforming two points on the vector line

* The transpose is needed because we want to take a point known in the A coordinate system and determine its coordinates in the B coordinate system.

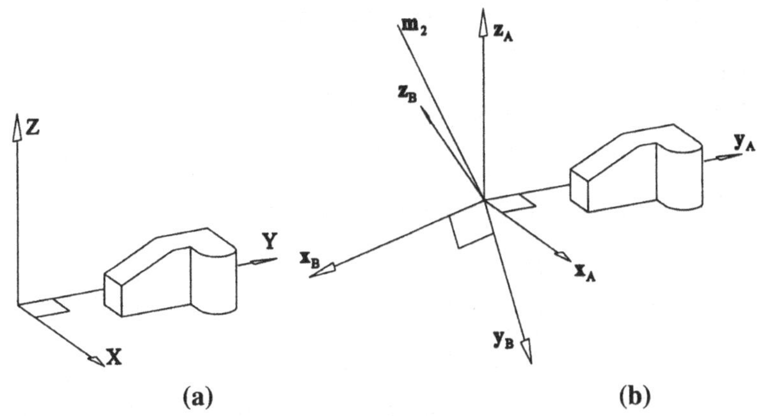

Figure 12.7. Rotated coordinate system.

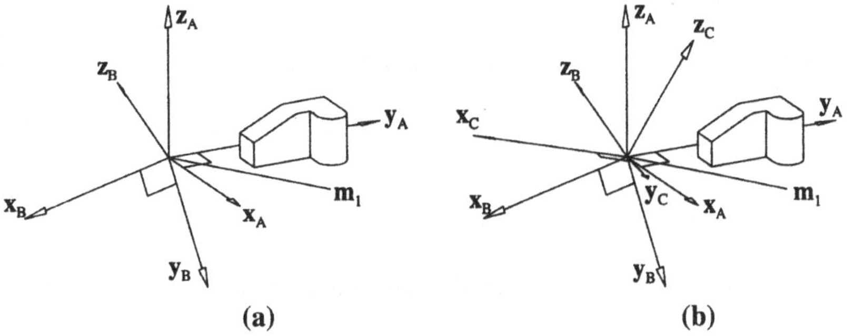

Figure 12.8. Two successive rotations.

from the A coordinate system to the B coordinate system and then calculating the vector direction as the difference between these two points. The points on the axis of rotation to be transformed are arbitrarily selected as $[3, 2, 0]^T$ and $[0, 0, 0]^T$, and the transformation equation to be used is given in Eq. (12.76). These two points transform to $[-0.8013, 0.5979, 0.0216]^T$ and $[0, 0, 0]^T$. Thus, the next axis of rotation as measured in the B coordinate system is $-0.8013\mathbf{i} + 0.5979\mathbf{j} + 0.0216\mathbf{k}$.

Again, the coordinates of a point that is rotated can be calculated as being the coordinates of a stationary point as seen in a coordinate system that is rotated in the opposite direction. Figure 12.8(a) shows the $\mathbf{m}_1$ vector, which is the next axis of rotation, and Figure 12.8(b) shows a new coordinate system C, which was initially aligned with B and was then rotated minus sixty degrees about the axis of rotation. The transformation that relates the B and C coordinate systems can be calculated from the transpose of Eq. (12.75) using $-0.8013\mathbf{i} + 0.5979\mathbf{j} + 0.0216\mathbf{k}$ as the direction of the axis vector $\mathbf{m}$ and minus sixty degrees for the rotation angle θ. This transformation matrix can be written as

$$ {}^{C}_{B}\mathbf{R} = \begin{bmatrix} 0.8210 & -0.2582 & 0.5091 \\ -0.2208 & 0.6787 & 0.7004 \\ -0.5264 & -0.6875 & 0.5002 \end{bmatrix}. \tag{12.78} $$

The coordinates of a point after the two specified rotations may now be written as

$$ {}^{C}\mathbf{P}_1 = {}^{C}_{B}\mathbf{R}\,{}^{B}_{A}\mathbf{R}\,{}^{A}\mathbf{P}_1. \tag{12.79} $$

Substituting Eqs. (12.77) and (12.78) into this equation yields

$$ {}^{C}\mathbf{P}_1 = \begin{bmatrix} -0.3500 & -0.9196 & 0.1784 \\ 0.9274 & -0.3133 & 0.2044 \\ -0.1321 & 0.2370 & 0.9625 \end{bmatrix} {}^{A}\mathbf{P}_1. \tag{12.80} $$

Writing the vector ${}^{A}\mathbf{P}_1$ as $[x, y, z]^T$, the transformation becomes

$$ {}^{C}\mathbf{P}_1 = \begin{bmatrix} -0.3500x - 0.9196y + 0.1784z \\ 0.9274x - 0.3133y + 0.2044z \\ -0.1321x + 0.2370y + 0.9625z \end{bmatrix}. \tag{12.81} $$

It is necessary to show that the quaternion operator $q_2q_1(\,)q_1^{-1}q_2^{-1}$ transforms a general point $\mathbf{P}_1$ in the same manner as the transformation Eq. (12.81). The product q_2q_1 was calculated in Eq. (12.70) as

$$ q_2q_1 = 0.5699 + 0.0143i + 0.1362j + 0.8102k. \tag{12.82} $$

Because q_1 and q_2 are both unit quaternions, the inverse of this product will equal the conjugate of the product. Thus,

$$ q_1^{-1}q_2^{-1} = 0.5699 - 0.0143i - 0.1362j - 0.8102k. \tag{12.83} $$

Writing the vector $\mathbf{P}_1$ as $[x, y, z]^T$, the overall transformation may now be written as

$$ \begin{aligned} q_2q_1(xi + yj + zk)q_1^{-1}q_2^{-1} &= (0.5699 + 0.0143i + 0.1362j + 0.8102k)(xi + yj + zk) \\ &\quad \times (0.5699 - 0.0143i - 0.1362j - 0.8102k). \end{aligned} \tag{12.84} $$

Expanding the first product $q_2q_1(xi+yj+zk)$ yields

$$ \begin{aligned} q_2q_1(xi + yj + zk) = &(-0.0143x - 0.1362y - 0.8102z) \\ &+ i(0.5699x + 0.1362z - 0.8102y) \\ &+ j(0.5699y + 0.8102x - 0.0143z) \\ &+ k(0.5699z + 0.0143y - 0.1362x). \end{aligned} \tag{12.85} $$

Writing this product as

$$ q_2q_1(xi + yj + zk) = d_1 + a_1i + b_1j + c_1k \tag{12.86} $$

allows Eq. (12.84) to be expressed as

$$ \begin{aligned} q_2q_1(xi + yj + zk)q_1^{-1}q_2^{-1} = &(0.5699d_1 + 0.0143a_1 + 0.1362b_1 + 0.8102c_1) \\ &+ 0.5699(a_1i + b_1j + c_1k) \\ &+ d_1(-0.0143i - 0.1362j - 0.8102k) \\ &+ \begin{vmatrix} i & j & k \\ a_1 & b_1 & c_1 \\ -0.0143 & -0.1362 & -0.8102 \end{vmatrix}. \end{aligned} \tag{12.87} $$

Expanding this expression yields

$$\begin{aligned} q_2 q_1 (xi + yj + zk) q_1^{-1} q_2^{-1} = {} & i(-0.3500x - 0.9196y + 0.1784z) \\ & + j(0.9274x - 0.3133y + 0.2044z) \\ & + k(-0.1321x + 0.2370y + 0.9625z). \end{aligned} \tag{12.88}$$

Equation (12.88) shows that the point $[x, y, z]^T$ has been transformed to the point $[(-0.3500x - 0.9196y + 0.1784z), (0.9274x - 0.3133y + 0.2044z), (-0.1321x + 0.2370y + 0.9625z)]^T$. This is the same transformation as listed in Eq. (12.81).

12.8 Summary

It has been shown in this chapter that rigid-body rotations can be modeled by quaternions. It is a very simple procedure to determine the quaternion q that will rotate a point about an arbitrary axis vector by the quaternion operator $q(\ )q^{-1}$. Fewer mathematical operations are needed to compute q compared to the rotation matrix of Eq. (2.59). For this reason, quaternion algebra is often employed in many applications, as, for example, in computer graphics. Quaternions are introduced in this chapter because they elegantly quantify rigid-body rotations, which are the cornerstone of the spatial kinematics discussed throughout this book.

12.9 Problems

1. Prove that successive rotations by the angles ϕ, $\frac{\pi}{2}$ radians, and ϕ respectively about the x, y, and z axes are equivalent to a single rotation of $\frac{\pi}{2}$ radians about the y axis.

2. The quaternions q_1 and q_2 are given as

$$q_1 = \cos(30°) + \frac{\sin(30°)}{5}(3i + 4k),$$

$$q_2 = \cos(60°) + \frac{\sin(60°)}{\sqrt{29}}(5j - 2k).$$

Solve the following equation for the quaternion q_3:

$$q_1 q_3 = q_2 q_1.$$

3. q_1 is a unit quaternion, and p is a quaternion with no scalar component. Under what conditions will $p = q_1(p)q_1^{-1}$?

4. A box is moved from position 1 to position 2 and then to position 3 as shown in Figure 12.9. Determine the axis and angle of rotation that would move the box directly from position 1 to position 3.

5. A box has been rotated forty degrees about an axis parallel to $2\mathbf{i} + \mathbf{j} + \mathbf{k}$. It was then rotated sixty degrees about an axis parallel to $\mathbf{i} + 3\mathbf{j} - 2\mathbf{k}$ (measured with respect to the fixed coordinate system). You wish to return the box to its original orientation with

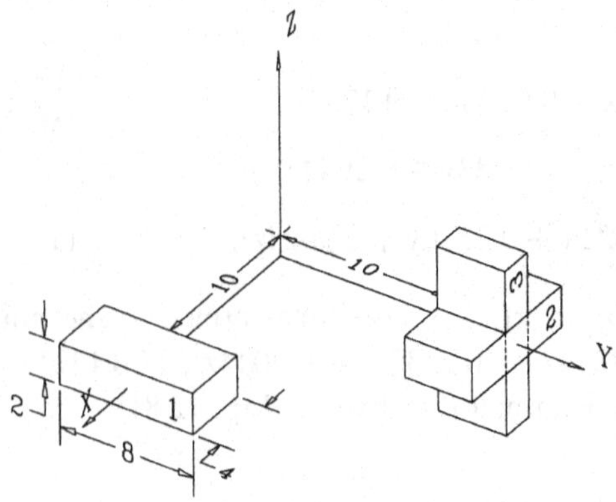

Figure 12.9. Two successive rotations.

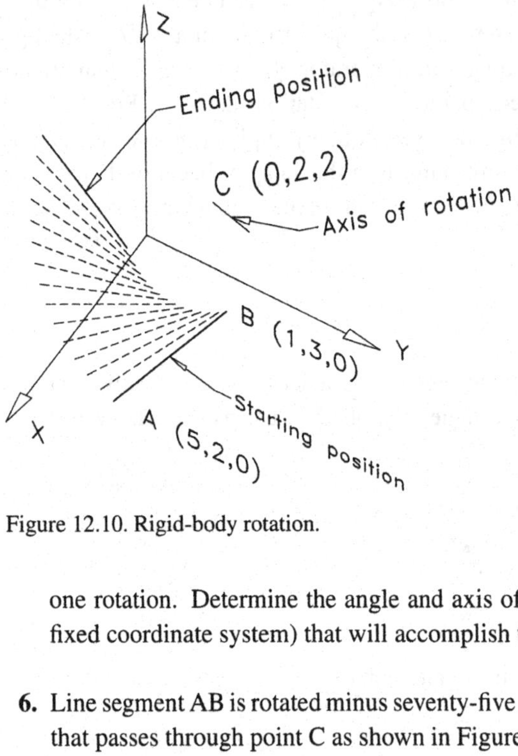

Figure 12.10. Rigid-body rotation.

one rotation. Determine the angle and axis of rotation (measured with respect to the fixed coordinate system) that will accomplish this.

6. Line segment AB is rotated minus seventy-five degrees about an axis parallel to $\mathbf{i}+\mathbf{j}+\mathbf{k}$ that passes through point C as shown in Figure 12.10. The coordinates of points A, B, and C are as follows:

A: $(5, 2, 0)$

B: $(1, 3, 0)$

C: $(0, 2, 2)$

Use quaternions to determine the coordinates of the endpoints of line segment AB after the rotation is accomplished.

7.

(a) Assume the quaternions p and q are given as follows:

$$p = (1, -1, 2, 3) \qquad q = (-3, 2, 1, 0.5)$$

What is the product pq? What is qp? What is $p^{-1}q$?

(b) Under what circumstances would the product pq equal qp? List all cases.

(c) What is the result of the quaternion product ijk?

Appendix

Spherical Equations
Polar Equations
Half-Tangent Laws
Direction Cosines

for a

triangle
quadrilateral
pentagon
hexagon
heptagon

Equations for a Spherical Triangle		
$X_1 = s_{23}s_2$	$Y_1 = s_{23}c_2$	$Z_1 = c_{23}$
$X_2 = s_{31}s_3$	$Y_2 = s_{31}c_3$	$Z_2 = c_{31}$
$X_3 = s_{12}s_1$	$Y_3 = s_{12}c_1$	$Z_3 = c_{12}$
$\bar{X}_1 = s_{23}s_3$	$\bar{Y}_1 = s_{23}c_3$	$\bar{Z}_1 = c_{23}$
$\bar{X}_2 = s_{31}s_1$	$\bar{Y}_2 = s_{31}c_1$	$\bar{Z}_2 = c_{31}$
$\bar{X}_3 = s_{12}s_2$	$\bar{Y}_3 = s_{12}c_2$	$\bar{Z}_3 = c_{12}$

Equations for a Polar Triangle		
$U_{12} = s_3s_{23}$	$V_{12} = s_3c_{23}$	$W_{12} = c_3$
$U_{23} = s_1s_{31}$	$V_{23} = s_1c_{31}$	$W_{23} = c_1$
$U_{31} = s_2s_{12}$	$V_{31} = s_2c_{12}$	$W_{31} = c_2$
$U_{21} = s_3s_{31}$	$V_{21} = s_3c_{31}$	$W_{21} = c_3$
$U_{32} = s_1s_{12}$	$V_{32} = s_1c_{12}$	$W_{32} = c_1$
$U_{13} = s_2s_{23}$	$V_{13} = s_2c_{23}$	$W_{13} = c_2$

Direction Cosines – Spatial Triangle

Set 1		
	$\underline{S}_1\,(0,\ 0,\ 1)$	$\underline{a}_{12}\,(1,\ 0,\ 0)$
	$\underline{S}_2\,(0,\ -s_{12},\ c_{12})$	$\underline{a}_{23}\,(c_2,\ s_2c_{12},\ U_{21})$
	$\underline{S}_3\,(\bar{X}_2,\ \bar{Y}_2,\ \bar{Z}_2)$	$\underline{a}_{31}\,(c_1,\ -s_1,\ 0)$
Set 2		
	$\underline{S}_2\,(0,\ 0,\ 1)$	$\underline{a}_{23}\,(1,\ 0,\ 0)$
	$\underline{S}_3\,(0,\ -s_{23},\ c_{23})$	$\underline{a}_{31}\,(c_3,\ s_3c_{23},\ U_{32})$
	$\underline{S}_1\,(\bar{X}_3,\ \bar{Y}_3,\ \bar{Z}_3)$	$\underline{a}_{12}\,(c_2,\ -s_2,\ 0)$
Set 3		
	$\underline{S}_3\,(0,\ 0,\ 1)$	$\underline{a}_{31}\,(1,\ 0,\ 0)$
	$\underline{S}_1\,(0,\ -s_{31},\ c_{31})$	$\underline{a}_{12}\,(c_1,\ s_1c_{31},\ U_{13})$
	$\underline{S}_2\,(\bar{X}_1,\ \bar{Y}_1,\ \bar{Z}_1)$	$\underline{a}_{23}\,(c_3,\ -s_3,\ 0)$
Set 4		
	$\underline{S}_1\,(0,\ 0,\ 1)$	$\underline{a}_{31}\,(1,\ 0,\ 0)$
	$\underline{S}_3\,(0,\ s_{31},\ c_{31})$	$\underline{a}_{23}\,(c_3,\ -s_3c_{31},\ U_{31})$
	$\underline{S}_2\,(X_3,\ -Y_3,\ Z_3)$	$\underline{a}_{12}\,(c_1,\ s_1,\ 0)$
Set 5		
	$\underline{S}_3\,(0,\ 0,\ 1)$	$\underline{a}_{23}\,(1,\ 0,\ 0)$
	$\underline{S}_2\,(0,\ s_{23},\ c_{23})$	$\underline{a}_{12}\,(c_2,\ -s_2c_{23},\ U_{23})$
	$\underline{S}_1\,(X_2,\ -Y_2,\ Z_2)$	$\underline{a}_{31}\,(c_3,\ s_3,\ 0)$
Set 6		
	$\underline{S}_2\,(0,\ 0,\ 1)$	$\underline{a}_{12}\,(1,\ 0,\ 0)$
	$\underline{S}_1\,(0,\ s_{12},\ c_{12})$	$\underline{a}_{31}\,(c_1,\ -s_1c_{12},\ U_{12})$
	$\underline{S}_3\,(X_1,\ -Y_1,\ Z_1)$	$\underline{a}_{23}\,(c_2,\ s_2,\ 0)$

Equations for a Spherical Quadrilateral

Fundamental Formulas:

$X_{12} = s_{34}s_3$	$Y_{12} = s_{34}c_3$	$Z_{12} = c_{34}$
$X_{23} = s_{41}s_4$	$Y_{23} = s_{41}c_4$	$Z_{23} = c_{41}$
$X_{34} = s_{12}s_1$	$Y_{34} = s_{12}c_1$	$Z_{34} = c_{12}$
$X_{41} = s_{23}s_2$	$Y_{41} = s_{23}c_2$	$Z_{41} = c_{23}$
$X_{21} = s_{34}s_4$	$Y_{21} = s_{34}c_4$	$Z_{21} = c_{34}$
$X_{32} = s_{41}s_1$	$Y_{32} = s_{41}c_1$	$Z_{32} = c_{41}$
$X_{43} = s_{12}s_2$	$Y_{43} = s_{12}c_2$	$Z_{43} = c_{12}$
$X_{14} = s_{23}s_3$	$Y_{14} = s_{23}c_3$	$Z_{14} = c_{23}$

Subsidiary Formulas:

$X_{12} = \bar{X}_3$	$-X^*_{12} = \bar{Y}_3$	$Z_1 = \bar{Z}_3$
$X_{23} = \bar{X}_4$	$-X^*_{23} = \bar{Y}_4$	$Z_2 = \bar{Z}_4$
$X_{34} = \bar{X}_1$	$-X^*_{34} = \bar{Y}_1$	$Z_3 = \bar{Z}_1$
$X_{41} = \bar{X}_2$	$-X^*_{41} = \bar{Y}_2$	$Z_4 = \bar{Z}_2$
$X_{21} = X_4$	$-X^*_{21} = Y_4$	$Z_2 = Z_4$
$X_{32} = X_1$	$-X^*_{32} = Y_1$	$Z_3 = Z_1$
$X_{43} = X_2$	$-X^*_{43} = Y_2$	$Z_4 = Z_2$
$X_{14} = X_3$	$-X^*_{14} = Y_3$	$Z_1 = Z_3$

Equations for a Polar Quadrilateral

Fundamental Formulas:

$U_{123} = s_4s_{34}$	$V_{123} = s_4c_{34}$	$W_{123} = c_4$
$U_{234} = s_1s_{41}$	$V_{234} = s_1c_{41}$	$W_{234} = c_1$
$U_{341} = s_2s_{12}$	$V_{341} = s_2c_{12}$	$W_{341} = c_2$
$U_{412} = s_3s_{23}$	$V_{412} = s_3c_{23}$	$W_{412} = c_3$
$U_{321} = s_4s_{41}$	$V_{321} = s_4c_{41}$	$W_{321} = c_4$
$U_{214} = s_3s_{34}$	$V_{214} = s_3c_{34}$	$W_{214} = c_3$
$U_{143} = s_2s_{23}$	$V_{143} = s_2c_{23}$	$W_{143} = c_2$
$U_{432} = s_1s_{12}$	$V_{432} = s_1c_{12}$	$W_{432} = c_1$

Subsidiary Formulas:

$U_{123} = U_{43}$	$U^*_{123} = -V_{43}$	$W_{12} = W_{43}$
$U_{234} = U_{14}$	$U^*_{234} = -V_{14}$	$W_{23} = W_{14}$
$U_{341} = U_{21}$	$U^*_{341} = -V_{21}$	$W_{34} = W_{21}$
$U_{412} = U_{32}$	$U^*_{412} = -V_{32}$	$W_{41} = W_{32}$
$U_{321} = U_{41}$	$U^*_{321} = -V_{41}$	$W_{32} = W_{41}$
$U_{214} = U_{34}$	$U^*_{214} = -V_{34}$	$W_{21} = W_{34}$
$U_{143} = U_{23}$	$U^*_{143} = -V_{23}$	$W_{14} = W_{23}$
$U_{432} = U_{12}$	$U^*_{432} = -V_{12}$	$W_{43} = W_{12}$

Half-Tangent Laws for a Quadrilateral

Set 1

$$x_1 = \frac{X_{34}}{Y_{34} + s_{12}} = -\frac{Y_{34} - s_{12}}{X_{34}} \qquad x_1 = \frac{X_{32}}{Y_{32} + s_{41}} = -\frac{Y_{32} - s_{41}}{X_{32}}$$

$$x_2 = \frac{X_{41}}{Y_{41} + s_{23}} = -\frac{Y_{41} - s_{23}}{X_{41}} \qquad x_2 = \frac{X_{43}}{Y_{43} + s_{12}} = -\frac{Y_{43} - s_{12}}{X_{43}}$$

$$x_3 = \frac{X_{12}}{Y_{12} + s_{34}} = -\frac{Y_{12} - s_{34}}{X_{12}} \qquad x_3 = \frac{X_{14}}{Y_{14} + s_{23}} = -\frac{Y_{14} - s_{23}}{X_{14}}$$

$$x_4 = \frac{X_{23}}{Y_{23} + s_{41}} = -\frac{Y_{23} - s_{41}}{X_{23}} \qquad x_4 = \frac{X_{21}}{Y_{21} + s_{34}} = -\frac{Y_{21} - s_{34}}{X_{21}}$$

Set 2

$$x_1 = \frac{X_4 - \bar{X}_2}{Y_4 - \bar{Y}_2} = -\frac{Y_4 + \bar{Y}_2}{X_4 + \bar{X}_2}$$

$$x_2 = \frac{X_1 - \bar{X}_3}{Y_1 - \bar{Y}_3} = -\frac{Y_1 + \bar{Y}_3}{X_1 + \bar{X}_3}$$

$$x_3 = \frac{X_2 - \bar{X}_4}{Y_2 - \bar{Y}_4} = -\frac{Y_2 + \bar{Y}_4}{X_2 + \bar{X}_4}$$

$$x_4 = \frac{X_3 - \bar{X}_1}{Y_3 - \bar{Y}_1} = -\frac{Y_3 + \bar{Y}_1}{X_3 + \bar{X}_1}$$

Direction Cosines – Spatial Quadrilateral

Set 1							
	$\underline{S}_1$ (0,	0,	1)	$\underline{a}_{12}$ (1,	0,	0)
	$\underline{S}_2$ (0,	$-s_{12}$,	c_{12})	$\underline{a}_{23}$ (c_2,	s_2c_{12},	U_{21})
	$\underline{S}_3$ ($\bar{X}_2$,	$\bar{Y}_2$,	$\bar{Z}_2$)	$\underline{a}_{34}$ (W_{32},	$-U^*_{321}$,	U_{321})
	$\underline{S}_4$ (X_{32},	Y_{32},	Z_{32})	$\underline{a}_{41}$ (c_1,	$-s_1$,	0)
Set 2							
	$\underline{S}_2$ (0,	0,	1)	$\underline{a}_{23}$ (1,	0,	0)
	$\underline{S}_3$ (0,	$-s_{23}$,	c_{23})	$\underline{a}_{34}$ (c_3,	s_3c_{23},	U_{32})
	$\underline{S}_4$ ($\bar{X}_3$,	$\bar{Y}_3$,	$\bar{Z}_3$)	$\underline{a}_{41}$ (W_{43},	$-U^*_{432}$,	U_{432})
	$\underline{S}_1$ (X_{43},	Y_{43},	Z_{43})	$\underline{a}_{12}$ (c_2,	$-s_2$,	0)
Set 3							
	$\underline{S}_3$ (0,	0,	1)	$\underline{a}_{34}$ (1,	0,	0)
	$\underline{S}_4$ (0,	$-s_{34}$,	c_{34})	$\underline{a}_{41}$ (c_4,	s_4c_{34},	U_{43})
	$\underline{S}_1$ ($\bar{X}_4$,	$\bar{Y}_4$,	$\bar{Z}_4$)	$\underline{a}_{12}$ (W_{14},	$-U^*_{143}$,	U_{143})
	$\underline{S}_2$ (X_{14},	Y_{14},	Z_{14})	$\underline{a}_{23}$ (c_3,	$-s_3$,	0)
Set 4							
	$\underline{S}_4$ (0,	0,	1)	$\underline{a}_{41}$ (1,	0,	0)
	$\underline{S}_1$ (0,	$-s_{41}$,	c_{41})	$\underline{a}_{12}$ (c_1,	s_1c_{41},	U_{14})
	$\underline{S}_2$ ($\bar{X}_1$,	$\bar{Y}_1$,	$\bar{Z}_1$)	$\underline{a}_{23}$ (W_{21},	$-U^*_{214}$,	U_{214})
	$\underline{S}_3$ (X_{21},	Y_{21},	Z_{21})	$\underline{a}_{34}$ (c_4,	$-s_4$,	0)
Set 5							
	$\underline{S}_1$ (0,	0,	1)	$\underline{a}_{41}$ (1,	0,	0)
	$\underline{S}_4$ (0,	s_{41},	c_{41})	$\underline{a}_{34}$ (c_4,	$-s_4c_{41}$,	U_{41})
	$\underline{S}_3$ (X_4,	$-Y_4$,	Z_4)	$\underline{a}_{23}$ (W_{34},	U^*_{341},	U_{341})
	$\underline{S}_2$ (X_{34},	$-Y_{34}$,	Z_{34})	$\underline{a}_{12}$ (c_1,	s_1,	0)
Set 6							
	$\underline{S}_4$ (0,	0,	1)	$\underline{a}_{34}$ (1,	0,	0)
	$\underline{S}_3$ (0,	s_{34},	c_{34})	$\underline{a}_{23}$ (c_3,	$-s_3c_{34}$,	U_{34})
	$\underline{S}_2$ (X_3,	$-Y_3$,	Z_3)	$\underline{a}_{12}$ (W_{23},	U^*_{234},	U_{234})
	$\underline{S}_1$ (X_{23},	$-Y_{23}$,	Z_{23})	$\underline{a}_{41}$ (c_4,	s_4,	0)
Set 7							
	$\underline{S}_3$ (0,	0,	1)	$\underline{a}_{23}$ (1,	0,	0)
	$\underline{S}_2$ (0,	s_{23},	c_{23})	$\underline{a}_{12}$ (c_2,	$-s_2c_{23}$,	U_{23})
	$\underline{S}_1$ (X_2,	$-Y_2$,	Z_2)	$\underline{a}_{41}$ (W_{12},	U^*_{123},	U_{123})
	$\underline{S}_4$ (X_{12},	$-Y_{12}$,	Z_{12})	$\underline{a}_{34}$ (c_3,	s_3,	0)
Set 8							
	$\underline{S}_2$ (0,	0,	1)	$\underline{a}_{12}$ (1,	0,	0)
	$\underline{S}_1$ (0,	s_{12},	c_{12})	$\underline{a}_{41}$ (c_1,	$-s_1c_{12}$,	U_{12})
	$\underline{S}_4$ (X_1,	$-Y_1$,	Z_1)	$\underline{a}_{34}$ (W_{41},	U^*_{412},	U_{412})
	$\underline{S}_3$ (X_{41},	$-Y_{41}$,	Z_{41})	$\underline{a}_{23}$ (c_2,	s_2,	0)

Equations for a Spherical Pentagon

Fundamental Formulas:

$X_{123} = s_{45}s_4$	$Y_{123} = s_{45}c_4$	$Z_{123} = c_{45}$
$X_{234} = s_{51}s_5$	$Y_{234} = s_{51}c_5$	$Z_{234} = c_{51}$
$X_{345} = s_{12}s_1$	$Y_{345} = s_{12}c_1$	$Z_{345} = c_{12}$
$X_{451} = s_{23}s_2$	$Y_{451} = s_{23}c_2$	$Z_{451} = c_{23}$
$X_{512} = s_{34}s_3$	$Y_{512} = s_{34}c_3$	$Z_{512} = c_{34}$
$X_{321} = s_{45}s_5$	$Y_{321} = s_{45}c_5$	$Z_{321} = c_{45}$
$X_{432} = s_{51}s_1$	$Y_{432} = s_{51}c_1$	$Z_{432} = c_{51}$
$X_{543} = s_{12}s_2$	$Y_{543} = s_{12}c_2$	$Z_{543} = c_{12}$
$X_{154} = s_{23}s_3$	$Y_{154} = s_{23}c_3$	$Z_{154} = c_{23}$
$X_{215} = s_{34}s_4$	$Y_{215} = s_{34}c_4$	$Z_{215} = c_{34}$

Subsidiary Formulas:

Set 1	$X_{123} = \bar{X}_4$	$-X^*_{123} = \bar{Y}_4$	$Z_{12} = \bar{Z}_4$
	$X_{234} = \bar{X}_5$	$-X^*_{234} = \bar{Y}_5$	$Z_{23} = \bar{Z}_5$
	$X_{345} = \bar{X}_1$	$-X^*_{345} = \bar{Y}_1$	$Z_{34} = \bar{Z}_1$
	$X_{451} = \bar{X}_2$	$-X^*_{451} = \bar{Y}_2$	$Z_{45} = \bar{Z}_2$
	$X_{512} = \bar{X}_3$	$-X^*_{512} = \bar{Y}_3$	$Z_{51} = \bar{Z}_3$
	$X_{321} = X_5$	$-X^*_{321} = Y_5$	$Z_{32} = Z_5$
	$X_{432} = X_1$	$-X^*_{432} = Y_1$	$Z_{43} = Z_1$
	$X_{543} = X_2$	$-X^*_{543} = Y_2$	$Z_{54} = Z_2$
	$X_{154} = X_3$	$-X^*_{154} = Y_3$	$Z_{15} = Z_3$
	$X_{215} = X_4$	$-X^*_{215} = Y_4$	$Z_{21} = Z_4$
Set 2	$X_{12} = X_{43}$	$Y_{12} = -X^*_{43}$	$Z_{12} = \bar{Z}_4$
	$X_{23} = X_{54}$	$Y_{23} = -X^*_{54}$	$Z_{23} = \bar{Z}_5$
	$X_{34} = X_{15}$	$Y_{34} = -X^*_{15}$	$Z_{34} = \bar{Z}_1$
	$X_{45} = X_{21}$	$Y_{45} = -X^*_{21}$	$Z_{45} = \bar{Z}_2$
	$X_{51} = X_{32}$	$Y_{51} = -X^*_{32}$	$Z_{51} = \bar{Z}_3$
	$X_{32} = X_{51}$	$Y_{32} = -X^*_{51}$	$Z_{32} = Z_5$
	$X_{43} = X_{12}$	$Y_{43} = -X^*_{12}$	$Z_{43} = Z_1$
	$X_{54} = X_{23}$	$Y_{54} = -X^*_{23}$	$Z_{54} = Z_2$
	$X_{15} = X_{34}$	$Y_{15} = -X^*_{34}$	$Z_{15} = Z_3$
	$X_{21} = X_{45}$	$Y_{21} = -X^*_{45}$	$Z_{21} = Z_4$

Equations for a Polar Pentagon

Fundamental Formulas:

$U_{1234} = s_5 s_{45}$	$V_{1234} = s_5 c_{45}$	$W_{1234} = c_5$
$U_{2345} = s_1 s_{51}$	$V_{2345} = s_1 c_{51}$	$W_{2345} = c_1$
$U_{3451} = s_2 s_{12}$	$V_{3451} = s_2 c_{12}$	$W_{3451} = c_2$
$U_{4512} = s_3 s_{23}$	$V_{4512} = s_3 c_{23}$	$W_{4512} = c_3$
$U_{5123} = s_4 s_{34}$	$V_{5123} = s_4 c_{34}$	$W_{5123} = c_4$
$U_{4321} = s_5 s_{51}$	$V_{4321} = s_5 c_{51}$	$W_{4321} = c_5$
$U_{3215} = s_4 s_{45}$	$V_{3215} = s_4 c_{45}$	$W_{3215} = c_4$
$U_{2154} = s_3 s_{34}$	$V_{2154} = s_3 c_{34}$	$W_{2154} = c_3$
$U_{1543} = s_2 s_{23}$	$V_{1543} = s_2 c_{23}$	$W_{1543} = c_2$
$U_{5432} = s_1 s_{12}$	$V_{5432} = s_1 c_{12}$	$W_{5432} = c_1$

Subsidiary Formulas:

Set 1	$U_{1234} = U_{54}$	$U^*_{1234} = -V_{54}$	$W_{123} = W_{54}$
	$U_{2345} = U_{15}$	$U^*_{2345} = -V_{15}$	$W_{234} = W_{15}$
	$U_{3451} = U_{21}$	$U^*_{3451} = -V_{21}$	$W_{345} = W_{21}$
	$U_{4512} = U_{32}$	$U^*_{4512} = -V_{32}$	$W_{451} = W_{32}$
	$U_{5123} = U_{43}$	$U^*_{5123} = -V_{43}$	$W_{512} = W_{43}$
	$U_{4321} = U_{51}$	$U^*_{4321} = -V_{51}$	$W_{432} = W_{51}$
	$U_{3215} = U_{45}$	$U^*_{3215} = -V_{45}$	$W_{321} = W_{45}$
	$U_{2154} = U_{34}$	$U^*_{2154} = -V_{34}$	$W_{215} = W_{34}$
	$U_{1543} = U_{23}$	$U^*_{1543} = -V_{23}$	$W_{154} = W_{23}$
	$U_{5432} = U_{12}$	$U^*_{5432} = -V_{12}$	$W_{543} = W_{12}$
Set 2	$U_{123} = U_{543}$	$V_{123} = -U^*_{543}$	$W_{123} = W_{54}$
	$U_{234} = U_{154}$	$V_{234} = -U^*_{154}$	$W_{234} = W_{15}$
	$U_{345} = U_{215}$	$V_{345} = -U^*_{215}$	$W_{345} = W_{21}$
	$U_{451} = U_{321}$	$V_{451} = -U^*_{321}$	$W_{451} = W_{32}$
	$U_{512} = U_{432}$	$V_{512} = -U^*_{432}$	$W_{512} = W_{43}$
	$U_{432} = U_{512}$	$V_{432} = -U^*_{512}$	$W_{432} = W_{51}$
	$U_{321} = U_{451}$	$V_{321} = -U^*_{451}$	$W_{321} = W_{45}$
	$U_{215} = U_{345}$	$V_{215} = -U^*_{345}$	$W_{215} = W_{34}$
	$U_{154} = U_{234}$	$V_{154} = -U^*_{234}$	$W_{154} = W_{23}$
	$U_{543} = U_{123}$	$V_{543} = -U^*_{123}$	$W_{543} = W_{12}$

Half-Tangent Laws for a Pentagon

Set 1

$$x_1 = \frac{X_{345}}{Y_{345} + s_{12}} = -\frac{Y_{345} - s_{12}}{X_{345}} \qquad x_1 = \frac{X_{432}}{Y_{432} + s_{51}} = -\frac{Y_{432} - s_{51}}{X_{432}}$$

$$x_2 = \frac{X_{451}}{Y_{451} + s_{23}} = -\frac{Y_{451} - s_{23}}{X_{451}} \qquad x_2 = \frac{X_{543}}{Y_{543} + s_{12}} = -\frac{Y_{543} - s_{12}}{X_{543}}$$

$$x_3 = \frac{X_{512}}{Y_{512} + s_{34}} = -\frac{Y_{512} - s_{34}}{X_{512}} \qquad x_3 = \frac{X_{154}}{Y_{154} + s_{23}} = -\frac{Y_{154} - s_{23}}{X_{154}}$$

$$x_4 = \frac{X_{123}}{Y_{123} + s_{45}} = -\frac{Y_{123} - s_{45}}{X_{123}} \qquad x_4 = \frac{X_{215}}{Y_{215} + s_{34}} = -\frac{Y_{215} - s_{34}}{X_{215}}$$

$$x_5 = \frac{X_{234}}{Y_{234} + s_{51}} = -\frac{Y_{234} - s_{51}}{X_{234}} \qquad x_5 = \frac{X_{321}}{Y_{321} + s_{45}} = -\frac{Y_{321} - s_{45}}{X_{321}}$$

Set 2

$$x_1 = \frac{X_{45} - \bar{X}_2}{Y_{45} - \bar{Y}_2} = -\frac{Y_{45} + \bar{Y}_2}{X_{45} + \bar{X}_2} \qquad x_1 = \frac{X_{32} - X_5}{Y_{32} - Y_5} = -\frac{Y_{32} + Y_5}{X_{32} + X_5}$$

$$x_2 = \frac{X_{51} - \bar{X}_3}{Y_{51} - \bar{Y}_3} = -\frac{Y_{51} + \bar{Y}_3}{X_{51} + \bar{X}_3} \qquad x_2 = \frac{X_{43} - X_1}{Y_{43} - Y_1} = -\frac{Y_{43} + Y_1}{X_{43} + X_1}$$

$$x_3 = \frac{X_{12} - \bar{X}_4}{Y_{12} - \bar{Y}_4} = -\frac{Y_{12} + \bar{Y}_4}{X_{12} + \bar{X}_4} \qquad x_3 = \frac{X_{54} - X_2}{Y_{54} - Y_2} = -\frac{Y_{54} + Y_2}{X_{54} + X_2}$$

$$x_4 = \frac{X_{23} - \bar{X}_5}{Y_{23} - \bar{Y}_5} = -\frac{Y_{23} + \bar{Y}_5}{X_{23} + \bar{X}_5} \qquad x_4 = \frac{X_{15} - X_3}{Y_{15} - Y_3} = -\frac{Y_{15} + Y_3}{X_{15} + X_3}$$

$$x_5 = \frac{X_{34} - \bar{X}_1}{Y_{34} - \bar{Y}_1} = -\frac{Y_{34} + \bar{Y}_1}{X_{34} + \bar{X}_1} \qquad x_5 = \frac{X_{21} - X_4}{Y_{21} - Y_4} = -\frac{Y_{21} + Y_4}{X_{21} + X_4}$$

Direction Cosines – Spatial Pentagon

Set 1

$\underline{S}_1\,(0,$	$0,$	$1)$	$\underline{a}_{12}\,(1,$	$0,$	$0)$
$\underline{S}_2\,(0,$	$-s_{12},$	$c_{12})$	$\underline{a}_{23}\,(c_2,$	$s_2c_{12},$	$U_{21})$
$\underline{S}_3\,(\bar{X}_2,$	$\bar{Y}_2,$	$\bar{Z}_2)$	$\underline{a}_{34}\,(W_{32},$	$-U^*_{321},$	$U_{321})$
$\underline{S}_4\,(X_{32},$	$Y_{32},$	$Z_{32})$	$\underline{a}_{45}\,(W_{432},$	$-U^*_{4321},$	$U_{4321})$
$\underline{S}_5\,(X_{432},$	$Y_{432},$	$Z_{432})$	$\underline{a}_{51}\,(c_1,$	$-s_1,$	$0)$

Set 2

$\underline{S}_2\,(0,$	$0,$	$1)$	$\underline{a}_{23}\,(1,$	$0,$	$0)$
$\underline{S}_3\,(0,$	$-s_{23},$	$c_{23})$	$\underline{a}_{34}\,(c_3,$	$s_3c_{23},$	$U_{32})$
$\underline{S}_4\,(\bar{X}_3,$	$\bar{Y}_3,$	$\bar{Z}_3)$	$\underline{a}_{45}\,(W_{43},$	$-U^*_{432},$	$U_{432})$
$\underline{S}_5\,(X_{43},$	$Y_{43},$	$Z_{43})$	$\underline{a}_{51}\,(W_{543},$	$-U^*_{5432},$	$U_{5432})$
$\underline{S}_1\,(X_{543},$	$Y_{543},$	$Z_{543})$	$\underline{a}_{12}\,(c_2,$	$-s_2,$	$0)$

Set 3

$\underline{S}_3\,(0,$	$0,$	$1)$	$\underline{a}_{34}\,(1,$	$0,$	$0)$
$\underline{S}_4\,(0,$	$-s_{34},$	$c_{34})$	$\underline{a}_{45}\,(c_4,$	$s_4c_{34},$	$U_{43})$
$\underline{S}_5\,(\bar{X}_4,$	$\bar{Y}_4,$	$\bar{Z}_4)$	$\underline{a}_{51}\,(W_{54},$	$-U^*_{543},$	$U_{543})$
$\underline{S}_1\,(X_{54},$	$Y_{54},$	$Z_{54})$	$\underline{a}_{12}\,(W_{154},$	$-U^*_{1543},$	$U_{1543})$
$\underline{S}_2\,(X_{154},$	$Y_{154},$	$Z_{154})$	$\underline{a}_{23}\,(c_3,$	$-s_3,$	$0)$

Set 4

$\underline{S}_4\,(0,$	$0,$	$1)$	$\underline{a}_{45}\,(1,$	$0,$	$0)$
$\underline{S}_5\,(0,$	$-s_{45},$	$c_{45})$	$\underline{a}_{51}\,(c_5,$	$s_5c_{45},$	$U_{54})$
$\underline{S}_1\,(\bar{X}_5,$	$\bar{Y}_5,$	$\bar{Z}_5)$	$\underline{a}_{12}\,(W_{15},$	$-U^*_{154},$	$U_{154})$
$\underline{S}_2\,(X_{15},$	$Y_{15},$	$Z_{15})$	$\underline{a}_{23}\,(W_{215},$	$-U^*_{2154},$	$U_{2154})$
$\underline{S}_3\,(X_{215},$	$Y_{215},$	$Z_{215})$	$\underline{a}_{34}\,(c_4,$	$-s_4,$	$0)$

Set 5

$\underline{S}_5\,(0,$	$0,$	$1)$	$\underline{a}_{51}\,(1,$	$0,$	$0)$
$\underline{S}_1\,(0,$	$-s_{51},$	$c_{51})$	$\underline{a}_{12}\,(c_1,$	$s_1c_{51},$	$U_{15})$
$\underline{S}_2\,(\bar{X}_1,$	$\bar{Y}_1,$	$\bar{Z}_1)$	$\underline{a}_{23}\,(W_{21},$	$-U^*_{215},$	$U_{215})$
$S_3\,(X_{21},$	$Y_{21},$	$Z_{21})$	$\underline{a}_{34}\,(W_{321},$	$-U^*_{3215},$	$U_{3215})$
$\underline{S}_4\,(X_{321},$	$Y_{321},$	$Z_{321})$	$\underline{a}_{45}\,(c_5,$	$-s_5,$	$0)$

Set 6

$\underline{S}_1\,(0,$	$0,$	$1)$	$\underline{a}_{51}\,(1,$	$0,$	$0)$
$\underline{S}_5\,(0,$	$s_{51},$	$c_{51})$	$\underline{a}_{45}\,(c_5,$	$-s_5c_{51},$	$U_{51})$
$\underline{S}_4\,(X_5,$	$-Y_5,$	$Z_5)$	$\underline{a}_{34}\,(W_{45},$	$U^*_{451},$	$U_{451})$
$\underline{S}_3\,(X_{45},$	$-Y_{45},$	$Z_{45})$	$\underline{a}_{23}\,(W_{345},$	$U^*_{3451},$	$U_{3451})$
$\underline{S}_2\,(X_{345},$	$-Y_{345},$	$Z_{345})$	$\underline{a}_{12}\,(c_1,$	$s_1,$	$0)$

Set 7

$\underline{S}_5\,(0,$	$0,$	$1)$	$\underline{a}_{45}\,(1,$	$0,$	$0)$
$\underline{S}_4\,(0,$	$s_{45},$	$c_{45})$	$\underline{a}_{34}\,(c_4,$	$-s_4c_{45},$	$U_{45})$
$\underline{S}_3\,(X_4,$	$-Y_4,$	$Z_4)$	$\underline{a}_{23}\,(W_{34},$	$U^*_{345},$	$U_{345})$
$\underline{S}_2\,(X_{34},$	$-Y_{34},$	$Z_{34})$	$\underline{a}_{12}\,(W_{234},$	$U^*_{2345},$	$U_{2345})$
$\underline{S}_1\,(X_{234},$	$-Y_{234},$	$Z_{234})$	$\underline{a}_{51}\,(c_5,$	$s_5,$	$0)$

Set 8

$\underline{S}_4\,(0,$	$0,$	$1)$	$\underline{a}_{34}\,(1,$	$0,$	$0)$
$\underline{S}_3\,(0,$	$s_{34},$	$c_{34})$	$\underline{a}_{23}\,(c_3,$	$-s_3c_{34},$	$U_{34})$
$\underline{S}_2\,(X_3,$	$-Y_3,$	$Z_3)$	$\underline{a}_{12}\,(W_{23},$	$U^*_{234},$	$U_{234})$
$\underline{S}_1\,(X_{23},$	$-Y_{23},$	$Z_{23})$	$\underline{a}_{51}\,(W_{123},$	$U^*_{1234},$	$U_{1234})$
$\underline{S}_5\,(X_{123},$	$-Y_{123},$	$Z_{123})$	$\underline{a}_{45}\,(c_4,$	$s_4,$	$0)$

Set 9						
	$\underline{S}_3$ (0,	0,	1)	$\underline{a}_{23}$ (1,	0,	0)
	$\underline{S}_2$ (0,	s_{23},	c_{23})	$\underline{a}_{12}$ (c_2,	$-s_2c_{23}$,	U_{23})
	$\underline{S}_1$ (X_2,	$-Y_2$,	Z_2)	$\underline{a}_{51}$ (W_{12},	U^*_{123},	U_{123})
	$\underline{S}_5$ (X_{12},	$-Y_{12}$,	Z_{12})	$\underline{a}_{45}$ (W_{512},	U^*_{5123},	U_{5123})
	$\underline{S}_4$ (X_{512},	$-Y_{512}$,	Z_{512})	$\underline{a}_{34}$ (c_3,	s_3,	0)
Set 10						
	$\underline{S}_2$ (0,	0,	1)	$\underline{a}_{12}$ (1,	0,	0)
	$\underline{S}_1$ (0,	s_{12},	c_{12})	$\underline{a}_{51}$ (c_1,	$-s_1c_{12}$,	U_{12})
	$\underline{S}_5$ (X_1,	$-Y_1$,	Z_1)	$\underline{a}_{45}$ (W_{51},	U^*_{512},	U_{512})
	$\underline{S}_4$ (X_{51},	$-Y_{51}$,	Z_{51})	$\underline{a}_{34}$ (W_{451},	U^*_{4512},	U_{4512})
	$\underline{S}_3$ (X_{451},	$-Y_{451}$,	Z_{451})	$\underline{a}_{23}$ (c_2,	s_2,	0)

Equations for a Spherical Hexagon

Fundamental Formulas:

$X_{1234} = s_{56}s_5$	$Y_{1234} = s_{56}c_5$	$Z_{1234} = c_{56}$
$X_{2345} = s_{61}s_6$	$Y_{2345} = s_{61}c_6$	$Z_{2345} = c_{61}$
$X_{3456} = s_{12}s_1$	$Y_{3456} = s_{12}c_1$	$Z_{3456} = c_{12}$
$X_{4561} = s_{23}s_2$	$Y_{4561} = s_{23}c_2$	$Z_{4561} = c_{23}$
$X_{5612} = s_{34}s_3$	$Y_{5612} = s_{34}c_3$	$Z_{5612} = c_{34}$
$X_{6123} = s_{45}s_4$	$Y_{6123} = s_{45}c_4$	$Z_{6123} = c_{45}$
$X_{4321} = s_{56}s_6$	$Y_{4321} = s_{56}c_6$	$Z_{4321} = c_{56}$
$X_{5432} = s_{61}s_1$	$Y_{5432} = s_{61}c_1$	$Z_{5432} = c_{61}$
$X_{6543} = s_{12}s_2$	$Y_{6543} = s_{12}c_2$	$Z_{6543} = c_{12}$
$X_{1654} = s_{23}s_3$	$Y_{1654} = s_{23}c_3$	$Z_{1654} = c_{23}$
$X_{2165} = s_{34}s_4$	$Y_{2165} = s_{34}c_4$	$Z_{2165} = c_{34}$
$X_{3216} = s_{45}s_5$	$Y_{3216} = s_{45}c_5$	$Z_{3216} = c_{45}$

Subsidiary Formulas:

Set 1	$X_{1234} = \bar{X}_5$	$-X^*_{1234} = \bar{Y}_5$	$Z_{123} = \bar{Z}_5$
	$X_{2345} = \bar{X}_6$	$-X^*_{2345} = \bar{Y}_6$	$Z_{234} = \bar{Z}_6$
	$X_{3456} = \bar{X}_1$	$-X^*_{3456} = \bar{Y}_1$	$Z_{345} = \bar{Z}_1$
	$X_{4561} = \bar{X}_2$	$-X^*_{4561} = \bar{Y}_2$	$Z_{456} = \bar{Z}_2$
	$X_{5612} = \bar{X}_3$	$-X^*_{5612} = \bar{Y}_3$	$Z_{561} = \bar{Z}_3$
	$X_{6123} = \bar{X}_4$	$-X^*_{6123} = \bar{Y}_4$	$Z_{612} = \bar{Z}_4$
	$X_{4321} = X_6$	$-X^*_{4321} = Y_6$	$Z_{432} = Z_6$
	$X_{5432} = X_1$	$-X^*_{5432} = Y_1$	$Z_{543} = Z_1$
	$X_{6543} = X_2$	$-X^*_{6543} = Y_2$	$Z_{654} = Z_2$
	$X_{1654} = X_3$	$-X^*_{1654} = Y_3$	$Z_{165} = Z_3$
	$X_{2165} = X_4$	$-X^*_{2615} = Y_4$	$Z_{216} = Z_4$
	$X_{3216} = X_5$	$-X^*_{3216} = Y_5$	$Z_{321} = Z_5$
Set 2	$X_{123} = X_{54}$	$Y_{123} = -X^*_{54}$	$Z_{123} = \bar{Z}_5$
	$X_{234} = X_{65}$	$Y_{234} = -X^*_{65}$	$Z_{234} = \bar{Z}_6$
	$X_{345} = X_{16}$	$Y_{345} = -X^*_{16}$	$Z_{345} = \bar{Z}_1$
	$X_{456} = X_{21}$	$Y_{456} = -X^*_{21}$	$Z_{456} = \bar{Z}_2$
	$X_{561} = X_{32}$	$Y_{561} = -X^*_{32}$	$Z_{561} = \bar{Z}_3$
	$X_{612} = X_{43}$	$Y_{612} = -X^*_{43}$	$Z_{612} = \bar{Z}_4$

	$X_{432} = X_{61}$	$Y_{432} = -X^*_{61}$	$Z_{432} = Z_6$
	$X_{543} = X_{12}$	$Y_{543} = -X^*_{12}$	$Z_{543} = Z_1$
	$X_{654} = X_{23}$	$Y_{654} = -X^*_{23}$	$Z_{654} = Z_2$
	$X_{165} = X_{34}$	$Y_{165} = -X^*_{34}$	$Z_{165} = Z_3$
	$X_{216} = X_{45}$	$Y_{216} = -X^*_{45}$	$Z_{216} = Z_4$
	$X_{321} = X_{56}$	$Y_{321} = -X^*_{56}$	$Z_{321} = Z_5$
Set 3	$X_{123} = X_{54}$	$-X^*_{123} = Y_{54}$	$Z_{12} = Z_{54}$
	$X_{234} = X_{65}$	$-X^*_{234} = Y_{65}$	$Z_{23} = Z_{65}$
	$X_{345} = X_{16}$	$-X^*_{345} = Y_{16}$	$Z_{34} = Z_{16}$
	$X_{456} = X_{21}$	$-X^*_{456} = Y_{21}$	$Z_{45} = Z_{21}$
	$X_{561} = X_{32}$	$-X^*_{561} = Y_{32}$	$Z_{56} = Z_{32}$
	$X_{612} = X_{43}$	$-X^*_{612} = Y_{43}$	$Z_{61} = Z_{43}$
	$X_{432} = X_{61}$	$-X^*_{432} = Y_{61}$	$Z_{43} = Z_{61}$
	$X_{543} = X_{12}$	$-X^*_{543} = Y_{12}$	$Z_{54} = Z_{12}$
	$X_{654} = X_{23}$	$-X^*_{654} = Y_{23}$	$Z_{65} = Z_{23}$
	$X_{165} = X_{34}$	$-X^*_{165} = Y_{34}$	$Z_{16} = Z_{34}$
	$X_{216} = X_{45}$	$-X^*_{216} = Y_{45}$	$Z_{21} = Z_{45}$
	$X_{321} = X_{56}$	$-X^*_{321} = Y_{56}$	$Z_{32} = Z_{56}$

Equations for a Polar Hexagon

Fundamental Formulas:

$U_{12345} = s_6 s_{56}$	$V_{12345} = s_6 c_{56}$	$W_{12345} = c_6$
$U_{23456} = s_1 s_{61}$	$V_{23456} = s_1 c_{61}$	$W_{23456} = c_1$
$U_{34561} = s_2 s_{12}$	$V_{34561} = s_2 c_{12}$	$W_{34561} = c_2$
$U_{45612} = s_3 s_{23}$	$V_{45612} = s_3 c_{23}$	$W_{45612} = c_3$
$U_{56123} = s_4 s_{34}$	$V_{56123} = s_4 c_{34}$	$W_{56123} = c_4$
$U_{61234} = s_5 s_{45}$	$V_{61234} = s_5 c_{45}$	$W_{61234} = c_5$
$U_{54321} = s_6 s_{61}$	$V_{54321} = s_6 c_{61}$	$W_{54321} = c_6$
$U_{43216} = s_5 s_{56}$	$V_{43216} = s_5 c_{56}$	$W_{43216} = c_5$
$U_{32165} = s_4 s_{45}$	$V_{32165} = s_4 c_{45}$	$W_{32165} = c_4$
$U_{21654} = s_3 s_{34}$	$V_{21654} = s_3 c_{34}$	$W_{21654} = c_3$
$U_{16543} = s_2 s_{23}$	$V_{16543} = s_2 c_{23}$	$W_{16543} = c_2$
$U_{65432} = s_1 s_{12}$	$V_{65432} = s_1 c_{12}$	$W_{65432} = c_1$

Subsidiary Formulas:

Set 1	$U_{12345} = U_{65}$	$U^*_{12345} = -V_{65}$	$W_{1234} = W_{65}$
	$U_{23456} = U_{16}$	$U^*_{23456} = -V_{16}$	$W_{2345} = W_{16}$
	$U_{34561} = U_{21}$	$U^*_{34561} = -V_{21}$	$W_{3456} = W_{21}$
	$U_{45612} = U_{32}$	$U^*_{45612} = -V_{32}$	$W_{4561} = W_{32}$
	$U_{56123} = U_{43}$	$U^*_{56123} = -V_{43}$	$W_{5612} = W_{43}$
	$U_{61234} = U_{54}$	$U^*_{61234} = -V_{54}$	$W_{6123} = W_{54}$
	$U_{54321} = U_{61}$	$U^*_{54321} = -V_{61}$	$W_{5432} = W_{61}$
	$U_{43216} = U_{56}$	$U^*_{43216} = -V_{56}$	$W_{4321} = W_{56}$
	$U_{32165} = U_{45}$	$U^*_{32165} = -V_{45}$	$W_{3216} = W_{45}$
	$U_{21654} = U_{34}$	$U^*_{21654} = -V_{34}$	$W_{2165} = W_{34}$
	$U_{16543} = U_{23}$	$U^*_{16543} = -V_{23}$	$W_{1654} = W_{23}$
	$U_{65432} = U_{12}$	$U^*_{65432} = -V_{12}$	$W_{6543} = W_{12}$
Set 2	$U_{1234} = U_{654}$	$V_{1234} = -U^*_{654}$	$W_{1234} = W_{65}$
	$U_{2345} = U_{165}$	$V_{2345} = -U^*_{165}$	$W_{2345} = W_{16}$
	$U_{3456} = U_{216}$	$V_{3456} = -U^*_{216}$	$W_{3456} = W_{21}$
	$U_{4561} = U_{321}$	$V_{4561} = -U^*_{321}$	$W_{4561} = W_{32}$
	$U_{5612} = U_{432}$	$V_{5612} = -U^*_{432}$	$W_{5612} = W_{43}$
	$U_{6123} = U_{543}$	$V_{6123} = -U^*_{543}$	$W_{6123} = W_{54}$
	$U_{5432} = U_{612}$	$V_{5432} = -U^*_{612}$	$W_{5432} = W_{61}$
	$U_{4321} = U_{561}$	$V_{4321} = -U^*_{561}$	$W_{4321} = W_{56}$
	$U_{3216} = U_{456}$	$V_{3216} = -U^*_{456}$	$W_{3216} = W_{45}$
	$U_{2165} = U_{345}$	$V_{2165} = -U^*_{345}$	$W_{2165} = W_{34}$
	$U_{1654} = U_{234}$	$V_{1654} = -U^*_{234}$	$W_{1654} = W_{23}$
	$U_{6543} = U_{123}$	$V_{6543} = -U^*_{123}$	$W_{6543} = W_{12}$
Set 3	$U_{1234} = U_{654}$	$U^*_{1234} = -V_{654}$	$W_{123} = W_{654}$
	$U_{2345} = U_{165}$	$U^*_{2345} = -V_{165}$	$W_{234} = W_{165}$
	$U_{3456} = U_{216}$	$U^*_{3456} = -V_{216}$	$W_{345} = W_{216}$
	$U_{4561} = U_{321}$	$U^*_{4561} = -V_{321}$	$W_{456} = W_{321}$
	$U_{5612} = U_{432}$	$U^*_{5612} = -V_{432}$	$W_{561} = W_{432}$
	$U_{6123} = U_{543}$	$U^*_{6123} = -V_{543}$	$W_{612} = W_{543}$

$U_{5432} = U_{612}$	$U^*_{5432} = -V_{612}$	$W_{543} = W_{612}$
$U_{4321} = U_{561}$	$U^*_{4321} = -V_{561}$	$W_{432} = W_{561}$
$U_{3216} = U_{456}$	$U^*_{3216} = -V_{456}$	$W_{321} = W_{456}$
$U_{2165} = U_{345}$	$U^*_{2165} = -V_{345}$	$W_{216} = W_{345}$
$U_{1654} = U_{234}$	$U^*_{1654} = -V_{234}$	$W_{165} = W_{234}$
$U_{6543} = U_{123}$	$U^*_{6543} = -V_{123}$	$W_{654} = W_{123}$

Half-Tangent Laws for a Hexagon

Set 1

$$x_1 = \frac{X_{3456}}{Y_{3456} + s_{12}} = -\frac{Y_{3456} - s_{12}}{X_{3456}} \qquad x_1 = \frac{X_{5432}}{Y_{5432} + s_{61}} = -\frac{Y_{5432} - s_{61}}{X_{5432}}$$

$$x_2 = \frac{X_{4561}}{Y_{4561} + s_{23}} = -\frac{Y_{4561} - s_{23}}{X_{4561}} \qquad x_2 = \frac{X_{6543}}{Y_{6543} + s_{12}} = -\frac{Y_{6543} - s_{12}}{X_{6543}}$$

$$x_3 = \frac{X_{5612}}{Y_{5612} + s_{34}} = -\frac{Y_{5612} - s_{34}}{X_{5612}} \qquad x_3 = \frac{X_{1654}}{Y_{1654} + s_{23}} = -\frac{Y_{1654} - s_{23}}{X_{1654}}$$

$$x_4 = \frac{X_{6123}}{Y_{6123} + s_{45}} = -\frac{Y_{6123} - s_{45}}{X_{6123}} \qquad x_4 = \frac{X_{2165}}{Y_{2165} + s_{34}} = -\frac{Y_{2165} - s_{34}}{X_{2165}}$$

$$x_5 = \frac{X_{1234}}{Y_{1234} + s_{56}} = -\frac{Y_{1234} - s_{56}}{X_{1234}} \qquad x_5 = \frac{X_{3216}}{Y_{3216} + s_{45}} = -\frac{Y_{3216} - s_{45}}{X_{3216}}$$

$$x_6 = \frac{X_{2345}}{Y_{2345} + s_{61}} = -\frac{Y_{2345} - s_{61}}{X_{2345}} \qquad x_6 = \frac{X_{4321}}{Y_{4321} + s_{56}} = -\frac{Y_{4321} - s_{56}}{X_{4321}}$$

Set 2

$$x_1 = \frac{X_{456} - \bar{X}_2}{Y_{456} - \bar{Y}_2} = -\frac{Y_{456} + \bar{Y}_2}{X_{456} + \bar{X}_2} \qquad x_1 = \frac{X_{432} - X_6}{Y_{432} - Y_6} = -\frac{Y_{432} + Y_6}{X_{432} + X_6}$$

$$x_2 = \frac{X_{561} - \bar{X}_3}{Y_{561} - \bar{Y}_3} = -\frac{Y_{561} + \bar{Y}_3}{X_{561} + \bar{X}_3} \qquad x_2 = \frac{X_{543} - X_1}{Y_{543} - Y_1} = -\frac{Y_{543} + Y_1}{X_{543} + X_1}$$

$$x_3 = \frac{X_{612} - \bar{X}_4}{Y_{612} - \bar{Y}_4} = -\frac{Y_{612} + \bar{Y}_4}{X_{612} + \bar{X}_4} \qquad x_3 = \frac{X_{654} - X_2}{Y_{654} - Y_2} = -\frac{Y_{654} + Y_2}{X_{654} + X_2}$$

$$x_4 = \frac{X_{123} - \bar{X}_5}{Y_{123} - \bar{Y}_5} = -\frac{Y_{123} + \bar{Y}_5}{X_{123} + \bar{X}_5} \qquad x_4 = \frac{X_{165} - X_3}{Y_{165} - Y_3} = -\frac{Y_{165} + Y_3}{X_{165} + X_3}$$

$$x_5 = \frac{X_{234} - \bar{X}_6}{Y_{234} - \bar{Y}_6} = -\frac{Y_{234} + \bar{Y}_6}{X_{234} + \bar{X}_6} \qquad x_5 = \frac{X_{216} - X_4}{Y_{216} - Y_4} = -\frac{Y_{216} + Y_4}{X_{216} + X_4}$$

$$x_6 = \frac{X_{345} - \bar{X}_1}{Y_{345} - \bar{Y}_1} = -\frac{Y_{345} + \bar{Y}_1}{X_{345} + \bar{X}_1} \qquad x_6 = \frac{X_{321} - X_5}{Y_{321} - Y_5} = -\frac{Y_{321} + Y_5}{X_{321} + X_5}$$

Set 3

$$x_1 = \frac{X_{56} - X_{32}}{Y_{56} - Y_{32}} = -\frac{Y_{56} + Y_{32}}{X_{56} + X_{32}}$$

$$x_2 = \frac{X_{61} - X_{43}}{Y_{61} - Y_{43}} = -\frac{Y_{61} + Y_{43}}{X_{61} + X_{43}}$$

$$x_3 = \frac{X_{12} - X_{54}}{Y_{12} - Y_{54}} = -\frac{Y_{12} + Y_{54}}{X_{12} + X_{54}}$$

$$x_4 = \frac{X_{23} - X_{65}}{Y_{23} - Y_{65}} = -\frac{Y_{23} + Y_{65}}{X_{23} + X_{65}}$$

$$x_5 = \frac{X_{34} - X_{16}}{Y_{34} - Y_{16}} = -\frac{Y_{34} + Y_{16}}{X_{34} + X_{16}}$$

$$x_6 = \frac{X_{45} - X_{21}}{Y_{45} - Y_{21}} = -\frac{Y_{45} + Y_{21}}{X_{45} + X_{21}}$$

Direction Cosines – Spatial Hexagon

Set 1

$\underline{S}_1$ (0,	0,	1)	$\underline{a}_{12}$ (1,	0,	0)
$\underline{S}_2$ (0,	$-s_{12}$,	c_{12})	$\underline{a}_{23}$ (c_2,	s_2c_{12},	U_{21})
$\underline{S}_3$ ($\bar{X}_2$,	$\bar{Y}_2$,	$\bar{Z}_2$)	$\underline{a}_{34}$ (W_{32},	$-U^*_{321}$,	U_{321})
$\underline{S}_4$ (X_{32},	Y_{32},	Z_{32})	$\underline{a}_{45}$ (W_{432},	$-U^*_{4321}$,	U_{4321})
$\underline{S}_5$ (X_{432},	Y_{432},	Z_{432})	$\underline{a}_{56}$ (W_{5432},	$-U^*_{54321}$,	U_{54321})
$\underline{S}_6$ (X_{5432},	Y_{5432},	Z_{5432})	$\underline{a}_{61}$ (c_1,	$-s_1$,	0)

Set 2

$\underline{S}_2$ (0,	0,	1)	$\underline{a}_{23}$ (1,	0,	0)
$\underline{S}_3$ (0,	$-s_{23}$,	c_{23})	$\underline{a}_{34}$ (c_3,	s_3c_{23},	U_{32})
$\underline{S}_4$ ($\bar{X}_3$,	$\bar{Y}_3$,	$\bar{Z}_3$)	$\underline{a}_{45}$ (W_{43},	$-U^*_{432}$,	U_{432})
$\underline{S}_5$ (X_{43},	Y_{43},	Z_{43})	$\underline{a}_{56}$ (W_{543},	$-U^*_{5432}$,	U_{5432})
$\underline{S}_6$ (X_{543},	Y_{543},	Z_{543})	$\underline{a}_{61}$ (W_{6543},	$-U^*_{65432}$,	U_{65432})
$\underline{S}_1$ (X_{6543},	Y_{6543},	Z_{6543})	$\underline{a}_{12}$ (c_2,	$-s_2$,	0)

Set 3

$\underline{S}_3$ (0,	0,	1)	$\underline{a}_{34}$ (1,	0,	0)
$\underline{S}_4$ (0,	$-s_{34}$,	c_{34})	$\underline{a}_{45}$ (c_4,	s_4c_{34},	U_{43})
$\underline{S}_5$ ($\bar{X}_4$,	$\bar{Y}_4$,	$\bar{Z}_4$)	$\underline{a}_{56}$ (W_{54},	$-U^*_{543}$,	U_{543})
$\underline{S}_6$ (X_{54},	Y_{54},	Z_{54})	$\underline{a}_{61}$ (W_{654},	$-U^*_{6543}$,	U_{6543})
$\underline{S}_1$ (X_{654},	Y_{654},	Z_{654})	$\underline{a}_{12}$ (W_{1654},	$-U^*_{16543}$,	U_{16543})
$\underline{S}_2$ (X_{1654},	Y_{1654},	Z_{1654})	$\underline{a}_{23}$ (c_3,	$-s_3$,	0)

Set 4

$\underline{S}_4$ (0,	0,	1)	$\underline{a}_{45}$ (1,	0,	0)
$\underline{S}_5$ (0,	$-s_{45}$,	c_{45})	$\underline{a}_{56}$ (c_5,	s_5c_{45},	U_{54})
$\underline{S}_6$ ($\bar{X}_5$,	$\bar{Y}_5$,	$\bar{Z}_5$)	$\underline{a}_{61}$ (W_{65},	$-U^*_{654}$,	U_{654})
$\underline{S}_1$ (X_{65},	Y_{65},	Z_{65})	$\underline{a}_{12}$ (W_{165},	$-U^*_{1654}$,	U_{1654})
$\underline{S}_2$ (X_{165},	Y_{165},	Z_{165})	$\underline{a}_{23}$ (W_{2165},	$-U^*_{21654}$,	U_{21654})
$\underline{S}_3$ (X_{2165},	Y_{2165},	Z_{2165})	$\underline{a}_{34}$ (c_4,	$-s_4$,	0)

Set 5

$\underline{S}_5$ (0,	0,	1)	$\underline{a}_{56}$ (1,	0,	0)
$\underline{S}_6$ (0,	$-s_{56}$,	c_{56})	$\underline{a}_{61}$ (c_6,	s_6c_{56},	U_{65})
$\underline{S}_1$ ($\bar{X}_6$,	$\bar{Y}_6$,	$\bar{Z}_6$)	$\underline{a}_{12}$ (W_{16},	$-U^*_{165}$,	U_{165})
$\underline{S}_2$ (X_{16},	Y_{16},	Z_{16})	$\underline{a}_{23}$ (W_{216},	$-U^*_{2165}$,	U_{2165})
$\underline{S}_3$ (X_{216},	Y_{216},	Z_{216})	$\underline{a}_{34}$ (W_{3216},	$-U^*_{32165}$,	U_{32165})
$\underline{S}_4$ (X_{3216},	Y_{3216},	Z_{3216})	$\underline{a}_{45}$ (c_5,	$-s_5$,	0)

Set 6

$\underline{S}_6$ (0,	0,	1)	$\underline{a}_{61}$ (1,	0,	0)
$\underline{S}_1$ (0,	$-s_{61}$,	c_{61})	$\underline{a}_{12}$ (c_1,	s_1c_{61},	U_{16})
$\underline{S}_2$ ($\bar{X}_1$,	$\bar{Y}_1$,	$\bar{Z}_1$)	$\underline{a}_{23}$ (W_{21},	$-U^*_{216}$,	U_{216})
$\underline{S}_3$ (X_{21},	Y_{21},	Z_{21})	$\underline{a}_{34}$ (W_{321},	$-U^*_{3216}$,	U_{3216})
$\underline{S}_4$ (X_{321},	Y_{321},	Z_{321})	$\underline{a}_{45}$ (W_{4321},	$-U^*_{43216}$,	U_{43216})
$\underline{S}_5$ (X_{4321},	Y_{4321},	Z_{4321})	$\underline{a}_{56}$ (c_6,	$-s_6$,	0)

Set 7

$\underline{S}_1$ (0,	0,	1)	$\underline{a}_{61}$ (1,	0,	0)
$\underline{S}_6$ (0,	s_{61},	c_{61})	$\underline{a}_{56}$ (c_6,	$-s_6c_{61}$,	U_{61})
$\underline{S}_5$ (X_6,	$-Y_6$,	Z_6)	$\underline{a}_{45}$ (W_{56},	U^*_{561},	U_{561})
$\underline{S}_4$ (X_{56},	$-Y_{56}$,	Z_{56})	$\underline{a}_{34}$ (W_{456},	U^*_{4561},	U_{4561})
$\underline{S}_3$ (X_{456},	$-Y_{456}$,	Z_{456})	$\underline{a}_{23}$ (W_{3456},	U^*_{34561},	U_{34561})
$\underline{S}_2$ (X_{3456},	$-Y_{3456}$,	Z_{3456})	$\underline{a}_{12}$ (c_1,	s_1,	0)

Set 8

$\underline{S}_6$ (0,	0,	1)	$\underline{a}_{56}$ (1,	0,	0)
$\underline{S}_5$ (0,	s_{56},	c_{56})	$\underline{a}_{45}$ (c_5,	$-s_5c_{56}$,	U_{56})
$\underline{S}_4$ (X_5,	$-Y_5$,	Z_5)	$\underline{a}_{34}$ (W_{45},	U^*_{456},	U_{456})
$\underline{S}_3$ (X_{45},	$-Y_{45}$,	Z_{45})	$\underline{a}_{23}$ (W_{345},	U^*_{3456},	U_{3456})
$\underline{S}_2$ (X_{345},	$-Y_{345}$,	Z_{345})	$\underline{a}_{12}$ (W_{2345},	U^*_{23456},	U_{23456})
$\underline{S}_1$ (X_{2345},	$-Y_{2345}$,	Z_{2345})	$\underline{a}_{61}$ (c_6,	s_6,	0)

Set 9

$\underline{S}_5$ (0,	0,	1)	$\underline{a}_{45}$ (1,	0,	0)
$\underline{S}_4$ (0,	s_{45},	c_{45})	$\underline{a}_{34}$ (c_4,	$-s_4c_{45}$,	U_{45})
$\underline{S}_3$ (X_4,	$-Y_4$,	Z_4)	$\underline{a}_{23}$ (W_{34},	U^*_{345},	U_{345})
$\underline{S}_2$ (X_{34},	$-Y_{34}$,	Z_{34})	$\underline{a}_{12}$ (W_{234},	U^*_{2345},	U_{2345})
$\underline{S}_1$ (X_{234},	$-Y_{234}$,	Z_{234})	$\underline{a}_{61}$ (W_{1234},	U^*_{12345},	U_{12345})
$\underline{S}_6$ (X_{1234},	$-Y_{1234}$,	Z_{1234})	$\underline{a}_{56}$ (c_5,	s_5,	0)

Set 10

$\underline{S}_4$ (0,	0,	1)	$\underline{a}_{34}$ (1,	0,	0)
$\underline{S}_3$ (0,	s_{34},	c_{34})	$\underline{a}_{23}$ (c_3,	$-s_3c_{34}$,	U_{34})
$\underline{S}_2$ (X_3,	$-Y_3$,	Z_3)	$\underline{a}_{12}$ (W_{23},	U^*_{234},	U_{234})
$\underline{S}_1$ (X_{23},	$-Y_{23}$,	Z_{23})	$\underline{a}_{61}$ (W_{123},	U^*_{1234},	U_{1234})
$\underline{S}_6$ (X_{123},	$-Y_{123}$,	Z_{123})	$\underline{a}_{56}$ (W_{6123},	U^*_{61234},	U_{61234})
$\underline{S}_5$ (X_{6123},	$-Y_{6123}$,	Z_{6123})	$\underline{a}_{45}$ (c_4,	s_4,	0)

Set 11

$\underline{S}_3$ (0,	0,	1)	$\underline{a}_{23}$ (1,	0,	0)
$\underline{S}_2$ (0,	s_{23},	c_{23})	$\underline{a}_{12}$ (c_2,	$-s_2c_{23}$,	U_{23})
$\underline{S}_1$ (X_2,	$-Y_2$,	Z_2)	$\underline{a}_{61}$ (W_{12},	U^*_{123},	U_{123})
$\underline{S}_6$ (X_{12},	$-Y_{12}$,	Z_{12})	$\underline{a}_{56}$ (W_{612},	U^*_{6123},	U_{6123})
$\underline{S}_5$ (X_{612},	$-Y_{612}$,	Z_{612})	$\underline{a}_{45}$ (W_{5612},	U^*_{56123},	U_{56123})
$\underline{S}_4$ (X_{5612},	$-Y_{5612}$,	Z_{5612})	$\underline{a}_{34}$ (c_3,	s_3,	0)

Set 12

$\underline{S}_2$ (0,	0,	1)	$\underline{a}_{12}$ (1,	0,	0)
$\underline{S}_1$ (0,	s_{12},	c_{12})	$\underline{a}_{61}$ (c_1,	$-s_1c_{12}$,	U_{12})
$\underline{S}_6$ (X_1,	$-Y_1$,	Z_1)	$\underline{a}_{56}$ (W_{61},	U^*_{612},	U_{612})
$\underline{S}_5$ (X_{61},	$-Y_{61}$,	Z_{61})	$\underline{a}_{45}$ (W_{561},	U^*_{5612},	U_{5612})
$\underline{S}_4$ (X_{561},	$-Y_{561}$,	Z_{561})	$\underline{a}_{34}$ (W_{4561},	U^*_{45612},	U_{45612})
$\underline{S}_3$ (X_{4561},	$-Y_{4561}$,	Z_{4561})	$\underline{a}_{23}$ (c_2,	s_2,	0)

Equations for a Spherical Heptagon

Fundamental Formulas:

$X_{12345} = s_{67}s_6$	$Y_{12345} = s_{67}c_6$	$Z_{12345} = c_{67}$
$X_{23456} = s_{71}s_7$	$Y_{23456} = s_{71}c_7$	$Z_{23456} = c_{71}$
$X_{34567} = s_{12}s_1$	$Y_{34567} = s_{12}c_1$	$Z_{34567} = c_{12}$
$X_{45671} = s_{23}s_2$	$Y_{45671} = s_{23}c_2$	$Z_{45671} = c_{23}$
$X_{56712} = s_{34}s_3$	$Y_{56712} = s_{34}c_3$	$Z_{56712} = c_{34}$
$X_{67123} = s_{45}s_4$	$Y_{67123} = s_{45}c_4$	$Z_{67123} = c_{45}$
$X_{71234} = s_{56}s_5$	$Y_{71234} = s_{56}c_5$	$Z_{71234} = c_{56}$
$X_{54321} = s_{67}s_7$	$Y_{54321} = s_{67}c_7$	$Z_{54321} = c_{67}$
$X_{65432} = s_{71}s_1$	$Y_{65432} = s_{71}c_1$	$Z_{65432} = c_{71}$
$X_{76543} = s_{12}s_2$	$Y_{76543} = s_{12}c_2$	$Z_{76543} = c_{12}$
$X_{17654} = s_{23}s_3$	$Y_{17654} = s_{23}c_3$	$Z_{17654} = c_{23}$
$X_{21765} = s_{34}s_4$	$Y_{21765} = s_{34}c_4$	$Z_{21765} = c_{34}$
$X_{32176} = s_{45}s_5$	$Y_{32176} = s_{45}c_5$	$Z_{32176} = c_{45}$
$X_{43217} = s_{56}s_6$	$Y_{43217} = s_{56}c_6$	$Z_{43217} = c_{56}$

Subsidiary Formulas:

Set 1	$X_{12345} = \bar{X}_6$	$-X^*_{12345} = \bar{Y}_6$	$Z_{1234} = \bar{Z}_6$
	$X_{23456} = \bar{X}_7$	$-X^*_{23456} = \bar{Y}_7$	$Z_{2345} = \bar{Z}_7$
	$X_{34567} = \bar{X}_1$	$-X^*_{34567} = \bar{Y}_1$	$Z_{3456} = \bar{Z}_1$
	$X_{45671} = \bar{X}_2$	$-X^*_{45671} = \bar{Y}_2$	$Z_{4567} = \bar{Z}_2$
	$X_{56712} = \bar{X}_3$	$-X^*_{56712} = \bar{Y}_3$	$Z_{5671} = \bar{Z}_3$
	$X_{67123} = \bar{X}_4$	$-X^*_{67123} = \bar{Y}_4$	$Z_{6712} = \bar{Z}_4$
	$X_{71234} = \bar{X}_5$	$-X^*_{71234} = \bar{Y}_5$	$Z_{7123} = \bar{Z}_5$
	$X_{54321} = X_7$	$-X^*_{54321} = Y_7$	$Z_{5432} = Z_7$
	$X_{65432} = X_1$	$-X^*_{65432} = Y_1$	$Z_{6543} = Z_1$
	$X_{76543} = X_2$	$-X^*_{76543} = Y_2$	$Z_{7654} = Z_2$
	$X_{17654} = X_3$	$-X^*_{17654} = Y_3$	$Z_{1765} = Z_3$
	$X_{21765} = X_4$	$-X^*_{26715} = Y_4$	$Z_{2176} = Z_4$
	$X_{32176} = X_5$	$-X^*_{32176} = Y_5$	$Z_{3217} = Z_5$
	$X_{43217} = X_6$	$-X^*_{43217} = Y_6$	$Z_{4321} = Z_6$

Set 2	$X_{1234} = X_{65}$	$Y_{1234} = -X^*_{65}$	$Z_{1234} = \bar{Z}_6$
	$X_{2345} = X_{76}$	$Y_{2345} = -X^*_{76}$	$Z_{2345} = \bar{Z}_7$
	$X_{3456} = X_{17}$	$Y_{3456} = -X^*_{17}$	$Z_{3456} = \bar{Z}_1$
	$X_{4567} = X_{21}$	$Y_{4567} = -X^*_{21}$	$Z_{4567} = \bar{Z}_2$
	$X_{5671} = X_{32}$	$Y_{5671} = -X^*_{32}$	$Z_{5671} = \bar{Z}_3$
	$X_{6712} = X_{43}$	$Y_{6712} = -X^*_{43}$	$Z_{6712} = \bar{Z}_4$
	$X_{7123} = X_{54}$	$Y_{7123} = -X^*_{54}$	$Z_{7123} = \bar{Z}_5$
	$X_{5432} = X_{71}$	$Y_{5432} = -X^*_{71}$	$Z_{5432} = Z_7$
	$X_{6543} = X_{12}$	$Y_{6543} = -X^*_{12}$	$Z_{6543} = Z_1$
	$X_{7654} = X_{23}$	$Y_{7654} = -X^*_{23}$	$Z_{7654} = Z_2$
	$X_{1765} = X_{34}$	$Y_{1765} = -X^*_{34}$	$Z_{1765} = Z_3$
	$X_{2176} = X_{45}$	$Y_{2176} = -X^*_{45}$	$Z_{2176} = Z_4$
	$X_{3217} = X_{56}$	$Y_{3217} = -X^*_{56}$	$Z_{3217} = Z_5$
	$X_{4321} = X_{67}$	$Y_{4321} = -X^*_{67}$	$Z_{4321} = Z_6$
Set 3	$X_{1234} = X_{65}$	$-X^*_{1234} = Y_{65}$	$Z_{123} = Z_{65}$
	$X_{2345} = X_{76}$	$-X^*_{2345} = Y_{76}$	$Z_{234} = Z_{76}$
	$X_{3456} = X_{17}$	$-X^*_{3456} = Y_{17}$	$Z_{345} = Z_{17}$
	$X_{4567} = X_{21}$	$-X^*_{4567} = Y_{21}$	$Z_{456} = Z_{21}$
	$X_{5671} = X_{32}$	$-X^*_{5671} = Y_{32}$	$Z_{567} = Z_{32}$
	$X_{6712} = X_{43}$	$-X^*_{6712} = Y_{43}$	$Z_{671} = Z_{43}$
	$X_{7123} = X_{54}$	$-X^*_{7123} = Y_{54}$	$Z_{712} = Z_{54}$
	$X_{5432} = X_{71}$	$-X^*_{5432} = Y_{71}$	$Z_{543} = Z_{71}$
	$X_{6543} = X_{12}$	$-X^*_{6543} = Y_{12}$	$Z_{654} = Z_{12}$
	$X_{7654} = X_{23}$	$-X^*_{7654} = Y_{23}$	$Z_{765} = Z_{23}$
	$X_{1765} = X_{34}$	$-X^*_{1765} = Y_{34}$	$Z_{176} = Z_{34}$
	$X_{2176} = X_{45}$	$-X^*_{2176} = Y_{45}$	$Z_{217} = Z_{45}$
	$X_{3217} = X_{56}$	$-X^*_{3217} = Y_{56}$	$Z_{321} = Z_{56}$
	$X_{4321} = X_{67}$	$-X^*_{4321} = Y_{67}$	$Z_{432} = Z_{67}$
Set 4	$X_{123} = X_{654}$	$Y_{123} = -X^*_{654}$	$Z_{123} = Z_{65}$
	$X_{234} = X_{765}$	$Y_{234} = -X^*_{765}$	$Z_{234} = Z_{76}$
	$X_{345} = X_{176}$	$Y_{345} = -X^*_{176}$	$Z_{345} = Z_{17}$
	$X_{456} = X_{217}$	$Y_{456} = -X^*_{217}$	$Z_{456} = Z_{21}$
	$X_{567} = X_{321}$	$Y_{567} = -X^*_{321}$	$Z_{567} = Z_{32}$
	$X_{671} = X_{432}$	$Y_{671} = -X^*_{432}$	$Z_{671} = Z_{43}$
	$X_{712} = X_{543}$	$Y_{712} = -X^*_{543}$	$Z_{712} = Z_{54}$

$X_{543} = X_{712}$	$Y_{543} = -X^*_{712}$	$Z_{543} = Z_{71}$
$X_{654} = X_{123}$	$Y_{654} = -X^*_{123}$	$Z_{654} = Z_{12}$
$X_{765} = X_{234}$	$Y_{765} = -X^*_{234}$	$Z_{765} = Z_{23}$
$X_{176} = X_{345}$	$Y_{176} = -X^*_{345}$	$Z_{176} = Z_{34}$
$X_{217} = X_{456}$	$Y_{217} = -X^*_{456}$	$Z_{217} = Z_{45}$
$X_{321} = X_{567}$	$Y_{321} = -X^*_{567}$	$Z_{321} = Z_{56}$
$X_{432} = X_{671}$	$Y_{432} = -X^*_{671}$	$Z_{432} = Z_{67}$

Equations for a Polar Heptagon

Fundamental Formulas:

$U_{123456} = s_7 s_{67}$	$V_{123456} = s_7 c_{67}$	$W_{123456} = c_7$
$U_{234567} = s_1 s_{71}$	$V_{234567} = s_1 c_{71}$	$W_{234567} = c_1$
$U_{345671} = s_2 s_{12}$	$V_{345671} = s_2 c_{12}$	$W_{345671} = c_2$
$U_{456712} = s_3 s_{23}$	$V_{456712} = s_3 c_{23}$	$W_{456712} = c_3$
$U_{567123} = s_4 s_{34}$	$V_{567123} = s_4 c_{34}$	$W_{567123} = c_4$
$U_{671234} = s_5 s_{45}$	$V_{671234} = s_5 c_{45}$	$W_{671234} = c_5$
$U_{712345} = s_6 s_{56}$	$V_{712345} = s_6 c_{56}$	$W_{712345} = c_6$
$U_{654321} = s_7 s_{71}$	$V_{654321} = s_7 c_{71}$	$W_{654321} = c_7$
$U_{543217} = s_6 s_{67}$	$V_{543217} = s_6 c_{67}$	$W_{543217} = c_6$
$U_{432176} = s_5 s_{56}$	$V_{432176} = s_5 c_{56}$	$W_{432176} = c_5$
$U_{321765} = s_4 s_{45}$	$V_{321765} = s_4 c_{45}$	$W_{321765} = c_4$
$U_{217654} = s_3 s_{34}$	$V_{217654} = s_3 c_{34}$	$W_{217654} = c_3$
$U_{176543} = s_2 s_{23}$	$V_{176543} = s_2 c_{23}$	$W_{176543} = c_2$
$U_{765432} = s_1 s_{12}$	$V_{765432} = s_1 c_{12}$	$W_{765432} = c_1$

Subsidiary Formulas:

Set 1	$U_{123456} = U_{76}$	$U^*_{123456} = -V_{76}$	$W_{12345} = W_{76}$
	$U_{234567} = U_{17}$	$U^*_{234567} = -V_{17}$	$W_{23456} = W_{17}$
	$U_{345671} = U_{21}$	$U^*_{345671} = -V_{21}$	$W_{34567} = W_{21}$
	$U_{456712} = U_{32}$	$U^*_{456712} = -V_{32}$	$W_{45671} = W_{32}$
	$U_{567123} = U_{43}$	$U^*_{567123} = -V_{43}$	$W_{56712} = W_{43}$
	$U_{671234} = U_{54}$	$U^*_{671234} = -V_{54}$	$W_{67123} = W_{54}$
	$U_{712345} = U_{65}$	$U^*_{712345} = -V_{65}$	$W_{71234} = W_{65}$
	$U_{654321} = U_{71}$	$U^*_{654321} = -V_{71}$	$W_{65432} = W_{71}$
	$U_{543217} = U_{67}$	$U^*{}_{543217} = -V_{67}$	$W_{54321} = W_{67}$
	$U_{432176} = U_{56}$	$U^*_{432176} = -V_{56}$	$W_{43217} = W_{56}$
	$U_{321765} = U_{45}$	$U^*_{321765} = -V_{45}$	$W_{32176} = W_{45}$
	$U_{217654} = U_{34}$	$U^*_{217654} = -V_{34}$	$W_{21765} = W_{34}$
	$U_{176543} = U_{23}$	$U^*_{176543} = -V_{23}$	$W_{17654} = W_{23}$
	$U_{765432} = U_{12}$	$U^*_{765432} = -V_{12}$	$W_{76543} = W_{12}$

Set 2	$U_{12345} = U_{765}$	$V_{12345} = -U^*_{765}$	$W_{12345} = W_{76}$
	$U_{23456} = U_{176}$	$V_{23456} = -U^*_{176}$	$W_{23456} = W_{17}$
	$U_{34567} = U_{217}$	$V_{34567} = -U^*_{217}$	$W_{34567} = W_{21}$
	$U_{45671} = U_{321}$	$V_{45671} = -U^*_{321}$	$W_{45671} = W_{32}$
	$U_{56712} = U_{432}$	$V_{56712} = -U^*_{432}$	$W_{56712} = W_{43}$
	$U_{67123} = U_{543}$	$V_{67123} = -U^*_{543}$	$W_{67123} = W_{54}$
	$U_{71234} = U_{654}$	$V_{71234} = -U^*_{654}$	$W_{71234} = W_{65}$
	$U_{65432} = U_{712}$	$V_{65432} = -U^*_{712}$	$W_{65432} = W_{71}$
	$U_{54321} = U_{671}$	$V_{54321} = -U^*_{671}$	$W_{54321} = W_{67}$
	$U_{43217} = U_{567}$	$V_{43217} = -U^*_{567}$	$W_{43217} = W_{56}$
	$U_{32176} = U_{456}$	$V_{32176} = -U^*_{456}$	$W_{32176} = W_{45}$
	$U_{21765} = U_{345}$	$V_{21765} = -U^*_{345}$	$W_{21765} = W_{34}$
	$U_{17654} = U_{234}$	$V_{17654} = -U^*_{234}$	$W_{17654} = W_{23}$
	$U_{76543} = U_{123}$	$V_{76543} = -U^*_{123}$	$W_{76543} = W_{12}$
Set 3	$U_{12345} = U_{765}$	$U^*_{12345} = -V_{765}$	$W_{1234} = W_{765}$
	$U_{23456} = U_{176}$	$U^*_{23456} = -V_{176}$	$W_{2345} = W_{176}$
	$U_{34567} = U_{217}$	$U^*_{34567} = -V_{217}$	$W_{3456} = W_{217}$
	$U_{45671} = U_{321}$	$U^*_{45671} = -V_{321}$	$W_{4567} = W_{321}$
	$U_{56712} = U_{432}$	$U^*_{56712} = -V_{432}$	$W_{5671} = W_{432}$
	$U_{67123} = U_{543}$	$U^*_{67123} = -V_{543}$	$W_{6712} = W_{543}$
	$U_{71234} = U_{654}$	$U^*_{71234} = -V_{654}$	$W_{7123} = W_{654}$
	$U_{65432} = U_{712}$	$U^*_{65432} = -V_{712}$	$W_{6543} = W_{712}$
	$U_{54321} = U_{671}$	$U^*_{54321} = -V_{671}$	$W_{5432} = W_{671}$
	$U_{43217} = U_{567}$	$U^*_{43217} = -V_{567}$	$W_{4321} = W_{567}$
	$U_{32176} = U_{456}$	$U^*_{32176} = -V_{456}$	$W_{3217} = W_{456}$
	$U_{21765} = U_{345}$	$U^*_{21765} = -V_{345}$	$W_{2176} = W_{345}$
	$U_{17654} = U_{234}$	$U^*_{17654} = -V_{234}$	$W_{1765} = W_{234}$
	$U_{76543} = U_{123}$	$U^*_{76543} = -V_{123}$	$W_{7654} = W_{123}$
Set 4	$U_{1234} = U_{7654}$	$V_{1234} = -U^*_{7654}$	$W_{1234} = W_{765}$
	$U_{2345} = U_{1765}$	$V_{2345} = -U^*_{1765}$	$W_{2345} = W_{176}$
	$U_{3456} = U_{2176}$	$V_{3456} = -U^*_{2176}$	$W_{3456} = W_{217}$
	$U_{4567} = U_{3217}$	$V_{4567} = -U^*_{3217}$	$W_{4567} = W_{321}$
	$U_{5671} = U_{4321}$	$V_{5671} = -U^*_{4321}$	$W_{5671} = W_{432}$
	$U_{6712} = U_{5432}$	$V_{6712} = -U^*_{5432}$	$W_{6712} = W_{543}$
	$U_{7123} = U_{6543}$	$V_{7123} = -U^*_{6543}$	$W_{7123} = W_{654}$

$$\begin{array}{lll}
U_{6543} = U_{7123} & V_{6543} = -U^*_{7123} & W_{6543} = W_{712} \\
U_{5432} = U_{6712} & V_{5432} = -U^*_{6712} & W_{5432} = W_{671} \\
U_{4321} = U_{5671} & V_{4321} = -U^*_{5671} & W_{4321} = W_{567} \\
U_{3217} = U_{4567} & V_{3217} = -U^*_{4567} & W_{3217} = W_{456} \\
U_{2176} = U_{3456} & V_{2176} = -U^*_{3456} & W_{2176} = W_{345} \\
U_{1765} = U_{2345} & V_{1765} = -U^*_{2345} & W_{1765} = W_{234} \\
U_{7654} = U_{1234} & V_{7654} = -U^*_{1234} & W_{7654} = W_{123}
\end{array}$$

Half-Tangent Laws for a Heptagon

Set 1

$$x_1 = \frac{X_{34567}}{Y_{34567} + s_{12}} = -\frac{Y_{34567} - s_{12}}{X_{34567}} \qquad x_1 = \frac{X_{65432}}{Y_{65432} + s_{71}} = -\frac{Y_{65432} - s_{71}}{X_{65432}}$$

$$x_2 = \frac{X_{45671}}{Y_{45671} + s_{23}} = -\frac{Y_{45671} - s_{23}}{X_{45671}} \qquad x_2 = \frac{X_{76543}}{Y_{76543} + s_{12}} = -\frac{Y_{76543} - s_{12}}{X_{76543}}$$

$$x_3 = \frac{X_{56712}}{Y_{56712} + s_{34}} = -\frac{Y_{56712} - s_{34}}{X_{56712}} \qquad x_3 = \frac{X_{17654}}{Y_{17654} + s_{23}} = -\frac{Y_{17654} - s_{23}}{X_{17654}}$$

$$x_4 = \frac{X_{67123}}{Y_{67123} + s_{45}} = -\frac{Y_{67123} - s_{45}}{X_{67123}} \qquad x_4 = \frac{X_{21765}}{Y_{21765} + s_{34}} = -\frac{Y_{21765} - s_{34}}{X_{21765}}$$

$$x_5 = \frac{X_{71234}}{Y_{71234} + s_{56}} = -\frac{Y_{71234} - s_{56}}{X_{71234}} \qquad x_5 = \frac{X_{32716}}{Y_{32716} + s_{45}} = -\frac{Y_{32716} - s_{45}}{X_{32716}}$$

$$x_6 = \frac{X_{12345}}{Y_{12345} + s_{67}} = -\frac{Y_{12345} - s_{67}}{X_{12345}} \qquad x_6 = \frac{X_{43217}}{Y_{43217} + s_{56}} = -\frac{Y_{43217} - s_{56}}{X_{43217}}$$

$$x_7 = \frac{X_{23456}}{Y_{23456} + s_{71}} = -\frac{Y_{23456} - s_{71}}{X_{23456}} \qquad x_7 = \frac{X_{54321}}{Y_{54321} + s_{67}} = -\frac{Y_{54321} - s_{67}}{X_{54321}}$$

Set 2

$$x_1 = \frac{X_{4567} - \bar{X}_2}{Y_{4567} - \bar{Y}_2} = -\frac{Y_{4567} + \bar{Y}_2}{X_{4567} + \bar{X}_2} \qquad x_1 = \frac{X_{5432} - X_7}{Y_{5432} - Y_7} = -\frac{Y_{5432} + Y_7}{X_{5432} + X_7}$$

$$x_2 = \frac{X_{5671} - \bar{X}_3}{Y_{5671} - \bar{Y}_3} = -\frac{Y_{5671} + \bar{Y}_3}{X_{5671} + \bar{X}_3} \qquad x_2 = \frac{X_{6543} - X_1}{Y_{6543} - Y_1} = -\frac{Y_{6543} + Y_1}{X_{6543} + X_1}$$

$$x_3 = \frac{X_{6712} - \bar{X}_4}{Y_{6712} - \bar{Y}_4} = -\frac{Y_{6712} + \bar{Y}_4}{X_{6712} + \bar{X}_4} \qquad x_3 = \frac{X_{7654} - X_2}{Y_{7654} - Y_2} = -\frac{Y_{7654} + Y_2}{X_{7654} + X_2}$$

$$x_4 = \frac{X_{7123} - \bar{X}_5}{Y_{7123} - \bar{Y}_5} = -\frac{Y_{7123} + \bar{Y}_5}{X_{7123} + \bar{X}_5} \qquad x_4 = \frac{X_{1765} - X_3}{Y_{1765} - Y_3} = -\frac{Y_{1765} + Y_3}{X_{1765} + X_3}$$

$$x_5 = \frac{X_{1234} - \bar{X}_6}{Y_{1234} - \bar{Y}_6} = -\frac{Y_{1234} + \bar{Y}_6}{X_{1234} + \bar{X}_6} \qquad x_5 = \frac{X_{2176} - X_4}{Y_{2176} - Y_4} = -\frac{Y_{2176} + Y_4}{X_{2176} + X_4}$$

$$x_6 = \frac{X_{2345} - \bar{X}_7}{Y_{2345} - \bar{Y}_7} = -\frac{Y_{2345} + \bar{Y}_7}{X_{2345} + \bar{X}_7} \qquad x_6 = \frac{X_{3217} - X_5}{Y_{3217} - Y_5} = -\frac{Y_{3217} + Y_5}{X_{3217} + X_5}$$

$$x_7 = \frac{X_{3456} - \bar{X}_1}{Y_{3456} - \bar{Y}_1} = -\frac{Y_{3456} + \bar{Y}_1}{X_{3456} + \bar{X}_1} \qquad x_7 = \frac{X_{4321} - X_6}{Y_{4321} - Y_6} = -\frac{Y_{4321} + Y_6}{X_{4321} + X_6}$$

Set 3

$$x_1 = \frac{X_{567} - X_{32}}{Y_{567} - Y_{32}} = -\frac{Y_{567} + Y_{32}}{X_{567} + X_{32}} \qquad x_1 = \frac{X_{432} - X_{67}}{Y_{432} - Y_{67}} = -\frac{Y_{432} + Y_{67}}{X_{432} + X_{67}}$$

$$x_2 = \frac{X_{671} - X_{43}}{Y_{671} - Y_{43}} = -\frac{Y_{671} + Y_{43}}{X_{671} + X_{43}} \qquad x_2 = \frac{X_{543} - X_{71}}{Y_{543} - Y_{71}} = -\frac{Y_{543} + Y_{71}}{X_{543} + X_{71}}$$

$$x_3 = \frac{X_{712} - X_{54}}{Y_{712} - Y_{54}} = -\frac{Y_{712} + Y_{54}}{xX_{712} + X_{54}} \qquad x_3 = \frac{X_{654} - X_{12}}{Y_{654} - Y_{12}} = -\frac{Y_{654} + Y_{12}}{X_{654} + X_{12}}$$

$$x_4 = \frac{X_{123} - X_{65}}{Y_{123} - Y_{65}} = -\frac{Y_{123} + Y_{65}}{X_{123} + X_{65}} \qquad x_4 = \frac{X_{765} - X_{23}}{Y_{765} - Y_{23}} = -\frac{Y_{765} + Y_{23}}{X_{765} + X_{23}}$$

$$x_5 = \frac{X_{234} - \bar{X}_{76}}{Y_{234} - Y_{76}} = -\frac{Y_{234} + Y_{76}}{X_{234} + X_{76}} \qquad x_5 = \frac{X_{176} - X_{34}}{Y_{176} - Y_{34}} = -\frac{Y_{176} + Y_{34}}{X_{176} + X_{34}}$$

$$x_6 = \frac{X_{345} - \bar{X}_{17}}{Y_{345} - Y_{17}} = -\frac{Y_{345} + Y_{17}}{X_{345} + X_{17}} \qquad x_6 = \frac{X_{217} - X_{45}}{Y_{217} - Y_{45}} = -\frac{Y_{217} + Y_{45}}{X_{217} + X_{45}}$$

$$x_7 = \frac{X_{456} - \bar{X}_{21}}{Y_{456} - Y_{21}} = -\frac{Y_{456} + Y_{21}}{X_{456} + X_{21}} \qquad x_6 = \frac{X_{321} - X_{56}}{Y_{321} - Y_{56}} = -\frac{Y_{321} + Y_{56}}{X_{321} + X_{56}}$$

Direction Cosines – Spatial Heptagon

Set 1

$\underline{S}_1$	$(0,$	$0,$	$1)$	$\underline{a}_{12}$	$(1,\quad 0,$	$0)$
$\underline{S}_2$	$(0,$	$-s_{12},$	$c_{12})$	$\underline{a}_{23}$	$(c_2,\quad s_2c_{12},$	$U_{21})$
$\underline{S}_3$	$(\bar{X}_2,$	$\bar{Y}_2,$	$\bar{Z}_2)$	$\underline{a}_{34}$	$(W_{32},\quad -U^*_{321},$	$U_{321})$
$\underline{S}_4$	$(X_{32},$	$Y_{32},$	$Z_{32})$	$\underline{a}_{45}$	$(W_{432},\quad -U^*_{4321},$	$U_{4321})$
$\underline{S}_5$	$(X_{432},$	$Y_{432},$	$Z_{432})$	$\underline{a}_{56}$	$(W_{5432},\quad -U^*_{54321},$	$U_{54321})$
$\underline{S}_6$	$(X_{5432},$	$Y_{5432},$	$Z_{5432})$	$\underline{a}_{67}$	$(W_{65432},\quad -U^*_{654321},$	$U_{654321})$
$\underline{S}_7$	$(X_{65432},$	$Y_{65432},$	$Z_{65432})$	$\underline{a}_{71}$	$(c_1,\quad -s_1,$	$0)$

Set 2

$\underline{S}_2$	$(0,$	$0,$	$1)$	$\underline{a}_{23}$	$(1,\quad 0,$	$0)$
$\underline{S}_3$	$(0,$	$-s_{23},$	$c_{23})$	$\underline{a}_{34}$	$(c_3,\quad s_3c_{23},$	$U_{32})$
$\underline{S}_4$	$(\bar{X}_3,$	$\bar{Y}_3,$	$\bar{Z}_3)$	$\underline{a}_{45}$	$(W_{43},\quad -U^*_{432},$	$U_{432})$
$\underline{S}_5$	$(X_{43},$	$Y_{43},$	$Z_{43})$	$\underline{a}_{56}$	$(W_{543},\quad -U^*_{5432},$	$U_{5432})$
$\underline{S}_6$	$(X_{543},$	$Y_{543},$	$Z_{543})$	$\underline{a}_{67}$	$(W_{6543},\quad -U^*_{65432},$	$U_{65432})$
$\underline{S}_7$	$(X_{6543},$	$Y_{6543},$	$Z_{6543})$	$\underline{a}_{71}$	$(W_{76543},\quad -U^*_{765432},$	$U_{765432})$
$\underline{S}_1$	$(X_{76543},$	$Y_{76543},$	$Z_{76543})$	$\underline{a}_{12}$	$(c_2,\quad -s_2,$	$0)$

Set 3

$\underline{S}_3\,(0,$	$0,$	$1)$	$\underline{a}_{34}\,(1,$	$0,$	$0)$
$\underline{S}_4\,(0,$	$-s_{34},$	$c_{34})$	$\underline{a}_{45}\,(c_4,$	$s_4c_{34},$	$U_{43})$
$\underline{S}_5\,(\bar{X}_4,$	$\bar{Y}_4,$	$\bar{Z}_4)$	$\underline{a}_{56}\,(W_{54},$	$-U^*_{543},$	$U_{543})$
$\underline{S}_6\,(X_{54},$	$Y_{54},$	$Z_{54})$	$\underline{a}_{67}\,(W_{654},$	$-U^*_{6543},$	$U_{6543})$
$\underline{S}_7\,(X_{654},$	$Y_{654},$	$Z_{654})$	$\underline{a}_{71}\,(W_{7654},$	$-U^*_{76543},$	$U_{76543})$
$\underline{S}_1\,(X_{7654},$	$Y_{7654},$	$Z_{7654})$	$\underline{a}_{12}\,(W_{17654},$	$-U^*_{176543},$	$U_{176543})$
$\underline{S}_2\,(X_{17654},$	$Y_{17654},$	$Z_{17654})$	$\underline{a}_{23}\,(c_3,$	$-s_3,$	$0)$

Set 4

$\underline{S}_4\,(0,$	$0,$	$1)$	$\underline{a}_{45}\,(1,$	$0,$	$0)$
$\underline{S}_5\,(0,$	$-s_{45},$	$c_{45})$	$\underline{a}_{56}\,(c_5,$	$s_5c_{45},$	$U_{54})$
$\underline{S}_6\,(\bar{X}_5,$	$\bar{Y}_5,$	$\bar{Z}_5)$	$\underline{a}_{67}\,(W_{65},$	$-U^*_{654},$	$U_{654})$
$\underline{S}_7\,(X_{65},$	$Y_{65},$	$Z_{65})$	$\underline{a}_{71}\,(W_{765},$	$-U^*_{7654},$	$U_{7654})$
$\underline{S}_1\,(X_{765},$	$Y_{765},$	$Z_{765})$	$\underline{a}_{12}\,(W_{1765},$	$-U^*_{17654},$	$U_{17654})$
$\underline{S}_2\,(X_{1765},$	$Y_{1765},$	$Z_{1765})$	$\underline{a}_{23}\,(W_{21765},$	$-U^*_{217654},$	$U_{217654})$
$\underline{S}_3\,(X_{21765},$	$Y_{21765},$	$Z_{21765})$	$\underline{a}_{34}\,(c_4,$	$-s_4,$	$0)$

Set 5

$\underline{S}_5\,(0,$	$0,$	$1)$	$\underline{a}_{56}\,(1,$	$0,$	$0)$
$\underline{S}_6\,(0,$	$-s_{56},$	$c_{56})$	$\underline{a}_{61}\,(c_6,$	$s_6c_{56},$	$U_{65})$
$\underline{S}_7\,(\bar{X}_6,$	$\bar{Y}_6,$	$\bar{Z}_6)$	$\underline{a}_{71}\,(W_{76},$	$-U^*_{765},$	$U_{765})$
$S_1\,(X_{76},$	$Y_{76},$	$Z_{76})$	$\underline{a}_{12}\,(W_{176},$	$-U^*_{1765},$	$U_{1765})$
$S_2\,(X_{176},$	$Y_{176},$	$Z_{176})$	$\underline{a}_{23}\,(W_{2176},$	$-U^*_{21765},$	$U_{21765})$
$S_3\,(X_{2176},$	$Y_{2176},$	$Z_{2176})$	$\underline{a}_{34}\,(W_{32176},$	$-U^*_{321765},$	$U_{321765})$
$\underline{S}_4\,(X_{32176},$	$Y_{32176},$	$Z_{32176})$	$\underline{a}_{45}\,(c_5,$	$-s_5,$	$0)$

Set 6

$\underline{S}_6\,(0,$	$0,$	$1)$	$\underline{a}_{67}\,(1,$	$0,$	$0)$
$\underline{S}_7\,(0,$	$-s_{67},$	$c_{67})$	$\underline{a}_{71}\,(c_7,$	$s_7c_{67},$	$U_{67})$
$\underline{S}_1\,(\bar{X}_7,$	$\bar{Y}_7,$	$\bar{Z}_7)$	$\underline{a}_{12}\,(W_{17},$	$-U^*_{176},$	$U_{176})$
$S_2\,(X_{17},$	$Y_{17},$	$Z_{17})$	$\underline{a}_{23}\,(W_{217},$	$-U^*_{2176},$	$U_{2176})$
$S_3\,(X_{217},$	$Y_{217},$	$Z_{217})$	$\underline{a}_{34}\,(W_{3217},$	$-U^*_{32176},$	$U_{32176})$
$S_4\,(X_{3217},$	$Y_{3217},$	$Z_{3217})$	$\underline{a}_{45}\,(W_{43217},$	$-U^*_{432176},$	$U_{432176})$
$\underline{S}_5\,(X_{43217},$	$Y_{43217},$	$Z_{43217})$	$\underline{a}_{56}\,(c_6,$	$-s_6,$	$0)$

Set 7

$\underline{S}_7$	$(0,$	$0,$	$1)$	$\underline{a}_{71}$	$(1,$	$0,$	$0)$
$\underline{S}_1$	$(0,$	$-s_{71},$	$c_{71})$	$\underline{a}_{12}$	$(c_1,$	$s_1c_{71},$	$U_{17})$
$\underline{S}_2$	$(\bar{X}_1,$	$\bar{Y}_1,$	$\bar{Z}_1)$	$\underline{a}_{23}$	$(W_{21},$	$-U^*_{217},$	$U_{217})$
S_3	$(X_{21},$	$Y_{21},$	$Z_{21})$	$\underline{a}_{34}$	$(W_{321},$	$-U^*_{3217},$	$U_{3217})$
S_4	$(X_{321},$	$Y_{321},$	$Z_{321})$	$\underline{a}_{45}$	$(W_{4321},$	$-U^*_{43217},$	$U_{43217})$
S_5	$(X_{4321},$	$Y_{4321},$	$Z_{4321})$	$\underline{a}_{56}$	$(W_{54321},$	$-U^*_{543217},$	$U_{543217})$
$\underline{S}_6$	$(X_{54321},$	$Y_{54321},$	$Z_{54321})$	$\underline{a}_{67}$	$(c_7,$	$-s_7,$	$0)$

Set 8

$\underline{S}_1$	$(0,$	$0,$	$1)$	$\underline{a}_{71}$	$(1,$	$0,$	$0)$
$\underline{S}_7$	$(0,$	$s_{71},$	$c_{71})$	$\underline{a}_{67}$	$(c_7,$	$-s_7c_{71},$	$U_{71})$
$\underline{S}_6$	$(X_7,$	$-Y_7,$	$Z_7)$	$\underline{a}_{56}$	$(W_{67},$	$U^*_{671},$	$U_{671})$
$\underline{S}_5$	$(X_{67},$	$-Y_{67},$	$Z_{67})$	$\underline{a}_{45}$	$(W_{567},$	$U^*_{5671},$	$U_{5671})$
$\underline{S}_4$	$(X_{567},$	$-Y_{567},$	$Z_{567})$	$\underline{a}_{34}$	$(W_{4567},$	$U^*_{45671},$	$U_{45671})$
$\underline{S}_3$	$(X_{4567},$	$-Y_{4567},$	$Z_{4567})$	$\underline{a}_{23}$	$(W_{34567},$	$U^*_{345671},$	$U_{345671})$
$\underline{S}_2$	$(X_{34567},$	$-Y_{34567},$	$Z_{34567})$	$\underline{a}_{12}$	$(c_1,$	$s_1,$	$0)$

Set 9

$\underline{S}_7$	$(0,$	$0,$	$1)$	$\underline{a}_{67}$	$(1,$	$0,$	$0)$
$\underline{S}_6$	$(0,$	$s_{67},$	$c_{67})$	$\underline{a}_{56}$	$(c_6,$	$-s_6c_{67},$	$U_{67})$
$\underline{S}_5$	$(X_6,$	$-Y_6,$	$Z_6)$	$\underline{a}_{45}$	$(W_{56},$	$U^*_{567},$	$U_{567})$
$\underline{S}_4$	$(X_{56},$	$-Y_{56},$	$Z_{56})$	$\underline{a}_{34}$	$(W_{456},$	$U^*_{4567},$	$U_{4567})$
$\underline{S}_3$	$(X_{456},$	$-Y_{456},$	$Z_{456})$	$\underline{a}_{23}$	$(W_{3456},$	$U^*_{34567},$	$U_{34567})$
$\underline{S}_2$	$(X_{3456},$	$-Y_{3456},$	$Z_{3456})$	$\underline{a}_{12}$	$(W_{23456},$	$U^*_{234567},$	$U_{234567})$
$\underline{S}_1$	$(X_{23456},$	$-Y_{23456},$	$Z_{23456})$	$\underline{a}_{71}$	$(c_7,$	$s_7,$	$0)$

Set 10

$\underline{S}_6$	$(0,$	$0,$	$1)$	$\underline{a}_{56}$	$(1,$	$0,$	$0)$
$\underline{S}_5$	$(0,$	$s_{56},$	$c_{56})$	$\underline{a}_{45}$	$(c_5,$	$-s_5c_{56},$	$U_{56})$
$\underline{S}_4$	$(X_5,$	$-Y_5,$	$Z_5)$	$\underline{a}_{34}$	$(W_{45},$	$U^*_{456},$	$U_{456})$
$\underline{S}_3$	$(X_{45},$	$-Y_{45},$	$Z_{45})$	$\underline{a}_{23}$	$(W_{345},$	$U^*_{3456},$	$U_{3456})$
$\underline{S}_2$	$(X_{345},$	$-Y_{345},$	$Z_{345})$	$\underline{a}_{12}$	$(W_{2345},$	$U^*_{23456},$	$U_{23456})$
$\underline{S}_1$	$(X_{2345},$	$-Y_{2345},$	$Z_{2345})$	$\underline{a}_{71}$	$(W_{12345},$	$U^*_{123456},$	$U_{123456})$
$\underline{S}_7$	$(X_{12345},$	$-Y_{12345},$	$Z_{12345})$	$\underline{a}_{67}$	$(c_6,$	$s_6,$	$0)$

Set 11

$\underline{S}_5\ (0,$	$0,$	$1)$	$\underline{a}_{45}\ (1,$	$0,$	$0)$
$\underline{S}_4\ (0,$	$s_{45},$	$c_{45})$	$\underline{a}_{34}\ (c_4,$	$-s_4c_{45},$	$U_{45})$
$\underline{S}_3\ (X_4,$	$-Y_4,$	$Z_4)$	$\underline{a}_{23}\ (W_{34},$	$U^*_{345},$	$U_{345})$
$\underline{S}_2\ (X_{34},$	$-Y_{34},$	$Z_{34})$	$\underline{a}_{12}\ (W_{234},$	$U^*_{2345},$	$U_{2345})$
$\underline{S}_1\ (X_{234},$	$-Y_{234},$	$Z_{234})$	$\underline{a}_{71}\ (W_{1234},$	$U^*_{12345},$	$U_{12345})$
$\underline{S}_7\ (X_{1234},$	$-Y_{1234},$	$Z_{1234})$	$\underline{a}_{67}\ (W_{71234},$	$U^*_{712345},$	$U_{712345})$
$\underline{S}_6\ (X_{71234},$	$-Y_{71234},$	$Z_{71234})$	$\underline{a}_{56}\ (c_5,$	$s_5,$	$0)$

Set 12

$\underline{S}_4\ (0,$	$0,$	$1)$	$\underline{a}_{34}\ (1,$	$0,$	$0)$
$\underline{S}_3\ (0,$	$s_{34},$	$c_{34})$	$\underline{a}_{23}\ (c_3,$	$-s_3c_{34},$	$U_{34})$
$\underline{S}_2\ (X_3,$	$-Y_3,$	$Z_3)$	$\underline{a}_{12}\ (W_{23},$	$U^*_{234},$	$U_{234})$
$\underline{S}_1\ (X_{23},$	$-Y_{23},$	$Z_{23})$	$\underline{a}_{71}\ (W_{123},$	$U^*_{1234},$	$U_{1234})$
$\underline{S}_7\ (X_{123},$	$-Y_{123},$	$Z_{123})$	$\underline{a}_{67}\ (W_{7123},$	$U^*_{71234},$	$U_{71234})$
$\underline{S}_6\ (X_{7123},$	$-Y_{7123},$	$Z_{7123})$	$\underline{a}_{56}\ (W_{67123},$	$U^*_{671234},$	$U_{671234})$
$\underline{S}_5\ (X_{67123},$	$-Y_{67123},$	$Z_{67123})$	$\underline{a}_{45}\ (c_4,$	$s_4,$	$0)$

Set 13

$\underline{S}_3\ (0,$	$0,$	$1)$	$\underline{a}_{23}\ (1,$	$0,$	$0)$
$\underline{S}_2\ (0,$	$s_{23},$	$c_{23})$	$\underline{a}_{12}\ (c_2,$	$-s_2c_{23},$	$U_{23})$
$\underline{S}_1\ (X_2,$	$-Y_2,$	$Z_2)$	$\underline{a}_{71}\ (W_{12},$	$U^*_{123},$	$U_{123})$
$\underline{S}_7\ (X_{12},$	$-Y_{12},$	$Z_{12})$	$\underline{a}_{67}\ (W_{712},$	$U^*_{7123},$	$U_{7123})$
$\underline{S}_6\ (X_{712},$	$-Y_{712},$	$Z_{712})$	$\underline{a}_{56}\ (W_{6712},$	$U^*_{67123},$	$U_{67123})$
$\underline{S}_5\ (X_{6712},$	$-Y_{6712},$	$Z_{6712})$	$\underline{a}_{45}\ (W_{56712},$	$U^*_{567123},$	$U_{567123})$
$\underline{S}_4\ (X_{56712},$	$-Y_{56712},$	$Z_{56712})$	$\underline{a}_{34}\ (c_3,$	$s_3,$	$0)$

Set 14

$\underline{S}_2\ (0,$	$0,$	$1)$	$\underline{a}_{12}\ (1,$	$0,$	$0)$
$\underline{S}_1\ (0,$	$s_{12},$	$c_{12})$	$\underline{a}_{71}\ (c_1,$	$-s_1c_{12},$	$U_{12})$
$\underline{S}_7\ (X_1,$	$-Y_1,$	$Z_1)$	$\underline{a}_{67}\ (W_{71},$	$U^*_{712},$	$U_{712})$
$\underline{S}_6\ (X_{71},$	$-Y_{71},$	$Z_{71})$	$\underline{a}_{56}\ (W_{671},$	$U^*_{6712},$	$U_{6712})$
$\underline{S}_5\ (X_{671},$	$-Y_{671},$	$Z_{671})$	$\underline{a}_{45}\ (W_{5671},$	$U^*_{56712},$	$U_{56712})$
$\underline{S}_4\ (X_{5671},$	$-Y_{5671},$	$Z_{5671})$	$\underline{a}_{34}\ (W_{45671},$	$U^*_{456712},$	$U_{456712})$
$\underline{S}_3\ (X_{45671},$	$-Y_{45671},$	$Z_{45671})$	$\underline{a}_{23}\ (c_2,$	$s_2,$	$0)$

References

Altmann, S.L. (1986), *Rotations, Quaternions, and Double Groups*, Clarendon Press, Oxford.

Brand, L. (1947), *Vector and Tensor Analysis*, John Wiley, New York.

Crane, C., Duffy, J., and Carnahan, T. (1991), "A Kinematic Analysis of the Space Station Remote Manipulator System," Journal of Robotic Systems, Vol. 8, No. 5, pp. 637–658.

Crane, C., Duffy, J., and Carnahan, T. (1992), "A Kinematic Analysis of the Modified Flight Telerobotic Servicer Manipulator System," Journal of Robotic Systems, Vol. 9, No. 3, pp. 461–480.

Denavit, J. (1958), "Displacement Analysis of Mechanisms Based on (2 × 2) Matrices of Dual Numbers," V.D.I. Berichte, Vol. 29, pp. 81–89.

Dimentberg, F.M. (1948), "A General Method for the Investigation of Finite Displacements of Spatial Mechanisms and Certain Cases of Passive Contraints," (Russian), Trudi Seminara po Teorii Mashin, Mekhanizmov, Akademiia Nauk, SSSR, Vol. 5, No. 17, pp. 5–39. Available as Translation No. 436, Purdue University, Indiana.

Duffy, J., and Habib-Olahi, H.Y. (1971), "A Displacement Analysis of Spatial Five-Link 3R-2C Mechanisms. Part 1. On the Closures of the RCRCR Mechanism," Journal of Mechanisms, Vol. 6, No. 3. pp. 289–301.

Duffy, J., and Habib-Olahi, H.Y. (1971a), "A Displacement Analysis of Spatial Five-Link 3R-2C Mechanisms. Part 2. Analysis of the RRCRC Mechanism," Journal of Mechanisms, Vol. 6, No. 4. pp. 463–473.

Duffy, J., and Habib-Olahi, H.Y. (1972), "A Displacement Analysis of Spatial Five-Link 3R-2C Mechanisms. Part 3. Analysis of the RCRRC Mechanism," Journal of Mechanisms, Vol. 7, No. 1. pp. 71–84.

Duffy, J., and Rooney, J. (1974), "A Displacement Analysis of Spatial Six Link 4R-P-C Mechanisms: Part 1: Analysis of RCRPRR Mechanism," Journal of Engineering for Industry, Trans. ASME, Vol. 96, Series B, No. 3, pp. 705–712. "Part 2: Derivation of Input-Output Displacement Equation for RCRRPR Mechanism," Journal of Engineering for Industry, Trans. ASME, Vol. 96, Series B, No. 3, pp. 713–717. "Part 3: Derivation of Input-Output Displacement Equation for RRRPCR Mechanism," Journal of Engineering for Industry, Trans. ASME, Vol. 96, Series B, No. 3, pp. 718–721.

Duffy, J., and Rooney, J. (1974a), "Displacement Analysis of Spatial Six-Link 5R-C Mechanisms," Journal of Applied Mechanics, Trans. ASME, Vol. 41, Series E, No. 3, pp. 759–766.

Duffy, J., and Rooney, J. (1975), "A Foundation for a Unified Theory of Analysis of Spatial Mechanisms," Journal of Engineering for Industry, Trans. ASME, Vol. 97, Series B, No. 4, pp. 1159–1164.

Duffy, J. (1977), "Displacement Analysis of Spatial Seven-Link 5R-2P Mechanisms," Journal of Engineering for Industry, Trans. ASME, Series B, Vol. 99, pp. 692–701.

Duffy, J. (1980), *Analysis of Mechanisms and Robot Manipulators*, John Wiley, New York.

Duffy, J., and Crane, C. (1980), "A Displacement Analysis of the General Spatial Seven Link, 7R Mechanism," Mechanism and Machine Theory, Vol. 15, pp. 153–169.

Freudenstein, F. (1973), "Kinematics: Past, Present, and Future," Mechanism and Machine Theory, Vol. 8, No. 2, pp. 151–161.

Hunt, K.H. (1978), *Kinematic Geometry of Mechanisms*, Clarendon Press, Oxford.

Keen, D.R. (1974), "A Unified Theory for the Analysis of Spatial Mechanisms by a Vector Loop and Matrix Method," Doctoral Dissertation, Liverpool Polytechnic, Liverpool, U.K.

Kotelnikov, A.P. (1895), "Screw Calculus and Some Applications to Geometry and Mechanics," Annals of the Imperial University of Kazan.

Lee, H.Y., and Liang, C.G. (1987), "Displacement Analysis of the Spatial 7-link 6R-P Linkages," Mechanism and Machine Theory, Vol. 22, No. 1, pp. 1–12.

Lee, H.Y., and Liang, C.G. (1988), "Displacement Analysis of the General Spatial 7-link 7R Mechanism," Mechanism and Machine Theory, Vol. 23, No. 3, pp. 219–226.

Lin, W. (1987), "Displacement Analysis of Two Special Cases of the TTT Manipulators," Master's Thesis, University of Florida.

Pelecudi, C. (1972), "Teoria Mecanismelor Spatiale," Editura Academiei Republicii Socialiste Romania.

Raghavan, M., and Roth, B. (1993), "Inverse Kinematics of the General 6R Manipulator and Related Linkages," Journal of Mechanical Design, Trans. ASME, Vol. 115, No. 3, pp. 502–508.

Reuleaux, F. (1876) (Translation by A. P. W. Kennedy), "The Kinematics of Machinery," Dover Pub. Inc., New York, 1963.

Rooney, J.J. (1974), "A Unified Theory for the Analysis of Spatial Mechanisms Based on Spherical Trigonometry," Doctoral Dissertation, Liverpool Polytechnic, Liverpool, U.K.

Soni, A.H., and Pamidi, P.R. (1971), "Closed Form Displacement Relations of a Five-Link RRCCR Spatial Mechanism," Journal of Engineering for Industry, Trans. ASME, Vol. 93, Series B, No. 1. pp. 221–226.

Study, E. (1901), "Geometrie der Dynamen," Verlag Teubner, Leipzig.

Todhunter, I. (1888), *An Elementary Treatise on the Theory of Equations*, Macmillan and Co., London, U.K.

Torfason, L.E., and Sharma, A.K. (1973), "Analysis of Spatial RRGRR Mechanisms by the Method of Generated Surfaces," Journal of Engineering for Industry, Trans. ASME, Series B, Vol. 95, No. 3, pp. 704–708.

Uicker, J.J., Denavit, J., and Hartenberg, R.S. (1964), "An Iterative Method for the Displacement Analysis of Spatial Mechanisms," Journal of Applied Mechanics, Trans. ASME, Vol. 86, Series E, No. 2, pp. 309–314.

Wallace, D.M. (1968), "Displacement Analysis of Spatial Mechanisms with More Than Four Links," Doctoral Dissertation, Columbia University, New York. University Microfilms, Ann Arbor, Michigan.

Wallace, D.M., and Freudenstein, F. (1970), "The Displacement Analysis of the Generalised Tracta Coupling," Journal of Applied Mechanics, Trans. ASME, Series E, Vol. 37, No. 3, pp. 713–719.

Wallace, D.M., and Freudenstein, F. (1975), "Displacement Analysis of the Generalised Clemens Coupling: The RRSRR Spatial Linkage," Journal of Engineering for Industry, Trans. ASME, Series B, Vol. 97, No. 2, pp. 575–580.

Weckert, M. (1952), "Analytische und Graphische Verfahren für die Untersuchung eigentlicher, Raumkurbelgetriebe, dargestellt an einem raumlichen viergliedrigen Kurbelgetriebe mit einem Drehgelenk und dre Drehschubgelenken," Doctoral Dissertation, Technische Hoschsule, München.

Wörle, H. (1962) "Untersuchungen Uber Kippelkurven Viergliedriger Raumlicher Kubelgetriebe," Konstruktion, Vol. 14, No. 10, pp. 390–392.

Yang, A.T. (1963), "Application of Quaternion Algebra and Dual Numbers to the Analysis of Spatial Mechanisms," Doctoral Dissertation, Columbia University, New York. University Microfilms, Ann Arbor, Michigan.

Yang, A.T., and Freudenstein, F. (1964), "Application of Dual-Number Quaternion Algebra to the Analysis of Spatial Mechanisms," Journal of Applied Mechanics, Trans. ASME, Vol. 31, Series E, pp. 300–308.

Yang, A.T. (1969), "Displacement Analysis of Spatial Five-Link Mechanisms Using (3 × 3) Matrices with Dual Number Elements," Journal of Engineering for Industry, Trans. ASME, Vol. 191, Series B, No. 1.

Yuan, M.S.C. (1970), "Displacement Analysis of the RRCCR Five-Link Spatial Mechanism," Journal of Applied Mechanics, Trans. ASME, Vol. 37, Series E.

Yuan, M.S.C. (1971), "Displacement Analysis of the RCRCR Five-Link Spatial Mechanism," Journal of Mechanisms, Vol. 6, No. 1.

Index